CORONA- AND RELATED VIRUSES

Current Concepts in Molecular Biology and Pathogenesis

ADVANCES IN EXPERIMENTAL MEDICINE AND BIOLOGY

Recent Volumes in this Series

Volume 375
DIET AND CANCER: Molecular Mechanisms of Interactions
Edited under the auspices of the American Institute for Cancer Research

Volume 376
GLYCOIMMUNOLOGY
Edited by Azita Alavi and John S. Axford

Volume 377
TISSUE RENIN–ANGIOTENSIN SYSTEMS: Current Concepts of Local Regulators in
Reproductive and Endocrine Organs
Edited by Amal K. Mukhopadhyay and Mohan K. Raizada

Volume 378
DENDRITIC CELLS IN FUNDAMENTAL AND CLINICAL IMMUNOLOGY, Volume 2
Edited by Jacques Banchereau and Daniel Schmitt

Volume 379
SUBTILISIN ENZYMES: Practical Protein Engineering
Edited by Richard Bott and Christian Betzel

Volume 380
CORONA- AND RELATED VIRUSES: Current Concepts in Molecular Biology
and Pathogenesis
Edited by Pierre J. Talbot and Gary A. Levy

Volume 381
CONTROL OF THE CARDIOVASCULAR AND RESPIRATORY SYSTEMS IN
HEALTH AND DISEASE
Edited by C. Tissa Kappagoda and Marc P. Kaufman

Volume 382
MOLECULAR AND SUBCELLULAR CARDIOLOGY: Effects of Structure and Function
Edited by Samuel Sideman and Rafael Beyar

Volume 383
IMMUNOBIOLOGY OF PROTEINS AND PEPTIDES VIII: Manipulation or Modulation
of the Immune Response
Edited by M. Zouhair Atassi and Garvin S. Bixler, Jr.

CORONA- AND RELATED VIRUSES

Current Concepts in Molecular Biology and Pathogenesis

Edited by

Pierre J. Talbot

Armand Frappier Institute
University of Québec
Laval, Québec, Canada

and

Gary A. Levy

The Toronto Hospital
University of Toronto
Toronto, Ontario, Canada

SPRINGER SCIENCE+BUSINESS MEDIA, LLC

Library of Congress Cataloging-in-Publication Data

Corona-- and related viruses   current concepts in molecular biology
and pathogenesis / edited by Pierre J. Talbot and Gary A. Levy.
      p.   cm. -- (Advances in experimental medicine and biology , v.
380)
   Includes bibliographical references and index.
   ISBN 978-1-4613-5775-9   ISBN 978-1-4615-1899-0 (eBook)
   DOI 10.1007/978-1-4615-1899-0
   1. Coronavirus infections--Congresses. 2. Coronaviruses-
-Congresses.   I. Talbot, Pierre, 1956-   . II. Levy, Gary A.
III. Series.
   QR399.C65 1995
   616'.0194--dc20   95-37889
                                                            CIP

Proceedings of the Sixth International Symposium on Corona and Related Viruses,
held August 27–September 1, 1994, in Quebec City, Quebec, Canada

ISBN 978-1-4613-5775-9

© 1995 Springer Science+Business Media New York
Originally published by Plenum Press, New York in 1995
Softcover reprint of the hardcover 1st edition 1995

10 9 8 7 6 5 4 3 2 1

PREFACE

Corona- and related viruses are important human and animal pathogens that also serve as models for other viral-mediated diseases. Interest in these pathogens has grown tremendously since the First International Symposium was held at the Institute of Virology and Immunobiology of the University of Würzburg, Germany. The Sixth International Symposium was held in Quebec City from August 27 to September 1, 1994, and provided further understanding of the molecular biology, immunology, and pathogenesis of corona-, toro-, and arterivirus infections. Lectures were given on the molecular biology, pathogenesis, immune responses, and development of vaccines. Studies on the pathogenesis of coronavirus infections have been focused mainly on murine coronavirus, and mouse hepatitis virus. Neurotropic strains of MHV (e.g., JHM, A59) cause a demyelinating disease that has served as an animal model for human multiple sclerosis. Dr. Samuel Dales, of the University of Western Ontario, London, Canada, gave a state-of-the-art lecture on our current understanding of the pathogenesis of JHM-induced disease.

Exciting new work with the identification of cellular receptors has proceeded in a number of laboratories around the world. This has resulted in the identification of the MHV receptor as members of the carcinoembryonic antigen family; the aminopeptidase-N for the 229E for the human coronavirus and porcine transmissible gastroenteritis virus and 9-O-acetylated neuraminic acid for bovine coronavirus. The counterpart domains on the viruses are now beginning to be identified. Work on coronavirus attachment and penetration is proceeding and should provide new insights into the molecular mechanisms involved in cell entry. It has become apparent, however, that virus binding to receptor alone is not sufficient for entry of the virus into the cell.

Dr. Kathryn Holmes, Uniformed Services of the Health Sciences, Bethesda, USA provided new insights into the binding domains on viral proteins, demonstrating that the presence of cellular receptors is not always sufficient to allow for infection.

Dr. Hubert Laude, Institut Nationale de la Recherche Agronomique, Jouy-en-Josas, France, provided a stimulating lecture on the structure and function of the structural and non-structural proteins of coronaviruses. The nonstructural proteins have been recently identified and one of these proteins, sM, has been discovered to be present on the virion. However, at present its function is unknown, but it may act as a channel protein similar to the influenza M2 protein.

Dr. Emil Skamene, Montreal General Hospital, Canada, gave an introductory lecture into the genetics and immunobiology of viral host interactions. Although at present the genetic basis for susceptibility and resistance to coronavirus infection is not clearly established, work in this area will provide important new insights into the mechanisms of viral immunopathogenesis. A number of additional lectures provided insights into the role of cellular and humoral immune mechanisms in the resistance and pathogenesis of disease.

Dr. Michael Buchmeier from La Jolla, California, USA, provided new information on the role of cytokines in the pathogenesis of MHV-induced disease.

Dr. Willy Spaan from Leiden, The Netherlands, gave a lecture on the common and distinctive features about corona-, toro-, and arteriviruses. The complete genomic sequences of four coronavirus species are now known and further sequence data was provided for porcine transmissible gastroenteritis virus. Dr. Michael Lai from the Departments of Microbiology and Virology at the University of Southern California, USA, provided new important information on the transcription, replication, recombination and engineering of coronavirus genes. A great deal of research activity has proceeded on characterization of the complex of proteins encoded by the 5'-end of the coronavirus genome which is involved in forming the RNA-dependent RNA polymerase complex. The mechanisms of RNA transcription and its regulation in particular, and of discontinuous transcription of subgenomic mRNAs are still not completely understood; however, a number of lectures provided new insights into these areas of research.

Dr. Luis Enjuanes of Madrid, Spain, discussed the development of vaccines against corona- and related virus diseases. These groups of viruses are of major economical importance and thus the development of vaccines is a priority. Various biotechnological approaches are now being explored, including the use of transgenic methodology.

The meeting was attended by over 140 scientists and was the largest meeting devoted to this topic to date.

The book is organized into eight sections covering pathogenesis, immune responses, vaccine development, viral proteins, cellular receptors, characterization of viral replicase, transcription and replication of viral RNA, recombination and mutation of viral RNA. The book includes proceedings of both podium and poster presentations.

The organizers of the meeting wish to thank all of those individuals who worked tirelessly to make the meeting a great success. We are especially grateful to Nathalie Arbour, Fanny Chagnon, Charmaine Mohamed, Mathilde Yu, Aurelio Bonavia, Robert Fingerote, Laisum Fung, Alain Lamarre and Yannic Salvas. We also wish to thank the members of the Scientific Advisory Committee for their help in preparing the scientific program: Robert Anderson, Lorne Babiuk, Samuel Dales, Serge Dea, J. Brian Derbyshire, Youssef Élazhary, Lucie Lamontagne and Peter Tijssen. We also wish to take the opportunity to acknowledge the financial support of Sandoz Canada, Inc., SmithKline Beecham Animal Health, Multiple Sclerosis Society of Canada, Université du Québec, Institut Armand-Frappier, Medical Research Council of Canada, Rhône Mérieux, International Union of Microbiological Societies, Connaught Laboratories Ltd., National Multiple Sclerosis Society (USA), Fort Dodge Laboratories, Ministère de l'Éducation du Québec, Biochem Thérapeutique, Langford Laboratories, Vetrepharm Canada, Inc., Boehringer Mannheim Canada, and Société de Promotion Économique du Québec Métropolitain, without which this meeting would have not been possible.

During the banquet that was held during the meeting, a special award and a plaque were presented to Professor Samuel Dales, Department of Microbiology and Immunology, University of Western Ontario, London, Ontario, Canada, for his invaluable lifetime scientific contributions to this field of research. On behalf of all of us, we wish to express to him our appreciation, gratitude, and pride.

The VIIth International Symposium on Corona- and Related Viruses will be held in Spain in 1997 and will be hosted by Dr. Luis Enjuanes.

Pierre Talbot   Gary A. Levy
Professor of Virology   Professor of Medicine
The Virology Research Centre   University of Toronto
Institut Armand-Frappier
Université du Québec

CONTENTS

Pathogenesis

Immune Responses

Vaccine Development

Viral Proteins

Cellular Receptors

Characterization of Viral Replicase

Transcription and Replication of Viral RNA

Recombination and Mutation of Viral RNA

Pathogenesis

GENETIC RESISTANCE TO CORONAVIRUS INFECTION

A Review

Ellen Buschman and Emil Skamene

McGill Centre for the Study of Host Resistance
Montreal General Hospital
Montreal, Quebec
Canada

GENETIC RESISTANCE TO CORONAVIRUS INFECTION

Introduction

Much information on the mechanisms of host genetic resistance to viral infections has come from research on the coronaviruses, particularly on the mouse hepatitis viruses (MHV). One of the fundamental observations made by Bang and co-workers some thirty years ago was that MHV infection of the host proceeds in a series of stages, which can be seen as three sequential barriers of host resistance[1-3]. These stages have also become the key to dissecting the genetic control of host resistance to coronaviruses. The first stage is the presence or absence of a specific cellular receptor which controls viral entry. Once the virus has gained entry, factors expressed by the host cells will then restrict or permit viral growth and acute disease. Finally, the humoral and cellular defenses of the host's immune system will determine whether the virus is eliminated or disseminated and a chronic disease is established. In this chapter, we have organized our review of genetic resistance to coronaviruses according to these three host resistance mechanisms: genetic control at the level of cellular receptors, genetic control at the level of the macrophage, and genetic control at the level of acquired immunity. However, we would like to stress that these 'levels' are purely operational boundaries. In reality, a host can be infected with a virus several times during its lifetime, and thus all available innate and immune resistance mechanisms will be called into play at once. In addition, we have included a general outline of the methods used to identify host resistance genes in mouse models of infection. Those interested in a more complete explanation of genetic analysis can refer to recent articles on this subject[4,5].

Corona- and Related Viruses, Edited by P. J. Talbot and G. A. Levy
Plenum Press, New York, 1995

Genetic Analysis of Host Resistance to Infections- Brief Outline

Host genetic resistance to viral infections, as for bacterial and parasitic infections, is usually expressed as a complex genetic trait. The initial approach to mapping, cloning and determining the function of the genes regulating resistance to infections is to dissect complex traits, such as disease susceptibility into simpler phenotypes, such as viral replication, that may be under single gene control[5]. The basic procedure is first to develop an animal model of infection, usually in the mouse, that has a clearly defined trait of resistance and susceptibility. Next, the genetic variation of the selected trait is analysed in a large panel of inbred strains. A pattern of resistant, susceptible and intermediate phenotypes (continuous variation) is suggestive of a complex trait controlled by multiple genes, whereas a pattern of clearly delineated susceptibility or resistance (discontinuous variation) suggests a single locus with two alternative alleles[6]. A Mendelian analysis is then undertaken on F1 and segregating backcross populations derived from resistant and susceptible progenitors to determine the mode of inheritance and to give an estimate of the number of genes involved[6]. Should the results indicate that more than one gene is acting then further genetic investigation may require the use of recombinant congenic strains[7] or multiple-locus linkage analysis[6,8]. Should single gene control be confirmed, one of the most frequently used gene mapping methods is linkage analysis in recombinant inbred strains of mice (RIS). The chromosomal location of the unknown locus is deduced by the concordance in the strain distribution pattern in the RIS panel with markers for previously mapped genes. Once the chromosomal location of the pertinent gene is known, gene cloning may be undertaken by positional cloning and/or by the candidate gene approach[4,5].

RESISTANCE TO CORONAVIRUS EXPRESSED AT THE LEVEL OF CELLULAR RECEPTORS

MHV-JHM and MHV-A59

Three classical genetic studies represent the major source of knowledge regarding the genetic control of infection with the neurotropic JHM strain of MHV which causes an acute fatal encephalitis in susceptible mice, and the sero-related, hepatotropic strain MHV-A59 which is considerably less virulent. Stohlman and Frelinger analysed the genetic control of acute encephalitis following JHM inoculation in susceptible B10.S and resistant SJL mice[9]. Their results from backcross and F2 generations supported the hypothesis that resistance to acute disease is under the control of two (unmapped) genes; one, termed *Rhv-1* is dominant and the second, *rhv-2*, is recessive. The genetic analysis also excluded any major effect of the H-2 complex on the resistance to acute JHM-induced disease. Secondly, Knobler et al. analyzed the genetic control of resistance to JHM virus in vivo in resistant SJL and susceptible BALB/C mice[10-12]. They concluded that a single recessive locus termed *Hv2* on chromosome 7, near *Svp-2* (seminal vesicle protein), determined resistance in vivo and in explanted macrophages. In contrast to later studies, these authors concluded that resistance to JHM infection was expressed by macrophages as an ability to restrict viral spread. Finally, Smith and co workers studied the ability of peritoneal macrophages from resistant SJL and susceptible mouse strains to support the growth of MHV- A59 in vitro[13]. They identified and mapped a locus, called *Mhv-1*, to chromosome 7, 41.5 cM from the albino locus, and showed that resistance was inherited in a recessive fashion. However, this group found that resistance was most likely expressed at the level of the viral receptor on the macrophage. Several common threads run through these genetic studies. First, *rhv-2*, *Hv2* and *Mhv-1* are most likely identical genes, as at least the latter two genes were shown to map to chromosome 7. Secondly, while the exact mechanism(s) of resistance to MHV-JHM and A59 expressed by SJL macrophages is unclear, it seems fair to say that it can be expressed both at the level of the viral receptor and at the level of viral synthesis. Discrepancies between the different studies could have arisen from different doses of virus used, the traits analysed and the treatments employed to obtain peritoneal macrophages. However, the two-gene hypothesis offered by Stohlman and Frelinger may eventually be shown to be correct when later results concerning viral binding to the MHV receptor (CEA/Bgp family of glycoproteins) on SJL tissues are considered.

The receptor for MHV-A59 (termed MHVR) was originally identified by a virus overlay protein blot assay for virus-binding activity as a 110- to 120-kDa glycoprotein on plasma membranes of intestinal epithelium or liver from susceptible BALB/c mice[14,15]. Interestingly, virus binding activity was only detected in membranes from hepatocytes or enterocytes of susceptible BALB/c and semisusceptible C3H mice, but not in comparable preparations of resistant SJL/J mouse membranes[14]. However, membrane preparations from MHV-resistant SJL/J mice were shown to express a homologue of MHVR in studies with antibodies directed against MHVR[15]. Therefore, it was reasoned that SJL/J mice may be resistant to MHV-A59 infection because they lack a functional virus receptor, a hypothesis which is now rejected with the cloning and expression of the MHVR isoforms (see below). Using the monoclonal antibody CC1 which recognizes the N-terminal 25-amino acid sequence of immunoaffinity-purified MHVR, Holmes and colleagues later demonstrated that the MHV receptor was identical to the predicted mature N termini of two mouse genes related to human carcinoembryonic antigen (CEA) and was strongly homologous to the N termini of members of the CEA family in humans and rats[16,17].

Subsequently, several variants of the MHV receptor (MHVR) have been cloned and sequenced[18,19]. MHVR1 is closely related to the murine CEA-related clone mmCGM1 (Mus musculus carcinoembryonic antigen gene family member). The cDNA sequence of this clone can encode a 424-amino-acid glycoprotein with four immunoglobulin like domains, a transmembrane domain, and a short intracytoplasmic tail. A second receptor, mmCGM$_2$, contains two immunoglobulin like domains and encodes a glycoprotein of 42kDa[20]. Additionally, Holmes and coworkers isolated two splice variants of MHVR, one containing two immunoglobulin-like domains [MHVR(2d)] and the other with four domains as in MHVR but with a longer cytoplasmic domain [MHVR(4d)L][18]. Alternative splicing mechanisms could explain how these CEA transcripts are derived from the same gene. All these variants have been recently identified as members of the biliary glycoprotein (Bgp) subfamily of the CEA family[21]. Somatic cell hybrid analysis suggests that the *Bgp* gene is located on chromosome 7 in the mouse[22].

The question of how all the different different MHVR isoforms are related to genetic susceptibility to MHV-A59 infection was then approached by asking whether variants isolated from resistant versus susceptible mouse strains could function as receptors for MHV-A59[18,19,23]. Phenotypically susceptible inbred mouse strains BALB/c, C3H, and C57BL/6 were found to express transcripts and proteins of the MHVR1 (mmCGM$_1$) isoform and/or its splice variants but not the mmCGM$_2$ isoform. In contrast, adult SJL/J mice, which are resistant to infection with MHV-A59, express transcripts and proteins only of the mmCGM2-related isoforms, not MHVR[18]. This strain distribution is compatible with the hypothesis that the MHVR and mmCGM$_2$ glycoproteins may be encoded by different alleles of the same gene. Surprisingly, especially in view of the structural differences between MHVR and mmCGM$_2$, both of the groups of Holmes and Lai have shown that transfection and expression of either mmCGM$_1$ or mmCGM$_2$ from SJL mice into MHV-resistant Cos 7 or BHK cells rendered the cells susceptible to MHV infection[18,19]. The ability of the SJL-related isoforms molecules to serve as MHV receptors was comparable to that of those from

C57BL/6. Thus, another factor must explain the genetic resistance of the SJL mouse in spite of its functional MHV receptor. Several possibilities have been put forward such as post-translational modification, but the most intriguing idea brought out by Yokomori and Lai[19], and one that fits well with the earlier two-gene hypothesis of Stohlman and Frelinger[9], is that the mechanism of genetic resistance of the SJL mouse lies in the product of a second gene whose product may be associated with the MHV receptor. This second factor of resistance has been envisioned as a protease that acts very early during virus entry to interfere with other step(s) in virus replication.

Cellular Receptor for Human 229E and Pig TGEV Coronaviruses

A second receptor, aminopeptidase N (APN), which in humans is identical with the CD13 differentiation antigen[24], has been identified as a receptor for both human and pig-specific coronaviruses[25,26]. However, genetic control in this case has only been demonstrated in humans[27]. Yeager et al. showed that human APN, located on chromosome 15[28] is a receptor for one strain of human coronavirus 229E, that is an important cause of upper respiratory tract infections[26]. A study of somatic cell hybrids demonstrated that a gene for susceptibility to the coronavirus 229E is located on chromosome 15 in the region q11-q12[27]. Secondly, APN has been identified as a receptor for transmissible gastroenteritis virus (TGEV), a coronavirus which causes fatal diarrhea in the newborn pig[25]. TGEV replicates selectively in the differentiated enterocytes covering the villi of the small intestine. In the small intestine, APN plays a role in the final digestion of peptides generated from hydrolysis of proteins by gastric and pancreatic proteases. Two lines of evidence supported the view that APN itself acts as a receptor. First, virions bound specifically to APN that was purified to homogeneity. Second, transfection of the APN gene into an otherwise non-permissive cell line conferred susceptibility to TGEV.

APN is thought to be involved in the metabolism of regulatory peptides by diverse cell types, including macrophages, granulocytes, and synaptic membranes of the central nervous system (CNS). In the mouse, the APN enzyme has not been identified as a coronavirus receptor, but the Lap-1 locus, (for leucine arylaminopeptidase-1 also called APN) has been linked to chromosome 9 in recombinant inbred lines and in intraspecific backcrosses[29]. Interestingly, mouse APN has also been identified in antigen-presenting cells and is co-expressed with H-2 class II molecules[30]. It is not known if the strain distribution for Lap-1 would fit any genetic model for MHV, for example the enteric strain MHV-Y, but it may be worth considering as a candidate gene/receptor especially considering that APN is expressed in macrophages.

GENETIC RESISTANCE TO MHV EXPRESSED BY MACROPHAGES

The MHV virulent type 2 (MHV-2) and 3 (MHV-3) strains induce acute hepatic failure in susceptible mouse strains following intraperitoneal injection. Death from MHV-2 infection occurs 2-3 days after inoculation and probably results from the destruction of hepatic parenchyma and the formation of large necrotic foci[1]. In acute MHV-3 infection, susceptible mice die within 5 days with the occurrence of hepatic necrosis, fibrin deposition and a heavy infiltrate of inflammatory cells[31]. Resistance to acute infection with MHV-2 and MHV-3 has been shown to be dependent on innate macrophage factors[1,2,31-33] as well as immune defense mechansims[32,34,35] rather than on cellular receptors. Below, we discuss the macrophage resistance mechanisms described in acute MHV-2 and MHV-3 infections which appear to be controlled by different, H-2-unlinked genes.

MHV-2

The *Hv1* locus, described by Bang and coworkers in the 1960's, controls susceptibility to lethal infection by MHV-2[1-3]. In this model, the dominant, susceptible allele of *Hv1* occurs in the PRI (Princeton) strain and the C3HSS congenic strain whereas the resistant, recessive *Hv1* allele occurs in the C3H strain. Resistance or susceptibility in vivo was shown to correlate almost perfectly with the permissivity of peritoneal macrophages to MHV-2 in vitro. The *Hv1* gene is probably not expressed as a viral receptor, since virus was shown to be equally well adsorbed by both resistant and susceptible macrophages[2]. Resistant macrophages were shown to block some aspect of viral RNA synthesis, thus affecting viral replication. Moreover, genetically susceptible mice could be rendered phenotypically resistant by treatment with the lectin concanavalin A (Con A), similarly, macrophages harvested from the Con A treated animals also manifested resistance in vitro[3]. Con A had no effect when administered directly to the macrophages, suggesting that reversal of susceptibility occurred following the induction of lymphokines. The chromosomal location of *Hv1* is unknown, but it segregates independently of the *Hv2* locus on chromosome 7, the *Bcg* locus on chromosome 1 and the locus for resistance to flavivirus on chromosome 5[36].

MHV-3

Genetic resistance to acute MHV-3 infection is comprised of at least two different host defense strategies. The acute phase of MHV-3 infection is characterized by a fulminant hepatic necrosis which kills susceptible strains of mice such as C57BL/6 and BALB/c within 3-5 days. During this phase, the host cells for viral replication are both macrophages and lymphoid cells[32], and, consequently, both of these cell types have adapted equally important resistance mechanisms. The group of Levy has extensively studied the role of the macrophage in the resistance to hepatitis caused by MHV-3[31,33,37-39]. Here, the inheritance of the trait of resistance/susceptibility to the fulminant, acute hepatitis and death caused by MHV-3 infection was analyzed in a set of recombinant inbred (RI) strains of mice derived from the resistant A/J and susceptible C57B1/6J progenitors[37]. The strain distribution pattern (SDP) showed a discontinuous variation ranging from fully resistant (no liver disease), to fully susceptible (death from fulminant hepatitis), with 16 RI strains showing intermediate degrees of susceptibility. This SDP was consistent with a two-recessive-gene model of resistance. These results were in contrast to earlier results (1979) of Levy-Leblond et al.[40], who found evidence for a single, recessive gene for resistance to acute MHV-3 disease in a panel of F1, F2 and backcross generation mice. However, this study [40] clearly showed that the gene(s) for resistance express an age-dependent penetrance. No evidence of linkage between acute MHV-3 sensitivity and genes of the H-2 complex was found in either of the two studies. The SDP for susceptibility/resistance to MHV-3 was also found to correlate with the induction of procoagulant activity (PCA), a prothrombinase[37]. PCA mediates inflammatory responses by virtue of its direct prothrombin cleaving activity [38]. The cellular nature of the production of the 74 kDa PCA molecule was also examined in RI strains and a restriction for induction of PCA was observed at the level of the macrophage[39]. Peritoneal macrophages from resistant parental A/J and RI strains could not be induced to express PCA when stimulated by MHV-3 alone or in the presence of lymphocytes from susceptible and H-2 compatible RI mice. In contrast, macrophages from susceptible RI strains of mice expressed a similar increase in PCA after stimulation with MHV-3 in the presence of L3T4+ (T helper cells, CD4+)

lymphocytes. In addition, T cells from MHV-3 immunized resistant RI mice, but not from unimmunized mice, were able to suppress induction of PCA. This suppressor activity could be detected in resistant mice even after 28 days of infection. The requirement for both T helper cells and macrophage PCA production for the full expression of MHV-3-induced acute hepatitis is consistent with the hypothesis of a two-gene model for susceptibility.

Two treatments have been found to inhibit the severe liver disease and procoagulant activity induced by MHV-3. In one, 16,16 dimethyl prostaglandin E2, dmPGE2, has been shown to specifically inhibit the activity of the PCA molecule by a posttranslational mechanism[33]. However, PGE2 treatment did not block MHV-3 viral replication or prevent death in susceptible animals. PCA activity could not be inhibited by other eicosanoids including prostacyclin (PGI2), PGF2a and leukotriene B4 (LTB4). In the second treatment, passive transfer of the anti-PCA monoclonal antibody 3D4.3 attenuated the development of hepatic necrosis and enhanced survival in a dose-dependent manner that correlated with the rates of metabolism of immunoglobulin[31]. Interestingly, the passive transfer treatment, in contrast to the E2 treatment, did decrease replication of MHV-3 in livers of susceptible animals to levels equivalent with those seen in resistant A/J mice. Taking the results of these two treatments together, the authors suggested that PCA has multiple pathogenic roles in MHV-3 infection[31]. First, PCA may activate the coagulation system that enhances hepatic necrosis, and secondly, they speculated that PCA could promote the proteolytic cleavage of the structural spike (S) protein into S1 and S2 subunits, thereby activating the cell fusion properties of the virus[41].

GENETIC RESISTANCE EXPRESSED AT THE LEVEL OF ACQUIRED IMMUNITY

According to our initial scheme, we have now arrived at the last 'level' of genetic resistance: that of acquired immunity to MHV. From a genetic point of view, this is the most difficult phase to analyse. The role of neutralizing antibodies, cellular immunity (including T, B, NK and antigen presenting cells) and the importance of the different antigenic epitopes of the virus are all elements which clamor for attention and surely would require an entire chapter to be discussed adequately. For our review, we have chosen to discuss the relationship between immunodeficiency and MHV-3 disease, as studied by Lamontagne and co workers[34,35,42], and secondly to review studies describing the role of CD4 and CD8 T cells during chronic JHM infection[43,44].

MHV-3

During the acute phase of MHV-3 infection, the role of immune defense mechanisms was initially revealed by evidence that resistant A/J mice become susceptible to the hepatic necrosis following immunosuppressive treatments[45]. More recently, Lamontagne et al have shown in vitro that the pathogenic L2-MHV3 strain could infect macrophages as well as lymphocytes, including thymocytes of both susceptible (C57BL/6) and resistant (A/J) mice[35]. In resistant A/J mice, however, a reduction in the level of infectious viral particles by the infected cell occurred at 48 hrs postinfection. This cellular resistance mechanism was observed to act as a recessive phenotype which acts to prevent viral replication very early following infection. Furthermore, these authors have shown that the pathogenic outcome of MHV-3 disease in susceptible mice correlated with the ability of the virus to cause lysis in vitro of Thy1.2$^+$ and surface IgM$^+$ cells. Susceptible mice also demonstrated significant atrophy of the thymic and splenic lymphoid follicles by 72 hours post infection, whereas no such atrophy or cell lysis occurred in resistant A/J animals. Moreover, it was determined by a virus interference assay that the virus could attach to the cell surface, and that the resistance mechanism acts to interfere with the activity of the viral RNA polymerase. These findings thus demonstrated that T and B lymphocytes can express intrinsic, H-2 unlinked[40] genetic resistance to the MHV-3 virus.

Lamontagne and co-workers have also begun to explore the consequences of the MHV3-induced immunodeficiency on the chronic phase of the disease[34]. Chronic MHV-3 infection results in aged F1 animals (bred from a cross of resistant A/J and susceptible C57BL/6 mice) who survive the acute hepatitis[40]. The disease is characterized by viral persistency in various organs, including the brain, spleen, and thymus and the occurrence of hindlimb paralysis at approximately 2 months after infection. It was determined that T and B cell depletions arose in F1 animals within a few days after infection, and low levels of splenic T and B cells were maintained for up to 3 months, until death of the mice[34]. The reduction of splenic lymphoid cells was determined to originate from depletion of all T cell subpopulations in the thymus and of pre-B and B cells in the bone marrow. However, infectivity studies performed in vitro revealed that the virus established a non-productive replication in thymic stromal cells with a low level of viral transmission to complexed thymocytes, whereas pre-B and B cells supported a productive, lytic viral replication.

The trait of hindlimb paralysis in the chronic form of MHV-3 disease has been genetically analysed by Levy-Leblond et al[40]. The analysis was performed in F1, F2 and backcross animals as well as in a panel of A strain mice congenic at the H-2 complex. The development of paralysis in the genetic crosses suggested that the chronic disease was controlled by one or two recessive genes. However, when the panel of A strain H-2 congenics was analysed, it appeared that the genetic control was linked to the H-2 complex, since the A.CA strain (H-2^f) was the only one which displayed complete resistance to the chronic disease. Using H-2 recombinant strains of mice, it was later determined that both class I H-2K and H-2D regions controlled the resistance to the development of paralysis[46]. Further analysis of the panel of H-2 recombinants for the trait of T and B cell depletions described above[34] could reveal if and how H-2 genes, or closely linked genes such as tumor necrosis factor (TNF)[47] or the transporter for antigen processing genes (TAP)[48], are involved in the development of MHV-3-induced immunodeficiencies.

MHV-JHM

In the first section of this review, we discussed how resistance to JHM-induced acute encephalitis is controlled by the MHVR (CEA) receptor and a second, unknown resistance factor. There is considerably less information concerning the basis of genetic resistance and susceptibility to the chronic phase of disease induced by the JHM virus, which is considered to be a model of central nervous system (CNS) disease. Susceptibility to the chronic disease is characterized clinically by the development of hindlimb paralysis 3-8 weeks post infection and histologically by evidence of demyelination in infected brains[44,49]. One model of resistance or susceptiblility to the late onset disease is called the maternal antibody protection model where the strains BALB/c and C57BL/6, which are fully susceptible to acute encephalitis, are protected from acute disease when immunized intranasally and nursed by immunized dams[49]. However, from 40-60% of the C57BL/6 mice will go on to develop late onset demyelinating disease whereas BALB/c mice are resistant and do not develop chronic symptoms. Neutralizing antibodies do not protect susceptible mice from the late onset disease, signifying that T cells may have the crucial role in viral clearance and in preventing CNS disease.

Table 1. Mouse loci controlling resistance to Coronavirus MHV infection

Locus	Trait	Chromosome	Inheritance of resistance	Reference
MHV-2				
Hv1	survival/replication in macrophages	?	recessive	1-3
MHV-3				
nd	survival/macrophage procoagulant activity	? [H-2 unlinked]	2 genes, recessive	37
nd	development of chronic disease[paralysis]	17, H-2K, H-2D	1 or 2 genes, recessive	40
MHV-JHM/A59				
Hv2 [? rhv2, Mhv1, Bgp]	infection of macrophages, neurons, liver cells	7	recessive	10-13, 22
Rhv1	survival/acute encephalitis	?	dominant	9
Pj1	chronic disease[paralysis]	7	?recessive	50
nd	chronic disease[paralysis]	17, H-2D^d	?	43,44

nd, not designated

It was determined by Stohlman's group in adoptive transfer experiments that class I H-2D restricted T cells of the helper phenotype (CD4+) and cytotoxic phenotype (CTL; CD8$^+$) were required for viral clearance[43]. This group also detected a strain difference correlating with resistance to viral elimination in that BALB/c mice mounted a CD8$^+$ response against the nucleocapsid (N) protein, whereas C57BL/6 mice did not. Castro et al pursued this question by determining, in H-2 congenic mice, whether replacement of the C57BL/6 type D^b locus by the BALB/c type D^d locus was sufficient to protect mice from the late onset disease[44]. The results showed that B10.A(18R) mice, which contain the d alleles of the D and L loci, exhibited a CTL response against the N protein as did the BALB/c mice, however, approximately 16% of the B10.A(18R) mice developed late onset symptomatic disease. These results suggested that the CD8$^+$, N-specific response is only partially protective against the development of the demyelinating disease. Additional results from this study showed that acutely and chronically JHM-infected C57BL/6 mice manifested an S protein specific CTL response in brains and spinal cords. BALB/c mice, however, did not contain anti-S CTL activity. Thus, the anti-S CTL activity present in the CNS of chronically infected animals could be involved in the pathogenesis of demyelination. Although the relationship of the CD4$^+$/CD8$^+$ interaction to CNS demyelination remains to be determined, it is apparent that several factors governing T cell function are likely to be involved[44].

The involvement of several factors is also supported by the data of Kyuwa et al, who have examined the genetic control of acute and late disease induced by MHV-JHM in the strains BALB/cHeA, STS/A, F1 hybrids and 13 (BALB/cHeA X STS/A) RI strains[50]. In this model, following intracerebral inoculation with JHM, all the BALB/cHeA mice died within 2 weeks from acute encephalitis whereas only 30% of STS/A mice died and were termed semisusceptible. The genetic analysis of the acute phase suggested the involvement of multiple genes, in agreement with the earlier studies of Stohlman and Frelinger[9]. For the chronic disease, the paralysis of delayed onset developed in 36% of STS/A, 40% of the F1 hybrids and eight of the 13 RI strains. However, the incidence varied widely among the RI strains, indicating that JHM-induced late disease is under multifactorial control. An alternative hypothesis is that variability of trait expression is large. This is supported by analogous results in the model of acute sensitivity to MHV-3 infection where results in RI and F1 mice suggested a highly variable, age-dependent penetrance of the resistance allele(s)[40]. Indeed, by ordering the mouse strains into two disease categories and evaluating their data considering a single gene hypothesis, Kyuwa et al found that the SDP indicated a putative gene effect located on chromosome 7, near Ly-15, which they designated *Pj1*[50]. The authors speculated that since Ly-15 is known to facilitate T cell recognition [51] and to be involved in the CTL response[52] it is possible that Ly-15 is involved in the CTL response during CNS disease. Therefore, the Ly-15 antigen could be a candidate for one of the additional factors governing CTL function in the maternal antibody protection model.

CONCLUDING REMARKS

It is apparent from reviewing these studies that host resistance to infection with MHV is under multigenic control (summarized in Table 1). As we said earlier, although the three levels of host resistance we discussed are artificial boundaries, the advantage of setting such boundaries is that one can functionally dissect a complex trait into parameters that are under single gene control. In this way, genetic susceptibility to MHV-3 acute fulminant hepatitis has already been disassociated into several traits, such as macrophage PCA production and lymphoid cell depletion. Once the responsible genes have been mapped , it may be possible to analyze the genetic control of MHV-3-induced hepatitis (or JHM-induced chronic CNS disease) by the use of multiple linked marker loci[6]. We await with interest as the story of genetic resistance to coronavirus infections continues to unfold.

REFERENCES

1 Kantoch M, Warwick A, Bang FB The cellular nature of genetic susceptibility to a virus J Exp Med 1963,117 781-798

2 Shif I, Bang FB In vitro interaction of mouse hepatitis virus and macrophages from genetically resistant mice I Adsorption of virus and growth curves J Exp Med 1970,131 843-850

3 Weiser WY, Bang FB Blocking of in vitro and in vivo susceptibility to mouse hepatitis virus J Exp Med 1977,146 1467-1472

4 Copeland NG, Jenkins NA, Gilbert DJ, et al A genetic linkage map of the mouse Current applications and future prospects Science 1993,262 57-66

5 Malo D, Skamene E Genetic control of host resistance to infection Trends in Genetics 1994,10 365-371

6 Risch N, Ghosh S, Todd JA Statistical evaluation of multiple-locus linkage data in experimental species and its relevance to human studies application to nonobese diabetic (NOD) mouse and human insulin-dependent diabetes mellitus (IDDM) Am J Hum Genet 1993,53 702-714

7 Demant P, Hart AAM Recombinant congenic strains-a new tool for analyzing genetic traits determined by more than one gene Immunogenetics 1986,24 416-422

8 Cornall RJ, Prins JB, Todd JA, et al Type 1 diabetes in mice is linked to the interleukin-1 receptor and Lsh/Bcg/Ity genes on chromosome 1 Nature 1991,353 262-265

9 Stohlman SA, Frelinger JA Resistance to fatal central nervous system disease by mouse hepatitis virus, strain JHM I Genetic analysis Immunogenetics 1978,6 277-281

10 Knobler RL, Harpel MV, Oldstone MBA Mouse hepatitis virus type 4 (JHM strain) induced fatal central nervous system disease I Genetic control and the murine neuron as the susceptible site of disease J Exp Med 1981,153.832-843

11 Knobler RL, Taylor BA, Woodell MK, Beamer WG, Oldstone MBA Host genetic control of mouse hepatitis virus type 4 (JHM strain) replication II The gene locus for susceptibility is linked to the Svp-1 locus on mouse chromosome 7 Exp Clin Immunol 1984,1 217-222

12 Knobler RL, Tunison LA, Oldstone MBA Host genetic control of mouse hepatitis virus type 4 (JHM strain) replication I restriction of virus amplification and spread in macrophages from resistant mice J Gen Virol 1984,65 1543-1548

13 Smith MS, Click RE, Plagemann BPGW Control of mouse hepatitis virus replication in macrophages by a recessive gene on chromosome 7 J Immunol 1984,133 428-432

14 Boyle JF, Weismiller DG, Holmes KV Genetic resistance to mouse hepatitis virus correlates with absence of virus-binding activity on target tissues J Virol 1987,61 185-189

15 Williams RK, Jiang GS, Snyder SW, Frana MF, Holmes KV Purification of the 110-kilodalton glycoprotein receptor for mouse hepatitis virus (MHV)-A59 from mouse liver and identification of a nonfunctional, homologous protein in MHV-resistant SJL/J mice J Virol 1990,64 3817-3873

16 Williams RK, Jiang GS, Holmes KV Receptor for mouse hepatitis virus is a member of the carcinoembryonic antigen family of glycoproteins Proc Natl Acad Sci USA 1991,88 5533-5536

17 Williams RK, Jiang GS, Holmes KV, Dieffenbach CW Cloning of the mouse hepatitis virus (MHV) receptor expression in human and hamster cell lines confers susceptibility to MHV J Virol 1991,65 6881-6891

18 Dveksler GS, Dieffenbach CW, Cardellichio CB, et al Several members of the mouse carcinoembryonic antigen-related glycoprotein family are functional receptors for the coronavirus mouse hepatitis virus-A59 J Virol 1993,67 1-8

19 Yokomori K, Lai MM The receptor for mouse hepatitis virus in the resistant mouse strain SJL is functional Implications for the requirement of a second factor for viral infection J Virol 1992,66 6931-6938

20 Turbide C, Rojas M, Stanners CP, Beauchemin N A mouse carcinoembryonic antigen gene family member is a calcium-dependent cell adhesion molecule J Biol Chem 1991,266 309-315

21 McCuaig K, Rosenberg M, Nedellec P, Turbide C, Beauchemin N Expression of the Bgp gene and characterization of mouse colon biliary glycoprotein isoforms Gene 1993,127 173-183

22 Robbins J, Robbins PF, Kozak CA, Callahan R The mouse biliary glycoprotein gene (Bgp) partial nucleotide sequence, expression, and chromosomal assignment Genomics 1991,10 583-587

23 Yokomori K, Lai MMC Mouse hepatitis virus utilizes two carcinoembryonic antigens as alternative receptors J Virol 1992,66 6194-6199

24 Look AT, Ashmun RA, Shapiro LH, Peiper SC Human myeloid plasma membrane glycoprotein CD13 (gp150) is identical to aminopeptidase N J Clin Invest 1989,83 1299-1307

25 Delmas B, Gelfi J, L'Haridon R, et al Aminopeptidase N is a major receptor for the enteropathogenic coronavirus TGEV Nature 1992,357 417-420

26 Yeager CL, Ashmun RA, Williams RK, et al Human aminopeptidase N is a receptor for human coronavirus 229E Nature 1992,357 420-422

27 Sakaguchi AY, Shows TB Coronavirus 229E susceptibility in man-mouse hybrids is located on human chromosome 15 Somat Cell Genet 1982,8 83-94

28 Kruse TA, Bolund L, Grzeschik K-H, et al Assignment of the human aminopeptidase N (peptidase E) gene to chromosome 15q13-qter FEBS Lett 1988,239 305-308

29 Womack JE, Lynes MA, Taylor BA Genetic variation of an intestinal leucine arylaminopeptidase (Lap-1) in the mouse and its location on chromosome 9 Biochem Genet 1975,13 511-518

30 Hansen AS, Noren O, Sjostrom H, Werdelin O A mouse aminopeptidase N is a marker for antigen-presenting cells and appears to be co-expressed with major histocompatibility complex class II molecules Eur J Immunol 1993,23 2358-2364

31 Li C, Fung LS, Chung S, et al Monoclonal antiprothrombinase (3D4 3) prevents mortality from murine hepatitis virus (MHV-3) infection J Exp Med 1992,176 689-697

32 Dupuy JM, Lamontagne L Genetically-determined sensitivity to MHV3 infections is expressed in vitro in lymphoid cells and macrophages Adv Exp Med Biol 1987,218 455-463

33 Chung SW, Sinclair SB, Fung LS, Cole EH, Levy GA Effect of eicosanoids on induction of procoagulant activity by murine hepatitis virus strain 3 in vitro Prostaglandins 1991,42 501-513

34 Jolicoeur P, Lamontagne L Impaired T and B cell subpopulations involved in a chronic disease induced by mouse hepatitis virus type 3 J Immunol 1994,153 1318-1317

35 Lamontagne L, Decarie D, Dupuy JM Host cell resistance to mouse hepatitis virus type 3 is expressed in vitro in macrophages and lymphocytes Viral Immunol 1989,2 37-45

36 Weiser W, Vellisto I, Bang FB Congenic strains of mice susceptible and resistant to mouse hepatitis virus Proc Soc Exp Biol Med 1976,152 499-502

37 Dindzans VJ, Skamene E, Levy GA Susceptibility/resistance to mouse hepatitis virus strain 3 and macrophage procoagulant activity are genetically linked and controlled by two non-H-2-linked genes J Immunol 1986,137 2355-2360

38 Dindzans VJ, MacPhee PJ, Fung LS, Leibowitz JL, Levy GA The immune response to mouse hepatitis virus expression of monocyte procoagulant activity and plasminogen activator during infection in vivo J Immunol 1985,135 4189-4197

39 Chung S, Sinclair S, Leibowitz J, Skamene E, Fung LS, Levy G Cellular and metabolic requirements for induction of macrophage procoagulant activity by murine hepatitis virus strain 3 in vitro J Immunol 1991,146 271-278

40 Levy-Leblond E, Oth D, Dupuy JM Genetic study of mouse sensitivity to MHV3 infection influence of the H-2 complex J Immunol 1979,122 1359-1362

41 Frana LS, Behnke JN, Sturman LS, Holmes KV Proteolytic cleavage of the E2 glycoprotein of murine coronavirus host dependent differences in proteolytic cleavage and cell fusion J Virol 1985,556 912-916

42 Lamontagne L, Jolicoeur P Mouse hepatitis virus 3-thymic cell interactions correlating with viral pathogenicity J Immunol 1991,146 3152-3159

43 Sussman MA, Shubin RA, Kyuwa S, Stohlman SA T-cell mediated clearance of mouse hepatitis virus strain JHM from the central nervous system J Virol 1989,63 3051-3056

44 Castro RF, Evans GD, Jaszewski A, Perlman S Coronavirus-induced demyelination occurs in the presence of virus-specific cytotoxic T cells Virology 1994,733-743

45 Dupuy JM, Levy-Leblond E, Le Provost C Immunopathology of mouse hepatitis virus type 3 infection II Effect of immunosuppression in resistant mice J Immunol 1975,114 226-230

46 Oth D, Lussier G, Cainelli-Gebara VC, Dupuy JM Susceptibility to murine hepatitis virus (type 3) induced paralysis influenced by class I genes of the MHC Eur J Immunogenet 1991,18 405-410

47 Muller V, Jongeneel CV, Nedospasov SA, Lindahl KF, Steinmetz M Tumor necrosis factor and lymphotoxin genes map close to H-2D in the mouse major histocompatibility complex Nature 1987,325 265-267

48 Trowsdale J, Hanson I, Mockbridge I, Beck S, Townsend A, Kelly A Sequences encoded in the class II region of the major histocompatibility complex related to the 'ABC' superfamily of transporters Nature 1990,348 741-743

49 Perlman S, Schelper R, Bolger E, Ries D Late onset, symptomatic, demyelating encephalomyelitis in mice infected with MHV-JHM in the presence of maternal antibody Microb Pathog 1987,2 185-194

50 Kyuwa S, Yamaguchi K, Toyoda Y, Fujiwara K, Hilgers J Acute and late disease induced by murine coronavirus, strain JHM, in a series of recombinant inbred strains between BALB/cHeA and STS/A mice Microb Pathogen 1992,12 95-104

51 Hogarth PM, Walker ID, McKenzie IFC, Springer TA The Ly-15 alloantigenic system a genetically determined polymorphism of the murine lymphocyte function-associated antigen-1 molecule Proc Natl Acad Sci USA 1985,82 525-530

52 Davignon D, Martz E, Reynolds T, Kurzinger K, Springer TA Lymphocyte function-associated antigen 1 (LFA-1) a surface antigen distinct from Lyt-2,3 that participates in T lymphocyte mediated killing Proc Natl Acad Sci USA 1981,78 4535-4539

2

FACTORS CONTROLLING CORONAVIRUS INFECTIONS AND DISEASE OF THE CENTRAL NERVOUS SYSTEM

A Review

S. Dales

Department of Microbiology and Immunology
University of Western Ontario
London, Ontario
N6A 5C1
Canada

INTRODUCTION

This review summarizes 18 years of investigations in this laboratory on factors controlling the infectious process in cells of the central nervous system (CNS). The model chosen by us which involves rodents challenged with neurotropic murine coronavirus (CV) JHMV, was selected on the presumption that pathogenesis due to viral infections can be a trigger for sequelae culminating in autoimmune disease like that presumed to cause multiple sclerosis (MS) in humans. Our model, based on the initial 1949 reports about initiation of neurologic disease in rats and mice by JHMV[1,2] has been refined and amplified upon[3]. More recent evidence, which includes detection of viral RNA and protein in neural cells some albeit circumstantial, has implicated both the rodent and ubiquitous human OC43 or 229E CVs in CNS pathologies including demyelinative disease[4,5]. The key interrelated parameters relevant to the model examined by us which will be considered here include the influence of age and development of the host animal, its immune responses, genetic constitution and variability of the infecting viral agent. For the sake of clarity, realizing that it may be an oversimplification, CNS infections associated primarily with neurons (NEU) are designated as grey matter (GM) disease while those involving glial cells, especially astrocytes (AS) and oligodendrocytes (OL), which cause demyelinative lesions are termed white matter (WM) disease.

Due to the brevity of this review, it was necessary to focus predominantly on data emanating from this laboratory, without intention to minimize the salient contributions of colleagues elsewhere. In most cases, the pertinent reference to studies of others can be found in bibliographies of the articles cited here.

Corona- and Related Viruses, Edited by P. J. Talbot and G. A. Levy
Plenum Press, New York, 1995

FACTORS IN THE DISEASE PROCESS

a) *age related resistance*: the intracerebral (IC) inoculation of JHMV into neonate rats induces an acute encephalitis due to rapidly and widely disseminated GM involvement. A postnatal interval between 1 week and 3 weeks of age, at the time weaning occurs in rats is a period or "window" of susceptibility involving the WM. This corresponds well with completion of CNS myelination and OL maturation[6]. By contrast, the more gradual, prolonged myelination and maturation in the murine CNS correlates with an extended period into the lifespan of susceptibility to WM disease induced by JHMV[7].

b) *immune responses*: the unambiguous evidence for the role of cellular immunity in controlling JHMV infections of the rat CNS contrasts with clear cut involvement of humoral immunity. Thus, in athymic nude (Nu/Nu) rats and mice or following immunosuppression by drugs such as cyclosporin A, the normal age-related resistance is abrogated[8,9] and a productive infection can be elicited in rats at any age post-weaning. However, as a consequence of impaired cellular immunity, the disease produced is predominantly of the GM, associated with extensive NEU involvement which results in an acute, rapidly fatal encephalitis[8,9]. In the case of humoral immunity, localized anti-JHMV immunoglobulin synthesis can be monitored in the cerebrospinal fluid without appearing to influence the progress of WM-associated paralytic disease[10].

c) *genetic constitution of the host*: the published information dealing with this complex subject is too voluminous for adequate review here, limiting the selected material to our own studies on rodents. Concerning the rat model, strain differences occur in the "window of susceptibility" interval during which WM disease can be elicited by JHMV. Wistar Furth (WF) rats are genetically deficient in circulating growth hormone. Presumably related to this deficiency is a slower postnatal CNS maturation. Wistar Lewis (WL) rats possess normal levels of growth hormone, earlier onset of WM maturity and easily provoked, vigorous inflammatory cell responses in the CNS. Comparisons on susceptibility to WM disease showed that in WF the paralytic, demyelinative process can be induced by JHMV inoculation ic for significantly longer periods than in WL rats[11]. Genetic analysis by cross-breeding and backcrossing revealed that the heritable phenotype for resistance (R) is controlled by a single gene and that in WF strain this trait segregates as a homozygous recessive rr while in WL rats it occurs in a heterozygous form RR, Rr and rr.

Genetically determined host resistance *vs* susceptibility to CV can be defined unambiguously in specific strains of mice. Comparison between strain SJL/J which is resistant and CD.1 (or BALB/c) which are sensitive to challenge with neurotropic CVs JHMV or A59 showed by crossbreeding in the F1 generation that resistance is recessive, i.e., susceptibility is dominant[12]. These data were reflected exactly by correlative results from virus replication in primary cell explants from the CNS of embryonic mice[13], as illustrated in Table I.

Analysis of JHMV adsorption and expression revealed that arrest of this CV strain in cells of SJL/J mice was most probably due to an arrest in cell-to-cell spread, not to the lack of either receptors or virus-specified expression. It is possible that SJL/J cells lack a protease activity which is required to activate the S fusogenic component of the virus.

d) *genetic variability of the virus*: once again the variability and complexity of the evidence gathered on this subject precludes inclusion of an authoritative review in this article. To explain virulence, tropism and pathogenesis of CVs, variability among the components, especially of the S glycoprotein has been invoked, sometimes based on contradictory data (see[12] for several citations). Our studies have been concerned with neural cell tropism of the closely related viscerotropic MHV_3 and neurotropic JHMV CVs. Challenge of primary neural rat cultures enriched for specific cell types demonstrated that MHV_3 had preferential tropism for type 1AS (1AS) and JHMV for OL[14]. JHMV produces WM

Table I. Comparison of CV replication in glial cultures[a] from purebred and hybrid mice

Mouse strain from which cultures originated	Time after inoculation (h)	Titer 10^2 (PFU/ml)[b]		
		JHMV[c]	MHV$_3$[c]	A59[c]
SJL	12	0 17 ± 0 29 (3)	590 ± 114 (3)	6 ± 2 (6)
	24	23 ± 12 (14)	435 ± 292 (14)	13 ± 0 5 (6)
	48	48 ± 20 (14)	886 ± 62 (10)	2 ± 0 6 (3)
	72	22 ± 9 (12)	475 ± 119 (6)	ND[d]
CD 1	12	350 ± 216 (5)	77 ± 12 (4)	ND
	24	3260 ± 2110 (13)	778 ± 30 (11)	21,300 ± 19,600 (3)
	48	9440 ± 500 (9)	3340 ± 2300 (7)	95,800 ± 2700 (3)
(CD 1 × SJL) F$_1$	24	961 ± 34 (5)	1120 ± 730 (5)	ND
	48	6260 ± 4320 (5)	2400 ± 2000 (5)	ND
	72	943 ± 480 (5)[e]	213 ± 52 (5)[e]	ND

[a]Primarily oligodendrocytes and astrocytes (35)
[b]The values are means with standard deviations The number of cultures tested is shown in parentheses
[c]MOI, 1 PFU per cell
[d]ND, not determined
[e]Decrease in titers was attributed to rapid cell killing (from ref 13)

disease but MHV$_3$ does not evoke any CNS disease symptoms. This mutually exclusive tropism demonstrated with rat neural cells does not occur in murine 1AS and OL, both of which are infectable by both the viscerotropic and neurotropic agents[15].

CORRESPONDENCE BETWEEN THE IN-VIVO *VS* IN-VITRO DISEASE PROCESS

The IC inoculation of rats with JHMV at intervals post-partum, up to the age of weaning causes either the acute encephalytic GM disease or a progressive demyelinative WM disease with paralysis or paresis, depending on the age at challenge[6]. To determine age and disease related JHMV expression in the CNS presence of viral antigen was monitored by immunocytology and viral RNA by probing with specific cDNA in dot-blot assay. Virus materials were pinpointed within five functional zones of the CNS, defined anatomically in

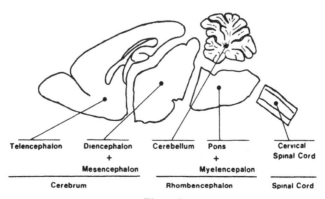

Figure 1

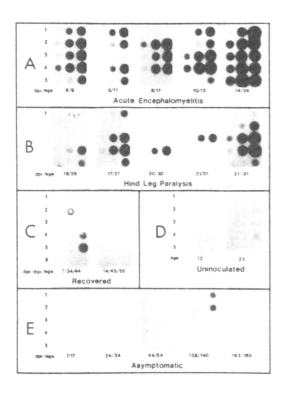

Figure 2. Dot-blot analysis of CNS tissue extracts prepared from samples taken from rats with either (A) acute encephalomyelitis, or (B) hind-leg paralysis, or (C) after recovery from paralysis, and (D) uninoculated controls as well as (E) asymptomatic animals Each row is composed of RNA extracts blotted at dilutions of 1/100, 1/10, and 1/1 The numbers of the ordinate represent the tissues sampled as follows 1· telencephalon, 2 diencephalon/mesencephalon, 3. cerebellum, 4 pons/myelencephalon, 5 cervical spinal cord, 6 lumbar spinal cord, dpr: days post recovery, dpi. days post inoculation *This result is anomalous since other blots in this dilution series were negative (Ref 10)

Figure 1. It is evident from the dot-blot analysis in Figure 2 that virus can be disseminated rapidly throughout the CNS. In the case of WM disease process, centers of infection are established prior to evidence of any tissue damage associated with the progressive form of demyelinative disease. JHMV specific RNA is detectable in the CNS of rats which have recovered from paralytic disease symptoms and in rats which have remained asymptomatic for 5 months and beyond. This and related studies demonstrated surprisingly the capacity of this neurotropic virus, possessing a plus sense single stranded RNA genome, to remain within the CNS for prolonged periods, perhaps indefinitely in a persistent or latent state, despite absence of any known mechanism for establishing a provirus replication strategy like that used by retroviruses.

We sought to define more precisely the parameters regulating JHMV infections of neural cells *in vivo* by examining virus-cell interactions under the tighter controlled conditions possible *in vitro*. For this purpose, explant cultures were established from embryonic or neonatal CNS tissue enriched in NEU originating in the hippocampus or OL lineage precursors from the telencephalon. The virus-cell interactions pertaining to either NEU or OL will now be considered in separate sections.

a) *studies on neurons*: our data, described in several published articles reveal that NEU involvement is extensive throughout the CNS GM, during acute encephalitis. When the progressive demyelinative, paralytic form of WM disease occur hippocampal and cerebellar neurons contain JHMV RNA and protein[16], illustrated in Figure 3. Since viral RNA persists for long periods in asymptomatic and recovered rats[10], we assume that NEU may provide the reservoir for virus persistence.

A specific, close association between JHMV nucleocapsids (N) and bundles of microtubules within neurites was observed in thinly sectioned cultured hippocampal neurons[17,18]. This association is perhaps established because there is protein "mimicry", uncov-

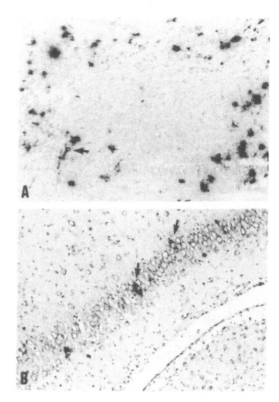

Figure 3. Coronavirus antigen and RNA in the hippocampus of JHMV-infected rats. (A) Immunohistochemical detection of N protein in the hippocampus of a JHMV-infected rat. Arrow points to an N+ neuron within Ammons's horn. Other N+ cells are likely glia. (B) Coronavirus RNA in the hippocampus detected by in situ hybridization using a JHMV-specific [35S} cDNA probe. Arrows point to two positive neurons. (Ref. 16).

ered by computer sequence matching between N and the microtubule-binding domain of the microtubule associated protein tau[18], illustrated in Figure 4. Relatedness of N to tau is further substantiated by immunological relatedness shown in a companion article[19]. These findings draw attention to: 1) a strategy of a CV for making use of the host's normal function on behalf of the virus. 2) involvement of microtubules in replication, trafficking of virus proteins and assembly of progeny particles[18].

b) *infections of oligodendrocytic lineage cells*: glial progenitors explanted from the rat optic nerve have the potential to differentiate into mature OL or 2A astrocytes[20]. These progenitors marked as of the O-2A phenotype are programmed in the same "time-clock" required for becoming mature, myelin producing OL as that which prevails *in vivo*. Throughout development and maturation phenotypic markers, many present at the cell surface, identify the various stages recognizable by means of specific antibodies, as shown in Figure 5. When such antibodies are applied, it is evident that soon after explantation of tissue from the telencephalon, most OL progenitors are in the mitotically active A2B5+ juvenile stage:

Figure 4. Evidence for amino acid sequence homology between tau and N. Identical residues are linked by vertical lines, related ones by two dots when homology is closer or one dot when it is less close. The tau sequence is the invariant 12-mer of the microtubule binding motif matched with N sequence from the carboxyl terminus.

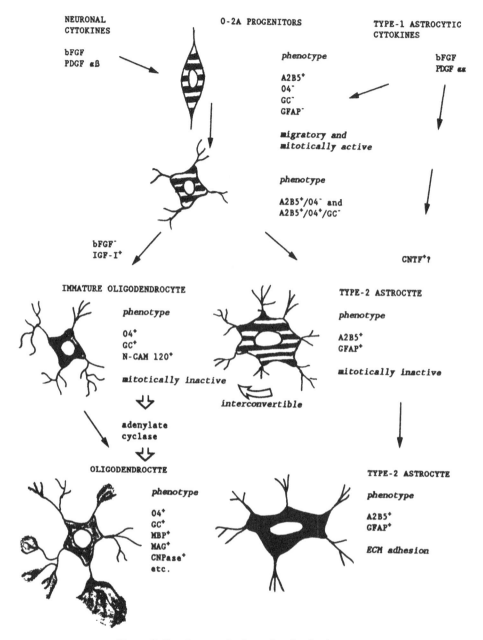

Figure 5. Developmental scheme for oligodendrocytes.

a phase during which they are amenable to mitotic arrest by application of ng amounts of cytokines bFGF and PDGFαα. The influence of cytokines is completely reversible. The A2B5+ cells which are refractory to infection by JHMV become susceptible when treated with bFGF and PDGF (Figure 6). Upon removal of the cytokines they regain their resistance[17].

Later, during terminal stages of development, the mature OL synthesize myelin specific and other constituents, including myelin basic protein (MBP), associated glycoprotein (MAG), galactocerebroside (GC) and 2':3'-cyclic nucleotide-3'-phosphohydrolase (CNPase) (Figure 5). During the terminal stages of differentiation, OL once again became

Figure 6. Effects of bFGF and PDGF on the replication of JHMV in enriched O-2A lineage cultures O-2A lineage cells derived from 9 DIV and P9 mixed telencephalic cultures were plated at 2 5 × 10⁵ cells per cm² and grown for 48 h in O1/T3 alone or O1/T3 supplemented with bFGF (10 ng/ml), PDGF (10 ng/ml), or bFGF plus PDGF before inoculation with JHMV at a MOI of 2 Data are the means of titers ± standard deviations from triplicate cultures and are representative of three independent experiments (Ref 17)

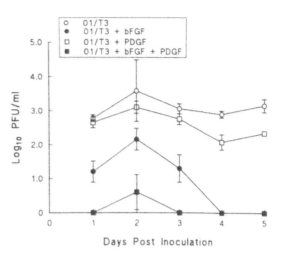

non-infectable by JHMV[14] Differentiation of OL can be deliberately accelerated by upregulating the adenylate cyclase metabolic pathway employing metabolites such as cyclic AMP[14 21] After treatment of purified progenitors with 1mM dbcAMP for 1 or 2 days the mature OL phenotype is expressed and coincidentally infectability by JHMV ceases The block occurs at a stage after adsorption but before any virus functions are expressed The arrested stage is likely to involve uncoating because 1) The 56 kDa protein N of inoculum virus is rapidly converted to the less phosphorylated 50 kDa form, prior to breakdown to low MW fragments, as shown in Figure 5 2) treatment with dbcAMP after replication has commenced does not suppress JHMV[14] 3) infection of mature OL can be initiated by

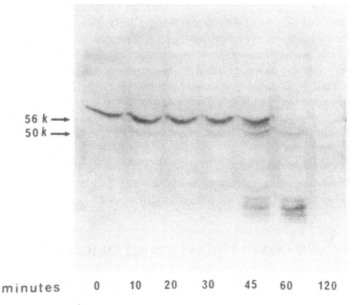

Figure 7. Modulation of N protein during 120 min of JHMV penetration into L-2 cells The antigen is identified by immunoblotting employing monoclonal anti N antibody (Mohandas, D , Wilson, G and Dales, S , unpublished)

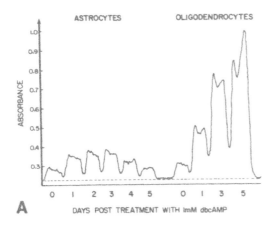

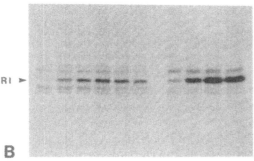

Figure 8. Modulation of RI and RII in primary rat astrocytes and oligodendrocytes during treatment with 1 mM dbcAMP. The concentrations of the regulatory subunits in cytosol (100,000 × g) fractions from astrocytes and oligodendrocytes were determined by binding of 8-azido-[^{32}P]cAMP, as described in Materials and Methods. A densitometer tracing (A) made from an autoradiogram (B), obtained after 10% SDS-polyacrylamide gel electrophoresis, enabled a comparison of the time-related changes in the RI regulatory subunit. Absorbance units have been normalized to the band of greatest density (oligodendrocytes, 5 days posttreatment). (Ref. 21)

transfecting with isolated genomic virus RNA (see[19] companion article). From the knowledge that the genome within inoculum virions is coated by the phosphorylated form of N[22] we hypothesize that deficiency of uncoating in mature OL might be due to suppression of dephosphorylation of N[21]. We also predicted that the dephosphorylating activity involved is a phosphoprotein phosphatase (PPPase) acting on N of penetrating inoculum virus. Evidence was obtained that a type 1 PPPase is associated with host cell endosomes, possesses a highly specific activity against phosphorylated N as the substrate[23] and is inhibitable by the regulatory subunit R of cAMP-dependent protein kinase I [24]. The upregulation of R expression in mature OL[21], shown in Figure 8, provided a rationale for the existence of a relationship between maturation and JHMV suppression in OL. This idea was supported by our demonstration of virus suppression upon transfection of the RI gene into non-neural L2 cells (see 19 companion article), but an absence of suppression when infectious genomic RNA is used in place of intact virions to initiate JHMV replication.

Experimental evidence obtained on the PPPase from assays in cell-free reactions supports the findings on intact cells[24] (and unpublished data).

SUMMARY AND CONCLUSIONS ABOUT OLIGODENDROCYTES

There is a correlation between specificity of tropism of JHMV for O-2A lineage cells from the rat and demyelination of white matter, associated with chronic disease. Susceptibility to infection, which can occur in O-2A cells before terminal differentiation may be influenced by cytokines. During the normal, age-related or rapidly induced maturation/dif-

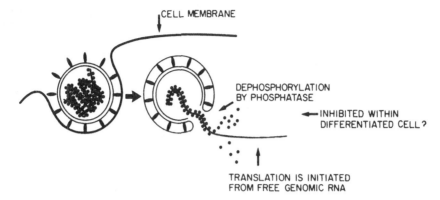

Figure 9. Early cell–virus interactions

ferentiation of rat oligodendrocytes, suppression of JHMV replication is correlated with upregulation of the subunit R1 of the cAMP-dependent protein kinase Virus inhibition occurs at a stage between penetration and initiation of genome expression Regulation over coronavirus infection of oligodendroglia is strictly controlled by the host cell There is evidence that induction of R1 subunit of protein kinase A influences uncoating, illustrated in Figure 9, by suppression of dephosphorylation during penetration Our former working hypothesis, now borne out by recent data predicts that the infection in mature oligodendrocytes is blocked because specific dephosphorylation of the capsid protein N, required for uncoating, etc is suppressed

REFERENCES

1 Bailey, O T , Pappenheimer, A M , Cheever, F S , Daniels, J B A murine virus (JHM) causing disseminated encephalomyelitis with extensive destruction of myelin II Pathology J Exp Med 1949,90 195-231

2 Cheever, F S , Daniels, J B , Pappenheimer, A M , Bailey, O T A murine virus (JHM) causing disseminated encephalomyelitis with extensive destruction of myelin I Isolation and biological properties of the virus J Exp Med 1949,90 181-194

3 Werner, L P Pathogenesis of demyelination induced by a mouse hepatitis virus (JHM virus) Arch Neurol 1973,28 298-303

4 Stewart, J N , Mounir, S , Talbot, P J Human coronavirus gene expression in brains of multiple sclerosis patients Virology 1992,191 502-505

5 Murray, R S , Brown, B , Brian, D , Cabirac, G F Detection of coronavirus RNA and antigen in multiple sclerosis brain Ann Neurol 1992,31 525-533

6 Sorensen, O , Percy, D , Dales, S In vivo and in vitro models of demyelinating disease III JHM virus infection of rats Arch Neurol 1980,37 478-484

7 Sorensen, O , Dugre, R , Percy, D , Dales, S In vivo and in vitro models of demyelinating disease caused by mouse hepatitis virus in rats and mice Inf and Immun 1982,37 1248-1260

8 Sorensen, O , Saravani, A , Dales, S In vivo and in vitro models of demyelinating disease XVII The infectious process in athymic rats inoculated with JHM virus Microb Path 1987,2 79-90

9 Zimmer, M J , Dales, S In vivo and in vitro models of demyelinating disease XXIV The infectious process in cyclosporin A treated Wistar Lewis rats inoculated with JHM virus Microb Path 1989,6 7-16

10 Sorensen, O , Coulter-Mackie, M B , Puchalski, S , Dales, S In vivo and in vitro models of demyelinating disease X Progression of JHM virus infection in the central nervous system of the rat during overt and asymptomatic phases Virology 1984,137 347-357

11 Sorensen, O , Beushausen, S , Coulter-Mackie, M , Adler, R , Dales, S In vivo and in vitro models of demyelinating disease In Kurstak, K , Lipowski, Z J , Morozow, P V, (eds) Viruses, immunity and mental disorders Plenum Med Book Co 1987 Chapter 18, pp 199-210

12 Pasick, J M M , Wilson, G A R , Morris, VL , Dales, S SJL/J resistance to mouse hepatitis virus JHM-induced neurologic disease can be partially overcome by viral variants of S and host immunosuppression Microb Path 1992,13 1-15

13 Wilson, G A R , Dales, S In vivo and in vitro models of demyelinating disease efficiency of virus spread and formation of infectious centers among glial cells is genetically determined by the murine host J Virol 1988,62 3371-3377

14 Beushausen, S , Dales, S In vivo and in vitro models of demyelinating disease XI Tropism and differentiation regulate the infectious process of coronavirus in primary explants of the rat CNS Virology 1985,141 89-101

15 Wilson, G A R , Beushausen, S , Dales, S In vivo and in vitro models of demyelinating diseases XV Differentiation influences the regulation of coronavirus infection in primary explants of mouse CNS Virology 1986,151 253-264

16 Sorensen, O , Dales, S In vivo and in vitro models of demyelinating disease JHM virus in the rat central nervous system localized by in situ cDNA hybridization and immunofluorescent microscopy J Virol 1985,434-438

17 Pasick, J M M , Dales, S Infection by coronavirus JHM of rat neurons and oligodendrocyte-type-2 astrocyte lineage cells during distinct developmental stages J Virol 1991,65 5013-5028

18 Pasick, J M M , Kalicharran, K , Dales, S Distribution and trafficking of JHM coronavirus structural proteins and virions in primary neurons and the OBL-21 neuronal cell line J Virol 1994,68 2915-2928

19 Kalicharran, K , Dales, S a) Dephosphorylation of the nucleocapsid protein of inoculum JHMV may be essential for initiating replication b) Involvement of microtubules and the microtubule-associated protein tau in trafficking of JHM virus and components within neurons These proceedings, p 57 and p 485

20 Raff, M C , Abney, E R , Miller, R H Two glial cell lineages diverge prenatally in rat optic nerve Dev Biol 1984,106 53-60

21 Beushausen, S , Narindrasorasak, S , Sanwal, B D , Dales, S In vivo and in vitro models of demyelinating disease activation of the adenylate cyclase system influences JHM virus expression in explanted rat oligodendrocytes J Virol 1987,61 3795-3803

22 Stohlman, S A , Fleming, J O , Patton, C D , Lai, M M C Synthesis and subcellular localization of the murine coronavirus nucleocapsid protein Virology 1983,130 527-532

23 Mohandas, D V , Dales, S Endosomal association of a protein phosphatase with high dephosphorylating activity against a coronavirus nucleocapsid protein FEBS Lett 1991,282 419-424

24 Wilson, G A R , Mohandas, D V , Dales, S In vivo and in vitro models of demyelinating disease Possible relationship between induction of regulatory subunit from cAMP dependent protein kinases and inhibition of JHMV replication in cultured oligodendrocytes In Cavanagh, D , Brown, T D K (eds) Coronavirus and their diseases Plenum Press, N Y

3

CHARACTERIZATION OF THE S PROTEIN OF ENTEROTROPIC MURINE CORONAVIRUS STRAIN-Y

Susan R. Compton and Satoshi Kunita[*]

Section of Comparative Medicine
Yale University School of Medicine
New Haven, Connecticut

ABSTRACT

The pathogenesis of enterotropic murine coronavirus strain MHV-Y differs extensively from that of prototypic respiratory strains of murine coronaviruses. The S protein of MHV-Y was characterized as a first step towards identifying viral determinants of enterotropism. Immunoblots of MHV-Y virions using anti-S protein specific antiserum revealed that the MHV-Y S protein was inefficiently cleaved. The MHV-Y S gene was cloned and sequenced. It encodes a protein predicted to be 1361 amino acids long. The presence of several amino acids changes within and surrounding the predicted cleavage site of the MHV-Y S protein may contribute to its inefficient cleavage.

INTRODUCTION

Mouse hepatitis virus (MHV), a singular name for several murine coronaviruses, causes a wide spectrum of diseases ranging from mild enteritis or rhinitis to fatal hepatitis or encephalitis. MHV strains can be divided into two biotypes, respiratory and enterotropic, on the basis of their initial site of replication. Following oronasal inoculation, respiratory MHV strains initiate replication in the upper respiratory tract and then disseminate to multiple organs if the mouse is sufficiently susceptible due to age, genotype or immune status[1,2]. On the other hand, replication of enterotropic MHV strains, such as MHV-Y, is largely restricted to the intestinal mucosa, with minimal or no dissemination to other organs[3,4]. All ages and genotypes of mice are susceptible to infection with enterotropic MHV-Y but only young mice develop disease in the form of enteritis[3,4]. Unlike respiratory MHV strains in which viral titers mirror

---

[*]Current address: Institute of Laboratory Animal Research, School of Science, Kitasato University, Kanagawa, Japan.

the severity of lesions, the titers of MHV-Y produced in the intestines do not reflect the level of lesion formation[1,4].

The molecular basis of MHV pathogenesis has been studied extensively for respiratory MHV strains such as MHV-4/JHM and MHV-A59. Although the molecular mechanisms of tissue tropism are still not clear, several studies have shown an important role for the MHV S protein in virulence and tissue tropism[2]. The S glycoprotein is synthesized as an 170-200 kD glycoprotein which is cleaved by trypsin-like host proteases into two 90 kD subunits. Cleavage is dependent on the virus strain and host cell type in which virus was grown. The S protein forms the characteristic viral peplomers and is believed to be responsible for the initiation and spread of infection, by mediating attachment of the virus to cell surface receptors and by inducing cell-cell fusion. The S protein also elicits neutralizing antibodies and cellular immune responses[2]. Variant viruses which possess mutations or deletions in the S protein have altered target cell specificity, rates of spread or virulence[5-11]. In contrast to respiratory MHV strains, the viral factors which determine the tropisms of enterotropic MHV strains have not been identified. The S protein of MHV-Y was characterized molecularly as a first step in determining the role of the S protein in the restricted tissue tropism of MHV-Y.

MATERIALS AND METHODS

Viruses and Cells

MHV-Y was originally isolated in NCTC 1469 cells from the intestine of a naturally infected infant mouse with acute typhlocolitis[12]. MHV-A59 was obtained from American Type Culture Collection. Virus stocks used in these studies were generated in J774A.1 cells. J774A.1 cells were obtained from American Type Culture Collection and maintained in RPMI medium 1640 supplemented with 10% fetal bovine serum.

Immunoblotting

Purified virion proteins were separated on 8% SDS-PAGE gels, electroblotted to nitrocellulose sheets and were probed with monospecific polyclonal goat antiserum to the S protein of MHV-A59 (gift of K. Holmes). Bound antibody was detected with peroxidase-conjugated Staphylococcal protein A and chemiluminescent reagents.

cDNA Cloning

Virion purified from MHV-Y infected J774A.1 supernatants were diluted in buffer containing RNasin and RNA was purified by SDS/Proteinase K treatment, phenol-chloroform extraction and ethanol precipitation. Virion RNA was reverse transcribed and PCR-amplified using the GeneAmp RNA PCR kit (Perkin Elmer) according to the manufacturers instructions. The MHV-Y S gene was PCR amplified in three overlapping fragments. Primer sets used to amplify the 5' and 3' end of the S gene were designed on the basis of the published S gene sequences of MHV-A59 and MHV-JHM and primer set used to amplify the central portion of the S gene was designed on the basis of sequences determined from clones of the 5' and 3' ends of the MHV-Y S gene[13,14]. The PCR products were cloned into the pCR™II cloning vector using the TA cloning kit (Invitrogen).

DNA Sequencing

The sequence of the MHV-Y S gene was determined in both directions from double stranded plasmid DNA. To exclude sequence errors based on misincorporation by the Taq polymerase, multiple clones from different RT-PCR reactions were sequenced and the consistency of the sequence was confirmed in at least 3 clones. Sequence comparisons were performed with the aid of the Translate, Gap, Bestfit, Pileup, Pretty and Peptidestructure programs in the Genetic Computer Group Sequence Analysis Software Package. Glycosylation sites were defined by the motif NXT or NXS where X does not equal P.

RESULTS AND DISCUSSION

As a first step in the characterization of the MHV-Y S protein, an immunoblot of proteins from MHV-Y virions grown in J774A.1 cells was performed using monospecific polyclonal antiserum specific for the S protein (Figure 1). The immunoblot showed negligible amounts of cleaved MHV-Y S protein subunits. A low to undetectable level of cleavage of the MHV-Y S protein was also seen in purified MHV-Y virions treated with trypsin, cell lysates from MHV-Y infected J774A.1 cells and homogenates of MHV-Y infected infant mouse intestines.

Given the lack of efficient cleavage of the MHV-Y S protein, we wanted to determine whether a functional cleavage site existed in the MHV-Y S protein. The entire MHV-Y S gene was cloned in three overlapping fragments and clones were characterized by restriction enzyme mapping. Cleavage patterns for the MHV-Y clones differed substantially from those predicted from MHV-A59 and MHV-4/JHM S sequences, so the entire MHV-Y S gene was sequenced[13-15].

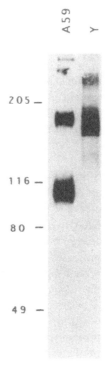

Figure 1. Immunoblots of MHV-A59 and MHV-Y virions. Virion proteins were separated on an 8% SDS-PAGE gel (Lanes: 1, MHV-A59 virions; 2, MHV-Y virions) and were probed with monospecific polyclonal goat antiserum to the S protein of MHV-A59. Bound antibody was detected with peroxidase-conjugated Staphylococcal protein A and chemiluminescent reagents. Positions of prestained protein markers are indicated on the margin.

Table 1. Summary of S gene characteristics

	MHV-4	MHV-A59	MHV-Y
Nucleotides	4128	3972	4083
Amino acids-total	1376	1324	1361
-S1	769	717	757
-S2	607	607	604
Pot. N-glycos. sites	22	21	22
Cleavage site	dysksRRARR/S	dysksRRAHR/S	nysttHRARR/S

The predicted MHV-Y S protein is 1361 amino acids long, 15 amino acid shorter than the MHV-4 S protein (Table 1). The 12 amino acids deleted in the S1 subunit of MHV-Y S protein were localized in 3 separate sites (amino acids 492, 515-522 and 574-576) within the region deleted in the MHV-JHM S protein[14]. The 3 amino acid deletion (amino acids 911-913) in the S2 subunit of the MHV-Y S protein is the first reported deletion in the S2 subunit of any MHV strain, though this deletion lies within a 7 amino acid region deleted from the S proteins of bovine coronaviruses and human coronavirus strain OC43[16-18]. Even though the S2 deletion maps to the region of the binding sites of monoclonal antibodies 5B19.2 and 5B170.3, both antibodies were capable of neutralizing MHV-Y infectivity in J774A.1 cells[19,20]. All four deletions found in the tissue culture-derived MHV-Y stocks were confirmed in cDNA clones derived from intestinal virus stocks of MHV-Y.

The nucleotide and amino acid sequence identities between MHV-Y and either MHV-4 or MHV-A59 (approximately 81% and 84% respectively) were lower than between MHV-4 and MHV-A59 (89% and 92% respectively). The MHV-Y S protein has 22 predicted glycosylation sites (Table 1), the same number predicted for the MHV-4 S protein, but the positions of only 20 predicted glycosylation sites for the MHV-Y S protein were conserved[14,15]. The predicted glycosylation sites at amino acids 582 and 709 of MHV-4 were absent from the MHV-Y S protein and were replaced by predicted glycosylation sites at amino acids 575 and 756 of the MHV-Y S protein. The N-terminal signal peptide located at amino acids 1 to 17 in the MHV-Y S protein was conserved relative to that of the MHV-4 S protein[14,15]. The hydrophobic membrane anchor sequence located at amino acids amino acids 1307 to 1322 of the MHV-Y S protein was also conserved. The KWPWYVWL motif found in all coronavirus S proteins was present in the MHV-Y S protein between amino acids 1299 and 1306[17].

The sequence adjacent to and including the cleavage signal sequence for the MHV-Y S protein (Table 1), differs from the sequence present in other MHV S proteins in several ways[13-15]. First, the MHV-Y S protein cleavage signal sequence differs from the consensus cleavage signal by the replacement of an arginine with a histidine (HRARR instead of RRARR). The presence of a histidine in this particular position within the cleavage signal sequence may alter interactions between the MHV-Y S protein and the active site of cellular proteases resulting in negligible MHV-Y S protein cleavage. Alternatively, the MHV-Y S protein signal cleavage signal (HRARR) may be functional but because of other changes in the region, this signal is not accessible to proteases. The KS to TT change at positions -6 and -7 relative to the cleavage site and/or the addition of a potential glycosylation site (NYS) at position -10 relative to the cleavage site may cause conformational changes within the protein resulting in the inaccessibility of the cleavage signal sequence to the trypsin-like proteases which cleave most MHV S proteins. It is interesting to note that even though the MHV-Y S protein is inefficiently cleaved, it still efficiently induces cell-cell fusion in many types of cells in culture and in vivo in the intestinal mucosa (unpublished data). This observation agrees with recent results indicating that cleavage of MHV S proteins may increase the efficiency of fusion but is not essential for fusion induction[17 26 28]

ACKNOWLEDGMENTS

We thank Frank Coyle for technical assistance and Lisa Ball-Goodrich for many helpful discussions This work was supported by NIH grant #RR02039

REFERENCES

1 Barthold S W , Smith A L Response of genetically susceptible and resistant mice to intranasal inoculation with mouse hepatitis virus JHM Virus Res 1987,7 225-239

2 Compton S R , Barthold S W , Smith, A L The cellular and molecular pathogenesis of coronaviruses Lab Anim Sci 1993,43 15-28

3 Barthold S W Host age and genotypic effects on enterotropic mouse hepatitis virus infection Lab Anim Sci 1987,37 36-40

4 Barthold S W , Beck D S , Smith A L Enterotropic coronavirus (mouse hepatitis virus) in mice influence of host age and strain on infection and disease Lab Anim Sci 1993,43 276-284

5 Dalziel R G , Lampert P W , Talbot P J , Buchmeier, M J Site-specific alteration of murine hepatitis virus type 4 peplomer glycoprotein E2 results in reduced neurovirulence J Virol 1986,59 463-471

6 Fleming J O , Trousdale M D , El-Zaatari F A K , Stohlman S A , Weiner, L A Pathogenicity of antigenic variants of murine coronavirus JHM selected with monoclonal antibodies J Virol 1986,58 869-875

7 Wege H , Winter J , Meyermann R The peplomer protein E2 of coronavirus JHM as a determinant of neurovirulence Definition of critical epitopes by variant analysis J Gen Virol 1988,69 87-98

8 Gallagher T M , Parker S E , Buchmeier, M J Neutralization-resistant variants of a neurotropic coronavirus are generated by deletions within the amino-terminal half of the spike glycoprotein J Virol 1990,64 731-741

9 Fazakerley J K , Parker S E , Bloom F , Buchmeier, M J The V5A13 1 envelope glycoprotein deletion mutant of mouse hepatitis virus type-4 is neuroattenuated by its reduced rate of spread in the central nervous system Virology 1992,187 178-188

10 Wang F-I , Fleming J O , Lai M M C Sequence analysis of the spike protein gene of murine coronavirus variants Study of genetic sites affecting neuropathogenicity Virology 1992,186 742-749

11 Hingley S T , Gombold J L , Lavi E , Weiss, S R MHV-A59 fusion mutants are attenuated and display altered hepatotropism Virology 1994,200 1-10

12 Barthold S W , Smith A L , Lord P F S , Bhatt P N , Jacoby R O , Main A J Epizootic coronaviral typhlocolitis in suckling mice Lab Anim Sci 1982,32 376-383

13 Luytjes W , Sturman L S , Breedenbeek P J , Charite J , van der Zeijst B A M , Horzinek M C , Spaan W J M Primary structure of the glycoprotein E2 of coronavirus MHV-A59 and identification of the trypsin cleavage site Virology 1987,161 479-487

14 Schmidt I , Skinner M A , Siddell, S G Nucleotide sequence of the gene encoding the surface projection glycoprotein of the coronavirus MHV-JHM J Gen Virol 1987,68 47-56

15 Parker S E , Gallagher T M , Buchmeier, M J Sequence analysis reveals extensive polymorphism and evidence of deletions within the E2 glycoprotein gene of several strains of murine hepatitis virus Virology 1989,173 664-673

16 Abraham S , Kienzle T E , Lapps W , Brian D A Deduced sequence of the bovine coronavirus spike protein and identification of the internal proteolytic cleavage site Virology 1990,176 296-301

17 Boireu P , Cruciere C , LaPorte J Nucleotide sequence of the glycoprotein S gene of bovine enteric coronavirus and comparison with the S proteins of two mouse hepatitis virus strains J Gen Virol 1990,71 487-492

18 Kunkel, F , Herrler, G Structural and functional analysis of the surface protein of human coronavirus OC43 Virology 1993,195 195-202

19 Luytjes W , Geerts D , Posthumus W , Meleon R , Spaan W Amino acid sequence of a conserved neutralizing epitope of murine coronaviruses J Virol 1989,63 1408-1412

20 Daniel C , Anderson R , Buchmeier M J , Fleming J O , Spaan W J , Wege H , Talbot, P J Identification of an immunodominant linear neutralization domain on the S2 portion of the murine coronavirus spike

glycoprotein and evidence that it forms part of a complex tridimensional structure J Virol 1993,67 1185-1194

21 Gombold J L , Hingley S T , Weiss, S R Fusion-defective mutants of mouse hepatitis virus A59 contain a mutation in the spike protein cleavage signal J Virol 1993,67 4504-4512

22 Stauber R , Pfleiderer M , Siddell S Proteolytic cleavage of the murine coronavirus surface glycoprotein is not required for fusion activity J Gen Virol 1993,74 183-191

23 Taguchi F Fusion formation by the uncleaved spike protein of murine coronavirus JHMV variant cl-2 J Virol 1993,67 1195-1202

4

BIOLOGICAL AND MOLECULAR DIFFERENTIATION BETWEEN CORONAVIRUSES ASSOCIATED WITH NEONATAL CALF DIARRHOEA AND WINTER DYSENTERY IN ADULT CATTLE

G. Millane, L. Michaud, and S. Dea

Centre De Recherche en Virologie
Institut Armand Frappier
Universite Du Québec
Laval, Québec
Canada, H7N 4Z3

ABSTRACT

Cytopathic coronaviruses were isolated in HRT-18 cells from bloody faecal samples collected from cows in Québec dairy herds with classical winter dysentery (WD). The formation of polykaryons in the infected cell cultures was found to be dependent on the presence of trypsin in the medium. Virus identification was confirmed by indirect immunofluorescence and indirect protein A-gold immunoelectron microscopy using rabbit hyperimmune serum, as well as monoclonal antibodies directed against the spike (S) and hemagglutinin-esterase (HE) glycoproteins of the prototype Mebus strain of bovine coronavirus (BCV-Meb). Four WD isolates differed from BCV-Meb by their ability to agglutinate rat erythrocytes at 4 and 37^0C, their higher receptor destroying enzyme activity, but lower acetylesterase activity. The WD isolates were serologically indistinguishable from the reference BCV-Meb strain by virus neutralization and Western immunoblotting, but could be differentiated by hemagglutination-inhibition. Sequence analysis of the PCR-amplified HE gene of a plaque-purified WD isolate (BCQ-2590) revealed sufficient number of nucleotide and amino acid substitutions which may explain this antigenic variability.

INTRODUCTION

Bovine coronavirus is known to cause significant economic losses throughout the world. It is an enteric virus which multiplies in the differentiated enterocytes of the small intestine and colon, causing severe diarrhoea in neonatal calf and chronic shedding in adult cattle.[1,2] Some recent reports have suggested that BCV also possesses a tissue tropism to the upper respiratory tract of calf causing pneumonia[3]. BCV is also associated to acute enteric infection in adult cattle during the winter season. The disease, known as winter dysentery (WD), is clinically characterized by an explosive apparition of acute diarrhoea in adult dairy and beef cattle, with dark bloody liquid diarrhoea, accompanied by decreased milk production and variable depression and anorexia[4]. The disease spreads rapidly to animal of all ages causing high morbidity but low mortality.

Although the etiological agent has not been conclusively identified, early investigation attributed the disease to *Campylobacter fetus* subspecies *jejuni*[5] but recent reports have described coronavirus particles in the faeces of WD affected cattle[2,4,6]. The coronavirus identified were isolated in cell cultures and shown to be serologically related to the BCV-Meb strain[7].

BCV is a well characterized hemagglutinating coronavirus. The viral particle is mostly spherical, enveloped and displays a double fringe of projections [8,9]. The viral genome consists of a large single-stranded RNA with positive polarity, approximately 30 kb in length, and encodes four major structural proteins: the nucleocapsid protein (N), the integral membrane protein (M), the spike glycoprotein (S) and the hemagglutinin/esterase glycoprotein (HE) which is associated to the acetylesterase activity (AE)[9,10].

Although WD isolates can be distinguished for their pathogenicity, several studies using different methods based on monoclonal and/or polyclonal antibodies[4,7], have shown that they are antigenically related to the BCV-Meb strain. However, possible antigenic difference between them was suggested[4] on the basis of virus neutralisation (VN). But there is still some controversy as to the existence of distinct BCV serotypes[4,11,12].

The purpose of this study was to characterize BCV strains isolated in cell cultures from bloody faecal samples of the WD affected cattle, in comparison to reference strains of neonatal calf diarrhoea in Québec (Canada). The cytopathogenicity on HRT-18 cells, infectivity titers, enzymatic activities and serological crossreactivities of Québec WD isolates were compared to those of the reference BCV-Meb strain. The HE gene of a Québec WD and two NCD isolates were also sequenced and their deduced amino-acid sequences were compared.

METHODOLOGY

The prototype Mebus strain of BCV[8], and the 67N strain of the porcine hemagglutinating encephalomyelitis virus (HEV) were obtained from the American Type Culture Collection (ATCC VR-874 and ATCC VR-740, respectively). The human HCV-OC43 respiratory coronavirus was provided to us by P. Talbot (IAF, Laval, Qc, Canada). The other coronavirus isolates were recovered from Québec herds experiencing clinical cases of Neonatal Calf Diarrhoea (NCD) during winter 1989 or typical outbreaks of WD during winters 1992 and 1993. All the BCV strains were propagated in HRT-18 cells and passaged not more than 5 times in the presence of 10 U/ml of bovine pancreatic trypsin[9]. The extracellular virions were purified from the supernatants of infected cell cultures by differential and isopycnic ultracentrifugation on sucrose gradients[9,13]. A rabbit hyperimmune serum was prepared against the prototype BCV-Meb strain according to immunization protocole described elsewhere[9]. Infectivity titers, and enzymatic activities were determined on purified BCV preparations, as previously described[9,14]. The HA activity was determined on rat erythrocytes at 4° and 37°C[9], and the antigenic relationships between WD and NCD strains were evaluated using VN and HAI tests[9].

Cloning and sequencing of the HE genes of two Québec NCD isolates (BCQ.3 and BCQ.571) and one WD isolate (BCQ.2590), were also done as previously described[15]. Their complete nucleotide and predicted amino-acid sequences were compared with those of reference BCV-Mebus strain. Sequence analyses were performed using the MacVector 3.5 (International Biotechnologies) and GeneWorks 2.2 (IntelliGenetics Inc) programs.

RESULTS AND DISCUSSION

The NCD and WD isolates of BCV were successfully propagated in HRT-18 cells in the presence of 10 U/ml of trypsin. Their biological and serological characteristics are summarised in Table 1. On the basis of cytopathogenicity on HRT-18 cells, Québec strains could be classified into weak fusogenic (induced production of small syncytia), highly fusogenic (syncytia increased in number and size) leading to complete destruction of the cell sheets 3 to 4 days after infection, and non fusogenic but highly cytolytic (no syncytia but intense degenerescence of the monolayers).

The hemagglutination (HA) titers obtained with rat erythrocytes were quite similar for the WD and NCD isolates after 1 hour incubation at 4^0C (results not shown). However a drastic drop in HA titers of WD isolates occurred if the incubation temperature was raised to 37^0C, whereas temperature did not seem to affect the HA activity of NCD isolates.

Although the Québec WD and NCD isolates were serologically indistinguishable from BCV-Meb strain by virus neutralisation (VN), 16 to 32 fold differences in HAI titer of the specific anti-BCV hyperimmune serum were obtained between the WD and the reference BCV-Meb strain. The reference anti-BCV serum reacted with same HAI titers to all NCD isolates tested, except BCQ.571 and BCQ.2070. No reactivity of the BCV hyperimmune

Table 1. Biological and serological characteristics of cell culture adapted-NCD and -WD bovine coronaviruses

Viral isolate	Disease	Type of CPE[a]	Infectivity titers [b]	Cross-reactivity [c]		Acetyl-esterase[d]	RDE titer[e]	
				VN	HAI		Rooster	Rat
BCQ 2439	WD	C	7 2	2560	80	3 292	16	8
BCQ 2442	WD	C	7 2	2560	80	3 193	ND	< 2
BCQ 2508	WD	C	8 7	1280	80	3 089	4	64
BCQ 2590	WD	C	9 4	2560	80	0 764	64	128
BCQ 7373	WD	C	8 7	NT	160	2 151	64	256
BCQ 3	NCD	B	8 4	1280	1280	0 492	32	< 2
BCQ 9	NCD	B	9 7	2560	1280	1 826	16	2
BCQ 189	NCD	A	8 9	1280	1280	3 388	64	< 2
BCQ 571	NCD	C	7 7	2560	80	3 210	8	< 2
BCQ 20	NCD	A	7 5	1280	40	1 782	2	< 2
BCQ 2070	NCD	C	7 9	2560	1280	1 877	32	< 2
BCV Meb	NCD	A	9 4	2560	1280	3 155	8	2

[a]On the basis of cytopathic changes induced in HRT-18 cells, Quebec BCV isolates could be classified into weak fusogenic (A), highly fusogenic (C), and non-fusogenic but highly cytolytic (B) strains
[b]$TCID_{50}$ 50% tissue culture infective dose
[c]Reciprocal of highest dilution of polyclonal anti-BCV serum (Mebus strain) that inhibited 100 $TCID_{50}$ or 4 hemagglutinating units of virus WD winter dysentery, NCD neonatal calf diarrhoea, NT non tested
[d]Optical density at 405 nm after 5 min of reaction with 1 mM p-nitrophenyl acetate

serum was demonstrated when tested against the heterologous hemagglutinating coronaviruses (HEV-67 and HCV-OC43).

To further investigate on the biological properties of WD isolates in comparison to
NCD isolates, different viral functions were assessed on purified BCV isolates containing
infectivities of $10^{7.2}$ to $10^{9.4}$ TCID$_{50}$/mL (Table 1). Although high variability was observed
amongst the infectivity titers of the BCV strains tested, they appeared to possess similar AE
activity with the exception of Québec isolates BCQ.3 and BCQ.2590 which possessed a
weak AE activity. The RDE (receptor destructive enzyme) titers determined for WD strains
with rat erythrocytes varied from 8 to 256, but the RDE activity was minimal or not detectable
for the NCD isolates. All the BCV tested demonstrated lower HA activity with rooster
erythrocytes (results not shown) than with rat erythrocytes, but elution through RDE activity
was detected with titers of 16 to 64 with rooster erythrocytes.

Previous studies suggested close resemblance between NCD and WD strains. The
present results provide additional data on the biological and serological properties of these
BCV isolates. As previously reported[7], the WD isolates were highly fusogenic. They induced
a severe cytopathic effect in comparison to the avirulent BCV strains. Interestingly two
virulent BCV strains (BCQ.571 and BCQ.2070) behaved also as highly fusogenic strains.
Changes in the amino acid sequences of the S glycoproteins between these two strains and
the reference BCV-Meb strain were recently reported[15]. Whether these differences at the
molecular level could be related to the BCV virulence remains to be elucidated. Our results
could also differentiate between WD and NCD strains on basis of their interaction with
different erythrocytes. Both NCD and WD isolates agglutinated rat erythrocytes with similar
titers at 4^0C, but a drastic drop was noticed in the HA titers of WD isolates at 37^0C. This
difference regarding HA activity could be related to receptor inactivation as demonstrated
by the RDE activity of HE protein which was more effective in tests involving WD than
NCD strains. The higher RDE activity could explain the high contagiousness of WD strains
and the short duration of the syndrome in the affected herd. On the other hand, the difference
in the RDE activity between rat and rooster erythrocytes could be due to a variation in
receptor binding properties or to structural differences in the receptors between the species
as it was already suggested[14]. If receptor binding and viral attachment to susceptible cells is
attributed to the S glycoprotein[16], the AE of the HE protein may play a role in virus release
from infected cells and viral spread.

Since differences identified between WD and NCD isolates appeared to be associated
to biological functions and antigenicity of their HE proteins, we investigated on a possible
explanation at the molecular level. The HE genes of two Québec NCD (BCQ.3 and BCQ.571)
and one WD (BCQ.2590) isolates, were cloned and sequenced. Their complete nucleotide
and deduced amino-acid sequences were compared with those of the reference BCV-Meb
strain.

At the nucleotide level, a high degree of similarity was demonstrated among Québec
BCV isolates BCQ.3, BCQ.571 and BCQ.2590. The only variation that appeared consisted
of 25 nucleotide substitutions which represented 5.8% of the entire HE gene sequence.
Among these substitutions, 11 appeared to be specific to the WD isolate and the deduced
amino acids sequence are interesting (results not shown). Three proline substitutions occurred between WD and the reference Mebus isolates. The first one is localized in the signal
peptide (at aa 5) and the second one at aa 377. These two substitutions were already identified
for the virulent BCV-LY138 strain[17]. The third substitution seemed to be specific to the WD
isolate. It was localized (at aa 53) in the vicinity of the sequences of the putative esterase
active domain (FGDS) which was conserved in all BCV strains analyzed.

Our results did not permit to identify major differences at the molecular level between
WD and NCD associated strains. Whether the proline substitutions in HE protein gene could
result in conformational changes[18] causing an alteration in the viral interaction with surface

receptors remain to be elucidated Further genetic characterisation of additional isolates from various geographical areas is necessary to determine the extent of variability at the genomic level

Monoclonal antibodies directed against WD strain BCQ 2590 were recently produced they were found to be able to differentiate between NCD and WD isolates using HAI tests Other studies consisting of identifying their respective antigenic determinants or epitopes are presently in progress

REFERENCES

1 Dea S , Roy R S , Elazhary Y La diarrhee neonatale due au coronavirus du veau Une revue de litterature Can Vet J 1981,22 51-58

2 Durham P J K , Hassard L E , Armstrong K R , Naylor J M Coronavirus-associated diarrhoea (winter dysentery) in adult cattle Can Vet J 1989,30 825-827

3 McNulty M S , Bryson D G , Allan G M , Logan E F Coronavirus infection of the bovine respiratory tract Vet Microbiol 1984,9 425-434

4 Saif L J, Brock, K V Redman D R , Kohler E M Winter dysentery in dairy herds electron microscopic and serological evidence for an association with coronavirus infection Vet Rec 1991,128 447-449

5 Campbell S G , Cookingham C A The enigma of winter dysentery Cornell Vet 1978,68 423-441

6 Takahashi E , Inaba Y , Sato K , Ito Y , Kuroci H , Akashi H , Satoda K , Omori T Epizootic diarrhoea of adult cattle associated with a coronavirus-like agent Vet Microbiol 1980,5 151-154

7 Benfield D A , Saif L J Cell culture propagation of a coronavirus isolated from cows with winter dysentery J Clin Microbiol 1990,28 1454-1457

8 Mebus C A , Stair E L , Rhodes M B , Twiehaus M J Neonatal calf diarrhoea propagation, attenuation, and characteristics of a coronavirus-like agent Am J Vet Res 1973, 34 145-150

9 Dea S , Verbeek A J , Tijssen P Antigenic and genomic relationships between turkey and bovine enteric coronavirus J Virol 1990,64 3112-3118

10 Cavanagh D , Brian D A, Enjuanes L , Holmes K V, Lai M M C , Laude H , Siddell S G , Spaan, W , Taguchi F , Talbot P Recommendation of the coronavirus study group nomenclature of the structural proteins, mRNAs and genes of coronaviruses Virology 1990 176 306-307

11 Reynolds D J , Debney T G , Hall G A , Thomas, L H , Parsons K R Studies on the relationship between coronaviruses from the intestinal and respiratory tracts of calves Arch Virol 1985,85 71-83

12 Michaud L , Dea S Characterization of monoclonal antibodies to bovine enteric coronavirus and antigenic variations among Quebec isolates Arch Virol 1993,131 455-465

13 Dea S , Roy R S , Begin M E Bovine coronavirus isolation in continuous cell lines Amer J Vet Res 1980,41 30-38

14 Storz J Zhang X M , Rott R Comparison of hemaglutinating, receptor destroying and acetylesterase activities of avirulent and virulent bovine coronavirus strains Arch Virol 1992,125 193-204

15 Rekik M R , Dea S Comparative sequence analysis of a polymorphic region of spike protein gene of bovine coronavirus isolates Arch Virol 1994,135 319-331

16 Spaan W , Cavanagh D , Horzinek M C Coronaviruses structure and genome expression J Gen Virol 1988,69 2939-2952

17 Zhang X , Kousoulas K G , Storz J The hemagglutinin/esterase glycoprotein of bovine coronavirus sequence and functional comparisons between virulent and avirulent strains Virology 1991,185 847-852

18 Yaron A , Naider F Proline-dependent structural and biological properties of peptide and proteins Critical Rev Biochem Mol Biol 1993,28 31-81

5

MOLECULAR DIFFERENTIATION OF TRANSMISSIBLE GASTROENTERITIS VIRUS AND PORCINE RESPIRATORY CORONAVIRUS STRAINS

Correlation with Antigenicity and Pathogenicity

D. J. Jackwood, H. M. Kwon, and L. J. Saif

Food Animal Health Research Program
The Ohio Agricultural Research and Development Center
The Ohio State University
Wooster, Ohio 44691

ABSTRACT

Transmissible gastroenteritis virus (TGEV) causes an economically important enteric disease of swine. Differences in the pathogenicity, antigenicity and tissue tropism have been observed among porcine coronaviruses. Although porcine respiratory coronavirus (PRCV) is antigenically similar but not identical to TGEV isolates, these respiratory coronaviruses differ markedly in pathogenicity and tissue tropism compared to TGEV isolates. Using a reverse transcriptase/polymerase chain reaction-restriction fragment length polymorphism (RT/PCR-RFLP) assay, TGEV and PRCV isolates were assigned to several distinct groups. By RFLP analysis of the 5' region of the S gene, TGEV strains were differentiated into 4 groups using the restriction enzyme Sau3AI. A fifth Sau3AI group contained the PRCV isolates. These 5 groups correlated with antigenic groups previously defined using mono-clonal antibodies in our laboratory. Several restriction enzymes could be used to differentiate the TGEV strains into Miller and Purdue types. Analysis of a PCR amplified product in the 3 and 3-1 genes indicated the RT/PCR-RFLP assay results for TGEV Miller strains could be correlated with lower virulence created by passage in cell culture.

INTRODUCTION

The antigenicity, pathogenicity and tissue tropism vary among the porcine coro-naviruses, transmissible gastroenteritis virus (TGEV) and porcine respiratory coronavirus (PRCV). Pathogenic differences among TGEV isolates have been documented[1, 2, 3, 4]. The

Corona- and Related Viruses, Edited by P. J. Talbot and G. A. Levy
Plenum Press, New York, 1995

molecular basis for virulence and tissue tropism was reported to reside in the region of RNA 3/ 3-1[2, 4] and in the region of the S glycoprotein gene[5, 6, 7], respectively. Britton *et al.*[2] reported that a deletion in RNA 3/3-1 (ORF-3a/3b) was observed in the attenuated TGEV strain 188-SG but not in the virulent D-52 strain of TGEV. Wesley *et al.*[4] reported that a similar deletion in RNAs 3/3-1 and 4 was present in an attenuated small-plaque variant of TGEV.

Certain MAbs to TGEV have been used to demonstrate antigenic differences among TGEV and PRCV strains[8, 9, 10, 11, 12, 13]. Mabs directed to a non-neutralizing epitope on the S glycoprotein of TGEV designated site B[8] or site D[9, 12, 13] did not bind to PRCV isolates but did identify TGEV strains. A panel of 12 MAbs generated against the virulent Miller TGEV strain was used to demonstrate differences in antigenicity among TGEV isolates[12, 13, 14 15, 16]. Five non-neutralizing MAbs were directed against the N protein and 4 neutralizing MAbs were directed against the S glycoprotein[12, 13, 14]. The neutralizing MAbs had different titers against heterologous (Purdue) and homologous (Miller) strains of TGEV indicating variability among these epitopes.

Differences in the tissue tropism of TGEV and PRCV isolates have been extensively characterized[7, 17, 18, 19, 20, 21]. Molecular differences in the S glycoprotein gene appear to affect the tissue tropism of these viruses[7, 17].

This study was initiated to determine if molecular differences observed using the reverse transcriptase/polymerase chain reaction-restriction fragment length polymorphism (RT/PCR-RFLP) assay could be correlated with differences in antigenicity, pathogenicity and tissue tropism among TGEV and PRCV viruses.

MATERIALS AND METHODS

Viruses

The M5C virulent strain was initially isolated from an outbreak of TGE in a local swine herd (Miller). It has been maintained by 5 serial passages in gnotobiotic pigs[16] and represents an intestinal suspension of the fifth passage in gnotobiotic pigs. The M6 virulent TGEV represents a plaque purified low cell-culture passage (6 times) of the Miller virulent strain of TGEV in swine testicular (ST) cells[16, 22]. The virus has retained its pathogenicity in gnotobiotic pigs (L.J. Saif, unpublished). The M60 Miller strain TGEV was passaged 60 times in ST cells[23] and has low pathogenicity for gnotobiotic pigs compared to the M5C and M6 strains (L. J. Saif, unpublished).

Nine field isolates of TGEV designated S387, T184, T232, T507, T517, T876, T988, U328, and Zy were described[12]. These viruses were obtained from Ohio, Canada, Nebraska, South Dakota, and Michigan (Table 1). Each isolate was obtained from swine with clinical signs of TGE and TGEV-positive immunofluorescence staining on gut tissue samples. The isolates which were confirmed to be virulent by passage in susceptible pigs include S387, T232, T876, U328, and Zy.

The ISU-1 (Ind/89) and ISU-3 (NC/89) strains of PRCV provided by Dr. H. Hill, Iowa State Univ., Ames, Iowa[24] were plaque-purified twice and passaged 8-14 times in ST cells. These viruses produce only subclinical infections when inoculated into gnotobiotic or conventional pigs and replicate exclusively in the upper respiratory tract[21, 25]. The PRCV strain designated DD312 was recently isolated from the respiratory tract of a pig in the United States (Saif, Weinau and Gadfield, unpublished).

Monoclonal Antibodies (MAbs)

TGEV-specific MAbs were produced and characterized in our laboratory[12, 13, 14, 15, 16]. These include neutralizing MAbs to sites A, B, and E of the S glycoprotein conserved on

Table 1. Isolation and passage history of porcine coronaviruses used in this study[12]

Isolate	Virulence[a]	Isolated		P#(PP)b	Source
		Data	State		
Reference Strains					
M5C Miller	Virulent	1965	Ohio	2(2)	E Bohl, OARDC, Wooster, OH
M6 Miller	Virulent	1965	Ohio	6(2)	L Saif, OARDC, Wooster, OH
M60 Miller	Attenuated	1987	Ohio	60(2)	R Woods, USDA Ames, Iowa
W184 Purdue	Virulent	1952	Indiana	4(1)	E Haelterman, Purdue Univ Purdue, Indiana
P115 Purdue	Attenuated	1965	Ohio	>115[c]	E Bohl, OARDC, Wooster, Ohio
CC1861	Attenuated	1972	Nebraska	?	M Welter, Ambico Vaccine Strain
TGEV Field Strains [d]					
S387	Virulent	1987	Ohio	3(0)	NG Herd, Bucyrus, Ohio
T184	?	1988	Canada	3(0)	P S Carman, Ontario Ministry of Agriculture and Food, Guelph, Ontario
T232	Virulent	1988	Ohio	6(0)	OARDC Swine Center, Wooster, Ohio
T507	?	1988	Nebraska	3(0)	R Moxley, Univ Nebraska
T517	?	1988	Nebraska	6(0)	R Moxley, Univ Nebraska
T876	Virulent	1988	Ohio	3(0)	FF Herd, Wauseon, Ohio
T988	?	1987	S Dakota	2(0)	D Benfield, S Dakota State Univ
U327	Virulent	1989	Michigan	6(0)	R Macs, Michigan State Univ
Zy	Virulent	1986	Ohio	6(0)	Zy Herd, Burbank, Ohio
PRCV Strains					
ISU-1		1990	Indiana	8(2)	H Hill, Iowa State Univ
ISU-3		1990	N Carolina	6(2)	H Hill, Iowa State Univ
DD312		1994	Illinois	4(0)	L Saif, OARDC, Wooster, Ohio

[a]Virulent = enteropathogenic, Attenuated = not enteropathogenic ? = pathogenicity was not confirmed although virus was isolated from a diarrheic pig from a typical TGE outbreak The virulence was determined by passage in susceptable or gnotobiotic pigs
[b]Number of times passaged in cell culture (Number of times plaque-purified) ? indicates the passage number in cell culture and the number of times the virus was plaque purifed is unknown
[c]P115 was plaque purified numerous times during passage in cell culture
[d]All field strains of TGEV were isolated from pigs with transmissible gastroenteritis confirmed using an immunofluoresence assay on intestinal samples

all strains of TGEV and PRCV tested and non-neutralizing MAbs (44C11 and 45A8) to site D of the S glycoprotein, reactive with TGEV strains but non-reactive with PRCV strains[12] Other non-neutralizing MAbs to the S glycoprotein include 11H8, 8G11 and 75B10 The latter two react with site V, conserved on Miller but not Purdue strains[12]

Reverse Transcriptase/Polymerase Chain Reaction-Restriction Fragment Length Polymorphism (RT/PCR-RFLP) Assay

Viral RNA was extracted and purified using proteinase K, acid phenol and the RNaid kit (BIO 101, La Jolla, CA) Briefly, ST cell culture solutions inoculated with the appropriate viruses were incubated for 5 minutes at 55°C in a solution containing 2% SDS and 250 µg/ml proteinase K Following incubation the samples were extracted with acid phenol and chloroform/isoamyl alcohol The RNaid kit was used to purify viral RNA according to the manufacturers instructions Purified RNA was suspended in diethyl-pyrocarbonate (DEPC) treated water and stored at -70°C before use in the reverse transcriptase (RT) reaction

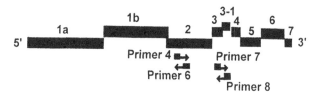

Figure 1. Schematic representation of the RNA genome of TGEV Miller strain, showing the location of the primers used for PCR. Primer set 4/6 amplifies a fragment of the S gene and primer set 7/8 amplifies RNAs 3 and 3-1.

The nucleotide sequences of the oligonucleotide primers were deduced and synthesized according to the published TGEV Miller strain sequence. The locations of primers used in this study are shown in Figure 1.

The RT reaction contained 2 μl of 10 X PCR buffer (500 mM KCl, 200 mM Tris, pH 8.4, 0.5 mg/ml nuclease-free bovine serum albumin), 2 μl of 10 mM each dNTP, 250 ng antisense primer, 40 units RNasin, 1.5 μl of 60 mM MgCl$_2$, and 3-5 μl of the purified viral RNA described above. A 20 μl total reaction volume was obtained by adding sterile DEPC-treated water. The reaction mixture was heated at 65°C for 10 minutes before 200 units of Moloney murine leukemia virus RT was added. The reaction was then incubated for 1 hour at 45°C. Following incubation, the reaction was stopped by heating to 95°C for 5 minutes.

The PCR reaction contained 8 μl of 10 X PCR buffer, 250 ng of the sense primer, 3.5 μl of 60 mM MgCl$_2$, and 2.5 units of Taq DNA polymerase. A 100 μl total reaction volume was obtained by adding sterile distilled water. Thirty-five cycles of denaturation at 94°C for 1 minute, annealing at 45°C for 2 minutes, and polymerization at 74°C for 5 minutes were conducted. The initial denaturation and polymerization steps were at 94°C for 5 minutes and 74°C for 6 minutes, respectively. The final polymerization step was conducted at 74°C for 10 minutes. The PCR products were analyzed on a 1% agarose gel containing ethidium bromide.

In most cases it was not necessary to gel purify PCR products before conducting the RE analysis. However, depending on the restriction enzyme used, purification was sometimes necessary. The total PCR reaction product was separated on a 1% agarose gel. The amplified DNA with the appropriate size was cut from the gel and purified using the Geneclean kit (Bio 101) according to the manufacturer's recommendation. The purified DNA was digested with selected REs according to the manufacturer's specification. The restriction fragment patterns were observed following electrophoresis on a 2% agarose gel.

RESULTS AND DISCUSSION

The TGEV primer pair 4/6 amplified a portion of the S glycoprotein gene which was approximately 1.5 kb. The size of the fragments from all TGEV and PRCV strains was similar indicating no major deletions or insertions. Using *Ssp*I to digest the PCR products amplified with primer pair 4/6 produced two RFLP patterns. Similar results were obtained using the enzyme *Dde*I. TGEV strains were placed into two groups using these enzymes. The first group was characterized by the Miller strain viruses and the second group by the Purdue strain viruses. When the enzyme *Sau*3AI was used, the viruses were placed into five groups (Table 2). Group 1 contained the Miller strain viruses and the PRCV strain DD312. The P115 Purdue strain, two TGEV field strains and the Ambico vaccine strain all contained similar RFLP patterns and were placed in group 2. Group 3 contained TGEV field strains and group

Table 2. Correlation of the RT/PCR-RFLP results using primer pair 4/6 and *Sau*3AI enzyme with the ELISA and CCIF results using monoclonal antibodies (MAb) previously described[12]

*Sau*3AI Groups	Viruses	MAb Defined[a] Antigenic Groups
1	Miller (M6), Miller (M5C) PRCV (DD312)[b]	8G11 Binding (Miller specific) 75B10 Binding 11H8 Binding 45A8/44C11 Binding
2	Purdue (P115), TGEV Field Isolates: (T876, Zy) Ambico strain (CC1861)	8G11 Non-binding 75B10 Non-binding 11H8 Non-binding 45A8/44C11 Binding
3	TGEV Field Isolates: (S387, T232, T507, T517, U328, T988, CC717)	8G11 Non-binding 75B10 Non-binding (except U328) 11H8 Binding (except U328) 45A8/44C11 Binding
4	TGEV Field Isolates: (T184, W184)	8G11 Weak Binding 75B10 Non-binding 11H8 Weak Binding 45A8/44C11 Binding
5	PRCV (ISU-1, ISU-3)	8G11 Non-binding 75B10 Non-binding 11H8 Binding 45A8/44C11 Non-binding

[a]Monoclonal antibodies (MAbs) were generated to the S glycoprotein of the virulent Miller TGEV strain and used to characterize viruses in ELISA and cell culture immunofluorescence assays[12].
[b]The PRCV strain DD312 has not been thoroughly tested with the MAbs listed.

4 contained the virulent Purdue strain (W184) and a field isolate from Canada (T184). Group 5 contained the two Indiana PRCV strains ISU-1 and ISU-3.

The five *Sau*3AI groups could not be correlated with virulence of the TGEV isolates because confirmed virulent and attenuated viruses were observed in three of the four TGEV groups. The viruses tested appeared to fall into groups with similar origins. For example, the Miller strain viruses were grouped together, the Purdue origin vaccine strains were in a separate group and the PRCV strains were grouped together. There were two exceptions; the PRCV strain DD312 was grouped with the Miller strains and the virulent Purdue strain was placed in group 4 with a field isolate from Canada (T184).

Although the five RFLP groups defined by the enzyme *Sau*3AI did not correlate with pathogenicity or tissue tropism, they did correlate with antigenicity as defined using MAbs. In our previous studies, MAbs prepared to the Miller strain TGEV were used to differentiate five reference strains and nine field strains of TGEV[12]. The panel of MAbs used differentiated the viruses into seven distinctly different antigenic groups. Five of these MAb defined groups correlated with the five *Sau*3AI groups (Table 2). Although the DD312 PRCV strain was placed in group 1 with the Miller TGEV strains using RFLP, data on the MAb reactivity of this virus has not been completed.

The RFLP and MAb results[12] indicated that strains of TGEV currently endemic in the U.S. are antigenically different from the Miller strain and Purdue strains of TGEV. The variability observed was in epitopes located on subsite V of the S glycoprotein gene[12].

The history of TGEV field isolates Zy and T876 is very interesting in the context of their RFLP (*Sau*3AI) and MAb (11H8, 8G11) assignment into a group with the Purdue attenuated TGEV strains These two virus isolates came from herds with a history of using commercial modified live TGEV vaccines (Ambico) However, the virulence of both strains has been confirmed by passage in gnotobiotic pigs

The region which spans RNA segments 3 and 3-1 was amplified using PCR primer pair 7/8 (Fig 1) The length of the 7/8 PCR products from PRCV strains was smaller than the TGEV strains with the one exception that the ISU-3 strain PCR product was similar in length to the Miller strain PCR product The virulent M6 Miller strain could be differentiated from the P115 strain because the M6 PCR product was 43 bases shorter than that observed for P115 (data not shown) The M60 strain did not amplify with the 7/8 primer pair due to a 531 base deletion in the 3-1 RNA region (data not shown) This region was reported to be important in defining the virulence of TGEV isolates[4] The NS3 open-reading-frame was non-functional in PRCV isolates[20] and an attenuated small plaque variant of the Miller TGEV suggesting a possible relationship with viral attenuation[4] Our results using the Miller virulent and attenuated strains would indicate that differences in the size of the RNA 3 and 3-1 regions can be correlated with virulence and attenuation of these viruses

REFERENCES

1 Cubero M J , Bernard S , Leon L , Berthon P and Contreras A Pathogenicity and antigen detection of the nouzilly strain of transmissible gastroenteritis coronavirus, in 1-week-old piglets J Comp Path 1994,106 61-73

2 Britton P , Kottier S , Chen C -M , Pocock D H , Salmon H , and Aynaud J M The use of PCR genome mapping for the characterisation of TGEV strains In Coronaviruses (H Laude, and J F Vautherot Eds) Plenum Press, New York 1994 pp29-34

3 Saif L J , and Wesley R Transmissible gastroenteritis In Diseases of Swine Eds A D Leman, B Straw, R D Glock, W L Mengeling, R H C Penny, and E Scholl Iowa State Univ Press, Ames, Iowa 1992 pp 362-386

4 Wesley R D , Woods R D , and Cheung A K Genetic basis for the pathogenesis of transmissible gastroenteritis virus J Virol 1990, 64 4761-4766

5 Enjuanes L , Sune C , Gebauer F , Smerdou C , Camacho A , Anton I M , Gonzalez S , Talamillo A , Mendez A , Ballesteros M L and Sanchez C Antigen selection and presentation to protect against transmissible gastroenteritis coronavirus Vet Microbiol 1992,33 249-262

6 Gebauer F , Posthumus W P A , Correa I , Sune C , Smerdou C , Sanchez C M , Lenstra J A , Meloen R H , and Enjuanes L Residues involved in the antigenic sites of gastroenteritis coronavirus S glycoprotein Virology 1991, 183 225-238

7 Sanchez, C M , Gebauer F , Sune C , Mendez A , Dopazo J , and Enjuanes L Genetic evolution and tropism of transmissible gastroenteritis coronavirus Virol 1992, 190 92-105

8 Callebaut P , Correa I , Pensaert M , Jimenez G , and Enjuanes L Antigenic differentiation between transmissible gastroenteritis virus of swine and a related porcine respiratory coronavirus J Gen Virol 1988,69 1725-1730

9 Delmas B , Gelfi J , and Laude H Antigenic structure of transmissible gastroenteritis virus II Domains in the peplomer glycoprotein J Gen Virol 1986, 67 1405-1418

10 Diego M D , Laviada M D , Enjuanes L , and Escribano J M Epitope specificity of protective lactogenic immunity against swine transmissible gastroenteritis virus J Virology 1992, 66 6502-6508

11 Hodatsu T , Eiguchi Y , Tsuchimoto M , Ide S , Yamagishi H , and Matumoto M Antigenic Variation of Porcine Transmissible Gastroenteritis Virus Detected by Monoclonal Antibodies Vet Microbiol 1987,14 115-124

12 Simkins R A , Weilnau P A , Bias J , and Saif L J Antigenic variation among transmissible gastroenteritis virus (TGEV) and porcine respiratory coronavirus (PRCV) strains detected with monoclonal antibodies to the S protein of TGEV Am J Vet Res 1992,53 1253-1258

13 Simkins R A , Weilnau P A , VanCott J , Brim T A , and Saif L J Competition ELISA, using monoclonal antibodies to the transmissible gastroenteritis virus (TGEV) S protein, for serologic differentiation of pigs infected with TGEV or porcine respiratory coronavirus Am J Vet Res 1993,54 254-259

14 Simkins R A , Saif L J , and Weilnau P A Epitope mapping and detection of transmissible gastroenteritis viral proteins in cell culture using biotinylated monoclonal antibodies in a fixed-cell ELISA Arch Virol 1989,107 179-190

15 Welch, S K W and Saif L J Production and characterization of monoclonal antibodies to transmissible gastroenteritis virus of pigs (Abstract) 67th Ann Mtg Conf Res Workers in Animal Disease, 1986 pp 9

16 Welch, S K W and Saif L J Monoclonal Antibodies to a Virulent Strain of Transmissible Gastroenteritis Virus Comparison of Reactivity Against the Attenuated and Virulent Virus Strains Arch Virol 1988,102 221-236

17 Enjuanes L , Sanchez C Gebauer F , Mendez A , Dopazo J , and Ballesteros M L Evolution and tropism of transmissible gastroenteritis coronavirus In Coronaviruses (H Laude, and J F Vautherot Eds) Plenum Press, New York 1994 pp35-42

18 O'Toole D , Brown I , Bridges A , and Cartwright S F Pathogenicity of experimental infection with pneumotropic porcine coronavirus Res Vet Sci 1989,47 23-29

19 Pensaert M B , Caillebaut P , and Vergote J Isolation of a new porcine respiratory, nonenteric coronavirus related to transmissible gastroenteritis Vet Q , 1986, 8 257-261

20 Rasschaert D , Duarte M , and Laude H Porcine respiratory coronavirus differs from transmissible gastroenteritis virus by a few genomic deletions J Gen Virol 1990,71 2599-2607

21 Wesley R D , Woods R D , Hill H T , Biwer J D Evidence for a porcine respiratory coronavirus, antigenically similar to transmissible gastroenteritis virus, in the United States J Vet Diagn Invest 1990,2 312-317

22 Bohl, E H and Saif L J Passive immunity in transmissible gastroenteritis of swine immunoglobulin characteristics of antibodies in milk after inoculating virus by different routes Infect Immunol 1975,11 23-31

23 Woods R D Humoral and cellular responses in swine exposed to transmissible gastroenteritis virus Am J Vet Res 1979, 40 108-110

24 Hill H , Biwer J , Wesley R D , and Woods R D Porcine respiratory coronavirus in the United States Abstr 70th Conf Res Workers in Animal Dis 1989, (242)

25 VanCott J L , Brim T A , Simkins R A , and Saif L J Isotype-specific antibody secreting cells to transmissible gastroenteritis virus and porcine respiratory coronavirus in gut- and bronchus-associated lymphoid tissues of suckling pigs J Immunol 1993,3990-4000

6

ORGAN SPECIFIC ENDOTHELIAL CELL HETEROGENEITY INFLUENCES DIFFERENTIAL REPLICATION AND CYTOPATHOGENICITY OF MHV-3 AND MHV-4

Implications in Viral Tropism

J. Joseph, R. Kim, K.Siebert, F.D. Lublin, C. Offenbach, and R.L. Knobler

Division of Neuroimmunology
Department of Neurology
Jefferson Medical College
Philadelphia, Pennsylvania 19107

INTRODUCTION

The various strains of mouse hepatitis virus exhibit distinct organ tropisms. MHV-4(JHM) and MHV-3 are predominantly neurotropic and hepatotropic respectively. Studies on the mechanisms involved in organ specific infection of mouse hepatitis virus (MHV) have focused on several factors such as dose, route of administration, age and strain of experimental animals[1]. Another potential mechanism for regulation of tissue specific spread of virus is the ability of organ specific endothelial cells to selectively support the replication of different strains of MHV. Endothelial cells form an interface between circulating blood and organs and thus could serve as a barrier to infection by viruses. Endothelial cells are also heterogeneous in different organs and may exhibit selectivity in permitting viral entry, thus influencing their tissue tropism[2].

In order to study the role of endothelial cells in regulating viral tropisms, primary cultures of hepatic endothelial cells, hepatocytes and cerebral endothelial cells derived from MHV susceptible BALB/c mice were infected with MHV-3 or MHV-4. Supernatants from infected cultures were collected, for performing plaque assays, at various time points after infection. Cytopathic effects were also monitored at these time points.

The data presented indicates that the endothelial cells but not hepatocytes, demonstrate differential responses to MHV-4 and MHV-3 infection. Endothelial cell heterogeneity thus could play an important role in regulating organ tropism of MHV-3 and MHV-4. The possible mechanisms of differential replication of MHV strains in organ specific endothelial cells are discussed.

Corona- and Related Viruses, Edited by P. J. Talbot and G. A. Levy
Plenum Press, New York, 1995

MATERIALS AND METHODS

Brain Derived Endothelial Cell Cultures

Cerebral microvascular endothelial cells (CEC), were isolated from the brains of BALB/c mice (2 months old) using the method described by Rupnick et al[3]. Briefly, cerebral microvessels were isolated by digestion of cerebral white matter in 0.5% collagenase (Sigma, St.Louis, MO), and density centrifugation of homogenized material in 15% dextran. The vascular pellet was further purified by gradient centrifugation on Percoll(45%). Capillaries were plated onto 0.1% gelatin-coated plates, and endothelial cells grew out in about 10 days. Endothelial cell lines were established from these initial outgrowths and were cultured in Medium 199 (GIBCO, Grand Island, NY), supplemented with 20% fetal bovine serum, 2mM L-glutamine, 90µg/ml heparin (porcine, Sigma, St. Louis, MO), and 20µg/ml endothelial cell growth factor (ECGF, Collaborative Research, Cambridge, MA).

Hepatic Endothelial Cell Cultures

Hepatic endothelial cells from BALB/c mice (2 months old) were isolated using a method described by Huber et al[4]. Liver tissues were minced and digested in 0.5% collagenase and homogenized. The homogenized material was spun at 160 x g to pellet the cells. The cell pellet was centrifuged at 400 x g for 15 minutes on a gradient consisting of 29%,37% and 55% isotonic Percoll (Pharmacia Fine Chemicals, Piscataway, NJ). Hepatic endothelial cells were found between the 37% and 55% Percoll interface. The endothelial cells were washed and plated in the medium utilized above for cerebral microvascular endothelial cells. Endothelial cell identity for all sources was established by the uptake of DiI-Ac-LDL (acetylated low density lipoprotein, Biomedical Technologies, Stoughton, MA), and specific binding of the lectin *Bandeiraea simplicifolia* BSI-B$_4$[5,6]. The endothelial cell lines used in these studies were between passages 8-12.

Hepatocyte Cultures

Primary hepatocyte cultures were established from BALB/c mice (2 months old) as described by Seglen[7]. Hepatocytes were isolated by collagenase perfusion (Type I, Sigma, St. Louis, MO, 130U/ml) through the portal vein and plated (5 X 10^5 cells/T25 flask) in complete Williams E medium (GIBCO Laboratories, Chagrin falls, OH) containing 10IU/ml penicillin, 10µg/ml gentamycin, 0.02U/ml insulin and 10% heat inactivated fetal bovine serum (Atlanta Biologicals, Atlanta, GA).

Virus Infection of Cultures

Endothelial cells and hepatocytes grown in T-25 flasks (5 x 10^5 cells/flask) were infected with MHV-3 or MHV-4 at an MOI of 0.1. MHV-4 (obtained from Dr. Robert Knobler) was grown to a stock containing 3 x 10^6 PFU/ml. MHV-3 (ATCC) was grown to a stock containing 6 x 10^6PFU/ml. At various times after infection (24, 48, 72 and 96 hours) culture supernatants were collected for measuring viral titers. At these time points cytopathic effects were also documented. Virus titers in the supernatants were measured by plaque assays on L-2 cells.

RESULTS

Comparison of MHV-3 and MHV-4 Effects on Hepatic Endothelial Cells

Fig. 1 compares the cytopathic effects on hepatic endothelial cells at 24, 48, 72 and 96 hours following treatment with MHV-3 or MHV-4. The hepatotropic MHV-3 causes syncitium formation and extensive cellular destruction on hepatic endothelial cells as early as 24 hours following infection (Panel A). In contrast the neurotropic MHV-4 infection does not result in syncytium formation or cause other cytopathic effects on hepatic endothelial cells at 24 hours (Panel E). Interestingly, the earliest evidence of any cytopathology (cellular granularity) was seen only at 96 hours after infection with MHV-4 (Panel D).

Comparison of MHV-3 and MHV-4 Effects on Hepatocytes

Fig. 2 compares the cytopathic effects on hepatocytes at 24, 48, 72 and 96 hours following infection with the hepatotropic MHV-3 or neurotropic MHV-4. Unlike the findings obtained with hepatic endothelial cells both virus strains caused equivalent cytopathic effects on hepatocytes. Evidence of granularity and cell death becomes apparent as early as 24 hours after infection with both MHV-3 and MHV-4 (Panels A and E). No differential effect was observed in the pattern of cytopathic effects on hepatocytes at later time points (48, 72, 96 hours) after infection with MHV-3 or MHV-4 (Compare panels B and F, C and G, D and H).

Comparison of MHV-3 and MHV-4 Effects on Brain Endothelial Cells

Fig. 3 compares the cytopathic effects on brain endothelial cells at 24, 48, 72 and 96 hours after infection with MHV-3 or MHV-4. In contrast to the findings obtained with hepatic endothelial cells treatment with the hepatotropic MHV-3 did not result in any cytopathic effects on brain endothelial cells at 24 hours (Panel A). The earliest evidence of cytopathology was at 96 hours after treatment of brain endothelial cells with MHV-3 (Panel D). Treatment of brain endothelial cells with the neurotropic MHV-4 resulted in extensive cellular destruction by 72 hours following infection (Panel G).

Two key finding are of interest. They are 1. Brain endothelial cells are not readily susceptible to the hepatotropic MHV-3 when compared to hepatic endothelial cells and 2. Brain endothelial cells are more susceptible to the neurotropic MHV-4 infection, as evidenced by the time course of cytopathic effects, when compared to the hepatotropic MHV-3.

Comparison of MHV-3 and MHV-4 Titers in Supernatants of Hepatic Endothelial Cells and Hepatocytes

The data presented in Table I demonstrate that the hepatotropic MHV-3 readily replicates in hepatic endothelial cells with viral titers peaking at 48 hours following infection (3.2×10^5 PFU/ml). In contrast, the hepatic endothelial cells did not support the replication of the neurotropic MHV-4 at any of the time points examined (0 PFU/ml at all time points). The selective ability to support MHV-3 replication was found to be restricted only to the endothelium. Hepatocytes were able to support the replication of both MHV-3 and MHV-4, with equivalent viral titers obtained at 96 hours after infection ($1.9-2 \times 10^5$ PFU/ml)(Table II).

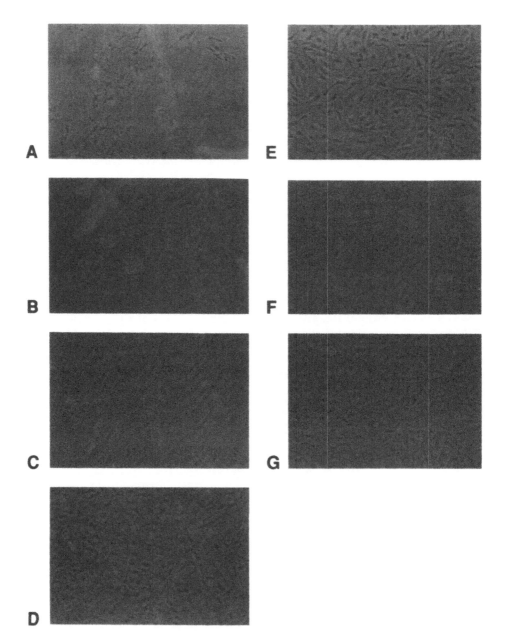

Figure 1. MHV-3 and MHV-4 induced cytopathology following infection of hepatic endothelial cells at an MOI of 0.1. Panels A. MHV-3, 24hrs, B. MHV-3, 48 hrs, C. MHV-3, 72 hrs, E. MHV-4, 24 hrs, F. MHV-4, 48 hrs, G. MHV-4, 72 hrs, D. MHV-4, 96 hrs. Syncytium formation as well as rapid cellular destruction occurs as early as 24 hours after infection with MHV-3. The cells remaining relatively intact following MHV-4 treatment as late as 96 hours after infection.

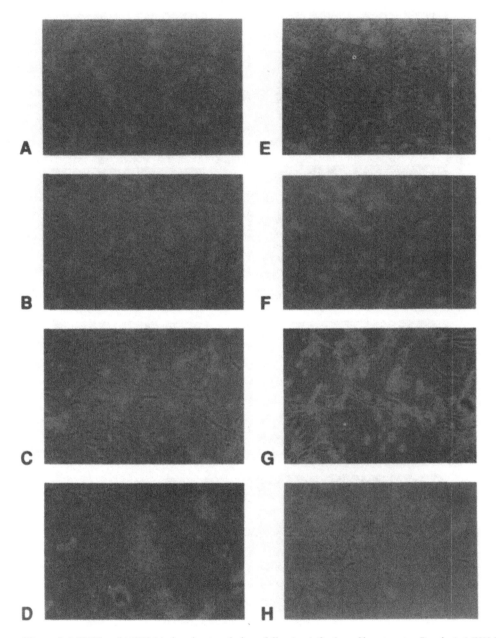

Figure 2. MHV-3 and MHV-4 induced cytopathology following infection of hepatocytes. Panels A. MHV-3, 24 hrs, B. MHV-3, 48 hrs, C. MHV-3, 72 hrs, D. MHV-3, 96 hrs, E. MHV-4, 24 hrs, F. MHV-4, 48 hrs, G. MHV-4, 72 hrs, H. MHV-4, 96 hrs. Comparable cytopathology is induced by both virus strains. Granularity and cell death starts at 24 hours with gradual progression to total destruction of the monolayer by 96 hours.

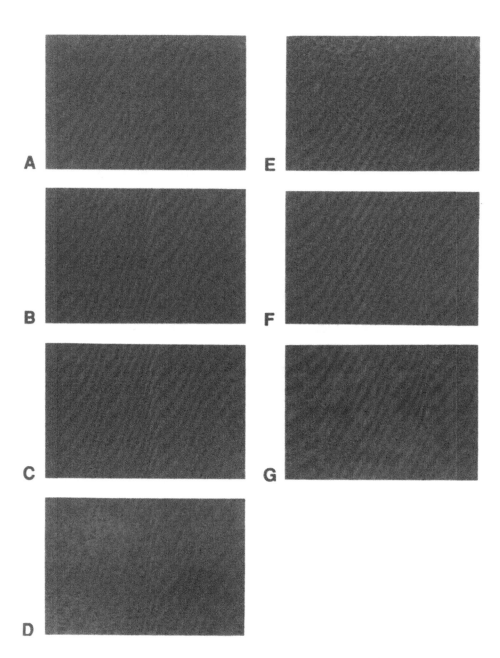

Figure 3. MHV-3 and MHV-4 induced cytopathology following infection of brain endothelial cells. Panels A. MHV-3, 24 hrs, B. MHV-3, 48 hrs, C. MHV-3, 72 hrs, D. MHV-3, 96 hrs, E. MHV-4, 24 hrs, F. MHV-4, 48 hrs, G. MHV-4, 72 hrs. MHV-4 is more cytopathic on these cells with total cellular destruction occuring by 72 hours while it takes up to 96 hours for this to occur with MHV-3.

DISCUSSION

The selective organ tropism of different strains of MHV are regulated by multiple factors such as age, dose, route of administration, and strain of experimental animals[1]. Additional factors that regulate MHV tropism are likely to be involved, depending on the

Table I. Comparison of neurotropic (MHV-4) and hepatotropic (MHV-3) virus replication in hepatic endothelium

Time (h)	MHV-4 (PFU/ML)	MHV-3 (PFU/ML)
0	0	0
24	0	$2\,9 \times 10^5$
48	0	$3\,2 \times 10^5$
72	0	4×10^4
96	0	0

manner of spread of virus within the body For instance hematogenous spread of virus occurs following infection of mice through the natural intranasal route of infection[8] The hematogenous spread can be selectively regulated by the endothelial cells, that form a barrier between blood and tissue, in different organs Endothelial cells are heterogeneous in different organs of the body and express organ specific cell surface properties as well as blood-brain barrier function in the brain[2]

The goal of the current study was to determine if endothelial cells from brain and liver demonstrate any heterogeneity in its response to the predominantly neurotropic MHV-4 (JHM) and hepatotropic MHV-3 strains The results obtained demonstrate an organ specific pattern of cytopathology and virus replication in endothelial cells, which mirrors the tropism of the virus strains In sharp contrast, the hepatocytes do not discriminate between the neurotropic and hepatotropic strains of the virus Hepatocytes support equivalent replication of both MHV-4 and MHV-3 These findings point toward an important regulatory role for the endothelial cells in organ specific virus infection that could ultimately influence their tissue tropism

Several possible mechanisms are likely to be involved in regulating differential effects of MHV strains on organ specific endothelial cells One possible difference between endothelial cells may lie in the heterogeneity of virus receptor expression Such differences have been previously described for murine cocksackievirus In this model it was found that the tropism of the virus correlated with their ability to infect endothelial cells from different organs as well as virus receptor expression on these cells[4] The MHV receptor has been identified as a member of the carcinoembryonic antigen family and several isoforms have been identified that show differences in organ and strain distribution[9 10] Heterogeneity in the expression of MHV receptor isoforms on organ specific endothelial cells could influence differential virus binding and replication of the various MHV strains This possibility is an area of further study

Recent studies in other laboratories have suggested that the presence of a functional virus receptor alone may not be sufficient to confer susceptibility to MHV infection Multiple cellular factors that influence virus internalization, virus uncoating or other steps in virus

Table II. Comparison of neurotropic (MHV-4) and hepatotropic (MHV-3) virus replication in hepatocytes

Time (h)	MHV-4 (PFU/ML)	MHV-3 (PFU/ML)
0	50	0
24	$3\,1 \times 10^3$	0
48	$2\,4 \times 10^4$	$4\,3 \times 10^3$
72	$4\,4 \times 10^5$	$5\,6 \times 10^4$
96	$2\,0 \times 10^5$	$1\,9 \times 10^4$

replication could determine virus susceptibilty in different endothelial cell populations[11][13] The *in vitro* endothelial cell system offers an opportunity to dissect out the various parameters in organ selective infection of the vascular beds, which could provide important insights into understanding the mechanisms of MHV tissue tropisms

REFERENCES

1 Lavi E , Gilden D H , Highkin M K , Weiss S R The organ tropism of mouse hepatitis virus A59 in mice is dependent on dose and route of inoculation Lab Anim Sci 1986,36 130-135

2 Auerbach R , Alby L , Morissey L , Tu M , Joseph J Expression of organ-specific antigens on capillary endothelial cells Microvasc Res 1985,29 401-411

3 Rupnick M A , Carey A , Williams S K Phenotypic diversity in cultured cerebral microvascular endothelial cells In Vitro 1988,24 435-444

4 Huber S A , Haisch C , Lodge P A Functional diversity in vascular endothelial cells Role in cocksackievirus tropism J Virol 1990,64 4516-4522

5 Voyta J C , Netland J C , Via D P, Zetter, B R Specific labeling of endothelial cells using fluorescent acetylated low density lipoproteins J Cell Biol 1984,99 81A

6 Sahagun G , Moore S A , Fabry Z , Schelper R L , Hart M N Purification of murine endothelial cell cultures by flow cytometry using fluorescein-labeled *Griffonia simplicifolia* agglutinin Am J Pathol 1989, 134 1227-1232

7 Seglen P O Preparation of isolated rat liver cells Methods Cell Biol 1976,13 29-33

8 Cabirac G F , Soike K F , Butunoi C , Hoel K , Johnson S , Cai G-Y , Murray R S Coronavirus JHM OMP1 pathogenesis in owl monkey CNS and coronavirus infection of owl monkey CNS via peripheral routes Adv Exp Med Biol 1993,342 347-352

9 Dveksler G S , Diffenbach C W , Cardellichio C B , McCuaig K , Pensiero M N , Jiang G S , Beauchemin N , Holmes K C Several members of the carcinoembryonic antigen-related glycoprotein family are receptors for the coronavirus mouse hepatitis virus-A59 J Virol 1993,67 1-8

10 Yokomori K , Lai M M C Mouse hepatitis virus utilizes two carcinoembryonic antigens as alternative receptors J Virol 1992, 66 6194-6199

11 Yokomori K , Lai M M C The receptor for mouse hepatitis virus in the resistant mouse strain SJL is functional Implication for the requirement of a second factor for viral infection J Virol 1992,66 6931-6938

12 Yokomori K , Asanaka M , Stohlman S A , Lai M M C A spike protein-dependent cellular factor other than the virus receptor is required for mouse hepatitis virus entry Virology 1993,196 45-56

13 Asanaka M , Lai M M C Cell fusion studies identified multiple cellular factors involved in mouse hepatitis virus entry Virology 1993,197 732-741

7

GENOMIC REGIONS ASSOCIATED WITH NEUROTROPISM IDENTIFIED IN MHV BY RNA-RNA RECOMBINATION

Ehud Lavi,[1] Qian Wang,[1] Susan T. Hingley,[2] and Susan R. Weiss[2]

[1] Division of Neuropathology
Department of Pathology and Laboratory Medicine
[2] Department of Microbiology
University of Pennsylvania School of Medicine
Philadelphia, Pennsylvania 19104-6079

ABSTRACT

Infection of mice with a neurotropic strain of MHV (MHV-A59), a non-neurotropic strain of MHV (MHV-2), and a set of recombinant viruses (kindly provided by Dr. Michael Lai) were used to map genetic determinants of viral neurotropism and demyelination. Following intracerebral (IC) inoculation of 4-week old C57Bl/6 mice, 1LD50 of MHV-A59 produced acute meningoencephalitis and hepatitis, and subsequently chronic CNS demyelinating disease. IC inoculation of 1LD50 of MHV-2 produced acute hepatitis without CNS disease. Recombinants ML-3, ML-11, ML-7, ML-8, ML-9 and ML-10 produced acute encephalitis similar to MHV-A59. According to previous oligonucleotide fingerprinting analysis the only common denominator of the neurotropic recombinant viruses was an M gene derived from MHV-A59. Sequencing of PCR-amplified viral S and M genes confirmed that the M genes of neurotropic viruses are derived from A59 while the S genes of neurotropic viruses are either derived from MHV-2 or from A59. In tissue culture, ML-11, ML-3 and MHV-2 are fusion negative, while A59, ML-7, ML-8 and ML-10 are fusion positive. Thus, neurotropism in MHVs is not linked to fusion or the S gene. Moreover, the M gene may be a significant determinant of neurotropism and acute encephalitis.

INTRODUCTION

Infection of mice by mouse hepatitis virus (MHV), a coronavirus, is a reliable and consistent experimental model system for chronic demyelinating diseases such as multiple sclerosis (MS). As in MS, the mechanism of MHV-induced inflammatory demyelination is not entirely clear, but there is evidence for direct cytolytic effect on oligodendrocytes, low level viral persistence, and immune-mediated pathology. We have previously used recombi-

Corona- and Related Viruses, Edited by P J Talbot and G A Levy
Plenum Press, New York, 1995

nant viruses derived from MHV strains JHM and A59, to map biological properties of the virus to viral genes[1]. Based on the correlation between the phenotypic properties and the oligonucleotide map of the recombinants, we concluded that the 3' end 25% of the MHV genome (genes 3-7), controls biological properties such as plaque morphology and replication in tissue culture, organ tropism and distribution of central nervous system pathology in the mouse[1]. This genomic region contains the genes encoding the viral structural proteins S, M, and N, as well as some non structural proteins. It has been previously shown that the envelope protein S is a determinant of fusion properties and may contribute to neuropathogenicity of MHV strains[2,3,4,5,6,7]. We have now used recombination between a neurotropic (MHV-A59) and a non-neurotropic (MHV-2) strain of MHV to further map the genetic determinants of viral neurotropism and demyelination.

EXPERIMENTAL DESIGN AND RESULTS

Recombinations between MHV-2 and MHV-A59. A set of recombinant viruses between MHV-2 and MHV-A59 were obtained from Dr. Michael Lai[6]. The recombinants, labeled ML-7, ML-8, ML-10, ML-11, ML-3 and ML-9 were prepared by crossing MHV-2 with a temperature sensitive mutant of MHV-A59. MHV-A59 is fusion positive while MHV-2 is fusion negative. The following is our preliminary characterization of biological and molecular properties of some the recombinant viruses as compared to parental viruses.

LD50 of MHV-2/A59 recombinant and parental viruses. Four week old C57Bl/6 mice (certified virus free mice purchased from Jackson Laboratories), were inoculated intracerebrally (IC) with 10 fold serial dilutions of MHV-A59, MHV-2 and recombinant viruses. Five mice were used per dilution; five 10 fold dilutions per virus were used. Mice were monitored daily for signs of disease and mortality. Disease signs were non specific and included ruffled fur, loss of appetite and weight, hunched position, lack of motility. The LD50 was calculated according to the standard Reed-Muench formula as previously described[8]. The LD50 for the viruses are listed in the Table 1 Only MHV-A59, MHV-2, ML-3 and ML-11 caused mortality at a dose equal or below the titer of the viral stock. The loss of hepatotropism in the non-lethal viruses may be the explanation for this phenomenon.

Viral growth and titer in tissue culture. MHV-2 reached titers of 2×10^7 in 24 hours in L2 cells, similar to MHV-A59. ML11 reached titers of 2×10^6 pfu/ml in 24 hours. ML-8 reached titers of 1.55×10^8 pfu/ml, ML-7: 1.32×10^5 pfu/ml and ML-10: 1.32×10^5 pfu/ml. While A59 produced fusion and cytopathic effect in infected L2 cells, both MHV-2 and ML-11 were fusion negative. Thus, the fusion property does not segregate with neurotropism, but as previous studies showed is controlled by the 3' portion of gene 3 (S protein).

Histopathology and organ tropism of recombinant and parental viruses. Following IC inoculation of 1LD50 into 4-week old C57Bl/6 mice, MHV-A59 produced acute meningoencephalitis and hepatitis, and subsequently chronic CNS demyelinating disease. IC inoculation of MHV-2 produced acute hepatitis without CNS pathology. Occasional inflammatory cells were found in the choroid plexus. Recombinants ML-11 and ML-3 produced acute encephalitis and hepatitis similar to MHV-A59. ML-7, ML-8, ML-9 and ML-10 produced mild acute encephalitis without hepatitis. ML-10 produced minimal hepatic inflammation but no evidence of hepatitis since there were no foci of necrosis or destructive changes in hepatocytes. MHV-A59 caused chronic spinal cord demyelination which was easily detected in all of the mice by H&E, LFB or toluidine-stained sections. However, H&E and LFB did not detect demyelination in the spinal cords of mice infected with 1 LD50 dose of ML-11, or undiluted ML-7, ML-8 and ML-10. Toluidine blue staining on Araldite-embedded sections revealed only rare demyelinated axons in multiple spinal cord sections of a mouse infected with ML-11. Thymic cortical depletion

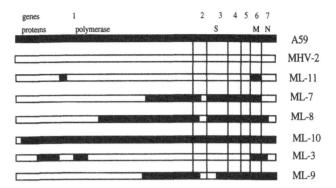

Figure 1. Recombinant viruses between MHV-2 and A59· The map is based on the published oligonucleotide fingerprinting map of recombinant viruses[6] and additional information derived from our sequencing data

of lymphocytes was detected in MHV-A59, ML-11, ML-3 and MHV-2, but not in ML-7, ML-8 and ML-10. It was always associated with hepatitis, but not with mild hepatic inflammation as seen in ML-10 infection.

Molecular characterization of recombinants. The comparative amino acid sequence analysis of M protein of MHV-A59, MHV-2 and recombinant viruses was deduced from nucleic acid sequencing of gene 6 of the viruses. RNA was purified from virus-infected L2 cells, and reversed transcribed using random hexamers (100pmol/20µl; Boehringer Mannheim Biochemicals) and Moloney MLV reverse transcriptase (400µ/20µl; Gibco BRL). PCR amplification of the cDNA templates was done with Taq polymerase (2.5 µ/100µl; BMB), using multiple primers, allowing overlapping sequencing of both strands. PCR products were purified using Promega's Magic PCR prep, then sequenced using the fmol DNA sequencing kit from Promega following the manufacturer's recommended protocol for the extension/termination reaction with an ^{35}S-dATP label.

The comparative analysis of the M proteins of MHV-2, A59 and ML-11 revealed that the M protein of ML-11, ML-7, ML-8, ML-3 and ML-9 were derived from MHV-A59, thus confirming the previous oligonucleotide fingerprinting analysis of gene 6 of these viruses. MHV-2 M protein sequence differs from that of MHV-A59 by 10 amino acids: asparagine

Table 1. Summary of biologic properties of recombinant viruses as compared to parental viruses MHV-A59 and MHV-2

Property	Virus							
	A59	MHV–2	ML–11	ML–7	ML–8	ML10	ML–3	ML–9
Peak morbidity (dpi)[1]	5–8	6–11	6–8	–	–	–	4–8	–
Mortality (dpi)[1]	6–8	7–11	7–8	–	–	–	5–9	–
LD50 (pfu).[2]	4x10³	4x10⁵	2x10⁴	>2 6x10³	>3 1x10⁶	>2 6x10³	1x10²	>6x10⁵
Fusion of L2 cells·[3]	+	–	–	+	+	+	–	
Hepatotropism[1]	+	+	+	–	–	–	+	–
Thymotropism[1]	+	+	+	–	–	–	+	–
Neurotropism.[1]								
Acute encephalitis	+	–	+	+	+	+	+	+
Demyelination	+	–	rare	–	–	–		

[1]Groups (2-5) of 4-week old C57Bl/6 mice were inoculated intracerebrally with 1-2 LD50 dose of virus assessed by LFB and toluidine blue stained sections

[2]Groups of 5 4-week-old C57Bl/6 mice were inoculated intracerebrally with 10 fold dilutions of the virus LD50 was calculated according to the standard Reed-Muench formula

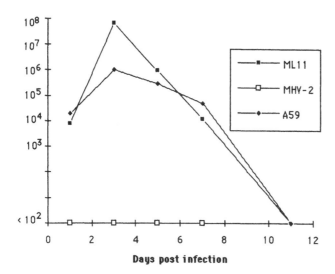

Figure 2. Growth curve of viruses (PFU/gm) in brains of mice infected with 1LD50 dose.

for serine in 2; glutamine for glutamic acid in 9; isoleucine for valine in 20; arginine (basic) for glutamine (uncharged) in 21; valine for isoleucine in 37; valine for isoleucine in 52; leucine for isoleucine in 59; isoleucine for valine in 96; cysteine (non polar) for serine (uncharged polar) in position 219, and the substitution of threonine (uncharged) for alanine (non polar) in position 221.

Sequencing of the S gene of ML-7, ML-8, two hepatotropism-deficient viruses revealed 4 identical amino acid substitution in the following positions: 98 (asparagine to serine), 375 (isoleucine to methionine), 652 (leucine to isoleucine), and 1067 (threonine to asparagine). Preliminary sequence analysis of the S gene of ML-11, and ML-3 suggests an MHV-2 origin of this gene. Partial characterization of the S gene of ML-9 confirms a 5' end derived from MHV-2 and a 3' end derived from A59.

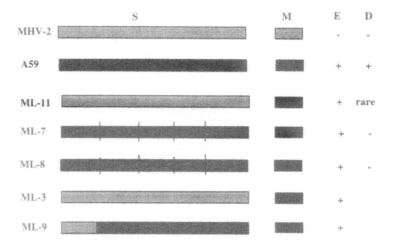

Figure 3. Preliminary sequence analysis of the S and M genes of parental and recombinant viruses and correlation with pathogenesis properties. S = Spike glycoprotein; M = membrane glycoprotein; E = acute encephalitis; D = demyelination.

DISCUSSION

Preliminary mapping of biologic properties to viral genes. The following conclusions concerning mapping of biologic properties can be suggested based on our preliminary studies:

1. Molecular determinants of hepatotropism and acute meningoencephalitis in MHV are not linked. This conclusion is based on several lines of evidence. Both wild type strains of MHV and recombinant viruses can exhibit either neurotropic properties alone (JHM, ML-7, ML-8, ML-10, ML-9), hepatotropism alone (MHV-2), or both (MHV-A59 ML-11 and ML-3).

2. Molecular determinants of acute hepatitis in MHV are linked to thymic cortical depletion of lymphocytes. Without exception, in every wild type MHV strain, recombinant variant or mutant MHV viruses that we have examined, these properties always segregated together. Thus, viral sequences that control these properties are linked.

3. Molecular determinants of acute meningoencephalitis are not linked to demyelination. Both processes are variations of affinity of MHV for the CNS, however, the extent of each one is not in direct proportion to the extent of the other. For example: Both JHM and A59 produce severe demyelination in mice surviving the acute disease, however, JHM produces severe encephalitis while A59 encephalitis is focal and restricted to specific regions of the brain which are part of the olfactory-limbic systems. The same dichotomy is seen in many variants and recombinant viruses. Moreover, manipulations and attenuation of acute encephalitis in JHM did not always affect the extent of demyelination[2,5]. The idea of separate control of these two aspects of neurotropism in MHV may support the speculation that different mechanisms exist for acute encephalitis vs chronic demyelination. While acute encephalitis may result primarily from acute neuronal infection, chronic demyelination may be related to chronic persistent infection of virus in oligodendrocytes with a possible contribution of an immune mediated phenomenon.

4. The M gene of MHV may contain molecular determinants of acute encephalitis. The major evidence for this conclusion is the phenotypic and molecular characterization of recombinants ML-11 and ML-3. These recombinants produces acute encephalitis like A59 while MHV-2 does not produce encephalitis. Molecular analysis of ML-11 by oligonucleotide fingerprinting and partial sequencing revealed that it was primarily derived from MHV-2 except for the entire gene 6, and a small portion of gene 1. Sequence analysis of S and M genes of ML-7, ML-8, and ML-3 help to substantiate this conclusion. An M gene derived from A59 correlates with acute encephalitis while acute encephalitis clearly does not correlate with the S gene. In both ML-11 and ML-3 sequencing revealed that the S gene is derived from MHV-2 but the viruses produce acute encephalitis similar to A59. Support for the possible contribution of the M gene to neurotropism also comes from a study that showed that treatment with two non-neutralizing monoclonal antibodies against M protected mice from encephalitis caused by JHM virus[9]. The fact that manipulation of the S protein with monoclonal antibodies modified the disease process does not contradict our results since the S gene controls many functions that are important for viral replication and viral-cell interaction, thus

modifications of these functions can indirectly modify pathogenesis. In other viral systems such as measles infection, a defective transmembrane protein M has been related to the difference in neurotropism, rendering this defective virus capable of producing subacute sclerosing panencephalitis (SSPE). Further studies are necessary to determine which of the 10 amino acid changes in MHV-2 M protein is/are associated with lack of neurotropism in MHV-2.

ACKNOWLEDGMENTS

This study was supported in part by a grant from the University of Pennsylvania Research Foundation (EL), by National Multiple Sclerosis Society grants PP-0284 and RG-2615A1/2 (EL) and by PHS grants NS-11037 and NS-21954 (SRW).

REFERENCES

1 Lavi E, Murray EM, Makino S, Stohlman SA, Lai MM, Weiss SR Determinants of coronavirus MHV pathogenesis are localized to 3' portions of the genome as determined by ribonucleic acid-ribonucleic acid recombination Lab Invest 1990,62 570-578
2 Buchmeier MJ, Lewicki HA, Talbot PJ, Knobler RL Murine hepatitis virus-4 (strain JHM) - induced neurologic disease is modulated in vivo by monoclonal antibody Virology 1984,132 261-270
3 Collins AR, Knobler RL, Powell H, Buchmeier MJ Monoclonal antibody to murine hepatitis virus-4 (strain JHM) define the viral glycoprotein responsible for attachment and cell-cell fusion Virology 1982,119 358-371
4 Dalziel RG, Lampert PW, Talbot PJ, Buchmeier MJ Site specific alteration of murine hepatitis virus type 4 peplomer glycoprotein S results in reduced neurovirulence J Virol 1986,59 463-471
5 Fleming JO, Trousdale MD, El-Zaatari FAK, Stohlman SA, Weiner LP Pathogenicity of antigenic variants of murine coronavirus JHM selected with monoclonal antibodies J Virol 1986,58 869-875
6 Keck JG, Soe LH, Makino S, Stohlman SA, Lai MMC RNA recombination of murine coronavirus recombination between fusion-positive mouse hepatitis virus A59 and fusion-negative mouse hepatitis virus 2 J Virol 1988,62 1989-1998
7 Makino S, Fleming JO, Keck JG, Stohlman SA, Lai MMC RNA recombination of coronaviruses localization of neutralizing epitopes and neuropathogenic determinants on the carboxyl terminus of peplomers Proc Natl Acad Sci USA 1987,84 6567-6571
8 Lavi E, Gilden DH, Wroblewska Z, Rorke LB, Weiss SR Experimental demyelination produced by the A59 strain of mouse hepatitis virus Neurology 1984,34 597-603
9 Fleming JO, Shubin RA, Sussman MA, Casteel N, Stohlman SA Monoclonal antibodies to the matrix (E1) glycoprotein of mouse hepatitis virus protect mice from encephalitis Virology 1989,168 162-167

8

INVOLVEMENT OF MICROTUBULES AND THE MICROTUBULE-ASSOCIATED PROTEIN TAU IN TRAFFICKING OF JHM VIRUS AND COMPONENTS WITHIN NEURONS[*]

Kishna Kalicharran[†] and Samuel Dales

Cytobiology Group
Department of Microbiology and Immunology
Health Sciences Centre
University of Western Ontario
London, Ontario, N6A 5C1
Canada

INTRODUCTION

The neurotropic coronavirus JHM (JHMV) is capable of inducing various forms of CNS disease in rodents, ranging from an acute encephalomyelitis to a delayed onset demyelination[1,2,3]. In rats, during the early stages of the disease process, neurons become cellular targets[3,4]. When introduced by intranasal inoculation, JHMV can invade the CNS of mice and rats by spreading along the olfactory neurons[5, 6, 7, 8]. Virus spread was shown to occur by the transneuronal route[5,8]. Subsequent spread within the CNS seems to involve specific neuronal populations and tracts[6, 8]. In particular, in Wistar Furth rats, Purkinje and hippocampal neurons are extensively involved[3,4]. Moreover, neurons have been shown to provide a repository site where both RNA and virions can persist for prolonged periods[9, 10, 11]. Recently, Pasick et al., (1994) demonstrated that trafficking of virus materials within neurons occurs asymmetrically along soma-todendritic and axonal processes, and appears to be dependent on the integrity of the microtubular network, as evident from analyses by light and electron microscopy which revealed that JHMV nucleocapsids (N) are closely associated with microtubules[12, 13]. This report seeks to define, more directly, by immunoprecipitation and immunoblotting, interactions between N and the microtubular arrays.

[*] Supported by a grant from the Medical Research Council of Canada.
[†] Recipient of a studentship from the Multiple Sclerosis Society of Canada.

RESULTS AND DISCUSSION

Previous work showed that neurons become infected early during establishment of the CNS disease by JHMV[8, 10, 11]. The predilection of N nucleoprotein for associating with neuronal microtubules was shown previously[12, 13]. This is evident in primary explants of rat hippocampal neurons infected for 24 hours with JHMV, then processed for electron microscopy. As illustrated in Figure 1, the dense granular nucleocapsid material is juxtaposed to microtubules, an observation consistent with previous findings[12, 13] on the colocalization of microtubules and nucleocapsid and determined by means of confocal microscopy following dual antibody marking with immunofluorescence (data not shown).

To explain the observed colocalization of N with microtubules, the database on microtubule-associated cellular proteins was searched for sequence homology which may exist with N. In fact, the microtubule-associated protein tau was found to possess a significant sequence match with N, where an overall 20% identity and 42% similarity was uncovered. It is significant that the closest homology lies within the microtubule-binding domain of the tau sequence which encodes a 31% identity and 54% similarity with N[12].

The phosphoprotein tau which has an mRNA generated by alternative splicing is developmentally regulated. Tau is differentially phosphorylated by several kinases and in its dephosphorylated state acts to stimulate formation of microtubule bundles[14]. Being a microtubule-associated protein, tau interacts with tubulin by a 18 amino acid repeat that constitutes the microtubule-binding domain[16]. Within neurons, tau is localized primarily to axons[15]. Considerable attention has been given to tau as a major pathologic feature in

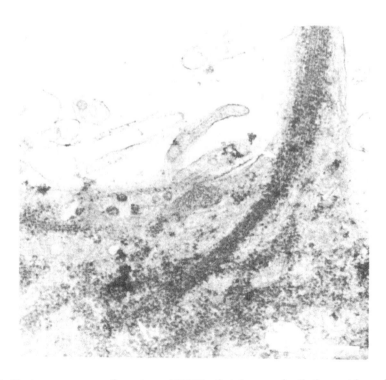

Figure 1. Electron microscopy of a portion of JHMV-infected neuronal cells from a telencephalic explant culture Arrows indicate a close association between the dense, granular nucleocapsid material and the microtubular network Magnification (Bar = 0 5µM) (from [13])

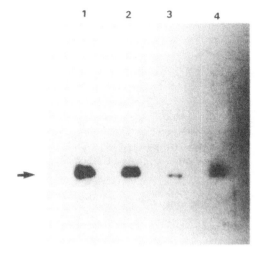

Figure 2. Immunoblot demonstrating cross reactivity between the N protein with anti-tau antibodies. Purified N was probed with two MAbs against N (lanes 1 and 2) or with MAb and polyclonal antibodies against tau (lanes 3 and 4) (from [12])

Alzheimer's disease because it self-assembles and is an integral component of the paired helical filaments constituting neurofibrillary tangles[17].

To determine whether N and tau are immunologically cross reactive, purified N was transferred to nitrocellulose and the blots were probed with anti-N antibodies as controls and both anti-tau monoclonal and polyclonal antibodies. It is evident from Figure 2, that both monoclonal (lane 3) and polyclonal (lane 4) anti-tau antibodies recognized purified N in an immunoblot assay albeit the monoclonal reacted with a lesser intensity.

To analyze the significance of the above findings further, immunoprecipitation and Western blotting were used to assess the interaction of with the microtubular protein. The OBL-21 is a neuronal cell line originating from olfactory neurons of CD.1 mice, which was immortalized by the myc gene of a replication defective avian retrovirus[18]. Following infection for 24 hours with JHMV, OBL-21 cell extracts were reacted with antibodies against tau, tubulin and N. The immunoprecipitates were subjected to SDS-PAGE and the resulting immunoblot probed with anti-N antibodies. As shown in lane 3 of Figure 3, anti-tau antibodies formed a precipitate with N This finding supports the presumption that amino acid sequence relatedness between N and tau is also one of immunological identity at the epitope(s) binding anti-tau antibodies. The ability of anti-tubulin antibodies to bring down N in the immunoprecipitates although in small amount (Figure 3, lane 2) indicates a reactivity with preexisting complexes with N and tubulin, which are established during infection, as

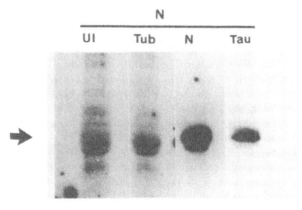

Figure 3. Immunoblot demonstrating the ability of antibodies against tau and tubulin to form immunoprecipitates with N present in JHMV-infected OBL-21 neuronal cells. JHMV-infected cell lysates were mixed with pre-immune serum (lane 1), anti-tubulin (lane 2), anti-N (lane 3) and anti-tau (lane 4) antibodies Immunoprecipitates were separated by SDS-PAGE then probed in immunoblots with anti- N antibodies

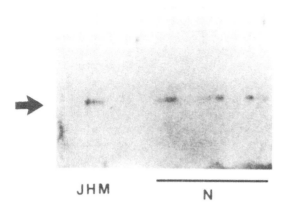

Figure 4. Blot overlay demonstrating the direct interaction between tubulin and N in a nitrocellulose matrix. Purified JHM virions (lane 1) and purified N protein (lanes 3-5) were transferred to nitrocellulose sheets and then purified tubulin was added. Following binding in buffer solution, the attached tubulin was detected by means of anti-tubulin antibodies.

suggested by the example in Figure 1. As a direct test of a specific interaction between N and tubulin, we performed a blot overlay according to procedures previously described[19]. For this we used N isolated and purified from cells and N existing in isolated JHM virions. The material was separated by SDS-PAGE, then transferred onto nitrocellulose membranes. The blots were exposed overnight to purified tubulin (and bovine serum albumin as a competitor) to allow specific tubulin attachment. The complexes formed were then subjected to Western blotting using anti-tubulin antibodies. As evident from Figure 4, the added tubulin appears to bind to the purified form of N (Figure 4, lanes 3-5) and to the N in virions (Figure 4, lane 1). These findings support the view that N is attached to tubulin inside the cell at the amino acid sequence which is homologous with that encoded by the microtubule-binding domain of tau.

In summary, JHMV nucleocapsids are closely associated with microtubules within neurites. The virus nucleocapsid protein (N) has a capacity to bind in vitro to purified tubulin. and evidently also inside infected cells. The existence of both an amino acid sequence and immunological relatedness between N and the microtubule-binding motif of tau suggests that there is a mimicry between the association of N and tau with microtubules .

REFERENCES

1. Sorensen, O., D. Percy and S. Dales. In vitro and in vivo models of demyelinating diseases III. JHM virus infection in rats. Arch. Neurol. 1980; 37:478-484.

2. Sorensen, O., R. Dugre, D. Percy, and S. Dales. In vivo and in vitro models of demyelinating disease: endogenous factors influencing demyelinating disease caused by mouse hepatitis virus in rats and mice. Infect. and Immun. 1982; 37:1248-1260.

3. Parham, D., A. Tereba, P.J. Talbot, D. Jackson, and V. Morris. Analysis of JHM central nervous system infection in rats. Arch. Neurol. 1986;43:702-708.

4. Sorensen, O., and S. Dales. In virtro and in vivo models of demyelinating disease: JHM virus in the rat central nervous system localized by in situ cDNA hybridization and immunofluorescent microscopy. J. Virol. 1985; 56:434-438.

5. Lavi, E., P. Fishman, M. Highkin., and S. Weiss. Limbic encephalitits after inhalation of a murine coronavirus. Lab. Invest. 1988; 58:31-36.

6. Perlman, S., G. Jacobsen, and S. Moore. Regional localization of virus in the central nervous system of mice persistenly infected with murine coronavirus JHM. Virology. 1988; 166:328-338.

7. Barnett, E.M., and S. Perlman. The olfactory and not the trigeminal nerve is the major site of CNS entry for mouse hepatitis virus, strain JHM. Virology, 1993; 194; 185-191.

8 Barnett, E M , M D Cassell, and S Perlman Two neurotropic viruses herpes simplex virus type 1 and mouse hepatitis virus, spread along different neural pathways from the main olfactory bulb Neurosci 1993, 57 1007-1025

9 Sorensen, O , M Coulter-Mackie, S Puchalski, and S Dales In vivo and in vitro models of demyelinating disease IX progression of JHM virus infection in the central nervous system of the rat during overt and asymptomatic phases Virol 1984, 137 347-357

10 Lavi, E , D Gilden, M Highkin , and S Weiss Persistence of mouse hepatitis virus A59 in a slow virus demyelinating infection in mice as detected by in situ hybridization J Virol 1984, 51 563-566

11 Fleming, J , J Houtman, H Alaca, H Hinze, D McKenzie, J Aiken, T Bleasdale, and S Baker Persistence of viral RNA in the central nervous system of mice inoculated with MHV-4 In Laude, H , and J Vautherot (eds) Coronaviruses Plenum Press New York 1994 327-332

12 Pasick, J M M , K Kalicharran, and S Dales Distribution and trafficking of JHM coronavirus structural proteins and virions in primary neurons and the OBL-21 neuronal cell line J Virol 1994, 68 2915-2928

13 Pasick, J M M , and S Dales Infection by coronavirus JHM of rat neurons and oligodendrocyte-type-2 astrocytes lineage cells during distinct developmental stages J Virol 1991, 65 5013-5028

14 Goedert, M , R A Crowther, and C C Garner Molecular characeterization of microtubule-associated proteins tau and MAP2 Trends Neurosci 1991, 14 193-199

15 Binder, L I , A Frankfurter, and L I Rebhun The distribution of tau in the mammalian central nervous system J Cell Biol 1985, 101 1371-1378

16 Lee, G , N Cowan, and M Kirschner The primary structure and heterogeneity of tau protein from mouse brain Science 1988, 239, 285-288

17 Goedert, M , C M Wischik, R A Crowther, J E Walker, and A Klug Cloning and sequencing of the cDNA encoding a core protein of the paired helical filament of Alzheimer disease identification as the microtubule-associated protein tau 1988, 85 4051-4055

18 Ryder, E F , E Y Snyder, and C L Cepko Establishment and characterization of multipotent neural cell line using retrovirus vector-mediated oncogene transfer J Neurobiol 1990, 2 356-375

19 Homann, H E , W Willenbrink, C J Buchholz, and W J Neubert Sendai virus protein-protein interactions studies by a protein blotting protein-overlay technique mapping of domains on NP protein required for binding to P protein J Virol 1991,65 1304-1309

9

EVOLUTION AND PERSISTENCE MECHANISMS OF MOUSE HEPATITIS VIRUS

Wan Chen and Ralph S. Baric

The Department of Epidemiology
The University of North Carolina at Chapel Hill
Chapel Hill, North Carolina

ABSTRACT

We established and characterized persistently-infected DBT cells with mouse hepatitis virus to study the molecular mechanisms of MHV persistence and evolution in vitro. Following infection, viral mRNA and RF RNA were coordinately reduced by about 70% as compared to acute infection suggesting that the reduction in mRNA synthesis was due to reduced levels of transcriptionally active full length and subgenomic length negative-stranded RNAs. Although the rates of mRNA synthesis were also reduced, the relative percent molar ratio of the mRNAs and RF RNAs were similar to those detected during acute infection. In contrast to the finding during BCV persistence, analysis of the MHV leader RNA indicated that the leader RNA and leader/body junction sequences were extremely stable. These data suggested that polymorphism and mutations resulting in intraleader ORFs was not required for MHV persistence. Conversely MHV persistence was significantly associated with a A to G mutation at nt 77 in the 5′ end untranslated region (UTR) of the genomic RNA.

INTRODUCTION

Although coronaviruses readily cause persistent infections, the mechanism by which these viruses establish and maintain a persistent infection in vitro and in vivo is unclear (1-6). Previous studies have suggested that virus evolution and mutation resulted in the production of temperature-sensitive, cold-sensitive, small plaque and fusion defective viral variants during coronavirus persistence (7, 8, 9, 10), but the role of these virus variants in persistence has not been established. During persistent bovine coronavirus (BCV) infection, mutations and evolution in the BCV leader RNA resulted in an intraleader open reading frame (ORF) which potentially attenuated the translation of downstream ORFs in each BCV mRNA (12). Since the intraleader mutation occurred after the establishment of viral persistence, these changes probably function in maintaining BCV persistence in vitro. Since little information is available concerning the molecular mechanisms by which MHV and other coronaviruses persist and evolve in vitro and in vivo, we have established persistently-infected cultures of DBT cells, examined virus transcription and gene expression, and studied the persistence and evolution mechanisms of the MHV-A59 in vitro. We demonstrate that viral transcription is reduced but that subgenomic negative strands and RF RNAs are present and transcriptionally active during MHV persistent infection. In contrast to finding in BCV, the MHV leader RNA sequences are extremely stable and do not evolve significantly. Rather, MHV persistence was significantly associated with mutation and evolution in the 5' (UTR) in the MHV genome length RNA.

MATERIAL AND METHODS

Virus and Cell Line

The MHV-A59 strain of mouse hepatitis virus and murine astrocytoma cells, DBT cells, were used throughout the course of these studies (13). Persistent and control cultures were simultaneously maintained and passaged under identical treatment conditions. Persistently- infected cultures were established by infection at a MOI 5. After acute infection, cells that survived infection (<95%) were cultured into stably-infected cell lines that continuously released infectious virus.

Plaque assays, infectious center assays and indirect immunofluorescence was performed as previously described (13,14).

Analysis of Viral RNA

Intracellular viral RNA was extracted from acutely and persistently infected cells using RNA STAT-60 (Tel-TEST "B", Inc.,) following the manufacturer's directions. Northern dot blot was performed with equivalent amounts of intracellular RNA using radiolabeled cDNA probes specific for genomic RNA (nt 6989-7527) or an N gene specific cDNA clone IBI76N for mRNA as previously described (13).

To metabolically radiolabel viral RNAs, cultures of cells were first grown overnight in 90% phosphate-free MEM (Gibco) containing 10% fetal calf serum (FCS). Following acute infection, control and persistently-infected cultures were maintained in phosphate free media containing 1% FCS for 5 hrs. Persistently- and acutely-infected cultures were radiolabeled with 200 µCi / ml ^{32}P-orthophosphate for 1 hr. Viral RNAs was extracted and analyzed as described by Sawiki and Sawiki (15).

Cloning and Sequencing of the 5' end of Genomic RNA and mRNA 3 and 7

The 5' RACE (Rapid Amplification of cDNA Ends) system (Gibco, BRL) was used for cloning and sequencing the 5' leader RNA of mRNA 3 and 7. Briefly, primer 1, which was complementary to nts 193-213 in the N gene, and primer 3, which was complementary to nts 511-531 in the S gene, were used for cDNA synthesis. After purification of the cDNAs and tailing with dCTP and TdT as described by the manufacturer, the products were mixed with either primer 2 (nt 26-47 in the N gene cording sequences), or primer 4 (nt 9-32 in S gene coding sequences), and 5' G tailed anchor primer for PCR amplification. Following 25-30 cycles, the PCR products were cloned into pAMP1 vector (BRL). The Sequenase Version 2.0 DNA sequencing kit (USB) and Sp6 primer were used for sequencing.

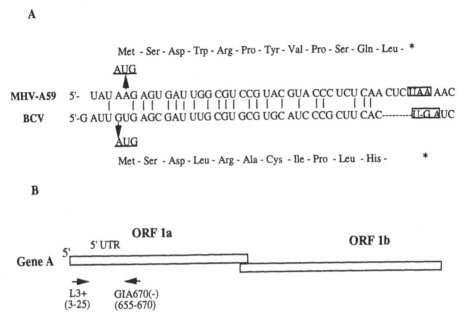

Figure 1. Nucleotide sequence comparisons of the BCV and MHV-A59 leader RNAs. The location of the intraleader ORF in BCV and the putative ATG mutation which could result in a intraleader ORF in MHV are shown in panel A. The cloning strategy of the 5' end of the MHV genome is shown in panel B.

To clone the 5' end of the genomic RNA, cDNA synthesis was accomplished using random primers and reverse transcriptase. Following cDNA synthesis, primers L3[+] (nt 3-25) and G1A 670 (-) (nt 655-670 in the gene 1) were used for PCR amplification and cloning of 5' end of the MHV genome (Fig. 1, B). The appropriately- sized PCR products were cloned into pGME-T vector (Promega) and sequenced.

RESULTS

Characterization of MHV-A59 Persistent Infection In Vitro

To identify the molecular basis for MHV persistence, we reasoned that establishing persistent infections in highly cytolytic cell lines would maximize the selection for mutation and evolution in the virus and cells that survived acute infection. Consequently, DBT cells, a murine astrocytoma cell line, were chosen since >95% of the cells lysed following acute infection. Within 4-6 days post infection, the surviving cells approached confluence and the persistent cultures were passaged every 3-4 days.

In agreement with previous findings (7,18), the viral titers were variable but generally reduced by 50-90% (Table 1) as compared to acute infection . The quantity of viral genomic RNA and mRNA levels were also reduced as much as 70-95% (Table 1).

The reduction in virus titers and RNA levels could be due to a reduction in the number of infected cells in the culture or due to a reduction in virus replication in each persistently-infected cell. To address this question, viral protein expression was monitored by immunofluorecence using monoclonal antibodies against the MHV S and M glycoproteins. Only 16-18% of persistently-infected cells expressed significant concentrations of viral antigens at 35 days post infection as compared to 81-83% during the acute infection.

Table 1. Virus titers and levels of genomic RNA and mRNA during acute and persistent infection

Time post infection	Viral titer (PFU/ml)	Genomic RNA (% of 24 hr P.I.)	mRNA (% of 24 hr P.I.)
12 hr	4.8×10^7	31.0	48.2
24 hr	9×10^7	100	100
3 days	7.5×10^2	3.7	3.3
7 days	3.5×10^4	5.1	2.6
11 days	2.2×10^4	4.0	7.2
27 days	5.2×10^5	3.7	12.2
35 days	3.6×10^5	5.5	27.2

Infectious center assays were also performed to determine the number of cells releasing infectious virus. Under identical conditions, both acutely and persistently-infected cells had equal numbers of infectious centers, suggesting that the reduction in virus titers and RNA synthesis was not due to a reduction in the number of infected cells (data not shown).

To detect if the generation of temperature sensitive (Ts) virus mutants occurred during MHV-A59 persistence, plaque assays were performed at 32°C and 39°C at different times post infection. In general, the Ts virus comprised only 20-25% of the total virus released at different time of MHV persistence (data not shown).

mRNA and Replicative form RNA Synthesis during Persistent Infection

Previous studies suggest the viral transcription was reduced during persistent infection (2). Although both full length and subgenomic length negative-stranded RNAs have

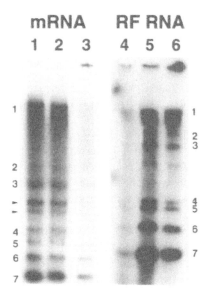

Figure 2. mRNA and RF RNA synthesis during acute and persistent infection. Lane 1, 2: mRNA from acute infection (6 hr P.I); lane 3: mRNA from persistent infection (30 days P.I); lane 4; RF RNA from persistent infection; lane 5, 6: RF RNA from acute infection. Arrows: Viral RNAs occasionally radiolabeled during MHV infection (16).

Table 2. Relative percent molar ratio of mRNA and RF RNA

Size	Acute infection		Persistent infection	
	mRNA	RF RNA	mRNA	RF RNA
1	0 7	8 8	0 1	11 9
2	4 2	8 0	4 3	5 4
3	6 5	7 9	7 7	5 4
4	11 1	8 1	7 6	8 9
5	13 6	10 9	8 4	15 0
6	18 3	20 2	22 4	23 5
7	45 6	36 0	50 5	34 2

been detected during BCV persistence, it was unclear whether these RNAs were present in transcriptionally-active replicative intermediate (RI) complexes (2) To address this question, cultures of persistently- or acutely-infected cells were radiolabeled with ^{32}P- orthophosphate for 1 hr , treated with RNase and DNase 1, and the products separated on 0 8% agarose gels (Fig 2) As in acute infection, all seven viral mRNAs were detected in persistently-infected cells DI RNAs were not detected and AMBIS scans indicated a coordinate approximate 70% reduction in viral mRNA synthesis during persistent infection Transcriptionally active full length and subgenomic length replicative form (RF) RNAs were clearly present at reduced levels, consistent with the notion that subgenomic negative strands were actively involved in the synthesis of the viral mRNAs (Fig 2) To determine whether the reduction in virus transcription specifically affected synthesis of specific mRNA or RF RNA subsets or whether all mRNAs were uniformly reduced, the relative percent molar ratio of mRNA and RF RNA were calculated and compared to acute infection As expected, the relative percent molar ratios were very similar during acute and persistent infection, suggesting that the block in virus transcription uniformly affected the transcription of all viral mRNAs

We also calculated rates of mRNA synthesis during acute and persistent infection During acute MHV infection at 37°C, the average RNA polymerase rates approached 1530±275 nt/min In contrast, polymerase rates in persistently infected cell cultures were approximately 541±275 nt/min, a significant ~ 65% reduction in the rate of transcription (p< 05)

Genetic Evolution of the Leader RNA and 5′ end of Genomic RNA

Recently, it has been demonstrated that a mutation resulting in a small ORF within the 5′ leader RNA of BCV mRNA was selected during persistent infection, and that this intraleader ORF probably attenuated translation of downstream ORFs in each viral mRNA (12) MHV and BCV have relatively similar 5′ end leader RNA sequences and a single A to U mutation at nt 5 could result in a similar intraleader ORF in MHV (Fig 1, A) To determine if the intraleader mutation occurred during MHV persistent infection, leader RNA sequences of mRNA 3 and 7 were cloned and sequenced by using 5′ RACE system The mRNA leader sequences were examined from acute infection (6 hr post infection), 35 days post infection and 105 days post infection In contrast to findings reported during BCV persistent infection, no 5′ terminal leader mutations were detected (Fig 3, A) Although several clones contained 1 or 2 nt truncations at the 5′ end which were probably caused by premature termination during cDNA synthesis, the extensive polymorphism and deletions at the 5′ end of BCV mRNAs were also not detected in MHV Rather, both the MHV leader sequence as well as

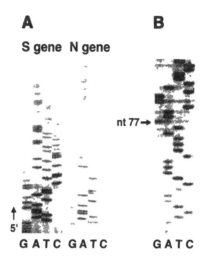

Figure 3. Sequence of the MHV leader RNA and 5' end UTR sequence during persistent infection Panel A: Leader RNA sequence of MRNA 3 and 7 at 105 days post infection. Panel B: The A to G mutation at nt 77 which results in new 5' end UTR ORF.

the leader/body junction were extremely stable. Only 2/30 clones contained point mutations at 105 days post infection, neither of which resulted in intraleader ORFs.

Since our data indicated that intraleader mutations and ORFs did not function in establishing or maintaining MHV persistence in DBT cells, we cloned and sequenced the 5' end of the genomic RNA because mutations in this region may affect the expression of the MHV polymerase genes and alter viral RNA synthesis. An A to G mutation at nt 77 was detected by 56 days post infection which resulted in a 16 aa ORF in the 5' UTR of the genominc RNA (Fig. 3, B; Fig. 4). The A to G mutation accumulated during MHV persistent infection until about 50-70% of the genome length molecules retained the mutation though 119 days post infection (data not shown).

To evaluate the mutation rate in the 5' UTR and p28 cording region during MHV persistent infection, the 5' end of the p28 protein was also sequenced and compared with 5' end UTR. The mutation rate in 5' UTR was significantly higher than the rate detected in p28 coding region (Table 3).

Significance of Mutations in MHV Persistence

As infectious clones are not available to evaluate the role of a particular mutation in MHV persistence, we used biostatistical significance test (Fisher's exact test), odds ratios and 95% confidence limits to determine if a particular mutation was significantly associated with MHV persistence in vitro (17).

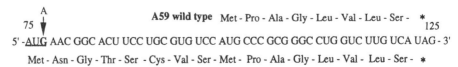

Figure 4. The putative protein sequence of the new 5' end UTR ORF generated by the A to G mutation at nucleotide 77

Table 3. Comparison of mutation rate at the 5' end UTR and p28 coding region

Passage	Mutation in UTR (%)	Mutation in p28 (%)
Acute infection	13	18
56 days P.I.	37	0
88 days P.I.	80	30
119 days P.I.	69	21

These data have clearly demonstrated a strong associations ($p<0.05$) between the mutation at nucleotide 77 and MHV persistence at days 88 and 119 P.I. Similar associations were not detected among the other mutations on the genomic length RNA.

DISCUSSION

In this study, we have initiated molecular genetic approaches to elucidate the mechanisms and complexities of coronavirus evolution, persistence, and replication during MHV persistent infection in vitro. In agreement with previous studies, yields of infectious virus, viral RNA and protein synthesis were generally suppressed during coronaviruses persistence (1, 7, 18). The number of antigen positive cells was also greatly reduced, yet the number of infectious centers closely approximated that seen during acute infection. Since these data indicate that all cells harbored infectious virus, virus replication was probably reduced in each infected-cell during the persistent infection. These findings would indicate that viral antigen is not a good indicator of MHV persistence and also explain why MHV can be cultured from antigen negative cells in vitro and in vivo (19).

Virus transcription was clearly reduced during persistent infection by about 50-70%. Radiolabeling experiments demonstrated that transcription of each viral mRNA and RF RNA were coordinately reduced but the relative percent molar ratio of each mRNA and RF RNAs were similar to the ratios detected during acute infection. Overall rates of mRNA synthesis were also reduced about 65%. These data suggest that reduced amounts of viral RNA polymerase and negative-stranded RNAs may be responsible for the reduction in viral transcription. Since both full length and subgenomic length RFs were radiolabeled, our results confirm previous findings that subgenomic negative strands are functional templates for mRNA synthesis (15).

Unlike the 5' end mutation, extensive hypervariability and polymorphism seen in leader RNA sequences during BCV persistence (12), surprisingly; the MHV leader RNA sequences were extremely stable. Intraleader ORFs did not evolve, yet the leader/body junction sites in both MHV and BCV remained stable and highly conserved in persistently-

Table 4. Biostatistical tests for the contribution of a particular mutation to persistence

Position of mutation	Passage (days P.I.)	Odds ratio	95% CL (cÔR)	Fisher's exact test (P)
NT 77	56	2.58	(.076, 87.86)	0.56
NT 77	88	66.42	(3.06, 1441)	<0.001
NT 77	119	26.87	(1.34, 539.92)	<0.05
NT 126	56	2.58	(.076, 87.86)	0.56
NT 119,126	88	4.89	(0.19, 131.83)	0.40

infected cells (20) These data demonstrate that MHV persistence does not require the presence of intraleader mutations and ORFs for establishing or maintaining viral persistence in vitro Rather these data support findings by Liao et al (21) and Zhang et al (22) that a highly stable MHV leader RNA sequence and leader / body junction are probably an absolutely essential cis / trans acting elements in virus transcription during acute infection and persistent infection in vitro The difference in the evolution of the BCV and MHV mRNAs is intriguing, but difficult to explain The most likely interpretation of these data is that different virus or host systems may provide different environments and selective pressures on the coronavirus genome, resulting in the evolution and accumulation of different mutations that initiate and maintain viral persistence (23)

While intraleader ORFs did not develop in mRNAs from persistently-infected DBT cells, MHV persistence was significantly associated with the evolution and accumulation of a specific mutation in the 5' UTR of the genomic RNA that resulted in the appearance of a new 16 amino acid ORF Such a mutation / ORF might result in reduced translation and expression of the MHV polygenic polymerase region readily explaining the reduction in virus transcription Alternatively, this mutation may contribute to MHV persistence by enhancing the maintenance, replication and expression of the genomic RNA Both mechanisms are likely since the 5' UTR of many positive strand RNA viruses modulate viral protein synthesis, or function in replication and transcription of viral positive- and negative-stranded RNAs (24-27) In the case of MHV, the 5' UTR contains both trans and cis acting elements that are critical for virus replication and the synthesis of subgenomic mRNAs (21), may interact with cellular proteins (28), or regulate translation of the 22 kb polygenic polymerase region Interestingly, the 5' UTR mutation resides within a 9 nt domain which may also regulate the initial synthesis of the viral subgenomic mRNAs and promote leader switching between RNA templates (28) While the exact function of this mutation in MHV is unclear and under study, the in vitro model for MHV persistence will provide a rationale framework for elucidating the mechanisms by which coronaviruses persist and evolve in humans and experimental animals (29-33)

ACKNOWLEDGMENTS

We thank Boyd Yount Jr for excellent technical assistance This work was supported by NIH grant AI-23946

REFERENCES

1 Baybutt HN, Wege H, Meulen VT, et al J Gen Virol 65 915-924, 1984
2 Hofmann MA, Sethna PB, Brian DA J Virol 64 4108-4114, 1990
3 Stohlman SA, Weiner LP Arch Virol 57 53-61, 1978
4 Knobler RL, Lampert PW, Oldstone MB Nature 298 279-280, 1982
5 Perlman S, Jacobsen G, Olson AL, Afifi A Virology 175 418-426, 1990
6 Fleming JO, Houtman JJ, Baker S, et al In Laude H and Vautherot JF (eds), Coronaviruses, Plenum Press, New York, 1994, pp327-332
7 Holmes KV, Behnke JN In Meulen VT, Siddell S, Wage H (eds), Biochemistry and Biology of Coronaviruses, 1981, pp287-299
8 Stohlman SA, Sakaguchi AY, Weiner LP Virology 98 448-455, 1979
9 Hirano N, Goto N, Makino S, Fujiwara K In Meulen VT, Siddell S, Wage H (eds), Biochemistry and Biology of Coronaviruses, 1981, pp301-308
10 Hingley ST, Gombold JL, Lavi E, Weiss SR Virology 200 1-10, 1994
11 Ahmed R, Stevens JG In Fields BN, Knipe DM (eds) Virology, Raven, New York, 1990, vol 1 pp241-265

12 Hoffman MA, Senanayake SD, Brian DA Proc Natl Acad Sci USA 90 11733-11737, 1993

13 Schaad MC, Stolhman SA, Baric RS, et al Virology 177 634-645, 1990

14 Fleming JO, Stohlman SA, Weiner LP, et al Virology 131 296-307, 1983

15 Sawicki SG, Sawicki DL J Virol 64 1050-1056, 1990

16 Makino S, Lai MMC J Virol 63 5285-5292, 1989

17 Bernard Rosner In Bernard Rosner (eds), Fundamentals of Biostatistics(3rd Ed), PWS-KENT Publishing Company, Boston, MA , 1990

18 Mizzen L Cheley S, Anderson R, et al Virology 128 407-417, 1983

19 Stohlman SA, Sakaguchi AY, Weiner LP Life Sci 24 1029-1036, 1979

20 Hofmann MA, Chang RY, Ku S, Brian DA Virology 196 163-171, 1993

21 Liao CL, Lai MMC J Virol 68 4727-4737, 1994

22 Zhang X, Liao CL, Lai MM J Virol 68 4738-4746, 1994

23 Domingo E, Holland JJ In Morse SS (eds), The evolutionary Biology of Viruses, Raven Press, New York, 1994, pp161-184

24 Scheper GC, Thomas AA, Voorma HO Biochim Biophys Acta 1089(2) 220-226, 1991

25 Elroy SO, Fuerst TR, Moss B Proc Natl Acad Sci USA 86 6126-6130, 1989

26 Martinez-Salas E, Sais JC, Domingo E, et al J Virol 67 3748-3755, 1993

27 Rohll JB, Percy N Barclay WS, et al J Virol 68 4384-4391, 1994

28 Furuya T, Lai MMC J Virol 67 7215-7222, 1993

29 Fleming JO, et al J Virol 58 869-875, 1986

30 Kyuwa S, et al Virology 65 1789-1795, 1991

31 Weismiller DG, Sturman LS, holmes KV, et al J Virol 64 3051-3055, 1990

32 Lucas A, Coulter M, Flintoff W, et al Virology 88 325-337, 1978

33 Murray RS, Cai GY, Cabirac GF, et al Virology 188 274-284, 1992

10

SPREAD OF MHV-JHM FROM NASAL CAVITY TO WHITE MATTER OF SPINAL CORD

Transneuronal Movement and Involvement of Astrocytes

S. Perlman, N. Sun, and E. M. Barnett

Departments of Pediatrics and Microbiology
University of Iowa
Iowa City, Iowa

ABSTRACT

C57Bl/6 mice infected intranasally with mouse hepatitis virus, strain JHM (MHV-JHM) develop hindlimb paralysis with histological evidence of demyelination several weeks after inoculation. Virus must spread from the site of inoculation, the nasal cavity, to the site of disease, the white matter of the spinal cord. It has been shown previously that after intranasal inoculation, virus enters the brain via the olfactory nerve and spreads to infect many of its neuroanatomic connections within the central nervous system (CNS). In this report, it is shown that virus infecting the spinal cord is first detected in the gray matter, with spread occurring to the white matter soon thereafter. Astrocytes are heavily infected during the process of spread from the gray to the white matter of the spinal cord. Since astrocytes are in intimate contact with neuronal synapses and are themselves connected via gap junctions, these results suggest that astrocytes may be a conduit for the spread of virus in these mice. Astrocytes provide factors for the proliferation and survival of oligodendrocytes, and widespread infection of these cells might contribute to the demyelinating process eventually observed in these mice. Additionally, since virus first appears at specific locations in the spinal cord, it should be possible to determine the source of the virus infecting the cord. While the results are not definitive, the data are most consistent with virus spreading from the ventral reticular formation to the gray matter of the cervical spinal cord.

INTRODUCTION

Mouse hepatitis virus (MHV), a member of the coronavirus family, causes hepatitis, enteritis and encephalitis in susceptible rodents[1]. MHV strain JHM (MHV-JHM) is highly neurotropic and causes acute and chronic encephalomyelitis[2]. The most virulent strains, characterized in part by encoding a full length (4139 nucleotides) surface glycoprotein (S)[3], cause an acute encephalitis in nearly all strains of mice, with death occurring in 5-7 days. This acute infection can be modified so that most mice survive, but instead develop histological evidence of demyelination. Modifications that result in this scenario include use of attenuated strains of MHV-JHM[4] or passive administration of either protective antibody or T cells[5,6]. Virus or viral products can be detected in the white matter of these mice, although mice are asymptomatic in most cases.

In the model developed in this laboratory, suckling mice are inoculated intranasally with MHV-JHM and are protected from the acute, fatal infection by nursing with dams previously immunized against the virus[7]. Several weeks later, a variable percentage (40-90%) develop hindlimb paralysis with histological evidence of a demyelinating encephalo-myelitis. Both the clinical and histological manifestations of this disease are primarily in the spinal cord suggesting that virus spreads to the spinal cord from the original site of inoculation (the nasal cavity). In several recent publications, the possible pathways used by the virus to spread within the brain have been described. First, the results indicate that virus spreads transneuronally, and presumably, trans-synaptically from the olfactory bulb. Surgical ablation of both olfactory bulbs or chemical destruction of the olfactory epithelium prevents virus entry into the brain, suggesting that virus spreads via infection of olfactory receptor neurons[8]. In contrast to what is observed with pseudorabies virus or herpes simplex virus[9,10], MHV-JHM does not enter the brain via the trigeminal nerve or via the sympathetic and parasympathetic nerves which also innervate the nasal cavity. Consistent with these observations, direct inoculation into the olfactory bulb labels precisely the same structures as does inoculation into the nasal cavity. The tropism for the olfactory bulb is so great that virus appears to enter the brain solely via the olfactory nerve even after inoculation into peripheral sites such as the tooth pulp or peritoneum (unpublished observations). Second, virus appears to spread primarily in a retrograde direction (from the axon to the cell body), although this is difficult to prove definitively since the olfactory bulb is reciprocally connected to its primary connections in most cases. Consistent with this, the olfactory tubercle, which receives projections from the bulb, but does not itself project to the bulb, only is rarely infected by MHV-JHM[10]. A recent publication suggests that MHV-JHM is released from both the dendritic and axonal surfaces of cultured neurons[11]. Release from the apical surface would make anterograde spread possible, but this direction of spread does not appear to occur in the infected animal to a significant extent.

Third, many, but not all of the primary connections of the main olfactory bulb are infected by MHV-JHM. Of particular note, the locus coeruleus is not infected by MHV-JHM and the hippocampus shows only minimal evidence of infection[10]. Both of these structures send prominent projections to the olfactory bulb. This distribution of infection is not shared by all viruses. Thus, after intranasal or intrabulbar infection, herpes simplex virus type I infects both of these structures, but does not infect other structures which are infected by MHV-JHM[10]. In a recent publication, rabies virus was shown to infect nearly the same structures as MHV-JHM, with sparing of the locus coeruleus[12]. The explanation for this sparing of the locus coeruleus by these two neurotropic RNA viruses, but not by a neurotropic herpesvirus remains to be determined.

RESULTS AND DISCUSSION

These previous experiments were all performed with young adult mice (6 weeks old) in which MHV-JHM was inoculated into the nasal cavity or olfactory bulb. These mice die before there is sufficient time for virus to reach the spinal cord. In previous reports, passive administration of neutralizing anti-MHV monoclonal antibody directed against the S glycoprotein fully protected mice from the acute encephalitis, but did not prevent demyelination[5]. Mice remained asymptomatic in that study. In preliminary studies, we delivered varying amounts of antibody at different times relative to the intranasal inoculation of virus. Administration of antibody prior to virus resulted in complete protection from the acute disease, but protection was so complete that we could not detect virus in the brain by in situ hybridization. We determined that 7 μL of antibody (1:1 mixture of two anti-S monoclonal antibodies-5A13.5 and 5B19.2, kindly provided by Dr. M. Buchmeier, The Scripps Research Institute) administered intraperitoneally 72 hours after infection fully protected mice from acute encephalitis, but virus could still be detected in the brain by in situ hybridization. By 72 hours, MHV-JHM has caused a significant infection of the olfactory bulb which would spread in the absence of antibody throughout the entire brain over the next 48-72 hours. The presence of relatively small amounts of protective antibody administered even after infection was well established in the olfactory bulb thus prevents an acute, fatal infection but not the transneuronal spread of virus to distal connections of the bulb. In our initial experiments we showed that the same brain structures were infected in the presence of antibody as we observed previously in the absence of antibody. We also showed that histological evidence of demyelination could be detected in the spinal cord 2 months after inoculation even though the mice remained asymptomatic, suggesting that virus had spread to the cord and persisted in this structure.

To determine the initial site of infection within the spinal cord, three week old C57Bl/6 mice were inoculated intranasally with virus and protected from the acute, fatal infection by the intraperitoneal administration of monoclonal antibody. At 6-7 days post infection, virus was readily detected in the spinal cord by both in situ hybridization and immunohistochemical techniques (Figure 1). Virus was first noted in the gray matter of the cervical spinal cord in laminae V-VII, although spread to the white matter occurred very soon thereafter. The initial site of labeling was always the same and suggested that spread to the spinal cord occurred from a single site or a few sites in the brain. In the next set of analyses, we attempted to determine the source of the virus which infected the cord. For these analyses, we assumed that virus spread solely in the retrograde direction, as this is consistent with previous studies.

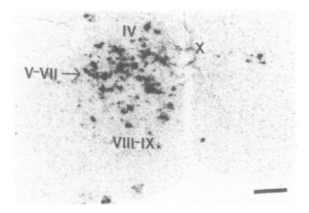

Figure 1. Virus appearance in the gray matter of the spinal cord. Brains and spinal cords were harvested at 6-7 days after intranasal inoculation with MHV-JHM and frozen in OCT. Sections were prepared and analyzed by in situ hybridization as described previously[10]. Virus was initially detected in laminae V-VII and then as shown in the figure, rapidly spread to involve other parts of the gray matter. A small amount of virus can also be detected in the white matter in this section. The approximate position of each lamina is shown in the figure. Magnification bar-200 μm.

The spinal cord has major projections to the reticular formation, the thalamus, the hypothalamus, the midbrain and the cerebellum, as well as less prominent connections to other CNS areas such as the ventral forebrain[13][17] Specific sites in the spinal cord project to each location For example, the spinothalamic tract originates primarily in the cervical cord, with a smaller projection from the lumbar spinal cord Within the rostral spinal cord, this tract originates from both dorsal and ventral sites Analysis of each of the above sites did not establish any single one as the definitive source for the virus infecting the spinal cord However, the cerebellum was not likely to be the source for virus, since the spinal cord projects to the cerebellar cortex and medial and interposed nuclei, none of which is infected by MHV-JHM to a significant extent On the other hand, the ventral reticular formation is likely to be the source for virus infecting the cord First, the ventral reticular formation receives many projections from the spinal cord, with a very high representation of fibers originating in the appropriate laminae (V-VII) of the cervical region Second, the ventral reticular formation is heavily infected by MHV-JHM in all cases, and this structure is labeled just prior to infection of the spinal cord Other structures which receive projections from the upper spinal cord, such as the midline thalamic nuclei, are not as heavily or consistently infected by MHV-JHM as is the reticular formation Thus, the specificity of labeling in the cervical spinal cord and reticular nuclei and the time course of the MHV-JHM infection all strongly suggest that these nuclei are the source for the spinal cord infection

A summary of the data is shown in Figure 2 In this figure, we assume that virus spreading to the cord originates from one of the four structures (ventral reticular formation, midline thalamic nuclei, lateral hypothalamus, central gray) which receives projections from the cord and is infected by MHV-JHM Likely pathways from the nasal cavity to these structures are shown, although connections to the pontine and medullary reticular formation are sufficiently ill-defined so that this approach is not possible for these structures

Once virus has spread to the gray matter of the spinal cord, it can be detected soon thereafter in the white matter In the next set of experiments, we defined more precisely the basis for this spread The presence of neutralizing antibody made it likely that virus moved to the white matter via cell-to-cell spread and not via the extracellular fluid In theory, virus could spread from the gray to the white matter without exposure to the extracellular fluid via either axons or astrocyte-astrocyte connections Astrocytes are extensively connected via gap junctions and have been postulated to form a giant functional syncytium[18] Astrocytes can be specifically identified using antibody to glial fibrillary acidic protein (GFAP) In the next set of experiments, infected spinal cords were simultaneously assayed for MHV-JHM

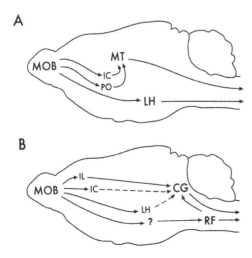

Figure 2. Pathways of virus spread from the site of inoculation, the nasal cavity, to the site of disease, the spinal cord Virus could spread from the brain to the spinal cord via the thalamus (A), the lateral hypothalamus (A), the reticular formation (B) or the midbrain (central gray) (B) The probable pathways involving these structures are shown in the figure and are indicated by solid lines Dotted lines show other pathways which are present, but less likely to be used by the virus to spread from the nasal cavity to the spinal cord Virus could spread to the reticular formation via many possible pathways, so for the sake of simplicity, no specific one is indicated Abbreviations-MOB-main olfactory bulb, PO-piriform cortex, MT-midline thalamic nuclei, IC-insular cortex, LH-lateral hypothalamus, IL-infralimbic cortex, CG-central gray, RF-reticular formation

RNA and astrocyte antigens The results showed that astrocytes were heavily infected during the initial stages (days 6-7) of virus spread to the white matter, with infection appearing to radiate throughout the ventral and lateral parts of the spinal cord These results are in general agreement with our previous results, in which we showed that a substantial fraction of infected cells in both symptomatic and asymptomatic mice were astrocytes[19]

Although these mice are asymptomatic, these results have implications for the pathogenesis of the demyelinating disease caused by MHV-JHM First, astrocytes are in close contact with oligodendrocytes, and are connected to them via gap junctions Virus could spread from astrocytes to oligodendrocytes at sites of gap junctions, since the two cellular membranes are juxtaposed at these locations MHV-JHM has a very fusogenic surface glycoprotein which is able to fuse cells lacking cellular receptor for the virus[20] Presumably, this protein would facilitate viral spread, even at sites lacking virus receptor Second, astrocytes are believed to produce factors critical for oligodendrocyte differentiation, proliferation and survival, as well as cytokines, such as a tumor necrosis factor, which are toxic for oligodendrocytes[21] Astrocyte dysfunction might contribute to the demyelinating process, either by lack of production of important trophic factors, or production of toxic agents In conclusion, infection of astrocytes may be the pathway by which virus spreads from neurons to oligodendrocytes and may also indirectly contribute to the demyelinating process

ACKNOWLEDGMENTS

This research was supported by grants from the N I H S P was supported by a Research Career Development Award from the N I H , N S by an N I H Institutional Training Grant and E B by a N I M H predoctoral fellowship

REFERENCES

1 Compton, S R , Barthold, S W , Smith, A L The cell and molecular pathogenesis of coronaviruses Lab Anim Sci 1993,43 15-28

2 Weiner, L P Pathogenesis of demyelination induced by a mouse hepatitis virus (JHM virus) Arch Neurol 1973,28 298-303

3 Parker, S E , Gallagher, T M , Buchmeier, M J Sequence analysis reveals extensive polymorphism and evidence of deletions within the E2 glycoprotein gene of several strains of murine hepatitis virus Virology 1989,173 664-673

4 Haspel, M V , Lampert, P W , Oldstone, M B A Temperature-sensitive mutants of mouse hepatitis virus produce a high incidence of demyelination Proc Natl Acad Sci 1978,75 4033-4036

5 Buchmeier, M J , Lewicki, H A , Talbot, P J , Knobler, R L Murine hepatitis virus-4 (strain JHM)-induced neurologic disease is modulated in vivo by monoclonal antibody Virology 1984,132 261-270

6 Stohlman, S A , Matsushima, G K , Casteel, N , Weiner, L P In vivo effects of coronavirus-specific T cell clones DTH inducer cells prevent a lethal infection but do not inhibit virus replication J Immunol 1986,136 3052-3056

7 Perlman, S , Schelper, R , Bolger, E , Ries, D Late onset, symptomatic, demyelinating encephalomyelitis in mice infected with MHV-JHM in the presence of maternal antibody Microbial Pathog 1987 2 185-194

8 Barnett, E M , Perlman, S The olfactory nerve and not the trigeminal nerve is the major site of CNS entry for mouse hepatitis virus, strain JHM Virology 1993,194 185-191

9 Sabin, A B Progression of different nasally instilled viruses along different nervous pathways in the same host Proc Soc Exp Med Biol 1938,38 270-275

10 Barnett, E M , Cassell, M , Perlman, S Two neurotropic viruses, herpes simplex virus type I and mouse hepatitis virus, spread along different neural pathways from the main olfactory bulb Neuroscience 1993,157 1007-1025

11 Pasick, J , Kalicharran, K , Dales, S Distribution and trafficking of JHM coronavirus structural proteins and virions in primary neurons and the OBL-21 neuronal cell line J Virol 1994,68 2915-2928

12 Astic, l , Saucier, D , Coulon, P , Lafay, F , Flamand, A The CVS strain of rabies virus as transneuronal tracer in the olfactory system of mice Brain Res 1993,619 146-156

13 Granum, S L The spinothalamic system of the rat I Locations of cells of origin J Comp Neurol 1986,247 159-180

14 Burstein, R , Cliffer, K D , Giesler, J , G J Cells of origin of the spinohypothalamic tract in the rat J Comp Neurol 1990,291 329-344

15 Cliffer, K D , Burstein, R , Giesler, J , G J Distributions of spinothalamic, spinohypothalamic and spinotelencephalic fibers revealed by anterorgrade transport of PHA-L in rats J Neurosci 1991,11 852-868

16 Villanueva, L , de Pommery, J , Menetrey, D , Le Bars, D Spinal afferent projections to subnucleus reticularis dorsalis in the rat Neurosci Lett 1991,134 98-102

17 Yezierski, R P , Mendez, C M Spinal distribution and collateral projections of rat spinomesencephalic tract cells Neuroscience 1991,44 113-130

18 Mugaini, E Cell junctions of astrocytes, ependyma and related cells in the mammalian central nervous system, withe emphasis on the hypothesis of a generalized functional syncytium of supporting cells In Fedoroff, S , Vernadakis, A (eds) Astrocytes Academic Press, New York 1986 pp329-362

19 Perlman, S , Ries, D The astrocyte is a target cell in mice persistently infected with mouse hepatitis virus, strain JHM Microbial Pathog 1987,3 309-314

20 Gallagher, T , Buchmeier, M , Perlman, S Cell receptor-independent infection by a neurotropic murine coronavirus Virology 1992,191 517-522

21 Gard, A L Astrocyte-oligodendrocyte interactions In Murphy, S , (eds) Astrocytes Pharmacology and Function Academic Press, San Diego 1993 pp 331-354

11

IN VITRO INTERACTION OF CORONAVIRUSES WITH PRIMATE AND HUMAN BRAIN MICROVASCULAR ENDOTHELIAL CELLS

G. F. Cabirac[1,2,3], R. S. Murray[2,4], L. B. McLaughlin[1], D. M. Skolnick[1], B. Hogue[5], K. Dorovini-Zis[6], and P. J. Didier[7]

[1] Rocky Mountain Multiple Sclerosis Center
[2] Colorado Neurological Institute
 Swedish Medical Center, Englewood, Colorado
[3] Department of Biochemistry, Biophysics and Genetics
 University of Colorado Health Sciences Center
 Denver, Colorado
[4] National Jewish Center for Immunology and Respiratory Medicine
[5] Department of Microbiology and Immunology
 Baylor College of Medicine, Houston, Texas
[6] Department of Pathology, University of British Columbia
 Vancouver, British Columbia, Canada
[7] Department of Pathology, Tulane Regional Primate Research Center
 Covington, Louisiana

ABSTRACT

Primary human and primate brain microvascular endothelial cells were tested for permissiveness to coronaviruses JHM and 229E. While sub-genomic viral RNAs could be detected up to 72 hours post-infection, primate cells were abortively infected and neither virus caused cytopathology. Human cells were non-permissive for JHM but permissive for 229E replication; peak production of progeny 229E and observable cytopathic effects occurred approximately 22 and 32 hour post-infection, respectively. Using the criterion of cytopathology induction in infected endothelial cells, 229E was compared to other human RNA and DNA viruses. In addition, virus induced modulation of intercellular adhesion molecule 1 (ICAM-1), vascular cell adhesion molecule 1 (VCAM-1) and HLA I was monitored by immunostaining of infected cells.

INTRODUCTION

While numerous studies of coronavirus infection in rodents has generated valuable information on the mechanisms of virus induced CNS disease[1-7], the question of coronavirus CNS disease in humans justifiably remains open. Many viruses with confirmed neurotropic potential can infect the CNS of the host following primary infection at extra neural sites. Studies on natural infections of humans and on both natural and experimental infections of animals indicate that viruses enter the CNS primarily through the vascular endothelium[8-10]. Infection of CNS tissue may occur after virus replicates in or is transported through endothelial cells[8,9]. Because of reports linking coronaviruses to human CNS disease[11-13] and the observation that peripheral inoculation of JHM into primates results in viral RNA/antigen expression in areas proximal to CNS blood vessels[14] we have chosen to characterize the *in vitro* interaction of coronaviruses with both primate and human brain microvascular endothelial cells.

MATERIALS AND METHODS

Cells and Viruses

DBT, WI-38, HCT, Vero and BSC-1 cells were grown in Dulbecco's modified Eagle's medium (DMEM; Gibco/BRL) supplemented with 10% fetal bovine serum (FBS; Gibco/BRL), 100 U/ml penicillin and 100μg/ml streptomycin. MHV JHM was obtained from Dr. Steve Stohlman[2]. JHM was assayed for hemagglutinating activity as previously described[15]. Human coronavirus 229E was obtained from the American Type Culture Collection (ATCC). Vaccinia (WR strain), herpes simplex type I (HSV-I, strain F) and a clinical isolate of varicella zoster virus (VZV) were generous gifts from Dr. Donald Gilden. Vaccinia, HSV-I, JHM, 229E and OC43 were propagated and titered on BSC-1, Vero, DBT, WI-38 and HCT cells, respectively. 229E was grown at 34°C and all other viruses at 37°C. Multiplicity of infection (m.o.i.) stated in this work, in reference to infection of endothelial cells, is calculated based on infectivity of the specific virus relative to its indicator cell line, e.g., 229E on WI38 cells.

UV-inactivation of JHM and 229E for use in RNA analysis was as follows. A volume of virus diluted in 5% FBS medium was placed into a plastic dish so that the depth of liquid was approximately 2mm. Dishes were placed in ice-H_2O and the inocula were exposed for 15 min to 15 watt 300nm UV-light bulbs placed a distance of 5cm from the surface of the liquid. Prior titration using this set-up showed that infectivity of either JHM on DBT or 229E on WI38 cells was completely eliminated after 5-10 minutes of UV exposure. Inactivated inocula were immediately used to infect cells used for RNA analysis.

Rhesus brain endothelial cells (RhBEC) were derived from animals in the breeding colony at Tulane Regional Primate Research Center (TRPRC) that were normally lost due to trauma and/or cachexia. Occasionally, control animals from other investigator projects were sacrificed. Animals of both sex and of all ages (none greater than 10 years of age) were used. No animals from studies involving infectious agents or drugs were used. Isolation and propagation of RhBEC was as follows. One-quarter to one-half of a monkey brain was generally used for this procedure. Brain tissue was stored in RPMI medium containing antibiotics and processed as soon as possible after collection. Meninges was removed from the brain and discarded then the tissue was put into approximately 100 ml of Hanks balanced salt solution in a beaker. The tissue was chopped into small pieces, then cut into fine pieces against the side of the beaker using a scapel. The resulting pieces were homogenized by drawing the suspension in and out of a 10 ml syringe. At this stage the tissue was processed using a Cellector (Bellco). If a large amount of tissue was used, processing was started with the 520µm screen, otherwise the 280µm screen was used. If the 520µm screen was used to start, the material on the screen was discarded and the suspension coming through the screen was saved for further use; on following screens the liquid coming through the screen was discarded and the material on screen was collected. When the homogenate became less dense and the liquid in the Cellector was reduced to approximately 10ml, the 190µm screen was used until the liquid coming through cleared and the small vasculature could be seen sticking to the screen. The vasculature was rinsed from the 190µm screen (total volume equal to 9 ml) and added to a small beaker. 1ml of 1% collagenase/dispase (Boehringer Mannheim) was added and the mixture incubated at 37°C with constant shaking for 1 hour. After incubation, 1ml of the suspension was placed into ten 60mm dishes. These were rotated until the liquid evenly coated the dishes and then the dishes were allowed to incubate a few minutes undisturbed. 1-2 ml of medium was slowly added to each dish then endothelial growth factor was added to a final concentration of 100µg/ml. The dishes were incubated overnight at 37°C then the media discarded and replaced with fresh medium and growth factor. All cultures were stained with rabbit polyclonal anti-factor VIII-related antigen (fVIIIRAg) antibody (Dako) to verify phenotype. All endothelial cells used in this study were grown in Iscove's modified Dulbecco's medium, 20% FBS, 100µg/ml heparin (Sigma; H-3149), 100 U/ml penicillin and 100µg/ml streptomycin (complete IMDM) supplemented with endothelial cell growth supplement (ECGS, Sigma; E2759) to a final concentration of 100µg/ml. Culture medium was changed every three days and confluent monolayers were passaged at a 1:3 to 1:6 split. Cultures were periodically stained for fVIIIRAg to monitor phenotype. All experiments with the RhBEC were done at or below a passage level of five.

The isolation and characterization of human brain endothelial cells (HBEC) has previously been described[16-18]. Cells were grown in M199 medium, 25 mM HEPES, 10 mM sodium bicarbonate, 10% FBS, 100µg/ml heparin, 100 U/ml penicillin and 100µg/ml streptomycin supplemented with 20 µg/ml ECGS on human fibronectin coated plastics (Corning). All virus infections were done on confluent HBEC monolayers. All experiments with the HBEC were done at or below a passage level of four.

One-Cycle Virus Growth Assays

Virus inocula were adsorbed to endothelial cell monolayers at 4°C for 30 minutes, unabsorbed virus was washed from monolayers by three rinses with cold medium then fresh medium added to monolayers; the zero time point samples were frozen immediately after addition of medium. At each time point duplicate or triplicate samples were frozen at -70°C. Time point samples were thawed and virus assayed on the appropriate indicator cell lines.

RNA Analysis

JHM and 229E inocula used for infection of RhBEC were in DME + 5% FBS. JHM and 229E were adsorbed at 37°C and 34°C, respectively, to monolayers for 1 hour. The monolayers were washed twice with warmed complete IMDM then complete IMDM + 100µg/ml endothelial cell growth supplement was added. All 229E infected cultures were incubated at 34°C. Total RNA was extracted from virus infected or mock infected cells by the single-step method of Chomczynski and Sacchi[19]. All reagents used for RNA analysis were prepared with water obtained from a PhotoCatalytics water system (PhotoCatalytics, Inc.). This instrument photoxidizes residual organic compounds in feed water not removed by conventional upstream water purification treatments; eluant water is free of RNases and therefore does not require diethylpyrocarbonate (DEPC) treatment before use in RNA manipulations[20]. Purified RNA was dissolved in 2mM EDTA and poly A$^+$ RNA prepared using Dynal Oligo (dT)$_{25}$ beads. Poly A$^+$ RNA samples were glyoxylated and electrophoresed on 1.0% agarose gels as described previously[21]. RNA was transferred to membranes (Nytran Plus, Schleicher & Schuell) by electroblotting in 40 mM Tris acetate, 1 mM EDTA (1X TAE buffer) at 4°C then immobilized by baking the membrane in a vacuum oven at 70-80°C for 1 hour. The cloned cDNA probes used to detect JHM and 229E RNA were cDNA clones G344 and L8, respectively[22,23]. cDNA was labeled by the random primer method[24] with α^{32}P dATP. Membranes were prehybridized in 50% deionized formamide, 10X Denhardt's, 2% SDS, 5X SSPE, and 200µg/ml single-stranded DNA at 44°C for 3 hours. Prehybridization buffer was removed, the membrane briefly rinsed in warmed hybridization buffer (without probe) then hybridization buffer containing denatured probe at 1 X 10^6 cpm/ml. Hybridization buffer was 50% deionized formamide, 5X Denhardt's, 0.2% SDS, 5X SSPE, and 100µg/ml DNA. Hybridization was at 44°C for 16-18 hours. Membranes were washed twice in 5X SSPE, 0.1% SDS for 15 minutes each, once in 1X SSPE, 0.2% SDS for 15 minutes, and once in 0.1X SSPE, 0.2% SDS for 15 minutes, all washes done at room temperature. A final wash was done in 1X SSPE, 1.0% SDS at 60°C for 30 minutes.

Immunohistochemical Staining

Monoclonal antibodies specific for human inter-cellular adhesion molecule 1 (ICAM-1), vascular cell adhesion molecule 1 (VCAM-1), and HLA-I, were obtained from commercial sources (Genzyme and Becton Dickinson). MAbs J.3.1 specific for JHM nucleocapsid and 5-11H.6 specific for 229E spike glycoprotein were gifts from Dr. John Fleming and Dr. Pierre Talbot, respectively. Cells used for staining were grown on either glass cover slips or 8-chambered plastic slides (Nunc) that were coated with fibronectin (described above) prior to seeding. Cells were briefly washed in cold serum-free medium, fixed in -20°C methanol for 10 minutes then washed three times in phosphate-buffered saline (PBS). Fixed monolayers were incubated with 5% normal goat serum (NGS) in PBS, 0.1% BSA for 20 minutes followed by primary antibody diluted in PBS, 0.1% BSA, 2% NGS for 1 hour. Monolayers were washed three times for 10 minutes each in PBS, incubated with gold-cojugated goat anti-mouse antibody (Amersham AuroProbeLM) for 1 hour, washed three times in PBS then three times in H$_2$O for 5 minute each. Bound, gold-conjugated antibody was detected by silver staining (Amersham IntenSEM) using the manufacturers recommended conditions.

RESULTS

Virus Replication on Brain Endothelial Cells and Cytopathic Effects

Standard one-cycle replication assays of JHM or 229E on RhBEC or HBEC showed that HBEC were permissive for 229E but not JHM and that RhBEC were non-permissive for both 229E and JHM (Figure 1). Each virus was tested for production of progeny virus from both RhBEC and HBEC on one animal/patient cell isolate. Peak titers of progeny 229E from infected HBEC occurred approximately 22 hr p.i.; input infectious JHM or 229E in RhBEC or JHM in HBEC decayed over the observed time course.

For the coronaviruses tested there was a correlation of CPE to productive infection, i.e., 229E but not JHM was cytopathic for HBEC and neither virus was cytopathic for RhBEC. Figure 2 shows mock infected HBEC and 229E infected HBEC, A and B, respectively. 229E induced cytopathology was observable on infected monolayers approximately 32 hr p.i.. To compare these coronaviruses to other common human viruses, the brain

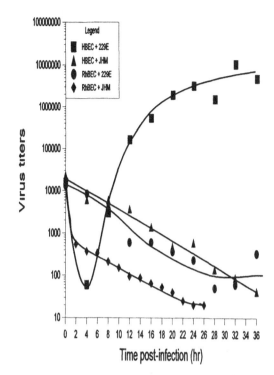

Figure 1. One-cycle virus replication assays on RhBEC and HBEC Titer values shown represent cell associated plus extracellular virus in cultures at each time point

endothelial cells were infected with a number of RNA and DNA viruses. Table 1 shows the tested viruses that were cytopathic or non-cytopathic for either RhBEC or HBEC. Three to five RhBEC isolates were used to test for CPE induction by each virus (left column); all virus infections of HBEC (right column) were done on one patient isolate.

Viral Antigen and RNA Production in Infected Cells

Immunostaining of 229E or JHM infected RhBEC monolayers with virus specic monoclonal antibodies showed that virus antigens were detectable up to 48 hr p.i.. Despite the presence of viral sub-genomic RNA in the infected cells (see below), the pattern of staining suggested that these antigens were from input virus and not from *de novo* synthesis in the cells; staining of JHM infected HBEC showed the same pattern of antigen as in the primate cells (data not shown). Staining with 229E specific MAb showed cytoplasmic viral

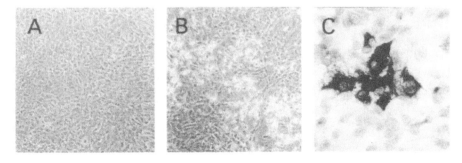

Figure 2. 229E CPE on HBEC and staining of viral antigen A, Mock infected HBEC, B, 229E infected HBEC ~38 hr p i ; C, 229E infected HBEC immunostained showing cytoplasmic viral antigen (~14 hr p i)

Table 1. Virus Induced Cytopathology in RhBEC and HBEC

RhBEC[a]	HBEC[b]
Cytopathic viruses tested:	Cytopathic viruses tested:
Vaccinia	Vaccinia
Herpes simplex type 1[c]	Herpes simplex type 1
	Varicella zoster virus
	Adenovirus types 1 & 7
	Echovirus type 9
	Coxsackievirus B5
	Coronavirus 299E
Non-cytopathic viruses tested:	Non-cytopathic viruses tested:
Coronaviruses:	Human cytomegalovirus[e]
JHM	Respiratory syncytial virus
SD[d]	Parainfluenza virus type 2
229E	Coronaviruses:
OC43	JHM
	SD
	OC43

[a]All viruses listed were tested on three to five different animal isolates.
[b]Viruses were tested on one patient isolate only.
[c]HSV-1 induced CPE was dependent on the animal isolate tested; some
isolates developed no CPE while other isolates were affected.
[d]SD designates putative MS tissue isolate[11].
[e]Towne strain.
All viruses not described in Materials and Methods were obtained from
the American Type Culture Collection and propagated on appropriate
cell lines.

antigen in productively infected HBEC (Fig. 2C). Staining of OC43 infected RhBEC or
HBEC with an OC43 cross-reactive HEV MAb (4E11.3) showed similar pattern of staining
to JHM or 229E non-productively infected cells (data not shown).

    Analysis of poly A$^+$ RNA extracted from infected RhBEC showed that subgenomic
RNAs were transcribed in both JHM and 229E infected endothelial cells. Figure 3 shows

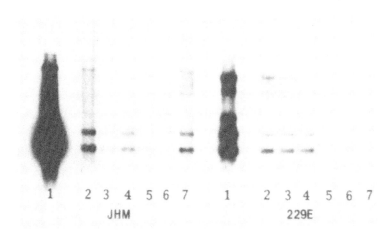

Figure 3. Northern analysis of RNA from JHM and 229E infected RhBEC. See text for description.

samples extracted 2, 8 and 24 hours p.i. from endothelial cells (JHM, left lanes 2, 4, 7;229E, right lanes 2-4). For both JHM and 229E the highest level of these transcripts appears to be at the 2 hour p.i. time point. Based on a comparison between the amount of RNAs loaded onto the gel and the autoradiographic signals, the quantities of JHM and 229E RNAs in the endothelial cells, at time of maximum production, are only 1-5% of the quantities in normal productively infected cells (DBT+JHM, left lane 1; WI38+229E, right lane 1). Despite the low levels produced in the endothelial cells, viral RNAs could be detected as late as 72 hours p.i. (data not shown). These results were repeatable for 229E and JHM with cells isolated from 2 and 5 different animals, respectively.

Infection of RhBEC with UV inactivated virus showed that production of detectable viral RNAs in cells was significantly reduced for JHM (lane 5) and 229E (lane 5). Using the same UV inactivated inocula and running approximately equivalent amounts of RNAs on the gel, JHM RNA could not be detected in infected DBT cells (JHM, lane 6) and a only a minor amount of 229E RNA could be detected in infected WI38 (2289E, lane 6). Viral RNAs could also be detected in JHM infected WI38 cells (JHM, lane 7) and 229E infected DBTs (229E, lane 7). RNA analysis was not done on virus infected HBEC.

ICAM-1, VCAM-1 AND HLA-I Expression on Infected HBEC

Infected HBEC monolayers were immunostained for ICAM-1, VCAM-1 and HLA-I to determine if virus can modulate expression of these cell surface antigens (Table 2). It appeared that there was an increase in ICAM-1 expression in both HSV-1 and 229E infected HBEC at 14 hr p.i.; vaccinia infected cells maintained base level of ICAM-1 expression. TNFα caused an expected large increase in ICAM-1 expression. No change in VCAM-1 expression was observed at 14 hr p.i. with the viruses tested. Similarly, HLA-I expression was not affected by any of the viruses tested or, unexpectedly, by TNFα.

DISCUSSION

Because the reticuloendothelial system clears hematogenous virus during most *in vivo* viremic infections the amount of cell-free virus in blood is usually low. Despite this, the vascular endothelium is the major route for natural CNS infection by viruses. Therefore to approximate one of the many factors involved in an *in vivo* CNS virus infection we have characterized the *in vitro* interaction of viruses with cultured brain endothelial cells. Cytopathology, virus replication and modulation of specific cell surface polypeptides were the criteria used to characterize coronavirus interaction with RhBEC and HBEC; the coronaviruses were also compared to other common human viruses.

Table 2. ICAM-1, VCAM-1 and HLA-I expression on virus infected HBEC

	ICAM-1	VCAM-1	HLA-I
Mock	+	±	±
TNFα	+++	++	±
Vaccinia	+	±	±
HSV-1	++	±	±
229E	++	±	±

Cells stained ~14 hr. post-infection/treatment

The rationale for using brain microvascular endothelial cells for this study is that there is evidence of endothelial cell phenotype heterogeneity that is dependent on tissue/organ origin. Endothelial cells isolated from specific vascular beds exhibit differences in expression of a number of markers when cultured *in vitro* and they maintain these differences over short-term passage[25-28]. In addition, there are also differences in response to virus infection when comparing tissue/organ specific endothelial cells[29-32]. Obviously, biological properties of cultured endothelial cells will not be identical to those of cells *in vivo* because of losing the modifying influence of the *in vivo* milieu. This has been shown for brain endothelial cells where extracellular matrix proteins, soluble factors released from astrocytes or even cell-cell contact with astrocytes can alter expression of certain markers[33]. Accurate *in vitro* determination of *in vivo* brain endothelial cell response to virus infections will probably require culturing the cells in the presence of these various modifying factors.

For JHM and 229E, induction of cytopathology in infected cells was predictive of a productive infection (see Fig. 1 and Table 1). Interestingly, when looking at the other viruses tested on HBEC there appeared to be a correlation between virus CPE and incidence of clinical CNS diseases[34]. Viruses that are known to cause encephalitis or meningitis produced cytopathology while viruses not identified as causative agents of CNS infection, e.g., RSV, did not induce CPE. Other than the coronaviruses, we did not confirm that CPE always indicated a productive infection or that lack of CPE was associated with a non-productive infection.

While neither JHM or 229E caused RhBEC cytopathology, the detection of viral sub-genomic RNA species indicates that a percentage of input virions uncoated in a manner that permitted some transcription of the virus genome. The results obtained from infection with UV-inactivated virus indicates that these RNAs were not present as a result of packaging in inoculum virions as demonstrated for BCV[35], TGEV[36], and IBV[37]. Preliminary immuno-precipitation experiments with JHM infected RhBEC showed that nucleocapsid was not synthesized (data not shown). This suggests that at least one block in virus replication in RhBEC occurs at the step of viral mRNA translation; this result also indicates that N detected by immunostaining of infected cells was from input virus. Since viral RNAs were detected in the non-permissive endothelial cells the question of receptor mediated versus non-specific entry of the viruses needs to be addressed. It is important to note that the stocks of JHM used for this work did not have hemagglutination activity.

We immunostained ICAM-1, VCAM-1 and HLA-I on infected HBEC as a prelimi-nary test to determine the effects of 229E on expression of endothelial cell surface antigens that are important in an immune response. ICAM-1 was the only one of the three cell-surface polypeptides affected at ~14 hr p.i.; the staining showed that the number of cells expressing elevated levels of ICAM-1was greater than the number of expected infected cells. This suggests that for 229E infected HBEC, ICAM-1 elevation was the result of secretion of some soluble mediator. This would be similar to the observation that CMV infection causes endothelial cells to secrete IL-1 resulting in upregulation of another adhesion molecule, E selectin[38]. Endothelium surface antigens may be differentially regulated by coronaviruses depending on the virus and host species since it has been reported that JHM causes a decrease in ICAM-1 expression on murine brain endothelial cells[39].

Endothelial cells play a crucial role in the control of inflammation, coagulation, leukocyte trafficking, tumor metastasis, and angiogenesis. The endothelium specific modu-lation of these processes is mediated by both endothelial cell associated molecules and secreted, soluble factors. During either a nonspecific, acute phase immune response or an antigen-specific, DTH response the expression of both cell associated and secreted bio-molecules by the affected endothelial cells can be altered by infecting virus. For coro-naviruses, a more thorough examination of virus interaction with endothelial cells is warranted. In addition to the data showing 229E infection of HBEC, the results with JHM suggest that other coronavirus strains have the potential to enter endothelial cells and undergo limited transcription, the possibility exist that endothelium specific factors are modulated despite a lack of virus replication It will be important to determine the effects of coronaviruses on the physiologic or pathophysiologic processes that occur at the CNS endothelium in humans

REFERENCES

1 Sorensen, O , Percy, D and Dales, S In vivo and in vitro models of demyelinating diseases III JHM virus infection of rats Arch Neurol 1980, 37 478-484

2 Stohlman, S A and Weiner, L P Chronic central nervous system demyelination in mice after JHM virus infection Neurol 1981, 31 38-44

3 Knobler, R , Haspel, M , and Oldstone, M Mouse hepatitis virus type-4 (JHM strain) - Induced fatal central nervous system disease I Genetic control and the murine neuron as the susceptible site of disease J Exp Med 1981, 153 832

4 Dubois-Dalcq, M E , Doller, E W , Haspel, M U , and Holmes,K Cell tropism and expression of mouse hepatitis viruses (MHV) in mouse spinal cord cultures Virol 1982, 119 317-331

5 Lavi, E , Gilden, D H , Highkin, M K and Weiss, S Persistence of mouse hepatitis virus A59 RNA in a slow virus demyelinating infection in mice as detected by in situ hybridization J Virol 1984, 51 563-566

6 Buchmeier, M , Lewicki, H , Talbot, P , and Knobler, R Murine hepatitis virus-4 (strain JHM) induced neurologic disease is modulated in vivo by monoclonal antibody Virol 1984, 132 261

7 Perlman, S , Jacobsen, G and Afifi, A Spread of a neurotropic murine coronavirus into the CNS via the trigeminal and olfactory nerves Virol 1989, 170 556

8 Johnson R T Viral Infections of the Nervous System New York Raven Press, 1982

9 Mims C A The Pathogeneis of Infectious Diseases 2nd ed London Academic Press, 1982

10 Johnson R T and Mims C A Pathogenesis of viral infections of the nervous system N Engl J Med 1968, 278 23-30,87-92

11 Burks, J S , DeVald, B L , Jankovsky, L D , and Gerdes, J C Two coronaviruses isolated from central nervous system tissue of two multiple sclerosis patients Science 1980, 209 933-934

12 Murray R S , Brown, B , Brian, D , and Cabirac, G F Detection of coronavirus RNA and antigen in multiple sclerosis brain Ann Neurol 1992, 31 525-533

13 Stewart, J N , Mounir, S , and Talbot, P J Human coronavirus gene expression in the brain of multiple sclerosis patients Virol 1992, 191 502-505

14 Cabirac, G F , Soike, K F , Hoel, K , Butunoi, C , Cai, G -Y, Johnson, S , and Murray, R S Entry of coronavirus into primate CNS following peripheral infection Micro Path 1994,

15 Hogue, B G and Brian, D A Structural proteins of human respiratory coronavirus OC43 Virus Res 1986, 5 131-144

16 Dorovini-Zis K , Prameya R , and Bowman P D Culture and characterization of microvascular endothelial cells derived from human brain Lab Invest 1991, 64 425-436

17 Wong D and Dorovini-Zis K Upregulation of intercellular adhesion molecule-1 (ICAM-1) expression in primary cultures of human brain microvessel endothelial cells by cytokines and lipopolysaccharide J Neuroimmunol 1992, 39 11-21

18 Huynh, H K and Dorovini-Zis K Effects of interferon-gamma on primary cultures of human brain microvessel endothelial cells Amer J Path 1993, 142 1265-1278

19 Chomczynski, P and Sacchi, N Single-step method of RNA isolation by acid guanidinium thiocyanate-phenol-chloroform extraction Anal Biochem 1987, 162 156-159

20 Cooper, G , Borish, L , Mascali, J , Watson, C , Kirkegaard, K , Morrissey, L , and Tedesco, J L The photocatalytic production of organic-free water for molecular biological and pharmaceutical applications J Biotech 1994, 33 123-133

21 Cabirac, G F , Mulloy, J J , Strayer, D S , Sell, S , and Leibowitz, J L (1986) Transcriptional mapping of early RNA from regions of the Shope fibroma and malignant rabbit fibroma virus genomes Virol 1986,153, 53-69

22 Budzilowicz, C J , Wilczynski, S P , and Weiss, S R Three intergenic regions of coronavirus mouse hepatitis virus strain A59 genome RNA contain a common nucleotide sequence that is homologous to the 3' end of the viral mRNA leader sequence Virol 1985, 53 834-840

23 Schreiber, S S , Kamahora, T , and Lai, M M C Sequence analysis of the nucleocapsid protein gene of human coronavirus 229E Virol 1989,169 141-151

24 Feinberg, A P, and Vogelstein, B A technique for radiolabeling DNA restriction endonuclease fragments to high specific activity Anal Biochem 1983, 132 6-13

25 Gumkowski, F, Kaminska, G , Kaminski, M , Morrissey, L W , and Auerbach, R Heterogeneity of mouse vascular endothelium In vitro studies of lymphatic, large blood vessel and microvascular endothelial cells Blood Vessels 1987, 24 11-23

26 Turner, R R , Beckstead, J H , Warnke, R A , and Wood, G S Endothelial cell phenotypic diversity In situ demonstration of immunologic and enzymatic heterogeneity that correlates with specific morphologic subtypes Amer J Clin Path 1987, 87 569-576

27 Belloni, P N and Nicolson, G L Differential expression of cell surface glycoproteins on various organ derived microvascular endothelia and endothelial cell cultures J Cell Physiol 1989, 136 398-410

28 Lodge, P A , Haisch, C E , Huber, S A , Martin, B , and Craighead, J C Biological differences in endothelial cells depending upon organ derivation Transplant Proc 1991, 23 216-218

29 Huber, S A , Haisch, C , and Lodge, P A Functional diversity in vascular endothelial cells role in coxsackievirus tropism J Virol 1990, 64 4516-4522

30 Goerdt, S and Sorg, C Endothelial heterogeneity and the acquired immunodeficiency syndrome a paradigm for the pathogenesis of vascular disorders Clin Investig 1992, 70 89-98

31 Lafon, M E , Gendrault, J L , Royer, C , Jaeck, D , Kirn, A , and Steffan, A M Human endothelial cells isolated from the hepatic sinusoids and the umbilical vein display a different permissiveness for HIV 1 Res Virol 1993, 144 99-104

32 Joseph, J , Kim, R , Siebert, K , Lublin, F D , Offenbach, C , and Knobler, R L Organ specific endothelial cell heterogeneity influences differential replication and cytopathogenicity of MHV-3 and MHV-4 Implications in viral tropism In Adv Expt Med Biol , (eds) Talbot P and Levy G 1995 Plenum Press, NY p 43

33 Abbott, N J , Revest, P A , and Romero, I A Astrocyte-endothelial interaction physiology and pathology Neuropath App Neurobiol 1992, 18 424-433

34 In Principles and Practice of Infectious Diseases (3rd Ed) (eds) Mandell, G L , Douglas, R G , and Bennett, J E 1990 Churchill Livingstone Inc

35 Hofmann, M A , Sethna, P B , and Brian, D A Bovine coronavirus mRNA replication continues through-out persistent infection in cell culture J Virol 1990, 64 4108-4114

36 Sethna, P B , Hofmann, M A , and Brian, D A Minus-strand copies of replicating coronavirus mRNAs contain antileaders J Virol 1991, 65 320-325

37 Cavanagh, D , Shaw, K , and Xiaoyan, Z Analysis of messenger RNA within virions of IBV In Adv Expt Med Biol , Vol 342 (eds) Laude H and Vautherot J-F 1993 Plenum Press, NY

38 Span, A H , Mullers, W , Miltenburg, A M , and Bruggeman C A Cytomegalovirus induced PMN adher-ence in relation to an ELAM-1 antigen present on infected endothelial cell monolayers Immunol 1991, 72 355-360

39 Joseph, J , Knobler, R L , Lublin, F D , and Burns, F R Regulation of the expression of intercellular adhesion molecule-1 (ICAM-1) and the putative adhesion molecule basigin on murine cerebral endothe-lial cells by MHV-4 (JHM) In Adv Expt Med Biol , Vol 342 (eds) Laude H and Vautherot J-F 1993 Plenum Press, NY

12

TREATMENT OF RESISTANT A/J MICE WITH METHYLPREDNISOLONE (MP) RESULTS IN LOSS OF RESISTANCE TO MURINE HEPATITIS STRAIN 3 (MHV-3) AND INDUCTION OF MACROPHAGE PROCOAGULANT ACTIVITY (PCA)

R. J. Fingerote,[1] J. L. Leibowitz,[2] Y. S. Rao,[1] and G. A. Levy[1]

[1] Department of Medicine
The Toronto Hospital
University of Toronto
Toronto, Ontario, Canada
[2] Department of Pathology
University of Texas Health Sciences Center
Houston, Texas

ABSTRACT

BALB/cJ mice die of fulminant hepatitis within 7 days of exposure to murine hepatitis virus strain 3 (MHV-3) whereas A/J mice are fully resistant to the lethal effects of MHV-3 infection. Previous studies have implicated macrophage activation with production of a unique macrophage prothrombinase (PCA) and lymphocyte cytokine secretion in the pathogenesis of MHV-3 susceptibility and have demonstrated that immunosuppression induces susceptibility in resistant mice. This study was undertaken to determine whether macrophages, derived from resistant A/J mice and treated *in vitro* with methylprednisolone sodium succinate (MP), elaborated PCA following MHV-3 exposure and whether therapy with MP altered resistance of A/J mice to MHV-3 infection *in vivo*.

Macrophages, incubated with MP *in vitro*, expressed dose dependent increases in PCA following infection with MHV-3. No induction of PCA occurred in macrophages treated with MHV-3 or MP alone. Analysis of mRNA transcripts for mouse fibrinogen like protein (musfiblp), the MHV-3 specific prothrombinase, in macrophages which were incubated with MP prior to exposure to MHV-3 demonstrated significantly increased mRNA levels as compared to macrophages not incubated with MP prior to MHV-3 exposure. *In vivo*, A/J mice treated for 3 days with 500 mg/kg/day of MP prior to infection with MHV-3 demonstrated extensive hepatocyte necrosis and fibrin deposition in hepatic sinusoids on histological examination of liver tissue, elevated serum transaminases and 100% mortality within 10 days of infection. These results therefore provide further support for the role of increased PCA in the pathogenesis of MHV-3 related liver necrosis.

INTRODUCTION

Infection of inbred mice by the coronavirus murine hepatitis virus strain 3 (MHV-3) causes a strain and age dependent spectrum of disease[1,2]. BALB/cJ mice are fully susceptible whereas mature A/J mice are fully resistant and develop no clinical disease[2]. We have shown that susceptibility to MHV-3 correlates with the induction of a unique monocyte/macrophage prothrombinase (PCA). High levels of PCA are produced *in vivo* and *in vitro* by macrophages from susceptible BALB/cJ mice following MHV-3 exposure whereas macrophages from resistant A/J mice fail to produce increased PCA under similar circumstances[2,3]. We have recently isolated and cloned a gene (musfiblp) that encodes a polypeptide with prothrombinase-like activity and is induced during MHV-3 infection[4]. MHV-3 infection of susceptible macrophages resulted in a marked increase in musfiblp mRNA which was detected as early as 6 hours p.i. In contrast, induction of musfiblp mRNA was seen in macrophages from A/J mice but was markedly less than that seen in macrophages from BALB/cJ mice and was not only seen until 12 hours p.i.

Treatment of resistant mice with corticosteroids results in the loss of resistance to MHV and the development of lethal acute hepatitis[5]. The mechanism for this loss of resistance is not known but may reflect impairment of cellular immunity which is a known consequence of administration of corticosteroids[6].

These present studies were initiated to determine whether corticosteroids affect viral replication, transcription of musfiblp and expression of its functional gene product (PCA) in macrophages from resistant A/J mice.

METHODS AND RESULTS

I. In Vivo Studies

Mature female A/J mice, infected with 1×10^6 plaque forming units (PFU) of MHV-3 administered intraperitoneally (IP), demonstrated no histologic or biochemical evidence of hepatitis and all mice survived. Animals pretreated for 3 days with Methylprednisolone Sodium Succinate (MP) (Solu-medrol[R], Upjohn Co, Don Mills, Ontario) 500/mg/kg daily with continuation of therapy following MHV-3 exposure demonstrated marked elevations of serum alanine transaminase, a marker of liver necrosis (1500 ± 450 vs 50 ± 10 IU/l in control mice), extensive hepatocyte necrosis with fibrin deposition in hepatic sinusoids on histological examination of the liver and 100% mortality within 10 days of infection. Peak hepatic viral titers of $1.12 \times 10^4 \pm 4.7 \times 10^3$ PFU/gm liver and $8.9 \times 10^3 \pm 3.8 \times 10^3$ PFU/gm liver were seen in MP treated and untreated, MHV-3 infected, animals respectively at 6 days post infection (p.i.). The differences in peak viral titers between these two groups were not statistically significant (p=0.244). At 9 days p.i., viral replication was undetectable in animals not receiving MP whereas high titers of MHV-3 persisted in MP treated mice.

II. In Vitro Studies

Peritoneal macrophages (2×10^6/ml) derived from A/J mice were preincubated with 0-30 μg of MP/ml 30 minutes prior to infection with 1000 PFU of MHV-3. Dose dependent

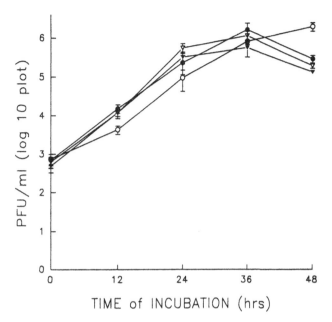

Figure 1. The effect of methyl-prednisolone (MP) on MHV-3 replication in A/J macrophages *in vitro* Macrophages from A/J mice were pretreated with 0 (O), 10 (●), 20 (▽) or 30 (▼) μg of MP/ml and infected with MHV-3 Viral titres were determined in standard plaque assay Results are mean ± 1 standard deviation of 3 experiments done in duplicate

enhanced viral replication was seen in the MP treated macrophages compared to MHV-3 infected and non MP treated macrophages in the first 24 hours of incubation. However, by 36 hours, there were no significant differences in viral titres in macrophages incubated with or without MP and, at 48 hours of incubation, viral titers were still increasing in MP untreated macrophages at a time when they were decreasing in the MP treated macrophages (Figure 1).

To assess the effect of MP on PCA expression, peritoneal macrophages (1×10^6/ml), harvested from A/J mice were preincubated for 30 minutes with 0 or 100 μg/ml of MP, infected with MHV-3 at a multiplicity of infection (MOI) of 1.0 and assayed for PCA. A/J macrophages infected with MHV-3 without prior *in vitro* MP exposure failed to express increased PCA above basal levels, as has been described previously[2,3]. In contrast, macrophages from A/J mice pretreated with MP *in vitro* expressed markedly increased levels of PCA following MHV-3 stimulation. An increase in PCA was seen as early as one hour p.i. reaching maximum levels at 6 hours p.i. and declining at 8 hours p.i. (Figure 2). Subsequent studies demonstrated that the PCA response of A/J derived macrophages to MP was dose dependent (Figure 3). Similar increases in PCA expression occurred in macrophages derived from A/J mice treated in vivo with 500 mg/kg/day of MP for 3 days prior to sacrifice (data not shown).

To determine the effect of MP on transcription of the recently identified PCA gene musfiblp, macrophages were preincubated with MP as described above and infected with MHV-3 at an MOI of 1.0. Macrophages were incubated for up to 8 hours following which total cellular RNA was isolated, resolved on a 1% agarose gel containing formaldehyde and transferred to nitrocellulose membranes. The membranes were subsequently hybridized with a musfiblp specific probe. Constitutive expression of musfiblp mRNA was not observed in MP treated or untreated macrophages. In macrophages which had been pretreated with MP and infected with MHV-3, significantly increased levels of musfiblp mRNA were detected as early as 2 hours p.i. which continued to increase at 8 hours p.i.. In contrast, in non MP treated but MHV-3 infected macrophages, significantly increased mRNA was only seen at 8 hours p.i. (Figure 4).

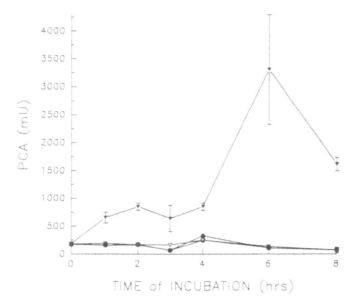

Figure 2. The effect of methylprednisolone (MP) on induction of procoagulant activity (PCA) by MHV-3 in A/J derived macrophages *in vitro*. Control macrophages (O), macrophages treated with 100 μg MP (●), MHV-3 infected macrophages (▽) and MHV-3 infected macrophages preincubated with 100 μgm MP (▼) were assessed for PCA in a one stage clotting assay. Results are mean ± 1 standard deviation of 3 experiments done in duplicate.

DISCUSSION

The mechanisms underlying the strain dependent spectrum of liver injury in mice following MHV-3 infection are unclear. Bang and Warwick suggested that susceptibility to MHV infection correlated with replication of MHV in isolated cultures of macrophages[7]. However, in more recent studies, MHV-3 replication has been demonstrated in macrophages[2] and hepatocytes[8] from both resistant and susceptible strains of mice. Thus, restriction of viral replication does not appear to explain resistance. Lamontagne et al have demonstrated that viral pathogenicity correlates with replication of MHV-3 in T and B lymphocytes and the subsequent consequences of loss of immunocompetence[9].

Susceptibility to MHV-3 in inbred strains of mice correlates with the ability of macrophages derived from these mice to produce PCA following MHV-3 exposure *in vitro*[2,3]. Induction of PCA in susceptible strains correlates with the MHV-3 related liver injury which

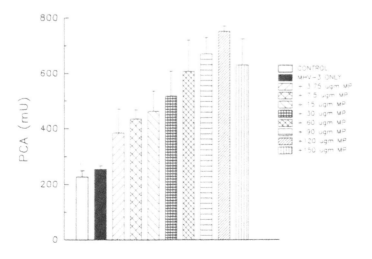

Figure 3. Enhancing effect of increasing concentrations of methylprednisolone (MP) on induction of procoagulant activity (PCA) by MHV-3 in A/J derived macrophages *in vitro*. Macrophages were preincubated with MP at 0 to 150 μg/ml and 30 minutes later infected with MHV-3. Following a 2 hour incubation, PCA was determined in a one stage clotting assay. Results are mean ± 1 standard deviation of 3 experiments done in duplicate.

1 2 3 4 5   6 7 8 9 10

Figure 4. Effect of methylprednisolone (MP) on transcription of musfiblp mRNA in A/J derived macrophages stimulated with MHV-3. 10 μg total RNA, isolated from A/J macrophages which were not pretreated with MP (Lanes 1-5) or were pretreated with MP (Lanes 6-10) and infected with MHV-3 for 0 hrs (Lanes 1 and 6), 2 hours (Lanes 1 and 7), 4 hours (Lanes 3 and 8), 6 hours (Lanes 4 and 9) or 8 hours (Lanes 5 and 10), were hybridized with a 1.3 kb random-primed musfiblp cDNA probe.

is characterized by sinusoidal thrombosis and abnormalities of microcirculatory flow[10]. Susceptible mice which were treated with high titered neutralizing antibody to PCA were protected from MHV-3 related liver injury *in vivo*[11]. Together, these studies support the concept that induction of procoagulant synthesis, as manifested by increased PCA, plays an important role in the pathogenesis of MHV-3 related liver disease.

Corticosteroids are known to inhibit immune function. They inhibit macrophage cytotoxic function and processing and presentation of antigen to T cells; decrease the activity of natural killer cells and induce apoptosis in immature B and T cell precursors and mature T cells[12]. Their effects on T lymphocytes include inhibition of production of IL-2 associated with a shift of the cytokine response from a TH1 to a TH2 profile[13]. We have recently reported that TH1 lymphocytes inhibit induction of PCA by macrophages in response to MHV-3 *in vitro* and also can prevent the lethality of MHV-3 infection *in vivo*[14]. Körner et al. have also reported the importance of TH1 cells in the resistance to MHV JHM infection[15].

We have confirmed that corticosteroid therapy *in vivo* results in loss of resistance to MHV-3 in A/J mice. We have now demonstrated that treatment of A/J macrophages with corticosteroids prior to infection with MHV-3 results in induction of PCA whereas non corticosteroid treated and MHV-3 infected A/J macrophages fail to express PCA. This was shown both by demonstration of increased transcription of the musfiblp gene thought to encode PCA and by expression of functional PCA in the one stage clotting assay in corticosteroid treated and MHV-3 infected macrophages. These results suggest that differences in expression of PCA following MHV-3 exposure in resistant and susceptible mice cannot be simply explained by differences in the coding sequence of musfiblp and suggests that differences in musfiblp gene transcription or stability of message may account for the disparate expression of functional PCA between resistant and susceptible mice.

These findings in combination with the fact that corticosteroid treated resistant A/J mice develop microvascular thrombosis and hepatic necrosis and die following MHV-3 infection similar to fully suscepible BALB/cJ, mice further supports the concept that induction of macrophage PCA is pivotal to the pathogenesis of MHV-3 disease.

ACKNOWLEDGMENTS

This work was supported by Medical Research Council of Canada program project grant PG 11810, NIH grant AI31069 and a grant from the Council of Tobacco Research We thank Mr L S Fung for his technical assistance and Ms Charmaine Mohamed for her help in the preparation of this manuscript

REFERENCES

1 LePrevost C , Levy-Leblond E , Virelizier J L , Dupuy J M , Immunopathology of mouse hepatitis virus type 3 infection I Role of humoral and cell-mediated immunity in resistance mechanisms J Immunol 1975,114 221-225

2 Levy G A , Leibowitz J L , Edgington T S Induction of monocyte procoagulant activity by murine hepatitis virus type 3 parallels disease susceptibility in mice J Exp Med 1981,154 1150-1163

3 Dindzans V J , Skamene E , Levy G A Susceptibility/ resistance to mouse hepatitis virus strain 3 and macrophage procoagulant activity are genetically linked and controlled by two non-H-2-linked genes J Immunol 1986,137 2355-2360

4 Parr R , Fung L S , Reneker J , Myers-Mason N , Leibowitz J , Levy G Association of mouse fibrinogen like protein (musfiblp) with murine hepatitis virus induced prothrombinase activity 1994 (submitted)

5 Gallily R , Warwick A , Bang F B Effect of cortisone on genetic resistance to mouse hepatitis virus in vivo and in vitro Proc Natl Acad Sci USA 1964,51 1158-1164

6 Behrens T W , Goodwin J S Glucocorticoids In Bray M A , Morley J , (eds) The pharmacology of lymphocytes Springer-Verlag, Berlin 1988 pp 425-439

7 Bang F B , Warwick A Mouse macrophages as host cells for the mouse hepatitis virus and the genetic basis of their susceptibility Proc Natl Acad Sci USA 1960,46 1065-1075

8 Arnheiter H , Baechi T , Haller O , Adult mouse hepatocytes in primary monolayer culture express genetic resistance to mouse hepatitis virus type 3 J Immunol 1982,129 1275-1281

9 Lamontagne L , Descoteaux J-P , Jolicoeur P Mouse hepatitis virus 3 replication in T and B lymphocytes correlate with viral pathogenicity J Immunol 1989,142 4458-4465

10 MacPhee P J , Dindzans V J , Fung L-S , Levy G A , Acute and chronic changes in the microcirculation of the liver in inbred strains of mice following infection with mouse hepatitis virus type 3 Hepatology 1985,5 649-660

11 Li C , Fung L S , Chung S , Crow A , Myers-Mason N , Phillips M J , Leibowitz J L , Cole E , Ottaway C A , Levy G Monoclonal antiprothrombinase (3D4 3) prevents mortality from murine hepatitis virus (MHV 3) infection J Exp Med 1992,176 689-697

12 Bateman A , Singh A , Kral T , Solomon S The immune-hypothalamic pituitary-adrenal axis Endocrine Rev 1989,10 92 112

13 Daynes R A , Araneo B A Contrasting effects of glucocorticoids on the capacity of T cells to produce the growth factors interleukin 2 and interleukin 4 Eur J Immunol 1989,19 2319-2325

14 Chung S , Gorczynski R , Cruz B , Fingerote R , Skamene E , Perlman S , Leibowitz J , Fung L , Flowers M , Levy G Characterization of murine hepatitis virus strain 3 (MHV-3) specific T cell lines effect on induction of macrophage procoagulant activity in-vitro and course of MHV-3 infection in-vivo Immunology 1994 (In Press)

15 Korner H , Schliephake A , Winter J , Zimprich F , Lassmann H , Sedgwick J , Siddell S , Wege H Nucleocapsid or spike protein-specific CD4[+] T lymphocytes protect against coronavirus-induced encephalomyelitis in the absence of CD8[+] T cells J Immunol 1991,147 2317-2323

ULTRASTRUCTURAL CHARACTERISTICS AND MORPHOGENESIS OF PORCINE REPRODUCTIVE AND RESPIRATORY SYNDROME VIRUS PROPAGATED IN THE HIGHLY PERMISSIVE MARC-145 CELL CLONE

S. Dea, N. Sawyer, R. Alain, and R. Athanassious

Centre de Recherche en Virologie
Institut Armand Frappier
Université du Québec
Laval, Québec
Canada, H7V 4Z3

ABSTRACT

A Québec reference strain of PRRSV (IAF-KLOP) was successfully propagated in MARC-145 cells, a highly permissive cell clone to PRRSV derived from the MA-104 cell line. Purified extracellular virions appeared as pleomorphic but mostly spherical enveloped particles, 50-72 nm in diameter, with an isometric core about 25-30 nm. By indirect immunofluorescence, detection of viral antigens within the cytoplasm was possible as soon as 6 h p.i. Nucleocapsids, budding at smooth endoplasmic reticulum (ER), and enveloped viral particles that tended to accumulate in the lumen of ER or Golgi vesicles, were the main features of the viral morphogenesis. The virus apparently was released by exocytosis.

INTRODUCTION

Porcine reproductive and respiratory syndrome virus (PRRSV) has recently been identified in North America and Europe as an important cause of reproductive failure in sows of any parities and respiratory disease in young pigs[1]. Preliminary morphological and molecular characterization of the Lelystad virus[2,3], as well as North american isolates[4,5], suggested that PRRSV belongs to the *Arterivirus* group, which includes equine arteritis virus (AEV), lactate dehydrogenase-elevating virus (LDV), and simian hemorrhagic fever virus (SHFV)[6].

Corona- and Related Viruses, Edited by P J Talbot and G A Levy
Plenum Press, New York, 1995

Virus replication has failed in many different kinds of primary and established cell lines[1]. Porcine alveolar macrophages (PAM) are the only known primary cells that support virus replication[1,2,5]. Recently, homogeneous high- and low-permissive cell clones to PRRSV were derived from the MA-104 monkey kidney cell line[7,8]. The purpose of the present study was to describe the ultrastructural characteristics and morphogenesis of a Québec reference strain of PRRSV that could be serially propagated in the MARC-145 cell line, previously reported to support the growth of North american isolates[8].

METHODOLOGY AND RESULTS

The IAF-Klop strain was initially isolated in PAM cells from lung homogenates of aborted fetuses from a Québec pig farm having experienced typical outbreak of PRRS[5]. Following two successive passages in PAM, it could be successfully propagated in MARC-145 cells, kindly provided to us by J. Kwang (U.S. Meat Animal Research Center, USDA, ARS, Clay Center, Nebraska). Infected cell cultures were monitored daily for the appearance of cytopathic effect. Subpassages were done at 5- to 6-day intervals. Serological identification of the virus was confirmed by indirect immunofluorescence using monoclonal antibody (Mab) SDOW17, directed against the nucleocapsid protein of the american reference strain ATCC-2332 (obtained from D.A. Benfield and E. Nelson, South Dakota State University, Brookings, SD). The fluorescence was restricted to the cytoplasm and could be observed as soon as 6 h p.i.

The virus was plaque-purified twice prior ultrastructural studies. For transmission electron microscopy, infected cells were fixed with 2.5% glutaraldehyde in 0.1M sodium phosphate buffer (pH 7.4) for 1 h at 4°C, and ultrathin sections were processed and stained with uranyl acetate and lead citrate[9]. Extracellular viral particles in clarified infected-cell culture fluid were negatively stained with 2% phosphotungstic acid at pH 7.0[5].

Protein-A immunogold labelling of PRRSV-infected MARC-145 cells, following incubation with porcine anti-PRRSV serum or MAb SDOW-17, allowed visualization of viral antigens within the cytoplasm as soon as 6 h p.i. By 12 to 14 h p.i., cytoplasmic inclusions mostly made of granular material could be observed (Figure 1a). Complete intracellular viral particles were observed by 12 to 18 h p.i. in the lumen of the cytoplasmic vesicles, but not in the nucleus (Figure 1b,c). Virions consist of empty (electron translucent) or complete (electron dense center) enveloped particles, 45-55 nm in diameter, with a central isometric core. By 18 to 24 h p.i., nucleocapsids budding at smooth endoplasmic reticulum (ER) (Figure 1 c), and enveloped viral particles that tended to accumulate in the lumen of ER or Golgi vesicles (Figure 1 b,d), were the main features observed. Thereafter, the virions appeared to escape from the cell within smooth-walled vesicles that progressively migrated to the plasma membrane and fused with it (Figure 1 e,f). No budding was demonstrated at the level of the cytoplasmic membrane. Anormal proliferation of cytoskeleton elements (microtubular system) could be observed as a result of PRRSV infection of MARC-145 cells.

Extracellular virions purified by isopycnic ultracentrifugation on CsCl density gradients appeared as pleomorphic but mostly spherical enveloped particles, 50-72 nm in diameter, with an isometric core about 25-30 nm. Specific protein-A immunogold labelling of aggregated extracellular virions present in the supernatant fluids of infected PAM cultures was obtained following incubation with porcine anti-PRRSV serum (data not shown).

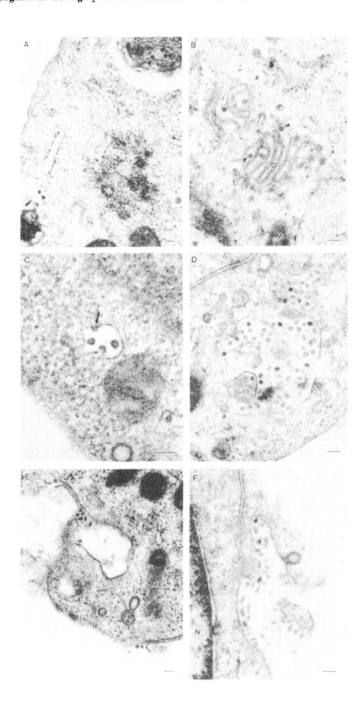

Figure 1. Transmission electron microscopy of PRRSV-infected MARC-145 cells. A) Cytoplasmic inclusions (Ci); B) Intracellular viral particles in the lumen of the Golgi vesicles; C) Budding of viral particles from the membrane of a smooth-walled vesicle; D) Accumulation of virions within smooth-walled vesicles; E) Smooth-walled vesicles containing viral particles migrating to the plasma membrane; F) Releasing of mature virions resulted after fusion of cytoplasmic vesicle with the plasma membrane. N = nuclei; M = Mitochondria; G = Golgi apparatus. Bars represent 100 nm.

CONCLUSION

The ultrastructural characteristics of the cell culture-adapted PRRSV isolate were apparently similar to that of *Togaviruses* Maturation of the intracellular virions occurred within smooth-walled vesicles derived from ER and Golgi apparatus, and as described for *Coronaviruses,* the mature virions appeared to escape by exocytosis Thus, PRRSV apparently depends on several host cell functions for virus maturation and release, including a cellular secretion pathway Cytoskeleton elements may also play a role in viral assembly and release The similarities between the morphogenesis of PRRSV and LDV supports the hypothesis, that PRRSV is a new member of the *Arteriviridae*[6]

REFERENCES

1 Goyal S M Porcine reproductive and respiratory syndrome, a review J Vet Diagn Invest 1993,5 656-664

2 Wensvoort G , Kluyver E P, Pol J M A , Wagenaar F , Moormann R J M , Hulst M M , Bloemraad R , den Besten A , Zetstra T , Terpstra C Lelystad virus, the cause of porcine epidemic abortion and respiratory syndrome a review of mystery swine disease research at Lelystad Vet Microbiol 1992,33 185-193

3 Meulenberg J J M , Hulst M , de Meijer E J, Moonen P L J M , den Besten A , de Kluyver E P , Wensvoort G , Moormann R J M Lelystad virus, the causative agent of porcine epidemic abortion and respiratory syndrome (PEARS), is related to LDV and EAV Virology 1993,192 62-72

4 Benfield D A, Nelson E , Collins J E , Harris L , Goyal S M , Robison D , Christianson W T , Morrison R B , Gorcyca D , Chladek D Characterization of swine infertility and respiratory syndrome virus (isolate ATCC VR-2332) J Vet Diagn Invest 1992,4 127-133

5 Mardassi H , Athanassious R , Bilodeau R , Dea S Porcine reproductive and respiratory syndrome virus morphological, biochemical and serological characteristics of Quebec isolates associated to acute or chronic outbreaks of PRRS virus Can J Vet Res 1993,58 55-64

6 Plagemann P G W , Moennig V Lactate dehydrogenase-elevating virus, equine arteritis virus, and simian hemorrhagic feber virus a new group of positive-strand RNA viruses Adv Virus Res 1992,41 99-192

7 Bautista E M , Goyal S M , Yoon I J , Joo H S , Collins J E Comparison of porcine alveolar macrophages and CL 2621 for the detection of porcine reproductive and respiratory syndrome (PRRS) virus and anti-PRRS antibody J Vet Diagn Invest 1993,5 163-165

8 Kim H S , Kwang J , Yoon I J , Joo H S , Frey M L Enhanced replication of porcine reproductive and respiratory syndrome (PRRS) virus in a homogeneous subpopulation of MA-104 cell line Arch Virol 1993,133 477-483

9 Dea S , Garzon S , Strykowski H , Tijssen P Ultrastructure and protein A-gold immunolabelling of HRT-18 cells infected with turkey enteric coronavirus Vet Microbiol 1989,20 21-33

14

ANTIGENIC AND GENOMIC VARIATIONS AMONG CYTOPATHIC AND NON-CYTOPATHIC STRAINS OF BOVINE ENTERIC CORONAVIRUS

S. Dea, L. Michaud, and R. Rekik

Centre de Recherche en Virologie
Institut Armand Frappier
Université du Québec
Laval, Québec
Canada, H7N 4Z3

INTRODUCTION

Bovine coronavirus (BCV) is a member of the *Coronaviridae* family known primarily as one of the major causative agent of neonatal calf diarrhea (NCD), a disease with substantial economic impact in the dairy and beef cattle industries[1]. The virus has been reported to be also responsible for acute hemorrhagic enteritis (Winter dysentery)[2] or mild chronic diarrhea in adult cattle, and for upper respiratory tract illness in growing calves[3]. Although there exist BCV strains associated to different pathological entities in the bovine species, there is still some controversy as to the existence of distinct BCV serotypes. In view of the probable occurrence of pathogenic together with non-pathogenic strains, studies of the determinants that distinguish clinical from subclinical infections are important to establish an adequate control on cattle coronaviral infections. The purpose of the present study was to define the biological, serological and genomic characteristics of BCV isolates associated with out-breaks of NCD and Winter dysentery in dairy herds in Québec.

METHODOLOGY

The cell culture-adapted Mebus strain of BCV (BCQ-Meb)[1] was obtained from the American Type Culture Collection (ATCC-VR874). The reference BCV strain was initially isolated in bovine fetal kidney cells from diarrhea fluid of a calf. Twenty other BCV isolates were recovered from clinical cases of epidemic diarrhea in newborn calves affecting dairy herds located in 4 different geographic areas in Québec[4,5]. Five additional isolates were recovered during winters of 1992 and 1993 from faecal samples collected from adult diarrheic cows from herds which were experiencing typical outbreaks of Winter dysentery (WD). No commercial BCV vaccine had been applied in these herds during the year preceeding emergence of clinical cases. The viruses were propagated in the human rectal tumor (HRT-18) cells in the presence of 10 U/ml of bovine pancreatic trypsin[6]. The extracellular virions were purified by differential and isopycnic ultracentrifugation on continuous 20 to 55% (V/V) sucrose gradients[6]. A rabbit hyperimmune serum was produced against purified tissue culture-adapted BCV-Meb strain. Specificity and cross-reactivity of this antiserum towards Québec BCV isolates were investigated by seroneutralization (SN) and hemagglutination-inhibition (HI) tests[7]. The reactivity of monoclonal antibodies (Mabs) directed against epitopes of the four major antigenic domains of the BCV spike (S) glycoprotein was tested by SN and indirect ELISA[4].

RT-PCR amplification of the hemagglutinin-esterase (HE) and S genes of the various BCV isolates were performed as previously described[5,8]. PCR products with A overhangs were ligated into TA cloning vector (pCR II vector; Invitrogen Co.), providing single 3'T overhangs at the insertion site. The ligation mixture was used to transform competent *E. coli* cells. Sequencing of cDNA clones was performed on both strands by the dideoxynucleotide chain-termination method by using T7 DNA polymerase[5]. Sequence analyses were performed using the MacVector 3.5 (International Biotechnologies) and GeneWorks 2.2 (IntelliGenetics Inc) programs.

RESULTS AND DISCUSSION

Upon their second and third passages in HRT-18 cells, most of clinical BCV isolates induced cytopathic changes within 48 to 72 h p.i. The various isolates could be differentiated into non-cytopathic (NCP) and cytopathic (CP) strains, the latters being classified into highly fusogenic and non-fusogenic strains. Nevertheless, the yield of viral production was similar for the different isolates (ranged between $10^{5\,5}$ to $10^{7\,00}$ $TCID_{50}/mL$), as revealed by immuno-peroxydase and calculation of infectivity titers. The five WD isolates behaved as highly fusogenic strains.

All BCV isolates tested were indistinguishable by SN and Western immunoblotting tests, using rabbit hyperimmune serum to the reference BCV-Meb strain. Interestingly, three CP isolates could be differentiated by their reactivity to a set of neutralizing Mabs (BCB1, BCF4, BCB5 and BCA3) directed against the S glycoprotein[4]. Based on competitive ELISA, three antigenic subgroups could be recognized among the BCV isolates studied[4]. These subgroups were identified by Mabs directed to neutralizing epitopes of antigenic domains A, B and C of the S glycoprotein. Antigenic domain D appeared to be highly conserved among Québec BCV isolates, as well as, non-neutralizing epitopes assigned to antigenic domain A and B. On the other hand, BCV isolates associated to Winter dysentery could be differentiated by their hemagglutinating activity at 37°C and 4°C, and absence of reactivity to the reference antiserum in the HAI tests.

Oligonucleotide primers, flanking genomic regions coding for the HE glycoprotein and the S1B immunodominant portion (nt 1185 to 2333) of the S glycoprotein, permitted amplification by PCR of all viruses tested[5]. Sequences of S1B gene fragment of Québec BCV isolates demonstrated a high degree of similarity. Frameshift, deletion, or insertion, and non sense mutations were not observed; the only variations among these sequences consisted of 40 nt substitutions, which represented 4% of the sequence[5]. The differing nucleotides were not distributed randomly over the entire sequence but rather were clustered in a highly polymorphic region (nt 1368 to 1776). This was also reflected in the deduced amino acid sequences; indeed, over the S1B sequence most of mutations were silent, whereas in the polymorphic region, most of the nt substitutions resulted in mutations of aa residues.

The fact, that these aa changes occurred in almost identical location suggests these differences may be significant. Comparative sequence analyses of the highly polymorphic region of the S1 subunit showed that highly cytopathic Québec BCV isolates are genetically divergent from the well characterized L9, F15 and LY-138 strains[5]. Four sporadic aa changes were located in antigenic domain II (aa residues 517 to 720) of their S1 subunits.

Sequence of the predicted proteolytic cleavage site (KRRSRR) was conserved within Québec BCV isolates studied. However, three isolates could be differentiated by their nt sequences proximal to the S proteolytic cleavage site. Phylogenetic analyses classify recent Québec clinical isolates in a distinct sublineage than other well characterized reference strains[5].

In conclusion, recent Québec BCV isolates could be differentiated from reference strains by serological and genomic studies, but, no correlation was found between nt and aa substitutions, the rate of viral replication and type of CPE induced in HRT-18 cells. A closer genomic relatedness was demonstrated between the virulent BCV-F15 strain and four highly cytopathogenic Québec isolates. At least 12 aa substitutions have been identified between the virulent and avirulent groups in the highly polymorphic region of the S1 peptide, suggesting that aa changes in this particular region of the S glycoprotein may be related to BCV virulence. Nevertheless, we cannot exclude that other regions of the BCV genome may be also involved.

REFERENCES

1 Mebus C A , Stair E L , Rhodes M B , Twiehaus, M J Neonatal calf diarrhoea propagation, attenuation, and characteristics of a coronavirus-like agent Am J Vet Res 1973,34:145-150

2 Benfield D A , Saif L Cell culture propagation of a coronavirus isolated from cows with winter dysentery J Clin Microbiol 1990,28.1454-1457

3. Reynolds D J , Debney T G , Hall G A , Thomas L H , Parsons K R Studies on the relationship between coronaviruses from the intestinal and respiratory tracts of calves Arch Virol 1985,85.71-83

4 Michaud L , Dea S Characterization of monoclonal antibodies to bovine ente-ric coronavirus and antigenic variations among Quebec isolates Arch Virol 1993, 131 455-465

5. Rekik M.R , Dea S Comparative sequence analysis of a polymorphic region of spike protein gene of bovine coronavirus isolates Arch Virol 1994 135 319-331

6 Dea S , Garzon S , Tijssen P Isolation and trypsin-enhanced propagation of turkey enteric (Bluecomb) coronaviruses in a continuous HRT-18 cell line Am J Vet Res 1989,50 1310-1318

7 Dea S , Tijssen P Antigenic and polypeptide structure of turkey enteric coronaviruses as defined by monoclonal antibodies J Gen Virol 1989,70 1725-1741

8 Zhang X , Kousoulas K G , Storz J Comparison of the nucleotide and deduced amino acid sequences of the S genes specified by virulent and avirulent strains of bovine coronaviruses Virology 1991,183 397-404

15

FRAGMENTATION AND REARRANGEMENT OF THE GOLGI APPARATUS DURING MHV INFECTION OF L-2 CELLS

Ehud Lavi,[1] Qian Wang,[1] Anna Stieber,[1] Youjun Chen,[1] Susan Weiss,[2] and Nicholas K. Gonatas[1]

[1] Division of Neuropathology
Department of Pathology and Laboratory Medicine
[2] Department of Microbiology
University of Pennsylvania School of Medicine
Philadelphia, Pennsylvania 19104-6079

INTRODUCTION

The Golgi apparatus-complex (GA) plays a seminal role in the transport, processing and targeting of polypeptides synthesized in the rough endoplasmic reticulum (RER). Important insight into the function of the GA has been gained with the use of viruses or their coat glycoproteins which are processed through the GA. Coronavirus mouse hepatitis virus (MHV), possesses a membrane protein (M) which is targeted to the trans-Golgi network (TGN) [1].

EXPERIMENTAL DESIGN AND METHODS

In this study, we examined the morphologic aspects of the GA in syncytia formation during infection of L-2 cells rat fibroblasts with MHV-A59. At 4 hour intervals cells were fixed in 2% paraformaldehyde, permeabilized and processed for immunohistochemistry with ABC Vector elite kit according to manufacturer's recommendations. Other cells were grown on therminox, fixed with Karnowski's fixative, post-fixed in osmium tetroxide and processed for electron microscopy as previously described [2].

RESULTS

Immunostaining with anti MG-160 antibodies, a GA specific marker[3], revealed fragmentation and translocation of the GA in the center of the syncytia 16-24 hours post infection. Electron microscopy confirmed the presence of a fragmented GA in the center of

Corona- and Related Viruses, Edited by P. J. Talbot and G. A. Levy
Plenum Press, New York, 1995

the syncytia. Antibodies against a RER protein [4], and against alpha and beta tubulin, revealed no significant changes in the distribution of the RER and cytoskeleton in MHV infected cells until the end stage of cell death (48 hours). Two fusion-defective, infectivity-competent, mutant MHVs, which contain an identical amino acid alteration in the cleavage signal sequence of the spike (S) glycoprotein [5] caused fragmentation of the GA but without complete aggregation of the GA in the center of the syncytia. Revertant viruses had fusion properties and GA staining as MHV-A59.

DISCUSSION

MHV-induced fragmentation of the GA is independent of the formation of syncytia. However, the translocation of the fragmented GA toward the center depends on syncytia formation. This translocation is associated with, and maps like fusion to the cleavage site of the S glycoprotein of the virus. Fragmentation and/or rearrangement of the GA has been previously shown to occur under several conditions: during mitosis[6], in certain cell lines infected with herpes simplex virus [7], and in motor neurons in the human disease amyotrophic lateral sclerosis [8]. The molecular mechanism(s) of the fragmentation of the GA induced by these conditions remains to be elucidated.

ACKNOWLEDGMENTS

This study was supported in part by a grant from the University of Pennsylvania Research Foundation (EL), by National Multiple Sclerosis Society grants PP-0284, RG-2615A1/2 (EL) and an NIH grant NS-05572 (NKG) and PHS grant NS-21954 (SRW).

REFERENCES

1 Swift AM, Machamer CE A Golgi retention signal in a membrane-spanning domain of coronavirus E1 protein J Cell Biol 1991,115:19-30

2 Lavi E, Wang Q, Stieber A, Gonatas NK Polarity of processes with Golgi apparatus in a subpopulation of type I astrocytes Brain Res 1994,647:273-285

3 Gonatas JO, Mezitis SGE, Stieber A, Fleischer B, Gonatas NK MG-160, a novel sialoglycoprotein of the medial cisternae of the Golgi apparatus J Biol Chem 1989,264:646-653

4 Chen Y, Hickey WF, Mezitis SGE, et al Monoclonal antibody 2H1 detects a 60-65 kD membrane polypeptide of the rough endoplasmic reticulum of neurons and stains selectively cells of several rat tissues J Histochem Cytochem 1991,39 635-643

5 Gombold JL, Hingley ST, Weiss SR Fusion-defective mutants of mouse hepatitis virus A59 contain a mutation in the spike protein cleavage signal J Virol 1993,67:4504-4512

6 Robbins E, Gonatas NK. The ultrastructure of a mammalian cell during the mitotic cycle J. Cell. Biol. 1964,21:429-463

7 Campadelli G, Brandimarti R, Di Lazzaro C, Ward PL, Roizman B, Torrisi MR Fragmentation and dispersal of Golgi proteins and redistribution of glycoproteins and glycolipids processed through the Golgi apparatus after infection with herpes simplex virus 1 Proc. Natl. Acad. Sci USA 1993,90:2798-2802.

8. Mourelatos Z, Yachnis A, Rorke L, Mikol J, Gonatas NK The Golgi apparatus of motor neurons in amyotrophic lateral sclerosis Ann Neurol 1993;33:608-615.

16

SEVERE COMBINED IMMUNODEFICIENCY (SCID) MOUSE HEPATITIS EXPERIMENTALLY INDUCED WITH LOW VIRULENCE MOUSE HEPATITIS VIRUS

Koji Uetsuka[1], Hiroyuki Nakayama[1], Naoaki Goto,[1] and Kosaku Fujiwara[2]

[1] Department of Veterinary Pathology
Faculty of Agriculture
The University of Tokyo
1-1-1 Yayoi, Bunkyo-ku, Tokyo 113, Japan
[2] Department of Veterinary Pathology
Nihon University School of Veterinary Medicine
1866 Kameino, Fujisawa 252, Japan

Severe combined immunodeficient (SCID) mice were experimentally infected with a low virulence strain of mouse hepatitis virus, MHV-2cc, and hepatic lesions of those mice were examined pathologically and compared with those of MHV-2cc hepatitis of athymic nude mice[1]

All mice died or were killed in moribund stage within 28 days

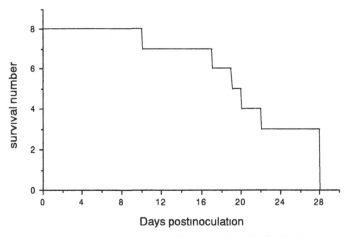

Figure 1. Survival number of SCID mice infected with MHV-2cc

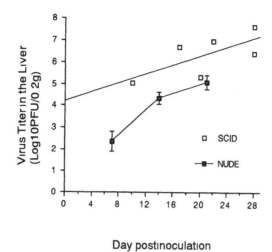

Day postinoculation

Figure 2. Virus titers in the liver of infected SCID mice compared to those of nude mice

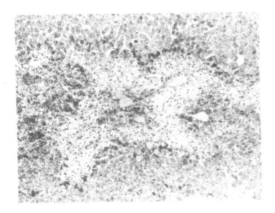

Figure 3. Hepatocytes disappeared leaving reticular fiber and cell debris 17day p i At 22 day p i many lesions were enlarged, confluenced and intermingled spreading diffusely

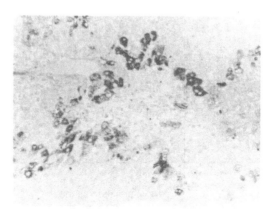

Figure 4. Viral antigen-positive hepatocytes at rims of the lesion 17day p i

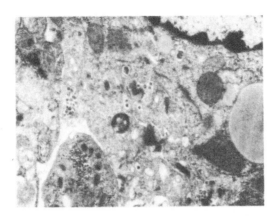

Figure 5. Virions in the cytoplasm of a degenerated hepatocyte of a SCID mice. 17day p.i.

Viral titers of infected SCID mice progressively increased until the mice died, being higher than those of infected nude mice (Fig. 2).

Histopathologically, at 10 day postinoculation (p.i.), focal small inflammatory lesions appeared, then these lesions became larger, spreaded infiltratively along with sinusoidal space. In the lesions, hepatocytes disappeared leaving reticular fiber and cell debris.

Immunostaining revealed some hepatocytes around the hepatic lesions showed viral antigen-positive at 10 day p.i., and then these viral antigen-positive hepatocytes were increased in number and were mainly observed at rims of the lesions even in much larger lesions.

By electron microscopy, in the cytoplasm of degenerated hepatocytes around the inflammatory areas, virions were seen in smooth endoplasmic reticulum as well as intercellular spaces among hepatocytes.

Infected SCID mice showed higher viral titer in the liver and died much earlier than nude mice, so viral replication in SCID mice was suggested not to be confined by the depletion of B lymphocytes. Rapid enlargement of hepatic lesions in infected SCID mice was well co-related to the increase of viral titer in the liver, and the destruction of hepatocytes was thought to be caused by the direct effects of viral proliferation in addition to the host resistance. For the persistency of MHV-2cc in hepatocytes, the host resistant conditions may be causative. The proliferation of connective tissues and the regeneration of hepatocytes in SCID mice were not so remarkable as in nude mice having B lymphocytes which might be participated in the regenerative courses of the liver.

In conclusion, depletion of B lymphocyte in SCID mice makes the lesion severe for establishing persistent infection.

REFERENCES

1. Goto, N., Inoue, T., Hirano, N., Sato, A., and Fujiwara, K. Pathology of active hepatitis in athymic nude mice caused by a mutant strain of mouse hepatitis virus, MHV-2-cc. Jpn. J. Vet. Sci. 1985. 47(6): 971-977.

17

MOUSE HEPATITIS VIRUS-SPECIFIC CD8+ CYTOTOXIC T LYMPHOCYTES INDUCE APOPTOSIS IN THEIR TARGET CELLS

Shinwa Shibata,[1] Shigeru Kyuwa,[2] Kosaku Fujiwara,[3] Yutaka Toyoda,[2] and Naoaki Goto[1]

[1] Laboratory of Veterinary Pathology, Faculty of Agriculture, University of Tokyo
1-1-1 Yayoi, Bunkyo-ku, Tokyo 113, Japan
[2] Department of Animal Pathology, Institute of Medical Science, University of Tokyo
4-6-1 Shirokanedai, Minato-ku, Tokyo 108, Japan
[3] Department of Pathobiology, Nihon University School of Veterinary Medicine
1866 Kameino, Fujisawa 252, Japan

CD8+ cytotoxic T lymphocytes (CTL) have been suggested to play an important role in virus clearance and development of demyelination during mouse hepatitis virus (MHV) infection in mice[1]. Production of antiviral cytokines such as γ-IFN and direct cytolysis of virus-infected cells have been considered to be the major role of CTL, but their precise mechanisms in MHV infection remain unclarified. We have previously established CTL line, P11D which specifically lyses cells infected with MHV, strain JHM (JHMV)[2] and saves mice from lethal JHMV infection[3]. To clarify the detailed mechanism of CTL during MHV infection, we examined morphological changes of JHMV-infected target cells in vitro cocultured with anti-MHV CTL line, P11D.

Cell culture and preparation for electron microscopy were described elsewhere[4]. Briefly, J774.1 target cells were infected with JHMV at a multiplicity of infection of 5 and incubated for 5 hours before starting coculture with CTL. Coculture was carried out at an effector-to-target cell (E/T) ratio of 2 for up to 6 hours. At various time intervals, cells were recovered and fixed with 2.5% glutaraldehyde for TEM, and 1% glutaraldehyde for SEM. Specimens for electron microscopy were prepared by conventional method. By the same way, virus titers were assayed from 0 to 4 hours after starting coincubation. Fragmentation of cellular DNA was also examined by agarose gel electrophoresis, whose DNA samples were obtained from a similar culture system but E/T ratio of 0.5 for 2 hours.

Morphological changes of JHMV-infected cells attacked by CTL were as follows. At 1 hour after starting coincubation with CTL, condensation and margination of chromatin, and degeneration of cytoplasm were found in target cells (Figure 1A). These cells were in

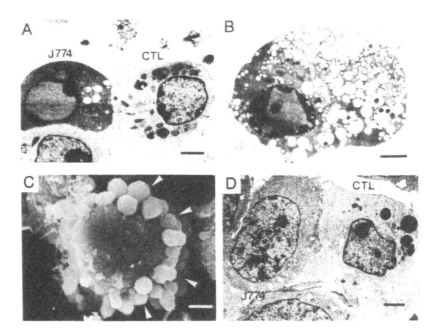

Figure 1. Coincubation of CTL and JHMV-infected target cells at 1 hour CTL was in contact with target cell, where we could observe chromatin condensation and cellular shrinkage (A) Sever vacuolar degeneration observed in target cell at 2 hours (B) Smooth of cell surface and bulbous formation (arrowheads) at 2 hours (C) Control incubation of CTL and uninfected target cells at 1 hour Despite of the contact with CTL, no degeneration were seen in the target cell (D) Each bars represent 2mm

contact with CTL but no degranulation could be seen. At 2 hours, severer shrinkage accompanying vacuolation (Figure 1B) and bulbous formation (Figure 1C) was found in target cells. These are characteristic morphological changes of apoptosis. Moreover, electrophoresis of cellular DNA showed DNA fragmentation in target cells (data not shown), indicating that MHV-specific CTL induce apoptosis in virus-infected cells. In the control culture of CTL and uninfected target cells, no degeneration was found in target cells, although cellular contact with CTL and target cells was observed (Figure 1D). At 4 to 6 hours, almost all the target cells showed sever shrinkage mentioned above, but no syncitium formation caused by virus infection was found, which was consistent with the inhibition of virus growth (data not shown).

Our present results show that CTL specific for MHV induce apoptosis in virus-infected cells, which results in virus clearance in vitro. We suggest that CTL-mediated apoptosis may be involved not only in virus clearance but also in development of demyelination in vivo, because T cells have reported to play an important role in virus clearance and demyelination during MHV infection in mice. This study may suggest a novel implication for understanding pathogenesis of MHV infection in mice.

REFERENCES

1 Kyuwa, S., and S A Stohlman. Pathogenesis of a neurotrophic murine coronavirus, strain JHM in the central nervous system of mice Semin Virol 1990 1 273-280

2 Yamaguchı, K , S Kyuwa, K Nagata, and M Hayamı Establıshment of cytotoxıc T-cell clones specıfic for cells ınfected wıth mouse hepatıtıs vırus J Vırol 1988 62 2505-2507

3 Yamaguchı, K , N Goto, S Kyuwa, M Hayamı, and Y Toyoda Protectıon of mıce from a lethal coronavırus ınfectıon ın central nervous system by adoptıve transfer of vırus-specıfic T cell clones J Neuroımmunol 1991 32 1-9

4 Shıbata, S , S Kyuwa, S -K Lee, Y Toyoda, and N Goto Apoptosıs ınduced ın mouse hepatıtıs vırus-ınfected cells by a vırus-specıfic CD8+ cytotoxıc T-lymphocyte clone J Vırol 1994 ın press

18

ECHOCARDIOGRAPHIC CHANGES FOLLOWING RABBIT CORONAVIRUS INFECTION

Lorraine K. Alexander, Bruce W. Keene, and Ralph S. Baric

The Department of Epidemiology
The University of North Carolina at Chapel Hill
Chapel Hill, North Carolina
The College of Veterinary Medicine
North Carolina State University
Raleigh, North Carolina

Much of our understanding of the mechanisms by which viruses cause myocarditis and/or dilated cardiomyopathy (DCM) is based on animal models of virus-induced heart disease. Information concerning cardiac function during acute and/or chronic viral infection in these models is limited (1). A well defined model in a species conducive to monitoring of cardiac function is needed to enhance our understanding of viral induced heart disease. We have previously demonstrated that rabbit coronavirus (RbCV) infection results in degeneration and necrosis of myocytes, myocarditis, and gross organ and histopathologic changes of DCM (2,3). We have also shown that electrocardiographic changes observed during RbCV infection mimic those in humans with myocarditis and DCM (submitted). This chapter describes the echocardiographic changes observed during RbCV infection.

Eleven male New Zealand white rabbits were sedated prior to echocardiography with a combination of xylazine (0.17 mg/kg) and ketamine (17 mg/kg). An electrocardiogram was monitored continuously during echocardiography and two-dimensional echocardiographic views were recorded with the animal in right lateral recumbancy from the right parasternal long and short-axis positions using a 7.5 MHz annular array transducer. Measurements of left ventricular (LV) size, systolic function, mitral valve motion, and aortic and left atrial diameter were made according to the American Society of Echocardiography standards for M-mode echocardiography. Briefly, M-mode measurements included LV end diastolic and systolic chamber dimensions and wall thickness obtained by guiding the M-mode cursor between the papillary muscles from a right parasternal short-axis imaging plane just ventral to the mitral valve leaflets at the level of the chordae tendinae. Aortic and left atrial dimensions were measured from an M-mode view obtained by guiding the cursor through the aorta and left atrium in a right parasternal short short axis view at the level of the aortic valve. The mitral valve motion and E-point - septal separation was observed and recorded from M-mode images obtained by guiding the cursor through a right parasternal

Table 1. Cardiac function values for 11 RbCV infected rabbits

Measurement	Uninfected[a] n = 11	Nonsurvivor[a,b] n= 6	Suvivor[a,b] n= 5
Left Ventricular (LV) diameter (d)[c] (cm)	1.42 ± 0.24	1.13 ± 0.44	1.14 ± 0.12
LV diameter (s)[d] (cm)	0.92 ± 0.17	0.93 ± 0.38	0.84 ± 0.17
% fractional shortening	35.5 ± 4.85	17.33 ± 6.19	26.17 ± 12
Septal wall thickness (d) (cm)	0.22 ± 0.07	0.25 ± 0.06	0.22 ± 0.05
Septal wall thickness (s) (cm)	0.38 ± 0.08	0.28 ± 0.09	0.33 ± 0.12
LV posterior wall thickness (d) (cm)	0.31 ± 0.11	0.32 ± 0.08	0.26 ± 0.03
LV posterior wall thickness (s) (cm)	0.50 ± 0.12	0.44 ± 0.13	0.42 ± 0.06
Left atrium diamter (cm)	0.88 ± 0.14	0.93 ± 0.15	0.86 ± 0.10
Aorta (cm)	0.66 ± 0.12	0.74 ± 0.13	0.68 ± 0.05
Left atrium/Ao	1.22 ± 0.20	1.36 ± 0.39	1.28 ± 0.14
E point septal separation (EPSS)	0.14 ± 0.04	0.22 ± 0.16	0.126± 0.09

a = Mean ± SD.
b = Day 3 after infection.
c = diastole.
d = systole.

short axis view at the level of the mitral valve. LV fractional shortening was calculated as an ejection phase index of systolic function. All values reported reflect the mean of 3 measurements made on sinus beats. Rabbits were infected with 0.3 ml of a $1X\ 10^3$ - $1X\ 10^4$ RID_{50} of RbCV and echocardiographic measurements were repeated using the same anesthetic and measurement protocol on days 3, 6, 9, 12 and 30 post-infection.

Two (18%) rabbits died during the acute phase of infection (day 3), 4 (36%) died in the early subacute phase (day 6), and 5 (46%) survived beyond day 12 into the chronic phase. Echocardiographic data is displayed in Table 1. The index of systolic ventricular function

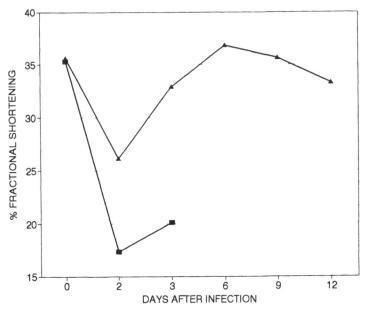

Figure 1. Percent fractional shortening in 11 RbCV infected rabbits.

chosen, % fractional shortening was depressed in all infected rabbits by day 3 post infection (Figure 1). Fractional shortening was more depressed in nonsurvivors (17.33 ± 6.19%, p= <.001 from controls) as compared to survivors (26.17 ± 12%, ns from control). Mean LV wall thickness, chamber dimensions, and left atrial dimensions were not significantly different from controls throughout the study in either survivors or nonsurvivors. These findings confirm our previous pathologic studies in which rabbits dying early in infection (days 2-5) did not have significantly different LV wall thickness, and chamber dimensions from control animals.

We conclude that RbCV infection depresses an ejection phase index of systolic LV function, that this depression precedes gross morphologic changes in the ventricle, and that severe systolic dysfunction correlates positively with mortality. These findings provide a direct link between the severity of virus-induced cardiac dysfunction and survival during RbCV infection, characterizing a reproducible model of cardiac dysfunction following viral infection of the heart.

REFERENCES

1. Woodruff JF. Am J Pathol 1980;101:427-84.
2. Edwards S. et al. J Infect Dis 1992;165:134-40.
3. Alexander LK. et al. J Infect Dis 1992;166:978-85

19

NEURONAL SPREAD OF SWINE HEMAGGLUTINATING ENCEPHALOMYELITIS VIRUS (HEV) 67N STRAIN IN 4-WEEK-OLD RATS

N. Hirano,[1] R. Nomura,[1] T. Tawara,[1] K. Ono,[1] and Y. Iwasaki[2]

[1]Department of Veterinary Microbiology, Iwate University
Morioka, Japan
[2]Department of Neurological Science, Tohoku University, School of
    Medicine
Sendai, Japan

INTRODUCTION

The HEV 67N strain causes encephalomyelitis or vomiting and wasting syndrome in piglets [1,2]. In experimental infection of piglets, the virus spreads along the nerve pathways to the central nervous system (CNS), and is restricted to the neurons. In our experimental studies of HEV 67N strain, the virus produced encephalitis in mice when inoculated by several routes, and propagated mainly in the neurons in the CNS [4]. However, 20-day-old or older mice were resistant to the virus inoculated by intravenous (i.v.), intraperitoneal (i.p.) or subcutaneous (s.c.) route. In contrast, 4-week-old rats died of encephalitis after i.v., i.p. and s.c. as well as intracerebral (i.c.) and intranasal (i.n.) inoculation. However, when rats were inoculated by s.c. route, they died a few days earlier than those by i.p. and i.v. routes, suggesting that the virus might be spread to the CNS by neural routes rather than blood stream. To see the virus spread from the peripheral nerve to the brain, the virus was directly inoculated in to the sciatic nerve of rats. These studies demonstrated that HEV spread from the sciatic nerve to the brain by the neural route, and that persistent infection of HEV was established in rats[6].

EXPERIMENTAL DESIGN AND RESULTS

In the present study, rats were inoculated with the virus under several treatments to follow the virus spread from the peripheral nerve to the CNS.

Plaque-purified 67N strain was propagated in SK-K cells and assayed for infectivity by plaque count method [5]. Four-week-old male Wistar rats were inoculated by several routes

Table 1. Effect of antiserum treatment on survival
and viral spread in rats at various times after
virus inoculation into sciatic nerve

Antiserum added at	CNS/Tested[a]	Dead/Tested[b]
−24 h	0/5	0/5
0	0/5	0/5
24	0/5	0/5
48	0/5	0/5
72	2/5	0/5
96	5/5	2/5
None	5/5	5/5

[a]No. of CNS positive/No. of tested.
[b]No. of dead/No. of tested.

as described previously [6]. In experiments, 5 rats were used in each group. For antiserum treatment, rats were administrated i.p. with 1 ml of rat antiserum against 67N strain (HI titer, 1:100 <).

In order to see the virus spread from the peripheral nerve to the CNS, the virus of 1 x 10⁴ PFU was inoculated into the foot pad. The virus was first detected in the posterior half of the spinal cord on day 2, and in the anterior half of the spinal cord and brain on day 4. The virus was never detectable from the spleen and liver. The virus specific antigen was first detected in the posterior horn of the spinal cord on day 3, and in the anterior horn of spinal cord and the brainstem on day 4. Antigen positive cells were found in the cerebral cortex on day 5.

To follow the virus spread via the peripheral nerve, 1 x10⁴ PFU of the virus was inoculated s.c. into the foot pad of the right leg. Rats were surgically operated to cut the sciatic nerve at 0, 6, 12, 24 and 48 hours (h) postinoculation (p.i.). The fatal infection of HEV was aborted by cutting the sciatic nerve within 6 to 12 h p.i.

Next, the virus (1 x10⁴ PFU) was directly inoculated into the sciatic nerve, and then the proximal or distal segment of the nerve was cut within 1 h after inoculation. The cutting of the proximal segment of the sciatic nerve protected the rats from encephalitis. Rats of other groups died of encephalitis.

To see the effect of antiserum on virus spread from the sciatic nerve to the CNS, rats were treated with antiserum after or before virus inoculation of 1 x 10⁴ PFU into the sciatic nerve. As shown in Table 1, with antiserum treatment within 72 h p.i. rats were protected from fatal encephalitis.

To see blocking of the virus spread in the CNS by antiserum, rats were treated with antiserum at -24, 0 and 24 h after virus inoculation by i.c. and intraspinal (i.s.) routes. Even with antiserum treatment 24 h before inoculation, all rats inoculated i.c. died of encephalitis, and 2 of 5 i.s. inoculated rats also died. All of other inoculated rats died of acute encephalitis.

From these results, the infection was aborted by cutting the ipsilateral sciatic nerve within 6 h after inoculation into the foot pad. In addition, when the virus was directly inoculated into the sciatic nerve, only cutting of the proximal but not distal segment of the nerve within 1 h protected the rats from encephalitis. These findings suggest that the virus spreads from the peripheral nerve to the CNS by neural route not by the blood stream.

When rats inoculated with the virus into the sciatic nerve were given antibody within 72 h p.i., all the inoculated rats survived for 14 days. However, when rats which had received antibody were challenged with the virus by i.c. and i.s. routes 24 h after treatment, all the rats inoculated i.c. and 2 of 5 rats inoculated i.s. died of encephalitis.

These findings suggest that the virus inoculated into sciatic nerve is easily neutralized or blocked to spread to the CNS by antibody and that spread of the virus inoculated directly into the CNS is not inhibited in the CNS by pretreatment with antiserum These results suggest that the inoculated virus might be neutralized by antibody when passing through the segment of the nerve or the synapse of the spinal cord preventing spread of the virus from the peripheral nerve to the CNS

Our previous and present studies strongly suggest the crucial role of neural spread of the virus in induction of fatal encephalomyelitis with HEV in rats

REFERENCES

1 Roe, C K and Alexander, T J L Canad Comp Med 22, 305-307 (1958)
2 Andries, K and Pensaert, M B Am J Vet Res 41, 1372-1385 (1980)
3 Andries, K and Pensaert, M B Adv Exp Med Biol 142, 399-408 (1981)
4 Yagami, K , Hirai, K and Hirano, N J Comp Med 96, 645-657 (1986)
5 Hirano, N , Ono, K , Takasawa, H , and Haga, S J Virol Methods 27, 91-100 (1990)
6 Hirano, N , Haga, S and Fujiwara, K Adv Exp Med Biol 342, 333-338 (1994)

20

CHARACTERIZATION OF HUMAN T CELL CLONES SPECIFIC FOR CORONAVIRUS 229E

J. S. Spencer,[1-3] G. F. Cabirac,[1,2,4] C. Best,[1] L. McLaughlin,[1] and
R. S. Murray[2,5]

[1] Rocky Mountain Multiple Sclerosis Center
[2] Colorado Neurological Institute
   Englewood, Colorado
[3] Department of Immunology
[4] Department of Biochemistry, Biophysics and Genetics
   University of Colorado Health Sciences Center
   Denver, Colorado
[5] National Jewish Center for Immunology and Respiratory Medicine
   Denver, Colorado

ABSTRACT

Coronaviruses (CV) are pleomorphic enveloped RNA viruses that are ubiquitous in nature, causing a variety of diseases in both man and domestic animals. In man, CV are generally associated with upper respiratory tract infections. The two prototype strains that are the best studied human CV isolates and which are thought to be responsible for most of the respiratory infections caused by CV are called 229E and OC43. Humoral responses consisting of neutralizing antibodies to CV are present in most individuals by six years of age. Although the cellular immune response to CV in man has not been characterized at all, it is known that the spike (S) and nucleocapsid (N) proteins elicit the major cell mediated immune responses in the mouse.

This report describes the production and characterization of eleven independently isolated T cell clones that are specific for the human CV(HCV) 229E. The T cell clones are CD4+ and presumably recognize a processed viral peptide presented by class II molecules on the surface of antigen presenting cells. Of six 229E-specific T cell clones tested against purified viral proteins, three recognize the 180 kD spike glycoprotein while the other three recognize the 55 kD nucleocapsid phosphoprotein. Analysis of the human T cell mediated response to HCV will provide information regarding which viral proteins elicit the immunodominant response, what the fine specificity of these T cell clones are (immuno-dominant peptides), and what the T cell receptor (TCR) and cytokine usage is of these virus specific clones.

Corona- and Related Viruses, Edited by P. J. Talbot and G. A. Levy
Plenum Press, New York, 1995

INTRODUCTION

Multiple sclerosis (MS) is a chronic demyelinating disease of the human central nervous system (CNS). The etiology of MS is unknown, although analysis of animal models of experimental allergic encephalomyelitis[1-3] as well as data from MS patients (reviewed in Ref. 4) indicate that both genetic and environmental factors contribute to this disease. Viruses are among the environmental factors that have been proposed as possible causative agents for MS. Although numerous viral agents have been implicated as candidates in the etiology of MS, no clear association between any particular virus and the disease has been confirmed[5].

Although CV infection in man is more frequently associated with upper respiratory tract infections, causing up to 35% of all cases of the common cold[6,7], there is evidence that CV are also involved in enteric infections[8,9] and childhood meningitis[10]. In the latter case, a coronavirus (Tettnang virus) was cultured from the cerebrospinal fluid of an 18 month old child with viral meningitis that followed an upper respiratory tract infection. The possible involvement of CV in MS was suggested initially by the observation of viral particles bearing the typical morphological features of CV in electron microscopic sections of brain taken from an MS patient[11], followed by a report describing the isolation of two separate CV from MS autopsy brain tissue[12,13]. Recently, our laboratory has identified murine-like coronavirus (MCV) RNA sequence and antigen in MS brain by in situ hybridization and immunohistochemical techniques[14]. Another group of investigators found human CV 229E sequences in MS brains by using the polymerase chain reaction[15], and also showed that the virus may be neurotropic due to its ability to infect a variety of human cell lines of CNS origin[16]. Following intracranial inoculation, CV can productively infect and disseminate in primate brains, resulting in encephalomyelitis and demyelination[17]. Similarly, MCV infections in rodents result in a panencephalitis accompanied by extensive demyelination[18,19]. Rodents infected with subacute levels of MCV JHM show evidence of chronic demyelination as a result of viral persistence within the CNS[20,21]. In mice both $CD4^+$ and $CD8^+$ T cells are required for CV clearance from the CNS and for prevention of encephalomyelitis[22,23]. Because the pathological changes in the CNS observed in rodents infected with JHM resemble those found in MS patients, this animal model has been used extensively to study both acute and chronic forms of demyelinating disease caused by viruses. We have chosen to characterize the human cell mediated response to 229E to determine whether there may be any relationship between CV and multiple sclerosis.

MATERIALS AND METHODS

Viruses and Cell Lines

The HCV strain 229E was grown on the human lung fibroblast cell line, WI-38. Both were obtained from American Type Culture Collection (Rockville, MD). The MCV strain JHM[24] was grown on the mouse tumor cell line DBT[25]. Both the virus and the cell line used for its propagation were originally obtained from Dr. Stephen Stohlman (University of Southern California, Los Angeles, CA). The cell lines were grown as monolayers in roller bottles using Dulbecco's modified Eagle medium (DMEM)(Gibco BRL, Grand Island, NY) supplemented with 10% fetal calf serum (FCS) in a humidified cell culture incubator at 37°C in an atmosphere of 5% CO_2. Virus infected cell monolayers were cultured at a lower temperature optimal for viral replication (34°C). Infected cell supernatants were harvested at 18 hr post-infection for JHM and 36 hr post-infection for 229E, with subsequent steps carried out at 4°C. Virus was purified from infected cell supernatants by precipitation with

0.5 M NaCl and 10% polyethylene glycol (mw 8,000; Sigma Chemical Co., St. Louis, MO), final concentration, followed by pelleting precipitated material at 10,000 x g. The virus was further purified by ultracentrifugation over two separate sucrose gradients, a 30%/50% discontinuous gradient centrifuged at 25,000 x g at 4°C for four hr followed by centrifugation of virus material through a 25%-55% continuous gradient at 25,000 x g for 18 hr. The purified virus band was removed from the gradient, pelleted by ultracentrifugation, and resuspended in sterile phosphate buffered saline (PBS) at a concentration of 1 mg/ml. Virus purity was determined by analyzing each preparation by polyacrylamide gel electrophoresis (SDS-PAGE) and protein concentrations were determined with the BCA protein assay (Pierce Chemical Co., Rockford, IL). The virus was completely inactivated by exposure to a combination of ultraviolet light (exposed for 15 min to a 15 watt 300 nm UV light source placed five cm from its surface), sonication and gamma irradiation (10,000 rad exposure from a cesium source IBL 437C cell irradiator; CIS-US, Inc., Bedford, MA) prior to being used in antigen presentation assays to prevent its replication in viable cells.

Autologous immortalized human B lines cell used as antigen presenting cells were obtained by transforming peripheral B cells using Epstein Barr virus (EBV). The EBV secreting transformed Marmoset lymphoblastoid cell line, B95-8 (originally obtained from American Type Culture Collection, Rockville, MD), was grown to high density, and the culture supernatant was centrifuged and filtered to remove cellular components. Peripheral blood lymphocytes (PBL) were obtained from heparinized blood by density gradient centrifugation over Histopaque-1077 solution (Sigma Chemical Co., St. Louis, MO). Approximately 1 x 10^7 PBL were incubated with 1 ml of EBV containing B95-8 supernatant for 1 hr at 37°C in a 10% CO_2 incubator. Cells were washed three times with RPMI 1640 medium (Gibco BRL, Grand Island, NY), resuspended to 2 x 10^6 cells per ml and plated into the first two rows of a flat bottomed 96 well plate at 200 µl per well. Subsequent rows contained serial two-fold dilutions of these cells. Ten µl of a 10 µg/ml stock of cyclosporin A (a generous gift of Sandoz Pharmaceuticals, East Hanover, NJ) was added to each well to inactivate the T cells in these cultures. The plate was fed weekly, with transformed B lymphoblastoid lines usually arising from three to five weeks after setting up the cultures. B cell lines were expanded as suspension cultures in Iscove's modified Dulbecco's medium (IMDM)(Gibco BRL, Grand Island, NY) supplemented with 5 x 10^{-5} M 2-mercaptoethanol, 10% FCS, 100 U/ml penicillin and 100 µg/ml streptomycin.

Isolation of Purified 229E Nucleocapsid (N) and Spike (S) Proteins

Whole inactivated 229E virus was electrophoresed on 10% SDS-PAGE gels under reducing conditions. The gels were fixed and stained briefly with Coomassie brilliant blue dye to locate the N and S proteins, destained and then rehydrated in distilled water. The protein bands were excised from the gel, minced into 2-3 mm^2 pieces and placed into an electro-elution concentrator block (CBS Scientific, Del Mar, CA) in 0.05 M Tris acetate, 0.1% SDS buffer. Proteins were electrophoretically eluted from gel pieces overnight at 50 V. The buffer in the unit was changed to PBS and electrophoresis continued for two hr prior to harvesting the purified proteins.

Preparation of 229E Specific Human T Cell Clones

Transformed human B cells were used as antigen presenting cells to generate 229E specific T cell lines from autologous PBL. The B cells were incubated overnight with whole gradient purified 229E to allow for antigen (Ag) processing to occur, with 1 x 10^7 B cells being incubated with 250 µg 229E in 5 ml of fresh X-VIVO 15 medium (Biowhittaker, Walkersville, MD). Cultures containing freshly isolated autologous PBL were set up the following day with irradiated (4000 rad) Ag pulsed B cells. PBL were isolated by gradient density centrifugation over Histopaque-1077. The buffy coat layer was transferred and the cells were washed three times in balanced salt solution. The PBL were mixed with the Ag-pulsed B cells resulting in final culture conditions of 5×10^5/ml PBL, 2×10^5/ml B cells and 5 µg 229E per well of a 24 well plate, with two plates being used. The cultures were incubated at 37°C in a humidified 10% CO_2 atmosphere for a total of ten days, being fed on day five with 1 ml of fresh medium. On day five (and on each day thereafter, up to day ten) eight wells were pooled and viable cells isolated by centrifugation over Histopaque-1077. The cells were cultured in fresh medium with 5% T-stim (contains IL-2, PHA and other T cell growth factors; Collaborative Biomedical Products, Bedford, MA) with 2×10^5/ml irradiated B cells as feeders. The cells were cloned in limiting dilution in 96 well plates using X-VIVO 15 medium with 8% T-stim and 2×10^4/well irradiated feeder B cells. Remaining bulk culture cells were frozen away for future use. The plates were fed 100 µl/well fresh medium with 20 U/ml recombinant human IL-2 (Collaborative Biomedical Products, Bedford, MA) on day five after cloning, and then fed every 5-6 days with fresh medium containing 8% T-stim. Wells that showed growth were expanded to 1 ml cultures using the same culture conditions and then tested in antigen proliferation assays five to seven days after expansion.

T Cell Proliferation Assay

T cells expanded to 24 well plates were tested in an Ag proliferation assay with Ag pulsed B cells irradiated just prior to setting up the assay. Generally, 5×10^3 to 1×10^5 viable T cells were cultured with 2×10^4 B cells with or without 1 µg/well 229E in flat bottomed 96 well plates in a final volume 200 µl. After two to three days, proliferation was measured during the final 18 hr of culture by the uptake of [^{3}H]-thymidine ([^{3}H]-TdR) and counted in a beta scintillation counter. Those cultures that responded well to whole 229E virus were reexamined after subcloning in a second proliferation assay with whole JHM virus (1 µg/well) and purified 229E N or S proteins (0.1-0.2 µg/well).

Subcloning of 229E-Specific T Cell Lines

T cell lines that showed a proliferative response towards 229E in the initial screening assay were expanded in fresh medium containing 20 U/ml human recombinant IL-2 for 3-5 days. The cultures were examined each day and the line was subcloned again in 96 well culture plates when it appeared that growth was optimal. Wells that showed positive growth were again expanded and tested for proliferation to whole 229E, with virus-specific subclones expanded for further study or to freeze down in reserve. Generally, 20 to 30 subclones were generated for each 229E-specific T cell line. The subclones were used in determining Ag dose response curves and to test Ag specific responses to the N and S proteins.

RESULTS AND DISCUSSION

This report describes the characterization of eleven independently isolated 229E-specific human T cell clones derived from two normal healthy donors. The 229E specific clones were obtained after culturing peripheral blood lymphocytes *in vitro* with virus pulsed APC. We assume that there are CV-specific T cells circulating in peripheral blood, considering the likelyhood of CV infections each year. A considerable amount of time was spent working out optimal conditions for the *in vitro* cultures and Ag presentation assays. Initially, primary *in vitro* cultures of PBL, APC and 229E were cloned from seven to fifteen days after onset.

In some cases after the primary stimulation with virus, cells were given a rest period by culturing in fresh medium with a small amount of growth factors (1% T-stim) for seven to ten days, followed by restimulation with 229E for an additional week. However, the majority of 229E-specific lines were derived from cells cloned from primary cultures that were five to seven days old (nine out of eleven). In our experience, reexposure of primary cultures to a second round of 229E stimulation did not increase the probability of generating 229E-specific lines.

The lines were selected on the basis of a high proliferative response to 229E virus pulsed APC relative to APC alone. Although the proliferative response was quantitated by the uptake of [³H]-TdR, lines that responded well showed an obvious difference in the numbers of viable cells, which often appeared as clumps of cell blasts that were discernable by visual inspection alone. Typical responses of the eleven 229E-specific T clones are shown in Figure 1.

Stimulation indices (SI) for these T cell clones ranged from an SI of 5 (clone 2A.3) to 91 (clone 2C.9-3H.11). Six of the lines were isolated from one individual after screening a total of 558 clones resulting from two separate *in vitro* cultures, while the remaining five lines were isolated from another individual after screening a total of 410 clones resulting from a single *in vitro* culture. The frequency of clones isolated that responded well and maintained their response after subcloning corresponds to a 1.1% and 1.2% efficiency, respectively, although the actual numbers of clones selected after the initial screening was closer to 5%. Numerous lines that were initially selected based on a stimulation index of three to five were retested or subcloned, but, with one exception, the results of retesting the lines and subclones of marginal responders were uniformly negative. It is unclear whether

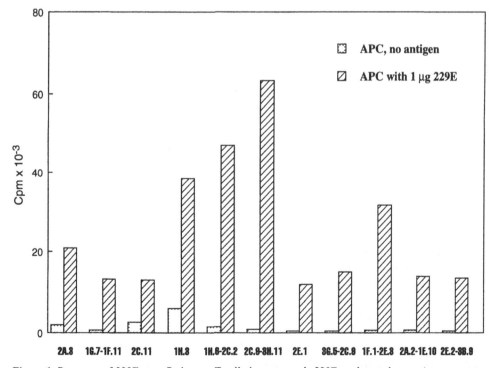

Figure 1. Response of 229E-specific human T cell clones towards 229E as detected in an Ag presentation assay. Approximately 2 x 10³ to 1 x 10⁵ T cells were cocultured with 2 x 10⁴ autologous irradiated transformed B cells with or without 1μg of inactivated 229E After two to three days, proliferation was measured by the uptake of [³H]-TdR during the final 18 hr period of culture and counted in a beta scintillation counter

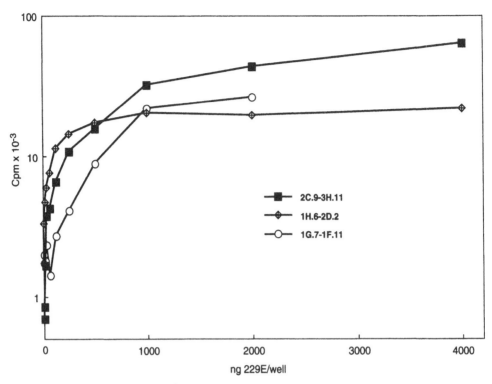

Figure 2. Dose response curve of 229E-specific T cell clones towards varying amounts of 229E. Autologous EBV transformed B cells lines were incubated overnight with serial two-fold dilutions of 229E ranging from 7 ng to 4000 ng per 100 ml. Approximately 2×10^4 Ag pulsed B cells per well were cocultured with T cells in triplicate. The number of viable T cells per well for each subclone was 2.0×10^4 for 1G.7-1F.11, 2.4×10^4 for 2C.9-3H.11, and 7.5×10^4 for 1H.6-2D.2. Proliferative responses were assessed by the uptake of $[^3H]$-TdR.

or not these weak responders consisted of multiple cell lines made up of a minor 229E-specific T cell population that was overgrown by a nonresponding line(s). All of the T cell clones characterized in this report are CD4[+] T cells, and therefore presumably recognize a processed viral peptide in the context of class II molecules on the APC.

To determine the limits of the response of 229E-specific T cell clones to antigen, a dose response curve was performed with three separate subclones by varying the concentration of 229E in the in the cultures from 7 ng to 4 μg per well, as illustrated in Figure 2. The half-maximal response for these subclones occurred with Ag amounts between 0.125 μg to 0.5 μg of 229E.

To determine the antigen specificity of the T cell clones, 229E N and S proteins were purified from SDS-PAGE gel slices by electroelution as shown in Figure 3.

We also wanted to determine if there was any crossreactivity of T cell epitopes on the MCV JHM, which is more closely related serologically to the HCV OC43 , but is antigenically unrelated to 229E[13,26]. The proliferative response of six representative T cell clones against whole virus and purified 229E N and S proteins is shown in Figure 4. As can be seen, three of the subclones tested responded to the N protein and three others responded to S. All of the T cell lines tested to date react with either N or S, indicating that these proteins probably elicit the immunodominant cell mediated responses in man, as had been previously reported for the mouse[23,27]. Currently, we are analyzing what peptides from the N and S proteins stimulate these 229E-specific T cell clones and what the TCR usage is on these clones.

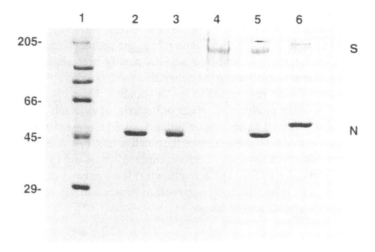

Figure 3. SDS-PAGE analysis of purified whole 229E and JHM coronaviruses and purified electroeluted 229E N and S proteins. Lane 1, molecular weight markers with numbers to the left indicating their relative mass in kDa; lanes 2 and 5, whole 229E; lane 3, 229E N protein; lane 4, 229E S protein; and lane 6, whole JHM. Approximately 1-2 µg protein was electrophoresed on a 10% polyacrylamide gel under reducing conditions. Lanes 1-3 were stained with Coomassie blue dye, while lanes 4-6 were developed using a silver stain to enhance the intensity of the S protein.

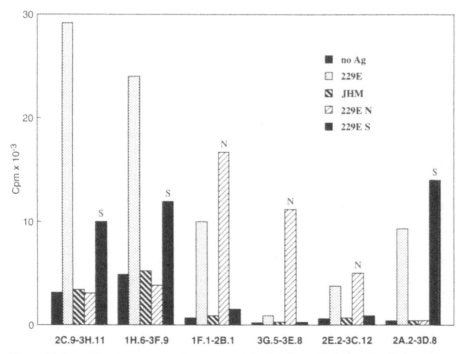

Figure 4. The proliferative response of six representative 229E-specific human T cell clones to a panel of antigens, including whole 229E and JHM virus (2 µg/well), and purified 229E N (0.2 µg/well) and S (0.1 µg/well) proteins. The reactivity patterns were confirmed by testing at least two individual subclones of each T cell line. The proliferation was determined by measuring the incorporation of [3H]-TdR during the final 18 hr of culture.

In several autoimmune or chronic inflammatory diseases in man, such as rheumatoid arthritis, insulin-dependent diabetes mellitus, multiple sclerosis, and Lyme disease, T cell mediated immunity is thought to play an important role in the pathogenesis of the disease. T cells isolated from the target organ involved have frequently been shown to express a restricted set of TCR molecules, supporting the notion that there is a preferential expansion of a subset of pathogenic T cells responding to an Ag stimulus. Knowing either what the offending antigens are (whether products of viruses or bacteria, or even self proteins) or the receptor usage of T cells involved in the disease process would allow for a specific target for pharmacologic intervention. In some instances, the putative target Ag and the TCR usage of T cells isolated from diseased tissue has been determined[28-30], while in other cases, an association between a particular TCR and an autoimmune disease exists, but the stimulating antigen is unknown[31,32]. The relationship between coronaviruses and MS remains an open question, but it certainly warrants further study.

ACKNOWLEDGMENTS

This work was supported in part by a grant from the Colorado Neurological Institute to J.S.S., support from Swedish Medical Center, and from the Nancy Davis Foundation. We are grateful to Dr. Judy van de Water (Division of Rheumatology, Allergy and Clinical Immunology, University of California, Davis, CA) for helpful discussions in generating the human 229E-specific T cell lines, and to Joanne Streib (National Jewish Center for Immunology and Respiratory Medicine, Denver, CO) for assisting us in generating the EBV transformed human B cell lines.

REFERENCES

1 Zamvil, S S , and L Steinman The T lymphocyte in experimental allergic encephalomyelitis Annu Rev Immunol 1990, 8 579

2 Acha-Orbea, H , D J Mitchell, L Timmermann, D C Wraith, and G S Tausch Limited heterogeneity of T cell receptors from lyphocytes mediating autoimmune encephalomyelitis allows specific immune intervention Cell 1988, 54 263

3 Ben-Nun, A , H Wekerle, and I R Cohen Vaccination against autoimmune encephalomyelitis with T-lymphocyte line cells reactive against myelin basic protein Nature 1981, 293 60

4 Martin, R , H F McFarland, and D E McFarlin Immunologic aspects of demyelinating diseases Annu Rev Immunol 1992, 10 153

5 Booss, J , and J H Kim Evidence for a viral etiology of multiple sclerosis In "Handbook of Multiple Sclerosis", S D Cook, ed Marcel Dekker, Inc , New York 1990, pp 41-61

6 McIntosh, K Coronaviruses In "Virology", B N Fields, et al , eds Raven Press, New York 1985, pp 1323-1330

7 Hamre, D , and M Beem Virologic studies of acute respiratory disease in young adults V Coronavirus 229E infections during six years of surveillance Am J Epidemiol 1972, 96 94

8 Resta, S , J P Luby, C R Rosenfeld, and J D Siegel Isolation and propagation of a human enteric coronavirus Science 1985, 229 978

9 Battaglia, M , N Passarini, A DiMatteo, and G Gerna Human enteric coronaviruses further characterization and immunoblotting of viral proteins J Inf Dis 1987, 155 140

10 Malkova, D , J Holubova, J M Kolman, F Lobkovic, L Pohlreichova, and L Zikmundova Isolation of Tettnang coronavirus from man? Acta Virol (Prague)(Eng Ed) 1980, 24 363

11 Tanaka, R , Y Iwasaki, and H J Koprowski Ultrastructural studies of perivascular cuffing cells in multiple sclerosis brain J Neurol Sci 1976, 28 121

12 Burks, J S , B L Devald, L D Jankovsky, and J D Gerdes Two coronaviruses isolated from central nervous system tissue of two multiple sclerosis patients Science 1980, 209 933

13 Gerdes, J C , I Klein, B L DeVald, and J S Burks Coronavirus isolates SK and SD from multiple sclerosis patients are serologically related to murine coronaviruses A59 and JHM and human coronavirus OC43, but not to human coronavirus 229E J Virol 1981, 38 231

14 Murray, R S , B Brown, D A Brian, and G F Cabirac Detection of coronavirus RNA and antigen in multiple sclerosis brain Ann Neurol 1992, 31 525

15 Stewart, J N , S Mounir, and P J Talbot Human coronavirus gene expression in the brains of multiple sclerosis patients Virology 1992, 191 502

16 Talbot, P J , S Ekande, N R Cashman, S Mounir, and J N Stewart Neurotropism of human coronavirus 229E In Adv in Exp Med and Biol , H Laude and J-F Vautherot, eds Plenum Press, New York 1993, 342 339

17 Murray, R S , G-Y Cai, K Hoel, J-Y Zhang, K F Soike, and G F Cabirac Coronavirus infects and causes demyelination in primate central nervous system Virology 1992, 188 274

18 Fleming, J O , M D Trowsdale, J Bradbury, S A Stohlman, and L P Weiner Experimental demyelination induced by coronavirus JHM (MHV-4) Molecular identification of a viral paralytic disease Microb Path 1987, 3 9

19 Wege, H , S G Siddell, and V Ter Meulen The biology and pathogenesis of coronaviruses Curr Top Microbiol Immunol 1982, 99 165

20 Nagashima, K , H Wege, R Meyermann, and V Ter Meulen Demyelinating encephalomyelitis induced by a long term coronavirus infection in rats Acta Neuropath 1979, 45 205

21 Kyuwa, S , and S A Stohlman Pathogenesis of a neurotropic murine coronavirus, strain JHM, in the central nervous system of mice Semin Vir 1990, 1 273

22 Williamson, J S P, and S A Stohlman Effective clearance of mouse hepatitis virus from the central nervous system requires both CD4$^+$ and CD8$^+$ T cells J Virol 1990, 64 4589

23 Korner, H , A Schiephake, J Winter, F Zimprich, H Lassmann, J Sedgwick, S Siddell, and H Wege Nucleocapsid or spike protein specific CD4$^+$ T lymphocytes protect against coronavirus-induced encephalomyelitis in the absence of CD8$^+$ T cells J Immunol 1991, 147 2317

24 Stohlman, S A , and L P Weiner Chronic nervous system demyelination in mice after JHM virus infection Neurology 1981, 31 38

25 Hirano, N , K Fugiwara, S Hino, and M Matumoto Replication and plaque formation of mouse hepatitis virus (MHV-2) in mouse cell line DBT culture Arch Gesamte Virusforsch 1974, 44 298

26 Weiss, S Coronavirus SD and SK share extensive nucleotide homology with murine coronavirus MHV-A59 more than that shared between human and murine coronaviruses Virology 1983, 126 669

27 Bergmann, C , M McMillan, and S Stohlman Characterization of the L^d-restricted cytotoxic T-lymphocyte epitope in the mouse hepatitis virus nucleocapsid protein J Virol 1993, 67 7041

28 Oksenberg, J R , M A Panzara, A B Begovich, D Mitchell, H A Erlich, R S Murray, R Shimonkevitz, M Sherritt, J Rothbard, C C A Bernard, and L Steinman Selection for T-cell receptor $V\beta$-$D\beta$-$J\beta$ gene rearrangements with specificity for a myelin basic protein peptide in brain lesions of multiple sclerosis Nature 1993, 362 68

29 Wucherpfennig, K W , K Ota, N Endo, J G Seidman, A Rosenzweig, H L Weiner, and D A Hafler Shared human T cell receptor usage to immunodominant regions of myelin basic protein Science 1990, 248 1016

30 Lahesmaa, R , M-C Shanafelt, A Allsup, C Soderberg, J Anzola, V Freitas, C Turck, L Steinman, and G Peltz Preferential usage of T cell antigen receptor V region gene segment Vβ5 1 by *Borrelia burgdorferi antigen-reactive T cell clones isolated from a patient with Lyme disease J Immunol 1993, 150 4125*

31 Paliard, X , S G West, J A Lafferty, J R Clements, J W Kappler, P Marrack, and B L Kotzin Evidence for the effects of a superantigen in rheumatoid arthritis Science 1991, 253 325

32 Conrad, B , E Weidmann, G Trucco, W A Rudert, R Behboo, C Ricordi, H Rodriquez-Rilo, D Finegold, and M Trucco Evidence for superantigen involvement in insulin-dependent diabetes mellitus aetiology Nature 1994, 371 351

21

STUDIES ON THE *IN VITRO* AND *IN VIVO* HOST RANGE OF PORCINE EPIDEMIC DIARRHOEA VIRUS

Anna Utiger,[1] Annekäthi Frei,[1] Ana Carvajal,[2] and Mathias Ackermann[1]

[1] Institute of Virology
Vet.- Med. Faculty, University of Zürich
Zürich, Switzerland
[2] Dept. de Patologia Animal
Fac. Vet., Universidad de Leon
Leon, Spain

One of the main reasons for the lack of information concerning the biological and molecular properties of porcine epidemic diarrhoea virus (PEDV) has been the difficulty to grow the agent in cell cultures. Interestingly, primary cells as well as cell lines of porcine origin have been found to support poorly, if at all, replication of PEDV. The same was true for cells and cell lines originating from other species such as mouse, bovine, hamster, and humans. Hofmann and Wyler were the first ones to report replication of PEDV in Vero cells.[1] Later, a Japanese strain of PEDV could be adapted to grow in a slightly broader range of cells.[2] Interestingly, it was observed that PEDV did not replicate in cell culture unless high amounts of trypsin were present in the culture medium throughout infection. Requirement for trypsin could be explained by two different hypothesis: (i) the receptor attachment protein of the virus had to be cleaved before adsorption or translocation into the cells was possible. Requirement of trypsin to enhance the infectivity in this way is known for many viruses, including corona-, rota-, and influenza viruses. (ii) Presence of trypsin was required to treat surface proteins of the cells, before attachment or translocation or even later events in virus replication were possible. Such treatment is likely to cause detachment of cells from plastic or glass support in conventional cell cultures. Before addressing the above questions, it was necessary to establish cell lines which are resistant to high amounts of trypsin during virus infection.

In order to adapt potential host cells of PEDV to the presence of increasing amounts of trypsin, confluent monolayers of diploid human lung cells (WI-38, ATCC#CCL75) were subjected to trypsin treatment as follows. The cell culture medium which contained 8% of fetal calf serum (FCS) was replaced by virus infection medium which was free of FCS but contained 0.3% tryptose phosphate broth and 2.5 μg/ml of trypsin.[3] The cells were kept for two hours at 4°C and for 24 hours at 37°C. Free floating and dead cells were then removed. The remaining cells were submerged in fresh cell culture medium containing FCS and left

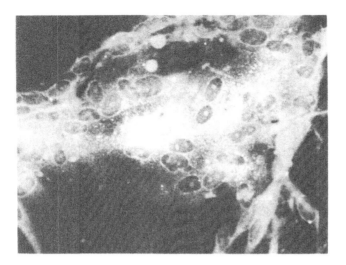

Figure 1. Immunofluorescence of PEDV infected WI-38 cells is shown. Monolayers of WI-38 cells were infected with PEDV (previously passaged already three times in WI-38 cells). 24 hours post infection, the cells were fixed with methanol and incubated consecutively with a monoclonal antibody specific for the N protein of PEDV (mAb2715, kindly supplied by L Johr[4]) and goat anti-mouse FITC

at 37°C until the monolayer was restored. Then, the selective treatment was repeated with 5 μg/ml and thereafter with 10 μg/ml trypsin.

As soon as the WI-38 cells appeared to tolerate 10 μg of trypsin for 24 hours, they were infected with our PEDV strain which had been adapted to replicate in Vero cells.[1] The virus could be passaged in WI-38 cells for at least 5 times. Characteristic syncytium formation was observed from the very first passage on WI-38. With time, the virus obviously adapted to the new cellular system, since more complete and more rapid formation of syncytia was observed. Immunofluorescent assays indicated that all the major structural proteins of PEDV (S, M, N) were synthesized in WI-38 infected cells. An example is shown in Fig. 1. Furthermore, coronavirus-like particles purified from the cell culture supernatant could be detected in the electron microscope (not shown).

These observations indicated that the host range of PEDV may be more extended than one has believed until now. It was never established, whether PEDV represented a genuine pig virus or whether it had been introduced to the porcine species from other animals or humans. Therefore, it was of interest to determine if PEDV or a related virus circulated in animal species other than porcines. In order to address this question, sera of humans and of cats were tested for antibodies to PEDV by using the PEDV specific ELISA described by Knuchel et al..[3] The details of these studies are reported elsewhere.[5] Interestingly, 6 out of 165 human sera and 9 out of 47 cat sera reacted positively in the PEDV ELISA. The interpretation of these observations is very tedious because many individuals, humans as well as cats, had also antibodies against other coronaviruses, e.g. human coronavirus 229E and feline infectious peritonitis virus. Although definitive proof is still lacking, we present for the first time serological evidence that PEDV or a closely related virus may circulate in humans and cats. Our observations are sustained by a report of Have et al.[6] who have described a PEDV-related virus in mink.

ACKNOWLEDGMENTS

These studies were supported by the Swiss Federal Institutions BAG and BVet.

REFERENCES

1. Hofmann and Wyler, J. Clin. Microbiol., 1988;26, 2235-2239.
2. Kusanagi et al., J. Vet. Med. Sci., 1992;54, 313-318.
3. Knuchel et al., Vet. Microbiol., 1992;32, 117-134.
4. Jöhr, Thesis, Vet. med. Faculty, University of Zürich (1989).
5. Frei, Thesis, Vet. med. Faculty, University of Zürich (in preparation).
6. Have et al., Vet. Microbiol., 1992;31, 1-10.

22

CORONAVIRUSES IN POLARIZED EPITHELIAL CELLS

J.W.A. Rossen,[1] C.P.J. Bekker,[1] W.F. Voorhout,[2] M.C. Horzinek,[1]
A. Van Der Ende,[3] G.J.A.M. Strous,[3] and P.J.M. Rottier[1]

[1] Institute of Virology
[2] Department of Functional Morphology
[3] Laboratory of Cell Biology
Utrecht University
Yalelaan 1
584 CL Utrecht
The Netherlands

ABSTRACT

Coronaviruses have a marked tropism for epithelial cells. In this paper the interactions of the porcine transmissible gastroenteritis virus (TGEV) and mouse hepatitis virus (MHV-A59) with epithelial cells are compared. Porcine (LLC-PK1) and murine (mTAL) epithelial cells were grown on permeable supports. By inoculation from the apical or basolateral side both TGEV and MHV-A59 were found to enter the polarized cells only through the apical membrane. The release of newly synthesized TGEV from LLC-PK1 cells occurred preferentially from the apical plasma membrane domain, as evidenced by the accumulation of viral proteins and infectivity in the apical culture fluid. In contrast, MHV was released preferentially from the basolateral membrane of mTAL cells. The apical release of TGEV and the basolateral release of MHV may explain the *in vivo* establishment of a local and systemic infection, respectively.

RESULTS AND DISCUSSION

Most coronaviruses cause enteric and/or respiratory infections. However, some coronaviruses spread systemically. The basis for these diseases appears to be the marked tropism of most coronaviruses for epithelial cells of the respiratory and intestinal tract. For example, TGEV which causes diarrhoea in pigs, infects intestinal epithelial cells[1]. MHV-A59 spreads systemically. This virus infects epithelial cells of the respiratory tract before being disseminated to other organs[2]. In epithelial plasma membranes two domains can be distinguished, the apical and basolateral, which are separated by tight junctions. In general, virus

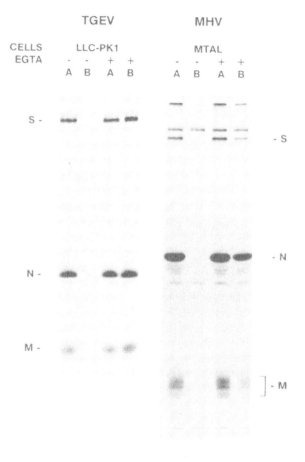

Figure 1. Entry of TGEV and MHV into LLC-PK1 and mTAL cells. Cells grown on permeable supports were inoculated from the apical (A) or basolateral (B) side with TGEV or MHV, labelled with 100 μCi ^{35}S-labelling mix (LLC-PK1:4-6 hpi; mTAL: 6-9 hpi) and lysed in lysis-buffer. Subsequently, viral proteins were precipitated from the cell lysates with MHV or TGEV specific antisera and separated in a SDS-10% polyacrylamide gel. The position of the spike (S), nucleocapsid (N) and membrane protein (M) are indicated.

entry into and release from epithelial cells are polarized, i.e. restricted to the apical or basolateral domain[3]. In this study we compared the side of TGEV and MHV-A59 entry into and the release of these viruses from polarized porcine LLC-PK1 and murine mTAL cells, respectively. Both cell lines are derived from the proximal tubule of a kidney.

LLC-PK1 and mTAL cells were grown on permeable supports to have separate access to their apical and basolateral plasma membrane domain. Subsequently, cells were inoculated with TGEV or MHV-A59 from the basolateral or apical side. Cells were labelled with ^{35}S-labelling mix, lysed and viral proteins were precipitated from the lysates with MHV or TGEV specific antisera. Figure 1 shows that the entry of both MHV and TGEV into the epithelial cells is restricted to the apical plasma membrane domain. To exclude the possibility that these findings resulted from the inability of the virus to pass the permeable support and reach the cells, cells were treated with EGTA before infection. This treatment opens the tight junctions between the cells. The cells could now be infected from both sides (Figure 1).

To determine the side of release of TGEV and MHV from these cells, apical and basolateral culture media of infected cells were collected and analyzed for the amount of infectious virus particles. In addition, media of infected, radiolabelled cells were collected and viral proteins were precipitated from these media using the virus specific antisera. Table 1 shows that infectious TGEV particles are preferentially released into the apical medium, whereas infectious MHV particles are preferentially released into the basolateral medium.

Table 1. Release of infective virus particles from
TGEV-infected LLC-PK1 and MHV-infected mTAL
cells into apical and basolateral medium

Virus	Titer[a] apical	Titer[a] basolateral	Ratio apical/basolateral
TGEV[b]	4.1×10^6	1.3×10^5	31.5
MHV[c]	7.1×10^4	5.8×10^5	0.12

[a]Mean value of six experiments.
[b]Titers in plaque forming units.
[c]Titers in $TCID_{50}$-units.

These data were confirmed by the results of the immunoprecipitation shown in Figure 2, as TGEV and MHV proteins were predominantly released into the apical and basolateral media, respectively.

In this paper it is shown that both TGEV and MHV enter epithelial cells through the apical side. However, in contrast to the apical release of TGEV, MHV is preferentially released from the basolateral membrane of epithelial cells. The results support the idea that the rapid lateral spread of the TGEV infection over the intestinal epithelia occurs by the preferential release of virus from infected epithelial cells into the gut lumen followed by efficient infection of nearby cells through the apical domain. In addition, the basolateral release of MHV from epithelial cells may be the first step in the establishment of a systemic disease.

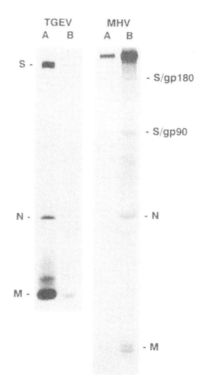

Figure 2. Release of TGEV and MHV from LLC-PK1 and mTAL cells. Cells grown on permeable supports were infected from the apical side with TGEV or MHV, labelled with 200 μCi ^{35}S-labelling mix (LLC-PK1:5-7 hpi followed by a two-hour chase; mTAL: 6-9 hpi). Viral proteins were precipitated from the apical (A) or basolateral (B) media using virus specific antisera and separated in a SDS-10% polyacrylamide gel.

REFERENCES

1 Pensaert, M , E O Haelterman, and E J Hinsman Transmissible gastroenteritis of swine virus-intestinal cell interactions 2 Electron microscopy of the epithelium in isolated jejunal loops Arch Gesamte Virusforsh 1990,31 335-351

2 Compton, S R , S W Barthold, and A L Smith The cellular and molecular pathogenesis of coronaviruses Laboratory Animal Science 1993,43 15-26

3 Tucker, S P, and R W Compans Virus infection of polarized epithelial cells Adv Virus Res 1993,42 187-247

23

ISOLATION AND EXPERIMENTAL ORAL TRANSMISSION IN PIGS OF A PORCINE REPRODUCTIVE AND RESPIRATORY SYNDROME VIRUS ISOLATE

R. Magar[1], Y. Robinson[1], C. Dubuc[2], and R. Larochelle[1]

[1] Laboratoire d'Hygiène Vétérinaire et Alimentaire, Agriculture Canada
Saint-Hyacinthe, Québec
[2] Institut de Recherches Vétérinaires
Nepean, Ontario
Canada

ABSTRACT

A virus inducing a cytopathic effect on porcine alveolar macrophages was isolated from the lungs of a pig with respiratory problems and lesions of proliferative and necrotizing pneumonia. The isolate was found to react with porcine reproductive and respiratory syndrome virus (PRRSV) monoclonal antibodies by indirect immunofluorescence and was designated LHVA-93-3. The virus could also be propagated on the MARC-145 cell line. The LHVA-93-3 macrophage-passaged isolate was inoculated orally or intranasally in four-week-old specific pathogen-free pigs. Histologically, focal to multifocal lesions of proliferative, necrotizing and interstitial pneumonia could be observed in the lungs of pigs inoculated orally or intranasally, 6 and 10 days post-inoculation. Virus could be reisolated from essentially the same tissues including serum following both routes of infection. The distribution of PRRSV antigens in fixed tissues as determined by immunogold silver staining (IGSS) was similar in orally or intranasally inoculated pigs. The results of this experimental transmission study indicate that pigs may become infected by PRRSV following oral as well as intranasal exposure.

Porcine reproductive and respiratory syndrome (PRRS) is a recently identified, economically important viral disease of swine in North America and in Europe. The causal agent of this disease is a new non-hemagglutinating, enveloped RNA virus[1,2]. The molecular characteristics of the PRRS virus (PRRSV) suggest it should belong to the Arterivirus group[3,4]. The virus is believed to be transmitted by contact or by aerosol exposure and the majority of studies on the pathogenesis of PRRSV have been carried out in swine using experimental intranasal inoculations. Little information exists on the transmission and infection by the virus *via* the oral route. The purpose of the present study was to report on

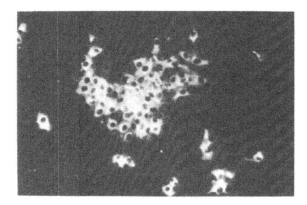

Figure 1. Indirect immunofluorescence of LHVA-93-3 infected MARC-145 cells using PRRSV monoclonal antibody SDOW17.

the identification of a PRRSV isolate and to evaluate if pigs could become infected following the experimental oral transmission of this PRRSV isolate.

The lung tissue specimen collected from a 12-week-old dyspneic pig demonstrated lesions corresponding to those of proliferative necrotizing pneumonia[5] complicated by a secondary bacterial infection. A clarified and filtered tissue homogenate was inoculated in embryonated chicken eggs and onto porcine alveolar macrophages (PAM). No haemagglutinating virus was identified following three passages in embryonated eggs. A cytopathic effect (CPE) was observed following the first passage on PAM and the CPE could be maintained for several passages. The fourth passaged isolate was inoculated on MARC-145 cells[6] and CPE could be observed during first passage and was characterized by a gradual formation of foci or small networks of clumped cells followed by rounding and detachment of these cells.

Imunofluorescence (IF) was performed in 96-well plates on acetone-fixed inoculated macrophages for virus identification. Using PRRSV monoclonal antibodies (MAbs) SDOW17, VO17 and EP147[7] bright cytoplasmic fluorescence could be observed in the PAM. The PAM were negative by IF to transmissible gastroenteritis virus, swine influenza virus, hemagglutinating encephalomyelitis virus, porcine parvovirus and bovine viral diarrhea virus. The isolate was designated LHVA-93-3. Cytoplasmic fluorescence was also noted in MARC-145 cells inoculated with this virus isolate using the SDOW17 (Figure 1), the VO17 and EP147 MAbs.

Immunogold electron microscopy (IEM) performed on aliquots of supernatants of inoculated PAM[8] demonstrated the presence of small aggregates of viral particles surrounded by gold granules (Figure 2). The virus isolate LHVA-93-3 propagated in MARC-145 cells was also examined by electron microscopy following purification on caesium chloride

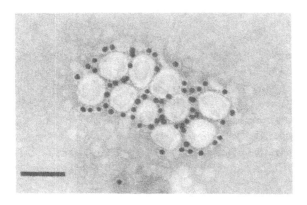

Figure 2. Immunogold electron microscopy of LHVA-93-3. Aggregate of viral particles densely labelled with gold granules. Uranyl acetate stained. Bar = 100 nm.

density gradients. Virus particles (density of 1.18 g/mL) were generally spherical but often oblong in shape averaging (average of 100 particles) 64 ± 8 nm.

Oral or intranasal inoculations of four-week-old specific pathogen-free pigs were performed using the clarified and filtered LHVA-93-3 isolate propagated in PAM. Two control pigs were orally inoculated with 2 mL each of clarified and filtered supernatant of uninfected PAM. Two pigs were orally inoculated with 2 mL of virus (approximately $10^7 TCID_{50}$) and two pigs were inoculated intranasally (1 mL per nostril) with the same preparation. One animal of each group was euthanized 6 and 10 days post-inoculation (pi). Tissue samples from tonsils, lungs, spleen, mesenteric lymph nodes and small intestine were collected for virus isolation. Samples from these tissues were also collected for histopathology and immunohistochemistry and were fixed in 10% neutral buffered formalin. Serum samples were collected prior to inoculations and at necropsy for virus isolation and serology. Detection of PRRSV antigens in formalin-fixed tissues was performed by immunogold silver staining (IGSS) using the SDOW17 MAb as previously described[8,9].

No major clinical signs were observed throughout the experiment. All pigs inoculated with the PRRSV isolate demonstrated some fever during the experimental period. Sera were tested for antibodies to PRRSV by IF performed in 96-well plates using MARC-145 cells inoculated with the Québec LHVA-93-3 PRRSV isolate. Only sera collected 10 days pi from pigs inoculated with virus were positive for PRRSV antibodies demonstrating titers of 1:2560 and 1:1280.

No major gross lesions were observed in any of the infected or control pigs and microscopic lesions were essentially limited to the lungs of the virus-inoculated pigs. Focal to multifocal areas of proliferative, necrotizing and intestitial pneumonia were noted in the lungs of pigs inoculated with PRRSV both orally and intranasally (Figures 3 and 4). Lesions were more marked in cranial lobes and in the lungs of pigs 10 days pi; they were characterized by the accumulation of necrotic cells admixed with macrophages, lymphocytes and occasionally neutrophils in the alveolar lumen, a mild to moderate hyperplasia of type II pneumocytes and focal areas of moderate thickening of alveolar septa by macrophages and lymphocytes. In some sections occasional syncytial cells were also noted in the alveolar lumen.

No PRRSV could be reisolated from tissues of control pigs including serum samples. PRRSV was isolated from tonsils, lungs, lymph nodes, spleen and sera collected at 6 days pi from both orally and intranasally inoculated pigs. Virus was also isolated from the intestinal tissue homogenate and from the intestinal content collected from the intranasally inoculated pig but not from the pig infected orally. PRRSV was recovered from all tissue samples and serum samples collected at 10 days pi from the infected pigs, except the intestinal contents from both pigs and the intestinal tissue from the orally inoculated pig.

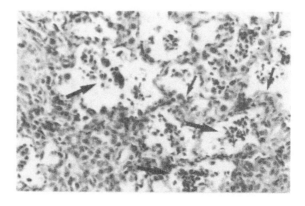

Figure 3. Lung of orally inoculated pig with LHVA-93-3 isolate of PRRSV, at 10 days pi, showing lesions of proliferative, necrotizing and intersitial pneumonia. Necrotic debris in alveoli (large arrows) and thickening of alveolar septa (small arrows) can be seen. Hematoxylin phloxin safran stain.

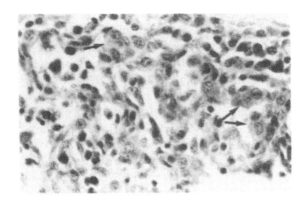

Figure 4. Lung of orally inoculated pig with LHVA-93-3 isolate of PRRSV, at 10 days pi, showing lesions of proliferative, necrotizing and intersitial pneumonia Necrotic debris in alveoli and hyperplasia of type II pneumocytes (arrows) can be seen. Hematoxylin phloxin safran stain

When the formalin-fixed paraffin-embedded lung tissue of the original field sample was tested for the presence of PRRSV antigens, dark specific staining of cells and necrotic debris could be observed (Figure 5), confirming the results obtained by virus isolation. In experimental studies, PRRSV antigens were demonstrated in tonsils, lungs, lymph nodes, spleen and intestine collected 6 days pi from both orally and intranasally inoculated pigs. In the lungs, labelled cells were scattered in the alveolar lumen and in the septa (Figure 6). Generally the number of labelled cells was greater in cranial and middle lobes than in caudal lobes. In the intestine, labelled cells were observed in the lamina propria of intestinal villi (Figure 7). In addition darkly stained cells including multinucleate giant cells were seen in the Peyer's patches (Figure 8). The distribution of labelled cells was similar in tissues from orally and intranasally inoculated pigs. At day 10 pi, PRRSV antigens were detected in tonsils, lungs and lymph nodes but not in spleen or intestine. No labelling could be noted in tissues of control pigs. Results of antigen detection in tissues were similar to those of virus reisolation.

In the present study the histopathological lung lesions noted in pigs 6 and 10 days following oral and intranasal inoculation with the LHVA-93-3 Québec isolate of PRRSV were similar to those described in proliferative and necrotizing pneumonia (PNP)[5]. Although necrotizing bronchiolitis described in PNP is not reported to be a characteristic lesion of PRRS affected lungs, these two conditions (PRRS and PNP) appear to overlap to a greater extent than originally reported. To date these conditions have been considered as separate entities. In the study described here the isolation of PRRSV from the lungs of a pig with

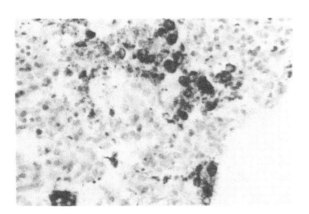

Figure 5. Immunogold silver staining of PRRSV infected lung tissue Original formalin-fixed paraffin-embedded field sample from which LHVA-93-3 was isolated. Numerous darkly stained macrophages and debris can be seen in alveoli. Hematoxylin counterstain.

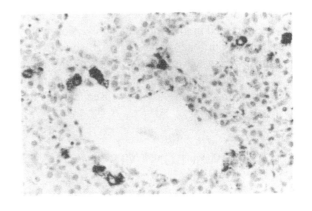

Figure 6. Immunogold silver staining of lung tissue collected from experimental pig orally inoculated with LHVA-93-3 isolate of PRRSV Stained mononuclear cells showing cytoplasmic labelling in septa Hematoxylin counterstain

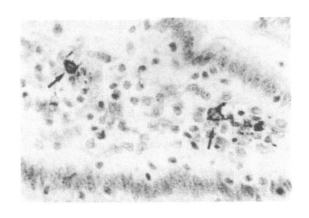

Figure 7. Immunogold silver staining of PRRSV antigens in the small intestine of experimental pig orally inoculated with LHVA-93-3 isolate of PRRSV (6 days pi) Stained mononuclear cells (arrows) in the lamina propria of intestinal villi can be seen Hematoxylin counterstain

lesions of PNP is consistent with the results and observations of recent studies[8][9][10] demonstrating an association of PRRSV with PNP

The results of the experimental transmissions in the present study indicate that pigs can become infected by PRRSV following oral as well as intranasal inoculation of virus Histological lung lesions were induced following both routes of inoculation and the pneumonic lesions observed were similar Virus could be isolated from several tissues and viral antigens could be detected in these tissues as well Labelled cells observed in the lungs were

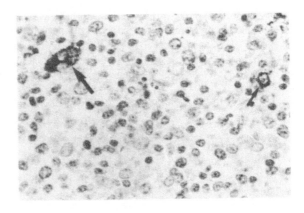

Figure 8. Immunogold silver staining of PRRSV antigens in Peyer's patches of experimental pig inoculated intranasally with LHVA-93-3 isolate of PRRSV (6 days pi) A mononuclear cell (small arrow) showing cytoplasmic labelling as well as a labelled multinucleate giant cell (large arrow) can be seen Hematoxylin counterstain

typical of alveolar macrophages and other mononuclear cells Stained cells were demonstrated in the Peyer's patches and in the lamina propria of the small intestine but no staining was observed in the villous enterocytes or the crypt cells The labelled cells in the intestine as well as in other organs (tonsils, spleen, lymph nodes) were cells of the mononuclear phagocyte system as demonstrated by the presence of labelled multinucleate giant cells These results would suggest that PRRSV can infect a pig through the oro-pharynx, probably by infecting the tonsils and a subsequent viremia may then insure virus dispersal to several tissues or organs

ACKNOWLEDGMENTS

The authors are grateful to Drs D Benfield and E Nelson (South Dakota State University, Brookings, SD) for monoclonal antibodies, K Eernisse (National Veterinary Services Laboratory, Ames, IA) for supplying the MARC-145 cells, Dr R Desrosiers for submitting sample, L Lessard, D Longtin and L Delorme for technical assistance

REFERENCES

1 Wensvoort G , Terpstra C , Pol J M A , Ter Laak E A , Bloemraad M , De Kluyver E P *et al* Mystery swine disease in the Netherlands the isolation of Lelystad virus Vet Quart 1991,13 121-130

2 Benfield D A , Nelson E , Collins J E , Harris L , Goyal S M , Robison D *et al* Characterization of swine infertility and respiratory syndrome (SIRS) virus (isolate ATCC VR-2332) J Vet Diagn Invest 1992,4 127-133

3 Meulenberg J J M , Hulst M M , De Meijer E J , Moonen P L J M , Den Besten A , De Kluyver E P *et al* Lelystad virus, the causative agent of porcine epidemic abortion and respiratory syndrome (PEARS), is related to LDV and EAV Virology 1993,192 62-72

4 Conzelmann K -K , Visser N , Van Woensel P , Thiel H -J Molecular characterization of porcine reproductive and respiratory syndrome virus, a member of the Arterivirus group Virology 1993,193 329-339

5 Morin M , Girard C , Elazhary Y , Fajardo R , Drolet R , Lagace A Severe proliferative and necrotizing pneumonia in pigs a newly recognized disease Can Vet J 1990,31 837-839

6 Kim H S , Kwang J , Yoon I J , Joo H S , Frey M L Enhanced replication of porcine reproductive and respiratory syndrome (PRRS) virus in a homogeneous subpopulation of MA-104 cell line Arch Virol 1993,133 477-483

7 Nelson E A , Christopher-Hennings J , Drew T , Wensvoort G , Collins J E , Benfield D A Differentiation of U S and European isolates of porcine reproductive and respiratory syndrome virus by monoclonal antibodies J Clin Microbiol 1993,31 3184-3189

8 Magar R , Larochelle R , Robinson Y , Dubuc C Immunohistochemical detection of porcine reproductive and respiratory syndrome virus using colloidal gold Can J Vet Res 1993,57 300-304

9 Larochelle R , Sauvageau R , Magar R Immunohistochemical detection of swine influenza virus and porcine reproductive and respiratory syndrome virus in porcine proliferative and necrotizing pneumonia cases from Quebec Can Vet J 1994,35 513-515

10 Magar R , Carman S , Thomson G , Larochelle R Porcine reproductive and respiratory syndrome virus identification in proliferative and necrotizing pneumonia cases from Ontario Can Vet J 1994,35 523-524

Immune Responses

24

THE PROTECTIVE ROLE OF CYTOTOXIC T CELLS AND INTERFERON AGAINST CORONAVIRUS INVASION OF THE BRAIN

Ehud Lavi and Qian Wang

Division of Neuropathology
Department of Pathology and Laboratory Medicine
University of Pennsylvania School of Medicine
Philadelphia, Pennsylvania 19104-6079

ABSTRACT

MHV-A59 causes focal acute encephalitis, acute hepatitis, and chronic demyelination while MHV-2 causes acute hepatitis and no brain involvement. The difference in organ tropism between these two closely related MHVs is not related to the ability of these viruses to grow in brain cells since both viruses grow equally well in primary glial cell cultures derived from neonatal mouse brains. We postulated therefore that the ability of the virus to stimulate certain host immunological factors may be important for protection of the brain against invasion and replication of the virus. In this study we performed preliminary experiments to investigate the potential role of two host factors in protection of the brain against MHV invasion: cytotoxic T cells and interferon.

Four week old $\beta2M^{(-/-)}$ mice, lacking $\beta2$ microglobulin, MHC class I expression and functional cytotoxic CD8+ T cells were inoculated intracerebrally (IC) with MHV-2 and analyzed at various intervals post infection for histopathology and viral titers in organs. Histology revealed both acute hepatitis and acute encephalitis. Acute encephalitis was observed in periventricular areas. Mononuclear lymphocytic infiltration involved the choroid plexus, the ependyma and in the surrounding brain parenchyma. There was no involvement of other areas of the brain including areas that are typically involved in A59 infection of C57Bl/6 mice. By contrast, C57Bl/6 mice infected with MHV-2 showed no involvement of the brain parenchyma and only slight inflammation of the choroid plexus was present. High titers of infectious virus was detected by plaque assay in both brains and livers of $\beta2M^{(-/-)}$ mice infected with MHV-2 in contrast to only liver titers in C57Bl/6 mice infected with a similar dose of MHV-2.

Polyclonal rabbit-anti mouse IFN α/β or anti IFN β (Lee Biomolecular Research LAb.) was given to groups of 4-week-old C57Bl/6 mice at a dose of 10,000 U per one I.P. treatment, 24 hours prior to I.C. inoculation of 1LD50 of MHV-2 or MHV-A59. At various intervals post inoculation virus titers from brains and livers were determined by plaque assay, and the histopathology of all the internal organs was analyzed by H&E staining. Treatment

Corona- and Related Viruses, Edited by P. J. Talbot and G. A. Levy
Plenum Press, New York, 1995

with preimmune serum from the same rabbit was used as control with no effect on disease outcome in either one of the viruses. While IFN antibodies had little or no effect on the outcome of disease in MHV-A59 infection, mice treated with either anti IFN α /β or anti IFN β had high titers of virus recovered from the brain and histopathological evidence of acute meningoencephalitis. Thus cytotoxic T cells and interferon may have a protective role against brain invasion of the virus in MHV-2 infection in mice.

INTRODUCTION

The ability of viruses to replicate and invade the brain is a function of the availability of cellular receptors for the virus, the ability of the virus to gain access into neural tissues (neuroinvasiveness) and a variety of host immunological barriers. Some coronaviruses contain properties of neurovirulence and neuroinvasiveness while other closely related strains are lacking these properties. For example MHV-A59 causes focal acute encephalitis, acute hepatitis, and chronic demyelination while MHV-2 causes acute hepatitis and no brain involvement. The difference in neurotropism in these viruses provides an excellent tool for the study of molecular control of neurotropism and protection of the brain from viruses. The difference in organ tropism between these two closely related MHVs is not related to the ability of these viruses to grow in brain cells since both viruses grow equally well in primary glial cell cultures derived from neonatal mouse brains. We postulated therefore that the ability of MHV to stimulate certain host immunological factors may be important for protection of the brain against invasion and replication of the virus. In this study we performed preliminary experiments to investigate the potential role of two host factors: cytotoxic T cells and interferon.

Cytotoxic T Cells

Cytotoxic T cells, CD8[+], are important for clearance of MHV infection[1]. Our previous experiments with acute MHV-A59 infection in β2M [(-/-)] mice revealed that inflammatory response in the brain is delayed in the absence of functional CD8+ cells, suggesting that cytotoxic T cells are an early inflammatory factor during MHV infection of the brain. Their protective action may be either by direct cytotoxic phenomenon or via the secretion of antiviral cytokines such as interferon.

Interferon

Using a comparative analysis between the neurotropic MHV (A59), a non-neurotropic MHV (MHV-2) and a set of recombinants between the two viruses, we are studying the molecular control of viral neurotropism. In our preliminary correlation between the phenotypic analysis and the nucleic acid sequence analysis of these viruses, it appeared that the M gene may be a determinant of acute encephalitis. A region at the 5' end of the M gene has been shown to control the ability of another coronavirus (TGEV) to induce interferon response[2]. Some coronaviruses such as TGEV are strong interferon inducers[3] while others such as MHV-JHM are not. Thus we wanted to test whether treatment with anti-interferon antibodies have an effect on the outcome of disease and organ tropism in MHV infection.

EXPERIMENTAL DESIGN AND METHODS

To investigate the role of CD8[+] cells in acute MHV-A59 infection we studied infection of β2 microglobulin negative transgenic mice. The expression of β2 microglobulin on the surface of cells is closely associated with the expression of the MHC class I protein, thus disruption of β2 microglobulin expression interferes with the normal expression of MHC class I. The cytotoxic T cell reaction is highly dependent upon recognition of antigens in the context of MHC class I. Furthermore, the development of CD8[+] cells is completely dependent upon MHC class I expression. The β2M [(-/-)] knockout mice are therefore devoid of MHC class I expression and therefore lacking functional CD8[+] cells.

Four week old β2M [(-/-)] mice, originally obtained from Dr. Koller[4], were inbred at the University of Pennsylvania, and were inoculated intracerebrally (IC) with MHV-2. Groups of 3-5 mice per time point (days 1,3,5,7,9, 11, days post inoculation) were sacrificed by anesthetic (methoxyflurane) overdose and perfused intracardially with PBS and 10% phosphate buffered formalin. Organs were removed and fixed in 10% phosphate buffered formalin. Tissues were embedded in paraffin, sectioned and stained with H&E for light microscopy[5].

Polyclonal rabbit-anti mouse IFN α/β or anti IFN β (Lee Biomolecular Research LAb.) was given to groups of 4-week-old C57Bl/6 mice at a dose of 10,000 U per one I.P. treatment, 24 hours prior to I.C. inoculation of 1LD50 of MHV-2 or MHV-A59[6]. At various intervals after inoculation virus titers from brains and livers were determined by plaque assay, and the histopathology of all the internal organs was analyzed by H&E staining.

RESULTS

Infection of β2M [(-/-)] Knockout Mice with MHV-2

Light microscopy histopathologic examination of H&E stained sections of β2M [(-/-)] mice infected I.C. with $1LD_{50\ dose}$ of MHV-2 revealed involvement of both brain and liver. During the acute stage of disease both acute hepatitis and acute encephalitis developed. Acute encephalitis was centered around the periventricular areas around the lateral ventricles, third and fourth ventricles. In all of these locations mononuclear lymphocytic infiltration involved the choroid plexus, the ependyma and slightly infiltrating in the surrounding parenchyma in the form of perivascular inflammation of the Virchow-Robin spaces and pericapillary. There was no involvement of other areas of the brain including areas that are typically involved in

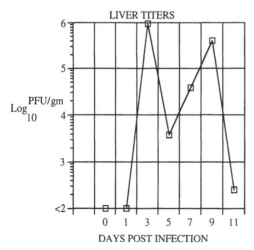
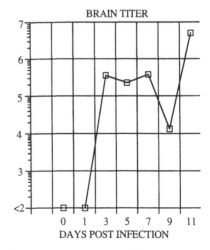

Figure 1. Organ titers following infection of β2M [(-/-)] mice with $1LD_{50}$ dose of MHV-2.

Table 1. Histopathological changes in C57Bl/6 mice receiving MHV-2 following treatment with rabbit polyclonal antibodies against interferon alpha/beta. Alternatively mice were inoculated intracerebrally with MHV-2 following treatment with preimmune rabbit serum

		PID					
		1	3	5	7	9	11
MHV-2	H	–	+	+++	+++	+	
pre-immune sera	T			+	+++	–	
	M	–	–	–	–	–	
	E	–	–	–	–	–	
	C	–	–	+	+	+	
MHV-2	H	–	+	+++	++	++	+
IFNδ/β abs	T			–		+++	–
	M	+	+	+	–	–	–
	E	–	–	++	+	+	–
	C	–	–	+	+	+	+

Key: PID - post inoculation day; H - hepatitis; T - thymic cortical depletion of lymphocytes; M - meningitis; E - encephalitis; C- choroiditis (choroid plexus inflammation).

A59 infection of C57Bl/6 mice. By contrast, C57Bl/6 mice infected with MHV-2 showed no involvement of the brain parenchyma and only slight inflammation of the choroid plexus was present.

To examine the growth of MHV-2 in the organs of infected mice and to rule out that inflammation of the brain was in this case a reactive non specific process we examined the organ titers at various time points after infection. The average of 2 mice per time point was assessed by plaque assay on L2 monolayers in six well plates inoculated with ten fold dilutions of the samples and is expressed in Log_{10} of pfu/gm tissue. High titers of infectious virus observed in both brain and liver in contrast to only liver titers in C57Bl/6 mice infected with a similar dose of MHV-2.

Anti-Interferon Treatment

While IFN antibodies had little or no effect on the outcome of disease in MHV-A59 infection, in mice treated with either anti IFN α/β or anti IFN β prior to infection with MHV-2, high titers of virus was recovered from the brain and histopathological evidence of acute meningitis and encephalitis was seen in H&E stained sections of brains (Table 1). Treatment with preimmune serum from the same rabbit was used as control with no significant effect on disease outcome in either MHV-2 or A59.

DISCUSSION

MHV-2 and A59 are closely related strains of MHV and share a high degree of molecular homology. However, the two strains differ substantially in pathogenesis and biologic properties. The main difference is in neurotropism. While A59 is neurotropic and produces acute encephalitis and chronic demyelination, MHV-2 is restricted in its ability to invade the brain even when the virus is introduced directly into the brain by I.C. inoculation. The protection of the brain in MHV-2 is not due to lack of receptors on brain cells for the virus since glial cells in culture can be infected equally with both viruses. Thus we considered the possibility that MHV-2 is capable of induction of an immunological response that protects the brain from further invasion of the virus. A59 may be defective in its ability to produce that immunological response. To begin to analyze the possible role of immune cells and cytokines in this phenomenon we asked whether mice lacking CD8+ cells or treated with anti interferon alpha/beta express more susceptibility to MHV-2 infection. In both cases we found that the brain was more susceptible to infection with MHV-2 although $\beta 2M^{(-/-)}$ mice did not express the same distribution of viral infection as C57Bl/6 mice infected with A59. Thus other factors of the immune system may play a additional role in protecting the brain against MHV-2 invasion.

ACKNOWLEDGMENTS

This study was supported in part by a grant from the University of Pennsylvania Research Foundation, and by National Multiple Sclerosis Society grants PP-0284 and RG-2615A1/2 (EL).

REFERENCES

1 Sussman MA, Shubin RA, Kyuwa S, Stohlman SA Cell mediated clearance of mouse hepatitis virus strain JHM from the central nervous system J Virol 1989,63:3051-3056

2 Laude H, Gelfi J, Lavenant L, Charley B Single amino acid changes in the viral glycoprotein M affect induction of interferon by the coronavirus transmissible gastroenteritis virus J Virol 1992,66:743-749

3 Laude H, Rasschaert D, Delmas B, Godet M, Gelfi J, Charley B Molecular biology of transmissible gastroenteritis virus Vet Microbiol 1990,23 147-154

4 Koller BH, Marrack P, Kappler JW, Smithies O Normal development of mice deficient in β2 M, MHC class I proteins, and CD8+ T cells Science 1990,248 1227-1230

5 Lavi E, Gilden DH, Highkin MK, Weiss SR The organ tropism of mouse hepatitis virus A59 is dependent on dose and route of inoculation Lab Anim Sci 1986,36 130-135

6 Su Y-H, Oakes JE, Lausch RN Ocular avirulence of a herpes simplex virus type 1 strain is associated with heightened sensitivity to alpha/beta interferon J Virol 1990,64 2187-2192

25

MHV-3 INDUCED PROTHROMBINASE IS ENCODED BY *MUSFIBLP*

Rebecca L. Parr,[1] Laisum Fung,[3] Jeffrey Reneker,[1] Nancy Myers- Mason,[3]
Julian L. Leibowitz,[1,2] and Gary Levy[3]

[1] Department of Pathology and Laboratory Medicine
[2] Department of Microbiology and Molecular Genetics
University of Texas Health Sciences Center
Houston, Texas
[3] Department of Medicine
The Toronto Hospital
University of Toronto
Toronto, Canada.

ABSTRACT

Previously, we demonstrated induction of a unique macrophage prothrombinase, PCA, in MHV-3 infected BALB/cJ mice. By immunologic screening, a clone representing PCA was isolated from a cDNA library and sequenced. The sequence identified this clone as representing part of a gene, *musfiblp*, that encodes a fibrinogen-like protein. Six additional clones were isolated, and one clone, p11-3-1, encompassed the entire coding region of *musfiblp*. Murine macrophages did not constituitively express *musfiblp*, but when infected with MHV-3, synthesized *musfiblp*-specific mRNA. *Musfiblp* mRNA induction was earlier and significantly greater in BALB/cJ than A/J macrophages. Prothrombinase activity was demonstrate when *musfiblp* was expressed from p11-3-1 in RAW 264.7 cells. These data suggest that *musfiblp* encodes the MHV-induced prothrombinase.

INTRODUCTION

MHV-3, a member of the coronavirus family, replicates in all strains of mice tested, but causes a strain-dependent spectrum of disease in inbred strains of mice[1]. This suggests that additional factors other than viral infection may be required for hepatic injury.

Several lines of evidence suggest that stimulation of the immune coagulation system by MHV-3 participates in the disease process. First, induction of monocyte/macrophage procoagulant activity (PCA) during MHV-3 infection correlates with the severity of the disease[2]. Second, there is a genetic linkage between induction of PCA in response to MHV-3, both *in vitro* and *in vivo*, and to susceptibility to liver disease[3]. Finally, treatment of mice with a monoclonal antibody to the MHV-3 induced PCA, a direct prothrombinase, prevents the lethality associated with MHV-3 infection[4].

In this study, we report the molecular cloning and sequencing of a cDNA isolated from a library prepared from macrophages which had been infected with MHV-3. The sequence of this cDNA is essentially identical to a previously described sequence corresponding to a gene encoding a mouse fibrinogen-like protein (*musfiblp*)[5]. *Musfiblp* was originally described as a cytotoxic T cell (CTL) specific gene which was constitutively expressed. In contrast, in macrophages, *musfiblp* is not normally expressed, but following MHV-3 infection its expression is upregulated. When the cDNA containing the entire coding region of *musfiblp* was expressed in the RAW 264.7 cell line, a prothrombinase activity was detected by both a one stage clotting assay and by cleavage of ^{125}I-labelled prothrombin.

MATERIAL AND METHODS

Mice

Female BALB/cJ and A/J mice, 6-8 weeks of age were purchase from Jackson Laboratories (Bar Harbor, ME.) and were housed in the animal facilities at the University of Toronto and UTHSC-Houston.

Cells and Viruses

Peritoneal macrophages were harvested from A/J and BALB/cJ mice as previously described[4]. RAW 264.7, BSC1, and RK13 were obtained from ATCC.

MHV-3 was grown and titrated as described previously[1]. Vaccinia virus expressing the T3 polymerase was generously provided by Dr. M. Esteban, Department of Biochemistry, SUNY Downstate Medical Center, Brooklyn, New York. Vaccinia virus was propagated in RK13 cells and titered in BSC1 cells[6].

Production and Screening of the cDNA Library

BALB/cJ peritoneal macrophages were infected with MHV-3 and at 6 hours p.i. RNA was extracted as previously described[7]. A random primed cDNA library was constructed in lambda ZAP II and screened by plaque immunoassay with the monoclonal antibody, 3D4.3. Plasmids were excised and purified as previously described[8]. Inserts were sequenced by automated cycle sequencing.

RNA Isolation and Analysis

Total RNA was isolated from both MHV-infected and uninfected macrophages and cell lines using guanidium hydrochloride, electrophoresed on 0.8% agarose gel in MOPS-formaldehyde, transferred to nylon membranes, and hybridized to a random-primed PCA-specific probe as previously described[7]. The membrane was autoradiographed and quantitation was performed using a Betagen scanner.

Transient Expression of *musfiblp*

Replicate cultures of 1 X 10⁶ RAW 264.7 cells were either infected with MHV-3 at an multiplicity of infection (MOI) of 5, infected with VVT$_3$pol at an MOI of 10, co-infected with both viruses, or mock infected. Immediately after viral infection, the cells were washed and transfected using the lipofectin procedure (GIBCO BRL) with 10 µg of plasmid DNA.

Procoagulant Activity (PCA)

The cell lysates were evaluated for PCA expression in a one stage clotting assay as previously described[9].

Prothrombin was radioiodinated and used in a prothrombin cleavage assay that has been described[10].

RESULTS

Identification of PCA-Specific cDNA clones

Plaques expressing immunoreactive protein were selected and repeatedly retested until a clonally isolated phage was obtained. The sequence of this clone, p1360-23 (1.25 kb) was compared with the GenBank and EMBL databases and was identical to a portion of exon 2 of *musfiblp* with the exception of one nucleotide. A series of six additional phagemids were identified by plaque hybridization, clonally isolated and propagated as plasmids (Figure 1). The clones were sequenced and arranged in relationship to each other, and the largest clone, p11-3-1, contained the entire coding sequence of *musfiblp*.

Northern Blot Analysis of PCA mRNA

In macrophages from BALB/cJ, *musfiblp*-specific RNA was first detected at 3 hours post infection, peaked at 6 hours, decreased therafter, and returned to baseline by 24 hours (Figure 2). The predominant RNA observed was approximately 4 kb in size, consistent with the previously reported size of *musfiblp* transcripts in CTL. Quantitation by betascanning indicated that an approximate eighty fold induction of the 4 kb transcript was obtained at 6 hours post infection. In A/J macrophages a significant but lesser and delayed induction of

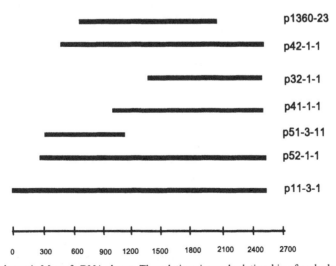

Figure 1. Schematic Map of cDNA clones. The relative size and relationship of each clone is shown.

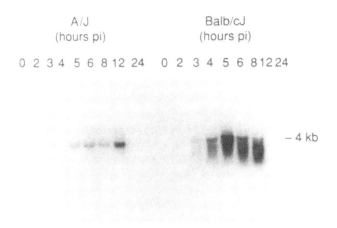

Figure 2. *Musfiblp* RNA expression in macrophages from A/J and BALB/cJ mice following MHV-3 infection. Thioglycollate elicited peritoneal macrophages (>96% MAC-1 positive) from A/J and BALB/cJ mice were mock infected (time 0 hours) or infected (times 2-24 hours) with MHV-3 at an M.O.I. of 1. Total RNA was extracted at the various time points post infection, as indicated in the figure, and 15 μg of total RNA was resolved on a formaldehyde gel, blotted onto nytran, and hybridized with the *musfiblp* cDNA probe, p1360-23, as described in experimental procedures. The relative equivalence of the amount and integrity of mRNA in each sample was verified by hybridization with an α-tubulin cDNA (data not shown).

musfiblp specific transcripts were observed. The maximum level of induction, 39-fold, was obtained at 12 hours post infection (Figure 2).

PCA Activity of Transfected Cell Lysates

To express the *musfiblp* protein, we used a macrophage cell line, RAW 264.7, which contained very low levels of *musfiblp* transcripts (data not shown) and which expressed no functional PCA, either constitutively, or following stimulation with LPS, or following infection with MHV-3 (Figure 3). For expression of the *musfiblp* protein, we exploited the presence of an upstream T3 promoter in the p11-3-1 plasmid. RAW 264.7 cells were transfected with the plasmid p11-3-1 and infected with a recombinant vaccinia virus expressing the T3 bacteriophage RNA polymerase (VVT$_3$pol). PCA was measured in a one

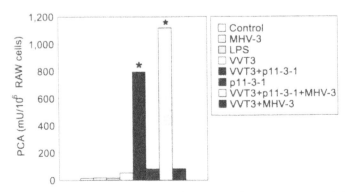

Figure 3. Induction of Procoagulant activity in RAW cells by murine hepatitis virus strain 3. Data represents the mean ± standard deviation of 4 experiments done in triplicate. Macrophages are the RAW cell line.

Figure 4. Expression of *Musfiblp* correlate: with prothrombinase activity. 0.05 ml of ho mogenates of replicate cultures of 1 X 10 RAW 264.7 cells either infected with MHV-: at an M.O I. of 5 alone, infected with VV-T: pal at an M.O.I. OF 10, co-infected with botl viruses, or mock infected in normal salin were incubated for 60 minutes at 37°C witl 0.01 ml of ^{125}I-prothrombin (100nM) an 0.01 ml of CaI$_2$ (25mM) SDS and EDTA were added and the samples electrophorese on a SDS-10% polyacrylamide slab gel an displayed by autoradiography Lane 1, ^{125}I prothrombin alone; lane 2, ^{125}I-prothrombii incubated with RAW cells infected witl MHV-3; lane 3, ^{125}I-prothrombin incubate with RAW cells stimulated with LPS for hours, lane 4, ^{125}I-prothrombin incubated

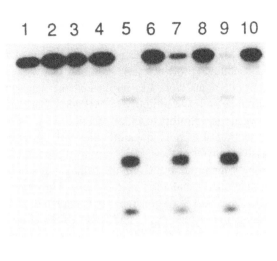

with RAW cells infected with VV-T3 pal, lane 5, ^{125}I-prothrombin incubated with RAW cells infected with both VV-T3 and p11-3-1, lane 6, ^{125}I-prothrombin incubated with RAW cells infected with p11-3-1, lane 7, ^{125}I-prothrombin incubated with RAW cells infected with VV-T3 pal, MHV-3 and p11-3-1; lane 8, ^{125}I-prothrombin incubated with VV-T3 pal and MHV-3; lane 9, ^{125}I-prothrombin incubated with Russel's Viper venom, lane 10, ^{125}I-prothrombin incubated with mock infected RAW cells

stage clotting assay and by the ability of the gene product to cleave prothrombin directly to the active moiety thrombin[10]. Uninfected and non-transfected RAW cells expressed low basal PCA (Figure 3). No augmentation was observed even following stimulation with MHV-3 or LPS. Infection with VVT$_3$pol by itself or in combination with MHV-3 resulted in a modest, although statistically significant, augmentation in PCA. Transfection of p11-3-1 in the absence of VVT$_3$pol to drive the T3 promoter resulted in a similar low level of activity. However, the combination of VVT$_3$pol and p11-3-1 induced a marked increase in PCA, suggesting that *musfiblp* expression is necessary for functional PCA. The combination of VVT$_3$pol, p11-3-1, and MHV-3, further increased PCA, suggesting that MHV infection may further upregulate PCA, although the mechanism for this is not clear.

To confirm that the PCA was indeed a prothrombinase, the cell lysates were added to ^{125}I-prothrombin and cleavage to thrombin was monitored by SDS-PAGE (Figure 4). Only RAW cells in which the expression of *musfiblp* was driven by VVT$_3$pol (lanes 5 and 7) demonstrated prothrombin cleavage. Neither infection with VVT$_3$pol (lane 4) nor transfection with p11-3-1 by itself (lane 6) resulted in prothrombin cleavage. The pattern of cleavage was identical to the physiologic cleavage observed following addition of factor Xa (lane 9).

DISCUSSION

In this study we have isolated a molecular clone corresponding to a mouse fibrinogen like gene (*musfiblp*)[5][11]. Originally *musfiblp* was isolated from a CTL cell line and was shown to be constitutively expressed in CTL but not in B cell lines, T helper cells, keratinocytes, mouse brain, or fibroblasts. Our data show that *musfiblp* RNA was not constitutively expressed in macrophages, but following MHV infection was markedly upregulated. Although we did not examine CTL specifically, lymphocytes recovered from popliteal lymph nodes only showed small amounts of constitutively expressed *musfiblp* RNA and MHV infection did not result in further augmentation (data not shown). Although macrophages from resistant A/J mice transcribe *musfiblp* in response to an MHV-3 infection, they express no functional PCA. There are several possibilities that could explain this observation. First, a post translational modification may be required for functional activity and this may not occur in A/J mice. Second, functional activity may require additional factors yet to be identified. Third, it is possible that the sequence of *musfiblp* in A/J mice precludes translation of a functional protein.

Previous work has implicated the macrophage as being of central importance in resistance and susceptibility to MHV-3[12]. Furthermore, we have established a correlation between susceptibility to lethal infection and induction of PCA by macrophages[13]. The recent production of monoclonal antibodies to PCA in our laboratory and the demonstration that passive transfer of these antibodies to susceptible mice confers resistance to MHV-3 argues strongly for a role for PCA in the pathogenesis of MHV-3[4]. The sequelae of prothrombinase production is generation of thrombin. Thrombin is the central bioregulatory enzyme in hemostasis and is responsible for conversion of fibrinogen to fibrin and for activation of coagulation factors V, VII, and XIII[14]. In addition, it is a potent activator of platelets either alone or in synergism with other agents, and may interact directly with endothelial cells causing vasoconstriction[15] and mediating leukocyte adherence[16], all of which are prominent features of MHV infection [17].

In conclusion, we have demonstrated that following MHV-3 infection, *musfiblp* RNA is markedly upregulated in macrophages from susceptible BALB/cJ mice, and that expression of *musfiblp* results in prothrombinase activity. Future experiments on the mechanism of regulation of this gene will lead to further understanding of the role of *musfiblp* in the pathogenesis of MHV-3 infection.

ACKNOWLEDGMENTS

This work was supported by a program project grant (PPG11810) from the Medical Research Council of Canada, NIH grant AI31069 and a grant from the Council for Tobacco Research. We would like to thank Dr. M. Estaban for kindly providing us with the VVT$_3$pol.

REFERENCES

1 Levy, G A , Leibowitz, J L , and Edgington, T S The induction of monocyte procoagulant activity by murine hepatitis virus MHV-3 parallels disease susceptibility in mice J Exp Med 1981,154 1150-1163

2 MacPhee, P J , Dindzans, V J , Fung, L S , and Levy, G A Acute and Chronic changes in the microcirculation of the liver in inbred strains of mice following infection with mouse hepatitis virus type 3 Hepatology 1985,5 649-660

3 Dindzans, V J , Skamene, E , and Levy, G A Susceptibility/resistance to mouse hepatitis virus strain 3 and macrophage procoagulant activity are genetically linked and controlled by two non-H-2 linked genes J Immunol 1986,137, 2355-2360

4 Li, C Y, Fung, L S , Chung, S , Crow, A , Myers-Mason, N , Phillips, M J, Leibowitz, J L , Cole, E , Ottaway, C A , and Levy G Monoclonal antiprothrombinase (3D4 3) prevents mortality from Murine Hepatitis Virus (MHV-3) infection J Exp Med 1992,176 689-697

5 Koyama, S , Hall, L R , Haser, W G , Tonegawa, S , and Saito, H Structure of a cytotoxic T-lymphocyte-specific gene shows a strong homology to fibrinogen β and γ chains Proc Natl Acad Sci USA 1987, 84 1609-1613

6 Rodriguez, D , Zhou, Y , Rodriguez, J-R , Durbin, R K , Jimenz, V , McAllister, W T , and Esteban, M Regulated expression of nuclear genes by T3 RNA polymerase and *lac* repressor, using recombinant vaccinia virus vectors J Virol 1990,64(10) 4851-4857

7 Evans, R , and Kamdar, S J Stability of RNA isolated from macrophages depends on the removal of an RNA-degrading activity early in the extraction procedure Biotechniques 1990,8(4) 357-360

8 Sambrook, J , Fritsch, E F , and Maniatis, T In Molecular Cloning A Laboratory Manual, Second Edition (Cold Spring Harbor, New York Cold Spring Harbor Laboratory Press) 1989

9 Levy, G A , and Edgington, T S Lymphocyte Cooperation is required for amplification of macrophage procoagulant activity J Exp Med 1980,151 1232-1244

10 Ottaway, C A , Warren, R E , Saibil, F G , Fung, L S , Fair, D S , and Levy, G A Monocyte procoagulant activity in Whipple's disease J Clin Immunol 1984,4(5) 348-358

11 Fung, L S , Neil, G , Leibowitz, J , Cole, E H , Chung, S , Crow, A , and Levy, G A Monoclonal antibody analysis of a unique macrophage procoagulant activity induced by murine hepatitis virus strain 3 infection J Biol Chem 1991,226(3) 1789-1795

12 Chung, S, Sinclair, S , Leibowitz, J, Skamene, E , Fung, L S , and Levy, G A Cellular and metabolic requirements for induction of macrophage procoagulant activity by murine hepatitis virus strain 3 in vitro J Immunol 1991,145 271-278

13 Dindzans, V J , MacPhee, P J , Fung, L S , Leibowitz, J L , and Levy, G A The immune response to mouse hepatitis virus Expression of monocyte procoagulant activity and plasminogen activator during infection in vivo J Immunol 1985,135(6) 4189-4197

14 Fenton, J W Thrombin Ann N Y Acad Sci 1986,485 5-15

15 Haver, V M and Namm, D H Characterization of the thrombin induced contraction of vascular smooth muscle Blood Vessels 1984,21 53-63

16 Jungi, T W , Spycher, M O , Nydegger, V E , and Banandum, S Platelet-leukocyte interaction Selective binding of thrombin stimulated platelets to human monocytes, polymorphonuclear leukocytes and related cell lines Blood 1986,67 629-636

17 Levy, G A , Leibowitz, J , and Edgington, T S Lymphocyte-instructed monocyte induction of the coagulation pathways parallels the induction of hepatitis by the murine hepatitis virus Prog Liver Dis 1982, 7 393-409

26

DEMYELINATION INDUCED BY MURINE CORONAVIRUS JHM INFECTION OF CONGENITALLY IMMUNODEFICIENT MICE

J. J. Houtman,[1] H. C. Hinze,[1] and J. O. Fleming[1,2,3]

[1] Departments of Medical Microbiology and Immunology
[2] Department of Neurology
   University of Wisconsin School of Medicine
[3] William S. Middleton Memorial Veterans Hospital
   Madison, Wisconsin

ABSTRACT

Mouse hepatitis virus JHM (JHMV or MHV-4) induces demyelination in rodents and has been studied as a model for the human disease, multiple sclerosis (MS). As is proposed in MS, the mechanism of subacute demyelination induced by JHMV appears to be primarily immunopathological, since demyelination in JHMV-infected mice is abrogated by immunosuppressive doses of irradiation and restored by adoptive transfer of splenocytes. Thy-1$^+$ cells play a critical role in transmitting disease to these recipient mice. To further characterize cells which may mediate JHMV-induced immunopathology, we inoculated congenitally immunodeficient mice with JHMV. By 12 days post-inoculation, both immunocompetent C57BL/6J controls and athymic nude C57BL/6 mice had severe paralysis and demyelination. In marked contrast, C57BL/6 mice with the severe combined immune deficiency (SCID) mutation had little or no paralysis or demyelination. Adoptive transfer of immune spleen cells from nude mice to infected SCID mice produced paralysis and demyelination. These findings suggest that a cell population present in immunocompetent C57BL/6J and nude mice but absent or non-functional in irradiated and SCID mice is essential for JHMV-induced demyelination. Identification of cells which mediate demyelination in this experimental system may have implications for our understanding of coronavirus pathogenesis and human demyelinating diseases.

INTRODUCTION

Infection of rodents with the neurotropic murine coronavirus JHM (MHV-4) produces an acute, often lethal encephalitis. Survivors exhibit a subacute or chronic paralytic-demyelinating disease which has been proposed as a model for the human demyelinating disease, multiple sclerosis.

Two mechanisms for JHMV-induced demyelination have been proposed. Early studies suggested that myelin damage was due to cytolytic viral infection of the myelin-producing oligodendrocytes[1,2]. Support for the oligodendrocyte lysis hypothesis included the localization of virions within oligodendrocytes and the occurrence of some demyelination in immunosuppressed mice. More recently, however, evidence has accumulated supporting a mechanism whereby myelin damage is caused by the immune response to viral infection. Immunosuppressive irradiation up to six days post-inoculation (PI) can prevent demyelination and adoptive transfer of immune splenocytes restores demyelination to infected irradiated recipients[3]. In addition, depletion of cells bearing the Thy-1 marker from the adoptively transferred cell population prevents restoration of demyelination, suggesting a role for T lymphocytes in demyelination[4]. An immunopathological mechanism for JHMV-induced demyelination has also been demonstrated in rats, with both CD4[+] and CD8[+] T lymphocytes contributing to disease[5].

To characterize cells which may participate in immune-mediated demyelination, we infected congenitally immunodeficient mice with JHMV. Athymic nude mice developed paralysis and demyelination. In contrast, mice possessing the severe combined immune deficiency (SCID) mutation showed minimal paralysis and little or no demyelination. In addition, demyelination was adoptively transferred to infected SCID mice with immune splenocytes from nude mice. This supports an immunopathological mechanism for demyelination and suggests that a cell population present in immunocompetent and nude mice but absent or non-functional in SCID and irradiated mice is essential for JHMV-induced demyelination.

MATERIALS AND METHODS

Male C57BL/6J, C57BL/6J-*nu* (nude) and C57BL/6J-*scid*/SzJ (SCID) mice were obtained from The Jackson Laboratory (Bar Harbor, ME) and used at 5-7 weeks of age. Mice were housed in microisolators and handled in a biosafety hood. The neuroattenuated JHMV antigenic variant 2.2-V-1 has been described previously and produces demyelination in immunocompetent mice with little or no encephalitis[6,7]. Mice were infected with 10^3 plaque forming units (pfu) of virus in 30 μl of Dulbecco's modified essential medium (DMEM) by the intracerebral (i.c.) route. Selected mice were irradiated 3 days after intracerebral inoculation with 850 rads of gamma-irradiation from a Cobalt-60 source[4]. Donor nude mice were immunized with 10^6 PFU intraperitoneally 6 days prior to transfer. Recipient SCID mice were infected 3 days prior to transfer. Donor splenocytes (4 x 10^6 cells) were transferred by the intravenous route.

Mice were monitored for clinical signs of disease until 12 days PI, when they were sacrificed. Brains and spinal cords were removed and subjected to virus isolation on DBT cells[8] or histopathological analysis[9]. A combined hematoxylin and eosin/luxol fast blue stain was used to visualize myelin. Selected mice were perfused for electron microscopic analysis[6].

RESULTS

As has been demonstrated previously, immunocompetent C57BL/6J mice infected with 2.2-V-1 undergo severe paralysis accompanied by marked demyelination, which can be prevented by immunosuppressive irradiation 3 days post-inoculation (Table 1, Groups 1

Table 1. Outcome at 12 days post-inoculation of immunocompetent and immunodeficient mice, and after adoptive transfer

Group	Experiment[a]	Paralysis[b]	Demyelination[b]	Virus[c]
1	C57BL/6	+++	+++	−
2	Irradiated	−	−	+++
3	Nude	+++	+++	++
4	SCID	+	+/ −	++
5	Nude to SCID Transfer	++	++	++

[a]Experimental groups of 6 to 8 mice. Mice were inoculated with 10^3 PFU of 2 2-V-1 by the intracerebral route "C57BL/6" refers to immunocompetent mice "Irradiated" refers to 850 rads at 3 days PI. "Nude to SCID transfer" refers to infected SCID recipients of adoptively transferred immune nude splenocytes.
[b]Paralysis and demyelination are indicated as severe (+++), moderate (++), minimal (+) or none (-) "+/-" indicates demyelination in one of six mice examined
[c]Recovery of infectious virus from brain homogenates at 12 days PI The sensitivity of the assay was 10^2 pfu per gram of brain

and 2)[4]. Nude mice infected with JHMV were also severely affected; most of these mice developed clinical paralysis, and both light microscopic and ultrastructural studies showed demyelination (Table 1, Group 3). In marked contrast, SCID mice showed only minimal clinical effects, mostly mild to moderate paraparesis (Table 1, Group 4). With the exception of one animal, no demyelination was evident by light microscopy or ultrastructural analysis in SCID mice studied. Whereas immunocompetent C57BL/6 mice were able to clear virus from the brain by day 12 PI, both nude and SCID mice had high titers of virus remaining in the brain at day 12 (Table 1).

Since the above results suggest that a cell population present in nude mice but absent in SCID mice is essential for JHMV-induced demyelination, immune splenocytes from nude mice were transferred into infected SCID mice. The results of this experiment are depicted in Table 1 (Group 5). Clinically, seven out of eight of these mice showed marked hindlimb weakness, and histopathologic analysis revealed plaques of demyelination. Disease in the SCID recipients of nude splenocytes was more severe than in normal SCID mice, but less severe than in nude or immunocompetent C57BL/6J mice.

DISCUSSION

Paralysis and demyelination have been previously reported in both nude mice and nude rats infected with JHMV[10,11]. These authors used non-neuroattenuated JHMV to infect athymic rodents and observed paralysis, demyelination and rapidly fatal encephalitis with destruction of neurons, making interpretation of these findings difficult. In the experiments reported here we used a neuroattenuated strain of JHM which allows us to study demyelination in immunodeficient mice with little or no confounding encephalitis.

The findings presented here support an immune-mediated mechanism for JHMV-induced demyelination rather than a viral cytolytic mechanism. Mice with the lowest level of immunocompetence (SCID and irradiated mice) showed the least demyelination. Thus, demyelination is correlated with the degree of immune function. The level of demyelination in SCID mice was augmented by the adoptive transfer of splenocytes from nude mice, suggesting that cells contributing to demyelination are present in the transferred splenocytes. SCID and irradiated mice showed little or no demyelination despite the presence of high titers of infectious virus in the CNS at 12 days PI. This argues against a viral cytolytic mechanism for JHMV-induced demyelination.

At 12 days PI, we isolated infectious virus from the brains of irradiated C57BL/6 mice, nude and SCID mice, and SCID recipients of immune nude splenocytes. Only immunocompetent C57BL/6 mice were able to clear infectious virus by 12 days PI. This is consistent with reports demonstrating a requirement for CD4[+] and CD8[+] T lymphocytes for viral clearance[12,13]. Since nude mice developed severe paralysis and demyelination, yet were unable to clear the virus, distinct cell populations may be involved in demyelination and viral clearance. Although T lymphocytes are essential for demyelination[4], thymically educated T lymphocytes do not appear to be required, since athymic nude mice develop severe demyelination.

In conclusion, we have demonstrated paralysis and demyelination in JHMV-infected immunocompetent and nude mice. In contrast, little or no demyelination was evident in SCID and irradiated mice. Adoptive transfer of immune splenocytes from nude mice resulted in paralysis and demyelination in SCID recipients. Our findings support an immune-mediated mechanism for demyelination. and suggest that a cell population present in immunocompetent and nude mice, but deficient in SCID and irradiated mice, is essential for JHMV-induced demyelination. Identification of this cell population may lead to new insights into the pathogenesis of human demyelinating diseases.

ACKNOWLEGEMENTS

This work was supported by the National Multiple Sclerosis Society (grant RG-2153-A-2). J.J.H. is a trainee on NIH Cellular and Molecular Biology Training Grant GM07215.

REFERENCES

1 Lampert, P W , Sims, J K , Kniazeff, A J Mechanism of demyelination in JHM virus encephalomyelitis Electron microscopic studies Acta Neuropath (Berlin) 1973,24 76-85

2 Weiner, L P Pathogenesis of demyelination induced by a mouse hepatitis virus (JHM virus) Arch Neurol 1973,28 298-303

3 Wang, F -I , Stohlman, S A , Fleming, J O Demyelination induced by murine hepatitis virus JHM strain (MHV-4) is immunologically mediated J Neuroimmunol 1990,30 31-41

4 Fleming, J O , Wang, F -I , Trousdale, M D , Hinton, D R , Stohlman, S A Interaction of immune and central nervous systems Contribution of anti-viral Thy-1[+] cells to demyelination induced by coronavirus JHM Regional Immunol 1993,5 37-43

5 Schwender, S , Hein, A , Imrich, H , Dorries, R On the role of different lymphocyte subpopulations in the course of coronavirus MHV IV (JHM)-induced encephalitis in Lewis rats In Laude, H , Vautherot, J F (eds) Coronaviruses Plenum Press, NY 1994 pp 425-430

6 Fleming, J O , Trousdale, M.D , El-Zaatari, F A K , Stohlman, S A , Weiner L P Pathogenicity of antigenic variants of murine coronavirus JHM selected with monoclonal antibodies J Virol 1986,58 869-875

7 Fleming, J O , Trousdale, M D , Bradbury, J , Stohlman, S A , Weiner, L P Experimental demyelination induced by coronavirus JHM (MHV-4) Molecular identification of a viral determinant of paralytic disease Microb Pathogen 1987,3 9-20

8 Stohlman, S A , Matsushima, G K , Casteel, N , Weiner, L P In vivo effects of coronavirus-specific T cell clones DTH inducer cells prevent a lethal infection but do not inhibit virus replication J Immunol 1986,136 3052-3056

9 Wang, F -I , Hinton, D R , Gilmore, W , Trousdale, M D , Fleming, J O Sequential infection of glial cells by the murine hepatitis virus JHM strain (MHV-4) leads to a characteristic distribution of demyelination Lab Invest 1992,66 744-754

10 Sorensen, O , Dugre, R , Percy, D , Dales, S In vivo and in vitro models of demyelinating disease Endogenous factors influencing demyelinating disease caused by mouse hepatitis virus in rats and mice Infect Immun 1982,1248-1260

11 Sorensen, O , Saravani, A , Dales, S *In vivo* and *in vitro* models of demyelinating disease XVII The infectious process in athymic rats inoculated with JHM virus Microb Pathogen 1987,2 79-90

12 Sussman, M A , Shubin, R A , Kyuwa, S , Stohlman, S A T-cell-mediated clearance of mouse hepatitis virus strain JHM from the central nervous system J Virol 1989,63 3051-3056

13 Williamson, J S P , Stohlman, S A Effective clearance of mouse hepatitis virus from the central nervous system requires both CD4+ and CD8+ T cells J Virol 1990,64 4589-4592

27

INDUCTION OF A PROTECTIVE IMMUNE RESPONSE TO MURINE CORONAVIRUS WITH NON-INTERNAL IMAGE ANTI-IDIOTYPIC ANTIBODIES

Mathilde Yu and Pierre J. Talbot

Laboratory of Neuroimmunovirology
Virology Research Center
Institut Armand-Frappier
Université Du Québec
Laval, Québec, Canada H7N 4Z3

ABSTRACT

Neurotropic murine coronaviruses (MHV) provide an excellent animal model to study experimental modulation of the immune response to a viral pathogen with anti-idiotypic antibodies. It is known that among the various types of anti-idiotypic antibodies (anti-Id), those designated beta (β) or internal image can molecularly mimic the antigen and induce biological activities such as anti-viral protection and neutralization. We have recently shown that polyclonal non-internal image anti-idiotypic antibodies of the γ-type could induce protective anti-coronavirus immunity[1].

In the present study, a polyclonal anti-Id (Ab2) was induced against a neutralizing murine monoclonal antibody (MAb1), designated 5B170.11. Mice immunized with this affinity-purified rabbit $Ab2_\alpha$, a non-internal image antibody, were partially protected against lethal infection by the JHM strain of MHV. However, other polyclonal and monoclonal non-internal image Ab2 induced against another neutralizing MAb1, designated 4-11G.6, were not able to protect mice against lethal infection with the A59 strain of MHV.

These results demonstrate that anti-viral protection by altering the idiotypic network with non-internal image-bearing anti-idiotype reagents can be achieved even with some anti-Id of the α-type.

Corona- and Related Viruses, Edited by P. J. Talbot and G. A. Levy
Plenum Press, New York, 1995

INTRODUCTION

Anti-idiotypic antibodies are potentially involved in the regulation of the immune response to a given antigen (Ag)[2]. Moreover, anti-Id have been used to manipulate the immune response *in vivo* in several experimental systems[3,4].

Anti-Id (Ab2) are classified in three different categories based on the idiotopes they recognize on Ab1. $Ab2_\alpha$ recognize idiotopes far from the antigen combining site (paratope) of Ab1. $Ab2_\gamma$ recognize idiotopes near the paratope of Ab1 and can compete with antigen for the binding site of Ab1. $Ab2_\beta$ are referred to as internal image anti-Id. They have the capacity to mimic the antigen used to generate the Ab1 and can substitute for antigen in inducing an anti-antigen response because they recognize an idiotype at the level of the paratope. However, the idiotypic network is complex because some non-internal image anti-Id can also induce an immune response to antigen, which can be described as biological rather than a structural mimicry. For example, $Ab2_\alpha$ could induce an anti-hepatitis B surface Ag response[5] and neutralizing antibodies against HIV[6]. Also, we have described an $Ab2_\gamma$-induced protective immune response against MHV-A59[1]. Therefore, non-internal image Ab2 can have interesting biological properties, although we do not yet understand their mechanisms of action.

Neurotropic murine coronaviruses (MHV strains JHM and A59) provide an excellent animal model to study the manipulation of the idiotypic network and its effect on a viral infection.

In the present study, we describe the production and characterization of non-internal image Ab2s against *in vitro* neutralizing and *in vivo* protective monoclonal antibodies (MAb1) specific to MHV-A59 and MHV-JHM. MAb1 4-11G.6 recognizes a discontinuous epitope on the spike (S) glycoprotein of MHV-A59, whereas MAb1 5B170.11 recognizes a linear epitope on the homologous protein of MHV-JHM. These two MAb1 had both been characterized previously and shown to neutralize virus infectivity *in vitro* and passively protect mice *in vivo* against a lethal MHV infection[7,10]. This suggested that the S protein has biological importance in immune protection against MHV infection, which was confirmed by the demonstration that affinity-purified S glycoprotein could vaccinate against lethal coronavirus infection[7].

Administration of these Ab2s to BALB/c mice showed that some non-internal image $Ab2_\alpha$ could partially mimic interesting biological properties of internal image Ab2, such as *in vivo* protection against viral infection. This emphasizes the complexity of the idiotypic network and how little is known on the mechanisms of induction of a protective immune response by anti-Id.

MATERIALS AND METHODS

Animals. New-Zealand white female rabbits of 2.5 to 3 kg were purchased from *Ferme de sélection Cunipur*, Stukely Sud, Québec, Canada. Four to 5 week-old female BALB/c mice were purchased from Charles River, St-Constant, Québec, Canada.

Antibodies. The production and characterization of the hybridoma secreting mouse neutralizing MAb 4-11G.6, specific for a discontinuous epitope on the S glycoprotein of MHV-A59, has been previously described[7]. Monoclonal antibody 5B170 is specific for a continuous epitope on the S protein of MHV-JHM[8]. All antibodies including normal rabbit immunoglobulins (NRIg) were purified by standard Protein-A-Sepharose chromatography.

Polyclonal anti-Id and antibody assays. The immunization protocols for generating polyclonal anti-Id and their characterization have been described elsewhere[1], including virus neutralization and protection assays and ELISA for detection of Ab3 against MHV-A59 in mice sera. ELISA for detection of idiotype in antiviral sera and of inhibition of binding of idiotype to antigen by anti-Id were also described previously[1].

Monoclonal anti-Id. The production of MAb was described previously[7]: BALB/c mice were immunized with 100 µg of affinity purified MAb 4-11G.6 and given two booster injections of 50 µg.

ELISA for detection of MAb2. Microtiter plates were coated with affinity purified F(ab')$_2$ MAb1 [1.25 µg/mL in phosphate buffered saline (PBS)] and incubated for 16 h at room temperature. The plates were blocked with PBS containing 10% (v/v) fetal calf serum and 0.2% (v/v) Tween-20 for 30 min at 37°C. Hybridoma culture supernatants (4 days of growth) diluted 1/2 were added to the wells and incubated for 90 min at room temperature. The plates were then washed five times with PBS containing 0.1% (v/v) Tween-20. Peroxidase-labeled goat anti-mouse Fc antibody (ICN Biologicals, Miles) was then added and incubated for 90 min at room temperature. The plates were washed five times and the reaction was developed with O-phenylenediamine and hydrogen peroxide. The reaction was stopped with 1N HCl and the absorbance read at 492 nm using an SLT EAR 400 AT plate reader.

Virus and cells. MHV-A59 and MHV-JHM were obtained from the American Type Culture Collection (Rockville, MD), plaque-purified twice, and passaged four times at a multiplicity of infection of 0.01 on DBT cells as described previously[9].

Plaque assays with brain homogenates of mice immunized with anti-Id. Brains were collected 5 days after virus challenge and plaque assays performed as described elsewhere[10].

Statistics. Results of *in vivo* protection assays were analyzed with the Kaplan-Meier survival curve[11]. Antiviral Ab3 antibody responses were evaluated at a 1/500 dilution and analyzed with the Mann-Whitney test[12]. Brain viral titers observed after NRIg or anti-Id treatment and the repeated experiments were first analyzed with a Manova test[13]. This test included an interaction test which in our case was shown to be significant, so the two treatments were then compared separately by a Student t-test[14].

RESULTS AND DISCUSSION

Polyclonal Anti-5B170.11

A polyclonal anti-Id 5B170.11 was purified by affinity chromatography from the serum of a rabbit immunized with MAb1 5B170.11. To determine if the anti-Id bound to the paratope of MAb1, inhibition of virus-binding by anti-Id was tested. This inhibition of attachment assay can discriminate α-type from β and γ-types anti-Id. The polyclonal anti-Id produced against MAb1 5B170.11 did not inhibit the interaction between MAb1 and antigen, consistent with it being an Ab2$_\alpha$ (Fig. 1A). The anti-Id was also tested for it capacity to inhibit the neutralizing ability of MAb1 in an inhibition of virus-neutralization assay. As much as 10 µg of anti-Id could not reduce the neutralization titer of MAb1 (data not shown), which confirmed the results of the inhibition of attachment assay.

On the basis of previous studies showing that non-internal image anti-Id could trigger an antigen-specific immune response[1,16,17] like internal image anti-Id, we examined the *in vivo* modulation capability of the rabbit polyclonal Ab2$_\alpha$ anti-5B170.11. Two groups of 6 BALB/c mice were immunized with anti-Id or NRIg three times at two-week intervals. After the third injection, the MHV-specific Ab3 response was examined by ELISA. No detectable antiviral antibodies were produced in mice immunized with Ab2$_\alpha$ (data not shown). To verify whether the Ab2$_\alpha$ could nevertheless induce a protective immune response, mice were challenged intracerebrally with 10 LD$_{50}$ of MHV-JHM, 10 days after the last booster anti-Id injection. All mice showed clinical signs of MHV infection. Animals in control groups died

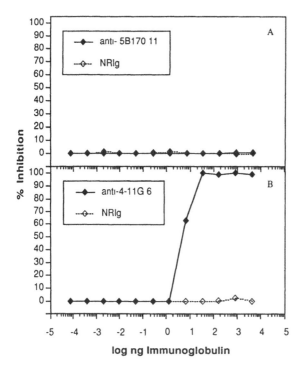

Figure 1. Inhibition of attachment assay for discrimination of α-type from β and γ-type anti-Id Microtiter plates were coated with 5 μg/mL of viral antigen Biotinylated MAb1 and dilutions of purified Ab2 or NRIg were pre-incubated together and transferred onto the viral antigen-coated plates. The residual binding of MAb1 to viral antigens was detected using peroxidase-labeled streptavidin

from MHV-JHM infection within 6 to 10 days, whereas 16% to 33% of mice in the $Ab2_\alpha$ group lived longer (until day 41) (Fig. 2A). In a repeat experiment some mice survived in both control and $Ab2_\alpha$ treated group (Fig. 2B). In the third repeat experiment, anti-viral Ab3s were detected by ELISA at a dilution of 1/500 (data not shown). The presence of specific Ab3s at this dilution was significant for a p value of 0.0014 in the Mann-Whitney test. However, these Ab3s could not neutralize viral infection *in vitro* (data not shown). No specific reactivity was observed in control groups (NRIg) or pre-immune sera. We then evaluated whether the apparent partial protection of mice correlated with reduced viral titers

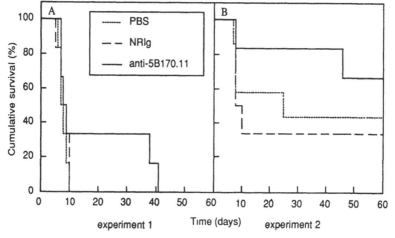

Figure 2. Survival curves of mice immunized with polyclonal Ab2 anti-5B170.11. Groups of 6 BALB/c mice were immunized with Ab2, NRIg or PBS and challenged with 10 LD_{50} of MHV-JHM. A: experiment 1; B: experiment 2

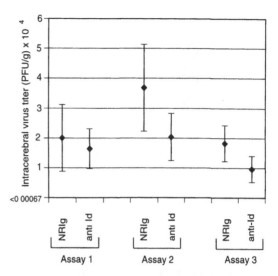

Figure 3. Brain viral titers observed after anti-Id and NRIg treatments. Virus titers from brains of mice immunized with Ab2 anti-5B170 11 or NRIg were quantitated by plaque assay

in the brain (Fig 3) The first plaque assay did not show a significant reduction of viral titers but two other plaque assays from the same brain aliquots did show significant reductions in viral titers between mice treated with Ab2$_\alpha$ and NRIg The reduction in viral titer was significant at a p value of 0 014 with the second assay and at a p value of 0 004 for the third assay Such reduced viral titers could explain the observed apparent protection

We observed either a weak or non-existent antiviral Ab3 response (data not shown) and variable protection after treatment with polyclonal Ab2$_\alpha$ anti-5B170 11 (Fig 2) This suggests either a need to optimize the conditions for antiviral Ab3 induction, or an involvement of cellular protective responses The induction of specific immune responses by Ab2$_\alpha$ has been studied in different experimental systems[15 16 17 18 19] and the activation of the cellular component of the immune response by anti-Id was reported[20 21 22]

Further experiments are needed to understand the mechanisms involved in the induction of protective immunity by non-internal image anti-Id For example, very little is known on the interactions between anti-idiotypic antibody and immune cells

The reasons why some Ab2$_\alpha$ induce protection and others do not are unclear Ab2$_\alpha$-induced immune responses might be the result of the induction of a regulatory pathway of idiotypes Anti-Id could induce a different series of immunological reactions within an idiotypic network than those induced by the antigen

Table 1. Ab2 anti-4-11G 6

		anti-4-11G 6			
	Polyclonal	MAb2 3-2A 1	MAb2 8-11G 1	MAb2 2-10F 1	MAb2 7-11E 1
Anti-Id	γ	α	α	α	α
Ab3[a]	<100	<100	<100	<100	<100
Neutralizing Ab3[b]	<50	<50	<50	<50	<50
Protection[c]	—	—	—	—	—

[a]Highest dilution where Ab3 anti-virus is detectable by ELISA
[b]Reciprocal of the highest dilution of serum that neutralized 50% of input virus
[c]BALB/c mice immunized with Ab2 and challenged intracerebrally with MHV-A59

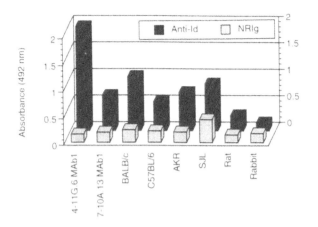

Figure 4. Detection of idiotype by Ab2. Microtiter plates were coated with 1.5 μg/mL of purified polyclonal Ab2 anti-4-11G.6 or NRIg. The binding of syngeneic (BALB/c), allogeneic (C57BL/6, AKR, SJL) and xenogeneic (rat and rabbit) anti-viral sera produced against MHV-A59 was determined by ELISA using peroxidase-labeled species-specific anti-Ig.

Polyclonal and Monoclonal Anti-4-11G.6

Polyclonal and monoclonal anti-Id were also produced in animals immunized with MAb1 4-11G.6. They are listed in Table 1.

Polyclonal Anti-4-11G.6. As a first approach towards determining whether the polyclonal anti-Id against MAb1 4B11.6 was an internal image (Ab2$_\beta$) anti-Id, its ability to bind to the paratope of MAb1 was tested. The results of the ELISA inhibition of attachment assay are shown in Fig.1B. Binding of MAb1 4-11G.6 to viral antigen could be inhibited by purified anti-Id in a dose-dependent manner. Twenty ng of anti-Id was enough to inhibit 100% of MAb1 binding to antigen, while the same amount of NRIg did not have any effect on this binding, indicating that the polyclonal anti-Id was not an Ab2$_\alpha$. The ability of the polyclonal Ab2 to abrogate the neutralization of virus infectivity was also tested. Ten μg of anti-Id could reduce the neutralization titer of MAb1 by 400-fold, whereas the same amount of NRIg did not have any effect (data not shown). These results confirmed the ELISA inhibition of attachment assay and suggested that this anti-Id binds at or near the paratope of MAb1 and was therefore an Ab2$_\beta$ or Ab2$_\gamma$. To distinguish between these two possibilities, we investigated the ability of this anti-Id to be recognized by antisera from different animal species raised against the initial antigen. An Ab2$_\beta$ should bind to all anti-MHV hyperimmune sera because of its internal image properties. As shown in Fig.4, our polyclonal anti-Id recognized a share idiotype in hyperimmune sera from BALB/c, C57BL/6, AKR and SJL mice. It also bound weakly to rat sera, but not to rabbit sera. These results demonstrated that this anti-Id could not induce antibody responses to the antigen across species barrier. Therefore, this polyclonal anti-Id must only bind near the antigen-binding site of MAb1. This gamma-type reaction was confirmed by the absence of both specific antiviral Ab3 induction and protection (Table 1).

Interestingly, we previously demonstrated that Ab2$_\gamma$ against MAb1 7-10A specific for a related epitope could vaccinate mice against infection by this coronavirus[1]. We now show that Ab2$_\gamma$ anti-4-11G.6 did not induce a protective immune response. MAbs1 7-10A and 4-11G.6 were previously shown to recognize two overlapping conformational epitopes by an ELISA competition assay[7]. Therefore, the mechanisms of protection induced by the Ab2$_\gamma$ anti-7-10A remain to be investigated.

Monoclonal Anti-4-11G.6. Anti-4-11G.6 MAb2 were also generated in BALB/c mice. They were all of IgG2b isotype and, as shown in Table 1, did not compete with antigen for binding of MAb1, which suggests that they recognize framework idiotopes and can be classified as Ab2$_\alpha$ These MAb2, coupled to KLH to enhance their immunogenicity, did not induce specific antiviral Ab3 nor induced protection against MHV-A59 (Table 1) Previous studies have shown that such Ab2s could induce Ab3s in other viral systems[16 17] However, in the present work, no MAb2$_\alpha$ were able to induce an antiviral Ab3 response

The use of monoclonal anti-Id with interesting biological activities should help clarify the mechanisms of protection induced by non-internal image anti-Id Moreover, the production of monoclonal internal image Ab2$_\beta$, although potentially interesting for characterization of molecular determinants involved in viral pathogenesis, identification of cellular receptors and vaccination, is technically difficult Molecular cloning of the antibody repertoire could overcome this technical problem[23] Such studies are in progress

ACKNOWLEDGMENTS

We are grateful to Eduardo Franco and Marie Desy for expert assistance with statistical analyses, Francine Lambert for excellent technical assistance, Alain Lamarre for assistance in the early stages of this work and Michael J Buchmeier (The Scripps Research Institute, La Jolla, CA, U S A) for his generous gift of MAb 5B170 Mathilde Yu acknowledges studentships from the Institut Armand-Frappier and the *Fonds pour la Formation de Chercheurs et l'Aide a la Recherche* (FCAR) and Pierre J Talbot scholarship support from the *Fonds de la Recherche en sante du Quebec* (FRSQ) This work was supported by a grant from the Medical Research Council of Canada (MRC) to P J T

REFERENCES

1 Lamarre, A , Lecomte, J , Talbot, P J 1991 Antiidiotypic vaccination against murine coronavirus infection J Immunol 147 4256-4262

2 Jerne, N K 1974 Towards a network theory of the immune system Ann Immunol 125c 373-389

3 UytdeHaag, G C M , Bunschoten, H , Weijer, K , Osterhaus, A D M E 1986 From Jenner to Jerne Towards idiotypic vaccines Immunol Rev 90 93-113

4 Zhou, E -M , Chanh, T C , Dreesman, G R , Kanda, P , Kennedy, R C 1987 Immune response to human immunodeficiency virus *In vivo* administration of anti-idiotype induces an anti gp160 response specific for a synthetic peptide J Immunol 139 2950-2956

5 Kennedy, R C , Eichberg, J W , Lanford, R E , Dreesman, G R 1986 Anti-idiotypic antibody vaccine for type B viral hepatitis in chimpanzees Science 232 220-223

6 Fung, M S C , Sun, C R Y , Liou, R S , Gordon, W , Chang, N T , Chang, T -W , Sun, N -C 1990 Monoclonal anti-idiotypic antibody mimicking the principal neutralization site in HIV-1 gp120 induces HIV-1 neutralizing antibodies in rabbits J Immunol 145 2199-2206

7 Daniel, C , Talbot, P J 1990 Protection from lethal coronavirus infection by affinity-purified spike glycoprotein of murine hepatitis virus, strain A59 Virology 174 87-94

8 Collins, A R , Knobler, R L , Powell, H , Buchmeier, M J 1982 Monoclonal antibodies to murine hepatitis virus-4 (strain JHM) define the viral glycoprotein responsible for attachment and cell-cell fusion Virology 199 358-371

9 Daniel, C , Talbot, P J 1987 Physico-chemical properties of murine hepatitis virus, strain A59 Arch Virol 96 241-248

10 Buchmeier, M J , Lewicki, H A , Talbot, P J , Knobler, R L 1984 Murine hepatitis virus-4 (strain JHM)-induced neurologic disease is modulated *in vivo* by monoclonal antibody Virology 132 261-270

11 Armitage, P , Berry, G Statistical methods in medical research Second edition, 1987, Blackwell Scientific Publications, Oxford pp 428-433

12 Armitage, P , Berry, G Statistical methods in medical research Second edition, 1987, Blackwell Scientific Publications, Oxford pp 411-412

13 Tabachnick, B G , Fidell, L S Using Multivariate Statistics Second edition, 1989, Harper Collins Publishers, Inc , New York p 376

14 Armitage, P , Berry, G Statistical methods in medical research Second edition, 1987, Blackwell Scientific Publications, Oxford pp 107-111

15 Francotte, M , Urbain, J 1984 Induction of anti-tobacco mosaic virus antibodies in mice by rabbit antiidiotypic antibodies J Exp Med 160 1485-1494

16 Schick, M R , Dreesman, G R , Kennedy, R C 1987 Induction of an anti-hepatitis B surface antigen response in mice by noninternal image (Ab2$_\alpha$) anti-idiotypic antibodies J Immunol 138 3419-3425

17 Zhou, E -M , Lohman, K L , Kennedy, R C 1990 Administration of noninternal image monoclonal anti-idiotypic antibodies induces idiotype-restricted responses specific for human immunodeficiency virus envelope glycoprotein epitopes Virology 174 9-17

18 Suñe, C , Smerdou, C , Anton, I M , Abril, P , Plana, J , Enjuanes, L 1991 A conserved coronavirus epitope, critical in virus neutralization, mimicked by internal-image monoclonal anti-idiotypic antibodies J Virol 65 6979-6984

19 Kang, C -U , Nara, P , Chamat, S , Caralli, V , Chen, A , Nguyen, M -L , Yoshiyama, H , Morrow, W J W , Ho, D D , Kholer, H 1992 Anti-idiotype monoclonal antibody elicits broadly neutralizing anti-gp120 antibodies in monkeys Proc Natl Acad Sci USA 89 2546-2550

20 Rees, A D M , Praputpittaya, K , Scoging, A Dobson, N , Ivanyi, J , Young, D , Lamb, J R 1987 T-cell activation by anti-idiotypic antibody evidence for the internal image Immunology 60 389-393

21 Huang, J -H , Ward, R E , Kohler, H 1986 Idiotope antigens (Ab2$_\alpha$ and Ab2$_\beta$) can induce *in vitro* B cell proliferation and antibody production J Immunol 137 770-776

22 Zhou, S -R , Whitaker, J N 1993 Specific modulation of T cells and murine experimental allergic encephalomyelitis by monoclonal anti-idiotypic antibodies J Immunol 150 1629-1642

23 Marks, J D , Hoogenboom, H R , Griffiths, A D , Winter , G 1992 Molecular evolution of proteins on filamentous phage J Biol Chem 267 16007-16010

28

TRANSCRIPTION AND TRANSLATION OF PROINFLAMMATORY CYTOKINES FOLLOWING JHMV INFECTION

Stephen A. Stohlman[1,2] Qin Yao[2] Cornelia C. Bergmann[1,2]
Stanley M. Tahara[1,2] Shigeru Kyuwa,[3] and David R. Hinton[4]

[1] Department of Microbiology
[2] Department of Neurology
University of Southern California
Los Angeles, California
[3] Department of Animal Pathology
University of Tokyo
Tokyo, Japan
[4] Department of Pathology
University of Southern California
Los Angeles, California

ABSTRACT

Infection with JHMV results in the transcriptional activation of two host cell genes encoding proinflammatory cytokines, tumor necrosis factor (TNF)-α and interleukin (IL)-1β. Analysis of irradiated mice showed that IL-1β mRNA accumulation in the central nervous system was predominantly derived from the mononuclear infiltrate. By contrast, accumulation of TNF-α mRNA was unaffected by immunosuppression, suggesting that resident cells were the source of this cytokine. Infected mice were treated with anti-TNF antibody to determine if TNF-α contributed to either the encephalomyelitis or demyelination associated with JHMV infection. Surprisingly, neither the cellular infiltrate nor demyelination were affected. In vitro analysis showed that IL-1β but not TNF was secreted from JHMV infected macrophages. The absence of TNF secretion is due to a block in translation of the TNF mRNA which accumulates during infection.

INTRODUCTION

The immune system is a critical component in the acute and chronic forms of central nervous system (CNS) infection by the JHMV strain of mouse hepatitis virus (1,2). All components of the innate and adaptive immune response, with the possible exception of NK cells, play a role in the preventing death of mice following CNS infection. For example, mice can be protected by the passive transfer of both neutralizing and non-neutralizing monoclonal antibodies (mAb). In addition, both virus-specific Th1 CD4[+] and cytotoxic T lymphocyte (CTL) CD8[+] populations can provide protection from acute disease (1,2).

Primary demyelination, the loss of myelin with relative sparing of axons is a hallmark of both the acute and chronic forms of JHMV-induced disease in the CNS (2,3). Protection from death, is however, not correlated with protection from the acute form of viral-induced demyelination. Although neutralizing antibodies provide protection from death and demyelination, protection from death in mice receiving either non-neutralizing mAb or CD4[+] T cells was not associated with a concomitant protection from demyelination (1,2). Total elimination of the immune response by whole body lethal irradiation early in infection prevents demyelination (4), suggesting that replication of virus within the oligoendroglia, a predominantly lethal event, was not by itself able to result in the pathological changes associated with demyelination. By contrast, primary demyelination is associated with the active removal of the myelin sheaths by activated macrophages, a cell type absent from the CNS of these immunosuppressed mice. Similarly, the protection afforded by the adoptive transfer of JHMV-CNS, predominantly virus associated with astrocytes, microglia and monocytes, and also a decrease in the amount of demyelination (5). Whether or not this reduction in demyelination was due to limited virus replication in the oligodendroglia or was mediated via an effect of the infected cells of the monocyte lineage, i.e. macrophage or microglia, or possibly cytokines released from astrocytes is unclear.

We recently demonstrated that JHMV infection results in an overall decrease in host protein synthesis, a concomitant decrease in the ability of host cell mRNA to be translated (6) and a loss of some, but not all of the host cell mRNA (7). Two of the host mRNAs which were not degraded following JHMV infection were the proinflammatory cytokines, TNF-α and IL-1β. Both of these cytokines are associated with fever, endotoxin mediated shock and have modulatory effects on the induction of immunity (8). TNF is of particular interest relative to CNS infections since it is associated with both the loss of blood brain barrier permeability and is directly toxic for in vitro cultures of oligodendroglia derivd from neonates (9,10,11).

MATERIALS AND METHODS

Macrophages: BALB/cBy mice were purchased from the Jackson Laboratory, (Bar Harbor, ME) at 6 wk of age. Peritoneal exudate cells (PEC) were induced by intraperitoneal injection of 3.0 ml of thioglycollate broth (Difco, Detroit, MI) as previously described (12). Mice were sacrificed 72 hours later by CO_2 asphyxiation and PEC were harvested aseptically with 6 ml of Joklik's-modified minimum essential medium (Gibco, Grand Island, NY) supplemented with 1.0 U/ml heparin. After overnight incubation at 37°C nonadherent cells were removed by three washes with RPMI medium.

Cytokine Induction and Assay: PEC were incubated in RPMI medium, 5% Fetal Calf Serum (FCS), supplemented with 25 ng/ml lipopolysaccharide (LPS) (E. coli 005:B5; Sigma Chemicals Company, St. Louis, MO. Human recombinant TNF-α or IL-1β (Collaborative Biomedical Products, Bedford, MA) or test supernatants were incubated with the target cells for 18 h at 37°C. Supernatants were UV-inactivated prior to analysis to prevent cytopathology induced by JHMV. TNF-α cytotoxicity was measured in a microassay using BC10ME target cells treated with 1 µg/ml actinomycin D as described (13). Viability was assessed using the 3-(4,5-dimethyl thiozol-2yl)-2-5-phenyl tetrazolium bromide dye reduction and absorbance read at 560 nm. IL-1β was measured using the IL-1 responsive LBRM TG6 cell line (14).

Histology: For histopathological analysis mice were sacrificed at 12 days post infection by CO_2 asphyxiation. Brains were removed, fixed for 3 hours in Clark's solution (75% ethanol and 25% glacial acetic acid) and embedded in paraffin. Sections were stained with either hematoxylin and eosin or luxol fast blue for routine examination.

RESULTS

In vivo induction of mRNAs specific for TNF-α and IL-1β were examined in the brains of mice following JHMV infection. Total RNA was prepared from the brains as previously described, separated by electrophoresis and probed for the expression of TNF-α and IL-1β mRNAs as previously described (7). Figure 1 shows that no mRNA for either cytokine was detected in brains from uninfected mice or in infected mice for the first 2 days post infection. By contrast, the mRNAs of both TNF-α and IL-1β were detected at 3, 4, 5 and 6 days post infection. To determine the role of mRNAs in CNS cells versus the immune mediated inflammation, TNF-α and IL-1β mRNA levels were examined at 6 days post infection in the CNS of infected mice and infected mice immunosuppressed with 400R whole body-irradiation administrated on the day of virus infection. The level of TNF-α mRNA was only reduced slightly by immunosuppression suggesting that the majority of the mRNA was transcribed in resident cells of the CNS. By contrast, the level of IL-1β mRNA was significantly reduced suggesting that the inflammatory cells recruited into the CNS provided a major source of IL-1β mRNA. Histological examination of the CNS of the immunosuppressed prior to infection showed no evidence of cellular infiltration. These data indicate that in vivo infection with JHMV also induces mRNAs for the proinflammatory cytokines, consistent with recently published data (15).

To determine if TNF-α played a role in JHMV-induced demyelination, mice were treated with 1 mm of rat anti-TNF neutralizing mAb MP6-XT3 (16) 1 day prior to infection and 1 day post infection. No differences in clinical score or mean day of death were noted between the anti-TNF and control mice treated with the same concentration of an irrelevant isotype matched mAb (GL-1113) (16). Surprisingly, histological comparisons of the brains of the JHMV-infected mice treated with the anti-TNF and control antibodies showed no

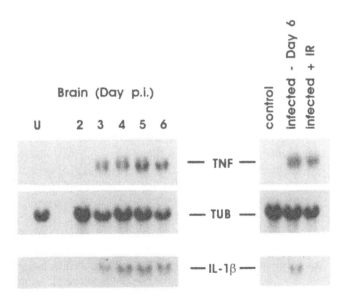

Figure 1. Induction of TNF-α and IL-1β mRNA in the brains of mice infected with JHMV. RNA was prepared from the brains of control and infected mice at the days post infection as indicated. Irradiated mice received 400 Rad whole body irradiation on the day of infection and were analyzed at day 6.

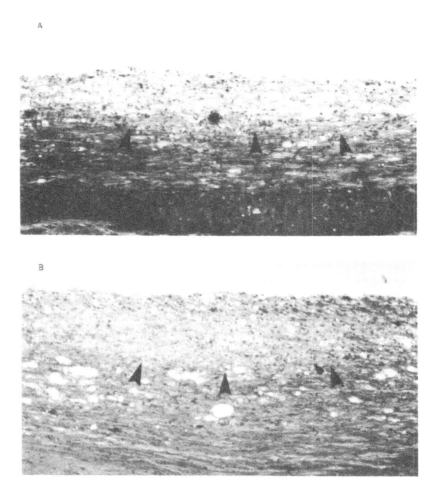

Figure 2. Encephalomyelitis and demyelination in anti-TNF treated JHMV-infected BALB/c mice at 12 days post infection Spinal cord tissues stained with luxol fast blue Magnification x 115 Untreated (A) and anti-TNF treated (B) Areas of demyelination are outlined by arrows

differences in either the extent of cellular infiltration or demyelination (Fig.2). Histochemical analysis also revealed that there was abundant rat immunoglobulin within the CNS of both groups of mice, indicating that the neutralizing anti-TNF mAb had gained access to the CNS during infection due to disruption of the blood brain barrier.

These data suggested that TNF-α did not play a primary role in JHMV-induced demyelination. To determine the effects of JHMV infection on the secretion of these two proinflammatory cytokines, BALB/c-derived PEC were infected with JHMV and the supernatants examined for the presence of TNF-α and IL-1β. TNF-α synthesis was measured at 6 hours following infection, LPS treatment or a combination of infection followed by LPS treatment. Although JHMV infection rapidly induces transcription of TNF-α mRNA (data not shown), no TNF-α was released following JHMV infection (Fig. 3, Panel A). Further infection, prevents the subsequent LPS-induced secretion of TNF-α. Interestingly, this effect was also apparent following infection with UV-inactivated virus. Infection also resulted in the rapid activation of IL-1β mRNA (7). However, in contrast to TNF-α, IL-1β was secreted following JHMV infection (Fig 3, Panel B).

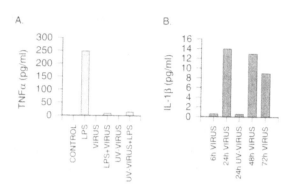

Figure 3. Cytokine release from JHMV and LPS treated macrophages derived from BALB/c mice. Samples were obtained from macrophages infected with JHMV at a multiplicity of infection of 5.0 or following treatment with 25 ng/ml LPS. TNF samples were obtained at 6 hours (panel A). IL-1β samples were obtained at various times post infection (Panel B).

Both of these proinflammatory cytokines are synthesized as precursors polypeptides which are cleaved to the active form prior to release. Recent data has shown preferential translation of mRNA containing the MHV leader sequence (6). To determine if the differential ability of these closely related peptides to be released from infected cells was due to a block in proteolytic processing of TNF-α, lysates of JHMV-infected cells were compared to cells treated with LPS (alone) by Western blot using a rabbit polyclonal antibody specific for TNF. Figure 4 shows an absence of both the precursor and the final TNF product in JHMV infected cells, while both can clearly be seen in the cells treated with LPS. Similar analysis of cells for the presence of IL-1β show the presence of the precursor protein in both infected and LPS treated cells (data not shown) consistent with the release of biologically active IL-1β from JHMV-infected cells.

DISCUSSION

The transcription of both TNF-α and IL-1β mRNA is increased in the CNS of mice following JHMV infection. In contrast to the increased level of IL-1β mRNA, which appears to be due to the recruitment of the inflammatory infiltrate, a substantial portion of the TNF-α mRNA is apparently due to active transcription in CNS cells. Treatment of JHMV-infected mice with neutralizing anti-TNF monoclonal antibodies demonstrated that neither JHMV-induced encephalomyelitis or its accompanying primary demyelination appear to be due to the pleotrophic effects of TNF-α. Surprisingly, in vitro analysis showed that although the mRNA levels for both these cytokines was increased following JHMV-infection (7) only IL-1β was released from infected cells. Analysis of the defect in TNF-α secretion demonstrated that similar to other host cell mRNA which lack the MHV leader RNA (6), the mRNA for TNF-α was not translated following infection. Examination of the 5' untranslated regions

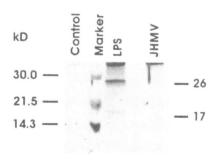

Figure 4. Western blot demonstrating the synthesis of TNF-α in LPS but not JHMV-infected macrophages. Lane A (Control); Control uninfected cells; Lane B (Marker); molecular wt markers; Lane C (LPS); PEC treated with LPS; Lane D (JHMV); PEC infected with JHMV. Lysates were prepared at 6 hours post infection.

of the TNF-α and IL-1β mRNAs showed no obvious differences or similarities to the 16 nucleotides present at the 3' end of the MHV leader RNA sequence required for translational competence in infected cells (6)

TNF-α has been suggested to have an in vivo role in demyelination based on its ability to produce death of oligodendroglia in vitro Indeed, the transcription of TNF-α mRNA is increased in the CNS of JHMV-infected mice However, analysis of the potential role of TNF-α to contribute to demyelination during JHMV infection suggests that only does it not contribute to demyelination, but also plays little or no role in the ability to immune cells to respond to a CNS viral infection This contrasts with the reports that anti-TNF is able to prevent the CD4⁺ T cell mediated induction of experimental allergic encephalomye-litis (17) Whether this is due to the ability of CD4⁺ T cells to effect the viability of oligodendroglial cells or to recruit activated macrophages into the CNS is not clear Similarly, the role of IL-1β, which has may properties in common with TNF-α, including the ability to induce fever and influence the activation of immune effectors, but not the ability to directly induce target cytolysis, is presently not clear

ACKNOWLEDGMENTS

This work was supported by Public Health Grants NS18146 and NS30880 and California Universitywide AIDS Research Program Grant USC 105 We wish to thank Dr Robert Coffman, DNAX corp for supplying the mAb, and acknowledge the technical assistance of Wengiang Wei and the assistance of Sonia Q Garcia and Charmaine Mohamed in manuscript preparation

REFERENCES

1 Kyuwa S and SA Stolman Sem Virol 1 273 (1990)
2 Fazakerley JK and MJ Buchmeier Adv Virus Res 42 249 (1993)
3 Lampert PW, JK Sims, AJ Kniazeff Acta Neuropathol 24 76 (1973)
4 Wang FI, Stohlman SA, JO Fleming J Neuroimmunol 30 31 (1990)
5 Stohlman S, Bergmann CC, van der Veen R, DR Hinton Submitted 1994
6 Tahara S, Dietlin T, Bergmann C, Nelson G, Kyuwa S and S Stohlman Virology 202 621 (1994)
7 Kyuwa S, Cohen M, Nelson G, Tahara SM, Stohlman S J Virol In press (1994)
8 Benvenisti EN Am J Physiol Cell Physiology 263 32 (1992)
9 Gutierrez EG, Banks WA, and AJ Kastin J Neuroimmunol 47 169 (1993)
10 Robbins DS, Shirazi Y, Drysdale BE, Lieberman A, Shin HS, ML Shin J Immunol 139 2593 (1987)
11 Selmaj KW and CS Raine Ann Neurol 23 339 (1988)
12 Matsushima GK and SA Stohlman J Immunol 146 3322 (1991)
13 Wang J, Stohlman SA, G Dennert J Immunol 152 3824 (1994)
14 Larrick J, Brindley L and M Doyle J Immunol Methods 79 39 (1985)
15 Pearce B,Hobbs M, McGraw T and Buchmeier M J Virol 68 5483 (1994)
16 Abrams JS, Roncarolo MG, Yssel H, Andersson U, Gleich GJ, Silver JE Immunol Rev 127 5 (1992)
17 Ruddle NH, Bergman CM, McGrath KM, Lingenheld EG, Grunnet ML, Padula SJ, Clark RB J Exp Med 172 1193 (1990)

29

PATHOLOGY OF MHV-A59 INFECTION IN ß2 MICROGLOBULIN NEGATIVE MICE

Ehud Lavi,[1] Qian Wang,[1] James Gombold,[2] Robyn Sutherland,[2]
Yvonne Paterson,[2] and Susan Weiss[2]

[1] Division of Neuropathology
Department of Pathology and Laboratory Medicine
[2] Department of Microbiology
University of Pennsylvania School of Medicine
Philadelphia, Pennsylvania 19104-6079

INTRODUCTION

Cytotoxic T cells, CD8[+], are important for clearance of MHV infection[1]. To investigate the role of CD8[+] cells in acute MHV-A59 infection we studied infection of ß2 microglobulin negative transgenic mice. These mice are devoid of MHC class I expression and therefore lacking functional CD8[+] cells.

EXPERIMENTAL DESIGN AND METHODS

Four week old β2M [(-/-)] mice were originally obtained from Dr. Koller[2] , were inbred at the University of Pennsylvania, and were inoculated intracerebrally (IC) with MHV-A59. Groups of 3-5 mice per time point (days 1,3,4, 5,7,8, 11,13, 21, 30 days post inoculation) were sacrificed by anesthetic (methoxyflurane) overdose and perfused intracardially with PBS and 10% phosphate buffered formalin. Organs were removed and fixed in 10% phosphate buffered formalin. Tissues were embedded in paraffin, sectioned and stained with H&E for light microscopy[3]. Immunohistochemistry was performed according to the ABC elite immunoperoxidase technique (Vector) with diaminobenzidine as substrate4. The effect of viral replication and demyelination in these mice is the subject of another study (Gombold et al. submitted for publication).

RESULTS

Following I.C. inoculation with 10 pfu of MHV-A59 (2 LD50), mice developed clinical signs of acute disease similar to C57Bl/6 mice infected with 1-2 LD50 (3000-5000

Table 1. Pathology of β2M$^{(-/)}$ transgenic mice infected with MHV-A59

Pathology	PID					
	4	7-8	11	13	21	30
Meningitis	3/6	6/6	2/2	4/4	1/3	0/3
Encephalitis	3/6	6/6	2/2	3/4	1/3	0/3
Hepatitis	0/6	5/6*	2/2#	3/4†	2/3¥	0/3
Thymic hypoplasia	0/2	2/5	1/1	2/2	0/2	0/3

*Mild-moderate
#Mild-severe
†Mild - moderate
¥Chronic persistent inflammatory form

pfu). These included ruffled fur, hunched position, decreased motility and appetite, lethargy, limb weakness, and eventually death. Histopathology revealed organ involvement during the acute disease similar to C57Bl/6 mice. Between 7-13 days post inoculation (PID) 100% of the mice had pathological changes in the brain, liver and thymus. There was no evidence of acute disease in the heart, lung, gastrointestinal tract, spleen, pancreas, adrenals, genital organs, salivary glands, bone marrow, or muscle. Acute meningoencephalitis was present with similar predilection for olfactory-limbic structures as in C57Bl/6 mice but with more extensive involvement of the basal ganglia and thalamus and more extensive proliferation of rod-shaped microglial cells. Immunohistochemistry revealed viral antigen in neurons in a similar anatomic distribution to C57Bl/6 mice. However, there was increased viral antigen in microglia, lymphocytes, endothelial and meningial cells. Acute hepatitis and thymus cortical hypoplasia in β2M $^{(-/-)}$ mice was delayed in onset and clearance, but otherwise similar to C57Bl/6 mice.

DISCUSSION

There was delayed onset and clearance of MHV-A59 in β2M $^{(-/-)}$ mice as compared to C57Bl/6 mice. Although the pattern of anatomic involvement of the brain was similar to C57Bl/6 infection, there was more viral replication in inflammatory, endothelial and meningial cells. Thus the increase in viral titers recovered from the brain, and the increased susceptibility of the β2M $^{(-/-)}$ mice to MHV-A59 may be attributed to the increase of number of infected cells in the brain.

ACKNOWLEDGMENTS

This study was supported in part by a grant from the University of Pennsylvania Research Foundation, by National Multiple Sclerosis Society grants PP-0284 and RG-2615A1/2 (EL) and PHS grant NS-11037 (SRW).

REFERENCES

1 Sussman MA, Shubin RA, Kyuwa S, Stohlman SA. Cell mediated clearance of mouse hepatitis virus strain JHM from the central nervous system J Virol 1989,63:3051-3056

2. Koller BH, Marrack P, Kappler JW, Smithies O. Normal development of mice deficient in ß2 M, MHC class I proteins, and CD8+ T cells. Science 1990;248:1227-1230.

3. Lavi E, Gilden DH, Highkin MK, Weiss SR The organ tropism of mouse hepatitis virus A59 is dependent on dose and route of inoculation Lab. Anim. Sci 1986;36:130-135

4. Lavi E, Fishman SP, Highkin MK, Weiss SR. Limbic encephalitis following inhalation of murine coronavirus MHV-A59 Lab Invest 1988,58:31-36

30

PRIMARY MURINE CORONAVIRUS INFECTION IN MICE

A Flow Cytometric Analysis

S. Kyuwa,[1] K. Machii,[2] A. Okumura,[1] and Y. Toyoda[1]

[1] Department of Animal Pathology
Institute of Medical Science
University of Tokyo
Tokyo, Japan
[2] Department of Veterinary Public Health
Institute of Public Health
Tokyo, Japan

T cell- mediated immune responses play a pivotal role in both viral clearance and immunopathology in mice infected with murine coronavirus, strain JHM (JHMV)[1,2.] In the present study, we attempted to characterize T cells induced during primary JHMV infection by flow cytometric analysis.

Female, 6 to 8 week old C57BL/6 (B6) mice were infected intraperitonealy with 10^6 PFU of JHMV. Although JHMV replicated for the first 3 days but was eliminated from spleens of B6 mice at 7 days postinfection (pi). Flow cytometric analysis was caried out to characterize spleen cells from JHMV-infected B6 mice.[3] Most drastic changes were noted as an increased number of CD8[+] T cells and their decreased CD8 intensity at 7 days pi. Time course study showed that intensity of $\alpha\beta$ T cell receptors declined with the CD8 intensity, while intensity of the lymphocyte function antigen-1 (LFA-1) and CD43 on CD8[+] T cells increased. Two-color analysis demonstrated that CD8[dull]LFA-1[bright] T cells were induced transiently in both C57BL/6 and BALB/c mice following JHMV infection (Figure 1). At 7 days pi a half of CD8[+] T cells were partitioned into CD8[dull]LFA-1[bright] T cells. Forward and side scatter profiles of CD8[dull]LFA-1[bright] T cells indicated that the population appeared to be activated T cells. Although CD45RB[dull]CD8[+] and CD44[bright]CD8[+] T cells were observed in JHMV-infected mice, expansion of CD25[+]CD8[+] and CD11b[+]CD8[+] T cells, which were reported as markers of cytotoxic T lymphocytes in choriomeningitis virus infection in mice[4,5], was not observed. Since the kinetics of the expansion of CD8[dull]LFA-1[bright] T cells was correlated with the viral elimination *in vivo*, we measured fresh cytotoxic activities of spleen cells from JHMV-infected B6 mice against syngeneic JHMV-infected macrophage-like cell line (IC-21 cells). The Ig[-] splenocytes from mice 7 days pi but neither those from uninfected or 14 days pi showed a weak, but significant cytotoxic activity against JHMV-infected H-2-matched cells *in vitro*. Therefore, these results suggest that the T cell population

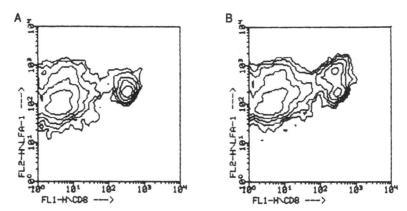

Figure 1. Two- color analysis of spleen cells from uninfected B6 mouse (A) and that at 7 days pi (B). Spleen cells were stained with anti-CD8 (FL1) and anti-LFA-1 (FL2) (antibodies, and analyzed by a FACScan™ flow cytometer.

may mediate the cytotoxicity against virus- infected cells *in vivo*, and thus the flow cytometric analysis is applicable to monitor coronavirus- induced primary cytotoxic T lymphocytes population *in vivo*.

REFERENCES

1. Kyuwa, S., Stohlman, S.A. Pathogenesis of a neurotropic murine oronavirus, strain JHM, in the central nervous system of mice. Semin Virol 1990;1:273-280.
2. Fleming, J.O., Wang, F.I., Trousdale, M.D., Hinton, D.R., Stohlman, S.A. Interaction of immune and central nervous system; contribution of anti-viral Thy-1[+] cells to demyelination induced by coronavirus JHM. Reg Immunol 1993;5:37- 43.
3. Kyuwa, S., Machii, K., Okumura, A., Toyoda, Y. Induction of LFA-1[high]CD8[+] T cells associated with antiviral cytotoxicity during primary murine
   coronavirus infection in mice. Submitted.
4. Saron, M.F., Colle, J.H., Dautry- Varsat, A., Truffa- Bachi, P. Activated T lymphocytes from mice infected by lymphocytic choriomeningitis virus display high affinity IL-2 receptors but do not proliferate in response to IL-2. J Immunol 1991;147:4333-4337.
5. McFarland, H.I., Nahill, S.R., Maciaszek, J.W., Welsh, R.M. CD11b (Mac-1): a marker for CD8[+] cytotoxic T cell activation and memory in virus infection. J Immunol 1992; 149:1326-1333.

31

NEUROVIRULENCE FOR RATS OF THE JHMV VARIANTS ESCAPED FROM NEUTRALIZATION WITH THE S1-SPECIFIC MONOCLONAL ANTIBODIES

Fumihiro Taguchi, Hideka Suzuki, Hiromi Takahashi, and Hideyuki Kubo

National Institute of Neuroscience, NCNP
4-1-1 Ogawahigashi, Kodaira
Tokyo 187, Japan

ABSTRACT

We have studied the neurovirulence for rats of the MAb-resistant variants isolated from a highly neurovirulent JHMV, cl-2. The variants, MM6 and MM13, with point mutation located within the N terminal 100 amino acids (aa) of the S1 protein showed no alteration in neurovirulence in comparison with cl-2, showing high neurovirulence. The variants, MM65 and MM85, with a deletion composed of about 150 aa located in the middle of the S1 subunit were revealed to be non-neurovirulent. A variant MM78 with one aa deletion, asparagic acid at number 543 from the N terminus of the S1, was shown to be low-virulence. The neurovirulence of these viruses paralleled with the viral growth potential in the rat brain. However, all of these variants as well as parental cl-2 showed high neurovirulence for mice. These results suggest that the domain composed of about 150 aa in the middle of the S1 is critical for high-neurovirulence of JHMV for rats.

INTRODUCTION

The spike (S) protein is suggested to be a major determinant of the neurovirulence of JHMV for mice[1,2]. The variant viruses escaped from neutralization of parental JHMV by the monoclonal antibodies (MAbs) specific for the S protein (MAb-resistant variants) have been revealed to have altered neurovirulence for mice[3,4]. The comparative nucleotide analyses of such variants and parental JHMV have identified the region on the S protein critical for the high neurovirulence for mice, although it is to be determined whether a single or multiple sites of the S gene are implicated with the neurovirulence[5,6]. However, it is not yet well studied on the viral factor influencing upon the neurovirulence of JHMV for rats, although it has been reported that JHMV variants with a large S protein are highly neurovirulent, while those with a small S protein are not[7 8] To analyze the role of the S protein in neurovirulence for rats, we have obtained MAb-resistant variants after neutralization by MAbs specific for the S protein By using those variants, we have tested whether the S protein is implicated with the neurovirulence for rats

EXPERIMENTAL DESIGN AND RESULTS

Variants obtained after neutralization by MAbs No 6, 13, 56, 78 and 85[9] were designated MM6, MM13, MM56, MM78 and MM85, respectively The nucleotide sequencing analyses showed that MM6 and MM13 have point mutations at aa numbers 83 (Gly to Glu) and 26 (Asn to His) from the N terminus of the S protein, respectively MM56 and MM85 were shown to contain a deletion of a stretch composed of 153 and 151 aa, respectively MM78 contained only one aa deletion at the position of 543, asparagic acid

We studied the neurovirulence of these variant viruses selected with various MAbs After inoculation of 10^5 PFU of variant viruses into the brain of 4 week old Lewis rats, MM6 and MM13 were demonstrated to be highly neurovirulent for rats as was their parental cl-2 About 80 % of rats infected these viruses showed clinically central nervous symptoms within 12 days postinoculation (p i) and most of these succumbed within 15 days The rats inoculated with MM65 and MM85 showed no clinical symptoms nor died during the observation period MM78 was shown to be intermediately virulent Highly virulent viruses, cl-2, MM6 and MM13 were revealed to grow well (10^3 to 10^4 PFU/g) as compared with the avirulent MM56 or MM85 (<10 PFU/g) MM78 with intermediate virulence showed their titer in the brain intermediate (10^2 to 10^3 PFU/g) between high- and non-neurovirulent viruses These results clearly showed that the neurovirulence of cl-2 and its variants correlated well with the growth potential in the brain of rats

We then examined the neurovirulence of our MAb-resistant viruses for BALB/c mice After inoculation of 10^3 PFU of cl-2, MM6, MM85 or MM78 with different neurovirulence for rats, there was no remarkable difference in the lethality and time to death of mice inoculated with each of these variants, showing clearly that there was no difference in neurovirulence for mice The difference in neurovirulence of JHMV observed in different laboratory[5 6] may result from the some point mutations in the S proteins of MAb-resistant viruses isolated different laboratories or it may due to the unidentified viral genetic differences

The present study indicates that the neurovirulence of JHMV for rats is determined by the S protein of the virus, the JHMV variants with the larger S protein are highly virulent and those with the deleted S protein are low- or non-neurovirulent The neurovirulence of the viruses correlates with the viral growth potential in the rat brain Identification of cell types in the rat brain to support the replication of JHMVs with the larger S protein but not that of JHMVs with the deleted S protein is important to analyze the mechanisms underlining the neurovirulence of JHMV for rats

REFERENCES

1 Holmes, K V, E W Doller and J N Behnke Analysis of the function of coronavirus glycoprotein by differential inhibition of synthesis with tunicamycin Adv Exp Med Biol 1981,142 133-142

2 Spaan, W, D Cavanagh, and M C Horzinek Coronaviruses structure and genome expression J Gen Virol 1988,69 2939-2952

3 Dalziel, R G, P W Lampert, P J Talbot, and M J Buchmeier Site-specific alteration of murine hepatitis virus type 4 peplomer glycoprotein E2 results in reduced neurovirulence J Virol 1986,59 463- 471

4 Fleming, J O , M D Trousdale, F A K El-Zaatari, S A Stohlman, and L P Weiner Pathogenicity of antigenic variants of murine coronavirus JHM selected with monoclonal antibodies J Virol 1986,58 869-875

5 Parker, S E , T M Gallagher, and M J Buchmeier Sequence analysis reveals extensive polymorphism and evidence of deletions within the E2 glycoprotein gene of several strains of murine hepatitis virus Virology 1989,173 664-673

6 Wang, F-I , J O Fleming and M M C Lai Sequence analysis of the spike protein gene of murine coronavirus variants Study of genetic sites affecting neuropathogenicity Virology 1992,186 742-7491

7 Taguchi, F , S G Siddell, H Wege, and V ter Meulen Characterization of a variant virus selected in rat brain after infection by coronavirus mouse hepatitis virus JHM J Virol 1985,54 429-435

8 Matsubara, Y , R Watanabe, and F Taguchi Neurovirulence of six different murine coronavirus JHMV variants for rats Virus Res 1991,20 45-58

9 Kubo, H , S Y Takase, and F Taguchi Neutralization and fusion inhibition activities of monoclonal antibodies specific for the S1 subunit of the spike protein of neurovirulent murine coronavirus JHMV cl-2 variant J Gen Virol 1993,74 1421-1425

32

T CELL IMMUNODEFICIENCY INVOLVED IN PATHOGENICITY OF ATTENUATED MHV3 MUTANTS

L Lamontagne,[1] C Page,[1] and J P Martin[2]

[1] Departement des Sciences Biologiques
Universite du Quebec a Montreal
Montreal, Quebec, Canada
[2] Laboratoire de Virologie, Unite INSERM 74
Universite Louis Pasteur
Strasbourg, France

ABSTRACT

Viral pathogenicity results when there is an imbalance between viral replication and the host's immune defenses The immune system plays and important role in the outcome of an acute disease induced by the mouse hepatitis virus type 3 (MHV3) Of use in the study of the role of viral properties involved in its pathogenicity is the attenuated escape mutants We reported that two MHV3 escape mutants were attenuated in their ability to deplete T cell subpopulations in the spleen in BALB/c mice according to inoculation route and time postinfection The highly attenuated CL12 mutant cannot induce depletion in T cells following intraperitoneal (i p) or intranasal (i n) inoculations, at three days postinfection (p i) The less attenuated 51 6 mutant, however, maintained the ability to deplete T cells following i p inoculation, as described for the pathogenic MHV3 In contrast, no depletion of T cells following i n inoculation was induced with this mutant The use of such mutants enables us to dissect the role of each compartment of the immune system

INTRODUCTION

Cellular resistance of mice to infection causes by a particular virus may indicate the absence of a specific genetically controlled viral permissivity factor in target cells or host factors, including the integrity of cellular and humoral immune defenses[1] Helper and cytotoxic T lymphocytes are involved in the viral elimination process and are dependent upon the thymus for normal T lymphopoiesis Mouse hepatitis virus type 3 (MHV3), a hepatotropic strain, is a coronavirus Its inoculation in mice is usually followed by a generalized infection characterized by an acute hepatic necrosis, killing the animal within a couple of days[2]. Resistant A/J mice support a subclinical infection, whereas other strains, such as BALB/c are fully susceptible to a fulminant hepatitis. Splenic T cells or thymocyte subpopulations were depleted in L2-MHV3 infected susceptible mice[3,4]. The susceptibility of BALB/c mice is associated with an impaired early activation of the immune system, demonstrated by the failure of lymphocyte proliferation resulting from the decrease in IL-2 production[5]. A useful tool in the study of the role of viral properties in pathogenicity is the use of attenuated escape mutants. The S protein is central to cellular tropism and MHV virulence[6,7]. We have selected MHV3 mutants by virtue of their resistance to neutralization by anti-S monoclonal antibodies (mAb) in order to study their pathogenic properties.

RESULTS AND DISCUSSION

To verify if mAb escape mutants were attenuated because they have lost the ability to induce lymphoid depletion in the spleen, favoring the development of an inflammatory response in infected organs, percentage and absolute number of splenic CD4+CD8-, and CD4-CD8+ subsets were recorded in groups of BALB/c mice i.p. or i.n. infected with either pathogenic L2-MHV3, or the attenuated mutants. Mutants MHV3-51.6 and MHV3-CL12 were selected from pathogenic MHV3 virus in the presence of S protein specific A51 and A37 mAbs, able to neutralize MHV3, and to slightly inhibit cellular fusion, but did not protect the sensitive mice against MHV3 infection[8]. Splenic T (CD4+CD8-, CD4-CD8+) were labelled with the following mAb: fluorescein isothiocyanate conjugated (FITC) anti-CD4 (mAb RM 4-5, Cedarlane, Hornby, Ontario, Canada), and phycoerythrin conjugated (PE) anti-CD8a (mAb 53-6.7, Cedarlane). For flow cytometry analysis, using a FACSCAN flow cytometer (Becton-Dickinson, Mountain View, CA), cells were gated according to forward and side angle light scatter. Five thousand cells were analyzed per sample. Thymocytes were similarly labelled, and CD4+CD8+, CD4+CD8-, or CD4-CD8+ subpopulations were recorded.

As shown in Table 1, depletion of splenic CD4+CD8- cells was higher in i.p. or i.n. L2-MHV3 infected mice ($p<0.001$), whereas there was no decrease, in absolute numbers, in those of mice i.n. infected with attenuated mutants. Similarly, splenic CD4-CD8+ subset depleted in i.p. or i.n. L2-MHV3-infected mice ($p<0.001$), but remained at normal level in mice i.n. infected with attenuated mutants at three or six days p.i. The thymuses were strongly depleted after i.p. infection with L2-MHV3 or MHV3-51.6, but not during MHV3-CL12 infection. The levels of CD4+CD8+ cells, however, as measured by the absolute number, were markedly decreased during L2-MHV3 or MHV3-51.6 infections ($p<0.001$), but not in mice infected with the MHV3-CL12 mutant at three days p.i. In contrast, this thymocyte subset was strongly depleted in mice i.n. infected with the attenuated mutants after six days ($p<0.001$) (results not shown).

We have previously demonstrated a correlation between viral pathogenicity and the depletion of cells in lymphoid organs, resulting from either a productive or abortive viral replication[3,4]. The use of neutralization escape mutants enlightens the role of T cell depletions in the pathogenic process during acute hepatitis. Pathogenic L2-MHV3 induces the depletion of splenic, thymic T cells following both i.p. or i.n. inoculations whereas splenic T and thymocyte subpopulations were less affected in mice infected with attenuated mAb escape mutants. The highly attenuated MHV3-CL12 mutant did not induce T lineage cell depletion, at least, before six days p.i. This work suggests a relationship between the pathogenic properties of attenuated MHV3 mutants and the depletion of T lineage cells in the spleen and thymus.

Table 1. Percentage and absolute numbers of splenic CD4+CD8- and CD4-CD8+ subpopulations in BALB/c mice infected i p or i n with pathogenic L2-MHV3 and attenuated mutants 51 6-MHV3 or CL12-MHV3

Viruses	Time post infection (days)	Inoculation route	Percentages of splenic T cells (abs num x 10^6)	
			CD4+CD8-	CD4-CD8+
Control	3	i p [a]	14 7 ± 1 2 (7 8 ± 1 0)	6 83 ± 0 5 (3 63 ± 0 2)
CL12-MHV3	3	i p	10 5 ± 1 0 (5 9 ± 0 6)	3 44 ± 0 3[b] (1 94 ± 0 1)[b]
51 6-MHV3	3	i p	8 0 ± 0 7[b] (2 1 ± 0 3)[b]	3 06 ± 0 2[b] (0 78 ± 0 1)[b]
L2-MHV3	3	i p	6 8 ± 0 7[b] (3 6 ± 0 5)[b]	1 76 ± 0 1[b] (0 92 ± 0 1)[b]
Control	3	i n	21 6 ± 1 8 (12 9 ± 0 7)	7 56 ± 0 6 (3 36 ± 0 3)
CL12-MHV3	3	i n	13 1 ± 1 0[b] (9 1 ± 0 6)	5 52 ± 0 5 (3 80 ± 0 4)
51 6-MHV3	3	i n	15 3 ± 1 2[b] (10 7 ± 0 7)	6 50 ± 0 5 (4 55 ± 0 6)
L2-MHV3	3	i n	7 6 ± 0 8[b] (4 0 ± 0 2)[b]	3 50 ± 0 3[b] (1 82 ± 0 2)[b]
Control	6	i n	11 5 ± 0 8 (9 1 ± 0 6)	4 34 ± 0 3 (3 40 ± 0 3)
CL12-MHV3	6	i n	5 8 ± 0 4[b] (6 9 ± 0 4)[b]	1 92 ± 0 2[b] (2 30 ± 0 2)
51 6-MHV3	6	i n	5 4 ± 0 4[b] (5 4 ± 0 4)[b]	2 10 ± 0 2[b] (1 47 ± 0 2)[b]
L2-MHV3	6	i n	N A	N A

[a] i p intraperitoneal, i n intranasal
[b] $p < 0001$
N A not applicable

REFERENCES

1 Brinton M A , Nathanson N Genetic determinants of virus susceptibility epidemiologic implications of murine models Epidemiol Rev 1981, 3 115-154
2 Piazza M , Piccinino F , Mutano F Hematologic changes in viral (MHV3) murine hepatitis Nature 1965, 205 1034-1035
3 Lamontagne L , Descoteaux J P , Jolicoeur P Mouse hepatitis virus 3 replication in T and B lymphocytes correlates with viral pathogenicity J Immunol 1989, 142 4458-4466
4 Lamontagne L , Jolicoeur P Mouse hepatitis virus 3-thymic cell interactions correlating with viral pathogenicity J Immunol 1991, 146 3152-3160
5 Dindzans V J , Zimmerman B , Sherker A , Levy, G A Susceptibility to mouse hepatitis virus strain 3 in Balb/cJ mice failure of immune cell proliferation and interleukin 2 production Adv Exp Med Biol 1987, 218 411-420
6 Makino S , Fleming J O , Keck J G , Stohlman S A , Lai M M C RNA recombination of coronavirus localization of neutralizing epitopes and neuropathogenic determinants on the carboxyl terminus of peplomeres Proc Natl Acad Sci USA 1987, 84 6567-6571
7 Talbot P J , Salmi A A , Knobler R L , Buchmeier M J Topographical mapping of epitopes on the glycoproteins of murine hepatitis virus 4 (strain JHM) correlation with biological activities Virology 1984, 132 250-256
8 Martin J P , Chen W , Obert G , Koehren F Characterization of attenuated mutants of MHV3 importance of the E2 protein in organ tropism and infection of isolated liver cells Adv Exp Med Biol 1990, 276 403-410

33

IMPAIRMENT OF BONE MARROW PRE-B AND B CELLS IN MHV3 CHRONICALLY-INFECTED MICE

P. Jolicoeur and L. Lamontagne

Département des Sciences Biologiques
Université du Québec à Montréal
Montréal, Québec, Canada

ABSTRACT

Mouse hepatitis virus type 3 (MHV3) appears to be an excellent model for the study of the relationship between viral-induced immunodeficiency and the development of chronic disease. Animal surviving acute hepatitis develop a chronic disease characterized by viral persistency in various organs, by a humoral immunodeficiency, and eventually die within the next three months postinfection. To verify if B cell immunodeficiency occurs during the chronic disease, percentage and absolute number of bone marrow B lineage cell subpopulations were recorded at various times postinfection (p.i.) in pathogenic L2-MHV3-infected (C57BL/6 x A/J) F1 mice. Absolute numbers of B ($c\mu+$ $s\mu+$) cells decreased as early as three days p.i. up to 15 days p.i., and then gradually returned toward normal values in L2-MHV3-infected mice during the chronic disease. In contrast, pre-B ($c\mu+$ $s\mu-$) cells were less significantly decrease during the chronic disease. In addition, abnormally enlarged cells (> 13 μm) were detected either in bone marrow pre-B or B cells from L2-MHV3-infected mice.

INTRODUCTION

Mouse hepatitis virus type 3 (MHV3) appears to be an excellent model to study the relationship between viral-induced immunodeficiency and the development of a chronic disease, since humoral and cellular immunodeficiency occur in chronically-MHV3 infected (C57BL/6 x A/J) F1 mice. Animals which survive to the acute hepatitis develop a chronic disease characterized by a wasting syndrome including weight loss, alopecia, and oily hair. In later stages, the mice develop incoordination and hindlimb paralysis, and eventually die within the next three months postinfection[1]. We have previously observed significant decreases of thymocytes, splenic cells, and macrophages in chronically-infected (C57BL/6 x A/J) F1 mice [2]. Low antibody titers against MHV3 virus were found for up to 60 to 70 days postinfection, mostly IgM[3]. In spite of the production of anti-MHV3 antibodies, all immunoglobulin types gradually decrease during the first three months, followed by a return to normal values in the paralyzed animals[3]. We propose that the humoral immunodeficiency observed in MHV3 chronically-infected (C57BL/6 x A/J) F1 mice results primarily from depletions of B cell subpopulations in the lymphoid organs during outcome of the chronic disease.

RESULTS AND DISCUSSION

To verify the integrity of bone marrow B lymphopoiesis during the MHV3 diseases, groups of three (C57BL/6 x A/J) F1 mice were i.p. injected with 1000 $TCID_{50}$ of L2-MHV3. Mock-infected mice received i.p. a similar volume of PBS. A double immunofluorescent labelling method was used to identify pre-B ($c\mu+ s\mu-$) or B ($c\mu+ s\mu+$) cell subpopulations[4]. Briefly, bone marrow samples (100 µl of 4 x 10^7 nucleated cells/ml suspension) were incubated for 30 min on ice with an optimal dilution of FITC anti-µ chain (Cappel Biochemical, Malvern, PA) for surface µ labelling, washed twice by centrifugation through FCS, cytocentrifuged, and fixed in ethanol-acetic acid. Cytoplasmic µ-chains were labelled with an optimal dilution of TRITC anti-µ directly on the cell spots, incubated for 30 min at room temperature in a humidified chamber. Slides were examined for $c\mu+ s\mu-$ and $c\mu+ s\mu+$ B lineage cell subpopulations under a fluorescence microscope (Leitz Dialux 22, Midland, Ontario, Canada) equipped with a mercury lamp and phase contrast optics. Percentage of labelled cells was determined by counting a total of 1000 cells and the absolute numbers were calculated by the percentage of positive labelled cells and the total bone marrow count.

As shown in Figure 1, the absolute number of B cells decreased as early as three days p.i. up to 15 days p.i., and then gradually returned toward normal values in L2-MHV3-infected mice during the chronic phase of the disease. In contrast, pre-B cells were less

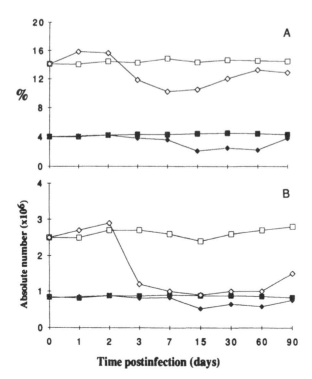

Figure 1. Percentage (A) and absolute numbers (B) at various times postinfection of bone marrow pré-B (■,r) and B (□,e) cells from L2-MHV3 infected F1 (C57BL/6 x A/J) (e ,r) and uninfected (□,■) mice.

significantly depleted during the chronic phase of the disease We have previously reported that abnormally enlarged pre-B or B cells occurred in bone marrow of L2-MHV3 acutely-infected parental susceptible C57BL/6 mice[5] To verify if the same maturation defect occurs in the B lineage cells of (C57BLI/6 x A/J) F1 mice, we analyzed, at various p i times, cell diameter distribution in bone marrow from pathogenic L2-MHV3-infected mice Small pre-B or B cells (six to nine μm) significantly decreased whereas the larger cells (10 to 12 μm), normally corresponding to a mitotic cell compartment,occurred up to 105 days p i (results not shown) In addition, abnormally enlarged cells (> 13 μm) were detected either in bone marrow pre-B or B cells from L2-MHV3-infected mice Such B cell maturation defects occurred as early as three days p i (results not shown) The results suggest that B lymphopoiesis disorder occurs rapidly in pathogenic L2-MHV3-infected (C57BL/6 x A/J) F1 mice, and is maintained during the chronic phase of the disease

We have previously demonstrated in parental C57BL/6 mice that B lineage cell depletions in spleen and bone marrow correlated with viral pathogenicity, and resulted from a lytic viral replication only in mature B and pre-B cells [5] In F1 (C57BL/6 x A/J) mice, B cells remained depleted in animals surviving the acute phase of the disease, for up to three mo until the occurrence of wasting syndrome, neurological signs and death Our results indicate a rapid onset of B cell immunodeficiency in L2-MHV3-infected mice, suggesting, at least, the inability of humoral immunity to eliminate viral infection

REFERENCES

1 Le Prevost C , Virelizier J L , Dupuy J M Immunopathology of mouse hepatitis virus type 3 infection III Clinical and virologic observation of a persistent infection J Immunology 1975, 115 640-646

2 Lamontagne L , Dupuy C , Leray D , Chausseau J P , Dupuy J M Coronavirus-induced immunosuppression role of mouse hepatitis virus 3-lymphocyte interaction 1985, Prog Leuk Biol 1 29-35

3 Leray D , Dupuy C , Dupuy J M Immunopathology of mouse hepatitis virus type 3 infection IV MHV-3 induced immunosuppression Clin Immunol Immunopathol 1982, 23 1457-1463

4 Park Y H , Osmond D G Phenotype and proliferation of early B lymphocyte precursor cells in mouse bone marrow 1987, J Exp Med 165 444-452

5 Jolicoeur P , Lamontagne L Mouse hepatitis virus 3 pathogenicity expressed by a lytic viral infection in bone marrow 14 8+μ+ B lymphocyte subpopulations 1989, J Immunology 143 3722-3728

Vaccine Development

34

DEVELOPMENT OF PROTECTION AGAINST CORONAVIRUS INDUCED DISEASES

A Review

Luis Enjuanes,[1] Cristian Smerdou,[1] Joaquín Castilla,[1] Inés M. Antón,[1]
Juan M. Torres,[1] Isabel Sola,[1] José Golvano,[2] Jose M. Sánchez,[1] and
Belén Pintado[1]

[1] Department of Molecular and Cellular Biology
Centro Nacional de Biotecnología, CSIC
Campus Universidad Autónoma
Cantoblanco, 28049 Madrid, Spain
[2] Universidad de Navarra
Facultad de Medicina
31080 Pamplona, Spain

INTRODUCTION

Current coronavirus vaccines are classical vaccines, i. e., are based on live attenuated coronavirus. These vaccines provide protection against certain strains of infectious bronchitis virus (IBV) and mouse hepatitis virus (MHV), and with variable results against bovine coronavirus (BCV) and transmissible gastroenteritis virus (TGEV).

Chicken immunized with whole inactivated IBV were protected against virulent challenge[15]. In contrast, chicken immunized with the purified spike (S) glycoprotein, nucleoprotein (N), or membrane (M) proteins did not induce protection[17]. Immunization with the S1 glycoprotein prevented replication of nephropathogenic IBV in kidney, but not in trachea of immunized chicken[36]. Protection did not correlate with the presence of virus neutralizing (VN) and haemagglutination inhibiting (HI) antibodies.

In feline infectious peritonitis virus (FIPV), humoral immunity to S protein is not protective, and cell mediated immune responses may be important in immunity[66]. The development of a safe and effective vaccine against FIPV has been very problematic. A variety of approaches have been unsuccessful, including the administration of inactivated FIPV, avirulent FIPV or sublethal doses of virulent FIPV, heterologous live virus vaccines [canine coronavirus (CCV), human coronavirus (HCV)-229E, and TGEV], and a recombinant vaccinia virus expressing FIPV S protein[98, 66]. Most recently, a temperature-sensitive mutant of FIPV has been developed as an FIPV vaccine. Unfortunately, while testing by the manufacturer provided 78% protection, independent testing gave from 0 to 50% protection, depending on the strain and dose exposure[78].

Lactogenic immunity is of primary importance in providing newborn piglets immediate protection against TGEV infection. Effective immunity has only been provided by the administration of virulent virus. Unfortunately, in this case the virus will become enzootic in the farm, preventing its use. Licensed TGEV vaccines have met limited success[74, 26], CCV and BCV vaccines provide partial protection and have to be improved. In summary, effective protection only to two coronaviruses can be elicited using current vaccines, implying the need of a new vaccine design.

Targets in Protection

The role in protection of the different viral proteins has a variable support. While there are many reports indicating that the S protein elicits protection against coronaviruses of different species[29, 17, 58, 43] the role of M protein, and particularly of the N protein, is less obvious. There is very little information on the role in protection of the small membrane (sM) protein[29, 17, 58, 43, 31], and of the non-structural proteins. Proteins exposed in virus or in virus infected cells have a higher probability of being involved in protection. Therefore, it is of interest to identify them.

S and M proteins have been detected on the surface of infected cells and these proteins have been identified as targets in immune protection in MHV[29, 35, 17, 48, 64, 82, 27, 58, 105, 43, 51, 70, 32, 89, 31, 20, 44, 102, 34]. S protein also induces protection against IBV and MHV. These proteins are frequently detected in small quantities in the surface of virus infected cells associated to the presence of virions, while they are detected in the cytoplasm of infected cells in large quantities[29, 35, 17, 48, 64, 82, 27, 58, 105, 43, 51, 70, 32, 89, 31, 20, 44, 102, 34]. These data are in agreement with the concept that coronaviruses assemble at rough endoplasmic reticulum and Golgi membranes. Consequently, most viral proteins are transported through intracellular compartments and only a small proportion, which is not incorporated into virions, is transported to the plasma membrane[86]. Expression of recombinant S, M, sM, haemagglutinin-esterase (HE), and hydrophobic (HP) proteins using poxvirus[83, 98, 97, 44], adenovirus[83, 98, 97, 108, 44], or baculovirus[105, 67, 32, 106, 107, 31, 108] has shown that these proteins are found in small amounts on the surface of transformed cells, in the absence of other coronavirus components[105]. This result does not necessarily reflect the situation after coronavirus infection, in which, as mentioned above, viral proteins are mainly found associated to virus bound to cell membranes.

M protein induces complement-dependent[48, 104] and possibly complement-independent neutralizing antibodies[101, 27] in assays performed *in vitro*. The antibodies elicited *in vivo* failed to protect against MHV challenge in some cases[102] but not in others[27]. According to the current topological model of coronavirus M protein, only the amino-terminus of the M protein is exposed on the virus surface[2, 9, 47, 50, 38]. In contrast, presented data support that the carboxy-terminus domain of M protein is exposed on the external virus surface both in TGEV and in MHV[90, 72] providing a new target for virus or virus-infected cells clearance.

The role of N protein in protection is being understood[102]. Antibodies to N protein have no virus neutralizing activity *in vitro*. An N protein fragment has been detected in the surface of infected cells (S. Stohlman, personal communication), and it has been shown that a monoclonal antibody (MAb) to the N protein prevented the cytopathic effect of MHV in L cell[52]. This antibody, as well as another non-neutralizing MAb, protected mice from acute disease [52]. In contrast, other anti-N protein antibodies did not provide *in vivo* protection[33, 13, 88, 52]. The mechanism by which the N protein protects is unclear. It may involve a cell-mediated defense mechanism, e.g. cytotoxic T lymphocyte (CTL) activity[85], T helper cells[103], or T cell-derived antiviral cytokines[102]. N-reactive helper T cells supporting virus-specific B cell responses have been demonstrated in TGEV[1]. This type of response might account for the enhancement of protection observed in MHV after combined immunization with the N and S proteins[102].

HE protein is also exposed on the surface of both BCV particles and infected cells[108]. HE expressed using recombinant Ad5 induced haemagglutination. Small membrane protein, an integral membrane protein of TGEV having Cexo-Nendo orientation is exposed on the surface of virions. The protein is also exposed on the surface of virus infected cells or insect cells transformed with baculovirus coding for sM[31]. The 9 kDa HP protein encoded by TGEV ORF7, or the equivalent protein of FIPV[21] is another accessible membrane-associated protein. The cellular location of HP suggests that it may play a role in the membrane association of replication complexes or in virus assembly[92].

In summary, there is clear evidence that S, M, HE, sM, and HP proteins are exposed on the surface of virus infected cells, mostly virion associated, while there is limited evidence on the exposure of N protein. Nevertheless, all these proteins constitute targets in protection against coronaviruses induced diseases.

B and T Cell Epitopes Involved in Protection

Antigenic sites in the major structural coronavirus proteins, S, M, N, and HE have been previously described[82]. Extensive studies on B cell epitopes have been done on the S protein of MHV, TGEV, IBV, FIPV, and BCV[13, 86, 23, 25, 59, 19, 18, 30, 53, 92, 20]. The location of antigenic sites inducing virus neutralizing antibodies has been summarized for TGEV, MHV, BCV, and IBV (Figure 1). Most of the information known on FIPV is coincident with that of TGEV[75]. Several S protein antigenic sites have been involved in neutralization: 3 in TGEV, 8 in MHV, 2 in BCV, and 3 in IBV. Most antigenic sites are located on the amino-terminal

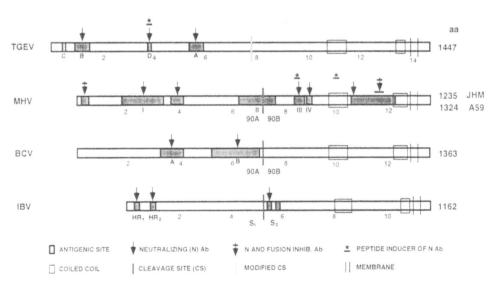

Figure 1. Antigenic sites on aligned coronavirus peplomer protein sequences. The location of dominant antigenic sites (dark squares), of antigenic sites recognized by neutralizing antibodies (arrows) and fusion inhibiting antibodies (arrows with plus sign), and of peptides eliciting protection (lines with star) is shown approximately to scale (small numbers below bars indicate amino acid number in hundreds). TGEV, capital letters below bar indicate the names of the antigenic sites [24, 30]. MHV, roman numbers indicate names of defined antigenic sites [59, 58, 73, 99, 20, 91]. BCV, capital letters below bar indicate defined antigenic sites [107]. IBV, HR$_1$ and HR$_2$ indicate the location of hypervariable regions[16, 54].

half domain of S protein. TGEV, MHV, and IBV have sites inducing neutralizing antibodies in the first 150 amino acids of S protein. Other sites inducing neutralizing antibodies on the amino-terminal half of S protein are approximately located between amino acids (aa) 200 and 400 (TGEV, MHV, BCV) and between aa 500 and 700 (TGEV, MHV, and BCV) (see Figure 1 for a more precise location). In the S2 half of IBV a site involved in neutralization has been defined next to the right of the trypsin cleavage site. In the S2 half of MHV two domains have been involved in virus neutralization. One contains several epitopes located between aa 800 to 900, and the other one is located between amino acids 1000 to 1220. Generally, monoclonal antibodies specific for S1 are more neutralizing than those S2 specific[83]. No antigenic site critical in TGEV neutralization has been identified in the stem portion of TGEV S protein.

Two hypervariable regions identified in IBV S protein at amino acid positions 38-51 and 99-115 from the N-terminus (excluding the signal sequence) are involved in the induction of neutralizing antibodies[61, 43]. In MHV a 13-mer synthetic peptide derived from S protein (aa 846 to 858) coupled to an influenza virus T-cell epitope protects against lethal MHV infection[43, 12]. A second peptide of 10-mer (aa 993 to 1002) of S protein, coupled to KLH elicited neutralizing antibodies and a protective immune response in BALB/c mice[87]. One MHV neutralizing MAb which selected escaping mutants with changes at amino acids 255 and 1116, also inhibited fusion activity[65, 41]. The information accumulated on the antigenic and functional structure of S protein in all coronaviruses, indicates that it has a complex three-dimensional structure in native state with interactions between distant domains in the primary structure[48, 30, 20, 41]. S protein fragments extended from the amino terminus, provided recombinant antigens inducing TGEV neutralizing antibodies[76].

Universal coronavirus vaccines based on a conserved domain of 124 aa found in the C terminal portions of TGEV, CCV, and FCV S protein have been patented. The report claims that this fragment is capable of eliciting a protective immune response against FIPV, FECV, TGEV, BCV, HCV, and IBV[81]. Further work is required to confirm these findings.

Less information is available on T-cell epitopes (Figure 2). In MHV virus-specific MHC class II-restricted cytotoxic CD4+ cells recognize an immunodominant epitope com-

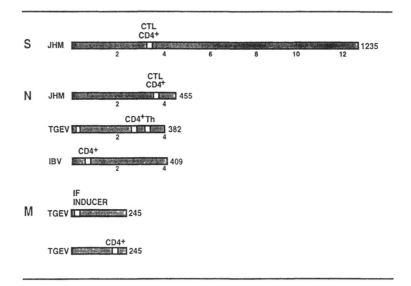

Figure 2. T cell epitopes precisely located in coronavirus proteins. The approximate location of T cell epitopes on the S, N, and M proteins is indicated by white boxes. IF, interferon inducer epitope; CTL, Cytotoxic T cell determinant; Th, T helper cell epitope[45, 7, 8, 49, 1, 34, 100].

prising the amino acids 329-343 of the viral S-glycoprotein. The epitope is recognized by polyclonal and virus specific cytotoxic T cell clones[34]. No evidence for the activation of the classical CD8[+] MHC class I-restricted CTL was found by these authors. MHV-JHM immune CD4[+] Th1 T cells transferred to MHV-JHM lethally infected mice results in protection from death, but provides little or no protection from demyelination[45]. CD8[+] cytotoxic T cells specific for MHV-JHM are involved in clearance of virus from CNS and thereby provide specific immunological protection. These CTL are specific for both S and N protein, but not for M or HE proteins. A subset of CD4[+] cells is also involved in virus clearance (S. Stohlman, personal communication). Most (15 out of 20) cell lines established from the brains of infected animals were specific for the N protein[45, 84]. MHV-JHM infection of rats induced immunodominant CD4[+] T-cells which recognized with high prevalence amino acids 388 to 563 of the carboxy-terminal domain of N protein. The same results were obtained from rats immunized with bacterially expressed N-protein or from animals vaccinated with a stable N-protein expressing vaccinia recombinant. Transfer of N specific CD4[+] T-cells to MHV-JHM infected rats mediated protection against acute disease[100] .

T cell responses to the three major TGEV structural proteins: S, M, and N have been identified in haplotype defined TGEV immune miniswine[1]. Four immunodominant T cell epitopes were identified using 15-mer synthetic peptides, three located in the nucleoprotein (N46, aa 46 to 60; N272, aa 272 to 286; and N321, aa 321 to 335) and one on the membrane protein (M196, aa 196 to 210). T lymphocytes from N321 peptide-immune miniswine reconstituted the in vitro synthesis of TGEV-specific antibodies by CD4[-] TGEV-immune B cells. The antibodies synthesized were directed at least against all three major structural proteins and neutralized TGEV infectivity, indicating that N321 peptide defines a functional T helper epitope which elicits T cells capable of collaborating with B cells specific for different proteins of TGEV. An interesting epitope has been defined on the M protein of TGEV which induces interferon in naive lymphocytes, and may be relevant in protection[49].

In IBV, protection is induced by generating cytotoxic and helper T cell responses that augment the activity of B cells, in producing virus neutralizing antibody. MHC class II-restricted CD4[+] T -cell hybridomas responded to IBV N protein amino acid sequence 71 to 78[6,8]. In summary, N protein seems to be a dominant antigen for T cell responses, although S and M proteins are also targets for cellular immunity.

Recombinant Vaccines to Protect against Coronavirus

Vaccinia virus (VV) recombinants expressing coronavirus proteins have been successfully used in preventing MHV induced disease, but not in diseases induced by other coronaviruses as FIPV or TGEV. A strong protection against acute encephalomyelitis was mediated in Lewis rats which were immunized with recombinant VV expressing a fusogenic S-protein[28]. By contrast, a VV recombinant encoding a non fusogenic S-protein variant or the N-protein was not capable to confer protection. In addition, MHV-JHM S-specific IgG antibodies elicited before MHV-JHM challenge modulated the disease process, changing it from an acute disease to subacute demyelinating encephalomyelitis. Immunization with recombinant VV expressing spike (S) protein of FIPV or CCV, induced early death syndrome (EDS)[98]. The epitope responsible for EDS in CCV is found on the first 700 amino-acids of the S protein.

Empty parvovirus capsids expressing linear epitopes of MHV have also been used to protect against MHV infections. The MHV protective epitope designated site A of strain A59 S protein[54] was inserted in the major structural protein (VP2) of human parvovirus and expressed using baculoviruses[12]. The peptide (H2N-846-SPLLGCIGSTCAE-858-COOH) has 13 amino acids. The 58 kDa protein assembled into capsids resembling native B19 virion, when expressed in insect cells. Immunoelectron microscopy indicated that the epitopes inserted in the loop were exposed on the surface of the chimeric particles. Antibodies specific for the inserted sequences were induced, which recognized the native virus. Mice immunized with the chimeric capsids were partially protected against a lethal challenge infection with MHV. It will be interesting to determine whether expression of linear epitopes from other coronavirus will also provide protection against the homologous virus.

Vectors to Induce Mucosal Immunity

Coronavirus often infect respiratory or enteric mucosal areas. Then, a strategy that could elicit protection against coronavirus infections is the induction of mucosal immunity. Extensive studies that span more than 20 years from many laboratories have shown that the precursor for mucosal IgA plasma cells originate in the lymphoepithelial structures in the gastrointestinal and respiratory tracts. This precursor, which switch or commits to IgA in gut, or bronchus-associated lymphoepithelial tissues (BALT) migrate from gut associated lymphoid tissues (GALT) to mesenteric lymph nodes, upper respiratory tracts or to exocrine tissues such as the mammary gland, where terminal differentiation into IgA secreting plasma cells takes place. Prokaryotic and eukaryotic vectors with tropism for GALT have been used to elicit immune response to coronavirus.

Human and bovine enteropathogenic strains of *Escherichia coli* have a multimeric surface protein CS31A that has been used as a carrier for foreign antigenic determinants[10]. Site D (named site C by H. Laude's group) of TGEV S protein includes a continuous epitope which can be represented by a synthetic peptide able to induce virus neutralizing antibodies[69, 30]. Sequences coding for antigenic site D[69, 30, 68] have been cloned in the V3 region of CS31A protein. The recombinant hybrid protein was purified and used to immunize mice with 20 μg of antigen per dose. Significant antibody titers and neutralization of the virus was demonstrated. In contrast, antibody titers to TGEV induced by the recombinant bacteria were low[10]. Site D from TGEV S protein fused to *E. coli* heat-labile toxin B (LT-B) subunit was expressed in attenuated *Salmonella* providing a collection of 6 recombinant antigens with a variable number of site D[80]. LT-B-site D fusion products were purified and subcutaneously inoculated into rabbits, in order to determine their immunogenicity. One of the hybrid proteins induced TGEV neutralizing antibodies. The corresponding plasmid was used to transform the vaccine strain *S. typhimurium* Δcya Δcrp Δasd χ3987. Constitutive expression of LT-B-site D protein led to the selection of bacteria expressing a recombinant product that formed pentamers and showed high stability *in vitro*. The recombinant bacteria also elicited TGEV specific antibodies *in vivo*[80]. Expression of TGEV S protein in avirulent *S. typhimurium* yield recombinant antigens with variable toxicity in bacteria. Transformed bacteria expressing a 53 kDa amino-terminal fragment of the S protein and the whole protein (144 kDa), respectively, showed an acceptable stability. Recombinant bacteria expressing the amino-terminal fragment of S protein was selected as a potential bivalent vector to induce both immunity to *Salmonella* and to TGEV. The bacteria colonized swine and induced an antibody response to Salmonella, but the response against TGEV was weak. Prokaryotic vectors developed so far, do not seem the ideal candidates to induce protective immunity against coronaviruses, possible because S protein is an essential antigen in protection against coronaviruses, and the conformation of this antigen is highly dependent on glycosylation[23, 19].

Among the eukaryotic vectors with tropism for GALT or BALT, adenoviruses are nowadays widely used. Two types of adenovirus vectors have been developed: non-defective (vectors with substitutions in E3) which can replicate in many permissive cells, and induce high level expression of the foreign insert, and defective vectors (with E1 gene deleted) which only replicate in 293 cells that provide E1 gene functions *in trans*. Non-defective and defective adenovirus vectors allow the expression of heterologous genes with up to 4,5 and 8 kb, respectively. These vectors are highly stable and induce strong responses[5, 76].

MHV S, N, and M proteins have been expressed using human adenovirus serotype 5 (Ad5). Mice intraperitoneally inoculated with these recombinants elicited serum antibodies which specifically recognized the respective proteins. Only antibodies to S protein neutralized MHV *in vitro*, but titers were low. N protein also induced detectable antibodies as determined by ELISA. By contrast, upon intracerebral challenge with a lethal dose of MHV, a significant fraction of animals vaccinated with adenovirus vectors expressing either the S or the N protein were protected. This protective effect was significantly stronger when the animals were immunized with recombinants expressing both S and N protein, as compared to survival after single immunizations[102].

TGEV spike protein has been expressed using Ad5. The whole protein, or fragments of different lengths: 378, 529, and 1109 amino acids extended from the amino-terminus have been expressed. The recombinant antigens included different sets of the antigenic sites C, B, D, and A, previously defined in the TGEV S protein[76]. These Ad5-TGEV-S recombinants infect hamsters and swine[91], and induce TGEV neutralizing antibodies. Furthermore, Ad5-TGEV-S immune hamsters secrete TGEV neutralizing antibodies in the milk during lactation[76], indicating that these recombinants might have a potential in protection against enteric diseases caused by coronaviruses. Vectors based on porcine adenoviruses are being developed[62, 71].

HE of BCV has been expressed using Ad5. The recombinant HE polypeptide existed in monomeric (65K) and dimeric form (130k). Mice inoculated intraperitoneally with live recombinant Ad5-HE elicited a significant level of BCV-neutralizing antibodies[108]. Protection experiments are being performed with live recombinant virus.

Development of Coronavirus Based Vectors

To immunize a determined animal species against a coronavirus it might be advisable to use a vector based on coronaviruses isolated from the corresponding specie. This would provide several advantages: i) The coronavirus is a multiantigenic system which could elicit a polyvalent immune response that may prevent the appearance of escaping mutants; ii) virus vectors can be selected with a tropism identical to the one exhibited by virulent wild type (*wt*)virus, inducing a local immune response where required; and, iii) vector approval for use in a given species by regulatory agencies may be easier when vaccines are derived from viruses affecting the same species. Then, it may be convenient to develop vectors based in coronaviruses. The ultimate goal is to develop a replication competent coronavirus RNA vector (Figure 3, C). From this, safer vectors could be derived by introducing a deletion removing an essential virus protein, which could be provided *in trans* by a packaging cell line (Figure 3, C). Also suicide vectors could be derived by deleting the encapsidation signal from the helper virus. Unfortunately, the coronavirus genome (around 30 kb) is too large to be cloned and expressed as a single infectious RNA for technical reasons. The construction of these vectors is based on genetic manipulation of single-strand positive-sense RNA genomes, which today can be done by recombination between *wt* genomes and genetically engineered defective interfering (DI) RNAs of small size.

At least two types of vectors based on coronaviruses can be anticipated as intermediate steps towards the construction of a non-defective coronavirus vector: i) a helper-dependent RNA-replication defective virion, which requires the cooperation of a helper virus for its replication (Figure 3, A); ii) a helper independent (replicase +) defective virus, capable of self replicating the RNA, but dependent for assembly on structural proteins provided *in trans* by a second RNA molecule derived from a DI genome (Figure 3, B). The replicase, as defined today, is coded by a 20 kb fragment, which makes difficult the construction of an

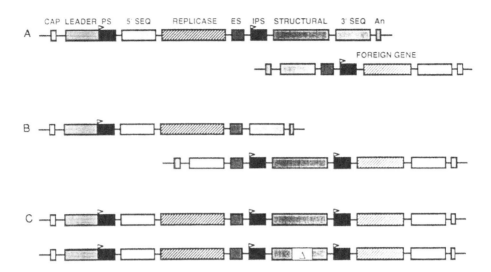

Figure 3. Models of potential coronavirus-derived expression constructs. A. Summarized structure of helper virus (upper part) and a DI genome engineered as a vector to express foreign genes (lower bar). B. Structure of a potential coronavirus based replicon (upper bar) and an engineered DI carrying information to code for structural proteins and foreign antigens (lower construct). C. Single RNA genome with the information for the replicase, essential proteins, and heterologous genes (upper bar). A defective genome can be derived by the introduction of a deletion (Δ) in an essential gene (lower bar). PS, promoter sequences; ES, encapsidation signal; IPS, internal promoter sequences; An, poly A. Triangles above bars indicate promoter position.

RNA replicon, to be expressed from a DNA vector, as shown for nodavirus, alphaviruses, and other small RNA viruses[56, 11, 4]. These vectors are safer, but their immunogenicity usually is lower than that of vectors which replicate and encapsidate. These vectors could be passaged several times before their ability to replicate is lost. The replication is maintained as far as both the replicon containing RNA and the RNA coding for structural proteins are passed together.

Key contributions to progress in this area have been the studies on the minimum *cis*-signal requirements for DI RNA replication[61, 60, 93, 39]. Most advanced work is coming from laboratories dealing with MHV. An MHV DI RNA-based expression system (Figure 4A)[57] represents one of the first MHV-based expression vectors derived from a coronavirus. This vector has low packaging efficiency, although it includes the packaging signal described so far.

Expression in these vectors relies on an internal ribosome entry site (IRES) before the heterologous gene (Figure 4A), or on the transcription of a subgenomic RNA by introducing a coronavirus derived internal promoter sequence (IPS) at the 5′ end of the heterologous gene (Figure 4B). Chloramphenicol acetyltransferase (CAT), luciferase, and MHV hemagglutinin-esterase HE proteins have been expressed using this type of vectors (Figure 4B)[55]. Both the wild-type and recombinant HE proteins are incorporated into viral particles, thus generating a pseudo-recombinant virus. The recombinant DI RNA can be passaged at least four times. This DI RNA is thus an effective expression vector for delivering foreign genes into virus-infected cells. Another RNA vector based on MHV DI particles has been obtained (Figure 4C)[94]. Studies on the replication of this vector have revealed that the MHV DI RNAs require a functional open reading frame (ORF) for efficient propagation[22]. The DI-vector could be used to modify the helper virus genome by homologous recombination to express heterologous genes. Site directed recombination has been shown between full length RNA genomes containing a *ts* mutation on the N gene, and a DI RNA genome,

Figure 4. Coronavirus derived expression constructs. A. Structure of a MHV DI genome used as a vector to express a foreign gene (CAT) using an IRES[57]. B. DI genome engineered to express a heterologous gene (HE) under the control of an internal promoter sequence (IPS)[55]. C. Recombination between a helper virus and a MHV DI genome (MIDI-C) to modify N protein of helper virus. This technique can be used to introduce foreign genes that could be expressed by coronavirus based vectors[96]. pT7, phage T7 promoter, ES, encapsidation signal; IRES, internal ribosome entry site; CAT, chloramphenicol acetyl transferase reporter gene; IPS, internal promoter sequence; HE, haemagglutinin-esterase; N, N protein; ts, temperature sensitive mutation.

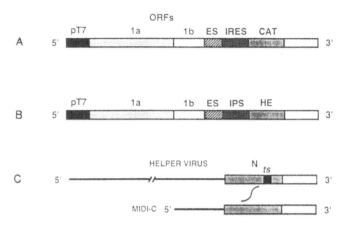

containing a wt N gene[40]. The frequency of site directed homologous recombination was significantly increased with the use of encapsidated co-replicating RNAs[96].

The aims are directed to the development of a system which obviates the need of a helper virus. A vector containing the gene encoding the RNA dependent RNA polymerase, and the replication signals will be required. Mapping the replication signals and polymerase gene reconstruction are priority goals. Such a vector can only be rescued into virus particles in the presence of the structural proteins. Assembly of DI particles in the absence of helper virus using transient expression of the structural proteins S, M, N, and sM, and the synthetic DI RNA has been achieved (W. Spaan, personal communication).

Optimization of the expression in coronavirus requires information on transcription regulation, which nowadays is under debate. The extent of base pairing between the promoter and the leader complementary promoter sequence has been proposed as the control of subgenomic RNA abundance[82, 46]. In fact, certain nucleotide substitutions in the 3' consensus sequence impair promoter activity. However, other data from MHV, IBV, and FIPV indicates that there is no correlation between potential base pairing and mRNA abundance[42,95]. Mutations disrupting the RNA duplex do not reduce promoter activity more than conservative substitutions in the same position[95]. One interpretation of all the data is that successful transcription initiation requires a promoter-leader duplex of minimal stability. Once this condition is met, extending the base pairing does not increase the rate of transcription initiation[95].

Coronavirus mRNAs are synthesized in MHV infected cells at a rate inversely related to their length[79]. In contrast, in FIPV and TGEV infected cells the shortest mRNAs (RNAs 7 and 6, respectively) are produced in much smaller quantities than the next larger mRNAs encoding the nucleocapsid protein. These data suggest that vectors based on different coronaviruses will have to use various promoters for optimum expression. Fusion of leader and body is imprecise. To explain the generation of this heterogeneity it has been proposed that leader RNAs are trimmed by a nuclease to remove the single-stranded RNA sequence from the 3' end after the leader complementary promoter sequence is annealed to the promoter[3]. Alternatively, it has been proposed[37] that a sequence-specific nuclease cleaves the leader RNA at position 1 or 2 of the consensus sequence. More recently it has been proposed[95] that transcription is initiated at multiple, possibly random, sites within the IPS. A 3' to 5' nuclease activity present in many RNA polymerases could remove 3' residues in the IPS causing the heterogeneity.

Resistance to Virus Infections: Intracellular Interference and Transgenic Animals

Two strategies in the construction of transgenic animals resistant to viral infection could be distinguished: i) "intracellular immunity", which could have been more appropriately named intracellular interference, and ii) development of transgenic animals providing immunoprotection. Intracellular interference includes: i) anti-sense RNA and ribozyme approaches; ii) trans-dominant negative mutant viral proteins; iii) molecular decoys; and, iv) intracellular expression of antibody to viral proteins. Antisense oligonucleotides containing sequences complementary to leader RNA induced a significant effect on the multiplication of MHV-JHM. A vector which expressed the antisense mRNA7 of MHV decreased virus multiplication in cell cultures. In transfected cell lines expressing a ribozyme against the 5' end of MHV genome, MHV-multiplication was inhibited by more than 98% at 12 h post infection[77]. Transgenic mice expressing antisense RNA against the nucleocapsid protein gene of MHV, under the control of RSV LTR, were more resistant to the lethal challenge of MHV than non-transgenic mice[63].

Interference in plant and mammalian viruses is provided by expression of the coat protein or the replicase of a virus in the cells susceptible of infection. Little has been done in coronaviruses in this subject. cDNAs engineered using DI RNAs might be used to study the interference in coronavirus replication, mediated by defective products of different ORFs.

Single-chain antibody molecules specific for surface S glycoprotein of MHV have been constructed to interfere with coronavirus induced diseases[41]. The single chain antibody was derived from a murine MAb that has a high binding affinity and neutralizes virus infectivity. Single chain antibodies have been expressed in CRFK and DBT cells using eukaryotic expression vectors with human cytomegalovirus or metallothionein promoters[41]. The resistance of these cells to virus infection and S protein-mediated membrane fusion is being evaluated.

Transgenic pigs secreting TGEV neutralizing antibodies in the milk, to passively protect the progeny, are being constructed. The project is based on the protection of newborn animals by oral administration of IgG or IgA antibodies purified from immune sows, previously reported. The genes coding for the light and heavy chain of TGEV neutralizing MAb 6A.C3, which recognizes a conserved epitope of S protein[75] have been expressed under the control of the whey acid protein (wap) promoter. This promoter is activated by peptidic and steroid hormones present in stimulating concentration in the mammary gland during lactation. Activation of wap promoter results in the synthesis of WAP protein very abundant in the milk. The recombinant antibody has been expressed under the control of both human cytomegalovirus and wap promoters in cell culture[14]. Recombinant antibody had titers of 10^4 as determined by RIA, and neutralization index of 4 log units, using TGEV antigen or infectious virus, respectively. Transgenic mice carrying the genes coding for both the heavy and light chains of 6A.C3 MAb have been developed. The presence of TGEV specific antibody in the milk of transgenic mice during lactation is being determined. Constructs providing optimum levels of TGEV neutralizing antibody will be used to produce transgenic pigs.

ACKNOWLEDGMENTS

This work has been supported by grants from the Consejo Superior de Investigaciones Científicas, the Comisión Interministerial de Ciencia y Tecnología, Instituto Nacional de Investigaciones Agrarias, La Consejería de Educación y Cultura de la Comunidad de Madrid, and Laboratorios Sobrino (Cyanamid) from Spain, and the European Communities (Projects Science and Biotech)

REFERENCES

1 Anton, I M, Gonzalez, S, Bullido, M J, Suñe, C, Meloen, R H, Borras-Cuesta, F, Enjuanes, L Immunodominant T cell epitopes of transmissible gastroenteritis virus major structural proteins J Virol 1994 In press

2 Armstrong, J, Niemann, H, Smeekens, S, Rottier, P, Warren, G Sequence and topology of a model intracellular membrane protein Nature 1984, 308 751-752

3 Baker, S C, Lai, M M C An In vitro system for the leader-primed transcription of coronavirus messenger RNAs EMBO J 1990, 9 4173-4179

4 Ball, A L cis-acting requirements for the replication of flock house virus RNA2 J Virol 1993, 67 3544-3551

5 Bett, A J, Prevec, L, Graham, F L Packaging capacity and stability of human Adenovirus type 5 vectors J Virol 1993, 67 5911-5921

6 Boots, A M, Van-Lierop, M J, Kusters, J G, Van-Kooten, G J, Van-der-Zeijst, B A, Hensen, E J MHC ClassII-restricted T-cell hybridomas recognizing the nucleocapsid protein of avian coronavirus IBV Immunology 1991, 72 10-14

7 Boots, A M H, Benaissatrouw, B J, Hesselink, W, Rijke, E, Schrier, C, Hensen, E J Induction of anti-viral immune responses by immunization with recombinant-DNA encoded avian coronavirus nucleo-capsid protein Vaccine 1992, 10 119-124

8 Boots, A M H, Kusters, J G, Vannoort, J M, Zwaagstra, K A, Rijke, E, Vanderzeijst, B A M, Hensen, E J Localization of a T-cell epitope within the nucleocapsid protein of avian coronavirus Immunology 1991, 74 8-13

9 Boursnell, M E, Brown, T D K, Binns, M M Sequence of the membrane protein gene from avian coronavirus IBV Virus Res 1984, 1 303-313

10 Bousquet, F, Martin, C, Girardeau, J P, Mechin, M C, Vartanian, M d CS31A capsule-like antigen as an exposure vector for heterologous antigenic determinants Infec Immun 1994, 62 2553-2561

11 Bredenbeeck, P J, Rice, C M Animal RNA virus expression systems Sem Virol 1992, 3 297-310

12 Brown, C S, Welling-Wester, S, Feijlbrief, M, Van Lent, J W M, Spaan, W J M Chimeric Parvovirus B19 capsids for the presentation of foreign epitopes Virology 1994, 198 477-488

13 Buchmeier, M J, Lewicki, H A, Talbot, P J, Knobler, R L Murine hepatitis virus-4 (strain JHM)-in-duced neurologic disease is modulated in vivo by monoclonal antibody Virology 1984, 132 261-270

14 Castilla, J, Sola, I, Pintado, B, Hennighausen, L, Enjuanes, L Expression of immunoglobulin genes in the mammary gland Resistance to virus infection 1994 Fundacion Juan March, Madrid, Spain

15 Cavanagh, D, Davis, P J, Derbyshire, J H, Peters, R W Coronavirus IBV virus retaining spike glycopolypeptide S2 but not S1 is unable to induce virus-neutralizing or haemagglutination-inhibiting antibody, or induce chicken tracheal protection J Gen Virol 1986, 67 1435-1442

16 Cavanagh, D, Davis, P J, Mockett, A P Amino acids within hypervariable region 1 of avian coronavirus IBV (Massachusetts serotype) spike glycoprotein are associated with neutralizing epitopes Virus Res 1988, 11 141-150

17 Cavanagh, D, Derbyshire, J H, Davis, P J, Peters, R W Induction of humoral neutralising and haemagglutination-inhibiting antibody by the spike protein of avian infectious bronchitis virus Avian Pathol 1984, 13 573-583

18 Correa, I, Gebauer, F, Bullido, M J, Suñe, C, Baay, M F D, Zwaagstra, K A, Posthumus, W P A, Lenstra, J A, Enjuanes, L Localization of antigenic sites of the E2 glycoprotein of transmissible gastroenteritis coronavirus J Gen Virol 1990, 71 271-279

19 Correa, I, Jimenez, G, Suñe, C, Bullido, M J, Enjuanes, L Antigenic structure of the E2 glycoprotein from transmissible gastroenteritis coronavirus Virus Res 1988, 10 77-94

20 Daniel, C, Anderson, R, Buchmeier, M J, Fleming, J O, Spaan, W J M, Wege, H, Talbot, P J Identification of an immunodominant linear neutralization domain on the S2 portion of the murine coronavirus spike glycoprotein and evidence that it forms part of a complex tridimensional structure J Virol 1993, 67 1185-1194

21 De Groot, R J, Andeweg, A C, Horzinek, M C, Spaan, W J M Sequence analysis of the 3' end of the feline coronavirus FIPV 79-1146 genome comparison with the genome of porcine coronavirus TGEV reveals large insertions J Virol 1988, 167 370-376

22 De Groot, R J , Vandermost, R G , Spaan, W J M The fitness of defective interfering murine
    coronavirus-DI-a and its derivatives is decreased by nonsense and frameshift mutations J Virol 1992,
    66 5898-5905

23 Delmas, B , Gelfi, J , Laude, H Antigenic structure of transmissible gastroenteritis virus II Domains in
    the peplomer glycoprotein J Gen Virol 1986, 67 1405-1418

24 Delmas, B , Rasschaert, D , Godet, M , Gelfi, J , Laude, H Four major antigenic sites of the coronavirus
    transmissible gastroenteritis virus are located on the amino-terminal half of spike protein J Gen Virol
    1990, 71 1313-1323

25 Deregt, D , Babiuck, L A Monoclonal antibodies to bovine coronavirus characteristics and topographi-
    cal mapping of neutralizing epitopes on the E2 and E3 glycoproteins Virology 1987, 68 41-420

26 Enjuanes, L , Van der Zeijst, B A M Molecular basis of transmissible gastroenteritis coronavirus
    (TGEV) epidemiology In Siddell S G , (eds), Coronaviruses Plenum Press, New York 1995 In press

27 Fleming, J O , Shubin, R A , Sussman, M A , Casteel, N , Stohlman, S A Monoclonal antibodies to the
    matrix (E1) glycoprotein of mouse hepatitis virus protect mice from encephalitis Virology 1989, 168
    162-167

28 Flory, E , Pfleiderer, M , Stuhler, A , Wege, H Induction of protective immunity against coronavirus-in-
    duced encephalomyelitis evidence for an important role of CD8+ T cells in vivo Eur J Immunol 1993,
    23 1757-1761

29 Garwes, D J , Lucas, M H , Higgins, D A , Pike, B V , Cartwright, S F Antigenicity of structural
    components from porcine transmissible gastroenteritis virus Vet Microbiol 1978, 3 179-190

30 Gebauer, F , Posthumus, W A P , Correa, I , Suñe, C , Sanchez, C M , Smerdou, C , Lenstra, J A , Meloen,
    R , Enjuanes, L Residues involved in the formation of the antigenic sites of the S protein of transmissible
    gastroenteritis coronavirus Virology 1991, 183 225-238

31 Godet, M , L'Haridon, R , Vautherot, J F , Laude, H TGEV coronavirus ORF4 encodes a membrane
    protein that is incorporated into virions Virology 1992, 188 666-675

32 Godet, M , Rasschaert, D , Laude, H Processing and antigenicity of entire and anchor-free spike
    glycoprotein-S of coronavirus TGEV expressed by recombinant baculovirus Virology 1991, 185
    732-740

33 Hasony, H J , MacNaughton, M R Antigenicity of mouse hepatitis virus strain 3 subcomponents in C57
    strain mice Arch Virol 1981, 69 33-41

34 Heemskerk, M H M , Schoemaker, H M , Spaan, W J M , Boog, C J P Induction of MHC class
    II-restricted CD4+ cytotoxic T cells by MHV-A59 European Immunology Meeting 1994 Barcelona,
    Spain W29

35 Holmes, K V , Doller, E W , Behnke, J N Analysis of the functions of coronavirus glycoproteins by
    differential inhibition of synthesis with tunicamycin Adv Exp Med Biol 1981, 142 133

36 Ignatovic, J , McWaters, P G Monoclonal antibodies to three structural proteins of avian infectious
    bronchitis virus characterization of epitopes and antigenic differentiation of Australian strains J Gen
    Virol 1991, 72 2915-2922

37 Joo, M , Makino, S Mutagenic analysis of the coronavirus intergenic consensus sequence J Virol 1992,
    66 6330-6337

38 Kapke, P A , Tung, F Y T , Hogue, B G , Brian, D A , Woods, R D , Wesley, R The amino-terminal
    signal peptide on the porcine transmissible gastroenteritis coronavirus matrix protein is not an absolute
    requirement for membrane translocation and glycosylation Virology 1988, 165 367-376

39 Kim, Y N , Lai, M M C , Makino, S Generation and selection of coronavirus defective interfering RNA
    with large open reading frame by RNA recombination and possible editing Virology 1993, 194 244-253

40 Koetzner, C A , Parker, M M , Ricard, C S , Sturman, L S , Masters, P S Repair and mutagenesis of
    the genome of a deletion mutant of the coronavirus mouse hepatitis virus by targeted RNA recombination
    J Virol 1992, 66 1841-1848

41 Kolb, A , Grosse, B , Siddell, S G Immunological prevention of coronavirus infection Resistance to
    viral infection 1994 Fundacion Juan March, Madrid, Spain

42 Konings, D A M , Bredenbeek, P J , Noten, J F H , Hogeweg, P , Spaan, W J M Differential premature
    termination of transcription as a proposed mechanism for the regulation of coronavirus gene expression
    Nuc Ac Res 1988, 16 10849-10860

43 Koolen, M J M , Borst, M A J , Horzinek, M C , Spaan, W J M Immunogenic peptide comprising a
    mouse hepatitis virus A59 B-cell epitope and an influenza virus T-cell epitope protects against lethal
    infection J Virol 1990, 64 6270-6273

44 Kubo, H , Taguchi, F Expression of the S1 and S2 subunits of murine coronavirus JHMV spike protein
    by vaccinia virus transient expression system J Gen Virol 1993, 74 2372-2383

45 Kyuwa, S , Stohlman, S Advances in the study of MHV infection of mice Adv Exp Med Biol 1990, 276 555-556

46 Lai, M M C Coronavirus - organization, replication and expression of genome Ann Rev Microbiol 1990, 44 303-333

47 Lapps, W , Hogue, B G , Brian, D A Sequence analysis of the bovine coronavirus nucleocapsid and matrix protein genes Virology 1987, 157 47-57

48 Laude, H , Chapsal, J M , Gelfi, J , Labiau, S , Grosclaude, J Antigenic structure of transmissible gastroenteritis virus I Properties of monoclonal antibodies directed against virion proteins J Gen Virol 1986, 67 119-130

49 Laude, H , Gelfi, J , Lavenant, L , Charley, B Single amino acid changes in the viral glycoprotein M affect induction of alpha interferon by the coronavirus transmissible gastroenteritis virus J Virol 1992, 66 743-749

50 Laude, H , Rasschaert, D , Huet, J C Sequence and N-terminal processing of the transmembrane protein E1 of the coronavirus transmissible gastroenteritis virus J Gen Virol 1987, 68 1687-1693

51 Laviada, M D , Videgain, S P , Moreno, L , Alonso, F , Enjuanes, L , Escribano, J M Expression of swine transmissible gastroenteritis virus envelope antigens on the surface of infected cells epitopes externally exposed Vir Res 1990, 16 247-254

52 Lecomte, J , Cainelli-Cebera, V , Mercier, G , Mansour, S , Talbot, P , Lussier, G , Oth, D Protection from mouse hepatitis virus type 3-induced acute disease by an anti-nucleoprotein monoclonal antibody Arch Virol 1987, 97 123-130

53 Lenstra, J A , Erkens, J H F , Langeveld, J G A , Posthumus, W P A , Meloen, R H , Gebauer, F , Correa, I , Enjuanes, L , Stanley, K K Isolation of sequences from a random-sequence expression library that mimic viral epitopes J Immunol Meth 1992, 152 149-157

54 Lenstra, J A , Kusters, J G , Koch, G , van der Zeijst, B A M Antigenicity of the peplomer protein of infectious bronchitis virus Molec Immunol 1989, 26 7-15

55 Liao, C -L , Lai, M M C The requirement of 5'-end genomic sequence as an upstream cis-acting element for coronavirus subgenomic mRNA transcription J Virol 1994, In Press

56 Liljestrom, P , Garoff, H A new generation of animal cell expression vectors based on the Semliki forest virus replicon Biotechnology 1991, 9 1356-1361

57 Lin, Y J , Lai, M M C Deletion mapping of a mouse hepatitis virus defective interfering RNA reveals the requirement of an internal and discontiguous sequence for replication J Virol 1993, 67 6110-6118

58 Luytjes, W , Geerts, D , Posthumus, W , Meloen, R , Spaan, W J M Amino acid sequence of a conserved neutralizing epitope of murine coronaviruses J Virol 1989, 63 1408-1415

59 Makino, S , Fleming, J O , Keck, J G , Stohlman, S T , Lai, M M C RNA recombination of coronaviruses localization of neutralizing epitopes and neuropathogenic determinants on the carboxyl terminus of peplomers Proc Natl Acad Sci 1987, 84 6567-6571

60 Makino, S , Joo, M , Makino, J K A System for study of coronavirus messenger RNA synthesis - a regulated, expressed subgenomic defective interfering RNA results from intergenic site insertion J Virol 1991, 65 6031-6041

61 Makino, S , Yokomori, K , Lai, M M C Analysis of efficiently packaged defective interfering RNAs of murine coronavirus - localization of a possible RNA-packaging signal J Virol 1990, 64 6045-6053

62 Mengeling, W L Porcine coronaviruses co-infection of cell cultures with transmissible gastroenteritis virus and hemagglutinating encephalomyelitis Am J Vet Res 1973, 34 779-783

63 Mizutani, T , Hayashi, M , Maeda, A , Sasaki, N , Yamashsita, T , Kasai, N , Namioka, S Inhibition of mouse hepatitis virus multiplication by antisense oligonucleotide, antisense RNA, sense RNA, and ribozyme Adv Exp Med Biol 1994, 276 129-135

64 Nakanaga, K , Yamanouchi, K , Fujiwara, K Protective effect of monoclonal antibodies on lethal mouse hepatitis virus infection in mice J Virol 1986, 59 168-171

65 Niesters, H G M , Lenstra, J A , Spaan, W J M , Zijderveld, A J , Bleumink-Pluym, N M C , van der Zeijst, B A M The peplomer protein sequence of the M41 strain of coronavirus IBV and its comparison with Beaudette strains Vir Res 1986, 5 253-263

66 Olsen, C W A review of feline infectious peritonitis virus molecular biology, immunopathogenesis, clinical aspects, and vaccination Vet Microbiol 1993, 36 1-37

67 Parker, M M , Masters, P S Sequence comparison of the N genes of five strains of the coronavirus mouse hepatitis virus suggests a three domain structure for the nucleocapsid protein Virology 1990, 179 463-468

68 Posthumus, W P A , Lenstra, J A , van Nieuwstadt, A P , Schaaper, W M M , van der Zeijst, B A M , Meloen, R H Immunogenicity of peptides simulating a neutralizing epitope of transmissible gastroenteritis virus Virology 1991, 182 371-375

69 Posthumus, W P A, Meloen, R H, Enjuanes, L, Correa, I, van Nieuwestadt, A, Koch, G Linear neutralizing epitopes on the peplomer protein of coronaviruses Adv Exp Med Biol 1990, 276 181-188

70 Pulford, D J, Britton, P Expression and cellular localisation of porcine transmissible gastroenteritis virus N and M proteins by recombinant vaccinia viruses Vir Res 1990, 18 203-218

71 Reddy, P S, Nagy, E, Derbishire, J B Restriction endonuclease analysis and molecular cloning of porcine adenovirus type 3 Intervirology 1993, 36 161-168

72 Risco, C, Anton, I M, Suñe, C, Pedregosa, A M, Martin-Alonso, J M, Parra, F, Carrascosa, J L, Enjuanes, L The membrane protein of transmissible gastroenteritis coronavirus exposes the carboxy-terminal region on the external surface of the virion 1994 Submitted

73 Routledge, E, Stauber, R, Pfleiderer, M, Siddell, S G Analysis of murine coronavirus surface glycoprotein functions by using monoclonal antibodies J Virol 1991, 65 254-262

74 Saif, L J, Wesley, R D Transmissible gastroenteritis In Leman A D, Straw, B, Mengeling, W L, D' Allaire, S, Taylor, D J, (eds), Diseases of swine Iowa State University Press, Ames, Iowa 1992, pp 362-386

75 Sanchez, C M, Jimenez, G, Laviada, M D, Correa, I, Suñe, C, Bullido, M J, Gebauer, F, Smerdou, C, Callebaut, P, Escribano, J M, Enjuanes, L Antigenic homology among coronaviruses related to transmissible gastroenteritis virus Virology 1990, 174 410-417

76 Sanchez, C M, Torres, J M, Suñe, C, Smerdou, C, Graham, F L, Enjuanes, L Transmissible gastroenteritis coronavirus spike protein expressed by adenovirus vectors elicited virus neutralizing antibodies in hamsters 1994 Submitted

77 Sasaki, N, Hayashi, M, Aoyama, S, Yamashita, T, Miyoshi, I, Kasai, N, Namioka, S Transgenic mice with antisense RNA against the nucleocapsid protein mRNA of mouse hepatitis virus J Vet Med Sci 1993, 55 549-554

78 Scott, F W, Corapi, W V, Olsen, C W Evaluation of the safety and efficacy of Primucell-FIPR vaccine Feline Health Topics 1992, 7 6-8

79 Sethna, P B, Hung, S -L, Brian, D A Coronavirus subgenomic minus-strand RNAs and the potential for mRNA replicons Proc Natl Acad Sci USA 1989, 86 5626-5630

80 Smerdou, C, Anton, I M, Plana, J, Curtiss, R, Enjuanes, L Expression of a continuos epitope from Transmissible gastroenteritis coronavirus S protein fused to *E coli* heat-labile toxin B subunit in attenuated *Salmonella* for oral immunization 1994 Submitted

81 Smith-Kline, B Universal coronavirus vaccine, spike protein cloning and expression for use as a recombinant vaccine Vaccine 1994, 12 671

82 Spaan, W, Cavanagh, D, Horzinek, M C Coronaviruses structure and genome expression J Gen Virol 1988, 69 2939-2952

83 Spaan, W, Cavanagh, D, Horzinek, M C Coronaviruses In van Regenmortel M H V, Neurath, A R, (eds), Immunochemistry of viruses II The basis for serodiagnosis and vaccines Elsevier, 1990, pp 359-379

84 Stohlman, S A, Bergmann, C, Cua, D, Wege, H Location of antibody epitopes within the mouse hepatitis virus nucleocapsid protein Virology 1994, 202 146-153

85 Stohlman, S A, Kyuwa, S, Cohen, M, Bergmann, C, Polo, J M, Yeh, J, Anthony, R, Keck, J G Mouse hepatitis virus nucleocapsid protein-specific cytotoxic T lymphocytes are L^d restricted and specific for the carboxy terminus Virology 1992, 189 217-224

86 Sturman, L, Holmes, K, V The novel proteins of coronaviruses Trends Biochem Sci 1985, 10 17-20

87 Talbot, P J, Dionne, G, Lacroix, M Vaccination against lethal coronavirus-induced encephalitis with a synthetic decapeptide homologous to a domain in the predicted peplomer stalk J Virol 1988, 62 3032-3036

88 Talbot, P J, Salmi, A A, Knobler, R L, Buchmeier, M J Topographical mapping of epitopes on the glycoprotein of murine hepatitis virus-4 (Strain JHM) correlation with biological activities Virology 1984, 132 250-260

89 To, L T, Bernard, S, Lantier, I Fixed-cell immunoperoxidase technique for the study of surface antigens induced by the coronavirus of transmissible gastroenteritis (TGEV) Vet Microbiol 1991, 29 361-368

90 Tooze, S A, Stanley, K K Identification of two epitopes in the carboxyterminal 15 amino acids of the E1 glycoprotein of mouse hepatitis virus A59 by using hybrid proteins J Virol 1986, 60 928-934

91 Torres, J M, Escribano, J A M, Enjuanes, L Induction of lactogenic immunity to transmissible gastroenteritis coronavirus with recombinant Adenovirus 5 expressing TGEV spike protein 1995 Submitted

92 Tung, F Y T, Abraham, S, Sethna, M, Hung, S L, Sethna, P, Hogue, B G, Brian, D A The 9-kDa hydrophobic protein encoded at the 3' end of the porcine transmissible gastroenteritis coronavirus genome is membrane-associated Virology 1992, 186 676-683

93 van der Most, R G , Bredenbeek, P J A domain at the 3' end of the polymerase gene is essential for encapsidation of coronavirus defective interfering RNAs J Virol 1991, 65 3219-3226

94 van der Most, R G , Bredenbeek, P J , Spaan, W J M A domain at the 3' end of the polymerase gene is essential for encapsidation of coronavirus defective interfering RNAs J Virol 1991, 65 3219-3226

95 van der Most, R G , De Groot, R J , Spaan, W J M Subgenomic RNA synthesis directed by a synthetic defective interfering RNA of mouse Hepatitis virus a study of Coronavirus transcription initiation J Virol 1994, 68 3656-3666

96 van der Most, R G , Heijnen, L , Spaan, W J M , Degroot, R J Homologous RNA recombination allows efficient introduction of site-specific mutations into the genome of coronavirus MHV-A59 via synthetic coreplicating RNAs Nuc Ac Res 1992, 20 3375-3381

97 Vennema, H , De Groot, R J , Harbour, D A , Horzinek, M C , Spaan, W J M Primary structure of the membrane and nucleocapsid protein genes of Feline infectious peritonitis virus and immunogenicity of recombinant vaccinia viruses in kittens Virology 1991, 181 327-335

98 Vennema, H , DeGroot, R J , Harbour, D A , Dalderup, M , Gruffydd-Jones, T , Horzinek, M C , Spaan, W J M Early death after feline infectious peritonitis challenge due to recombinant vaccinia virus immunization J Virol 1990, 64 1407-1409

99 Wang, F I , Fleming, J O , Lai, M M C Sequence analysis of the spike protein gene of murine coronavirus variants Study of genetic sites affecting neuropathogenicity Virology 1992, 186 742-749

100 Wege, H , Schliephake, A , Korner, H , Flory, E , Wege, H Coronavirus induced encephalomyelitis an immunodominant CD4+-T cell site on the nucleocapsid protein contributes to protection Adv Exp Med Biol 1994, 342 413-418

101 Welch, S K W , Saif, L J Monoclonal antibodies to a virulent strain of transmissible gastroenteritis virus comparison of reactivity with virulent and attenuated virus Arch Virol 1988, 101 221-235

102 Wesseling, J G , Godeke, G J Mouse hepatitis virus spike and nucleocapsid proteins expressed by adenovirus vector protect mice against a lethal infection J Gen Virol 1993, 74 2061-2069

103 Williamson, J S P , Stohlman, S A Effective clearance of mouse hepatitis virus from the central nervous system requires both CD4+ and CD8+ T cells J Virol 1990, 64 4589-4592

104 Woods, R D , Wesley, R D , Kapke, P A Complement-dependent neutralization of transmissible gastroenteritis virus by monoclonal antibodies Adv Exp Med Biol 1987, 218 493-500

105 Yoden, S , Kikuchi, T , Siddell, S G , Taguchi, F Expression of the peplomer glycoprotein of murine coronavirus JHM using a baculovirus vector Virology 1989, 173 615-623

106 Yoo, D , Parker, M D , Babiuk, L A The S2 subunit of the spike glycoprotein of bovine coronavirus mediates membrane fusion in insect cells Virology 1991, 180 395-399

107 Yoo, D , Parker, M D , Song, J , Graham, F L , Deregt, D , Babiuk, L A Structural analysis of the conformational domains involved in neutralization of Bovine Coronavirus using deletion mutants of the spike glycoprotein S1 subunit expressed by recombinant baculoviruses Virology 1991, 183 91 - 98

108 Yoo, D W , Graham, F L , Prevec, L , Parker, M D , Benko, M , Zamb, T , Babiuk, L A Synthesis and processing of the haemagglutinin-esterase glycoprotein of bovine coronavirus encoded in the E3 region of adenovirus J Gen Virol 1992, 73 2591-2600

35

PRODUCTION AND IMMUNOGENICITY OF MULTIPLE ANTIGENIC PEPTIDE (MAP) CONSTRUCTS DERIVED FROM THE S1 GLYCOPROTEIN OF INFECTIOUS BRONCHITIS VIRUS (IBV)

Mark W. Jackwood and Deborah A. Hilt

University of Georgia
College of Veterinary Medicine
Poultry Diagnostic and Research Center
Athens, Georgia 30602

ABSTRACT

Synthetic peptides were prepared as multiple antigenic peptide (MAP) constructs to the S1 glycoprotein of infectious bronchitis virus (IBV). The MAP system has been used in the production of anti-peptide and anti-protein antibodies. It has an advantage over linking peptides to a highly immunogenic carrier molecule because antibodies are not produced to the MAP core matrix of lysine residues. Two 25-residue peptides were synthesized to the Arkansas serotype and two were synthesized to the Massachusetts serotype of IBV. The peptide sequences correspond to amino acid residues 64 to 88 and to residues 117 to 141 for each of the IBV serotypes. A MAP construct for each peptide was prepared by linking 4 copies of a peptide to the immunogenetically inert core matrix of lysine residues. The MAP constructs were used to immunize specific pathogen free chickens. Anti-peptide ELISA titers and the dot immunobinding assay against the homologous peptide were positive for all of the sera tested whereas the anti-whole virus ELISA titers and virus neutralization titers were negative for all of the sera tested. Hyperimmune sera against whole virus did not cross react with synthetic peptides made to the heterologous virus suggesting a possible role for the MAP constructs in a serotype specific dot blot or ELISA test for IBV.

INTRODUCTION

Peptides have been used experimentally as vaccines to protect against several viral disease agents including foot and mouth disease virus[1], canine parvovirus[2], and mouse

Corona- and Related Viruses, Edited by P. J. Talbot and G. A. Levy
Plenum Press, New York, 1995

hepatitis virus[3,4]. They have also been used to study the structural features of viral immuno-genic proteins[5].

A number of algorithms have been developed to predict epitopes in proteins from the amino acid sequence[5,6,7,8]. In general, peptide antigenicity has been correlated with the structural features of the protein. Some of those features include hydrophilicity, amphipa-thicity, and surface accessibility[8]. In addition, sequence variability has been used to predict epitopes on viral proteins[8].

Several factors need to be considered when designing synthetic peptide vaccines. Since peptides do not generally stimulate an immune response by them selves, they must be conjugated to a carrier molecule to be immunogenic. The two most popular carrier molecules are bovine serum albumin (BSA) and keyhole limpet hemocyanin (KLH)[8].

Multiple antigenic peptide (MAP) constructs provide an alternative means of producing immunogenic peptides. First described by Tam[9] in 1988, MAP constructs consist of either four or eight peptides linked by the COOH-terminal amino acid to a core matrix of lysine residues. An advantage of using MAP constructs is that antibodies are not produced to the immunologically inert core matrix of lysine residues. In addition the need to conjugate the peptide to a carrier molecule is eliminated. A disadvantage of using MAP constructs is that it is not possible to use a T-cell stimulating carrier molecule which has been shown to be critical in developing a protective immune response using some peptides[1,10]. However, T-cell stimulating epitopes could be incorporated into the same MAP construct as B-cell epitopes since several peptides can be linked to the same core matrix of lysine residues.

In this study, we investigated the feasibility of developing a vaccine using MAP constructs of synthetic peptides against the S1 glycoprotein of infectious bronchitis virus (IBV).

METHODS

Synthetic Peptides

Two synthetic peptides to the Arkansas 99 (Ark 99) strain of IBV, designated Ark 99A and Ark 99B, and two synthetic peptides to the Mass 41 strain of IBV designated Mass 41A and Mass 41B, were designed based the sequence comparison of the S1 glycoproteins of the two viruses. The sequences of the 25 residue peptides were taken from variable regions between residues 64 to 88 and from residues 117 to 141 (Figure 1). Analysis of the peptides with the Hopp- Woods[7], and the Chou-Fasman[6] algorithms which predicts antigenic deter-minants based on the hydrophobicity of the sequence and secondary structure respectively showed that each peptide contained hydrophilic regions and a ß pleated sheet and or a α helix.

The 'Fmoc' method was used to synthesize the peptides. Then the peptides were linked to a 4-branch multiple antigenic peptide (MAP) resin (Applied Biosystems, Inc., Foster City, CA) following the manufacturer's instructions.

Immunization Schedule

Five one-day old specific pathogen free chicks per group were immunized with either Ark 99A, Ark 99B, Mass 41A, Mass 41B, Ark 99A and Ark 99B, Mass 41A and Mass 41B, whole Ark99 virus, or whole Mass 41 virus. A negative control group was given sterile saline.The immunization schedule and sera collection times are shown in Table 1. Prior to

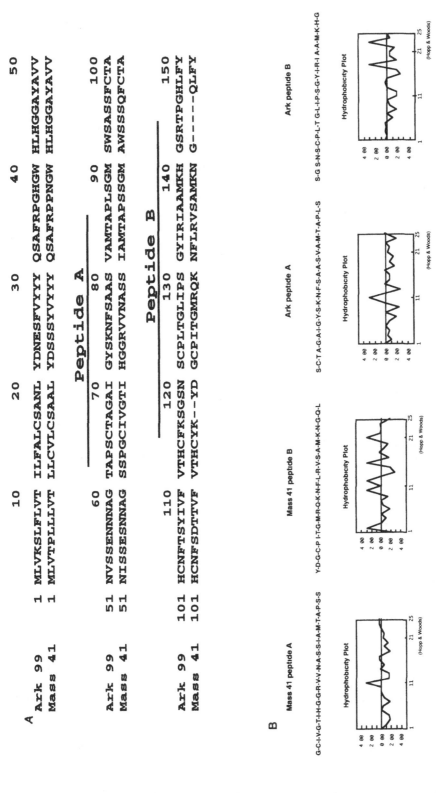

A

```
                 10         20         30         40         50
Ark 99   1  MLVKSLFLVT ILFALCSANL YDNESFVYYY QSAFRPGHGW HLHGGAYAVV
Mass 41  1  MLVTPLLLVT LLCVLCSAAL YDSSSYVYYY QSAFRPPNGW HLHGGAYAVV

                           Peptide A
                 60         70         80         90        100
Ark 99   51 NVSSENNNAG TAPSCTAGAI GYSKNFSAAS VAMTAPLSGM SWSASSFCTA
Mass 41  51 NISSESNNAG SSPGCIVGTI HGGRVVNASS IAMTAPSSGM AWSSSQFCTA

                           Peptide B
                120        130        140        150
Ark 99  101 HCNFTSYIVF VTHCFKSGSN SCPLTGLIPS GYIRIAAMKH GSRTPGHLFY
Mass 41 101 HCNFSDTTVF VTHCYK--YD GCPITGMRQK NFLRVSAMKN G-----QLFY
```

B

Mass 41 peptide A

G-C-I-V-G-T-I-H-G-G-R-V-V-N-A-S-S-I-A-M-T-A-P-S-S

Hydrophobicity Plot

(Hopp & Woods)

Ark peptide A

S-C-T A-G-A-I-G-Y-S-K-N-F-S-A-A-S-V-A-M-T-A-P-L-S

Hydrophobicity Plot

(Hopp & Woods)

Mass 41 peptide B

Y-D-G-C-P I-T-G-M-R-Q-K-N-F-L-R-V-S-A-M-K-N-G-Q-L

Hydrophobicity Plot

(Hopp & Woods)

Ark peptide B

S-G S-N-S-C-P-L-T G-L-I-P-S-G-Y-I-R-I A-A-M-K-H-G

Hydrophobicity Plot

(Hopp & Woods)

Figure 1. A The relative location of synthetic peptides in the S1 glycoprotein of the Ark 99 and Mass 41 strains of IBV B The peptide sequences and their hydrophobicity plot based on the Hopp and Woods algorithm

Table 1. The immunization[A] schedule and sera collection times for
chickens given synthetic peptides to IBV

Age in days	Immunization site	Adjuvant	Sera collection
1	SQ[B]	Complete Freund's	Yes
7	IM[C]	Incomplete Freund's	No
21	IV[D]	None	No
28	IV	None	Yes
35	—	—	Yes

[A]Five specific pathogen free chickens per group were immunized each time with
approximately 500 ng of peptide. The birds were immunized with either Mass
41A, Mass 41B, Ark 99A, Ark 99B, Mass 41A and Mass 41B, Ark 99A and
Ark 99B, Mass 41 whole virus, or Ark 99 whole virus.
[B]SQ= subcutaneous injection in the back of the neck.
[C]IM= intramuscular injection in the breast muscle.
[D]IV= intravenous injection.

immunization sera was collected from all of the birds and tested for maternal antibodies
against IBV by ELISA.

Analysis of Sera

All of the sera collected on days 28 and 35 were tested in a peptide specific ELISA
where the appropriate peptide was coated on the plates and by whole virus ELISA where the
homologous strain of the virus was coated on the plates.

Sera found to be positive in the peptide specific ELISA test were further tested
in immunobinding assays. The dot immunobinding assay was conducted by applying
each of the peptides or whole virus onto nitrocellulose filters. Allantoic fluid from
non-inoculated embryonating eggs was also dotted onto the filters and served as a negative
control.

Western blot analysis was conducted following the procedures of Sambrook *et al.* [11]
using sucrose gradient purified whole virus.

The virus neutralization assay was conducted in 10-day old embryonating eggs using
standard procedures[12].

RESULTS

Analysis of Sera

The Ark 99A peptide was not soluble and no antibodies were detected by ELISA in
birds given that peptide. Thus, it was dropped from the study. Geometric mean titers for the
sera from the birds given the other peptides are shown in Table 2. The anti-peptide ELISA
titers against the homologous peptide were positive for all of the sera tested and ranged from
400 for the Mass 41A peptide to 6400 for the Mass 41 A&B combination. The anti-whole
virus ELISA titers were negative for all of the sera tested.

The virus neutralization titers were negative for all of the sera tested.

Dot immunobinding analysis of the sera positive in the peptide specific ELISA test
revealed that the sera reacted with the homologous peptide but did not react with the whole
virus (Figure 2). Cross reactivity for some of the sera was observed with the other homolo-

Table 2. Geometric mean titers of sera collected from birds at 35 days post-inoculation with synthetic peptides to the IBV S1 glycoprotein

Group	Anti-peptide ELISA[A]	Anti-virus ELISA[B]	VN[C]
Mass 41A	400	≤ 200	≤2
Mass 41B	1600	≤200	≤2
Mass 41 A&B	6400	≤200	≤2
Ark B	800	≤200	≤2
Ark A&B	6400	≤200	≤2

[A]The homologous peptide was coated on the ELISA plate.
[B]The homologous whole virus was coated on the ELISA plate.
[C]VN= Virus neutralization. Conducted in 10-day old embryonating eggs.

gous peptide but not with the heterologous peptides. Hyperimmune antisera against whole virus reacted with the whole virus and the homologous peptides but not with the heterologous peptides in the dot immunobinding assay.

Western blot analysis (Figure 3) showed that the anti-peptide antibodies in the sera were directed against the S1 glycoprotein of the homologous virus. Only antibodies against the Mass 41B peptide reacted with the heterologous virus.

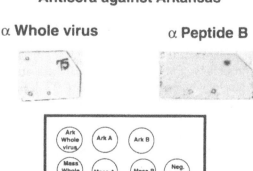

Figure 2. Binding of antibodies to whole virus and MAP constructs prepared to the S1 glycoprotein of IBV. Representative immunoblots for each sera are presented. Antigen placement on the filters is shown in the key at the bottom of the figure. Neg. Control= allantoic fluid from noninoculated embryonating eggs.

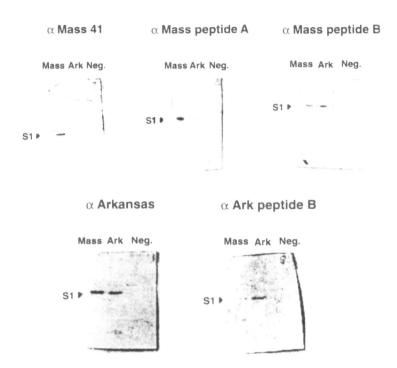

Figure 3. Binding of antibodies to the denatured S1 glycoprotein of IBV Representative Western blots are shown for each sera Neg = allantoic fluid from noninoculated embryonating eggs

DISCUSSION

In this study, synthetic peptides to IBV were prepared as MAP constructs and used to immunize SPF chickens. Sera collected at two different times following immunization contained anti-peptide antibodies and antibodies against the denatured S1 glycoprotein. No antibodies against the native protein or virus neutralizing (VN) antibodies were detected.

Although VN antibodies were not detected in our study, we cannot draw any conclusions regarding the immune status of those birds against challenge with IBV. Studies with foot and mouth disease virus have shown little or no correlation between the presence of *in vitro* neutralizing antibodies and protection following immunization of cattle with synthetic peptides[1]. Those researchers concluded that T-cells were responsible for the protection observed in cattle and that helper T-cells must be stimulated to get protection.

Empirical rules for the prediction of peptide antigenicity and T-cell epitopes from the amino acid sequence of a protein have been developed[8]. Unfortunately the methods do not take into consideration the individual's histocompatibility antigens (MHC) and thus, are not absolutely fool proof[4,8,13]. Researchers are still left with a trial and error approach to developing immunogenic peptides. We designed our peptides based on sequence variability between the Ark 99 and Mass 41 strains of IBV. Computer analysis of the hydrophobicity and secondary structure of the peptides indicated that each peptide contained hydrophilic regions and a ß pleated sheet and or a α helix. It was hoped that the MAP constructs of those peptides would generate VN antibodies in chickens. Perhaps a better immune response could have been generated with the peptides linked to a carrier molecule. Particularly a T-cell stimulating carrier such as ovalbumin or sperm whale myoglobulin which may be recognized by a greater number of histocompatibility types[1,10].

In our study it should be noted that hyperimmune sera against whole virus did not cross react with synthetic peptides made to the heterologous virus in the dot immunobinding assay This result suggests a possible role for synthetic peptides prepared as MAP constructs in a serotype specific dot blot or ELISA test for IBV

REFERENCES

1 Murphy, B R, and Chanock, R M Immunization against viruses In Fields, B N, and Knipe, D M (eds) Virology 2nd ed Raven Press, Ltd, New York 1990 pp 469-502

2 Langeveld, J P M, Casal, J I, Osterhaus, A D M E, Cortes, E, de Swart, R, Vela, C, Dalsgaard, K, Puijk, W C, Schaaper, W M M, and Meloen, R H First peptide vaccine providing prptection against viral infection in the target animal Studies of canine parvovirus in dogs J of Virol 1994,68 4506-4513

3 Koolen, M J M, Borst, M A J, Horzinek, M C, and Spaan, W J M Immunogenic peptide comprising a mouse hepatitis virus A59 B-cell epitope and an influenza virus T-cell epitope protects against lethal infection J of Virol 1990,64 6270-6273

4 Talbot, P J, Dionne, G, and Lacroix, M Vaccination against lethal coronavirus-induced encephalitis with a synthetic decapeptide homologous to a domain in the predicted peplomer stalk J of Virol 1988,62 3032-3036

5 Daniel, C, Lacroix, M, Talbot, P J Mapping of linear antigenic sites on the S glycoprotein of a neurotropic murine coronavirus with synthetic peptides A combination of nine prediction algorithms fails to identify relevant epitopes and peptide immunogenicity is drastically influenced by the nature of protein carrier Virology 1994,202 540-549

6 Chou, P Y, and Fasman, G P Prediction of protein conformation Biochemistry 1974,13 222-224

7 Hopp, T P, and Woods, K R Prediction of protein antigenic determinants from amino acid sequences Proc Natl Acad Sci USA 1981,78 3824-3828

8 van Regenmortel, M H V, Briand, J P, Muller, S and Plaue, S Synthetic polypeptides as antigens In Burdon, R H and van Knippenberg, P H (eds) Laboratory techniques in biochemistry and molecular biology vol 19 Elsevier, Amsterdam 1988 pp 1-75, 131-191

9 Tam, J P Synthetic peptide vaccine design Synthesis and properties of a high-density multiple antigenic peptide system Proc Natl Acad Sci USA 1988,85 5409-5418

10 Rothbard, J Synthetic peptides as vaccines Nature 1987,330 106-107

11 Sambrook, J, Fritsch, E F, Maniatis, T Molecular Cloning A laboratory Manual Cold Spring Harbor Laboratory Press, New York 1989, pp 18 60-18 65

12 Senne, D A Virus propagation in embryonating eggs In Purchase, H G, Arp, L H, Domermuth, C H, Pearson, J E (eds) A Laboratory Manual for the Isolation and Identification of Avian Pathogens 3rd ed The American Association of Avian Pathologists, Kennett Square 1989, pp 176-181

13 Livingstone, A Epitopes In Roitt, I M, Delves, P J (eds) Encyclopedia of Immunology Academic Press, San Diego 1992 pp 515-517

36

PROTECTION OF CATS FROM INFECTIOUS PERITONITIS BY VACCINATION WITH A RECOMBINANT RACCOON POXVIRUS EXPRESSING THE NUCLEOCAPSID GENE OF FELINE INFECTIOUS PERITONITIS VIRUS

T. L. Wasmoen, N. P. Kadakia, R. C. Unfer,[*] B. L. Fickbohm, C. P. Cook, H-J. Chu, and W. M. Acree

Biological Research and Development
Fort Dodge Laboratories
800 Fifth Street NW
Fort Dodge, Iowa, 50501

ABSTRACT

Feline Infectious Peritonitis Virus (FIPV) is a coronavirus that induces an often fatal, systemic infection in cats. Various vaccines designed to prevent FIPV infection have been shown to exacerbate the disease, probably due to immune enhancement mediated by virus-specific immunoglobulins against the outer envelope (S) protein. An effective vaccine would be one that induces cell-mediated immunity without disease enhancing antibodies. In this report, we describe the use of a recombinant raccoon poxvirus that expresses the gene encoding the nucleocapsid protein of FIPV (rRCNV-FIPV N) as an effective vaccine against FIPV-induced disease. Cats were parenterally or orally vaccinated twice, three weeks apart. Cats were then orally challenged with Feline Enteric Coronavirus (FECV), which induces a subclinical infection that can cause enhancement of subsequent FIPV infection. Three weeks later, cats were orally challenged with FIPV. The FIPV challenge induced a fatal infection in 4/5 (80%) of the controls. On the other hand, all five cats vaccinated subcutaneously with rRCNV-FIPV N showed no signs of disease after challenge with FIPV. Four of the five subcutaneous vaccinates survived an additional FIPV challenge. Vaccination with rRCNV-FIPV N induced serum IgG antibody responses to FIPV nucleocapsid protein, but few, if any, FIPV neutralizing antibodies. In contrast to the controls, protected vaccinates main-

---

[*] Current Address Rush Presbyterian St. Lukes Medical Center, Chicago, Il 60612.

Corona- and Related Viruses, Edited by P. J. Talbot and G. A. Levy
Plenum Press, New York, 1995

tained low FIPV serum neutralizing antibody titers after FIPV challenge. This suggests that the protective immune response involves a mechanism other than humoral immunity consisting of FIPV neutralizing antibodies.

INTRODUCTION

Feline infectious peritonitis virus (FIPV) is a member of the *Coronaviridae* family of enveloped viruses with a single-stranded RNA genome of positive polarity.[1,2] The virion is composed of three major structural proteins: a 200 kDa outer envelope peplomer or spike (S) glycoprotein, a 25-32 kDa transmembrane glycoprotein (E1, also known as the matrix (M) protein), and a 45 kDa phosphorylated nucleocapsid protein (N).[2] Antibodies against all three structural proteins are found during natural and experimental FIPV infections.[3,4] Virus neutralizing antibodies appear to be predominantly against the S protein.[5,6]

Various vaccines designed to prevent FIPV infection have also been shown to exacerbate the disease caused by the virus. These vaccines included attenuated-live or inactivated FIPV,[7,8] closely related coronaviruses,[9-12] and a vaccinia recombinant expressing the S glycoprotein.[13] All of these unsuccessful vaccines induced a strong humoral antibody response consisting of FIPV neutralizing antibodies directed against the S protein. In this report we describe the use of a recombinant raccoon poxvirus expressing the FIPV N protein that induces protective immunity against FIP in cats.

MATERIALS AND METHODS

Cells and Viruses. FIPV strain 79-1146 (ATCC VR-990), FIPV strain Fort Dodge Type II (FD-II) (Dr. John Black, American BioResearch, Seymour, TN), and FECV strain 79-1683 (ATCC VR-989) were propagated in Crandell feline kidney cells (CRFK cells, ATCC CCL 94). Raccoon poxvirus (RCNV) (ATCC VR-838) was grown in Vero cells (ATCC CCL 81). Cells were maintained in Eagle's minimum essential medium (MEM, GIBCO BRL, Bethesda, MD) containing 5 % heat-inactivated fetal bovine serum. Recombinant RCNV were propagated in Vero cells using a multiplicity of infection of 0.01 to 0.05.

Preparation of Recombinant RCNV Expressing the FIPV Nucleocapsid Protein. Cloning of the nucleocapsid gene from FIPV (FD-II) is described elsewhere.[14] The gene segment encoding the N gene was subcloned into the *Sma*I site of the poxvirus transfer vector pSC11 (a generous gift of Dr. Bernard Moss, NIH, Bethesda MD) to create pSC11/FIPV-N. Expression of the N gene is under the control of the vaccinia early/late promoter, p7.5. A recombinant raccoon poxvirus expressing the FIPV N gene (rRCNV-FIPV N) was generated by methods described previously.[15] Six individual rRCNV-FIPV N clones were selected and plaque purified five times. Expression of the FIPV N gene by the rRCNV was confirmed by radioimmunoprecipitation assays carried out as described previously.[6] Briefly, lysates of FIPV-infected CRFK cells or rRCNV-FIPV N infected Vero cells were prepared after metabolic labeling with L-[^{35}S]methionine and cysteine. Immunoprecipitation was done using a monoclonal antibody against FIPV N and Pansorbin cells (Calbiochem, San Diego CA). Precipitated proteins were analyzed by SDS-PAGE through 12 % gels. Expression was also confirmed by indirect immunofluorescence assay.[16] Briefly, Vero cells were infected with the various recombinant RCNVs and FIPV N expression was confirmed using a monoclonal antibody against FIPV N and a fluorescein-labeled goat anti-mouse IgG as the secondary antibody (Kirkegaard & Perry Laboratories, Gaithersburg MD).

Preparation of Vaccine. A single clone of rRCNV-FIPV N, showing significant N protein expression, was selected for large-scale expansion to serve as a vaccine virus. All recombinant and wild-type virus expansions and titrations were performed on Vero cells in MEM containing 2.5% fetal bovine serum (FBS). The titer of rRCNV-FIPV N was determined to be $10^{6.97}$ TCID$_{50}$/mL. The titer of the wild-type RCNV (negative control vaccine)was $10^{6.44}$ TCID$_{50}$/mL.

Vaccination of Cats. Nineteen, nine (9) month old cats (specific pathogen-free, Harlan Sprague Dawley, Madison, WI) were used. Cats were divided into three groups and vaccinated twice, 21 days apart as follows: Group 1 (n=5) was vaccinated subcutaneously with 3 mL of rRCNV-FIPV N ($10^{7.44}$ TCID$_{50}$/cat), Group 2 (n=4) was vaccinated subcutaneously with 3 mL of a 1:10 dilution of rRCNV-FIPV N ($10^{6.44}$ TCID$_{50}$/cat), and Group 3 (n=10) were vaccinated subcutaneously with wild type RCNV ($10^{6.92}$ TCID$_{50}$/cat) and served as the negative controls.

Challenge of Cats with FECV and FIPV. Two weeks following the second vaccination, all of the vaccinates and five of the control cats were orally inoculated with $10^{3.4}$ TCID$_{50}$ of FECV strain 79-1683. This virus induces a subclinical infection which can cause enhancement of subsequent FIPV infection. Three weeks later, these cats were orally challenged with $10^{3.4}$ TCID$_{50}$ of FIPV (strain 79-1146, ATCC VR-990). Cats were monitored daily for 64 days after challenge for signs of clinical disease including: fever, icterus, anorexia, depression, dehydration, and peritoneal swelling. Cats were monitored weekly after challenge for leukopenia, anemia, and weight loss. Cats deemed moribund were euthanized by the attending veterinarian and post-mortem pathological examination performed. Clinical disease signs were scored as shown in Table 1. The clinically healthy full dose rRCNV-FIPV N vaccinates and five additional control cats (exposed to FECV as described above) were challenged with FIPV a second time, 98 days after the first inoculation with 79-1146, using the same methods. Cats were monitored for clinical disease as described above.

Antibody Assays. FIPV-neutralizing activity in heat-inactivated cat sera was determined as described previously[17] using CRFK cells in a 96-well microplate assay. Titers were expressed as the reciprocal of the highest dilution of serum that inhibited 50% of viral cytopathic effects. Antibody responses to FIPV nucleocapsid were detected by Western blot against FIPV antigens using methods previously described[18] and anti-cat IgG substituted as the secondary antibody.

Table 1. Scoring system for clinical signs following FIPV challenge

Clinical Sign		Score/Observation
Fever	103.0 to 103.9 °F	1 point*
	104.0 to 104.9 °F	2 points
	≥105.0 °F	3 points
Dehydration		1 point
Depression		1 point
Anorexia		1 point
Peritoneal Swelling		1 point
Icterus		1 point
Weight Loss	20% to 29% decrease	1 point
	30% to 49% decrease	2 points
	>50% decrease	5 points
Leukopenia	Decrease of ≥50%	3 points
	Total Count <6000/μL	2 points
Anemia	Hematocrit <25%	3 points
Moribund/Death		25 points

*Fever was not scored until >1°F above baseline

RESULTS

Cats were immunized with two doses of rRCNV-FIPV N or wild-type RCNV (controls). The full dose rRCNV-FIPV N vaccinates and control cats were monitored for the presence of antibodies against the FIPV nucleocapsid protein by Western blot. Antibodies against the FIPV nucleocapsid protein were detected in vaccinated, but not control cats (data not shown). Serum antibodies capable of neutralizing FIPV virus were not detected in either recombinant rRCNV-FIPV N vaccinates or control cats after immunization.

Two weeks after the second vaccination, cats were inoculated with feline enteric coronavirus (FECV). This virus caused a subclinical infection in the cats. Prior infection with FECV has been shown to induce enhanced disease when cats are subsequently infected with FIPV (data not shown). Three weeks following FECV challenge, cats were orally inoculated with FIPV virus. Infection with FIPV induced a rapid onset of clinical disease in control cats characterized by fever, depression, dehydration, icterus, anorexia, weight loss, anemia, and leukopenia (Table 2). Four out of five control cats were deemed moribund by the attending veterinarian and were humanely euthanized between 14 and 64 days after virus inoculation. Post-mortem examination revealed both wet (effusive) and dry (non-effusive) forms of FIP with typical pathological lesions in multiple organs of the control cats (Table 3). Similar FIP disease was noted in the 1/10th dose rRCNV-FIPV N vaccinated cats after challenge (Tables 2 and 3, Figure 1) On the other hand, clinical disease was essentially absent in the full dose rRCNV-FIPV N vaccinates (Table 2). The full dose vaccinates showed a statistically significant reduction in clinical signs ($p<0.05$, by ANOVA) and death ($p<0.01$, by Chi Square Analysis) when compared to the control cats (Figure 1).

The five surviving, clinically healthy, vaccinates were rechallenged with FIPV 79-1146 at 98 days after the initial FIPV inoculation. Five cats, that had been RCNV vaccinated and exposed to FECV as described before, were included as controls in this phase of the study. Four out of five control cats developed severe clinical disease and were euthanized between 18 and 26 days after FIPV challenge (Table 4, Figure 1). Post-mortem

Table 2. Total clinical score following initial challenge with FIPV

CAT ID	Fever	Weight Loss	Leukopenia	Anemia	Clinical Signs*	Death	Total Score
Full Dose RCNV-FIPV N Vaccinates							
1260	0	0	0	0	0	0	0
1262	1	0	0	0	0	0	1
1264	1	0	0	3	0	0	4
1266	0	0	0	0	0	0	0
1268	2	0	0	0	0	0	2
1/10 Dose RCNV-FIPV N Vaccinates							
1313	29	5	9	3	36	25	107
1315	17	0	0	0	20	25	62
1317	3	0	3	0	9	25	40
1319	1	0	0	0	6	0	7
Controls (RCNV Vaccinated)							
1321	39	3	18	3	63	25	151
1323	5	0	2	0	14	25	46
1327	0	1	2	0	15	25	43
1329	5	1	6	0	11	25	48

Table 3. Type of clinical disease and onset of death following initial FIPV challenge

CAT ID	Disease form	Organs involved	Onset of death* (days)
1/10 Dose RCNV-FIPV N Vaccinates			
1313	Effusive	Lungs, Thoracic Cavity, Liver	42
1315	Effusive	Lungs, Spleen, Peritoneum, Kidneys	61
1317	Effusive	Lungs, Thoracic Cavity, Liver, Mesentery, Omentum	19
1319	N/A	N/A	N/A
Controls			
1321	Non-effusive	Lungs, Spleen, Kidney	64
1323	Non-effusive	Lungs, Spleen, Large Intestine, Mesentery, Liver, Spleen	22
1327	Effusive	Pericardium, Throacic Cavity, Lungs, Omentum, Mesentery, Kidney	22
1329	Effusive	Fascia, Lungs, Liver, Lymph Nodes	14
1337	N/A	N/A	N/A

N/A = Not Applicable
*Number of days between FIPV challenge and death

examination findings were consistent with systemic, pathological FIP disease. On the other hand, a second FIPV inoculation, 98 days after the first FIPV challenge, induced fatal disease in only 1/5 (20%) of the vaccinates. The ill vaccinate died 14 days after the second FIPV challenge of a non-effusive form of infectious peritonitis (Figure 1). Clinical disease signs were essentially absent in the other four vaccinates (Table 4).

The surviving vaccinates were held and monitored weekly for outward clinical signs of disease for a total of 158 days following the second FIPV challenge (256 days after the

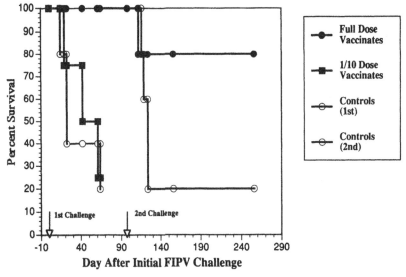

Figure 1. Survival curves following FIPV challenge The survival curves for full dose rRCNV-FIPV N vaccinates, 1/10th Dose rRCNV-FIPV N vaccinates, and controls are shown following the initial inoculation with virulent FIPV on day 0 Survival curves for the full dose rRCNV-FIPV N vaccinates and additional control cats are shown following a second FIPV challenge (on day 98) and monitoring through 256 days after the initial challenge

Table 4. Total clinical scores following second challenge with FIPV

CAT ID	Fever	Weight Loss	Leukopenia	Anemia	Clinical Signs*	Death	Total Score
Full Dose RCNV-FIPV N Vaccinates							
1260	0	0	0	0	0	0	0
1262	2	0	0	0	0	0	2
1264	0	0	0	0	0	0	0
1266	15	0	3	0	12	25	55
1268	0	0	0	0	0	0	0
Controls (RCNV Vaccinated)							
D016	26	0	0	0	12	25	65
D020	12	0	0	0	3	25	40
D028	6	0	3	0	0	0	9
D043	14	0	0	0	11	25	50
AMW2	5	0	0	0	5	25	35

*Clinical Signs Include Depression, dehydration, anorexia, peritoneal swelling, and icterus

first FIPV challenge) During this time, the cats showed no clinical signs of infectious peritonitis At the end of this extended observation period, the cats were euthanized and a gross pathological examination was performed No signs of tissue nor organ damage indicative of FIPV infection were found in these vaccinates at the time of necropsy

DISCUSSION

The development of an effective vaccine against FIPV has been difficult Vaccine formulations using attenuated-live and inactivated FIPV[7 8] and other closely related coronaviruses[9 12] have failed to provide protection against FIPV challenge and, in some cases, have actually enhanced the disease Unsuccessful protection and disease enhancement was also observed in cats vaccinated with a recombinant vaccinia virus expressing the FIPV S glycoprotein [13] It has been suggested that the accelerated disease progression and failed protection of these vaccines was related to the humoral immune response to the FIPV S protein [19 20] In particular, antibodies to the spike protein have been shown to enhance FIPV infection of macrophages [19] These studies suggest that exclusion of the S protein from a vaccine formulation, may avoid the induction of immune-enhancement that can be induced by this protein

This report demonstrates the ability of a recombinant RCNV expressing the FIPV nucleocapsid protein to induce protection against clinical disease induced by this coronavirus In the challenge model used, both effusive and non-effusive forms of FIP disease were induced and death was consistently found in 80% of the controls Control cats were vaccinated with wild type RCNV and, therefore, non-specific responses related to raccoon poxvirus alone did not play a role in the protection noted Further, the protection against FIP was correlated with the dose of rRCNV-FIPV N administered $10^{7\,44}$ $TCID_{50}$ induced protective immunity, whereas $10^{6\,44}$ $TCID_{50}$ did not In addition to the prevention of mortality, the vaccine appeared to prevent persistent, subclinical infection of tissues as demonstrated by a lack of detectable gross lesions in tissues at 256 after initial FIPV inoculation

Previous attempts to vaccinate cats with a recombinant vaccinia virus expressing the FIPV N protein failed to induce strong protective immunity to FIPV[21] There may be several reasons for the observed differences in efficacy between the recombinant vaccinia and raccoon poxviruses expressing FIPV N, which could include 1) differences in the physical properties of the expressed N proteins, 2) better replication of raccoon poxvirus compared to vaccinia virus in cats, 3) altered magnitude or form of immune response to raccoon poxvirus compared to vaccinia virus in cats, and 4) increased expression of the recombinant proteins by raccoon poxvirus *in vivo* Differences in the protection against FIPV may also be related to the route of vaccination and the type of challenge model used

Vaccination with the rRCNV-FIPV N failed to induce detectable antibodies capable of neutralizing FIPV These data suggest that vaccination with rrRCNV-FIPV N provided protection against FIPV by a mechanism other than a humoral immune response consisting of virus neutralizing antibodies Future studies will be necessary to elucidate the mechanism of protection induced by this virus vaccine

REFERENCES

1 Lai, M M C 1990 Coronavirus Organization, replication, and expression of genome Annu Rev Microbiol 44 303-333

2 Spaan, W, Cavanagh, D, and Horzinek, M C 1988 Coronaviruses Structure and genome expression J Gen Virol 69 2939-2952

3 Boyle, J F, Pedersen, N C, Everman, J F, McKeirnan, A J, Ott, R L, and Black, J W 1984 Plaque assay, polypeptide composition and immunochemistry of feline infectious peritonitis virus and feline enteric coronavirus Adv Exp Med Biol 173 133-147

4 Horzinek, M C, Ederveen, J, Egberink, H, Jacobse-Geels, H E L, Niewold, T, and Prins, J 1986 Virion polypeptide specificity of immune complexes in cats inoculated with feline infectious peritonitis virus Amer J Vet Res 47 754-761

5 De Groot, R J, Van Leen, R W, Dalderup, M J M, Vennema, H, Horzinek, M C, and Spann, W J M 1989 Stably expressed FIPV peplomer protein induces cell fusion and elicits neutralizing antibodies in mice Virol 171 493-502

6 Vennema, H, Rottier, P J M, Heijnen, L, Godeke, G J, Horzinek, M C, and Spaan, W J M 1990 Biosynthesis and function of the coronavirus spike protein Adv Exp Med Biol 276 9-19

7 Pedersen, N C 1987 Virologic and immunologic aspects of feline infectious peritonitis virus infection Adv Exp Med Biol 218 529-550

8 Pedersen, N C, and Black, J W 1983 Attempted immunization of cats against feline infectious peritonitis using a virulent live virus or sublethal amonts of virulent virus Am J Vet Res 44 229-234

9 Pedersen, N C, Boyle, J F, Floyd, K, Fudge, A, and Barker J 1981 An enteric coronavirus infection of cats and its relationship to feline infectious peritonitis Am J Vet Res 42 368-377

10 Pedersen, N C, Evermann, J F, McKeirnan, A J, and Ott, R L 1984 Pathogenicity studies of feline coronavirus isolates 79-1146 and 79-1683 Am J Vet Res 45 2580-2585

11 Stoddart, C A, Barlough, J E, Baldwin, C A, and Scott, F W 1988 Attempted immunization of cats against feline infectious peritonitis using canine coronavirus Res Vet Sci 45 383-388

12 Woods, R D, and Pedersen, N C 1979 Cross-protection studies between feline infectious peritonitis and porcine transmissible gastroenteritis viruses Vet Microbiol 4 11-16

13 Vennema, H, De Groot, R J, Harbour, D A, Dalderup, M, Jones, T G, Horzinek, M C, and Spaan, W J M 1990 Early death after feline peritonitis virus challenge due to recombinant vaccinia virus immunization J Virol 64 1407-1409

14 Dale, B, Yamanaka, M, Acree, W M, Chavez, L Felin infectious peritonitis virus diagnostic tools European Patent Application 0, 376,744, Published July 4, 1990

15 Mackett, M, Smith, G L, and Moss, B 1982 Vaccinia virus A selectable eukaryotic cloning and expression vector Proc Natl Acad Sci USA 79 7415-7419

16 Stoddart, C A, and Scott, F W 1988 Isolation and identification of feline peritoneal macrophages for in vitro studies of coronavirus-macrophage interactions J Leukocyte Biol 44 319-328

17 Ingersoll, J D, and Wylie, D E 1988 Comparison of serologic assays for measurement of antibody response to coronavirus in cats Am J Vet Res 49 1472

18. Chu, H-J., Chavez, L.G., Blumer, B.M., Sebring R.W., Wasmoen, T.L., and Acree, W.M. Immunogenicty and efficacy study of a commercial *Borrelia burgdorferi* bacterin. J. Am. Vet. Med. Assoc. 201:403-411, 1992.

19. Olson, C.W., Corapi, W.V., Ngichabe, C.K., Baines, J.D., and Scott, F.W. 1992. Monoclonal antibodies to the spike protein of feline infectious peritonitis virus mediate antibody-dependent enhancement of infection of feline macrophages. J. Virol. 66: 956-965.

20. Weiss, R.C., and Scott, F.W. 1981. Antibody-mediated enhancement of disease in feline infectious peritonitis: Comparison with dengue hemorrhagic fever. Comp. Immunol. Microbiol. Infect. Dis. 4: 175-189.

21. Vennema, H., De Groot, R.J., Harbour, D.A., Horzinek, M.C., and Spaan, W.J.M. 1991. Primary structure of the membrane and nucleocapsid protein genes of feline infectious peritonitis virus and immunogenicity of recombinant vaccinia viruses in kittens. Virol. 181: 327-335.

37

EFFICACY OF AN INACTIVATED VACCINE AGAINST CLINICAL DISEASE CAUSED BY CANINE CORONAVIRUS

R. Fulker, T. Wasmoen, R. Atchison, H-J. Chu, and W. Acree

Fort Dodge Laboratories
Fort Dodge, Iowa

ABSTRACT

Canine Coronavirus (CCV) is a causative agent of diarrhea in dogs. The reproduction of severe clinical disease with experimental CCV infection has been difficult. We have recently developed a CCV challenge model which reproduced clinical signs of disease in susceptible dogs. The following study was designed to determine whether immunization with an inactivated CCV vaccine would protect dogs from clinical disease induced using this model. Dogs (n=13) were vaccinated with an inactivated CCV vaccine. Vaccinates and controls (n=5) were orally inoculated with virulent CCV virus and treated with dexamethasone on days 0, 2, 4, and 6 after virus challenge. Control dogs developed clinical signs including diarrhea, dehydration, anorexia, depression, and nasal and ocular discharge. Diarrhea was noted in 80% of the controls and 60% progressed to a severe watery or bloody diarrhea that persisted for multiple days. Conversely, only 2/13 (15%) vaccinates developed mild diarrhea and none developed bloody diarrhea. The control dogs averaged 10.8 days of diarrhea compared to 1.4 days for vaccinates over the 21 day observation period. In addition to reduced clinical signs, the number of days of virus shedding and the level of CCV in feces was different for controls (100% shed virus) and vaccinates (38% shed virus). This study demonstrates that vaccination with an inactivated CCV vaccine can significantly reduce not only viral replication, but the occurrence of clinical disease following a virulent CCV infection.

INTRODUCTION

Coronavirus was first recognized as a disease agent in dogs in 1971 when it was isolated during an epizootic of diarrhea in Germany.[1] Canine coronavirus (CCV) was subsequently isolated from outbreaks of canine viral enteritis in the United States in 1978[2] and serological surveys have identified the prevalence of coronavirus exposure in dogs in numerous countries.[3-6] Although the disease noted in field cases of canine coronavirus infection may be severe,[1] it has been difficult to induce significant clinical signs using

Corona- and Related Viruses, Edited by P. J. Talbot and G. A. Levy
Plenum Press, New York, 1995

experimental models.[7,8] In this report, enhanced clinical disease was induced when gluco-corticoids were administered in conjunction with CCV inoculation of dogs. This laboratory model was used to demonstrate the effectiveness of an inactivated vaccine in preventing clinical disease in dogs.

MATERIALS AND METHODS

Animals. Mixed breed dogs, 16-20 weeks in age and of both sexes were used. Dogs were confirmed to lack serum antibodies against canine coronavirus (by serum neutralization assay as previously described[8]), as an indication of no prior virus exposure, before initiation of the study.

Vaccine. A commercially available inactivated CCV vaccine (Duramune® CvK, Fort Dodge Laboratories, Fort Dodge, IA) was used.

Viruses. The I-71 isolate[1] and the SA4 isolate (American BioResearch, Seymour, TN) of CCV at low culture passage were used for challenge. Viruses were grown on Crandell Feline Kidney (CRFK) cells and pooled in approximately equal proportions to yield a challenge pool. The challenge pool had a titer of $10^{6.3}$ $TCID_{50}$/mL on CRFK cells.

Detection of Virus Shedding. A minimum of 0.2 grams of feces was collected from each dog daily from -1 to 21 days after CCV inoculation. Fecal material was diluted with 1.8 mL of Hanks Balanced Salt Solution containing gentamicin (100 µg/mL). Fecal suspensions were vortexed and clarified by centrifugation (500 x g, 20 minutes). Supernatant fluids were inoculated onto Crandell Feline Kidney (CRFK) cells and incubated for 7 days a 37°C. Virus was detected by direct immunofluorescence using a fluorescein conjugated antibody (American BioResearch, Seymour, TN).

Experimental Design. Dogs were divided into vaccinated (n=13) and control (n=5) groups. Vaccinated dogs were immunized with two doses of vaccine given 21 days apart (Serial #147179A for first vaccination and Serial #147177A for second vaccination) by either subcutaneous injection in the scruff of the neck (n=6) or intramuscular injection in a rear leg (n=7). Fourteen days after the second vaccination, immunized and control dogs were inoculated orally with 6 mL of the CCV challenge pool ($10^{7.1}$ $TCID_{50}$ per dog). Dogs were injected intramuscularly with 2 mg of dexamethasone (Azium,® Schering Plough, Kenilworth, NJ; average of 0.12 mg/kg of body weight) on the day of virus inoculation and again on days 2, 4, and 6 after challenge. Dogs were monitored for fever (rectal temperature), weight, and clinical signs of disease from -3 to 21 days after CCV inoculation. Clinical signs were scored as outlined in Table 1. Clinical signs of disease in vaccinated animals were compared to controls using Mann Whitney Ranked Sum Analysis (Statview II, Abacus Concepts, Inc.) using a Macintosh PC.

RESULTS

Dogs were vaccinated with two doses of an adjuvanted, inactivated CCV vaccine administered by intramuscular (n=6) or subcutaneous (n=7) injection, 21 days apart. Fourteen days after the second immunization, vaccinated and control dogs were challenged orally with a pool of two virulent canine coronaviruses. A mildly immunosuppressive dose of dexamethasone was administered on the day of CCV challenge and 2, 4, and 6 days later. Clinical signs of CCV-induced disease including diarrhea, depression, dehydration, inappetance, nasal discharge, and ocular discharge were noted in control dogs between 2 and 21 days after virus inoculation. In previous studies, dexamethasone administration in the absence of CCV inoculation did not result in these types of clinical signs in dogs.[9] Canine coronavirus was detected in fecal samples from all control dogs with a duration of 1 to 6 days. Virus shedding was detected between days 3 and 8 after virus inoculation and preceded

Table 1. Scoring of clinical signs following virulent CCV inoculation

Sign		Score/Observation
Fever	103.5-103.9 °F	1 point
	104.0-104.9 °F	2 points
	≥105.0 °F	4 points
Diarrhea	Soft, Mucous	1 point
	Watery	2 points
	Bloody	3 points
Dehydration		1 point
Inappetance		1 point
Nasal/Ocular Discharge		1 point
≥10% Weight Loss		3 points

the days of peak clinical signs (days 5 through 12, see Figure 1). Significant diarrhea was noted in 80% of the control dogs and progressed to severe, watery diarrhea in three out of five of these animals. No significant weight loss was noted in control dogs during the observation period after CCV challenge.

Vaccinated animals, on the other hand, showed few clinical signs of disease (Figure 1). CCV shedding was only noted in 5/13 (38%) vaccinated animals compared to 5/5 (100%) controls. When clinical signs in vaccinates were compared to controls (Figure 2), statistically significant reductions were noted in the occurrence of dehydration (100% reduction, $p<0.001$), diarrhea scores (90% reduction, $p<0.02$), number of days of diarrhea (86% reduction, $p = 0.02$), and number of days of virus shedding (87% reduction, $p = 0.003$). Vaccinates had fewer signs of inappetance, nasal discharge, and ocular discharge than controls, although changes in these clinical signs were not statistically significant. The average cumulative clinical score for

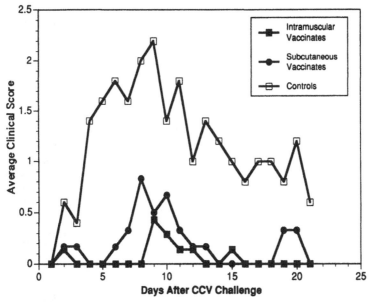

Figure 1. Average daily clinical scores following CCV challenge. The average daily clinical score (calculated as shown in Table 1) for vaccinated and control animals is shown for the 21 day period following CCV inoculation.

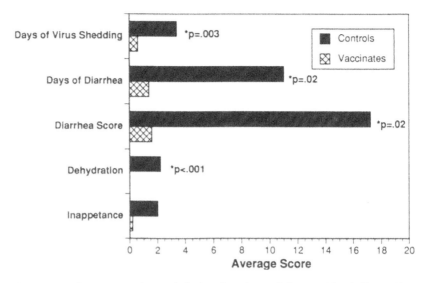

Figure 2. Comparison of vaccinates and controls for key clinical signs following CCV challenge The average scores (calculated as shown in Table 1) for vaccinates as compared to controls are shown for key clinical signs following CCV challenge including. Inappetance, dehydration, diarrhea The average number of days of diarrhea and virus shedding are also shown for vaccinates and controls *Statistical significance values are shown for vaccinates compared to controls as determined by Mann Whitney Ranked Sum Analysis

intramuscular vaccinates was 1.3 and for subcutaneous vaccinates was 4.2, which were both significantly lower than the 24.8 average score in controls (Figure 3).

DISCUSSION

Canine coronavirus was discovered as the causative agent of gastroenteritis in dogs.[1] However, the importance of this pathogen has been questioned because of difficulty in

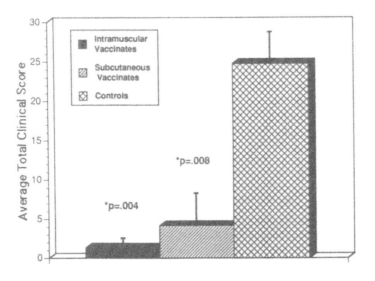

Figure 3. Average total clinical scores following CCV challenge The cumulative clinical score for each animal was summed and the group averages (± SE) are shown for intramuscular vaccinates, subcutaneous vaccinates, and controls *Statistical significance values are shown for vaccinates compared to controls as determined by Mann Whitney Ranked Sum Analysis

reproducing clinical disease in dogs under experimental conditions. Conditions causing stress in dogs, such as poor nutrition or overcrowding, and concurrent infections with other agents have been hypothesized to contribute to the severity of disease that has been documented with CCV in field cases. In support of the latter hypothesis, experimental inoculation of dogs with CCV in combination with canine parvovirus has been shown to be fatal under conditions where either virus by itself caused a self-limiting disease.[10] Because canine parvovirus can infect cells of the immune system, this study also suggested that CCV-induced disease may be augmented by immunosuppression.

In the study reported here, clinical disease was induced in dogs inoculated with CCV and the immunosuppressive drug dexamethasone. The clinical signs observed in this study included diarrhea, dehydration, depression, inappetance, nasal discharge and ocular discharge. In previous studies, where dexamethasone treatment was used in the absence of CCV inoculation, such clinical signs were not observed in dogs.[9] The clinical disease found in control dogs in this study resembles the findings of other investigators who have inoculated young puppies (<6 weeks old) with CCV alone.[1,8] However, this is the first report where significant clinical signs have been induced in older dogs. This model of experimental CCV infection supports the hypothesis that stress, which also can cause immunosuppression, may play an important role in the severity of CCV infection under natural conditions.

Previous vaccine studies have relied on the demonstration of reduced intestinal infection and CCV shedding as parameters of vaccine efficacy.[11] In this case, diminished infection rates were assumed to predict a reduced disease incidence in vaccinated animals. In the study reported here, the ability of an inactivated CCV vaccine to prevent disease was directly assessed. Using this CCV infection model, it was found that immunization resulted in a reduction in the occurrence and severity of diarrhea, inappetance, dehydration, depression, nasal discharge, ocular discharge, and virus shedding when controls were compared to vaccinates after CCV challenge. Therefore, this study shows that an inactivated vaccine is capable of inducing an immune response which will protect against clinical signs caused by CCV. This experimental model may provide a means to more fully assess the mechanism of immune protection against CCV infection and disease.

REFERENCES

1 Binn, LN, Lazar E C, Keenan K P, Huxsoll D L, Marchwicki R H, Strano A J 1974 Recovery and characterization of a coronavirus from military dogs with diarrhea Proceedings of the 78th Meeting of the U S Animal Health Association, pp359-366

2 Appel M, Meunier P, Pollock R, et al 1980 Canine Viral Enteritis Canine Pract 7:22-36

3 Miller J, Evermann J, Ott, R 1980 Immunofluorescence test for canine coronavirus and parvovirus West Vet 18 14-19

4 Tuchiya, K, Horimoto, T, Azetka, M, Takahashi, E, Konishi, S 1991 Enzyme-linked immunosorbent assay for the detection of canine coronavirus and its antibody in dogs Vet Microbiol 26 41-51

5 Herbst, W, Zhang, X M, and Schliesser, T 1988 The seroprevalence of coronavirus infections in the dog in West Germany Berl Munch Tierarztl Wochenschr 101 381-383

6 Rimmelzwann, G F, Groen, J, Egberink, H, Borst, G H, UtydeHaag, F G, Osterhaus, A D 1991 The use of enzyme-linked immunosorbent assay systems for serology and antigen detection in parvovirus, coronavirus and rotavirus infections in dogs in the Netherlands Vet Microbiol 26 25-40

7 Appel M 1987 Canine coronavirus In Virus Infections of Carnivores Ed M Appel, Elsevier, New York, pp 115-122

8 Tennant, B J, Gaskell, R M, Kelly, D F, Carter, S D 1991 Canine coronavirus infection in the dog following oronasal inoculation Res Vet Sci 51 11-18

9 Wasmoen, T L, Sebring, R W, Blumer, B M, Chavez, L G, Chu, H-J, Acree, W M 1992 Examination of Koch's postulates for Borrelia burgdorferi as the causative agent of limb/joint dysfunction in dogs with borreliosis J Am Vet Med Assoc 201 412-418

10 Appel, M J G 1988 Does canine coronavirus augment the effects of subsequent parvovirus infection? Vet Med 83 360-366

11 Acree, W M , Edwards, B , Black, J W Canine Coronavirus Vaccine International Application Published under the Patent Cooperation Treaty, International publication date 3rd January 1985, number WO 85/00014, World Intellectual Property Organisation

38

CLONING AND EXPRESSION OF FECV SPIKE GENE IN VACCINIA VIRUS

Immunization with FECV S Causes Early Death after FIPV Challenge

S. Klepfer,[1] A.P. Reed,[1] M. Martinez,[2] B. Bhogal,[2] E. Jones,[1] and T. J. Miller[3]

[1] Department of Molecular Biology, Smithkline Beecham Animal Health
P.O. Box 1539, King of Prussia, Pennsylvania
[2] Zonagene, Inc.
2408 Timberloch Place, B4, The Woodlands, Texas
[3] Smithkline Beecham Animal Health
601 W. Cornhusker Hwy., Lincoln, Nebraska

ABSTRACT

The spike gene of the feline enteric coronavirus (FECV), strain FECV-1683, was PCR amplified from total RNA extracted from FECV-infected cells and its sequence determined. A primary translation product of 1454 amino acids is predicted from the nucleotide sequence, containing a N-terminal signal sequence, a C-terminal transmembrane region and 33 potential N-glycosylation sites. The sequence shares 92% homology with the previously published feline infectious peritonitis virus, strain WSU-1146; however, several regions were identified that distinguished FECV from Feline Infectious Peritonitis virus, FIPV. The full length FECV S gene was cloned and expressed in vaccinia virus. Recombinants produced a 200 kD protein which was recognized by sera from cats infected with FIPV. When kittens were immunized with the vaccinia/FECV S recombinant, neutralizing antibodies to FIPV were induced. After challenge with a lethal dose of FIPV, the recombinant vaccinated animals died earlier than control animals immunized with vaccinia virus alone.

INTRODUCTION

Two laboratory strains of feline coronavirus, feline infectious peritonitis virus (FIPV WSU 79-1146 and a closely related Nor15/DF2) and feline enteric coronavirus FECV 1683 have been studied extensively in the literature [1-5]. WSU 79-1146, although antigenically similar to FECV 1683 exhibits a different type of pathogenesis. The hallmark of FIPV infection is an immune-mediated peritonitis which can be accelerated experimentally by previous exposure to the virus [2]. Cats with FIP disease typically have high circulating antibody titers as well as deposition of antigen-antibody complexes in many major organs. Susceptibility of macrophages/monocytes to virus infection is believed key to the widespread immunopathology observed following FIPV infection[6].

The spike protein appears to be largely the focus of the immune mediated pathology in that the other major structure proteins (nucleocapsid and membrane) cannot induce sensitization to early death syndrome [7]. Vennema et al., reported that a lethal FIPV challenge of kittens immunized with a vaccinia recombinant expressing the FIPV S protein developed FIP disease and died more rapidly than non-vaccinated animals following virulent FIPV challenge [8] More recently, Horsburgh et al., [9] showed that immunization with vaccinia recombinants expressing either the CCV or TGEV S protein accelerated the death of kittens challenged with a lethal dose of FIPV indicating a conserved feature of the spike protein immune response that predisposes cats to FIP.

The differences between FIPV-like and FECV-like virus may not be do to the genetic specificity of the virus, but rather, to the site of replication. Because the cell tropism of FECV 1683 differs dramatically from FIPV, the sequence of the FECV (1683) spike gene was determined and compared to that of the FIPV WSU 79-1146 and Nor15/DF2. FECV 1683 does not cause accelerated death nor FIP symptoms upon oral/nasal exposure of the virus to 12 week old kittens, in contrast Nor15/DF2 can cause acute symptoms with only a single exposure when administered by this route. To examine whether FECV spike protein retains the specificity of immune response pathology as its antigenically similar sister FIPV-DF2 the spike proteins of both were expressed using vaccinia virus vectors and compared side by side in a cat immunization study[7].

MATERIALS AND METHODS

Materials. All restriction enzymes were purchased from New England Biolabs (Beverly, MA) or Bethesda Research Labs (Gaithersburg, MD) and used according to manufacturer's specifications. Bluo-Gal was obtained from Sigma. All PCR reagents were produced by Perkin Elmer-Cetus (Norwalk, CT). Alkaline-phosphatase labeled antibodies and the BCIP/NBT substrate system were received from Kirkegaard-Perry Laboratories, Inc. Anti-FIPV serum was produced at SmithKline Beecham Animal Health (SBAH), Lincoln, Nebraska in cats that were vaccinated with a temperature sensitive DF2 strain of FIPV (Primucell[R] SmithKline Beecham Animal Health) and challenged with DF2 FIPV [10]. POTSKF33 was obtained from Dr. Christine DeBouck, SB Pharmaceuticals.

Cells and Virus Strains Norden Laboratories Feline Kidney (NLFK) cells and Type II FIPV Nor15/DF2 were grown and maintained as described previously [10,11]. A feline enteric coronavirus, FECV (WSU-1683), was obtained from Washington State University. Vaccinia virus strain WR was received from Dr. Bernard Moss, NIH.

For vaccinia virus infections, human thymidine-kinase-negative (HuTK-) and African green monkey kidney (CV-1) cell lines were used and maintained in Dulbecco's modified Eagle's medium (DMEM) containing 10% fetal bovine serum.

RNA purification. Roller bottles of confluent NLFK cells were infected with either DF2 FIPV or FECV virus at MOI = 0.1 in 50 ml of BME supplemented with 2% FBS. DF2 FIPV infections were performed in serum-free medium. The virus was absorbed for 2 hours and then 250 ml of growth medium added. The cultures were monitored for cytopathic effect (CPE) and typically harvested at 24 - 36 hours post-infection. Total cytoplasmic RNA was prepared from the infected monolayers by guanidine isothiocyanate extraction [12]

Oligonucleotide Design and Synthesis. Oligonucleotides were designed based on the nucleotide designation of the WSU 1146 FIPV S gene (GenBank Accession # X06170) using *Sma* I and *Stu* I sequences at the ends to facilitate cloning into pSC11. The primers were synthesized on an Applied Biosystem Model 380B DNA Synthesizer using the phosphoramidite method and were gel-purified prior to use.

PCR Amplification. Synthesis of cDNA from total RNA isolated from cells infected with a specific coronavirus was performed using standard procedures. Amplification of the cDNA was performed essentially according to the method of Saiki *et al.*, [12] using the *Taq* polymerase. The reaction was performed in the Perkin-Elmer Cetus thermal cycler for one cycle by denaturing at 95°C for 1', annealing at 37°C for 2' followed by an extension at 72°C for 40'. A standard PCR profile was then performed by a 95°C-1' denaturation, 37°C-2' annealing, 72°C-3' extension for 40 cycles. A final extension cycle was done by 95°C-1' denaturation, 37°C-2' annealing, 72°C-15' extension and held at 4°C until analyzed. PCR products were analyzed by electrophoresing 5.0 μl of the reaction on a 1.2 % agarose gel run 16-17 hours. Bands were visualized by ethidium bromide staining the gel and fluorescence by UV irradiation at 256 nm.

Cloning and sequencing of the FECV S gene. DNA sequence was determined from overlapping cloned regions of the FECV S gene. PCR-amplified DNAs were digested with *Xma*I and *Stu*I, excised and eluted from low-melting temperature gels and ligated into pBluescript vector (Stratagene). Insert-bearing clones were identified by restriction mapping. Nested set deletions were prepared and the sequence determined by Lark Sequencing Technologies, Houston TX, from both strands using the chain termination method [14.] DNA sequence analysis was performed using the University of Wisconsin GCG package of programs.

Cloning of Full-length FIPV and FECV S genes in pSC11. The full-length FECV S gene was amplified by PCR, digested with *Sma*I and *Stu*I, excised as above amd ligated into a bacterial expression plasmid, pOTSKF33, using standard procedures. Full-length DF2 FIPV and FECV spike gene 1-1454aa inserts were isolated from established pOTSKF33 plasmid clones by *Sma*I/*Stu*I digestion of plasmid DNA and the excised gene cloned into the *Sma*I site of the vaccinia recombination plasmid, pSC11.

Western Blots of virus expressed proteins. HuTK- monolayers infected with recombinant viruses were harvested at 2 days post-infection by scraping into the medium. Pelleted cells were washed with PBS and resuspended in 0.6 ml RIPA buffer (0.15 M NaCl, 0.1% SDS, 1.0% Na deoxycholate, 1.0% Triton X-100, 5 mM EDTA, 20 mM Tris, pH 7.4). RIPA lysates were frozen/thawed 3X and sonicated briefly. Non-reducing sample buffer (2% SDS, 80 mM Tris, pH 6.8, 10% glycerol, 0.02% bromophenol blue) was added and samples boiled 10 min prior to electrophoresis on 10% SDS-polyacrylamide gels as described by Laemmli. Proteins were transferred to Immobilon-P (Millipore) at 20-25 mA for 18 h in Tris/Glycine/Methanol buffer. Filters were blocked in 2% skim milk, 1% gelatin, and TBS (20 mM Tris, pH 7.5, 500 mM NaCl) for 1-2 hours (h) at room temperature (RT), rinsed with TTBS (TBS + 0.05% Tween-20) and incubated with anti-FIPV cat serum (from cats vaccinated twice with Primucell[R] licensed temperature sensative DF2 FIPV vaccine) at a 1:50 dilution in TTBS and 1% gelatin for 1-2 h at RT. Filters were washed in TTBS 3X for 10 min each and incubated with goat anti-cat alkaline-phosphatase labeled IgG at a 1:1000 dilution for 1 h. Filters were washed as before and incubated 5 - 15 min in BCIP/NBT substrate according to the manufacturer's instructions. Filters were then rinsed in water and air-dried.

RESULTS

The sequence comparison of the FECV 1683 spike gene (Genbank accession number) and WSU 79-1146 (accession number) are shown in Table 1. The gene is 4365 bp in length

Table 1. Percent homology comparison between FECV 1683
and WSU 1146 Comparison values represent
percent homology

Strain	WSU 1146	DF2
Nucleotide comparison		
FECV 1683	94 3(255)	94 3(250)
WSU 1146		99 8(15)
Amino Acid Comparison		
FECV 1683	97 3 (s)	97 2(s)
	95 5 (ı) (67)	95 4 (ı) (67)
WSU 1146		99 6 (s)
		99 6 (ı) (8)

Numbers in parentheses denote actual number of
nucleotide or amino acid differences (s) represents percent
amino acid homology based on Dayhoff similarity matrix
(ı) represents percent amino acid identity Predicted
number of glycosylation sites for each amino acid
sequence are FECV 1683=33 WSU 1146=35 DF2=35

and encodes an open reading frame of 1454 aa (~161,160 d molecular weight) As predicted
for a surface glycoprotein, the FECV S gene contained a signal sequence of 14 residues at
the amino terminus, a C-terminal transmembrane domain upstream of a 38 amino acid
cytoplasmic tail and 33 predicted N-glycosylation sites

The the S gene of FECV-1683 and the WSU-1146 strain share 94 3% homology at
the nucleotide level and 95 5% homology at the amino acid level Interestingly, the S gene
of FECV is more closely related to CCV at the gene sequence than to FIPV The FECV S
gene sequence contains 6 extra nucleotides (positions 351 - 356) when compared to the
published WSU-1146 FIPV S gene sequence These two additional amino acids at positions
119 and 120 increase the size of the FECV S gene to 1454 aa in contrast to the 1452 aa
reported for WSU-1146 Overall, only 66 amino acid differences were observed between the
FECV and WSU-1146 FIPV S sequences (95 5% identity) The greatest area of heterogeneity
is in the first 300 amino acids of the N terminal reagion, 89 6% identity, while in the
remainder of the gene the homology increased to 97 1%

A recombinant vaccinia virus expressing the full length FECV S gene was generated using standard transfection procedures [15] As a control, a recombinant containing
the full length S gene from WT FIPV virus, strain Nor15/DF2, was also constructed The
presence of coronavirus DNA in each recombinant was confirmed by hybridization
Expression of the spike gene in each recombinant was regulated by the 7 5K early-late
promoter such that S protein would be produced throughout the vaccinia replication cycle
Lysates prepared from cells infected with the vaccinia recombinants were probed for the
production of the spike protein by Western analysis using polyclonal sera to FIPV virus
(Figure 1)

Bands of 180-~200 kD were detected in lysates prepared from vaccinia/FECV or
FIPV infected lysates which co-migrated with spike protein in FIPV-infected cells These
bands were not observed in uninfected cells or in cultures infected with vaccinia alone Both
the size and the diffuse nature of the bands suggest that the FECV spike is glycosylated in
this expression system

Immunization of Kittens 14-week-old SPF kittens, eight per group, were immunized
subcutaneously with 2 X 10⁷ PFU of the vaccinia recombinants expressing feline coronavirus

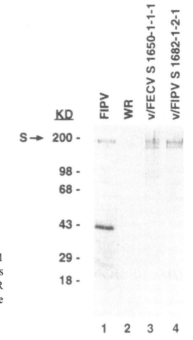

Figure 1. Western blots of recombinant S expressed genes. Lane 1 is purified DF2 virus, lane 2 is wt WR strain of vaccinia, lane 3 is wtWR expressing FECV 1683 spike protein, and lane 4 is wtWR expressing FIPV-DF2 spike protein. FIPV cat sera was used for the blot as described in Materials and Methods.

S genes. A group of eight kittens received the same amount of wild-type WR vaccinia virus (v/WR) as a negative control. Kittens were clinically examined; daily and rectal temperatures were taken. A second immunization with the same amount of virus was given after 3 weeks. Two weeks after the second immunization, kittens were challenged orally with 1 X 10^6 or 1 X 10^4 PFU of WT DF2 FIPV and survival monitored. Virus-neutralization titers were determined on the day of challenge and one and two weeks post-challenge. Serum samples were taken on the days of first and second vaccination, challenge, and post-challenge days 3, 7, 14, 21, and 28.

The kittens appeared uninfected by the immunization regimen and showed modest seroconversion to both vaccinia and to FECV by ELISA after the second immunization. Virus neutralizing titers were also present at 5 weeks post-vaccination (Table 2). Two weeks after the second immunization, kittens were challenged orally with either 1 X 10^6 PFU (high dose) or 1 X 10^4 PFU (low dose) of wild-type (WT) DF2 FIPV and survival monitored. A group of non-vaccinated kittens were also challenged with the low dose of WT DF2 FIPV to serve as infection controls. FIP disease progressed normally in this control group of animals with the first symptoms displayed at 17 -20 days post challenge and a 60% group survival rate after 27 days.

No cats were protected from WT DF2 FIPV challenge regardless of the immunogen but accelerated death was observed as compared to the non-vaccinated controls in groups immunized with vaccinia recombinants expressing either FECV S or DF2 FIPV S. Symptoms of FIP disease were first observed 9 - 11 days after challenge in the animals immunized with the vaccinia/coronavirus recombinants, regardless of the dose of challenge. 100% mortality was observed in the recombinant vaccinated groups as compared with in the non-immunized, challenge controls and in the group immunized with vaccinia alone and challenged with the low dose of WT DF2 FIPV, respectively. Post-mortem examinations were performed on all animals to confirm the diagnosis of FIP.

Table 2. Mortality and ELISA antibody titers of vaccinia virus
immunized FIPV-challenged kittens

Vaccine virus	50% Mortality (days)	Titers by ELISA		
		0 PCD	7 PCD	14 PCD
wtWR low	27	8	2,195	8,240
wtWR high	21	7	2,426	14,380
DF2-S low	18	55	4,963	48,441
DF2-S high	15	84	10,149	65,554
FECV-S low	15	42	3,185	56,661
FECV-S-high	14	53	1,775	11,990
Naive low	23	8	754	15,875

Groups challenged with 10^4 (low) and 10^6 (high) dose of virus is indicated
Post-challenge days (PCD) Mortality is reported as the number of days to 50%
deaths per group IgG serum antibody titers to DF2 FIPV were determined by
ELISA, Virus neutralization titers were also measured using conditions
previously described [10]

DISCUSSION

The FECV 1683 spike protein although isolated from a mildly pathogenic virus,
shares a high degree of homology with the WSU 79-1146 FIPV spike protein, and can equally
sensitize kittens to FIP disease when expressed in the systemic compartment. However, the
low incidence of naturally occuringFIP (0.1% in single cat households and 2-5% in multiple
cat households), yet high number of cats that are seropositive to coronavirus argues against
FECV as a predominant factor in causing FIP in the field. The low incidence of disease must
be attributed to more than the spike protein itself since spike gene sequences from CCV,
TGE and FECV have now been shown to cause accelerated death in cats. This fact coupled
with the high seroprevalence of coronavirus antibodies in cats would indicate that the several
other factors separately or acting together cooperate to precipitate disease. In the intestinal
tract the virus is evidently innocous, but allowed to replicate systemically (in macrophages)
may initiate the disease pathogenesis. The macrophage is the target cell for FIPV infection
and dissemination but interestingly, FECV replicates poorly in the macrophage 6. Perhaps
vaccinia, which can readily infect macrophages, delivers the FECV spike protein which
causes the cascade of immune response pathology induced by the spike protein expressed
in this cell type. Under this scenario the mechanism by which the virus can leave the intestinal
compartment and infect macrophages may be a worthy area of future study.

REFERENCES

1 H Lutz, B Hauser, and M C Horzinek 1986 Feline infectious peritonitis (FIP) - The present state of
knowledge, *J Small Animal Pract* 27·108

2 Pedersen, N C 1983 A review Feline Infectious Peritonitis and Feline Enteric Coronavirus Infections
Feline Practice **13** 13-19

3 N C Pedersen 1987 Virologic and immunologic aspects of feline infectious peritonitis virus infection,
Adv Exp Med Biol 218 529-550

4 N C Pedersen and J F Boyle 1980 Immunologic phenomena in the effusive form of feline infectious
peritonitis *Am J Vet Res* 41 868-876

5 N C Pedersen, J F Boyle, K Floyd, A Fudge and J Barker 1981 An enteric coronavirus of cats and its
relationship to feline infectious peritonitis virus *Am J Vet Res* **42**·368-377

6 Stoddart, C A and Scott, F W 1989 Intrinsic resistance of feline peritoneal macrophages to coronavirus infection correlates with in vivo virulence *J Virol* **63** 436-440

7 Vennema, H , De Groot, R J , Harbour, D A , Horzinek, M C and Spaan, W J M 1991 Primary structure of the membrane and nucleocapsid protein genes of Feline Infectious Peritonitis Virus and immunogenicity of recombinant vaccinia viruses in kittens *Virol* **181** 327-335

8 Vennema, H , De Groot, R J , Harbour, D A , Dalderup, M , Horzinek, M C and Spaan, W J M 1990 Early death after Feline Infectious Peritonitis Virus challenge due to recombinant vaccinia virus immunization *J Virol* **64** 1407-1409

9 Horsburgh

10 Gerber, J D , Ingersoll, J D , Gast, A M , Christianson, K K , Selzer, N L , Landon, R M , Pfeiffer, N E , Sharpee, R L and Beckenhauer, W H 1990 Protection against feline infectious peritonitis by intranasal inoculation of a temperature sensitive FIPV vaccine *Vaccine* **8** 536-542

11 Christianson, K K , Ingersoll, J D , Landon, R M , Pfeiffer, N E and Gerber, J D 1989 Characterization of a temperature sensitive feline infectious peritonitis virus *Arch Virol* **109** 185-196

12 R K Saiki, S Scharf, F A Faloona, K B Mullis, G T Horn, H A Erlich and N Arnheim 1985 Enzymatic amplification of Beta-Globin Genomic sequences and restriction site analysis for diagnosis of sickle cell anemia *Science* 230 1350-1354

39

THE USE OF ARMS PCR AND RFLP ANALYSIS IN IDENTIFYING GENETIC PROFILES OF VIRULENT, ATTENUATED OR VACCINE STRAINS OF TGEV AND PRCV

Chih-Hung Lai,[1] Mark W. Welter,[2] and Lisa M. Welter[1]

[1] Ambico West
USC School of Medicine
Los Angeles, California 90033
[2] Ambico, Inc.
Dallas Center, Iowa 50063

ABSTRACT

The use of ARMS (amplification refractory mutation system) PCR coupled with RFLP (restriction fragment length polymorphism) analysis has been used to identify a unique genetic marker on the Ambico oral vaccine strain. This method was also used to characterize the genetic profiles of a number of other TGEV strains. This procedure takes advantage of the nucleotide differences between the Ambico strain, and the Miller and Purdue strains. Within the S gene there are three nucleotide differences between the Ambico strain and the published Purdue sequence. There are additional nucleotide differences in the structural and non-structural gene sequences, but we have chosen to focus on the differences contained within the S gene. The Ambico strain has a closer sequence homology to the Purdue strain than to the Miller strain. The Ambico and Purdue strains contain a six nucleotide deletion at position 1122 that is not present in the Miller published sequence or the ISU-1 strain of PRCV (based on our PCR experiments). We have designed a 5' oligo whose sequence is homologous to a region located 80 nucleotides upstream of the TGEV and PRCV S gene initiation codons to be used in conjunction with either of two 3' oligos whose sequences are identical with the exception of the last six nucleotides of their 3' ends. When utilized with the appropriate PCR conditions, these oligos can differentiate between PRCV, Miller and Purdue prototype virus strains. These PCR products were then subjected to RFLP analysis using four separate restriction enzymes (*Bst*E II, *Alw*26 I, *Dra* III, or *Msp*A1 I). We have used this procedure to analyze six TGEV vaccine strains, intestinal derived virulent viruses, cell cultured viruses at different cell passage numbers, and field isolates of TGEV or PRCV.

Corona- and Related Viruses, Edited by P. J. Talbot and G. A. Levy
Plenum Press, New York, 1995

INTRODUCTION

TGEV was first identified in 1946 by Doyle and Hutchins [1]. TGEV causes gastroenteritis in pigs resulting in a mortality rate approaching 100% in piglets less than two weeks of age. TGEV has been shown to infect epithelial cells of the gut and respiratory tract. PRCV, a variant of TGEV with a predominate tropism for the respiratory tract, was first isolated in Europe in 1984 by Pensaert [2]. PRCV was first isolated in the United States in 1989 by Hill et. al. and designated ISU-1 [3]. Since then, four additional strains of PRCV have been isolated and characterized in the United States [4,5,6]. Based upon serology, these strains have been shown to be similar to the European strain, but based upon their nucleic acid sequence they are different. Through sequence analysis, attempts have been made to correlate characteristics of tropism and pathogenicity to genotypic changes among various TGEV and PRCV strains [7,8,9]. It is presently speculated that deletions in the 5' region of the S gene in PRCV are responsible for its altered tropism while deletions affecting the expression of mRNA 3a and 3b are responsible for altered pathogenicity [10,11,12].

MATERIALS AND METHODS

Viruses

Reference virus stocks were obtained from various sources and a description of these strains is listed in Table 1. Vaccine strains of TGEV were isolated from commercially available products from Grand Laboratories, Inc., Larchwood, Iowa; Schering-Plough Animal Health, Omaha, Nebraska; Oxford Veterinary Laboratories, Inc., Worthington, Minnesota; Diamond Laboratories, Des Moines, Iowa; Fort Dodge Laboratories, Fort Dodge, Iowa; and Ambico, Inc., Dallas Center, Iowa.

Isolation of Virus from Infected Tissue Samples.

Four TGEV seronegative piglets, two-to-three days of age were divided into two groups and housed in individual isolation boxes. Each group was inoculated either intranasally or orally with $10^{6.2}$ TCID$_{50}$ per pig of Ambico TGEV vaccine . One pig from each group was sacrificed at 24 hours and 48 hours post-inoculation. The following tissues were carefully collected using different instruments for each of the seven samples: lung; tonsils; small intestinal extract (SIE = actual tissue homogenate); nasal lavage and scrapings; mesenteric lymph nodes (MLN); small intestinal contents (SIC); and fecal samples. A sample of jejunum, ileum and duodenum was taken for staining by specific IFA of the frozen sections for TGEV. Infectious TGEV virus titers for the lung, tonsil, SIC, fecal, SIE, nasal, and MLN samples were determined by inoculation of confluent ST cell cultures and evaluation by plaque assay and TGEV specific IFA.

Isolation of Viral RNA and cDNA Synthesis

Frozen reference virus stocks consisted of either small intestinal extract (SIE) or ST cell culture passaged material. Small intestinal contents collected from infected piglets were diluted to 50% in PBS, and debris was removed by high speed centrifugation. An equal volume of 50% sucrose was added and the material was stored at -70°C as a 25% final SIE. This material was used to generate ST cell passaged stocks. Virus stocks were thawed and clarified of particulate matter by centrifugation at 10,000 xg. The supernatant was collected and used for viral RNA isolation.

TGEV vaccine strains from Grand Laboritories, Schering-Plough, Diamond Laboratories, and Ambico, Inc. were obtained as lyophilized powdered cakes. These samples were resuspended in 10 ml of sterile purified water. The Fort Dodge material was received from Dr. Ron Wesley already rehydrated (10 dose vial rehydrated with 20 ml). The rehydrated material was clarified of particulate matter by centrifugation. The supernant was collected for viral RNA isolation. The Oxford product was received as a liquid suspension and was first diluted two-fold with sterile water and then extracted twice with toluene, once with a 1:1 ratio of toluene/phenol and one time with a 1:1 ratio of chloroform/toluene. This material was then used for viral RNA isolation.

Tissue samples were collected and a 50% homogenized extract was prepared in PBS. Samples were clarified of debris by centrifugation at 10,000 x g. The supernatants were collected, aliquoted and quick frozen at -70°C. Frozen samples were thawed and diluted two-fold in detergent containing buffer and clarified of any debris by centrifugation. The supernatant was collected and used for viral RNA isolation.

Viral RNAs were isolated by digestion in 100 mM Tris (pH 7.5),12.5 mM EDTA, 150 mM NaCl, 1% SDS and 200 μg proteinase K (Boehringer Mannheim) at 37°C for 30 minutes. After proteinase K treatment, the samples were extracted twice with phenol and once with 1:1 ratio of phenol/chloroform. RNA was precipitated using LiCl and ethanol at -70°C. The RNA was pelleted by centrifugation and the pellet was rinsed two times with 70% ethanol and resuspended in sterile RNase free water. First strand cDNA was generated using reverse transcriptase and oligo WL23 (5' TGTGTACCATTACCACAG 3'). Oligo WL23 is complimentary to the 3' ends of TGEV and PRCV S genes (corresponding to nucleotides 3591 to 3608 of the Miller sequence). Briefly, the RNAs were denatured at 94°C for 2 minutes and then incubated at 42°C for 60 minutes in a 50 μl reaction containing 50 mM Tris (pH 8.0),6 mm $MgCl_2$, 40 mM KCl, 100μM dNTPs, 5U RNAsin (Promega), 2U of AMV reverse transcriptase (Seikagaku America, Inc.) and 500 pmol of oligo WL23. Final reactions were stored at -70°C until use.

ARMS PCR and RFLP Analysis

The cDNAs were then used in PCR type-specific reactions containing 100 μM dNTPs, 2U of Taq polymerase (Boehringer Mannheim), one tenth volume of supplied 10x Taq reaction buffer, one tenth volume of the cDNA material and 250 pmol of each 5' and 3' oligos. The 5' oligo WL5 (5' GGATTACTAAGGAAGGGTAAGTTG 3') is homologous to TGEV and PRCV viral genomic sequences located at positions -80 to -56 nucleotides upstream of their S gene initiation codons and either of two 3' oligos whose sequences are identical with the exception of the last six nucleotides of their 3' ends. These oligos, utilized with the proper PCR conditions, are able to differentially amplify Purdue-like or Miller-like virus strains. This is based on the property of Taq polymerase, which will not extend a mismatched 3' end [13,14,15]. The Tms (melting temperatures) of these 3' oligos are higher than that of oligo WL23. Therefore, the annealing temperatures used are too stringent for the WL23 oligo to interfere. The 3' oligo WL29 (5' TCGAGTCACTCACTGTATCATTAT 3') is complementary to nucleotides 1121 to 1144 of the Miller S gene sequence and contains an additional six nucleotides not present in the Purdue sequence, hence this oligo is specific for Miller and PRCV strains of coronavirus. The 3' oligo WL28 (5'CGAGT-CACTCACTGTATAACATG3') is complementary to nucleotides 1115 to 1137 of the Purdue strain and does not contain these additional six nucleotides, hence it is specific for the Purdue virus sequences.

Table 1. History and source of TGEV and PRCV virus strains

---

1 Miller MGV: virulent intestinal extract, Received from Dr Wesley, USDA, APHIS, NVSL, VS, Ames, Iowa

2 Miller PP3 · plaque purified three times on ST cells, Received from Dr Wesley, USDA, APHIS, NVSL, VS, Ames, Iowa

3 Miller TGEV APHIS 69-7· virulent intestinal extract; Received from Dr Tamogolia, Vet Biologics Div (VBD) - Animal Res Section (ARS) at USDA, Ames, Iowa.

4 Miller/ST-3. TGEV APHIS 69-7 Miller virulent strain passaged 3x in ST cells

5 Purdue TGEV Std Challenge Lot-4: virulent intestinal extract; Received from Dr. Haelterman, Purdue University, Indiana

6 Purdue/ST-4: TGEV Std Challenge Lot-4 Purdue virulent strain passaged 4x on ST cells

7 Purdue P-115 · Purdue avirulent strain; Received from Dr. Bohl, Ohio State University, Ohio

8 TGEV. Illinois Strain , Received from Dr Pat Gough, Vet Med Res Institute (V M R I), ISU, Ames, Iowa

9 PRCV ISU-I · passaged 3x in ST cells; Received from Dr Hill, Iowa State University, Ames, Iowa

---

We have demonstrated that oligos WL5 and WL28 amplify a PCR product of 1217 nt from the Ambico strain, whereas oligos WL5 and WL29 fail to amplify a PCR product. The reverse is true for oligos WL5 and WL29 which amplify a PCR product of 1224 nucleotides from the Miller strain and 543 nucleotides from the PRCV strain. Oligos WL5 and WL28 fail to amplify a PCR product from Miller or PRCV strains (data not shown). PCR conditions consisted of a hot start, whereby Taq was added after the tubes had reached a temperature of 94°C. The initial cycle consisted of 94°C for 3 minutes followed by 30 cycles of 94°C for 1 minute; 68°C for 1 minute; 72°C for 2 minutes; with a final cycle of 72°C for 10 minutes. One tenth volume of the PCR reactions were analyzed by gel electrophoresis. The PCR fragments were then subjected to RFLP analysis using the following restriction enzymes *Bst*E II, *Msp*A1 I, *Dra* III (New England Biolabs) and *Alw*26 I (Promega). All restriction enzyme reactions were carried out according to the manufacturer's specifications.

RESULTS AND DISCUSSION

We compared the published sequences for the Miller, Purdue-115, and PRCV (GenBank accession numbers S51223, D00118 and Z24675 respectively) with the Ambico vaccine strain, in order to choose restriction enzymes that would selectively recognize sequences within the Purdue, Miller, or Ambico generated PCR fragments. The predicted genetic profiles obtained by RFLP analysis are shown in Table 2.

Initial ARMS-PCR analysis was conducted on the reference viral material using the oligo combination of WL5-WL28 (specific for Purdue sequences) or WL5-WL29 (specific for Miller sequences). A summary of these results is presented in Table 3. All of the Miller virus strains amplified specifically with the WL5-WL29 oligos. RFLP analysis of this material showed that all of the Miller strains of virus have the predicted Miller genetic profile. Unexpectedly, the Purdue Std. Challenge SIE and the Purdue Std. Challenge ST/4 also amplified with the WL5-WL29 oligos. However, RFLP analysis of this material showed a mixed genetic profile, with a Miller profile for restriction enzyme *Dra* III and a Purdue profile for restriction enzymes *Bst*E II and *Msp*A1 I. It is unlikely that this represents a mixed viral population, because samples failed to amplify a PCR product with the WL5-WL28 combination. It more likely represents a precursor of the Purdue isolate prior to the selective

Table 2. Predicted RFLP pattern of digested ARMS-PCR amplified fragments

Virus Strain	Enzyme Pattern obtained with			
	*Bst*E II	*Alw*26 I	*Dra* III	*Msp*A1 I
Purdue-115	1000	1096	no cutting	no cutting
	217	121		
Miller	1056	1103	705	735
	110	121	519	489
Ambico	1000	615	no cutting	no cutting
	217	481		
		121		

pressures of tissue culture cell passage and plaque purification. It is interesting to note that this virus yields two plaque sizes of large and small phenotype. However, upon plaque purification of large or small plaques, the mixed phenotype of large and small plaques is still observed regardless of the size of the plaque picked (unpublished observations). We are presently attempting to isolate viral RNA from individual plaques for ARMS-PCR/RFLP analysis to determine if there are any discernible differences associated with the plaque phenotype. ARMS-PCR analysis of the Purdue-115 strain shows that it specifically amplified with the WL5-WL28 oligos and had the predicted RFLP genetic profile for the published Purdue-15 and Purdue-115 viral sequences. The Illinois strain (SIE of convalescent pigs) was the only sample that amplified a product with both sets of oligos, however it only amplified a very faint product using the WL5-WL28 oligos, whereas with the WL5-WL29 oligos the PCR product was more abundant. This most likely represents a mixed population of viruses in the Illinois sample since RFLP analysis of the WL5-WL28 PCR product with *Bst*E II and *Dra* III exhibited a Purdue profile, whereas RFLP analysis of the PCR product amplified with WL5-WL29 showed an identical genetic profile as the Purdue Std. Challenge strains. In previous studies we have shown that mixtures of Miller with Ambico virus strains will amplify PCR products from both combinations of oligos. However, RFLP analysis shows that the PCR product amplified with the WL5-WL28 oligos has the genetic profile of the Ambico strain and the product amplified with WL5-WL29 oligos has the genetic profile of the Miller strain. This indicates that the oligos are amplifying specifically. The same six nucleotide insertion is also present in a virulent British field isolate [9].

ARMS-PCR/RFLP analysis of Grand Laboritories, Schering-Plough and Diamond Laboratories vaccine strains showed that all amplified with the WL5-WL28 oligo combination and had identical RFLP genetic profiles as the Purdue-115 strain. Only the Fort Dodge vaccine strain amplified with the WL5-WL29 oligo combination and showed a RFLP genetic profile identical to the Miller strain. The Ambico strain amplified with the WL5-WL28 oligo combination. However, it posses a unique genetic profile for RFLP analysis with the enzyme *Alw*26 I. The RFLP analysis of the Ambico strain with *Bst*E II, *Dra* III, and *Msp*A1 I correlate with the genetic profile of Purdue-115. This data is in partial agreement with the S1 nuclease mapping studies published by Register and Wesley [16] who showed that the Solvay and Fort Dodge vaccine strains exhibited a similar S1 protection pattern as the Miller PP3. They also showed that the Diamond and Ambico vaccine strains showed a similar S1 nuclease protection pattern as the Purdue-115 strain. Our data contradicts theirs in that the S1 nuclease pattern for the Ambico strain should have an additional band as compared to the Diamond and Purdue-115 strains, due to the single base change contained in the Ambico strain. In question is the ability and sensitivity of S1 nuclease to detect single base pair mismatch versus the sensitivity of ARMS-PCR/RFLP analysis. They state that they have consistently

Table 3. ARMS-PCR and RFLP analysis of reference virus and vaccine strains

Virus Strain	WL28	WL29	*Bst*E II	*Alw*26I	*Dra* III	*Msp*AlI
			Amplification with WL5 and			
Mil.MGV	–	+	M	M/P	M	M
Mil.PP3	–	+	M	M/P	M	M
Mil.APHIS 69-7	–	+	M	M/P	M	M
Mil.APHIS 69-7/ST3	–	+	M	M/P	M	M
Pur. Std. Chall/ Lot-4	–	+	P	M/P	M	P
Pur. Std. Chall./ST-4	–	+	P	M/P	M	P
Purdue P-115	+	–	P	M/P	P	P
TGEV Illinois Strain	+/–	+	P	M/P	M	P
PRCV ISU-I*	–	+	NA	NA	NA	NA
Ambico Vaccine	+	–	P	A	P	P
Grand Labs	+	–	P	M/P	P	P
Schering-Plough	+	–	P	M/P	P	P
Diamond Labs	+	–	P	M/P	P	P
Fort Dodge	–	+	M	M/P	M	M
Oxford	+/–	–	ND	ND	ND	ND

Note: M = Miller pattern; P = Purdue pattern; A = Ambico pattern; NA = Not Applicable; ND = Not done due to lack of sufficient material. (*) PRCV amplified a fragment of 543 nucleotides.

detected deletions as small as 3-6 bp but they do not indicate if they have been successful in using the S1 nuclease protection assay to detect single base pair mismatches. ARMS-PCR analysis has been used to detect allele-specific differences based on a single nucleotide difference [13,14,15,17].

To determine if the Ambico vaccine strain retains its genetic marker while passaged through piglets, we generated tissue extracts for re-isolation of the Ambico vaccine strain. The results of the PCR amplification with WL5-WL28 and the re-isolation of virus titer evaluated by plaque assay and TGEV specific IFA are shown in Table 4. None of the tissue

Table 4. Sizes of PCR fragments and virus titers isolated from tissues of piglets inoculated with the ambico vaccine TGEV

	24 Hours Post-Inoculation				48 Hours Post-Inoculation			
	Orally		Intranasally		Orally		Intranasally	
	Pig-1		Pig-2		Pig-3		Pig-4	
Sample	PCR	PFU	PCR	PFU	PCR	PFU	PCR	PFU
Fecal	Neg.	(Neg.)	1217	(Neg.)	Neg.	(Neg.)	1217	(Neg.)
Lung	1217	(Neg.)	1217	(Neg.)	Neg.	(Neg.)	1217*	(Neg.)
MLN	1217	(Neg.)	1217	(Neg.)	Neg.	(Neg.)	Neg.	(Neg.)
Nasal	1217	(Neg.)	1217	(1.9)	Neg.	(Neg.)	Neg.	(2.4)
SIE	1217	(Neg.)	1217	(Neg.)	1217*	(Neg.)	1217*	(Neg.)
SIC	Neg.	(Neg.)	Neg.	(Neg.)	Neg.	(Neg.)	Neg.	(Neg.)
Tonsil	1217	(Neg.)	1217	(Neg.)	1217	(Neg.)	Neg.	(Neg.)

samples amplified a PCR product using the oligo combination WL5-WL29 All PCR products amplified from the tissue samples exhibited the genetic profile of the Ambico strain upon RFLP analysis (data not shown) This data indicates that the genetic marker is maintained subsequent to passage in pigs Infectious TGEV was reisolated at low titers from the nasal samples of the 2 pigs that were inoculated intranasally The remaining 26 samples were negative for TGEV by IFA evaluation of plaque assays IFA specific staining for TGEV of frozen sections of jejunum, ileum or duodenum were also negative for TGEV (data not shown)

Furthermore, ARMS-PCR/RFLP analysis of field samples obtained from pigs vaccinated with the Ambico vaccine strain showed that the Ambico strain maintained its genetic marker in the field (data not shown) We have used this method to evaluate a limited number of field samples and have been successful in identification of Miller, Purdue and PRCV strains We are presently utilizing a second set of primers located downstream of the WL5-WL28/29 oligo combinations to be used for further characterization of field isolates as compared to Miller or Purdue strains of TGEV and PRCV This method represents a rapid and sensitive way to monitor virus genetic profiles, which may have application in the diagnostics field A similar system has been used to identify single nucleotide mismatches in other virus systems [15 17] ARMS-PCR analysis coupled with RFLP represents an alternative to nucleic acid sequencing in correlating nucleic acid changes and the resulting altered tropism and pathogenicity

REFERENCES

1 Doyle L P , Hutchings L M A transmissible gastroenteritis in pigs J Am Vet Med Assoc 1946,108 257-259

2 Pensaert M , Callebaut P , Vergote J Isolation of a porcine respiratory, non-enteric coronavirus related to transmissible gastroenteritis Vet Q 1986,8 257-261

3 Hill H T , Biwer J D , Wood R D , Wesley R D Porcine respiratory coronavirus isolated from two U S swine herds Proc Am Assoc Swine Prac 1989,333-335

4 Paul P S , Vaughn E M , Halbur P G Characterization and pathogenicity of a new porcine respiratory coronavirus strain AR310 Proc Int Pig Vet Soc Congr 1992,12 92

5 Wesley R D , Woods R D , Hill H T , Biwer J D Evidence for a porcine respiratory coronavirus, antigenically similar to transmissible gastroenteritis virus J Vet Diagn Invest 1990,2 312-317

6 Vaughn E M , Halbur P G , Paul P S , Three new isolates of porcine respiratory coronavirus with pathogenicities and spike (S) gene deletions J Clin Micro 1994,32 1809-1812

7 Britton P , Kottier S , Chen C -M , Pocock D H , Salmon H , Aynaud J M The use of PCR genome mapping for the characterization of TGEV strains Adv Exp Med and Biol 1993,342 29-34

8 Sanchez C M , Gebauer F , Sune C, Mendez A , Dopazo J , Enjuanes L Genetic evolution and tropism of transmissible gastroenteritis coronaviruses Virology 1992,190 92-105

9 Britton P , Page K W Sequence of the S gene from a virulent British field isolate of transmissible gastroenteritis virus Vir Res 1990,18 71-80

10 Laude H , Van Reeth K , Pensaert M Porcine respiratory coronavirus molecular features and virus-host interactions Vet Res 1993,24 125-150

11 Wesley R D , Wood R D , Cheung A K Genetic basis for the pathogenesis of transmissible gastroenteritis virus J Virol 1990,64 4761-4766

12 Wesley R D , Wood R D , Cheung A K Genetic analysis of porcine respiratory coronavirus, an attenuated variant of transmissible gastroenteritis virus J Virol 1991,65 3369-3373

13 Newton C R , Graham A , Heptinstall L E , Powell S J , Summers C , Kalsheker N , Smith J C , Markham A F Analysis of any point mutation in DNA The amplification refractory mutation system (ARMS) Nuc Acids Res 1989,17 2503-2516

14 Huang M M , Arnheim N , Goodman M F Extension of base mispairs by Taq DNA polymerase implications for single nucleotide discrimination in PCR Nuc Acids Res 1992,20 4567-4573

15. Kwok S., Kellogg D.E., McKinney N., Spasic D., Goda L., Leverson C., Sninsky J.J. Effects of primer-template mismatches on the polymerase chain reaction: human immunodeficiency virus type 1 model studies. Nuc. Acids Res. 1990;18 999-1005.

16. Register K.B., Wesley R.D. Molecular characterization of attenuated vaccine strains of transmissible gastroenteritis virus. J. Vet. Diagn. Invest. 1994;6:16-22.

17. Ault G.S., Ryschkewitsch C.F., Stoner G.L. Type-specific amplification of viral DNA using touchdown and hot start PCR. J. Virological Meth. 1994;46:145-156.

Viral Proteins

40

INTRACELLULAR LOCALIZATION OF POLYPEPTIDES ENCODED IN MOUSE HEPATITIS VIRUS OPEN READING FRAME 1A

Weizhen Bi,[1] Pedro J. Bonilla,[2] Kathryn V. Holmes,[3] Susan R. Weiss,[2] and Julian L. Leibowitz[1]

[1] Department of Pathology and Laboratory Medicine and the Department of Microbiology and Molecular Genetics, University of Texas Health Sciences Center
Houston, Texas
[2] The Department of Microbiology, University of Pennsylvania
Philadelphia, Pennsylvania
[3] The Department of Pathology, Uniformed Services University of the Health Sciences
Bethesda, Maryland

ABSTRACT

We have investigated the intracellular localization of several of the proteolytic cleavage products derived from the 5' portion of mouse hepatitis virus (MHV) gene 1. Antisera UP1 recognizes the N-terminal ORF1a cleavage product p28. Immunofluorescent staining of cells with this antisera resulted in a diffuse punctate pattern of cytoplasmic staining, indicating that this protein is widely distributed in the cytoplasm. Immunofluorescent staining of infected cells with antisera which recognize polypeptides p240 and p290 stained discrete vesicular perinuclear structures suggesting that these proteins localized to the Golgi. This was confirmed by double immunofluorescent staining of BHK cells expressing the MHV receptor (BHK-R) with a Golgi specific antibody in addition to our anti-MHV ORF1a antibodies. Antisera UP102 recognizes p28 and the immediately downstream p65 gene product. Double immunofluorescent staining of MHV infected BHK-R cells with UP102 labeled discrete vesicular structures overlapping the Golgi complex. In addition there was punctate staining more widely distributed in the cytoplasm. The simplest explanation for this pattern is that p65 is also localized to the Golgi region of the cell, whereas p28 is more widespread. Plasmids containing the first 4.7 and 6.75 kb of ORF 1a have been expressed using the coupled vaccinia virus - T7 polymerase system. Images obtained by immunofluorescent staining of transfectants with our anti-ORF1a antisera are similar to those obtained during infection with A59. These studies indicate that the signals which direct p290 to the Golgi are likely contained between the C-terminus of p28 and ORF1a residue 1494.

INTRODUCTION

Mouse hepatitis virus, a murine coronavirus, contains a single - stranded positive - sense RNA genome. One of the initial steps in MHV replication is the translation of the 5'-most, approximately 21.8 kb of the genome. This region of the genome, designated gene 1, contains two overlapping large open reading frames[1,2]. The first open reading frame, ORF1a encodes 4488 amino acids. The second open reading frame ORF1b, overlaps ORF1a for 75 nucleotides and is 2771 amino acids long. *In vitro* transcription and translation studies of this region have demonstrated that these two ORFs are translated as an 800,000 kDa polyprotein by a ribosomal frameshifting mechanism[3]. Analysis of the nucleotide sequence and the predicted amino acid sequence revealed ORF1a contains several functional domains: two hydrophobic domains, three cysteine - rich domains, a picornaviral 3C-like protease domain, and two papain-like protease domains[2,3]. The relative positions of these ORF1a domains are depicted schematically in Figure 1. ORF1b is predicted to encode domains required for the MHV RNA-dependent RNA polymerase activity - the SDD polymerase motif, a nucleotide binding motif, a metal binding domain, and a helicase domain[4].

The presence of protease domains encoded by ORF1a, the extremely large size of the predicted gene 1 primary translation product, and genetic evidence that as many as six separate transacting MHV gene products are needed for RNA synthesis[5], all suggest that the gene 1 polyprotein is proteolytically processed into multiple polypeptides. Biochemical studies have recently begun to elucidate the proteolytic cleavage events in processing the gene 1 primary translation product. These have been most successful for ORF1a. An N-terminal 28 kDa protein, p28, is cotranslationally cleaved from the primary translation product by the first papain-like protease encoded in ORF1a[6-8]. Subsequently a 65 kDa protein

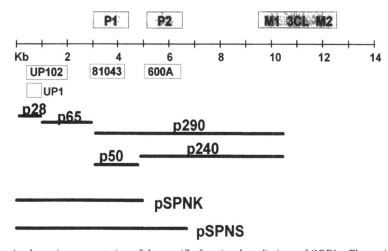

Figure 1. A schematic representation of the specific functional predictions of ORF1a. The positions of the predicted papain domains are indicated by the boxed P1 and P1; the position of the predicted picornavirus 3C-like protease domain and the two flanking hydrophobic domains are indicated by boxed M1, 3CL, M2. The approximate locations of those ORF1a polypeptide sequences used as immunogens and the names of the resulting antibodies are boxed below the kilobase scale. The approximate lengths and map positions of the plasmids pSPNK and pSPNS are shown.

(p65) is cleaved from a large precursor protein as are discrete polypeptides approximately 290 kDa (p290), 240 kDa (p240), and 50 kDa (p50) molecular mass[8,9]. The presumed relationships of these cleavage products to the MHV genome and to each other are depicted in Figure 1.

Enzymes capable of directing MHV RNA synthesis have been shown to sediment with the membrane fraction of the cytosol[10]. Thus we have begun an investigation into the intracellular localization of the various gene 1 protein products. Since the processing pathway for ORF1a gene products is much better understood than that for ORF1b, and we have in hand well characterized antibodies recognizing defined regions encoded by ORF1a, our initial efforts have been restricted to proteins encoded within this portion of the genome.

MATERIALS AND METHODS

Cells and Virus

The origin and growth of L-2 cells and BHK cells expressing the MHV receptor (BHK-R) have been described[11,12]. The origin and growth of MHV-A59 and vaccinia virus expressing the T7 RNA polymerase have been described[11,13].

Antibodies

The antibodies used in this study of been described[8,9]. These antisera were raised in rabbits using procaryote/viral fusion proteins as immunogens. The locations of viral sequences encoding the various polypeptides used to raise these antibodies, as well as the proteins immunoprecipitated by these antibodies, is shown in Table 1. The Golgi-specific anti-mannosidase II monoclonal antibody 53FC3[14] was purchased from Berkeley Antibody Company. Fluorescein conjugated goat anti-rabbit IgG and rhodamine conjugated goat anti-mouse IgG were purchased from Jackson Research.

Plasmids and Transfections

Plasmids pSPNK and pSPNS are derived from cDNAs and contain MHV-A59 gene 1 sequences from the *Nar*I site at nucleotide 182 to the *Kpn*I site at nucleotide 4664 and the *Spe*I site at nucleotide 6756, respectively[2]. In both cases the MHV-A59 cDNA is oriented in the plasmid pSP72 such that the coding sequences are under the control of the T7 promoter. pSPNK contains the coding sequence for p28, p65, and the first papain-like domain of MHV-A59; pSPNS extends past the second papain-like domain (Figure 1).

For transfection experiments 60 mm dishes containing sterile coverslips were seeded with 2.2 x 10^6 BHK-R cells and incubated overnight. The cell were then infected at a MOI=5 with recombinant vaccinia virus expressing T7 polymerase, vTF7.3. The cultures were

Table 1. Antisera used for Immunofluorescent localization of A59 ORF1a products

Antisera	Size of viral sequences used as immunogen (aa)	cDNA for fusion proteins spans nucleotides	Polypeptides immunoprecipitated from infected cells
UP1	228	461-1144	p28
81043	363	2879-3968	p290, p240
UT600	495	5000-6486	p290, p240
UP102	601	182-1984	p28, p65

incubated at 37°C for 2.5 hours and then transfected with 5 μg of plasmid using lipofectin (GIBCO-BRL) according to the manufacturer's instructions. After being returned to the incubator for $2\frac{1}{2}$ hours the transfection media was replaced with fresh media and the cells incubated for a further six hours, at which time they were washed with PBS and fixed with cold acetone.

Indirect Immunofluorescence

L-2 cells and BHK-R cells were grown for two days on glass coverslips and infected with MHV-A59 at an MOI of 10 or mock-infected. L-2 cells and BHK-R cells were fixed with cold acetone at 6-7 hours and at 8 hours post-infection, respectively. For transfection experiments, BHK-R cells were similarly fixed 8.5 hours after transfection. Cells were treated with 5% non-fat milk in phosphate buffered saline, pH 7.4, at 4°C, incubated on ice with rabbit anti-ORF1a antibody (1:100 dilution in PBS containing 3% BSA) for 90 minutes. Following incubation with primary antibody, the coverslips were washed extensively with PBS containing 0.2% Tween 20, incubated for 60 minutes with FITC-conjugated goat anti-rabbit IgG (1:100 dilution), and washed five times with PBS containing 0.2% Tween 20. For double labeling experiments a mixture of rabbit anti-ORF1a antibodies (diluted 1:100) and mouse monoclonal anti-mannosidase II antibody (diluted 1:600) were used as primary reagents, and a mixture of FITC-conjugated goat anti-rabbit IgG and rhodamine-conjugated goat anti-mouse IgG were used as secondary antibodies. The cells were mounted in elvanol mounting solution for examination with a Multi-probe 2001, CLSM confocal microscope (Molecular Dynamics). The CLSM produces simultaneous dual channel images. Colocalization of the two chromophores was revealed by the yellow color resulting from their overlapping emission. To obtain black and white prints from these images, the colors were remapped by computer. Yellow was remapped to white (100% gray), green was remapped to light gray (70% gray), and red was remapped to dark gray (30% gray).

RESULTS

Preliminary studies were performed using single color immunofluorescent staining of MHV-infected L2 cells to see if there was differential intracellular localization of various ORF1a polypeptide products. Antisera UP1 recognizes the N-terminal ORF1a cleavage product p28. Immunofluorescent staining of cells with this antisera resulted in a diffuse punctate pattern of cytoplasmic staining, indicating that this protein is widely distributed in the cytoplasm (data not shown). Immunofluorescent staining of infected cells with antisera which recognize polypeptides p240 and p290 (sera 81043 and UT600) resulted in intense staining of discrete vesicular perinuclear structures suggesting that these proteins localized to the Golgi (data not shown).

To investigate this further, two color immunofluorescent staining was performed using a Golgi-specific anti-mannosidase II monoclonal antibody in addition to the various anti-ORF1a antibodies. The anti-Golgi antibody is a mouse monoclonal antibody and stains murine cells poorly, if at all, but clearly stains hamster cells[14]. Thus we utilized BHK cells expressing the MHV receptor, BHK-R cells, for all subsequent experiments. MHV-infected cells were stained with anti-ORF1a antisera UP1, UP102, 81043, UT600 and Golgi - specific monoclonal antibody 53FC3. In the black and white print shown in Figure 2, white indicates coincident detection of FITC staining (anti-ORF1a antibodies) and rhodamine staining (anti-Golgi antibody); light gray indicates FITC staining only; and dark gray indicates rhodamine staining only. Staining of p28 with UP1, and the Golgi with 53FC3, demonstrated that most of the staining for p28 did not coincide with the Golgi complex (Figure 2, Panel

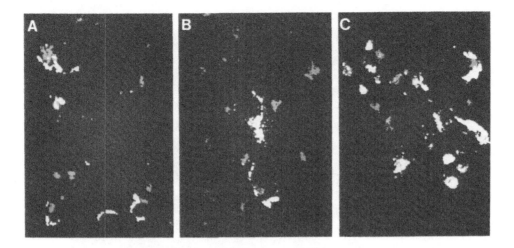

Figure 2. Immunofluorescent localization of ORF1a products in MHV infected cells. BHK-R cells were infected with MHV-A59, fixed, stained with anti-ORF1a and anti-Golgi antibodies, and visualized by CLSM as described in Material and Methods. Areas stained only by the Golgi-specific antibody 53FC3 are indicated by dark gray. Areas stained only by either UP1 (anti-p28, Panel A), UP102 (anti-p28+p65, Panel B), or 81043 (anti-p240/p290, Panel C) are light gray. Areas stained by anti-ORF1a and anti-Golgi antibodies are white.

A). As shown in Panel B, virtually all the staining of p240 and p290 proteins by antisera 81043 coincided with the staining of the Golgi complex by antibody 53FC3. Note that almost no light gray regions can be seen, indicating that virtually all of the polypeptide recognized by 81043 are localized to the Golgi region. A similar result was obtained with antisera UT600 (data not shown). None of the anti-ORF1a sera stained uninfected cells. Biochemical fraction of the cells into soluble and membrane fractions followed by immunoprecipitation with UP1, UT600, and 81043 were consistent with the immunofluorescent staining patterns. Staining with antibody UP102, which recognizes both p28 and p65, revealed a pattern that was a composite of the patterns observed with the anti-p28 and anti-p240/p290 sera. As shown in Figure 2, Panel C, approximately two thirds of the antigens recognized by UP102 (light gray + white) coincided with the Golgi complex stained with 53FC3 (dark gray + white). This suggested to us that p65 might also be localized to the Golgi.

Inspection of the hydrophobicity plot of the entire predicted ORF1a polypeptide reveals that the only obvious transmembrane domains are far downstream, flanking the picornavirus 3C-like protease domain. The sequences contained within p290 and p240 extend into this region of ORF1a (Figure 1). To investigate if these transmembrane domains are required for targeting p290 and p240 for localization in the Golgi region, we have utilized plasmids containing the first 4482 nucleotides, pSPNK, and 6574 nucleotides, pSPNS, of ORF1a under the control of the T7 promoter. The polypeptides encoded in both of these constructs contain a functional papain - like protease and cleave p28 from their primary *in vitro* translation products[7]. These two constructs have been transfected into BHK-R cells and expressed using the coupled vaccinia virus T7 polymerase system. Cells transfected with pSPNS and infected with vTF7.3 were fixed at 8.5 hours post transfection and stained with either the anti-p290/p240 antibodies UT600 or 81043 in combination with the anti-Golgi antibody 53FC3, stained with UP1 and 53FC3, and stained with UP102 and 53FC3. The images obtained by CLSM after transfection were very similar to those obtained during infection with MHV-A59 (data not shown). This experiment indicated that the predicted

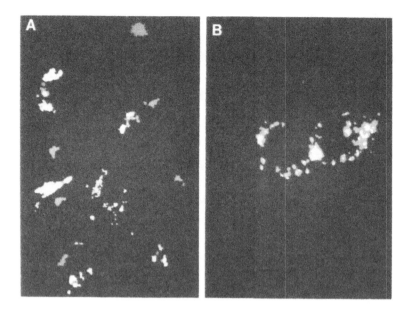

Figure 3. Immunofluorescent localization of ORF1a products in pSPNK transfected cells. BHK-R cells were infected with vFT7.3 and transfected with pSPNK, incubated, fixed, stained with anti-ORF1a and anti-Golgi antibodies, and visualized by CLSM as described in Material and Methods. Areas stained only by the Golgi-specific antibody 53FC3 are indicated by dark gray. Areas stained only by either UP102 (anti-p28+p65, Panel A), or 81043 (anti-p240/p290, Panel B) are light gray. Areas stained by anti-ORF1a and anti-Golgi antibodies are white. The enlargement for Panel B is approximately twice that of Panel A.

transmembrane domains flanking the 3C-like protease domains were not necessary for localization of p290/p240 in the Golgi region of the cell.

There is no obvious candidate for a transmembrane domain targeting p290/p240 for retention in the Golgi encoded within pSPNS. To further investigate where the Golgi localization signal(s) map, a further truncation of ORF1a was expressed. The plasmid pSPNK was expressed in a similar manner as pSPNS. The transfectants were stained with antibodies 81043, UP1, and UP102 in combination with the anti-Golgi antibody 53FC3. Since the sequences recognized by UT600 are not encompassed by this plasmid, it was not used for this experiment. The staining patterns observed were very similar to those obtained previously. Staining for p28 with UP1 again demonstrated that this polypeptide did not specifically localize to the Golgi region (not shown). As shown in Figure 3, Panel A, all the staining of p240 and p290 proteins by antisera 81043 coincided with the staining of the Golgi complex by antibody 53FC3; only white, indicating double staining structures, and dark gray regions, indicating Golgi not containing MHV ORF1a antigens, can be seen. The staining pattern obtained with UP102 was again similar in appearance to a composite of the pattern obtained with UP1 and 81043. As shown in Figure 3, Panel B, approximately one third of the antigens recognized by UP102 (light gray + white) coincided with the Golgi complex stained with 53FC3 (dark gray + white). These results indicate that the signals localizing p290/p240 to Golgi are encoded within the first 4664 nucleotides of the MHV-A59 genome. The staining pattern with UP1 is consistent with the biochemical *in vitro* data that p28 cleavage only requires the first papain protease domain, and that p28 is diffusely spread through the cytosol. The pattern obtained in transfected cells stained with UP102 is consistent with the hypothesis that p65 localizes to the Golgi region.

DISCUSSION

To better understand the functions of the MHV ORF1 encoded proteins we have initiated a series of experiments to determine their intracellular location As reported here, p28 appears to be free in the cytosol, it is more widely distributed than p290/240, as judged by immunofluorescent staining with UP1, and it partitions into the soluble fraction of the cytosol, suggesting that it is not membrane bound This was true in MHV-A59 infected cells and in transfection with experiments with both pSPNS and pSPNK, suggesting that p28 is cleaved normally by the protease domain contained within P1 (see Figure 1) during these transfections

We have demonstrated by double immunofluorescent staining that several ORF1a polypeptides localize to the Golgi region of the cell Both UT600 and 81043 antisera recognize the p290 and p240 ORF1a cleavage products and localize these proteins to the Golgi This is supported by the partition of these polypeptides into the membrane fraction of infected cells These proteins do not appear to be targeted to the Golgi region by the downstream transmembrane domains M1 and M2 (Figure 1) since staining cells transfected with pSPNK with 81043 still localized these sequences to the Golgi This suggests that the targeting sequences are located within the first 1494 amino acids of ORF1a, downstream of the p28 cleavage site Since no clear transmembrane domain is contained within this region it is possible that the protein is only loosely bound to Golgi membranes by a non-membrane spanning domain or by a protein-protein interaction Further experiments are needed to more precisely determine the Golgi targeting signal in these proteins

The localization of p65 is more uncertain at this time Since UP102 recognizes both p28 and p65, the simplest explanation for the pattern observed with UP102 is that p65 is also localized to the Golgi region of the cell, whereas p28 is more widespread Antibodies recognizing just p65 are needed to confirm this assignment

In conclusion these studies indicate that the MHV ORF1a cleavage products p290 and p240 localize to the Golgi complex, p28 is more widely distributed in the cytosol, and suggests that p65 may also localize to the Golgi region of the cell The localization signals for these proteins remain to precisely defined but appear to reside between the C-terminus of p28 and ORF1a residue 1494, the last amino acid encoded in pSPNK

ACKNOWLEDGMENTS

This work was supported in part by National Multiple Sclerosis Society Research Grant RG 2203-A-5, NIH grants AI17418 and AI25231

REFERENCES

1 Lee, H -J , Shieh, C -K , Gorbalenya, A E , Koonin, E V , LaMonica, N , Tuler, J , Bagdzhadzhyan, A , and Lai, M M C The complete sequence (22 kilobases) of murine coronavirus gene-1 encoding the putative proteases and RNA polymerase Virology 1989,180 567-582

2 Bonilla J , Gorbalenya, A E , and Weiss, S R Mouse hepatitis virus strain A59 RNA polymerase gene ORF 1a Heterogeneity among MHV strains Virology 1994,198 736-740

3 Breedenbeek, P J , Pachuk, C J , Noten, A F H , Charite, J , Luyjtes, W , Weiss, S R , and Spaan, W J M The primary structure and expression of the second open reading frame of the polymerase gene of the coronavirus MHV-A59 a highly conserved polymerase is expressed by an efficient ribosomal frameshifting mechanism Nucleic Acids Res 1990,18 1825-1832

4 Pachuk, C J , Breedenbeek, P J , Zoltick, P W , Spaan W J M , and Weiss, S R Molecular cloning of the gene encoding the putative polymerase of mouse hepatitis coronavirus strain A59 Virology 1989,171 141-148

5. Leibowitz, J.L., DeVries, J.R., and Haspel, M.V. Genetic analysis of murine hepatitis virus strain JHM. J. Virol. 1982;42:1080-1087.

6. Denison, M.R., and Perlman, S. Identification of a putative polymerase gene product in cells infected with murine coronavirus A59. Virology 1987;157:565-568.

7. Baker, S.C., Shieh, C.-K., Soe, L.H., Chang, M.-F., Vannier, D.M., and Lai, M.M.C. Identification of a domain required for autoproteolytic cleavage of murine coronavirus gene A polyprotein. J. Virol. 1989;63:3693-3699.

8. Denison, M.R., Zoltick, P.W., Hughes, S.A., Giangreco, B., Olson, A.L., Perlman, S., Leibowitz, J.L., and Weiss, S.R. Intracellular processing of the N-terminal ORF 1a proteins of the coronavirus MHV-A59 requires multiple proteolytic events. Virology 189:274-284, 1992.

9. Hughes, S.A., Bonilla, P., Denison, M.R., Leibowitz, J.L., Baric, R.S., and Weiss, S.R. A newly identified MHV-A59 ORF1a polypeptide p65 is temperature sensitive in two RNA negative mutants. Adv. Exp. Med. Biol. (H. Laude and J.-F. Vautherot, eds.), 1993;342:221-226, Plenum Press, New York, NY.

10. Brayton, P.R., Lai, M.M.C., Patton, C.D., and Stohlman, S.A. Characterization of two RNA polymerase activities induced by mouse hepatitis virus. 1982;42:847-853.

11. Leibowitz, J.L., Wilhelmsen, K,C., and Bond, C.W. The virus-specific intracellular RNA species of two murine coronaviruses: MHV-A59 and MHV-JHM. Virology 1981;114:39-51.

12. Dveksler, G.S., Pensiero, M.N., Cardelluchio, C.B., Williams, R.K., Jiang, G.S., Holmes, K.V., and Dieffenbach, C.W. Cloning of the mouse hepatitis virus (MHV) receptor: expression in human and hamster cell lines confers susceptibility to MHV. J. Virol. 1991;65:6881-6891.

13. Fuerst, T., Earl, P., and Moss, B. Use of a hybrid vaccinia virus-T7 RNA polymerase system for expression of target genes. Mol. Cell. Biol. 1987;7:2538-2544.

14. Burke, B., Griffiths, G., Reggio, H., Louvard, D., and Warren, G. A monoclonal antibody against a 135-K Golgi membrane protein. EMBO J. 1982;1:1621-1628.

41

BOVINE CORONAVIRUS NUCLEOCAPSID PROTEIN PROCESSING AND ASSEMBLY

Brenda G. Hogue

Department of Microbiology and Immunology and Division of Molecular
Virology
Baylor College of Medicine
Houston, Texas

ABSTRACT

The coronavirus nucleocapsid protein (N) encapsidates the genomic RNA to form a helical nucleocapsid. The requirements for coronavirus nucleocapsid assembly are being studied. Two forms (~50 kDa and 55 kDa) of the bovine coronavirus (BCV) N protein were detected in infected cells. However, only one form, a 50 kDa species, was detected in extracellular virions. After treatment with calf intestinal alkaline phosphatase (CIAP), the 55 kDa intracellular form increased in mobility to comigrate with the 50 kDa form; whereas, the 50 kDa intracellular species and N from extracellular virions was not sensitive to CIAP treatment. The data indicate that specificity exists with regard to assembly of N into the mature virion. The data suggests that processing of N may take place during assembly of either nucleocapsids or virions and that the processing may be a dephosphorylation event.

INTRODUCTION

The coronavirus nucleocapsid protein is a multifunctional protein. It is a major virion structural protein that interacts with the genomic RNA to form a helical nucleocapsid [1]. In addition, the protein is also involved in viral replication and pathogenesis. N interacts with RNA sequences contained within the leader RNA found at the 5' ends of both the genomic and subgenomic mRNAs and it has been suggested that this interaction may be involved in viral transcription [2]. The protein is also important with regard to viral pathogenesis [3,4].

The first step in the assembly of the coronavirus particle most likely involves a specific interaction between the N protein and the genomic RNA. The exact requirements for this initial step are not fully understood. The BCV N protein, like other coronavirus N proteins, is phosphorylated [5]. It is generally hypothesized that phosphorylation plays a role in the interaction of N and RNA and therefore, plays an important role in assembly.

In the present study, analysis of the requirements for the N protein to be assembled into virions has been initiated. Two forms of N were detected in BCV infected cells. Both forms were phosphorylated. Only one form of N was detected in extracellular virions. Surprisingly, the extracellular virion N comigrated with the faster migrating (50 kDa) intracellular form.

MATERIALS AND METHODS

HCT (HRT-18) cells and BHK cells (both from ATCC) were infected with BCV at an MOI of 5. Subconfluent cells in 60mm plates were infected at 24 hpi. For ^{32}P labeling, cells were rinsed with phosphate free medium and then labeled with 50 µCi ^{32}P orthophosphate in phosphate free medium supplemented with 10% normal DMEM and 10% dialyzed fetal calf serum. For methionine labeling, infected cells were labeled with 100 µCi EXPRE^{35}S^{35}S label (NEN) and chased in medium containing an excess of methionine and cysteine. At the end of labeling, the medium was removed from the cells, clarified and extracellular virus pelleted at 35K rpm for 3 hrs using an SW50.1 rotor. Cells were lysed and the cytoplasmic fraction collected. Virion pellets were resuspended in lysis buffer and both cell and virion lysates were precleared by incubating with protein A Sepharose for 1 hr. Precleared lysates were immuno-precipitated with an anti-BCV polyclonal antibody. Immune complexes were recovered by incubating with protein A Sepharose for 1 hr, eluted in 2X PAGE sample buffer and analyzed by SDS-PAGE. The ^{32}P labeled Sepharose-immune complex pellets were washed, resuspended in buffer without detergent and divided into three aliquots. One third of each immunoprecipi-tation was treated with either 40 units/ml RNase or 23 units of calf intestinal alkaline phosphatase for 30 min at 37°C. One third was treated in the same way except without RNase or CIAP. After digestion, Sepharose-immune complexes were pelleted in a microfuge and eluted with 2X PAGE sample buffer and analyzed SDS-PAGE.

RESULTS

Comparison of Intracellular and Extracellular Virions N

The BCV N protein, like other coronavirus N proteins, is phosphorylated [5]. As studies have been initiated on the assembly of coronaviruses we noticed a difference in the migration of the N protein when expressed alone and the N protein from purified BCV virions. This led us to compare the proteins from both the intracellular and extracellular fractions of infected cells. Cells infected with BCV were labeled with EXPR^{35}S^{35}S for 30 min and either lysed immediately or after a 9 hour chase in medium containing an excess of methionine and cysteine. After the pulse two forms of N were detected, a 50 kDa and 55 kDa species (Fig 1, lane 4). Both intracellular forms were detected after the 9 hr chase (lane 5). However, when pelleted virions were immunoprecipitated from medium off the cells after the 9 hr chase, only one form of N was detected in the extracellular virions (lane 6). Labeling with ^{32}P orthophosphate showed that both forms of the intracellular protein were phosphorylated (lane 3). The 50 kDa form from extracellular virions was also phosphorylated (lane 2).

Effect of Calf Intestinal Alkaline Phosphatase on N

When we labeled for a time shorter than discussed above we noted that the 55 kDa form of N was detected after the pulse and it appeared to chase to give rise to the 50 kDa form (data not shown). This suggested that some type of processing might be taking place. To begin investigating the nature of the different forms of N, the proteins were treated with calf intestinal alkaline

Figure 1. Expression of intracellular and extracellular forms of N in BCV infected BHK cells Infected cells were labeled for 30 min with EXPRE³⁵S³⁵S (lanes 4-6) or ³²P orthophosphate (lane 3) for 30 min and the intracellular proteins immunoprecipitated immediately (lane 3-4) or chased for 9 hrs at which time both the intracellular fraction (lane 5) and extracellular virions (lane 6) were immunoprecipitated The intracellular and extracellular fractions in lanes 1 and 2 were immunoprecipitated from BCV infected cells that were labeled for 12 hrs with ³²P orthophosphate

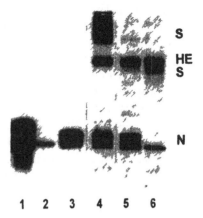

phosphatase (CIAP) after immunoprecipitation The slower migrating 55 kDa form of N was susceptible to phosphatase treatment (Fig 2, lane 3) However, the faster migrating intracellular N and extracellular virion N were not susceptible to CIAP (lanes 3 and 6)

A higher molecular weight ³²P labeled species which may represent an oligomeric form of N was also detected (lane 2) We and others previously noted such a species in purified virions, however, the nature and any function of this species remains to be determined [6][7]

DISCUSSION

This report shows that multiple forms of the BCV N protein are expressed in virion infected cells The data suggest that the two forms represent different phosphorylation states

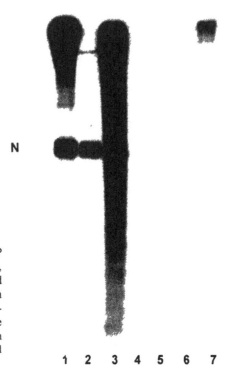

Figure 2. BCV infected HCT cells were labled with ³²P orthophosphate for 19 hrs Cell lysates (lanes 1-3), extracellular pelleted virions (lanes 4-6) and uninfected cell lysates (lane 7) were immunoprecipitated with a polyclonal anti-BCV serum Immune complexes isolated with protein A Sepharose were divided into three parts and not treated (lanes 1, 2 and 7), treated with RNase (lanes 2 and 5) or treated with calf intestinal alkaline phosphatase (lanes 3 and 6)

of the protein. This differs from what was previously reported for mouse hepatitis coronavirus (MHV) JHM since only one phosphorylated species was detected in virion infected cells and the mature virion N comigrated with the phosphorylated intracellular N [8]. Faster migrating forms of both MHV JHM and A59 have been detected intracellularly; however, these have generally been hypothesized to be cleavage or degradation products of N [7,9]. MHV N is phosphorylated at only one or two serines, even though there are a large number of potential sites in the protein [10]. Experiments to determine which sites are phosphorylated on the BCV N are currently underway.

In contrast to the intracellular pool of N in BCV infected cells, only one form of N was present in extracellular virions. It was surprising that the virion form is what appears to be the less phosphorylated form seen in the intracellular fraction. At this point we have not demonstrated which form of N initially interacts with the RNA to be assembled into the mature virions. However, the data indicate that specificity exists with regard to what is incorporated into virions since only one form is seen in the mature virion. If the more phosphorylated form actually initially interacts with the virion RNA, it is apparently processed in some way to give rise to a form that comigrates with the faster migrating intracellular form. Assuming that two populations of N molecules are not phosphorylated differently during translation, two types of processing could account for the different forms of N, a dephosphorylation or proteolytic processing. It is possible that the 55 kDa form of N is converted to the phosphorylated 50 kDa form by dephosphorylation. Data not shown here indicates that the slower migrating form, the 55 kDa species, is initially seen during a short pulse and appears to give rise to the faster migrating form, the 50 kDa species during a chase. Taken together this suggests that proteolytic processing may not be responsible for the faster migrating form seen in the mature virion. If so, dephosphorylation of the 55 kDa species may generate the 50 kDa species.

The functional role of phosphorylation of the coronavirus N protein is not known; however, it is generally hypothesized that phosphorylation plays a role in the interaction between the protein and the viral RNA. MHV N binds elements within the 5' leader sequence found on all mRNAs and the genomic RNA [2]. It was suggested that the interaction may play a role in transcription. Our data also indicates that the BCV N is associated with virion RNAs (Fig. 2 and data not shown). In addition to any role in transcription, the protein must also be involved in assembly. It must specifically encapsidate the genomic RNA and as part of the nucleocapsid, most likely interacts with other viral proteins to assemble the mature virion [11]. These multiple roles of N may be regulated by different phosphorylation states of the protein.

Previously it was shown that MHV JHM N is quickly phosphorylated after synthesis and concomitantly becomes associated with the cell membrane fraction [8]. It was impossible to distinguish multiple phosphorylated forms of N in the MHV [8]. Since we detect the two forms of N in the intracellular fraction of BCV infected cells this may allow us to study the role(s) this modification may play in replication and assembly.

REFERENCES

1. Sturman, L.S. and Holmes, K.V., The molecular biology of coronaviruses. Adv. Virus Res. 1983, 28:35-112.
2. Baric, R.S., Nelson, G.W., Fleming, J.O., Deans, R.J., Keck, J.G., Casteel, N. and Stohlman, S.A., Interactions between coronavirus nucleocapsid protein and viral RNAs; implications for viral transcription, J. Virol. 1988, 62:4280-4287.
3. Nakanaga, K., Yamanouchi, K. and Fujiwara, K., Protective effect of monoclonal antibodies on lethal mouse hepatitis virus infection in mice. J. Virol. 1986, 59:168-171.

4 Stohlman, S A , Kyuwa, S , Polo, J M Brady, D , Lai, M M C and Bergmann, C C , Characterization of mouse hepatitis virus-specific cytotoxic T cells derived from the central nervous system of mice infected with the JHM strain J Virol 1993, 67 7050-7059

5 King, B and Brian, D A , Bovine coronavirus structural proteins J Virol 1982, 42 700-707

6 Robbins, S G , Frana, M F , McGowan, J J , Boyle, J F and Holmes, K V , RNA-binding proteins of MHV detection of monomeric and multimeric N protein with an RNA overlay-protein blot assay Virology 1986, 150 402-410

7 Hogue, B G , King, B and Brian, D A , Antigenic relationships among proteins of bovine coronavirus, human respiratory coronavirus OC43, and mouse hepatitis coronavirus A59 J Virol 1984, 51 384-388

8 Stohlman, S A , Fleming, J O , Patton, Ç D , and Lai, M M C , Synthesis and subcellular localization of the murine coronavirus nucleocapsid protein, Virology, 1983, 130 527-532

9 Cheley, S , and Anderson, R , Cellular synthesis and modification of murine hepatitis virus polypeptides J Gen Virol 1981, 54 301-311

10 Wilbur, S M , Nelson, G W , Lai, M M C , McMillan, M , and Stohlman, S A , Phosphorylation of the mouse hepatitis virus nucleocapsid protein, Biochem biophys Res Commun 1986, 141 7-12

11 Sturman, L S , Holmes, K V , and Behnke, J , Isolation of coronavirus envelope glycoproteins and interaction with the viral nucleocapsid, J Virol 1980, 33 449-462

42

EXPRESSION AND IMMUNOGENICITY OF THE SPIKE GLYCOPROTEIN OF PORCINE RESPIRATORY CORONAVIRUS ENCODED IN THE E3 REGION OF ADENOVIRUS

P. Callebaut and M. Pensaert

Laboratory of Virology
Faculty of Veterinary Medicine
University of Gent
Casinoplein 24
B-9000 Gent, Belgium

ABSTRACT

The full length spike (S) gene of porcine respiratory coronavirus (PRCV) was inserted into the genome of human adenovirus type 5 downstream of the early transcription region 3 promoter. The recombinant virus replicated in cultures of the swine testicle ST cell line and directed the synthesis of S antigen to an amount of approximately 33 µg per 10^6 cells, as determined by ELISA. The antigen was cell-associated except in the late phase of the infection, when a low amount (4 µg per 10^6 cells) was released in the culture supernatant. The cell-associated antigen consisted of 2 polypeptides of 160 K and 175 K, respectively. The 160 K polypeptide comigrated with the authentic S' precursor from PRCV-infected cells. The 175 K polypeptide had the same mobility as the authentic mature S protein from PRCV-infected cells and from PRCV released in the supernatant. The extracellular recombinant antigen corresponded with the 175 K mature protein. Immunofluorescent staining gave evidence that some recombinant S protein was exposed on the cell surface; it also showed that the protein was recognized by conformation-specific anti-S monoclonal antibodies. Piglets, immunized oronasally with the recombinant adenovirus vector developed PRCV-neutralizing serum antibodies and were partially protected against PRCV - challenge.

INTRODUCTION

The use of human adenovirus type 5 (Ad5) for the construction of recombinant expression vectors is well established[1]. Previous work in our laboratory has shown that this

Corona- and Related Viruses, Edited by P. J. Talbot and G. A. Levy
Plenum Press, New York, 1995

virus causes a subclinical respiratory infection in swine[2]. Therefore, Ad5 is a candidate as a vaccine vector for the induction of an immune response in the porcine respiratory tract.

Porcine respiratory coronavirus (PRCV) can serve as a well known model suitable to study the vaccine potential of recombinant Ad5-based vectors. The major virus neutralization mediating antigenic sites are located on the spike (S) glycoprotein[3,4]. In PRCV-infected cells the S protein is initially synthesized as a precursor species S' of 1255 amino acids. By posttranslational processing the mature S protein is produced which is incorporated in the viral membrane[5].

In the present communication we report the development of an infectious Ad vector (AdgpS) which carries the entire PRCV S gene in the early transcription region 3 (E3) of its genome and directs a high level expression in porcine cells. We also describe the biochemical and biological properties of the expressed product and demonstrate that the vector has immunogenic and protective potential in piglets.

MATERIAL AND METHODS

Viruses and Cells

The PRCV isolates TLM 83[6] and 91V44[7] were used as the source of viral genomic RNA and for challenge of piglets, respectively. Both isolates were grown in cultures of the swine testicle (ST) cell line. The same cell line was used for titration of PRCV from nasal swabs of piglets, for growth and titration of AdgpS and for Ad5 neutralization assays. DNA transfection was performed in the human embryonal cell line 293. Swine kidney (SK6) cells were used for PRCV/TGEV neutralization tests[8].

Construction of the Recombinant Vector

The PRCV S gene insert was prepared starting from genomic RNA, extracted from purified PRCV virions, as described[9]. A cDNA was synthesized using the oligonucleotide primer 5'TCTGCTAGCTTAAATTTAATGGACGTGCAC; the Supercript Preamplification system (BRL) was used in the conditions prescribed by the manufacturer except for the incubation time with reverse transcriptase, which was 90 min. The DNA was amplified by PCR using the above oligonucleotide and the oligonucleotide 5' TAGCTAGCCACACCAT-GAAAAAATTATTTG as 5' and 3' amplimers, respectively. Thirty cycles (95°C for 1 min, 72°C for 15 min, 56°C for 2 min) were performed with 5 units of Taq DNA polymerase (Perkin Elmer Cetus).

The recombinant adenovirus AdgpS was constructed according to the procedure of Graham and Prevec[10]. Plasmids pFG144[11] and pFG173[12] were kindly provided by L. Enjuanes (Centro Nacional de Biotecnologia, Spain). Transfection of 293 cells was performed using the calcium phosphate coprecipitation method[13].

Immunologic Assays

To measure the level of expression by antigen-capture sandwich ELISA, cells were lysed in 0.05M Tris-HCl pH 8, 0.5M NaCl, 1% Triton X-100, 0.1% deoxycholate (lysis buffer). To improve detection of extracellular S antigen the culture supernatants were concentrated 10-fold by dialysis against 20% polyethylene glycol 20 000 (PEG) and subsequentely diluted 1/2 in 2x lysis buffer. The ELISA was performed essentially as described[14], using pig anti-TGEV IgG as the capture antibody and pig anti-TGEV horseradish peroxidase conjugate. Serial dilutions of samples were tested and ELISA absorbances were converted to S protein concentrations using a cell lysate with a known concentration of purified PRCV virions as a standard; the S protein content of this preparation was considered 1/4 of the total viral protein.

For immunoblot analysis of recombinant S protein, cells were lysed in 0.05 M Tris-HCl, pH 8, 1% SDS (gel-loading buffer). Cell culture supernatant was concentrated by dialysis against 20 % PEG and subsequently diluted 1/2 in 2x gel-loading buffer to give a final volume which was equal to that of the cell lysate. The proteins were resolved by electrophoresis on a 8% SDS-PAGE gel and were transferred onto nitrocellulose membrane (Bio-Rad) using a Bio-Rad Trans-Blot Cell. The membrane was probed with pig anti-TGEV serum and the reaction was developed using rabbit anti-pig IgG antibodies conjugated to horseradish peroxidase (Nordic).

Indirect immunofluorescence assays were performed using ST cell cultures, grown on glass coverslips and infected with AdgpS at a multiplicity of 10. After 24 h the cells were fixed in 0.1 % paraformaldehyde for 30 min at 4 °C[15] and probed with 2 pools of monoclonal antibodies (Mabs), each of which was directed against one of the antigenic sites A and D of the S protein. The MAbs were gifts of L. Enjuanes and were diluted 1/100. Bound antibody was detected with fluorescein -conjugated anti-mouse IgG antiserum (Nordic).

Serum PRCV neutralizing activity was measured by a microtiter virus neutralization (VN) assay described previously[8]. Ad5-neutralizing activity of the sera was detected by a similar microneutralization assay.

Inoculation and Challenge of Piglets

Four piglets of 4 weeks of age which were free of Ad5-neutralizing serum antibodies were obtained from a PRCV negative farm. Two of them were inoculated oronasally with each 2.5×10^{10} median tissue culture infectious doses ($TCID_{50}$) of AdgpS. The two remaining piglets were not inoculated and served as challenge controls. Challenge was performed four weeks after the AdgpS inoculation, using 10^7 $TCID_{50}$ of PRCV per animal, administered via aerosol as described[7]. In order to study the antibody response to PRCV, serum samples were collected on the day of inoculation with AdgpS, on the day of challenge and on post-challenge days (PCD) 2, 4, 6, 8 and 10, and tested in the VN assay. To estimate the degree of protection against challenge, the shedding of PRCV was determined in nasal swabs collected daily from the day of challenge until PCD10. Infectious virus titers in the secretions were determined as described previously[7].

RESULTS AND DISCUSSION

Construction of the Recombinant Vector

A full length DNA copy of the S gene coding sequence (from 6 nucleotides upstream to 3684 nucleotides downstream the initation codon) was obtained by reverse transcription of the viral genomic RNA, followed by PCR amplification. The amplimers were chosen to introduce a Nhe I site at each end, thus allowing cloning into the Ad5 transfer vector pFG144[11]. In the resulting plasmid pFG144gpS the PRCV gene replaced the largely deleted E3 region which was previously shown to be nonessential for Ad5 replication[2]. The insert was oriented parallel to the direction of transcription of the E3 promoter. The recombinant Ad5-gpS sequences in plasmid pFG144gpS were rescued in the Ad5 genome following cotransfection of human 293 cells along with the overlapping Ad5 subgenomic plasmid pFG173[12] by homologous overlap recombination.

Expression in Porcine Cells in Culture

The kinetics of the S expression as determined in cell lysates and culture supernatants of the porcine cell line ST by quantitative sandwich ELISA are shown in Fig.1. In the cell lysates the cumulative amount of recombinant S antigen increased at the highest rate between 18 and 24 h post-inoculation. This was before the appearance of cytopathic effect and probably consistent with transcription driven by the E3 promoter. The maximum level of 33 μg / 10^6 cells was produced at 48 h post-inoculation, when 50 % of the cells showed cytopathic effect. In the culture supernatant S antigen was detectable starting at 48 h post-inoculation. Late in the infection, at 72 h post-inoculation when cytopathic effect was complete, the yield of extracellular S antigen was 4.4 μg / 10^6 cells.

Immunoblot analysis was performed to identify the recombinant protein in the ST cell lysate and culture supernatant at the terminal phase of the infection. The cell-associated protein and extracellular protein, produced by equal numbers of cells, were analysed. In the lysate 2 polypeptides of 160 K and 175 K, respectively were identified (Fig. 2). The 160 K polypeptide was the predominant species and had the same mobility as the authentic precursor S′ protein present in PRCV-infected cell lysate but not in PRCV released in the medium. The minor 175 K polypeptide comigrated with the authentic mature S protein present in PRCV-infected cells as well as in extracellular PRCV. In the supernatant of AdgpS-infected cells the 175 K polypeptide was identified. These results indicated that the recombinant S is properly processed intracellularly. It was noted that late in the infection the majority of the processed protein is exported out of the cells.

Cells, fixed with paraformaldehyde and assayed by indirect immunofluorescence at 24 h after infection with AdgpS, showed membrane fluorescence when reacted with Mabs against antigenic sites A and D of the S protein. At that early time after infection it was unlikely that the appearance of cell surface fluorescence was caused by S protein, released from the cells and subsequently bound back to cell surface receptors. Therefore, the immunofluorescence suggests that some recombinant S is transported to the cell surface. In addition, the reactivity with the Mabs indicates that the recombinant S is folded in a conformation similar to that of the authentic protein and retains the important antigenic sites.

Immunogenicity in Piglets

Four weeks after the oronasal inoculation of 2 piglets with AdgpS, both piglets showed seroconversion against PRCV and against Ad5. The PRCV-neutralizing activity was

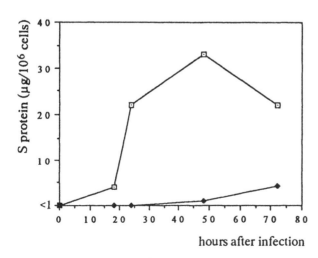

Figure 1. Kinetics of recombinant S production in ST cells. Cells were inoculated with AdgpS at a multiplicity of 10. At the indicated times post-infection cell extracts (□) and culture supernatants (♦) were assayed by antibody sandwich ELISA. The amounts of S protein were expressed in μg per 10^6 cells, calculated as described in Methods.

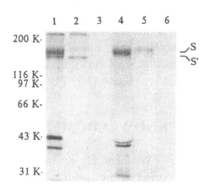

Figure 2. Immunoblot analysis of recombinant S synthesized in ST cells. Cells were inoculated with AdgpS (lanes 2 and 5), with PRCV (lanes 1 and 4) and with Ad5 (lanes 3 and 6) at a multiplicity of 10. At 72 h after infection with the adenoviruses and 48 h after infection with PRCV, cells were lysed (lanes 1 to 3) and culture supernatants (lanes 4 to 6) were concentrated as described in Methods. Following electrophoresis under non-reducing conditions, proteins were transferred to a nitrocellulose sheet and probed by anti-TGEV porcine serum. Precursor S′ and mature S polypeptides are indicated.

detected at titers of 4 and 12; the Ad5 VN-titers were 8 and 32, respectively. Two non-inoculated control piglets remained negative in the VN-tests.

Following challenge the AdgpS inoculated piglets responded by 6 days with a marked increase in their VN-titer, typical of an anamnestic response (Fig. 3.A). The controls showed a response typical of immunologically naive pigs. As indicated in Fig. 3.B, both AdgpS-inoculated piglets showed similar patterns of shedding of PRCV in nasal secretions; there was a marked reduction in the quantity and duration of virus excretion, compared to that of the control piglets.

In spite of the low number of animals studied, the finding that AdgpS elicited a low level of PRCV-neutralizing antibodies indicates that recombinant S protein was expressed in the pigs.

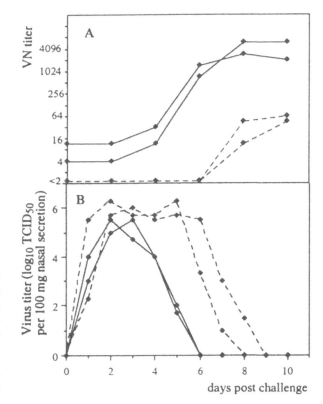

Figure 3. PRCV neutralizing antibody titers in serum (A) and PRCV-titers in nasal secretions (B) at the indicated times after PRCV-challenge of 2 piglets, inoculated with AdgpS 4 weeks before challenge (—) and of 2 control piglets, not inoculated with AdgpS (- - - -). PRCV titers are expressed in $\log_{10}$ median tissue culture infectious doses ($TCID_{50}$) per 100 mg of secretion.

This is confirmed by the observed partial protection and boosted serum antibody response upon challenge. The present data suggest that the immune response induced in the pigs by the recombinant S protein is boosted by the S antigen produced in the initial rounds of replication of the challenge virus, resulting in the shortened duration of challenge virus shedding. Work is in progress to corroborate these results and to study the suitability of the AdgpS vector as a live vaccine.

ACKNOWLEDGMENTS

We thank Dr. K. Van Reeth for help with the animal experiments. This work was supported by the Institute for the Encouragement of Scientific Research in Industry and Agriculture, Brussels.

REFERENCES

1. Berkner, K.L. (1992). Curr. Top. Microbiol. Immunol. 158 : 39.
2. Callebaut, P., Pensaert, M. and Enjuanes, L. (1994). Adv. Exp. Med. Biol. 342 : 469.
3. Correa, I., Gebauer, F., Bullido, M.J., Suné, C., Baay, M.F.D., Zwaagstra, K.A., Posthumus, W.P.A., Lenstra, J.A. and Enjuanes, L.(1990). J. Gen. Virol. 71 : 271.
4. Delmas, B., Rasschaert, D., Godet, M. and Laude, H.(1990). J. Gen. Virol. 71 : 1313.
5. Rasschaert, D., Duarte, M. and Laude, H.(1990). J. Gen. Virol. 71 : 2599.
6. Pensaert, M., Callebaut, P. and Vergote, J.(1986). Vet. Quart. 8 : 257.
7. Van Reeth, K. and Pensaert, M.B.(1994). Am.J.Vet. Res. (in press).
8. Voets, M.T., Pensaert, M. and Rondhuis, P.R.(1980). Vet.Quart. 2 : 211.
9. Callebaut, P. and Pensaert, M.(1992). Med. Fac. Landbouww. Univ. Gent, 57/4b : 2077.
10. Graham, F.L. and Prevec, L.(1991). Methods Mol. Biol. 7 : 109.
11. Ghosh-Choudhury, G., Haj-Ahmad, Y., Brinkley, P., Rudy, J. and Graham, F.L.(1986) Gene 50 : 161.
12. Zheng, B., Graham, F.L., Johnson, D.C., Hanke, T., McDermott, M.R. and Prevec, L.(1993). Vaccine 11 : 1191.
13. Graham, F.L. and van der Eb, A.J.(1973). Virology 52 : 456.
14. Callebaut, P. Debouck, P. and Pensaert , M.(1982). Vet. Microbiol. 7 : 295.
15. Tò, L.T. and Bernard, S.(1992). Res.Virol. 143 : 241.

43

CHARACTERIZATION OF STRUCTURAL PROTEINS OF LELYSTAD VIRUS

Janneke J.M. Meulenberg,[1*] Annelien Petersen-den Besten,[1]
Eric P. de Kluyver,[1] Rob J.M. Moormann,[1] Wim M.M. Schaaper,[2] and
Gert Wensvoort[1]

[1] Institute for Animal Science and Health (ID-DLO)
Department of Virology
Houtribweg 39, NL-8200 AJ Lelystad
[2] Institute for Animal Science and Health (ID-DLO)
Laboratory of Molecular Immunology
Edelhertweg 15, NL-8200 AB Lelystad
The Netherlands

ABSTRACT

The genome of Lelystad virus (LV), a positive-strand RNA virus, is 15 kb in length and contains 8 open reading frames that encode putative viral proteins. Synthetic polypeptides of 15 to 17 amino acids were selected from the amino acid sequences of ORFs 2 to 7 and anti-peptide sera were raised in rabbits. Using these anti-peptide sera and porcine anti-LV serum, we identified three structural proteins and assigned their corresponding genes. Virions were found to contain a nucleocapsid protein of 15 kDa (N), an unglycosylated membrane protein of 18 kDa (M), and a glycosylated membrane protein of 25 kDa (E). The N protein is encoded by ORF7, the M protein is encoded by ORF6, and the E protein is encoded by ORF5.

INTRODUCTION

Lelystad virus (LV) is a small enveloped virus containing a positive-strand RNA genome. It was first identified in 1991 in the Netherlands by Wensvoort et al.[1] and in the United states by Collins et al.[2] as the causative agent of porcine reproductive respiratory syndrome (PRRS). PRRS is mainly characterized by reproductive failure in sows and respiratory problems in pigs of all ages[3].

The genome of LV is a polyadenylated RNA molecule of about 15 kb, which contains eight ORFs that probably encode the replicase genes (ORFs 1a and 1b), the envelope proteins

---

[*] Address correspondence to Dr. Meulenberg.

Corona- and Related Viruses, Edited by P. J. Talbot and G. A. Levy
Plenum Press, New York, 1995

(ORFS 2 to 6) and the nucleocapsid protein (ORF7)[4,5]. ORFs 2 to 7 are most likely expressed from six subgenomic RNAs, which are synthesized during replication[4,6].

LV (also named PRRS virus) resembles equine arteritis virus (EAV), lactate dehydrogenase-elevating virus (LDV) and simian hemorrhagic fever virus (SHFV) in genome organization, replication strategy, amino acid sequence of the proteins and preference for infection of macrophages, both *in vivo* and *in vitro*[4,5]. Because of these similarities, proposals have been made to classify LV, EAV, SHFV, and LDV into a new virus family, tentatively named the *Arteriviridae*[4,5,7]. Arteriviruses have a genome organization and replication strategy similar to coronaviruses but the size of their genome is much smaller (12-15 kb) and they have different morphological and physicochemical properties.

Although the replication strategy of LV has been studied, and the complete nucleotide sequence of the viral genome has been determined, little is still known about the structural proteins of LV. In this paper these viral proteins were studied in more detail. The E, M, and, N proteins encoded by ORFs 5 to 7 respectively were shown to be structural proteins of LV.

MATERIALS AND METHODS

Cells and Viruses

LV was either grown on porcine alveolar macrophages or on CL2621 cells (courtesy of Boehringer-Ingelheim, St. Joseph, Mo.). Macrophages were maintained as described before[1]. CL2621 cells were maintained in Eagles basal medium supplemented with 5% fetal bovine serum, 100 IU/ml penicillin, and 100 µg/ml streptomycin. To prepare concentrated and purified virions, confluent monolayers of CL2621 cells were infected at a multiplicity of infection (MOI) of 0.1. At the beginning of cytopathic changes (48-56 h after infection), the medium was harvested and centrifuged for 20 min at 1200 x g. The virus in the medium was concentrated by precipitating it with 6% polyethylene glycol 20,000 overnight at 4 °C, and was then centrifuged at 10,000 x g for 45 min. The pellet was resuspended in TNE buffer (0.01 M Tris-HCl pH 7.2, 0.1 M NaCl, 1 mM EDTA) and layered on a 30-0% glycerol 0-50%-di-K-tartrate gradient[8].

Endoglycosidase Treatment

Purified LV preparations were resuspended in 25 µl Endoglycosidase buffer (1% NP40, 1mM phenylmethylsulfonyl fluoride and 1 µg/ml aprotinin, 1 µg/ml pepstatin A, and 1 µg/ml leupeptin in phosphate-buffered saline (PBS)). Then 800 mU of peptide N-glycosidase F (PNGaseF; Boehringer Mannheim) was added and the reaction mixture was incubated overnight at 37 °C. Controls were treated similarly, except the PNGaseF was omitted.

Preparation of Antisera

Polyvalent antiserum 21 directed against LV was obtained from a specific-pathogen-free (SPF) pig infected intranasally with 10^5 TCID$_{50}$ of a fifth cell culture passage of LV (CDI-NL-91; Institute Pasteur I-1102). Blood samples were taken 42 days after infection. Gene-specific rabbit sera directed against ORFs 2 to 7 were obtained by use of synthetic peptides of 15 to 17 residues containing an amino acid sequence specific for each ORF. The peptides were conjugated to keyhole limpet hemocyanin. SPF rabbits were immunized intramuscularly and subcutaneously with 1 mg peptide conjugated to keyhole limpet hemocyanin in complete Freund's adjuvant. After one month, the rabbits were given a booster injection of the same amount of conjugated peptide in incomplete Freund's adjuvant. The rabbits were bled at 12 weeks after the first immunization. Sera were tested for their reactivity with the various peptides in an enzyme-linked immuno sorbent assay (ELISA), using peptides coated to M96 plates[9]. Sera were also tested for their reactivity with viral antigen in an immunoperoxidase monolayer assay (IPMA) using LV infected alveolar lung macrophages, essentially as described by Wensvoort et al.[1].

Western Blot Analysis

Viral protein samples were suspended in Laemmli sample buffer[10], heated for 2 min at 100 °C, and separated by sodium dodecyl sulfate polyacrylamide gel electrophoresis (SDS-PAGE) on gels containing 12.5% polyacrylamide. The separated proteins were transferred to nitrocellulose paper by electroblotting[11]. Polyclonal antiserum 21 and anti-peptide sera were diluted 1:50 in PBS containing 2% NaCl, 0.05% Tween-80, and 5% horse serum. Nitrocellulose strips were incubated with these diluted antisera for 1 h at 37 °C. The strips were washed three times with PBS containing 2% NaCl and 0.05% Tween-80. They were then incubated with rabbit anti-swine IgG horseradish peroxidase (1:500) or goat anti-rabbit IgG horseradish peroxidase (1:1000) diluted in PBS containing 2% NaCl, 0.05% Tween-80, and 5% horse serum for 1 h at 37 °C. Finally, the strips were washed three times in PBS and stained in a solution of 0.6 mg/ml 4-chloro-1-naphtol, 20% (v/v) methanol, and 0.3 μl/ml H_2O_2 (30%).

RESULTS AND DISCUSSION

Identification of Structural Proteins

Gene-specific antisera, containing antibodies directed against peptides of ORFs 2 to 7 were raised in rabbits. Six sera - - 690 (anti-ORF2), 694 (anti-ORF3), 698 (anti-ORF4), 704 (anti-ORF5), 710 (anti-ORF6), and 714 (anti-ORF7) - - were selected that reacted positively with the corresponding peptide in an ELISA (Table 1). Most of them also reacted positively in an IPMA with LV-infected alveolar macrophages and immunoprecipitated the in vitro translation products of their corresponding ORFs (Table 1). The generated gene-specific anti-peptide sera were used to identify the proteins incorporated in virus particles. Lelystad virus was purified on a glycerol-di-K-tartrate gradient, and infectious peak fractions, found at densities of 1.16-1.17 g/cm³, were subjected to Western blot analyses using convalescent serum 21 and the gene-specific anti-peptide sera. Serum 21 recognized three structural proteins with an apparent molecular weight (Mw) of 25, 18, and 15 kDa (Fig.1a). Besides these three proteins, two faint bands of 28 and 42 kDa were observed. These were not detected on the control strip

Table 1. Reactivity of sera raised against LV-specific peptides of ORFs 2 to 7, using different test-systems

Serum	ORF	Amino acids[a]	Sequence	IPMA[b]	IVT[c]	WB[d]
690	2	64-78	CTLPNYRRSYEGLLPN	–	+	–
694	3	75-92	CKIGHDRCEERDHDELLM	+	+	–
698	4	62-77	CQEKISFGKSSQCREAV	+	+	–
704	5	145-161	CNFIVDDRGRVHRWKSPI	+	+	+
710	6	154-171	CVLGGKRAVKRGVVNLVKY	+	–	+
714	7	43-60	CGGQAKKKKPEKPHFP	+	+	+

[a]Location of the peptide sequence in each ORF.
[b]Immunoperoxidase monolayer assay on macrophages infected with Lelystad virus.
[c]Immunoprecipitation of radiolabeled in vitro translation products of ORFs 2 to 7.

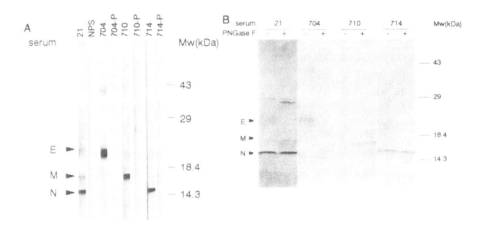

Figure 1. Western blot analysis of virions of LV. Virions were purified by isopycnic sedimentation on a glycerol-di-K-tartrate (A) Infectious fractions were resuspended in Laemmli sample buffer and were separated on a 12 5% polyacrylamide gel by SDS-PAGE. Proteins were transferred to nitrocellulose paper, and nitrocellulose strips were immunostained with porcine anti-LV serum 21, negative pig serum (NPS), gene-specific anti-peptide sera 704, 710, and 714, and their corresponding presera, 704-P, 710-P, and 714-P. The positions of the 15 kDa N protein, the 18 kDa M protein and the 25 kDa E protein are indicated with an arrow head. (B) Samples of LV virions were treated with PNGaseF (+) or left untreated (-) After SDS-PAGE and electrotransfer to nitrocellulose paper, proteins were immunostained with porcine anti-LV serum 21 and gene-specific anti-peptide sera 704, 710, and 714. The positions of the E, M, and N protein are indicated

incubated with a negative pig serum. Anti-peptide serum 704 (specific for ORF5), recognized the 25 kDa protein (E) and a faint band of 42 kDa. We concluded that the E protein is encoded by ORF5. The much fainter protein band observed at 42 kDa might be a dimer of the ORF5 encoded protein, still present to a limited extent under denaturing conditions. Anti-peptide serum 710 stained an 18 kDa protein (M), indicating that this structural protein is expressed from ORF6. Anti-peptide serum 714 reacted with a 15 kDa protein (major band) and a 28 kDa protein (minor band). This finding proves that the 15 kDa protein (N) is encoded by ORF7. The 28 kDa protein is probably a dimeric form of the N protein. No staining was observed when these Western blot strips were incubated with 704, 710, or 714 presera, or with anti-peptide sera specific for ORFs 2, 3, and 4.

Glycosidase Treatment of Purified LV

To establish which structural proteins of LV are glycosylated PNGaseF- treated and untreated virus preparations were analyzed on Western blots stained with convalescent serum 21 and the anti-peptide sera specific for ORFs 5, 6, and 7. As is shown in Figure 1b, after treatment with PNGaseF the apparent Mw of the E protein was reduced to approximately 17 to 18 kDa. The apparent Mw of the dimer of the E protein was reduced from 42 to 34 kDa. The size of the M and N proteins remained the same before and after treatment with PNGaseF. These results show that E is an N-glycosylated structural protein, whereas M and N are not. The size difference ($\pm$ 7 kDa) between the unglycosylated and glycosylated E protein suggested that both putative N-glycosylation sites are functional *in vivo*.

Comparison of Structural Proteins of LV, EAV, LDV, and SHFV

Previous studies have shown that LV resembles EAV, LDV, and SHFV in genome size, virion architecture, genome organization, gene expression strategy, and amino acid sequences of viral proteins. The identification of three major structural proteins, designated N, M, and E further confirms the relationship between LV and EAV, LDV, and SHFV. The amino acid sequence of the 15 kDa protein (N) encoded by ORF7 is extremely basic and is 41 and 20 % identical with the amino acid sequence nucleocapsid protein of LDV[12,13] and EAV[14] respectively. The identity of the nucleocapsid proteins of EAV and LDV was established by virus fractionation experiments. After the virus particles were treated with detergent only the N proteins of 15 to 16 kDa co-sedimented with the viral genome in the bottom fractions of a sucrose gradient[15]. In a sucrose gradient layered with NP40-treated LV-virions, the N protein was found in the bottom fraction of the gradient at a density of 1.18 g/cm^3, whereas the M and E protein cosedimented in the middle of the sucrose gradient at a density of 1.10 g/cm^3. This supplies further evidence for the assumption that the 15 kDa N protein of LV is the nucleocapsid protein.

The 18 kDa non-N-glycosylated envelope protein M encoded by ORF6 has the same hydrophobicity profile as the M protein of mouse hepatitis virus (MHV), infectious bronchitis virus (IBV), EAV, LDV, and the E protein of Berne torovirus[4]. These proteins are characterized by the presence of three hydrophobic segments at the N-terminus. Protease protection experiments have shown that the M proteins of MHV-A59 and IBV are type III integral membrane proteins[16,17]. They are anchored in the membrane by the three successive hydrophobic domains, whereas the C-terminal part is thought to be associated with the membrane surface.

The E protein of 25 kDa encoded by ORF5 was shown to be N-glycosylated, probably at two different sites. The E protein incorporated in virus particles was sensitive to PNGaseF (Fig. 1b), but partially resistant to EndoH (data not shown). These findings indicate that during virus maturation, E is transported through the Golgi apparatus and its N-linked oligosaccharides undergo Golgi-specific modifications. E is the counterpart of G$_l$, a structural envelope protein encoded by ORF5 of EAV[18]. G$_l$ migrated as a heterogeneous protein of 30 to 42 kDa on SDS-PAGE, because a variable number of lactosamine repeats were added to the N-linked core oligosaccharide. The E protein of LV, however, was not susceptible to Endo-ß-galactosidase (data not shown). Therefore the maturation of the N-linked oligosaccharide side chains of the E and G$_l$ protein is probably different.

Although the structural proteins of SHFV have not been studied in detail, a nucleocapsid protein of 12 kDa (N), an unglycosylated protein of 16-18 kDa (M), and a glycosylated glycoprotein of 50 kDa (counterpart of E and G$_l$) were identified in virus particles[7].

The N, M, and E proteins were also detected by Nelson *et al.*[19] in cell lysates of CL2621 cells infected with LV or with a United states isolate of LV. Although we have generated gene-specific anti-peptide sera, that recognized the N-glycosylated *in vitro* translation products of ORFs 2, 3, and 4 (Table 1) we were not able to detect the proteins encoded by these ORFs in cell lysates or purified LV. Perhaps these proteins were expressed only at very low levels in CL2621 cells and only small amounts of these proteins were incorporated in virus particles. Furthermore, the affinity of the polyvalent sera and monospecific peptide sera might not be high enough to detect such low amounts of protein. We do not yet know whether the gene products of ORFs 2 to 4 are structural proteins. The hydrophobicity profile of ORF2 of LV is similar to that of ORF2 of EAV and LDV. The gene product of ORF2, a 25 kDa glycoprotein designated G$_s$, was detected in a purified virus preparation of EAV[18]. It constituted only 1-2 % of the virion protein and was not recognized by a polyvalent anti-virion serum. Apparently this protein is only incidentally incorporated into virions.

ACKNOWLEDGMENTS

We thank J. Langeveld for the synthesis and analysis of the synthetic peptides. Part of this work was supported by Boehringer Ingelheim, Germany, and the Produktschap voor Vee en Vlees (PVV), the Netherlands.

REFERENCES

1 Wensvoort, G , Terpstra, C , Pol, J M A , Ter Laak, E A , Bloemraad, M , de Kluyver, E P, Kragten, C ,
   van Buiten, L , den Besten, A , Wagenaar, F , Broekhuijsen, J M , Moonen, P L J M , Zetstra, T , de Boer,
   E A , Tibben, H J , de Jong, M F , van 't Veld, P , Groenland, G J R , van Gennep, J A , Voets, M Th ,
   Verheijden, J H M , and Braamskamp, J Mystery swine disease in the Netherlands the isolation of
   Lelystad virus Vet Quart 1991,13 121-130
2 Collins, J E , Benfield, D A , Christianson, W T , Harris, L , Hennings, J C , Shaw, D P, Goyal, S M ,
   McCullough, S , Morrison, R B , Joo, H S , Gorcyca, D E , Chladek, D W Isolation of swine infertility
   and respiratory syndrome virus (Isolate ATCC-VR-2332) in North America and experimental reproduction of the disease in gnotobiotic pigs J of Vet Diagn Invest 1992,4 117-126
3 Wensvoort, G Lelystad virus and the porcine epidemic abortion and respiratory syndrome Vet Res
   1993,24 117-124
4 Meulenberg, J J M , Hulst, M M , de Meijer, E J , Moonen, P L J M , den Besten, A , de Kluyver, E P,
   Wensvoort, G , and Moormann,, R J M Lelystad virus, the causative agent of porcine epidemic abortion
   and respiratory syndrome (PEARS) is related to LDV and EAV Virology 1993,192 62-74
5 Conzelmann, K K , Visser, N , van Woensel, P, and Tiel, H J Molecular characterization of porcine
   reproductive and respiratory syndrome virus, a member of the Arterivirus group Virology 1993,193 329-
   339
6 Meulenberg, J J M , de Meijer, and Moormann, R J M Subgenomic RNAs of Lelystad virus contain a
   conserved junction sequence J Gen Virol 1993,74 1697-1701
7 Plagemann, P G W , and Moennig, V Lactate dehydrogenase-elevating virus, equine arteritis virus, and
   simian hemorrhagic fever virus a new group of positive-strand RNA viruses Adv in Virus Res
   1991,41 99-192
8 Purchio, A F , Larson, R and Collet, M S Characterization of bovine viral diarrhea proteins J Virol
   1984,50 666-669
9 Suter, M A Modified ELISA techniques for anti-hapten antibodies J Immunol Meth 1982,53 103-108
10 Laemmli, U K Cleavage of structural proteins during the assembly of the head of bacteriophage T4
   Nature (London) 1970,227 680-685
11 Towbin, H , Staehelin, T , and Gordon, J Electrophoretic transfer of proteins from polyacrylamide gels
   to nitro cellulose sheets Procedure and some applications Proc Natl Acad Sci 1979,76 4350-4354
12 Chen, Z , Kuo, L , Rowland, R R R , Even, C , Faaberg, K S , and Plagemann, P G W Sequences of 3'
   end of genome and of 5'end of open reading frame 1a of Lactate dehydrogenase-elevating virus and
   common junction motifs between 5'leader and bodies of seven sugenomic mRNAs J Gen Virol
   1993,74 643-660
13 Godeny, E K , Speicher, D W , and Brinton, M A Map location of lactate dehydrogenase-elevating virus
   (LDV) capsid protein (Vp1) gene Virology 1990,177 768-771
14 den Boon, J A , Snijder, E J , Chirnside, E D , de Vries, A A F , Horzinek, M C , and Spaan, W J M Equine
   arteritis virus is not a togavirus but belongs to the coronavirus superfamily J Virol 1991,65 2910-2920
15 Brinton-Darnell, M , and Plagemann, P G W Structure and chemical-physical characteristics of lactate-
   dehydrogenase-elevating virus and its RNA J Virol 1975,16 420-433
16 Rottier, P , Brandenburg, D , Armstrong, J , van der Zeijst, B A M , and Warren, G Assembly in vitro of
   a spanning membrane protein of the endoplasmic reticulum The E1 glycoprotein of coronavirus mouse
   hepatitis virus A59 Proc Natl Acad Sci 1984,81 1421-1425
17 Cavanagh, D , Davis, P J , and Pappin, D J C Coronavirus IBV glycopeptides locational studies using
   proteases and saponin, a membrane permeabilizer Virus Res 1986,4 145-156
18 De Vries, A A F , chinside, E D , Hozinek, M C , and Rottier, P J M Structural proteins of equine arteritis
   virus J Virol 1992,66 6294-6303
19 Nelson, E A , Chrisopher-Hennings, J , Drew, T , Wensvoort, G , Collins, J E , and Benfield, D A
   Differentiation of United states and european isolates of porcine reproductive and respiratory syndrome
   virus by monoclonal antibodies J of Clin Microbiol 1993,31 3184-3189

44

STRUCTURAL GENE ANALYSIS OF A QUEBEC REFERENCE STRAIN OF PORCINE REPRODUCTIVE AND RESPIRATORY SYNDROME VIRUS (PRRSV)

H. Mardassi, S. Mounir, and S. Dea

Centre De Recherche En Virologie
Institut Armand-Frappier
Université Du Québec
Laval, Québec
Canada, H7V 4Z3

ABSTRACT

The 3′ end genomic region of a Québec PRRSV reference strain (IAF-exp91), propagated in porcine alveolar macrophages (PAM), was sequenced and compared to the prototype European strain, the Lelystad virus (LV). The sequence, which represents the 3′-terminal 2834 nucleotides, encompassed 5 ORFs corresponding to ORFs 3 to 7 of LV. Extensive genomic variations resulting from an important rate of nucleotide additions, substitutions, and deletions were demonstrated between the two viruses. Indeed, the two corresponding sequences displayed a total of 66% and 63% identity at the nucleotide and amino acid levels, respectively. The predicted products of ORFs 5, 3, and 7, showed the highest rate of amino acid variations with percentages of identity of 52, 54, and 59, respectively. Sequence analysis of an additional Québec strain that could be propagated in a continious cell line (MARC-145), suggested that Québec PRRSV strains belong to a genotype distinct from that of LV, thus confirming previous serological results which allowed to divide PRRSV isolates into two distinct antigenic subgroups (U.S. and European).

Six viral major polypeptides with apparent M_rs of 14.5K, 15K, 19K, 24.5K, 29K, and 42K could be identified from lysates of viral infected cells, of which the 15, 19 and 24.5K species seemed to be structural. In vitro translation products of ORFs 7 and 6 comigrated with the 15 and 19 K viral proteins, whereas that of ORF 5 may be associated to the 24.5K when translated in presence of microsomes. Consequently, it is likely that ORFs 7 to 5, encode the three major structural proteins.

INTRODUCTION

It is now well established that porcine reproductive and respiratory syndrome (PRRS), is associated to a new porcine virus (PRRSV)[1,2,3] that has emerged, and which display strong morphological and biological similarities with members of the *Arterivirus* group, namely: equine arteritis virus (EAV), lactate dehydrogenase elevating virus (LDV), and simian hemorraghic fever virus (SHFV)[4]. Sequencing analysis of the whole genome of two reference strains isolated in the Netherlands[5,6], confirmed the relatedness between PRRSV and *Arteriviridae*. Effectively, the 15 kb positive-stranded and polyadenylated RNA molecule, contains eight ORFs similarly organized as EAV. The two first (ORF 1a and 1b), at the 5' end, were associated to the polymerase gene, whereas 5 ORFs (ORFs 2 to 6) identified at the 3' region, may encode membrane-associated proteins. The most 3' ORF (ORF 7) was assumed to encode the viral nucleocapsid protein[5,6]. As arteriviruses, PRRSV replicates in the cytoplasm of infected cells and generates multiple subgenomic mRNAs which form a 3' coterminal nested set[5].

Evidence for antigenic variations between North American and European isolates of PRRSV was previously reported[7], and lately confirmed by the extremely high genomic variations observed at the nucleocapsid gene[8]. The aim of the present study was to investigate the extent of this genomic variation and to analyse the products of ORFs 5 to 7, which by analogy to EAV may encode viral structural proteins.

METHODOLOGY

The origin, propagation, and identification of the Québec reference IAF-exp91 and IAF-Klop isolates of PRRSV, as well as polyclonal antisera directed against these two isolates were described elsewhere[9,10]. IAF-exp91 could be only propagated on PAM cells, whereas IAF-Klop which was initially isolated on PAM cells, was adapted to grow in the established monkey kidney cell line MARC-145[11], kindly provided to us by Dr J. Kwang (U.S Meat Animal Research Center, USDA, ARS, Clay Center, Nebraska).

Genomic RNA of both isolates was extracted by the method of Chomczinsky & Sacchi[12]. Specific viral cDNA clones representing the 3' end of the IAF-exp91 genome were obtained by setting up a cDNA library using an oligo (dT) primer as previously reported[8]. Plasmids containing viral cDNAs were sequenced on both strands according to the method of Sanger et al.[13]. Reverse transcription and polymerase chain reaction (RT/PCR) using three primer pairs designed according to the obtained IAF-exp91 sequence, was carried out in order to generate a cDNA copy of each of IAF-Klop ORFs 5 to 7[10]. The nucleotide and amino acid sequences were analysed using the MacVector 3.5 (International Biotechnologies) and GeneWork 2.2 (Intelligenetics) sequence analysis programs.

The products of ORFs 5 to 7 were analysed by in vitro translation experiments. The coding sequence of these ORFs were in vitro transcribed and then translated using wheat germ extracts or rabbit reticulocyte lysates in case where canine pancreatic microsomal membranes were added. The products were analysed by SDS-PAGE and compared to those obtained by radioimmunoprecipitation[9] assay (RIPA) of [^{35}S] methionine labelled IAF-Klop infected MARC-145 cells.

RESULTS AND DISCUSSION

A consensus sequence of 2834 nucleotides of the IAF-exp91 strain was derived from eight viral specific cDNAs clones, where 5 ORFs corresponding to ORFs 3 to 7 of

LV were identified. This sequence extends to the poly(A) tail which is separated from the ORF 7 stop codon by a 3' non coding region of 152 nucleotides long. Computer assisted analysis of this 3'-terminal region revealed a high rate of genomic variation randomly distributed along the sequence. Base substitutions, additions and deletions could be identified resulting in 66% nucleotide identity with the corresponding LV 3'-terminal region. As a consequence, considerable changes were introduced along the predicted product of each ORF. Indeed, amino acid identity with LV varied from 52% and 81% depending on the ORF (Table. 1), ORFs 5, 3, and 7 being the most variable. These findings are in agreement with previous results which divided PRRSV isolates into two distinct antigenic subgroups[7]. Thus, classification of PRRSV isolates into two distinct genotypes should be considered. Such a proposal is reinforced by the high relatedness observed between the two Québec PRRSV isolates, IAF-exp91 and IAF-Klop, despite difference in their tissue culture tropism. Indeed, amino acid identity between among both Québec strains , was above 90% for ORFs 7, 6, and 5 (Table 1). It is noteworthy, that PRRSV has been identified in North America at least four years before its recognition in Europe. This could represent one explanation for these extensive genomic variations between North American and European strains. Another characteristic which is shared by North American isolates, is the length of their 3' non coding region which exceeds that of European strains by 22 nucleotides. This makes North American isolates of PRRSV singular among the arteriviruses in that they contain the longest 3' non coding region. An RT/PCR based on this difference was developed in our laboratory allowing the differenciation between North American and European isolates[10].

It has been previously shown that LV is much more closely related to LDV, than to EAV[5,6]. This was also true in case of the two Québec isolates despite their high divergence from LV. Whether PRRSV has emerged from LDV due to continual genomic variation and host change remains questionable.

Among the features that characterize arteriviruses, is the resemblance of their structural polypeptides profiles[4]. Since PRRSV is similar to LDV and EAV, it is likely that the 3' most ORFs (ORFs 5 to 7) encode structural proteins. The characteristics of these ORFs products as well as those of ORFs 3 and 4 are presented in Table 2. [^{35}S] methionine labelling of intracellular viral proteins and immunoprecipitation experiments with rabbit or porcine polyclonal antisera permitted identification of at least 6 major viral specific polypeptides with M_rs of 14.5 K, 15 K, 19 K, 24.5 K, 29 K, and 42 K (Fig.1, lanes 8, 9 and 10). Among these polypeptides, the 15 K, 19 K, and 24.5 K seemed to be the major viral structural proteins, since they were also precitated from extracellular virions (data not shown). *In vitro* translation of ORFs 7 and 6, yielded one polypeptide each, which comigrated respectively with the viral 15 K and 19 K (Fig.1, lanes 3 and 4). The translated product of ORF 5 has a M_r of 18.5 K and could not be associated with any viral protein. Nonetheless, in presence of microsomes, ORF 5 yielded a protein which migrated slightly above the 24.5 K viral protein

Table 1. Percentage amino acid identity between predicted IAF-exp91 ORFs 3 to 7 products and those of IAF-Klop, LV, LDV, and EAV

ORF	IAF-klop	LV	LDV	EAV
7	99	59	49	23
6	98	81	51	17
5	94	52	46	18
4	ND	68	35	17
3	ND	54	24	11

ND: not determined

Table 2. IAF-exp91 ORFs 3 to 7 characteristics

ORF	Position	Number of amino acids encoded	Calculated size (kDa)		Potential N-Glycosylation sites
7	2311-2679	123 (128)		13.6 (13.8)	0 (1)
6	1797-2318	174 (173)		19.1 (18.9)	1 (2)
5	1210-1809	200 (201)	19.0	22.4 (22.4)	2 (2)
4	663-1196	178 (183)	17.4	19.6 (20.0)	4 (4)
3	118-879	254 (265)	26.3	29.0 (30.6)	7 (7)

The corresponding values for LV are indicated within the parenthesis. In case of ORFs 3, 4, and 5, the lower values correspond to the molecular mass of the proteins after removal of the predicted N-terminal signal sequence.

(data not shown). It is worthy to note that all these in vitro translated products were recognized by antiserum directed against the virion (data not shown). Given the position in the genome of ORFs 5 to7, the molecular weight of their translated products, and finally, the fact that all these in vitro products were recognized by antivirion sera, we are allowed to speculate that they encode for the viral 15 K, 19 k and 24.5 K proteins, respectively. This preliminary analysis awaits for confirmations that should be obtained when reagents such as monoclonal antibodies or monospecific antisera directed against each of these viral proteins will be available.

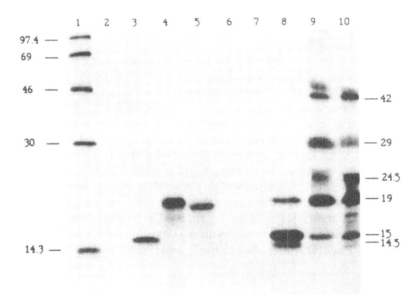

Figure 1. SDS-PAGE analysis of the *in vitro* translation products of ORFs 7, 6 and 5, and comparison with the viral specific polypeptides. Five µl of the translation reaction product of IAF-exp91 ORF 7, and IAF-Klop ORFs 6 and 5 were analyzed on 15% polyacrylamide gel in lanes 3, 4, and 5. Non recombinant plasmid was used as negative control (lane 2). For comparison, IAF-Klop viral polypeptides were precipitated from lysates of infected MARC-145 cells, using rabbit polyclonal antiserum directed against IAF-exp91 (lane 8), porcine hyperimmune serum directed against the homologous strain (lane 9) or a pooled neutralizing anti-PRRSV porcine serum (lane 10). Negative controls, included the incubation of homologous porcine hyperimmune serum with lysates of mock-infected cells (lane 6), and incubation of serum from serologically-negative SPF pigs with lysates of PRRSV-infected cells (lane 7). Lane 1: [14]C methylated protein markers.

CONCLUSION

The extentive genomic variations observed between Quebec and European PRRSV isolates, confirms the previous serologic findings, and supports their classification into two distinct genotypes within the *Arteriviridae* This study provides additional data that make PRRSV more related to the other members of the *Arteriviridae* Indeed, as EAV[14], genes encoding structural proteins seemed to be located at the 3' end of the viral genome

REFERENCES

1 Wensvoort G , Terpstra C , Pol J M A , Ter Laak E A , Bloemraad M , De Kluyver E P , Kragten C , Van Buiten L , Den Besten A , Wagenaar F , Broekhuijsen J M , Moonen P L J M , Zetstra T , De Boer E A , Tibben H J , De Jong M F , Van't Veld P , Groenland G J R , Van Gennep J A , Voets M TH , Verheijden J H M , Braamskamp J Mystery swine disease in the Netherlands The isolation of Lelystad virus Vet Q 1991,13 121-130

2 Benfield D A , Nelson E , Collins J E , Harris L , Goyal S M , Robinson D , Christianson W T , Morrison R B , Gorcyca D , Chladek D Characterization of swine infertility and respiratory syndrome (SIRS) virus (isolate ATCC VR-2332) J Vet Diagn Invest 1992,4 127-133

3 Dea S , Bilodeau R , Athanassious R , Sauvageau R , Martineau G P Swine reproductive and respiratory syndrome in Quebec isolation of an enveloped virus serologically-related to Lelystad virus Can Vet J 1992,33 801-808

4 Plagemann P G W , Moening V Lactate dehydrogenase-elevating virus, equine arteritis virus, and simian hemorrhagic fever virus a new group of positive-strand RNA viruses Adv Virus Res 1992,41 99-192

5 Meulenberg J J M , Hulst M M , De Meijer E J , Moonen P L J M , Den Besten A , De Kluyer E P , Wenswoort G , Moormann R J M Lelystad virus, the causative agent of porcine epidemic abortion and respiratory syndrome (PEARS), is related to LDV and EAV Virology 1993,192 62-72

6 Conzelmann K K , Visser N , VanWoensel P , Thiel H-j Molecular characterization of porcine reproductive and respiratory syndrome virus, a member of the arterivirus group Virology 1993,193 329-339

7 Wensvoort G , De Kluyver E P , Luijtze E A , Den Besten A , Harris L , Collins J E , Christianson W T , Chladek D Antigenic comparison of Lelystad virus and swine and respiratory syndrome (SIRS) virus J Vet Diagn Invest 1992,4:134-138

8 Mardassi H , Mounir S , Dea S Identification of major differences in the nucleocapsid protein genes of a Quebec strain and European strains of porcine reproductive and respiratory syndrome virus J Gen virol 1994,75 681-685

9 Mardassi H , Athanassious R , Mounir S , Dea S Porcine reproductive and respiratory syndrome virus Morphological, biochemical and serological characteristics of Quebec isolates associated to acute and chronic outbreaks of PRRS Can J Vet Res 1994,58 55-64

10 Mardassi H , Wilson L , Mounir S , Dea S Detection of porcine reproductive and respiratory syndrome virus and efficient differenciation between canadian and european strains by reverse transcription and PCR amplification J Clin Microbiol 1994,32 2197-2203

11 Kim H S , Kwang J , Yoon I J , Joo H S , Frey M L Enhanced replication of porcine reproductive and respiratory syndrome (PRRS) virus in a homogeneous subpopulation of MA-104 cell line Arch of Virol 1993,133 477-483

12 Chomczynsky P , Sacchi N Single-Step Method of RNA Isolation by Acid Guanidium Thiocyanate-Phenol-Chlorophorm Extraction Anal Biochem 1987,162 156-159

13 Sanger F , Nicklen S , Coulson A R DNA sequencing with chain-terminating inhibitors *PNAS U S A 1977 74 5463-5467*

14 De vries A A F , Chirnside E D , Horzinek M C , Rottier J M Structural proteins of equine arteritis virus J Virol 1992,66 6294-6303

45

SITE DIRECTED MUTAGENESIS OF THE MURINE CORONAVIRUS SPIKE PROTEIN

Effects on Fusion

E. C. W. Bos, L. Heijnen, and W. J. M. Spaan

Institute of Virology, Faculty of Medicine
University of Leiden
The Netherlands

ABSTRACT

Mutations were introduced in the transmembrane region of the spike protein of the murine coronavirus A59. The maturation of these mutant S proteins was not affected, they were all expressed at the cell surface, and became acylated, however some mutant S proteins did not induce cell-to-cell fusion. An I→K change in the middle of the predicted transmembrane (TM) anchor and mutation of the first three cysteine residues of the TM domain resulted in a fusion-negative phenotype. We propose a model by which these data can be explained.

INTRODUCTION

The spike (S) protein of the murine coronaviruses has several important features. The glycoprotein has been shown to bind to the receptor on the host cell[1], and is capable of inducing cell-to-cell fusion[2]. Furthermore the spike protein can induce neutralizing antibodies[3].

The murine coronavirus Spike protein is synthesized as a 110 kDa core protein that is co-translationally glycosylated in the ER to become a 150 kDa glycoprotein. During transport from the ER to and through the Golgi stacks, the protein matures to its 180 kDa form. Subsequently, part of the molecules are cleaved into two subunits of 90 kDa (S1 and S2). In recent years several groups have shown that cleavage of S is not a prerequisite for cell-to-cell fusion. However, cleaved S induces fusion more rapidly[4, 5, 6].

Although it has been known for several years that the spike protein functions as a fusion protein, the fusion mechanism is still unknown. Gallagher et al.[7] showed that fusion could be induced at low pH when mutations occurred in the first heptad repeat. Furthermore, Grosse and Siddell[8] found that spike proteins that were resistant to neutralizing antibody

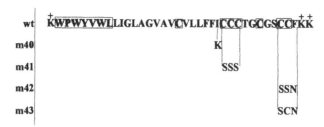

Figure 1. Schematic representation of the transmembrane anchor of the coronavirus spike protein. The amino acid sequence of the wild type protein is depicted at the upper line. For the mutants, only the changed amino acids are depicted. Charged residues at the N- and C-terminus of the TM anchor are marked +. Conserved residues (the WYV domain and the cysteine residues) are in boxes.

11F, did contain mutations in the second heptad repeat of the S2 sub-unit. This monoclonal antibody 11F had previously been shown to inhibit cell fusion, but not the virus-receptor interaction[9]. Thus, the S2 sub-unit of the S protein is very likely to be involved in the pH-related conformational changes that are required for fusion. In this study we have focused on the role of the transmembrane region in cell fusion.

RESULTS

The TM anchor is defined as the region between charged residues that is potentially inserted into the membrane. The charged residue at the C-terminus of the TM anchor functions as a stop translocation signal. From a comparison of the TM anchor of all coronavirus S genes sequenced to date, some general features can be deduced. Firstly, at the N-terminus of the TM seven large hydrophobic residues are located, which are conserved in all S proteins. Secondly, this domain is followed by 11 to 22 hydrophillic amino acids, containing several cysteines residues. The cysteine residues in this region are candidates for palmitoylation. Finally, the length of the TM domain is unusual large.

We have introduced several mutations in the TM domain as indicated in Fig.1 and analyzed the expression and the fusogenicity of the mutant S protein.

First, transport of the different mutants was studied. The proteins were expressed in L-cells, using the vaccinia T7 expression system. In a pulse-chase experiment, the endo-H profiles of the mutants was determined. All mutants were transported to the medial Golgi. Transport of the mutants to the trans Golgi stacks was not affected, since all mutants were partially cleaved into the two 90 kDa sub-units (Table 1).

By using immunofluorescence it was demonstrated that all mutants were transported to the cell surface. However, not all mutants were fusogenic. M43 was as fusogenic as the

Table 1. Characteristics of the spike mutants.

	Cleavage	Cell surface expression	Acylation	Cell-to-cell fusion
wt	+	+	+	+
m40	+	+	+	–
m41	+	+	+	–
m42	+	+	+	+
m43	+	+	+	+

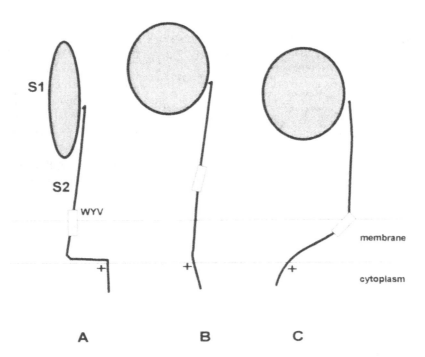

Figure 2. Possible models of the spike protein of MHV before (A) and after conformational changes have occurred (B and C). The S1 and S2 sub-units of the monomeric form of the spike protein are indicated. The highly conserved WYV domain is depicted as a box. The charged residue that functions as transfer stop signal is indicated by + at the cytoplasmic side of the membrane.

wild type spike protein, the fusion induced by m42 was delayed and the syncytia remained smaller, and m40 and m41 did not induce syncytia (Table 1).

For influenza virus, it has been shown that the degree of palmytoilation of the fusion protein can affect fusogenicity. The HA protein of some influenza strains is not fusogenic when palmytoilation is blocked[10, 11]. For other strains, however, block of acylation had no effect on fusion[12, 13]. To study the effect of the cysteine mutations on acylation of the murine coronavirus S protein, L-cells were labelled with ^{3}H-palmitic acid, and subsequently subjected to immunoprecipitation. Unfortunately all mutants became acylated, albeit to a different extent. Mutants m41 and m42 were less acylated, suggesting that some of the cysteine residues that were changed in these mutants are involved in acylation (Table 1).

We propose a model, by which our data can be explained, Fig.2. We hypothesize that after binding of S to the receptor, some conformational changes in both S1 and S2 occur as indicated in Figure 2. Either, the entire protein is pulled out of the membrane, thus exposing the highly conserved WYV domain (Fig.2B). Or as depicted in Figure 2C, the TM region of S could be twisted in the membrane, such that it becomes inserted in a different angle. Both proposed conformational changes require a large TM domain. Due to acylation the cysteine residues become hydrophobic, and can therefore be inserted into the lipid bilayer. The proposed changes in the S2 domain are not possible for our mutants m40 and m41. The charged residue in m40 may serve as a stop-transfer signal during translocation of the S protein across the ER membrane. As a result the TM domain of m40 is fixed in the membrane. In the absence of acylation of the cysteine residues which have been mutated in m41 the hydrophobic feature of the cysteine rich domain is less pronounced resulting in again a fixed

TM domain The phenotype of our mutants indicates that both the length of the TM and a position effect of acylated cysteines might be involved in fusion

Finally, we have developed an assay by which we can test the effect of mutations in S or other structural proteins on infectivity To this end, we transfected vaccinia virus infected L cells with four different constructs encoding for S, M, N and sM As RNA, the DI-RNA[14] was also introduced in the cells The release of DI RNA containg particles was analyzed by adding helpervirus MHV A59 to the medium and infecting a new monolayer of L-cells using this mixture Replication of the DI-RNA was analyzed following several undiluted passages

It was demonstrated that DI-RNA was only transferred to helpervirus-infected L-cells when DI-RNA was co-expressed with all structural proteins In the absence of the structural proteins no DI-RNA was observed following undiluted passage Although not proven directly, we assume that the DI-RNA which was released into the medium of the transfected cells was packaged into a virus like particle

REFERENCES

1 Collins A R , Knobler R L , Powell H , Buchmeier M J Monoclonal antibodies to murine hepatitis virus-4 (strain JHM) define the viral glycoprotein responsible for attachment and cell-cell fusion Virology 1982,119 358-371

2 Sturman L S , Holmes K V The molecular biology of coronaviruses Adv Virus Res 1983,28 35-112

3 Fleming J O , Stohlman S A , Harmon R C , Lai M M C , Frelinger J A Weiner L P Antigenic relationships of murine coronaviruses analysis using monoclonal antibodies to JHM (MHV-4) virus Virology 1983,131 296-307

4 Stauber R , Pfleiderera M , Siddell S Proteolytic cleavage of the murine coronavirus surface glycoprotein is not required for fusion activity J Gen Vir 1993,74 183-191

5 Taguchi F Fusion formation by the uncleaved spike protein of murine coronavirus JHMV variant cl-2 J Vir 1993,67 1195-1202

6 Gombold J L , Hingley S T , Weiss S R Fusion-defective mutants of mouse hepatitis virus A59 contain a mutation in the spike protein cleavage signal J Vir 1993,67 4504-4512

7 Gallagher T M , Escarmis C , Buchmeier M J Alteration of the pH dependence of coronavirus-induced cell fusion effect of mutations in the spike glycoprotein J V 1991, 65 1916-1928

8 Grosse B , Siddell S G Single amino acid changes in the S2 subunit of the MHV surface glycoprotein confer resistance to neutralization by S1 subunit-specific monoclonal antibody Virology 1994, 202 814-824

9 Routledge E , Stauber R , Pfleiderer M , Siddell S G Analysis of murine coronavirus surface glycoprotein functions by using monoclonal antibodies J V 1991, 65 254-262

10 Naeve C W , Williams D Fatty acids on the A/Japan/305/57 influenza virus hemagglutinin have a role in membrane fusion EMBO J 1990,9 3857-3866

11 Lambrecht B , Schmidt M F G Membrane fusion induced by influenza virus hemnagglutinin requires protein bound faaty acids FEBS 1986,202 127-132

12 Veit M , Kretzschmar E , Kuroda K , Garten W , Schmidt M F G , Klenk H D , Rott R Site-specific mutagenesis identifies three cysteine residues in the cytoplasmic tail as acylation sites of influenza virus hemagglutinin J V 1991, 65 2491-2500

13 Steinhauer D A , Wharton S A , Wiley D C , Skehel J J Deacylation of the hemagglutinin of influenza A/Aichi/2/68 has no effect on membrane fusion properties Virology 1991, 184 445-448

14 Van der Most R G , Bredenbeek P J , Spaan W J M A domain at the 3' end of the polymerase gene is essential for encapsidation of coronavirus defective interfering RNAs J V 1991, 65 3219-3226

46

IDENTIFICATION OF PROTEINS SPECIFIED BY PORCINE EPIDEMIC DIARRHOEA VIRUS

Anna Utiger, Kurt Tobler, Anne Bridgen,[*] Mark Suter, Mahender Singh, and Mathias Ackermann

Institute of Virology, Vet. -med. Faculty
University of Zurich, Zurich, Switzerland

ABSTRACT

Up to now, little was known about the proteins of porcine epidemic diarrhoea virus (PEDV). Using (i) metabolic labelling, (ii) antisera directed against synthetic peptides and monoclonal antibodies, and (iii) eukaryotic and prokaryotic expression systems, we have started to identify and characterize these proteins. The nucleocapsid protein (N) of PEDV was identified as a phosphoprotein with a relative mobility (Mr) of 57 k. No additional phospho-proteins were detected. At least three glycoproteins were virion associated. The 180/200 k band most probably represented the surface glycoprotein (S), whereas the 27 k protein was the membrane protein (M). The 21 k species could not yet be identified. A rabbit anti-M protein antiserum was generated by using synthetic peptides corresponding to the C-terminus of M. The specificity of this serum was reconfirmed by transient expression of the M-gene in Vero cells followed by immunofluorescence. In order to generate antisera against the putative gene products of ORF3 and of the internal N reading frames (I-1, I-2, I-3), additional anti-peptide sera were raised. The putative ORF3 product which had been tagged by a 6 Histidine tail, was expressed in *E. coli* and purified by nickel chelate affinity chromatography before 2-dimensional polyacrylamide gel electrophoresis and immunostaining with a rabbit anti-peptide serum directed against the N-terminus of the ORF3 product. Transient expression of the I-1 and I-3 reading frames was used to confirm the specificity of the corresponding anti-peptide sera. The immunological tools presented in this paper are now being used to identify the putative corresponding gene products specified by PEDV.

INTRODUCTION

The nucleotide sequence of approximately 3.300 bases nearest to the 3' end of the porcine epidemic diarrhoea virus (PEDV) genome has been published recently.[1,2] Yet little

---

[*] Present address: Institute of Virology, University of Glagow, Glasgow G11 5JR.

Corona- and Related Viruses, Edited by P. J. Talbot and G. A. Levy
Plenum Press, New York, 1995

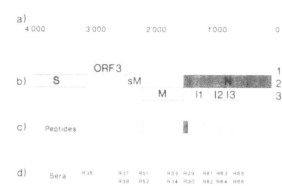

Figure 1. Overview of PEDV genome and the synthetic peptides and antisera used throughout this study. (a) Ruler, indicating nucleotides counted from the poly-A tail. (B) The genomic organization of PEDV nearest to the 3' end of the genome is shown, including the reading frames (1, 2, 3 on the right). The designation of each gene is indicated within the boxes which represent the genes. (c) Approximate map location of the peptides used to generate antisera. (d) Designation of the resulting rabbit anti-peptide sera.

is known about the viral proteins encoded by these genes. Using (i) antisera directed against synthetic peptides, and (ii) eukaryotic and prokaryotic expression systems, we have started to identify and characterize the PEDV proteins.

MATERIALS AND METHODS

PEDV strain CV777 was propagated in Vero cells essentially as described previously.[3] Infected and mock infected cell proteins were metabolically labelled by supplementing the cell culture medium with either ^{35}S-methionine, ^{32}P-orthophosphate, or ^{14}C-glucosamine.

For two dimensional (2D) electrophoresis, the samples were solubilized in a urea-ampholine solution and separated in the first dimension on a non equilibrium pH gradient as described previously.[4] SDS polyacrylamide gel electrophoresis, immunoblotting, and radioimmunoprecipitation (RIP) were done according to conventional protocols. 12% acrylamide gels crosslinked with bis-acrylamide and nitrocellulose sheets (BA85) were used. Binding of the antibodies to the transferred viral proteins was visualized either with goat anti-rabbit peroxidase conjugate or with unlabelled rabbit anti-mouse IgG and protein-A peroxidase followed by incubation in a substrate containing chloronaphtol and peroxide. For RIP, the soluble antigens were incubated with antibodies as indicated for each experiment before the immune complexes were precipitated using protein-A sepharose.

The deduced amino acid sequences of the PEDV genes were analyzed with the program Peptidestructure of the University of Wisconsin Genetics Computer Group in order to select theoretically immunogenic peptides of the putative proteins. Peptide synthesis and coupling to ovalbumin were performed by a commercial supplier (Neosystem S.A., Strasbourg, France). New Zealand White rabbits were subcutaneously injected with the peptide-carrier conjugate in Freund's complete or incomplete adjuvant as described previously.[5] The locations of the selected peptides and the designations of the corresponding rabbit sera are shown in Fig. 1.

To construct transient expression vectors, the genes of interest were excised from cloned plasmids and ligated into pSCT GAL X-556 [6] using appropriate restriction sites. This resulted in plasmids termed pSCT-"gene of interest" with the gene of interest under the control of the CMV immediate early promoter. For transient expression, Vero cells were transfected with DNA of pSCT-"gene of interest" and incubated for 36 to 48 hours, fixed with methanol, and incubated in succession with rabbit anti-peptide sera and goat anti-rabbit FITC.

Similar strategies were used to clone the genes of interest into a transfer vector containing the polyhedrin promoter in order to generate recombinant baculoviruses to express these genes.

In addition, the ORF3 gene was inserted into an *E. coli* expression vector. A histidine tag was added to the carboxy terminus of the putative ORF3 product. Due to the cloning procedure into pRBS,[7] both the N- and the C-terminus of the ORF3 product expressed by *E. coli* were slightly modified. The N-terminus contained Met-Arg-Gly-Ser in front of the original methionine, whereas the C-terminal addition consisted of Arg-Ser-His-His-His-His-His-His.

RESULTS AND DISCUSSION

The phosphoproteins encoded by PEDV were determined first. For this, PEDV infected and mock infected cell cultures were incubated in the presence of carrier free ^{32}P-orthophosphate. Total cellular proteins were harvested and subjected to polyacrylamide gel electrophoresis, Western transfer, and autoradiography. One viral phosphoprotein migrating with Mr 57 k could be detected unambiguously (Fig. 2A, lane 2). The same blot was immunostained with serum R29 raised against a synthetic peptide representing the NH$_2$-terminus of nucleocapsid protein (N). A protein band migrating at the same position as the 57 k phosphoprotein was detected (Fig. 2A, lane 5). The 57 k band was also observed with antigen obtained from a recombinant baculovirus which expressed the N open reading frame (Fig. 2A, lane 6).

These results show that the N open reading frame encodes a protein of 57 k which is phosphorylated and rabbit serum R29 (as well as R30, data not shown) is specific for N. We could only detect one viral phosphoprotein, even though several had been predicted by computer analysis of the sequencing data.[1,2]

The glycoproteins of PEDV were determined next. For this, PEDV infected and mock infected cell cultures were incubated in the presence of ^{14}C-glucosamine. Virions were purified from the harvested culture supernatants and subsequently subjected to polyacrylamide gel electrophoresis, Western transfer, and autoradiography. At least three species of glycoproteins could be identified (Fig. 2B). A double band of Mr 180/200 k which may represent the spike glycoprotein (S) was seen. A broad band around 27 k most probably

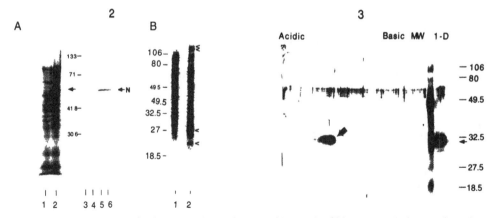

Figure 2. (A) Autoradiographs (lanes 1 and 2) and immunoblots with rabbit serum R29 (lanes 4 through 6) are shown. Lane 1: phosphoproteins in mock infected Vero cells. Lane 2: phosphoproteins in PEDV infected Vero cells. Lane 3: molecular weight marker. Lane 4: mock infected cell lysate. Lane 5: PEDV infected cell lysate. Lane 6: lysate of Sf9 cells infected with a recombinant baculovirus expressing N of PEDV. The arrow points to the major viral phosphoprotein N. (B) Autoradiographs of ^{14}C-glucosamine labelled virion preparations, purified from mock infected (lane 1) and PEDV infected (lane 2) cells are shown. The open arrows point to the putative glycoproteins specified by PEDV.

Figure 3. Immunostaining (R36) of ORF3-6His following one (arrow on the right) and two dimensional (bold arrow) gel electrophoresis and Western transfer.

represented the membrane protein (M). The third species, migrating around 21 k, may represent the small membrane protein (sM). The specificity of rabbit serum R34 for M was determined by transient expression of M in Vero cells followed by immunofluorescence (data not shown). Corresponding antisera for the detection of sM and S have not yet been raised.

A third series of experiments was aimed at the identification of the ORF3 and the internal N (I) gene products of PEDV. These putative polypeptides have stirred our interest because of two observations made during the determination of the genomic sequence of PEDV.[1,2] (i) Comparison of the nucleotide sequences obtained from two different viral isolates revealed almost complete sequence identity with the exception of variations and truncations within ORF3. (ii) The predicted open reading frame for the I-protein was interrupted by two STOP codons which truncated the reading frame and its putative protein to approximately one third of its potential length.

In order to generate antisera against the putative ORF3 product, two strategies were used: (i) to raise antisera against peptides representing the N- and C-terminus, respectively, of the putative protein. (ii) to express the ORF3 product including a 6-His tag (ORF3-6His) in E. coli, and to purify the polypeptide by nickel chelate affinity chromatography before immunizing rabbits.[7] According to the predictions made during sequencing, it was expected that the putative ORF3 product(s) specified by PEDV could be differentiated using 2D gel electrophoresis by specific pI and molecular weight. Fig. 3 shows the E. coli expressed ORF3-6His following one and two dimensional electrophoresis. The Western blot was immunostained with rabbit serum R36 which is specific for the N-terminus of the ORF3 product. Fast green staining of a similar blot revealed that ORF3-6His was more than 90% pure following nickel chelate affinity chromatography (not shown). These results indicate that R36 was able to recognize the E. coli expressed ORF3 polypeptide and should provide a good tool to test whether ORF3 protein is synthesized in PEDV infected cells. The same potential usefulness is anticipated for an antiserum which is being raised against the purified ORF3-6His.

In order to identify the putative I-1, I-2, and I-3 products, synthetic peptides which represented antigenic regions of I (see Fig. 1) were synthesized, coupled to ovalbumin, and used to immunize rabbits. Transient expression of I-1 and I-3 followed by immunofluorescence indicated that R61, R62, R65, and R66 recognized their antigen specifically (data not shown).

It remains to be tested whether or not the authentic products of ORF3 and/or I-1, I-2, and I-3 are synthesized in PEDV infected cells. The data presented in this paper, however, indicate that the strategy of employing antibodies to synthetic peptides in combination with eukaryotic and prokaryotic expression systems is extremely useful to identify and characterize so far unknown viral gene products.

ACKNOWLEDGMENTS

These studies were supported by the Swiss National Science Foundation, grant #31-37418.93.

REFERENCES

1. Bridgen et al., J. gen. Virol., 74, 1795-1804 (1993)
2. Duarte et al., Virology, 198, 466-476 (1994)
3. Knuchel et al., Vet. Microbiol., 32, 117-134 (1992)
4. Ackermann et al., J. Virol., 52, 108-118 (1984)
5. Fraefel et al., J. Virol. 68(5), 3154-3162 (1994)
6. S. Rusconi et al., Gene, 89, 211-221 (1990)
7. Stüber et al., Immunological Methods, Vol. IV, 122-152, Academic Press (1990)

47

COEXPRESSION AND ASSOCIATION OF THE SPIKE PROTEIN AND THE MEMBRANE PROTEIN OF MOUSE HEPATITIS VIRUS

Dirk-Jan E. Opstelten, Martin J. B. Raamsman, Karin Wolfs, Marian C. Horzinek, and Peter J. M. Rottier

Institute of Virology
Department of Infectious Diseases and Immunology
Faculty of Veterinary Medicine
Utrecht University
The Netherlands

ABSTRACT

The M and S envelope glycoproteins of mouse hepatitis virus associate in the process of virus assembly. We have studied the intrinsic properties of M/S heterocomplexes by coexpressing M and S in the absence of other coronaviral proteins. The formation of M/S complexes under these conditions indicates that M and S can interact independently of other coronaviral factors. Pulse-chase analysis revealed that M and S associate in a pre-Golgi compartment. M/S complexes are efficiently transported beyond the coronavirus budding compartment to the Golgi complex. The failure to detect complexes at the surface of coexpressing cells demonstrated that they are retained intracellularly. Thus, coexpression of the envelope glycoproteins drastically affects the intracellular transport of the S protein: instead of being transported to the cell surface, S is retained intracellularly by its association with M.

INTRODUCTION

The spike (S) protein and the membrane (M) protein of the mouse hepatitis virus strain A59 (MHV-A59) are involved in intermolecular interactions in the process of virus assembly. Using specific detergent conditions heterocomplexes consisting of M and S can be extracted from solubilized virions and from MHV-infected cells[1]. The interaction between M and S is likely to be essential for the incorporation of the S protein into virus particles since the latter is thought to be dispensable for virus assembly. Tunicamycin treatment of MHV-infected cells results in the secretion of spikeless virions[2,3] suggesting that the S protein is not required for the budding process. A specific interaction between the M and S proteins, therefore, could effect the incorporation of the latter into virus particles.

Coronaviruses are assembled by budding of the nucleocapsid into pre-Golgi membranes[4,5,6]. According to a prevailing hypothesis the envelope proteins of membrane viruses determine the site of budding[7,8]. In the case of coronaviruses, however, neither of the envelope glycoproteins contain targeting signals that specify their accumulation in the budding compartment. When expressed independently, the M protein localizes to the Golgi complex[6,9,10] whereas the S protein is transported to the cell surface (Vennema et al., in prep.). M/S complexes, however, might have acquired a signal that prevents their transport beyond the budding compartment. To investigate the effect of the association between M and S on their intracellular transport we have coexpressed the envelope glycoprotein genes in the absence of other coronaviral proteins.

MATERIALS AND METHODS

Cells, Virus and Antisera

OST7-1 cells, a kind gift of Dr. B. Moss, were maintained in Dulbecco's minimal essential medium (DMEM) containing 10% fetal calf serum, penicillin, and streptomycin (DMEM-10%FCS) supplemented with 400 µg/ml G-418 (Geneticin, GIBCO). The recombinant vaccinia virus vTF7-3 expressing the T7 RNA polymerase[11] was also obtained from Dr. B. Moss. The production of the rabbit polyclonal antiserum to MHV-A59 has been described previously[2]. The monoclonal antibodies (MAbs) J7.6 and J1.3 against S and M, respectively, were kindly provided by Dr. J. Fleming.

Infection, Transfection and Metabolic Labeling

Infection and transfection of OST7-1 cells: Subconfluent monolayers of OST7-1 cells in 35-mm dishes were washed with DMEM and inoculated with vTF7-3 at a multiplicity of infection (m.o.i.) of approx. 10 in DMEM for 45 min at 37°C. After inoculation the cells were washed with DMEM and transfected with the vector PTUM-M[12] and/or PTUM-S (Vennema et al., in prep.) which contain a cDNA copy of the MHV M and S protein, respectively, under the control of the T7 promotor. For this purpose, a mixture of 200 µl DMEM with plasmid DNA and 10 µl lipofectin reagent (Bethesda Research Laboratories, Life Technologies, Inc.) was added to the cells. After a 10-min incubation at room temperature 800 µl DMEM was added and cells were incubated at 37°C.

Labeling: 2 hr after inoculation, cells were transferred to 32°C. Starting at 4.5 hr after inoculation, the cells were starved for 30 min in MEM (GIBCO) without methionine. Cells were pulse-labeled with 100-200 µCi ^{35}S-in vitro labeling mix (Amersham) for the times indicated, then washed once with DMEM-10%FCS supplemented with 10 mM HEPES, 2 mM L-methionine, and 2 mM L-cysteine (chase medium) and chased for various times in chase medium. The cells were lysed on ice in 50 mM Tris (pH 8.0), 62.5 mM EDTA, 0.5% Nonidet P-40, 0.5% Na-deoxycholate (detergent solution) containing 2mM phenylmethyl-sulfonyl fluoride (PMSF). The lysates were spun for 3 min at 12,000 x g at 4°C to pellet nuclei and cell debris.

Immunoprecipitation and Gel Electrophoresis

Viral proteins were immunoprecipitated with the polyclonal MHV-A59 antiserum (10µl), the MAb J1.3αM (10µl), or with the MAb J7.6αS (20 µl). Serum was added to aliquots of cell lysates which had been diluted with detergent solution to a final volume of 600 µl. After overnight incubation at 4°C immune complexes were collected using 50 µl of a 10% (w/v) suspension of formalin-fixed *Staphylococcus aureus* cells (Bethesda Research Laboratories, Life Technologies, Inc.). After a 30 min incubation at 4°C they were washed three times with detergent solution and finally suspended in 25 µl 62.5 mM Tris-HCl (pH 6.8), 2% SDS, 10% glycerol, 20 mM DTT (sample buffer). The proteins were analysed in 10% SDS-polyacrylamide gels. The samples were heated for 2 min at 95°C before loading on the gel.

Surface Immunoprecipitation

Transfected cells were labeled with ^{35}S-in vitro labeling mix from 5-5.5 hr p.i. and subsequently chased for 3 hr in chase medium. The cells were put on ice and washed with PBS/5%FCS and incubated for 2 hr in 800 µl PBS/5%FCS containing the MAb J1.3αM (15µl) and/or MAb J7.6αS (30µl). Thereafter, cells were extensively washed with PBS/5%FCS and lysed with detergent solution containing 2 mM PMSF. The lysates were spun for 3 min at 12,000 x *g* and 4°C and 50 µl of a 10% (w/v) suspension of formalin-fixed *Staphylococcus aureus* cells was added to collect the immune complexes. After a 30 min incubation at 4°C the cells were pelleted by centrifugation. The supernatant was subjected to a second round of immunoprecipitation using the same antibodies. The bacterial cells were washed three times with detergent solution and finally suspended in sample buffer.

RESULTS AND DISCUSSION

Association of the Coexpressed M and S Proteins

The interaction between the M and S proteins of MHV-A59 can be detected under specific analytical conditions. When a combination of ionic and nonionic detergents is used for the solubilization of virions or infected cells M/S heterocomplexes can be immunoprecipitated using monospecific antisera against M or S[1,13]. To study whether M and S associate independently of other viral proteins we have coexpressed their respective genes in the absence of other coronaviral proteins using the vaccinia virus infection/transfection expression system[11].

Cells coexpressing M and S were labeled for 30 min with ^{35}S-methionine and chased for 90 min and subsequently lysed using a buffer containing 0.5% NP40 and 0.5% NaDOC. Equal fractions of the cell lysate were used for immunoprecipitation using a polyclonal anti-MHV serum or a monoclonal antibody (MAb) against S. As shown in Fig. 1, the material precipitated using the anti-MHV serum represents the total amount of labeled viral proteins. It mainly consists of the spike precursor S/gp150 and differentially glycosylated forms of M. The two other bands presumably represent vaccinia virus proteins which were precipitated nonspecifically. The S protein as well as a significant fraction of the M protein were also precipitated using the MAbαS. In addition, a fraction of labeled S was coprecipitated with the M specific antibodies. This indicates that M and S were associated under these conditions. Apparently, the formation of M/S complexes does not require other coronaviral proteins. To exclude the possibility that M and S had formed complexes after lysis we expressed the proteins independently in separate cell cultures and performed the immunoprecipitations using a mixture of the cell lysates (data not shown). Under these conditions no complexes between M and S were found indicating that M/S complexes derived from coexpressing cells were specific.

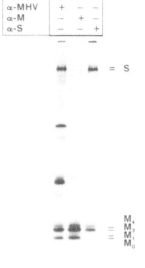

α-MHV	+	–	–
α-M	–	+	–
α-S	–	–	+

= S

M$_4$
M$_3$
M$_1$
M$_0$

Figure 1. Association of the coexpressed MHV M and S proteins. Cells coexpressing M and S were labeled for 30 min and chased for 90 min. The cell lysate was split into three equal fractions and the viral proteins were precipitated using a polyclonal anti-MHV serum, a MAbαM, or a MAbαS, respectively.

Formation and Intracellular Transport of M/S Complexes

To determine the kinetics of complex formation and to analyze the intracellular transport of M/S complexes we performed a pulse-chase labeling. Cells coexpressing M and S were pulse labeled for 15 min and chased for various time periods. The cell lysates were split into three equal portions from which the viral proteins were immunoprecipitated using the polyclonal anti-MHV serum, a MAbαM, and a MAbαS, respectively.

Again, the material precipitated using the polyclonal anti-MHV serum represents the total amount of labeled viral proteins (Fig. 2). After the pulse, predominantly precursor forms of M (M_0) and S (S/gp150) can be detected. The M protein becomes glycosylated posttranslationally when it is transported from the ER to the Golgi complex giving rise to the formation of different glycosylated forms (M_1-M_3-M_4). The addition of the first sugar, GalNAc, takes place in the intermediate compartment - which is identical to the budding compartment[5,14] - and M_3 and M_4 appear after the protein has reached the Golgi complex[10]. A fraction of S/gp150 was converted into S/gp180 during the chase as a result of modifications of its N-linked oligosaccharides. Cleaved forms of the S protein were hardly detected in this experiment.

Using the MAbαS we analyzed which fraction of the labeled M molecules was engaged in M/S complexes. After the pulse, only small amounts of labeled M protein were coprecipitated by the S specific antibodies. The amount of M that was coprecipitated with S increased during the chase and reached a maximum around 60 min of chase. The coprecipitation of the unglycosylated form of M (M_0) seen after short chase periods indicates that M/S complexes are formed in a pre-Golgi compartment. After longer chase periods, this form was efficiently processed into M_3 and M_4 demonstrating that the complexes were transported to the Golgi complex. Surprisingly, no accumulation of M_1 is observed indicating that M/S complexes are not retained in the budding compartment by themselves.

When M/S complexes were precipitated using the MAbαM we found that coprecipitation of labeled S started to appear after 15 min of chase. Apparently, newly synthesized M and S become engaged in heterocomplexes with different kinetics. The S protein needs more time to become association competent which is probably the result of its slow folding[12,13,15]. Like the M protein, the S protein present in M/S complexes underwent

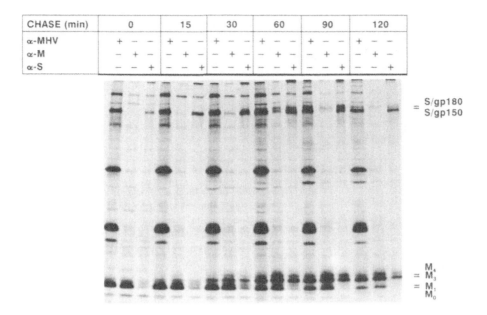

Figure 2. Formation and intracellular transport of M/S complexes Cells coexpressing M and S were labeled for 10 min and chased for the indicated periods. Each cell lysate was split into three equal fractions which were used for immunoprecipitations with the polyclonal anti-MHV serum, the MAbαM, or the MAbαS, respectively

processing during longer chase periods: its precursor form S/gp150 was converted into S/gp180 which indicates that it was transported from the ER to the Golgi complex. We conclude from this experiment that M and S associate in a pre-Golgi compartment and that they are transported as a complex beyond the site of virus budding to the Golgi region. This implies that additional factors, e.g. the nucleocapsid or the recently identified small membrane protein[16], determine pre-Golgi budding of MHV.

Intracellular Accumulation of S by Its Interaction with M

Knowing that M/S complexes are transported beyond the budding compartment and that the independently expressed M and S proteins are transported to different cellular locations, i.e. Golgi complex and plasma membrane, respectively, we were interested in the destination of M/S complexes. The previous experiment (Fig. 2) shows that M/S complexes reach the Golgi complex but the observation that only a very small fraction of S was cleaved into S/gp90 suggests that the complexes do not reach the cell surface. To investigate the cell surface expression of M/S complexes we have performed cell surface immunoprecipitations. Cells expressing S alone or together with M were labeled for 30 min and chased for 3 hr to allow ample time for the proteins to reach their final destinations. As described in the Methods section, we then first precipitated the viral proteins that were expressed at the plasma membrane and in a second round of immunoprecipitation we collected the proteins which were kept within the cells.

When expressed independently, a large fraction of the cleaved form of S as well as some uncleaved S/gp180 were found at the cell surface (Fig. 3). As expected, and at the same time illustrating the validity of the approach, the remaining precursor S/gp150 was only detected intracellularly. When coexpressed with M, however, hardly any S protein was

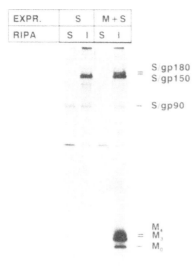

EXPR.	S		M + S	
RIPA	S	I	S	I

= S gp180
= S gp150

– S gp90

= M₁
= M₁
– M₀

Figure 3. Intracellular accumulation of M/S complexes. Cells expressing S alone or together with M were labeled for 30 min and chased for 3 hr. Viral proteins expressed at the cell surface (S) were collected in a first round of immunoprecipitation and the remaining fraction of intracellular proteins (I) were immunoprecipitated in a second round.

detected at the cell surface. Instead, we now observed an intracellular accumulation of S/gp180 and S/gp90. Although the M protein had reached the Golgi complex as judged by the appearance of M_3 and M_4 it could not be detected at the cell surface. We conclude that coexpression of M and S does not affect the localization of M. Rather, it specifically results in the intracellular retention of the S protein due to its interaction with M.

REFERENCES

1. Opstelten, D.-J. E., M. C. Horzinek, and P. J. M. Rottier. Complex formation between the spike protein and the membrane protein during mouse hepatitis virus assembly. Adv. Exp. Med. Biol. 1994;342:189-195.

2. Rottier, P. J. M., M. C. Horzinek, and B. A. M. van der Zeijst. Viral protein synthesis in mouse hepatitis virus strain A59-infected cells: effects of tunicamycin. J. Virol. 1981;40:350-357.

3. Holmes, K. V., E. W. Doller, and L. S. Sturman. Tunicamycin resistant glycosylation of a coronavirus glycoprotein: demonstration of a novel type of viral glycoprotein. Virology 1981;115:334-344.

4. Tooze, J., S. A. Tooze, and G. Warren. Replication of coronavirus MHV-A59 in sac⁻ cells: determination of the first site of budding of progeny virions. Eur. J. Cell Biol. 1984;33:281-293.

5. Krijnse-Locker, J., M. Ericsson, P. J. M. Rottier, and G. Griffiths. Characterization of the budding compartment of mouse hepatitis virus: evidence that transport from the RER to the Golgi complex requires only one vesicular transport step. J. Cell Biol. 1994;124:55-70.

6. Klumperman, J., J. Krijnse Locker, A. Meijer, M. C. Horzinek, H. J. Geuze, and P. J. M. Rottier. Coronavirus M proteins accumulate in the Golgi complex beyond the site of virion budding. J. Virol. 1994;in press.

7. Dubois-Dalcq, M., K. V. Holmes, and B. Rentier. Assembly of enveloped RNA viruses. Springer-Verlag, Vienna 1984.

8. Pettersson, R. F. Protein localization and virus assembly at intracellular membranes. Curr. Top. Microbiol. Immunol. 1991;170:67-106.

9. Rottier, P. J. M., and J. K. Rose. Coronavirus E1 glycoprotein expressed from cloned cDNA localizes to the Golgi region. J. Virol. 1987;61:2042-2045.

10. Krijnse Locker, J., G. Griffiths, M. C. Horzinek, and P. J. M. Rottier. O-glycosylation of the coronavirus M protein: differential localization of sialyltransferases in N- and O-linked glycosylation. J. Biol. Chem. 1992;267:14094-14101.

11 Fuerst, T R , E G Niles, F W Studier, and B Moss Eukaryotic transient-expression system based on recombinant vaccinia virus that synthesize bacteriophage T7 RNA polymerase Proc Natl Acad Sci USA 1986,83 8122-8126

12 Opstelten, D -J E , P de Groote, M C Horzinek, H Vennema, and P J M Rottier Disulfide bonds in folding and transport of mouse hepatitis coronavirus glycoproteins J Virol 1993,67 7394-7401

13 Opstelten, D -J E , P de Groote, M C Horzinek, and P J M Rottier Folding of the mouse hepatitis virus spike protein and its association with the membrane protein Arch Virol (Suppl) 1994,9 319-328

14 Tooze, S A , J Tooze, and G Warren Site of addition of N-acetyl-galactosamine to the E1 glycoprotein of mouse hepatitis virus-A59 J Cell Biol 1988,106 1475-1487

15 Vennema, H , P J M Rottier, L Heijnen, G J Godeke, M C Horzinek, and W J M Spaan Biosynthesis and function of the coronavirus spike protein Adv Exp Med Biol 1990,276 9-19

16 Yu, X , W Bi, S R Weiss, and J L Leibowitz Mouse hepatitis virus gene 5b protein is a new virion envelope protein Virology 1994,202 1018-1023

FUNCTIONAL DOMAINS IN THE SPIKE PROTEIN OF TRANSMISSIBLE GASTROENTERITIS VIRUS

H. Laude,[1] M. Godet,[1] S. Bernard,[2] J. Gelfi,[1] M. Duarte,[1] and B. Delmas[1]

[1] Unité de Virologie et Immunologie Moléculaires
Jouy-en-Josas
[2] Unité de Pathologie Infectieuse et Immunologie
Nouzilly
INRA, France

ABSTRACT

The coronavirus spike protein S is assumed to mediate essential biological functions, including recognition of target cells. Earlier studies from our and other groups identified two regions of the TGEV S (220K) protein possibly implicated in such functions. The first of these corresponds to the 224 amino acid N-terminal region which is deleted in PRCV, the respiratory variant of TGEV. We have examined the pathogenicity for the newborn piglet of a series of neutralization escape mutants encoding an S protein mutated in this region. Several amino acid changes were correlated with a dramatic loss of enterovirulence, thus indicating that crucial determinants are associated with this domain of S. The second region of potential relevance is the major neutralization domain. Baculovirus-vectored expression of 150 to 220 amino acid-long stretches encompassing this region, which is encoded by both TGEV and PRCV, was performed. The resultant recombinant proteins were shown to react with the cognate antibodies and to bind APN specifically, thus localizing the receptor-binding site on the S primary structure. Altogether these data lend support to the view that a domain of S protein structurally distinct from the receptor binding site is required for the virus to express its enteric tropism.

INTRODUCTION

Transmissible gastroenteritis virus (TGEV) is a porcine coronavirus which is responsible for an acute diarrhoea, often fatal for the swine neonate. The virus replicates preferentially in the differentiated enterocytes covering the small intestine mucosa[1]. It has been recently reported that TGEV uses an ectoenzyme abundantly expressed in the apical brush border, aminopeptidase N, as a receptor for gaining entry into cells[2]. However, the molecular basis of the viral enteric tropism remains poorly understood. In 1984, a naturally occuring variant of TGEV started to diffuse to a wide extent in the European swine population. This porcine respiratory coronavirus (PRCV), of which the replication was restricted to the respiratory tract, was essentially avirulent[3]. Partial analysis of its genomic sequence revealed two genetic defects, one in the spike protein gene, which results in a N-terminally truncated S protein, the other in ORF3, putatively encoding a non-structural protein[4]. However, the function(s) potentially impaired in PRCV are yet undetermined.

In the present paper, we report two studies which both pertain with the localization of functionally important domains in the amino-terminal half of the TGEV S protein. The results lead us to propose that the determinants involved in the recognition of the receptor and in the expression of enterovirulence are associated with two well-distinct regions of the S molecule.

MATERIAL AND METHODS

All the animal experiments were performed in the animal facilities of INRA, Nouzilly. Large White piglets were weaned at 3 day, then artificially reared. They were infected oronasally at 5 day with parental or mutant virus at the indicated dose.

Neutralization escape mutants were selected as previously described[5]. Stocks of mutant virus were produced in ST cells under selection with the relevant MAb diluted 1:100.

Direct RNA sequencing, immunoprecipitation and indirect immunofluorescence assays were performed as reported elsewhere[6]. Expression of S gene fragments and of porcine aminopeptidase N gene[2] in Sf9 cells using baculovirus were done following published procedures[7].

RESULTS

In Vivo Virulence of Site D Escape Mutants

As an approach to determine whether virulence determinants are located within the N-terminal region of the S protein, site D epitope mutants were generated and tested for their virulence in neonate piglets. Site D is a major antigenic site defined by four MAbs which bind within the amino acid region 82-212 of the mature S polypeptide[6]. MAb 40.1, the sole site D MAb exhibiting neutralizing activity, was used to select 7 independent escape mutants. Direct sequencing of genomic RNA was performed to determine the induced amino acid changes. All the mutants had a single nucleotide (nt) change leading to a one amino acid substitution, except mutant 10 which had a 12 nt deletion resulting in the loss of 4 residues. All the amino acid alterations clustered within the stretch 145-Pro to 155-Cys (Figure 1).

The animal model consisted of piglets weaned and infected at 5 days. Ninety seven animals were inoculated orally with mutant or parental virus (Purdue-115) at an infectious dose varying from 10^5 to 10^8 PFU, as specified. Three to six animals were infected per dose per mutant virus. The virulence was monitored according to daily weight increase, onset of diarrhoea, and morbidity and mortality rates, this over a 10 days period. Seventeen mock-infected animals remained healthy and showed a total weight increase of 1450 ± 250 g between J1 and J10. All the animals infected with 10^5 or 10^7 PFU of parental virus developed signes of anorexia, vomiting and severe diarrhoea, generally appearing at 3 days p.i., and died within 4 to 8 days p.i. Their total weight increase was - 110 ± 70 g (Figure 2). Similar findings were observed with 5 escape mutants selected towards site A or B MAbs (data not shown). Strikingly different data were obtained with the 40.1 escape mutants, as partially shown in Figure 2 for the mutants 6 and 8 (single mutation) and 10 (small deletion). The daily growth curves of animals inoculated with 10^5 PFU was comparable to that of the controls and no mortality was recorded. Mortality, never exceeding 50 per cent, was only

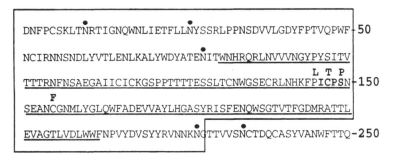

Figure 1. Amino acid sequence of the N-terminal region of the mature TGEV S protein (Purdue-115 strain[8]) Substitutions and deletion identified in the sequence of MAb 40 1 escape mutants are shown in bold characters The sequence deleted in PRCV S protein (European strains [4]) is boxed The sequence where the site D has been mapped[6] is underlined Putative Asn-linked glycosylation sites are dotted

observed with animals inoculated with 10^8 PFU. At intermediate doses (10^6-10^7), the animals suffered of only mild diarrhoea and all recovered. From these data, it was concluded that site D mutants exhibit a dramatically reduced virulence.

Functional Properties of the Isolated CO-26K Region

Our earlier studies led to the identification of a proteolytic fragment generated by collagenase cleavage of antibody-protected S protein, called CO-26K. This fragment expressed a group of epitopes mapped within the two closely related sites A and B, which together define the major neutralization-mediating domain of TGEV[6]. This observation was suggestive of the 26K region being an independent domain of the S molecule. It was thus of interest to examine functional properties of baculovirus-expressed, CO-26K-derived polypeptides. Three constructs, S223, S150 and S122, expressing amino-coterminal products covering the CO26K

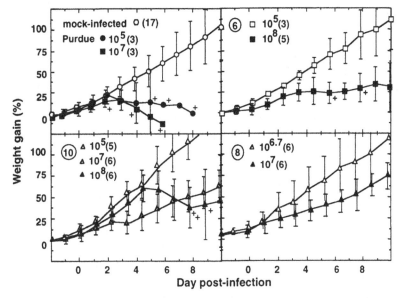

Figure 2. Pathogenicity of site D mutants for the newborn piglet The daily weight increase and mortality of animals infected by three different mutants (selector MAb 40 1) are compared with that of wild type (Purdue)- or mock-infected animals The infectious dose (10^5 to 10^8) and the number of animals within each group (brackets) are indicated Mutant 6 Pro145-Leu, mutant 8 Cys147-Arg, mutant 10 deletion 146 to 149

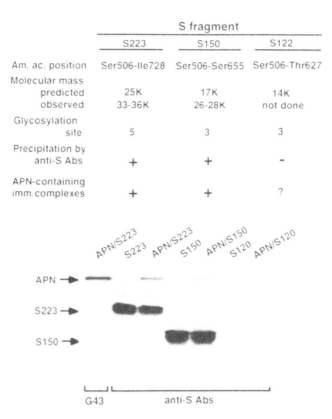

| | S fragment | | |
	S223	S150	S122
Am. ac. position	Ser506-Ile728	Ser506-Ser655	Ser506-Thr627
Molecular mass predicted	25K	17K	14K
observed	33-36K	26-28K	not done
Glycosylation site	5	3	3
Precipitation by anti-S Abs	+	+	-
APN-containing imm.complexes	+	+	?

Figure 3. Interaction between TGEV S-derived fragments and aminopeptidase N (APN). The characteristics of the baculovirus-expressed fragments are listed in the left panel. The constructs have been made so as to link the S signal peptide to the S body sequences via three extraneous residues (RSG). Bottom panel: extracts from labeled cell cultures singly or dually infected by recombinant baculoviruses expressing either an S fragment or APN were subjected to immunoprecipitation with an anti-APN (G43) or anti-S polyclonal antibodies as indicated. Immunoprecipitated material was analyzed by SDS-PAGE 15% gel and autoradiography.

sequence were made (Figure 3). The antigenicity of each S derivative was analyzed through immunoprecipitation of lysates from labeled cultures or indirect immunofluorescence of acetone-fixed cultures. The S223 and S150 products had the expected molecular mass and were secreted. Both were found to express all the site A and B epitopes, whereas the S122 products was only reactive to few MAb in indirect IF (data not shown).

As a next step, we sought to determine whether the S CO-26K domain was involved in binding to the APN receptor. To this end, porcine APN was expressed by using a recombinant baculovirus. Cell lysates from cultures infected by either APN- or S derivative-expressing baculoviruses, or coinfected by both, were subjected to immunoprecipitation with anti-S or anti-APN polyclonal antibodies. Baculovirus-expressed APN appeared as an uncleaved band of 120K. A similar band was coimmunoprecipitated with S223 or S150 proteins from dually infected cultures (Figure 3). S223 bound approximately one third of the material immunopre-cipitated by the most potent of our anti-APN MAbs. The S150 species bound consistently less APN material. No conclusion could be drawn about the S122 species since it was not immunoprecipitated by any of the anti-S antibody used. The anti-APN MAbs failed to immunoprecipitate S/APN complexes. This was consistent with the fact that they were selected on the basis of a receptor-blocking activity[2]. Altogether, these data led us to conclude that a recombinant polypeptide derived from the CO-26K domain, representing 10-15 per cent of the whole S polypeptide chain, retained a substantial capacity to bind APN.

DISCUSSION

By examining the pathogenicity of naturally occuring neutralization-resistant variants of TGEV in newborn piglets it was found that a small alteration within the N-terminal region of the S protein (position 145 to 155) may result in a dramatic attenuation of the virus. It is worth noting that all of the attenuated variants grew as the parental virus in cell culture and exhibited a normal plaque phenotype, thus indicating that neither virus-to-cell binding nor essential steps of the replication were impaired *in vitro*. While we cannot rule out the possibility that additional genetic changes may lie outside the S gene, this seems to be highly unlikely given that i) the selection pressure applied for the generation of the variants was due to a neutralizing MAb 40.1, directed to one epitope of the site D of S protein ; ii) variants obtained following the same approach but using MAbs directed to other epitopes (site A and B) exhibited a virulence phenotype indistinguishable of that of the parental virus ; iii) independent variants having the same mutation behaved identically in infected animals.

The above observation provides the first demonstration that a determinant crucial for enterovirulence is associated with the S protein of TGEV. Similar findings have been obtained with another coronavirus, mouse hepatitis virus (MHV), for which attenuation of the neurovirulence has been correlated with changes in the S protein gene product, most of them being located in its N-terminal half [9,10]. The present data lend strong support to the view that the large deletion present in the N-terminus of S protein is responsible for the nearly complete loss of replication of PRCV in the intestine. However, as pointed out elsewhere[11], the absence of enterovirulence does not explain the marked respiratory tropism of PRCV, which is not a common feature among the TGEV strains. In this respect, it would be of interest to examine the ability of our site D mutants to replicate in the respiratory tract of infected piglets.

Experiments using baculovirus-expressed polypeptides allowed us to demonstrate that an isolated domain of the S molecule, corresponding to the about 200 residues downstream of the serine 506, was able to recognize efficiently porcine aminopeptidase N. This region thus represents a major receptor-binding site of TGEV. Moreover, it is common to the S proteins of TGEV and PRCV, which is consistent with our earlier observation that PRCV also uses APN as a receptor, at least in cell culture.

These data, together with those of Kubo et al.[12] who identified a receptor-binding site on MHV-S protein, formally establish that S is an attachment protein for coronaviruses. S is the most polymorphic of the coronavirion proteins and can undergo a large deletion without alteration of its main functions. A pairwise alignment of TGEV and MHV S amino acide sequences reveals that the N-terminal region, which is absent in the respiratory viruses PRCV and HCV 229E relative to TGEV, is vis-à-vis the region containing the MHV receptor-binding site. Reciprocally, the MHV S polymorphic region is facing the TGEV receptor-binding site (Figure 4). This may indicate that both of these regions correspond to

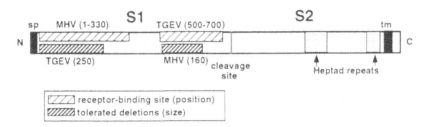

Figure 4. Relative positions of functional domains in TGEV and MHV S polypeptides. Data from [8] and this paper for TGEV and from [12] and [13] for MHV.

structural modules that are relatively independent from the rest of the molecule for their function

The picture emerging from our whole data is that the interaction S/APN, shown to be responsible for the species-specificity of TGEV/PRCV[14], is not a primary determinant of the tissue tropism, though it restricts the multiplication of the virus to epithelial cells covering the mucosa Nevertheless, the S protein carries determinant(s) that are crucially important for the enterotropism, and we propose that at least one such determinant is associated with the N-terminal region of the molecule In other words, this region would be necessary for the virus to infect the enterocytes but not other cells The precise function that is involved is still elusive It has been speculated by other authors, that different receptor-binding sites in TGEV S protein could be recognized by the enteric and respiratory tissues[15] This would imply that the N-terminal domain bind to a cell surface molecule different from APN and specifically expressed in the enterocytes An alternative hypothesis, however, is that this domain is required for a proper folding of the S molecule when it is faced up to the physico-chemical environment of the digestive tract, which might differently affect the conformation of the truncated (or mutated) and of the wild type S protein[11]

REFERENCES

1 Pensaert, M , Haelterman, E O & Burnstein, T Transmissible gastro-enteritis of swine virus-intestine cell-interactions Archiv fur die gesamte Virusforschung 1970,31 321-334

2 Delmas, B , Gelfi, J , L'Haridon, R , Vogel, L K , Sjostrom, H , Noren, O & Laude, H Aminopeptidase N is a major receptor for the entero-pathogenic coronavirus TGEV Nature 1992,357 417-419

3 Pensaert, M , Callebaut, P & Vergote, J Isolation of a porcine respiratory, non-enteric coronavirus related to transmissible gastroenteritis Veterinary Quaterly 1986,8 257-261

4 Rasschaert, D , Duarte, M , Laude, H Porcine respiratory coronavirus differs from transmissible gastro-enteritis virus by a few genomic deletions J gen Virol 1990,71 2599-2607

5 Delmas, B , Gelfi, J & Laude, H Antigenic structure of transmissible gastroenteris virus II Domains in the peplomer glycoprotein J gen Virol 1986,67 1405-1418

6 Delmas, B , Rasschaert, D , Godet, M , Gelfi, J , & Laude, H Four major antigenic sites of the coronavirus transmissible gastroenteritis virus are located on the amino-terminal half of spike glycoprotein S J gen Virol 1990,71 1313-1323

7 Godet, M , Rasschaert, D , & Laude, H Processing and antigenicity of entire and anchore-free spike glycoprotein S of coronavirus TGEV expressed by recombinant baculovirus Virology 1991,185 732-740

8 Rasschaert, D & Laude, H The predicted primary structure of the peplomer protein E2 of the porcine coronavirus transmissible gastroenteritis virus J Gen Virol ,1987 ,68 1883-1890

9 Fazakerley, J K , Parker, S E , Bloom, F & Buchmeier, M J The V5A13 1 envelope glycoprotein deletion mutant of mouse hepatitis virus type-4 is neuroattenuated by its reduced rate of spread in the central nervous system Virology 1992,187 178-188

10 Wang, F I , Fleming, J O & Lai, M C L Sequence analysis of the spike protein gene of murine coronavirus variants study of genetic sites affecting neuropathogenicity Virology 1992,186 742-749

11 Laude, H , Van Reth, K , & Pensaert, M Porcine respiratory coronavirus molecular features and virus-host interactions Vet Res 1993,24 125-150

12 Kubo, H , Yamada, Y K & Taguchi, F Localization of neutralizing epitopes and the receptor-binding site within the amino-terminal 330 amino acids of the murine coronavirus spike protein J Virol 1994,68 5403-5410

13 Parker, S E , Gallagher, T M & Buchmeier, M J Sequence analysis reveals extensive polymorphism and evidence of deletions within the E2 glycoprotein gene of several strains of murine hepatitis virus Virology 1989,173 664-673

14 Delmas, B , Gelfi, J , Kut, E , Sjostrom, H , Noren, & Laude, H Determinants essential for the transmissible gastroenteritis virus-receptor interaction reside within a domain of aminopeptidase-N that is distinct from the enzymatic site J Virol 1994,68 5216-5224

15 Sanchez, C M , Gebauer, F , Sune, C , Mendez, A , Dopazo, J , & Enjuanes, L Genetic evolution and tropism of transmissible gastroenteritis coronaviruses Virology 1992,190 92-105

49

RECOMBINANT EXPRESSION OF THE TGEV MEMBRANE GLYCOPROTEIN M

P. Baudoux, B. Charley, and H. Laude

Unite de Virologie et Immunologie Moléculaires
INRA
Jouy-en-Josas, France

ABSTRACT

We have previously shown that the membrane protein M of TGEV is involved in efficient induction of alpha interferon (IFNα) synthesis by non-immune peripheral blood mononuclear cells incubated with fixed, TGEV-infected cells or inactivated virions[1,2]. In order to determine whether M protein is able to induce interferon in the absence of other viral factors, we expressed the protein either stably in the porcine ST cells or transiently in the simian COS7 cells. Although showing no obvious difference in intracellular localization or glycosylation compared to its viral counterpart, the recombinant molecule failed to induce significant IFN activity.

INTRODUCTION

Transmissible gastroenteritis coronavirus (TGEV) is the causative agent of an acute diarrhea in swine, often fatal in newborn piglets[3]. Infection is associated with high interferon titers in the intestinal tract, lungs and serum[4,5]. *In vitro*, TGEV is able to induce an early and strong synthesis of IFNα by a minor subpopulation of nonimmune peripheral blood mononuclear cells (PBMC) from different species[6]. The IFN induction mechanism differs from the double-stranded RNA-mediated, classical pathway since induction is observed even with non replicating structures, such as glutaraldehyde-fixed, infected cells or U.V.-inactivated, purified virions[1,2]. Thus the inducer may be one or several structural viral component(s). In addition, the IFN-inducing capacity of both TGEV-infected cells and viral particles was markedly inhibited by 2 monoclonal antibodies (MAbs) specifically directed against the exposed amino-terminus of the membrane glycoprotein M[1,2], whereas MAbs specific for the other structural proteins had no effect. Finally, 2 mutant viruses which escaped complement-mediated neutralisation by the induction-inhibiting MAbs were about 100-fold less interferogenic than the parental Purdue virus[2]. Both mutations were shown to be localized at the M exposed amino-terminus and to result in alteration of glycosylation. Induction was also

Corona- and Related Viruses, Edited by P. J. Talbot and G. A. Levy
Plenum Press, New York, 1995

found to be reduced when oligosaccharides are enzymatically removed from the virion or if cells are infected in the presence of tunicamycin[7]. Together, these data strongly support the view that M protein plays a crucial role in IFN induction.

The aim of this study was to investigate whether the M protein exhibits interferogenic activity in the absence of any other viral factors. Its expression was achieved at relatively high level in two systems, either stably in the porcine ST cell line or transiently in the simian COS7 cells. In addition, subcellular localization, glycosylation, antigenicity and interferogenicity of the recombinant product were examined.

MATERIALS and METHODS

Viruses and Monoclonal Antibodies

Propagation of the high-passage Purdue-115 strain of TGEV in the PD5-cell line, virus titration on swine testis (ST) cells and the production of anti-M monoclonal antibodies (MAb) have been described[8]. Obtention and characterization of dm49-4 mutant virus has been already published[2].

cDNA Cloning

The M gene was amplified by PCR from the plasmid pTG2.15[9] with the upstream oligonucleotide 2218 (5′CAACCCCGCTCGAGCA-CTC3′) and the downstream primer 2219 (5′ATTTAGAATTCTAGTTATACC3′). The amplified product was restricted with *Xho*I and *Eco*RI (sites provided by the primers) and cloned in pTEJ4[10] digested with *Sal*I and *Eco*RI. The resulting plasmid, pM, was double-restricted with *Hind*III and *Bam*HI and the insert was subcloned into the corresponding sites of pBluescript SK- (Stratagene) to give pBM. This plasmid was restricted with *Hind*III, treated with Klenow enzyme to make blunt ends and cut with *Bam*HI. The insert was then cloned between the filled-in *Eco*RI and *Bam*HI sites of pUHD10-3[11]. The resulting plasmid, referred to as pUM, was constructed to allow expression of the M gene under the inducible promoter.

The plasmid pCM was constructed by inserting the *Hind*III-*Eco*RI fragment of pM into the corresponding sites of pcDNAI (Invitrogene Corp.). A cDNA of the M gene was amplified by RT-PCR from TGEV infected ST cells, with the oligonucleotides 2190 (5′AG-CACTCCAAGCTTGAACTA) and 2282 (5′GTTGGCGAATTCGAAGTTTAGTTATAC-CATATG3′). This cDNA was restricted with *Hind*III and *Eco*RI (both sites are provided by the primers) and inserted into the corresponding sites of pcDNAI to give pCM2. The plasmid pTM and pTM2 were derived from pCM and pCM2 respectively, by replacing the *Xba*I-*Hpa*I fragment which contains the small intron-polyadenylation signal with the *Xba*I-*Nae*I polyadenylation signal of pUHD10-3. In pCΔC, the second half of the M gene was deleted by self-ligating *Nsi*I-cut pCM. Its expression resulted in a fusion protein composed of the 103 first amino acids of the M protein followed by the peptide SRGPYSIVSPKC.

The plasmid pcDNAP$_2$ expressing the porcine aminopeptidase N was kindly provided by Dr Delmas.

Transfection

The ST cells and the COS7 cells were transfected with lipofectin- and lipofectamine™ (GIBCO BRL), respectively, according to the manufacturer's procedure. Stable ST cell clones were selected with hygromycin (350 µg/ml).

IFN Induction and Titration

In order to make accessible the recombinant product, M expressing cells as well as TGEV-infected cells were lysed by 3 freeze-thawing cycles. Interferon induction on porcine blood mononuclear cells and titration were then performed as previously described[1].

RESULTS AND DISCUSSION

Stable Expression of the M Protein in the Porcine ST Cells

TGEV-infected ST cells are routinely used to induce a high level IFN synthesis by leukocytes, with no detectable background in the absence of infection. These cells were thus chosen to stably express the M protein. The first attempt was made with the M gene placed under the strong constitutive ubiquitin promoter of the plasmid pTEJ4[10]. No expression was detected with 8 clones in the genome of which the plasmid pM was integrated. We next used the inducible system developped and kindly provided by Dr Gossen[11]. Eight clones showed strong reactivity toward the carboxy terminus-specific Mab 3.60, only under induced conditions. When analyzed by indirect immunofluorescence, the staining appeared to be restricted to a perinuclear area (data not shown). This observation was consistent with a Golgi localization, as previously reported for transiently expressed M proteins of IBV[12,13], MHV[14,15] and TGEV[16]. Clones were maintained for up to 4 weeks in induced conditions, without alteration of expression or cell growth, thus indicating that the recombinant product was not toxic.

Possible Incorrect Folding of the Stably Expressed M Protein

Following induction, the MAb 3.60 immunoprecipitated a product with an apparent molecular mass ranging from 29 to 36 kD (Figure 1). The diffuse migration pattern is as

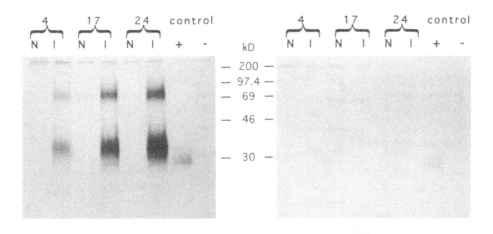

Figure 1. Stable expression of the M protein in ST cells The ST cell clones 4, 17 and 24 were cultured for one day in the presence (non induced N) or absence (induced I) of tetracyclin at 1μg/ml Radiolabeled extracts were immunoprecipitated in RIPA buffer (Tris pH8 0 10mM, NaCl 150mM, KCl 600mM, MgCl$_2$ 0,5mM, Triton X100 2%) with the anti-M MAbs 3.60 (left panel) or 25.22 (right panel) and analyzed in SDS-12% PAGE Controls are normal ST cells infected (+) or not (-)

expected for a protein bearing Golgi-modified, complexe carbohydrates. The 29 kD form co-migrated with the predominant mannose-rich form of the M protein from TGEV-infected cells[17]. Expression was also performed in the presence of either the N-glycosylation inhibitor tunicamycin or monensin which blocks the Golgi transport. In both cases, the recombinant product was shown to have the same electrophoretic mobility as its viral counterpart, thus confirming its glycosylated nature (data not shown). Unexpectedly, we failed to detect the recombinant polypeptide with 3 amino terminus-specific MAbs. This might be the result of misfolding. A strong label around 70 kD was also detected, although inconsistently, with the MAb 3.60 (Figure 1). It might correspond to a homo- or hetero- dimer of the recombinant product, possibly reflecting imperfectly folded material.

From the ST cell clone 30, a 18 kD product was immunoprecipitated by the amino-terminus specific MAbs but not by the MAb 3.60 (data not shown). This truncated protein seemed to accumulate in the perinuclear area. It was also glycosylated and dependent on induction, and thus was supposed to be expressed from a gene deleted of the 3' end. Together, these data suggest that the lack of recognition of the full length product by anti-amino terminus MAbs was not due to sequence alteration of the epitope.

Infection of the M-Expressing ST Cells

The M protein might require viral or virus-derived factor(s) to acquire a proper conformation. To address this point, induced cell clones were infected with dm49-4 virus. This escape mutant codes for a 26 kD, unglycosylated M protein which is no more recognized by the amino terminus-specific MAbs[2]. The expression level of the recombinant product was repeatedly enhanced by infection up to approximately the same amount as the viral counter-part (Figure 2). However, the recombinant protein expressed in dm49-4-infected cells still failed to react with the amino terminus-specific MAbs. The lack of a *trans*-acting factor which would be present only in infected cells might thus not be sufficient to explain the apparent misfolding.

The gene transfected in the ST cell clones showed 1 amino acid difference (Gly82-Cys) compared to the expected sequence (clone pTG2.15[9]). In addition, direct RNA sequencing of TGEV gene M revealed two amino acid changes compared to that deduced from the sequence of pTG2.15 clone: Asn144-Lys and Gly195-Asp[2]. All these mutations were

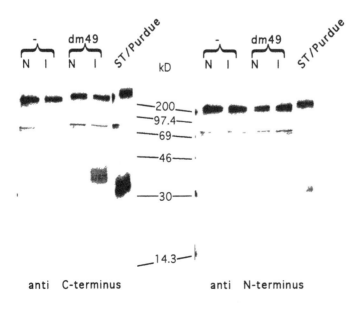

Figure 2. Infection of the M-expressing ST cells. The ST cell clone 37 was induced (I) or not (N), and either infected with the mutant virus dm49-4 (10 pfu/cell) or not infected (-). Cells were then labeled from 2 to 7 hours post adsorption and extracts were immunoprecipitated as in Figure 1.

localized downstream of the first transmembrane domain and were supposed to face the cytoplasm. M gene was therefore recloned from TGEV-infected cells and used in parallel with the mutated cDNA in subsequent experiments.

Transient Expression in COS7 Cells

COS7 cells were transfected with one of the following plasmids: pTM which expressed a cDNA identical to those previously transfected in ST cells, pTM2 coding for the wild type recloned gene, and pCΔC encoding a protein deleted of its carboxy half. The product of the latter construct was recognized solely by the amino-terminus specific MAbs, similar to the truncated form expressed by the ST cell clone 30 (Figure 3). The products expressed from both pTM and pTM2 were immunoprecipitated with the amino-terminus specific MAb 25.22 as efficiently as with the anti C-terminus MAb 3.60. In these conditions, the full-length recombinant protein seemed thus able to acquire a proper conformation.

In order to make possible the infection of the COS7 cells, the constructs were cotransfected with a plasmid expressing the TGEV receptor, aminopeptidase N (APN). Once again, infection enhanced the expression of the recombinant products up to approximately the same amount as the viral M protein.

Interferogenic Activity of the Recombinant Proteins

Purdue-infected cell lysates were highly interferogenic (about 10^4 IFN U/ml for ST cells, and up to 10^3 U/ml for the COS7 cells transiently expressing APN). By contrast, no significant IFN titers were found with the recombinant products in both cell systems, although the expression level of M protein should not be limiting.

Concluding Remarks

The M protein of TGEV was highly expressed either stably in ST cells or transiently in COS7 cells. The localization of the recombinant product appeared to be restricted to the Golgi apparatus, allowing complex glycosylation to occur, as described previously for IBV[12,13], MHV[14,15], and TGEV[16]. However, the ST cell clones showed a reduced ability to produce a correctly or stably folded recombinant M protein. The recombinant product expressed in both cell systems failed to induce IFN at detectable levels. This finding differs from those of two recent reports describing IFN induction by recombinant viral glycoprote-

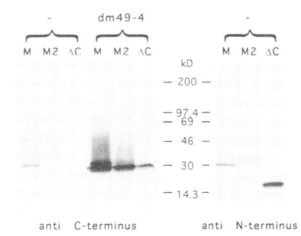

Figure 3. Transient expression in COS7 cells. The COS7 cells were cotransfected with pcDNAP$_2$ and either pTM (M), pTM2 (M2) or pCΔC. Two days post-transfection, cells were infected with dm49-4 (10 pfu/cell) or mock-infected (-). Immunoprecipitation were performed in lysis buffer (Tris pH8 50mM, EDTA 40mM, NaDOC 0.5%, NP40 0.5%).

ins[18 19] Thus, COS7 cells transiently expressing the HN protein of human parainfluenza type 4A were able to induce IFN as efficiently as infected cells[19] With TGEV, experiments are pursued to determine whether the lack of interferogenicity of the expressed M protein is related to the non-exposure at the cell surface or wether the interferogenic determinant is associated to another viral component

REFERENCES

1 Charley B , Laude H J Virol 1988, 62 8-11

2 Laude H , Gelfi J , Lavenant L , Charley B J Virol 1992, 66 743-749

3 Haelterman E D J Am Vet Med Assoc 1972, 160 534-540

4 La Bonnardiere C , Laude H Infect Immun 1981, 32 28-31

5 La Bonnardiere C , Laude H Ann Rech Vet 1983, 14 507-511

6 Charley B , Petit E , Laude H , La Bonnardiere C Ann Virol 1983, 134E 119-126

7 Charley B , Lavenant L , Delmas B Scand J Immunol 1991, 33 435-440

8 Laude H , Chapsal J M , Gelfi J , Labiau S , Grosclaude J J Gen Virol 1986, 67 119-130

9 Laude H , Rasschaert D , Huet J -C J Gen Virol 1987, 68 1687-1693

10 Johansen T E , Schøller M S , Tolstoy S , Schwartz T W FEBS 1990, 267 289-294

11 Gossen M and Bujard H Proc Natl Acad Sci USA 1992, 89 5547-5551

12 Machamer C E , Rose J K J Cell Biol 1987, 105 1205-1214

13 Machamer C E , Mentone S A , Rose J K , Farquhar M G Proc Natl Acad Sci USA 1990, 87 6944-6948

14 Rottier P J M , Rose J K J Virol 1987, 61 2042-2045

15 Armstrong J , McCrae M , Colman A J Cell Biochem 1987, 35 129-136

16 Pulford D J , Britton P Virus Res 1990, 18 203-218

17 Delmas B , Laude H Virus Res 1991, 20 107-120

18 Capobianchi M R , Ankel H , Ameglio F , Paganelli R , Pizzoli P , Dianzani F AIDS Res Human Retroviruses 1992, 8 575-579

19 Ito Y , Bando H , Komada H , Tsurudome M , Nishio M , Kawano H , Matsumura H , Kusagawa S , Yuasa T , Ohta H , Ikemura M , Watanabe N J Gen Virol 1994, 75 567-572

50

TRANSLATION OF THE MHV sM PROTEIN IS MEDIATED BY THE INTERNAL ENTRY OF RIBOSOMES ON mRNA 5

Volker Thiel and Stuart Siddell

Institute of Virology
University of Würzburg
Versbacher Str. 7
97078 Würzburg, Germany

INTRODUCTION

Mouse hepatitis virus (MHV) has a positive strand RNA genome of about 31 kilobases (1). In the infected cell, viral gene expression is mediated by translation from both genomic RNA and subgenomic mRNAs. These mRNAs form a 3' co-terminal set and they contain a common 5'leader sequence (2). Only the region of each mRNA absent from the next smallest mRNA, the so-called unique region, is thought to be translationally active (3,4). Most coronavirus mRNAs contain a single open reading frame (ORF) in their unique region and appear to be functionally mono-cistronic. One exception is the MHV mRNA 5 which contains two ORFs in its unique region, designated as ORF 5a and ORF 5b. Studies on the in vitro translation of synthetic mRNAs suggest that the MHV mRNA 5 is funtionally bicistronic (5). The ORF 5b gene product has been detected in MHV infected cells and virus particles and is equivalent to the small membrane (sM) proteins of infectious bronchitis virus (IBV) and transmissible gastroenteritis virus (TGEV)(6,7,8,9). Two mechanisms can be proposed for the expression of the MHV ORF 5b product. One possibility is based upon the leaky scanning model, as proposed by Kozak (10). In this case, the expression of ORF 5b would be mediated by ribosomes that scan from the 5' end of the mRNA, but fail to recognise the ORF 5a initiation codon. An alternative model is a cap-independent mechanism involving ribosome entry at an internal position on the MHV mRNA 5. Such a mechanism has been described for a variety of picornavirus RNAs and hepatitis C virus RNA (11,12). In the experiments reported here, we have analysed the in vitro translation products of synthetic mRNAs that contain the unique region of MHV mRNA 5, preceded by an ORF derived from the ß-galactosidase gene. The results show that the ß-galactosidase ORF prevents the movement of ribosomes from the 5' end of the mRNA but ORF 5b is, nevertheless, translated. We conclude that translation of the sM protein is mediated by an internal ribosome entry mechanism.

Corona- and Related Viruses, Edited by P. J. Talbot and G. A. Levy
Plenum Press, New York, 1995

METHODS

Cloning Recombinant Plasmids and in Vitro RNA Synthesis

To construct recombinant plasmids corresponding to the unique region of MHV mRNA 5, a 630 bp *Dde*I fragment of pJMS1010 (13) was blunt end cloned into *Sma*I linearised pGEM1. In the resulting construct, p5ab, the initiation codon of ORF 5a is 37 bp downstream of the cloning site. For the construction of a plasmid corresponding to ORF 5b alone, a *Taq*I - *Rsa*I fragment of pJMS1010 DNA, containing the coding region of ORF 5b, was blunt end cloned into *Sma*I linearised pGEM1 to produce the construct p5b. In order to place an ORF upstream of ORF 5a, a PCR-product containing an ORF comprised of the ß-galactosidase gene from nucleotide 6 to nucleotide 1125, was cloned into pGEM1 and p5ab. The resulting constructs are designated as pZ and pZ5ab, respectively. To increase the methionine content of the ORF 5b product, 8 AUG codons were engineered into the ORF 5b coding region, 10 nucleotides upstream of the ORF 5b termination codons in p5b, p5ab and pZab. The resulting constructs are designated as $p5b^{10}$, $p5ab^{10}$ and $pZ5ab^{10}$. The nucleotide sequences of all the plasmids described above were confirmed by chain-termination sequencing. Figure 1 shows the structure of these plasmids. For in vitro transcription, plasmid DNAs were linearised with restriction enzymes (as shown in Figure 1) and RNA was synthesised with T7 RNA polymerase in the presence of the cap structure m7G(5')ppp(5')G as described previously (14).

In Vitro Translation in an L Cell Lysate

The L cell lysate was prepared from L929S cells as previously described (4). The lysate was treated with micrococcal nuclease and 2.5 pmols (0.4 to 1.8 µg) of synthetic mRNA was added to each translation reaction. Aliquots of the translation mixture (15 µl) were electrophoresed on 17% SDS-polyacrylamide gels as described by Laemmli (15). Cytoplasmatic, polyadenylated RNA from MHV-infected cells was prepared as previously described (4).

RESULTS

In Vitro Translation of MHV mRNA 5 Derived Constructs

In order to identify the translation products of the MHV ORFs 5a and 5b, mRNAs were synthesised from *Bst*EII-linearised p5ab (mRNA 5a) and *Bam*HI-linearised p5b (mRNA 5b). In vitro translation of mRNA 5a directs the synthesis of a polypeptide of 12 kDa (figure 2a lane 3) and mRNA 5b directs the synthesis of a polypeptide of 14 kDa (figure 2a, lane 4). The mRNA $5b^{10}$ translation product has an apparent size of 11 kDa (Figure 2b, lane 4). The in vitro translation of a structurally bicistronic mRNA derived from *Bam*HI-linearised p5ab, i.e. mRNA 5ab, directs the synthesis of both the ORF 5a and ORF 5b products (Figure 2a, lane 5). This result is consistent with the idea that the MHV mRNA 5 is functionally bicistronic (16,5,6). To strengthen this conclusion, we translated mRNA derived from *Bam*HI-linearised $p5ab^{10}$, i.e. mRNA $5ab^{10}$. In this case, the detection of the ORF $5b^{10}$ product should be enhanced by the incorporation of additional radioactivity and, indeed, this result is clearly seen in Figure 2b, lane 5.

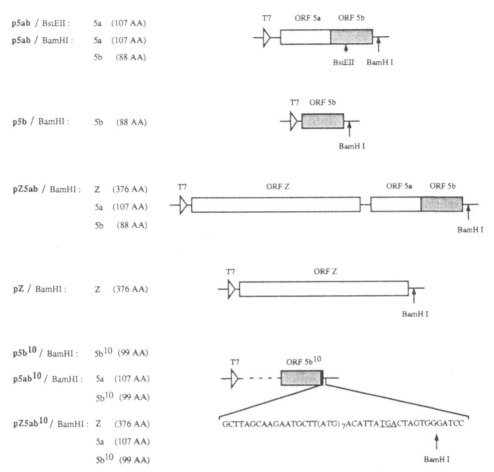

Figure 1. Structure of transcription plasmids. The transcription plasmids used in this study are illustrated. The position of the T7 promoter is shown and ORFs are indicated as boxes. The positions of relevant restriction enzyme recognition sites and the size of potential translation products are also indicated. The broken line represents sequences upsteam of ORF 5b in the plasmids p5b[10], p5ab[10] and pz5ab[10]

In Vitro Translation of ORF 5b but Not ORF 5a from a Tricistronic mRNA Containing an Additional Upstream ORF

To test whether ORF 5b can be expressed independently of ribosomes that enter from the 5' end of the mRNA, we translated mRNA derived from *Bam*HI-linearised pZ5ab, i.e. the tricistronic mRNA Z5ab. The result is shown in Figure 2a, lane 6. As expected, the upstream ORF Z is expressed, resulting in the synthesis of a polypeptide of 51 kDa. Importantly, no ORF 5a product can be detected in the translation reaction. This indicates that very few, if any, ribosomes scan through the upstream ORF Z and initiate the synthesis of an ORF 5a polypeptide. In contrast, the ORF 5b product is readily detected. The amount of ORF 5b product synthesised from the tricistronic mRNA Z5ab is similar to the amount of ORF 5b product expressed from an equimolar concentration of the bicistronic mRNA 5ab (compare Figure 2a, lanes 5 and 6). To rule out the possibility that a polypeptide of 14 kDa can be synthesised by the aberrant translation of ORF Z (for example, premature termination

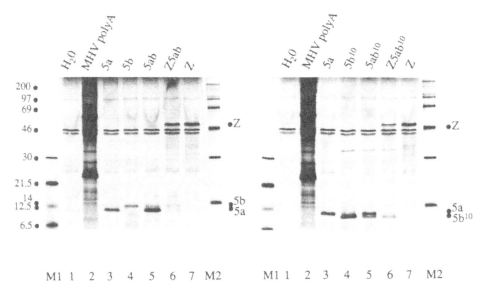

Figure 2. In vitro translation of mRNAs derived from transcription plasmids. The translation products of cell free protein synthesis were electrophoresed in 17% SDS polyacrylamide gels and detected by autoradiography

or internal initiation), we translated mRNA derived from *Bam*HI-linearised pZ, i.e. mRNA Z. As expected, this mRNA directs the synthesis of the ORF Z gene product and no 14 kDa polypeptide can be detected (Figure 2a, 2b, lane 7). To strengthen the conclusion that ORF 5b, but not ORF 5a, is translated from a tricistronic mRNA containing an additional upstream ORF, we carried out a further experiment. We translated mRNA derived from *Bam*HI-linearised pZ5ab[10], i.e. mRNA Z5ab[10]. The result is shown in Figure 2b, lane 6. In this translation reaction, the ORF Z product and the ORF 5b[10] product are easily identified. Again, using equimolar concentrations of mRNA, approximately equal amounts of ORF 5b[10] products are translated from the bicistronic and tricistronic mRNAs, mRNA5ab[10] and mRNA Z5ab[10] (compare Figure 2b, lanes 5 and 6). The ORF 5a product is not expressed from the tricistronic mRNA Z5ab[10].

DISCUSSION

The results presented in this study show that, in the context of the MHV mRNA 5 unique region, the initiation of ORF 5b protein synthesis occurs independently of ribosomes that enter from the 5′ end of the mRNA. This has been shown by the translation of the tricistronic mRNAs, Z5ab and Z5ab[10], where the 5′ proximal and 5′ distal ORFs, ORF Z and ORF 5b/b[10], are translated, whilst the internal ORF, ORF 5a, is not. Clearly, the upstream ORF Z, provides an effective barrier to scanning ribosomes but does not prevent the initiation of ORF 5b translation. Liu and Inglis (17) have concluded that the tricistronic mRNA 3 of IBV encodes three proteins, 3a, 3b and 3c, and that the translation of the most distal ORF 3c is mediated by a cap-independent mechanism involving internal initiation. Taken together, these data strongly suggest that the translation of the coronavirus sM proteins, i.e. the ORF 3c product of IBV and the ORF 5b product of MHV, involves the internal entry of ribosomes on a polycistronic mRNA. The initiation of protein synthesis by internal ribosome entry has been most extensively studied in picornavirus RNAs (18). In this case, ribosome entry is mediated by the so-called "internal ribosome entry site" (IRES) or "ribosome landing pad" (RLP) An obvious question is whether or not similar structures can be identified in the unique region of the MHV mRNA 5 Furthermore, it will be of interest to examine interactions between the putative MHV mRNA 5 IRES/RLP element and cellular proteins In the long term, the biological relevance of internal ribosome entry on the MHV mRNA 5 has to be explained The MHV ORF 5b product is thought to be an essential structural protein of the virus, however, its functional role(s) in the replication cycle is still unknown Why, in contrast to all other MHV subgenomic mRNAs, does the unique region of mRNA 5 encode two proteins? And why is the initiation of ORF 5b translation mediated by a complex mechanism such as internal ribosome entry? The answers to these and other questions must await further experiments

REFERENCES

1 Pachuk, C J, Bredenbeek, P J, Zoltik, P W, Spaan, W J M and Weiss, S R (1989) Virology 171 141-148

2 Lai, M M C, Patton, C D, Baric, R S, and Stohlman, S A (1983) Journal of Virology 46 1027-1033

3 Leibowitz, J L, Weiss, S R, Paavola, E, and Bond, C W (1982) Journal of Virology 43 905-913

4 Siddell, S (1983) Journal of General Virology 64 113-125

5 Budzilowicz, C J, and Weiss, S R (1987) Virology 157 509-515

6 Leibowitz, J L, Perlman, S, Weinstock, G, DeVries, J R, Budzilowicz, C, Weissemann, J M, and Weiss, S R (1988) Virology 164 156-164

7 Yu, X, Bi, W, Weiss, S R and Leibowitz, J L (1994) Virology 202 1018-1023

8 Liu, D X, and Inglis, S C (1991) Virology 185 911-917

9 Godet, M, L'Haridon, R, Vautherot, J F, and Laude, H (1992) Virology 188 666-675

10 Kozak, M (1989) Journal of Cell Biology 108 229-241

11 Jackson, R J, Howell, M T, and Kaminsky, A (1990) Trends in Biochemical Science 15 477-483

12 Brown, E A, Zhang, H, Ping, L, and Lemon, S M (1992) Nucleic Acids Research 20 5041-5045

13 Ebner, D, Raabe, T, and Siddell, S G (1988) Journal of General Virology 69 1041-1050

14 Contreras, R, Cheroutre, H, Degrave, W, and Fiers, W (1982) Nucleic Acids Research 10 6353-6362

15 Laemmli, U K (1970) Nature 227 680-685

16 Skinner, M A, Ebner, D, and Siddell, S G (1984) Journal of General Virology 66 581-592

17 Liu, D X, and Inglis, S C (1992) Journal of Virology 66 6143-6154

18 Meerovitch, K, and Sonenberg, N (1993) Seminars in Virology 4 217 227

51

IDENTIFICATION AND CHARACTERIZATION OF THE PORCINE REPRODUCTIVE AND RESPIRATORY VIRUS ORFS 7, 5 AND 4 PRODUCTS

S. Mounir, H. Mardassi, and S. Dea

Centre De Recherche En Virologie
Institut Armand-Frappier
Université du Québec
Laval, Québec
Canada, H7N 4Z3

ABSTRACT

Porcine reproductive and respiratory syndrome virus (PRRSV) causes abortions and respiratory diseases in pigs. The PRRSV genome is a positive-sense polyadenylated RNA molecule of about 15 kb. Along with the genes for structural proteins (envelope, matrix and nucleocapsid proteins) PRRSV genome contains a number of ORFs potentially encoding nonstructural proteins (ns). To investigate the nature of the PRRSV ORFs 7, 5 and 4 products, we have cloned the envelope (ORF5) nucleocapsid (ORF7) and ns4 (ORF4) protein genes in the bacterial expression vector pMAL™-c2 under control of the "tac" promoter and expressed the proteins in *E.coli*. The recombinant proteins were recognized by porcine and rabbit hyperimmune serums to PRRSV, suggesting their structural nature.

INTRODUCTION

Porcine reproductive and respiratory syndrome virus (PRRSV) is an enveloped positive-strand RNA virus. The morphological characteristics, genome structure and size of PRRSV are comparable to viruses of the *Arterivirus* group including equine arteritis virus, lactate dehydrogenase-elevating virus of mice, and simian hemorrhagic fever virus [1-4]. During PRRSV replication, six major subgenomic mRNAs are produced, ranging in size from 15 kb to 0.9 kb [6,7]. The subgenomic mRNAs have a 3'-coterminal nested structure, the sequence of each mRNA starts from the 3'-end of the genome, extending for various lenght into the 5'-end. Each mRNA contains a leader sequence at its 5'-end, which is derived from the 5'-end of the genome and which is joined to the bodies of the mRNAs during the transcription process[7]. Sequencing studies of cDNAs clones representing the 3' portion of PRRSV genomic RNA have shown the presence of 6 ORFs potentially encoding structural and nonstructural proteins. By analogy to the LDV and LV, it has been suggested that ORFs 7, 6 and 5 encode viral proteins, the nucleocapsid protein N, the matrix protein M and the surface glycoprotein G, respectively [5].

In order to characterize the nature of three PRRSV-ORFs products (ORF 7, 5 and 4) we have cloned in pMAL™-c2 expression vector and expressed in *Escherichia coli* the corresponding genes. Using antivirion hyperimmune serums in Western immunoblotting experiments and expression of individual ORFs, the genes encoding for the putative structural proteins were identified.

METHODS AND RESULTS

The origin and cultivation of the porcine alveolar macrophages (PAM) and the IAF-exp 91 strain of PRRSV were reported elsewhere[3]. Virus purification by isopycnic ultracentrifugation on CsCl density gradients, and, extraction, reverse transcription and polymerase chain reaction (PCR) amplification of viral RNA were also done as described previously [6,8]

Six specific oligonucleotides primers were designated for PCR amplification of the IAF-exp91- ORF7, ORF5 and ORF4 genes. These ORFs were amplified by PCR, using specific oligonucleotides containing *Bam*HI and *Eco*RI restriction sites. The PCR products were purified and cloned into the *Bam*HI *Eco*RI-digested vector pMAL™-c2 (New England Biolabs). In the resulting constructs, each ORF is fused at its 5'-end to the 3'-end of the *Escherichia coli* malE gene[9]. Protein expression was induced by addition of IPTG to a final concentration of 0.3 mM. Recombinant proteins were loaded onto an amylose column (New England Biolabs) and the fusion proteins were eluted with maltose at 10 mM. The MBP-ORF7, 5 and 4 fusion proteins were efficiently expressed and were 90% pure as judged from Coomassie blue stained gels. Their molecular weights were estimated to approximately 56K, 65K and 62K, respectively (Fig. 1).

To confirm the size of the ORF7 product, the recombinant protein was cleaved with factor Xa. After enzymatic cleavage, two bands could be observed after SDS-PAGE corresponding to the MBP protein having MW of 42.7K (Fig. 2, Lane 2) and the N protein with estimated MW of 13.6K. The latter was purified and injected to rabbits and mice to produce monospecific antibodies. The two other fusion proteins (ORFs 5 and 4) were injected to mice to produce monospecific antibodies, omitting factor Xa cleavage.

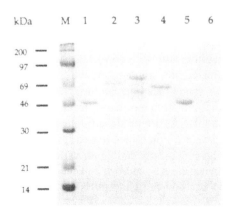

Figure 1. SDS-PAGE (10%) analysis of the PRRSV ORFs 7, 5 and 4 expressed recombinant proteins. (1, 5, 6) : pMAL™-c2 Vector without insert after IPTG induction. (2) : pMAL™-c2-ORF7 after IPTG induction and column purification. (3) : pMAL™-c2-ORF4 after IPTG induction and column purification. (4) : pMAL™-c2-ORF5 after IPTG induction and column purification. M : Molecular size marker in kilodaltons.

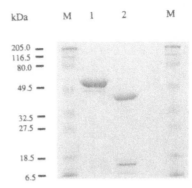

Figure 2. SDS-PAGE (10%) analysis of the expressed recombinant N protein before (1) and after (2) cleavage with factor Xa. The uncleaved and cleaved fusion proteins have estimated MW of 56K and 13.6K, respectively. M : Molecular size marker in kilodaltons.

To analyse the nature of the recombinant proteins produced in *E. coli*, they were subjected to Western blot analysis using hyperimmune serums from rabbits that have been inoculated with PRRSV purified virion. The anti-PRRSV serums recognized the three recombinant proteins (ORF7, ORF5 and ORF4) but failed to recognized the MBP protein (Fig. 3), thus confirming their viral specificity. The reactivity of the antisera towards ORF4 recombinant protein appeared very low, indicating probably the weak antigenicity of this ORF product in rabbits.

The recombinant proteins were also recognized by a serum obtained from a PRRSV-experimentaly infected pig[3]. This together with results obtained following incubation in the presence of rabbit anti-PRRSV serums indicated that these viral proteins are probably structural.

Interestingly, we observed that the ORF4 recombinant protein was more intensively recognized by the homologous antiserum produced in pigs rather than antisera produced in rabbits (Fig. 4). The low reactivity of the rabbit anti-PRRSV serums towards ORF4 products was also confirmed in RIPA experiments conducted with *in vitro* translation product (data not shown).

CONCLUSION

High level expression of the PRRSV ORF7, 5, and 4 products was obtained using the pMAL™-c2 prokaryotic expression vector.

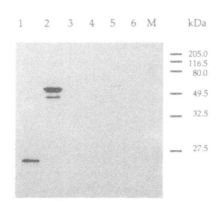

Figure 3. Western blot analysis using the rabbit anti-PRRSV serum. ORF7-fusion protein after factor Xa cleavage (Lane 1), purified ORF7-fusion protein (Lane 2), purified ORF4-fusion protein (Lane 3), purified ORF5-fusion protein (Lane 5). MBP controls (Lanes 4 and 6). Recombinant proteins were resolved by electrophoresis on 10% SDS-polyacrylamide gel, transferred onto a nitrocellulose membrane and reacted with a 1/1000 dilution of rabbit anti-PRRSV serum. M : Molecular size marker in kilodaltons.

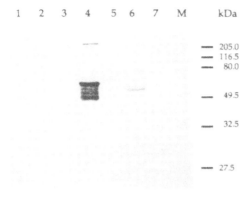

Figure 4. Western blot analysis using the porcine anti-PRRSV serum. ORF7-fusion protein after factor Xa cleavage (Lane 1), purified ORF7-fusion protein (Lane 2), purified ORF4-fusion protein (Lane 4), purified ORF5-fusion protein (Lane 6). MBP controls (Lanes 5 and 7). Proteins were resolved by electrophoresis on 10% SDS-polyacrylamide gel, transferred onto a nitrocellulose membrane and reacted with a 1/1000 dilution of porcine anti-PRRSV serum. M : Molecular size marker in kilodaltons.

The fusion proteins (ORF7, 5, and 4) were recognized by anti-PRRSV serums produced in rabbits and by the serum of a PRRSV-experimentally infected pig.

The results of the present study suggest that the PRRSV-ORF7, 5 and 4 encode structural proteins.

REFERENCES

1. Goyal S.M. Porcine reproductive and respiratory syndrome; a review. J Vet Diagn Invest 1993; 5: 656-664.
2. Wensvoort G., Kluyver E.P., Pol J.M.A., Wagenaar F., Moormann R.J.M., Hulst M.M., Bloemraad R., den Besten A., Zesta T., Terpstra C. Lelystad virus, the cause of porcine epidemic abortion and respiratory syndrome: a review of mystery swine disease research at Lelystad. Vet Microbiol 1992, 33: 185-193.
3. Mardassi H., Athanassious R., Bilodeau R., Dea S. Porcine reproductive and respiratory syndrome virus : morphological, biochemical and serological characteristics of Quebec isolates associated to acute or chronic outbreaks of PRRS virus. Can Vet Res 1993, 58: 55-64.
4. Conzelmann K.K., Visser N., Van Woensel P., Thiel H.-J. Molecular characterization of porcine reproductive and respiratory syndrome virus, a member of the arterivirus group. Virology 1993; 193: 329-339.
5. Meulenberg J.J.M., Hulst M.M., De Meijer E.J., Moonen P.L.J.M., Den Besten A., De Kluyer E.P., Wenswoort G., Moormann R.J.M. Lelystad virus, the causative agent of porcine abortion and respiratory syndrome (PEARS), is related to LDV and EAV. Virology 1993; 192: 62-72.
6. Mardassi H., Mounir S., Dea S. Identification of major differences in the nucleocapsid protein genes of a Quebec strain and european strains of porcine reproductive and respiratory syndrome virus. J Gen Virol 1994; 75: 681-685.
7. Meulenberg J.J.M., De Meijer E.J., Moormann R.J.M. Subgenomic RNAs of Lelystad virus contain a conserved leader-body junction sequence. J Gen Virol 1993; 74: 1697-1701.
8. Mounir S., Talbot P.J. Human coronavirus OC43 RNA 4 lacks two open-reading-frames located downstream of the S gene of bovine coronavirus. Virology 1993; 192: 355-360.
9. Maina C.V., Riggs P.D., Grandea III A.G., Slatko B.E., Moran L.S., Tagliamonte J.A., McReynolds L.A., di Guan C. An *Escherichia coli* vector to express and purify by fusion to and separation from maltose-binding protein. Gene 1988, 74: 365-373.

52

STRUCTURAL PROTEINS OF PORCINE REPRODUCTIVE AND RESPIRATORY SYNDROME VIRUS (PRRSV)

E. A. Nelson, J. Christopher-Hennings, and D. A. Benfield

Department of Veterinary Science
South Dakota State University
Brookings, South Dakota

INTRODUCTION

PRRSV is a small, enveloped, RNA virus, which is similar to the arteriviruses in morphology, physicochemical properties, nucleotide sequence, genomic organization and replication strategy.[1, 2, 3, 4] The arteriviruses include lactate dehydrogenase-elevating virus (LDV), equine arteritis virus (EAV) and simian hemorrhagic fever virus (SHFV).[5] Sequence information for the VR-2332,[6] Lelystad[3] and German isolates[2] of PRRSV suggests the presence of 6 open reading frames (ORFs) that may code for structural proteins. Three putative, structural proteins have been identified for the VR-2332 isolate of PRRSV.[7] However, additional structural proteins may exist and little information is available regarding the biochemical characteristics of these proteins. Therefore, the purpose of this study was to identify and conduct the initial characterization of the PRRSV structural proteins.

MATERIALS AND METHODS

The VR-2332 and Lelystad isolates of PRRSV were propagated on MA-104 cells. Virus from clarified culture supernatant was concentrated and purified using continuous sucrose density gradients.[1] Virus was labeled with Tran[35S]cysteine and [35S]methione (>1000 Ci/mmol; ICN) as previously described.[7] Alternatively, virus was labeled with 10 μCi of [6-3H]-glucosamine (40 Ci/mmol; ICN) per ml as for the Tran[35S]-label except a low glucose (450 mg/liter glucose) MEM replaced the cysteine- and methionine-free MEM.

Radioimmunoprecipitation (RIP) was done using a modified protein A-Sepharose technique.[7] Polyacrylamide gel electrophoresis (PAGE) was done on SDS-PAGE (12% acrylamide) gels[8] under reduced or nonreduced conditions. Immunoblotting was done with lysates from PRRSV infected and mock infected MA-104 cells and polyclonal antisera as previously described.[7]

Corona- and Related Viruses, Edited by P. J. Talbot and G. A. Levy
Plenum Press, New York, 1995

To identify PRRSV glycoproteins, [3H]glucosamine-labeled proteins from virus infected and uninfected cell lysates were separated by SDS-PAGE and incorporated [3H]glucosamine was detected by PPO fluorography. Gels were also sliced and incorporated [3H]glucosamine was quantitated by liquid scintillation. Tran[35S]-labeled PRRSV immunoprecipitates or purified virus preparations were also treated with endoglycosidase F/N-glycosidase F (glyco F; Boehringer Mannheim) prior to gel electrophoresis.[9]

Immuno-gold electron microscopy was done by floating collodion coated, carbon stabilized grids on droplets of PRRSV suspension for 2 h at room temperature. Grids were then floated on SDOW17 or NS95 MAb[7] diluted 1:100 in Tris buffer containing 1% BSA (BSA-Tris) for 1.5 h. After 2 washes in BSA-Tris buffer, grids were floated on a suspension of Protein A labeled gold beads (5 nm; Pelco) in BSA-Tris for 30 min. Grids were then washed, negatively stained and examined by electron microscopy as previously described.[10, 11]

RESULTS AND DISCUSSION

Three major proteins of 15-, 19-, and 26- to 30-kDa or 15-, 18-, and 26- to 30-kDa were identified in preparations of the VR-2332 and Lelystad isolates of PRRSV, respectively. These proteins were consistently identified by gel electrophoresis of purified Tran[35S]cysteine- and [35S]methionine-labeled virus, radioimmunoprecipitation and immunoblotting techniques. A less abundant, virus associated protein of 22-kDa was also detected. Protein bands of higher molecular mass were observed in both infected and mock-infected preparations, suggesting they were of cellular origin. The three major proteins likely correspond with the nucleocapsid protein, the nonglycosylated envelope protein and the envelope glycoprotein of LDV[12] and EAV.[9] A less abundant 25-kDa protein recently identified in EAV[9] may correspond with the 22-kDa protein observed in our study.

Immuno-gold electron microscopy using the VR-2332 isolate of PRRSV and the SDOW17 or NS95 MAbs indicated that the 15-kDa viral protein is present in the nucleocapsid of PRRSV. Monoclonal antibody and gold particles selectively bound to virus nucleocapsids and aggregates of nucleocapsids surrounded by antibody and gold particles were apparent in IEM preparations. Monoclonal antibody did not bind to intact, enveloped virus particles. Both LDV and EAV have a single nucleocapsid protein of similar relative molecular mass coded by ORF 7.[5]

To determine which virus proteins were glycosylated, PRRSV was labeled with [3H]glucosamine and gradient purified. Electrophoresis gel fractions were then analyzed for incorporated [3H] by liquid scintillation. A strong peak of incorporated [3H] was detected in gel fractions from SDS-PAGE gel lanes containing PRRSV but not in PAGE gel lanes containing a mock infected MA-104 cell preparation. The peak of incorporated [3H] corresponded with a relative molecular mass of approximately 26- to 30-kDa as determined from molecular mass standards and Tran[35S]-labeled virus included on the same PAGE gel. Electrophoresis of gradient purified [3H]glucosamine-labeled PRRSV followed by PPO fluorography demonstrated a single broad band of 26- to 30-kDa. No band was apparent in the lane containing a mock-infected preparation. Similar studies have indicated that the 21- to 44-kDa and 30- to 42-kDa proteins of LDV and EAV, respectively, are also glycoproteins.[5, 9, 13] The recently identified 25-kDa protein of EAV is glycosylated, but is present in the EAV particle in very low abundance.[9] A peak that corresponds with the 22-kDa protein of PRRSV was not identified in our [3H]glucosamine studies. If the 22-kDa protein of PRRSV is glycosylated, it may be difficult to verify due to the apparent low abundance of this protein.

To further examine glycosylations, purified Tran[35S]-labeled PRRSV was treated with glyco F prior to gel electrophoresis and no shifts in relative molecular mass for the 15- or 19-kDa proteins of the VR-2332 or the 15- or 18-kDa proteins of the Lelystad isolate of

PRRSV were apparent The broad 26- to 30-kDa protein band shifted downward and appeared as a sharp band at 16 5- or 22-kDa, suggesting that this protein is N-glycosylated Glyco F treatment of EAV by de Vries, et al [9] indicated that the 25- and 30- to 42-kDa proteins of EAV are N-glycosylated proteins while the remaining two EAV proteins lack N-glycans

REFERENCES

1 Benfield D A , Nelson E A , Collins J E , Harris L , Goyal S M , Robison D , Christianson W T , Morrison R B , Gorcyca D , Chladek D Characterization of swine infertility and respiratory syndrome (SIRS) virus (Isolate ATCC VR-2332) J Vet Diag Invest 1992,4 127-133

2 Conzelmann K K , Visser N , Van Woensel P , Thiel H J Molecular characterization of porcine reproductive and respiratory syndrome virus, a member of the Arterivirus group Virology 1993,193 329-339

3 Meulenberg J J M , Hulst M M , de Meuer E J , Moonen P L J M , den Besten A , de Kluyver E P Wensvoort G, Moormann R J M Lelystad virus, the causative agent of porcine epidemic abortion and respiratory syndrome (PEARS), is related to LDV and EAV Virology 1993,192 62-72

4 Wensvoort G , Terpstra C , Pol J M A , der Laak E A , Bloemrad M , deKluyver E P , Kragten C , van Buiten L , den Besten A , Wagenaar F , Broekhuijsen J M , Moonen P L J M , Zetstra T , de Boer E A , Tibben H J , de Jong M F , van't Veld P , Groenland G J R , van Gennep J A , Voets M T , Verheijden J H M , Braamskamp J Mystery swine disease in the Netherlands the isolation of Lelystad virus The Vet Quarterly 1991,13 121-130

5 Plageman P G W ,Moennig V Lactate dehydrogenase-elevating virus, equine arteritis virus, and simian hemorrhagic fever virus a new group of positive-strand RNA viruses Adv Virus Res 1992,41 99-183

6 Murtaugh M The PRRS virus Allen D Leman Swine Conference, St Paul, MN 1993 pp 43-45

7 Nelson E A , Christopher-Hennings J , Drew T , Wensvoort G , Benfield D A Differentiation of U S and European isolates of porcine reproductive and respiratory syndrome (PRRS) virus using monoclonal antibodies J Clin Micro 1993,31 3184-3189

8 Laemmli U K Cleavage of structural proteins during the assembly of the head of bacteriophage T4 Nature (London) 1970,227 680-685

9 de Vries A F , Chirnside E D , Horzinek M C , Rottier P J M Structural proteins of equine arteritis virus J Virol 1992,66 6294-6303

10 Benfield D A , Stotz I , Moore R , McAdaragh J P Shedding of rotavirus in the feces of sows before and after farrowing J Clin Micro 1982 16 186-190

11 Ritchie A E , Fernelius A L Direct immuno-electron microscopy and some morphological features of hog cholera virus Arch Virol 1968,23 292-298

12 Brinton-Darnell M , Plagemann P G W Structure and chemical-physical characteristics of lactate dehydrogenase-elevating virus and its RNA J Virol 1975,16 420-433

13 Zeegers J J W , van der Zeijst B A M , Horzinek M C The structural proteins of equine arteritis virus Virology 1976,73 200-205

Cellular Receptors

53

CELLULAR RECEPTORS FOR TRANSMISSIBLE GASTROENTERITIS VIRUS ON PORCINE ENTEROCYTES

Hana M. Weingartl and J. Brian Derbyshire

Department of Veterinary Microbiology and Immunology
University of Guelph
Guelph
Canada N1G 2W1

ABSTRACT

The activity of aminopeptidase-N (APN), reported to be a major receptor for porcine transmissible gastroenteritis virus (TGEV), in enterocyte fractions harvested from the jejunal villi and crypts of newborn and weaned piglets, did not correspond with the levels of saturable virus binding previously demonstrated for the same fractions. Plasma membranes prepared from enterocytes harvested from the jejunal villi of a newborn piglet were used in the preparation of a monoclonal antibody (MAb) which blocked the binding of TGEV, but not that of the porcine respiratory coronavirus (PRCV), to ST cells. This MAb immunoprecipitated a 200 kDa non-glycosylated protein from lysates of ST cells, which was not precipitated by an anti-APN MAb. The 200 kDa protein was shown by immunostaining and fluorescence activated cell scanning to be present on ST cells and on villous enterocytes from newborn piglets, but not on MDBK cells or enterocytes from weaned piglets. APN was demonstrated by the same techniques to be present on villous enterocytes from both newborn and weaned piglets, as well as on ST cells. It was concluded that the 200 kDa protein may be a second receptor for TGEV, contributing to the high susceptibility of newborn piglets to the virus.

INTRODUCTION

Transmissible gastroenteritis virus (TGEV) is an important cause of enteric disease in swine[1], associated with replication of the virus in the villous enterocytes of the small intestine, which leads to villous atrophy and malabsorption. The disease is most severe, and frequently fatal, during the first two weeks of life. We have been interested in the possible role of cellular receptors for TGEV in determining the tropism of the virus for villous enterocytes, and in the high susceptibility of newborn piglets to the virus. In earlier studies[2], we demonstrated high levels of saturable binding of TGEV to enterocytes from the villi of newborn piglets, while virus binding to enterocytes from older piglets was at a lower level, and non-saturable.

Aminopeptidase-N (APN) has been identified as a major receptor for TGEV on porcine enterocytes and ST cells[3]. Since APN is present in pigs of all ages, it could contribute to the intestinal tropism of the virus, but might not explain the preferential tropism of the virus to the villous enterocytes of the newborn, and the high level of virus binding to these cells. In the present study, further details of which have been published elsewhere[4], we identified a 200 kDa protein, in ST cells and in villous enterocytes from newborn piglets, which may be a second receptor for TGEV contributing to the high susceptibility to the virus of very young animals.

METHODS AND RESULTS

APN Enzymatic Activity in Porcine Enterocyte Fractions

Enterocytes were harvested by chelation[2] from the jejunum of two newborn and two 3 weeks-old piglets in a series of seven fractions from the tips of the villi to the crypts. The cells in each fraction were assayed for APN enzymatic activity[5], as well as for alkaline phosphatase (AP) activity[6], which was used as an indicator of the level of cell differentiation. Similar levels of APN activity were found in the newborn and weaned piglets, and there were no consistent differences in the levels of APN activity among the various fractions, while the levels of AP activity declined progressively in fractions I through IV as anticipated (Figure 1). Thus the distribution of APN activity failed to correspond to the previously reported[2] high level binding of TGEV, restricted to the villous enterocytes in fractions I and II of newborn piglets.

Preparation of Monoclonal Antibodies

Villous enterocytes were harvested from the jejunum of a newborn piglet, and monoclonal antibodies were prepared from plasma membranes harvested from the enterocytes as described[7]. Eighteen clones reacted in an enzyme immunoassay (EIA) against a lysate of ST cells, and four of these hybridomas blocked the replication of the Miller-6 strain

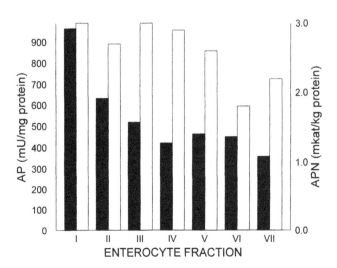

Figure 1. Distribution of alkaline phosphatase (AP - solid bars) and aminopeptidase-N (APN - open bars) activities in seven fractions of enterocytes from the tips of the villi (fraction I) to the crypts (fraction VII) of the jejunum of a newborn piglet.

of TGEV in a cytopathic effect reduction assay (CPERA) in ST cells. One of these antibodies (MAb 166) was selected for further study, and shown to block the replication of the Purdue, Diamond and Ambico strains of TGEV, in addition to the Miller-6 strain. However, MAb 166 failed to block the replication of the porcine respiratory coronavirus (PRCV) in ST cells in the CPERA, under conditions in which PRCV replication was blocked by a MAb (G43) against porcine APN, which was a generous gift from Dr. B. Delmas, I.N.R.A., Jouy-en-Josas, France. MAb 166 failed to react in an EIA against porcine kidney cytosol APN or porcine kidney microsomal APN, in which MAb G43 was used as a positive control.

Immunoprecipitation of ST Cell Lysates

In order to identify the specificity of MAb 166, it was used to immunoprecipitate radiolabeled ST cell lysate. Autoradiographs of the precipitated proteins, bound to recombinant protein G (rPG) beads and resolved by SDS-PAGE, indicated that a 200 kDA protein was precipitated from the ST cell lysate. The molecular weight of the 200 kDA protein was not changed by treatment with endoglycosidase H. The specificity of immunoprecipitation was confirmed by Western blotting of immunoprecipitates transferred to PVDF membranes after SDS-PAGE. The anti-APN MAb G43 precipitated a protein of 150 kDa molecular weight from the ST cell lysate. Labeled MDBK cell lysate was used as a negative control in these experiments.

Competitive EIA between TGEV and MAbs

This test was used to determine whether MAbs 166 or G43 would compete with TGEV for binding to ST cell lysates. MAb 310, which reacted with ST cell lysate in the EIA but failed to block TGEV replication in the CPERA, was used as a negative control. Incubation of the ST cell lysate-coated wells for 2 hours with the Miller-6 strain of TGEV blocked 30% of the subsequent binding of MAb 166 and 24% of the binding of MAb G43, while only 5.5% of the binding of MAb 310 was blocked by prior incubation of the ST cell lysate with virus.

Additive blocking of TGEV binding by MAbs 166 and G43

The CPERA was used to determine the combined effect of these MAbs in blocking the replication of TGEV in ST cells. When the ST cells were treated with dilutions of MAb G43 in the presence of MAb 166, the titer of the antibody which protected the ST cells against CPE was 16-fold higher than when the antibody was diluted in an irrelevant MAb directed against ovine interferon, kindly provided by Dr. R. L'Haridon, I.N.R.A., Jouy-en-Josas, France. Thus mixtures of MAbs 166 and G43 seemed to have an additive protective effect on ST cells challenged with TGEV.

Distribution of APN and 200 kDa Protein on Porcine Enterocytes

Sections of the jejunum of two newborn and two 3 weeks-old piglets were fixed in paraformaldehyde and immunostained with the Zymed streptavidin-biotin system for immunological staining, with MAbs G43 or 166 as the primary antibodies. When MAb G43 was used, uniform staining of the villous enterocytes was obtained in both the newborn and older piglets, while MAb 166 stained only groups of enterocytes located on the upper villi of newborn piglets, as illustrated in a previous publication[4]. Neither antibody stained the cryptal enterocytes, and no specific staining was observed when an irrelevant MAb directed against

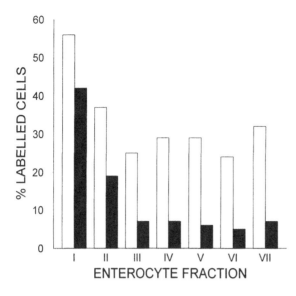

Figure 2. Labeling of enterocytes by MAb G43 (open bars) or MAb 166 (solid bars) in seven fractions from the tips of the villi (fraction I) to the crypts (fraction VII) of the jejunum of newborn piglets, as determined by FACS. The values are the means for two piglets.

Rhodococcus equi, kindly supplied by Dr. J.F. Prescott, University of Guelph, was used as the primary antibody, or when the primary antibody was omitted.

For analysis by fluorescence activated cell scanning (FACS), enterocyte fractions were collected from the jejunum of two newborn piglets, and the cells were labeled with either the anti-APN MAb G43, or MAb 166, directed against the 200 kDa protein. The percentage of cells in each fraction labeled by each antibody is shown in Figure 2. The highest percentage of labeled cells was found in fraction I, collected from the tips of the villi, and while few cells in fractions III to VII were labeled with MAb 166, relatively larger numbers of cells in each fraction were labeled with MAb G43. The distribution of the 200 kDA protein therefore corresponded with the previously described[2] binding of TGEV to the same enterocyte fractions.

DISCUSSION

While APN has been identified as a major receptor for TGEV[3], the present study has shown that the distribution of APN, as determined by assays of enzymatic activity, and by immunohistochemistry and FACS with an anti-APN MAb, failed to correspond with the previously described[2] high saturable virus binding to the villous enterocytes of newborn piglets. A MAb prepared against villous enterocytes collected from the jejunum of a newborn piglet blocked the replication of TGEV in ST cells, and immunoprecipitated a 200 kDa non-glycosylated protein from ST cell lysates. While the anti-APN MAb stained most of the villous enterocytes in both newborn and older piglets, only patches of enterocytes located on the upper villi of the jejunum of newborn piglets stained with the anti-200 kDa protein MAb. This corresponded with the distribution of fetal cells on the villi of newborn piglets[8], and with the distribution of cells stained with anti-TGEV antibodies in sections of the jejunum of piglets in the early stages of infection with the virus[9].

On the basis of the immunostaining and FACS results we postulate the existence of two receptors which contribute additively to the binding of TGEV, the previously described[3] APN, which seems to be widely distributed on enterocytes and probably on other tissues, irrespective of age, and a second 200 kDa protein which is restricted to the villous enterocytes of newborn piglets, and which may be a major factor in determining the high susceptibility of newborn piglets to TGEV. More than one receptor has been identified in relation to the tissue tropism of several other viruses, including mouse hepatitis virus[10], but TGEV may be the first virus for which a second receptor with a specific role in relation to age sensitivity has been identified.

It has been proposed[11] that the S protein of enteric strains of TGEV possesses two receptor binding sites, while the respiratory variant has only a single receptor binding site. Our finding that the binding of TGEV, but not PRCV, to ST cells was blocked by antibodies against the putative 200 kDa receptor for TGEV seems to support this hypothesis.

ACKNOWLEDGMENTS

This research was supported by the Natural Sciences and Engineering Research Council of Canada and by the Ontario Ministry of Agriculture, Food and Rural Affairs. We thank Dr. B. Delmas for kindly providing the monoclonal antibody G43.

REFERENCES

1 Saif L J , Wesley R D Transmissible gastroenteritis In Leman A D , Straw B E , Mengeling W L , D'Allaire S D , Taylor D J , (eds) Diseases of swine Iowa State University Press, Ames, Iowa 1992 pp 362-386

2 Weingartl H M , Derbyshire J B Binding of porcine transmissible gastroenteritis virus by enterocytes from newborn and weaned piglets Vet Microbiol 1993,35 23-32

3 Delmas B , Gelfi J , L'Haridon R , Vogel L K , Sjostrom H , Noren O , Laude H Aminopeptidase N is a major receptor for the enteropathogenic coronavirus TGEV Nature 1992,357 417-420

4 Weingartl H M , Derbyshire J B Evidence for a putative second receptor for porcine transmissible gastroenteritis virus on the villous enterocytes of newborn piglets J Virol 1994,68 In press

5 Sjostrum H , Noren O , Jeppeson L , Staun M , Svennson B , Christiansen L Purification of different amphiphilic forms of a microvillus aminopeptidase from pig small intestine using immunoadsorbent chromatography Eur J Biochem 1978,88 503-511

6 Weiser M M Intestinal epithelial cell surface membrane glycoprotein synthesis I An indicator of cell differentiation J Biol Chem 1973,248 2536-2541

7 Gratecos D , Knibiehler M , Benoit V , Semeriva M Plasma membranes from rat intestinal epithelial cells at different stages of maturation I Preparation and characterization of plasma membrane subfractions originating from crypt cells and from villous cells Biochim Biophys Acta 1978,512 508-524

8 Smith M W , Jarvis L G Growth and cell replacement in the newborn pig intestine Proc R Soc Lond B 1978,203 69-89

9 Jordan L T , Derbyshire J B Antiviral activity of interferon against transmissible gastroenteritis virus in cell culture and ligated intestinal segments in neonatal pigs Vet Microbiol 1994,38 263-276

10 Nedellec C , Dveksler G S , Daniels E , Turbide C , Chow B , Basile A A , Holmes K V , Beauchemin N Bgp2, a new member of the carcinoembryonic antigen-related gene family encodes an alternative receptor for mouse hepatitis viruses J Virol 1994,68 4525-4537

11 Sanchez C M , Gebauer F , Sune C , Mendez A , Dopazo J , Enjuanes L Genetic evolution and tropism of transmissible gastroenteritis coronaviruses Virology 1992,190 92-105

54

OVEREXPRESSION OF THE MHV RECEPTOR

Effect on Progeny Virus Secretion

T. M. Gallagher

Department of Microbiology and Immunology
Loyola University Medical Center
Maywood, Illinois 60153

ABSTRACT

The intracellular interaction of the coronavirus mouse hepatitis virus (MHV) with its cellular receptor (MHVR) was investigated. Overexpression of MHVR from vaccinia vectors during an ongoing MHV infection resulted in dramatic inhibition of virus production. Infectivity in both cytoplasmic extracts and supernatants was reduced by over three orders of magnitude relative to control cultures in which a truncated MHVR lacking virus binding activity was expressed. Complete MHV virions were not detectable in supernatants of MHVR expressing cells. In the presence of overexpressed MHVR, the coronavirus spike protein was not cleaved into posttranslation products S1 and S2, nor was it fully processed into a form resistant to endoglycosidase H digestion, indicating that intracellular engagement of spike with receptor prevented spike transport and consequent association with virions.

INTRODUCTION

The murine coronaviruses comprise a group of seven known strains, commonly known as strains of mouse hepatitis virus (MHV). Each MHV strain recognizes one or more members of the murine carcinoembryonic antigen (CEA) glycoprotein family as a cell surface receptor[1]. The primary CEA gene family member that is recognized by all strains tested thus far is known as CGM #1[2], or more simply as the MHV receptor (MHVR). This recognition event involves the binding of the protruding spike (S) glycoprotein of the invading enveloped virions to the most exposed distal portion of the MHVR[3, 4]. The fate of virus following this binding event is virus strain-specific. Some strains undergo a spike protein-mediated fusion of virion and cell membranes either at or very near the plasma membrane, while others are internalized into endosomes where declining pH of the endosome interior promotes a similar spike-mediated fusion of virion and endosome membranes[5, 6]. The outcome of both delivery pathways is virion uncoating and nucleocapsid delivery to the host cell cytosol.

The MHV A59 strain is one which fuses its virion envelope rapidly after binding to MHVR, with little or no requirement for acidic endosome exposure. The possibility therefore exists that MHVR : MHV A59 spike interaction *per se* is sufficient to promote fusion and consequent nucleocapsid delivery. This possibility is consistent with previous models of pH-independent fusion activation. For HIV[7], paramyxoviruses[8], and even alphaviruses[9] it is argued that interaction of receptor with cognate virion attachment protein results in conformational alterations that are prerequisite to fusion function. There has been no need to appeal to this model of "receptor-mediated activation of fusion" for those viruses requiring internalization. For example, low pH (c. 5.5) is well-known to convert the hemagglutinin spike of the endosome-dependent influenza virus into a fusion-competent conformation[10].

In the experiments reported here, MHVR was allowed to interact with spike glycoproteins by expressing both ligands at high levels in the same cell. Such co-expression allows the proteins to accumulate within the lumen of the endoplasmic reticulum (ER) and thereby enhances the likelihood of their interaction. MHVR was found to bind and thereby effect the nearly complete retention of the spikes in a pre-medial Golgi compartment.

METHODS AND RESULTS

Transient and Rapid Synthesis of MHVR from Vaccinia Vectors

To provide for high level expression of the MHV receptor in the face of an ongoing MHV infection, MHV - susceptible 17Cl1 cells were infected 3 hours prior to MHV infection with vaccinia virus (VV) recombinants expressing various forms of MHVR. Expression was dependent on co-infection of cells with the widely used recombinant VV-T7[11] along with VV recombinants harboring cDNA encoding the MHVR. In the prepared VV-MHVR recombinants, MHVR cDNA was juxtaposed downstream of the T7 promoter and was therefore readily transcribed by VVT7 - derived bacteriophage T7 RNA polymerase. Recombinant protein expression in this system initiates at 6 to 7 hours post VV infection and generally proceeds to levels exceeding 1 μg per 10^6 cells [12].

Two VV-MHVR recombinants were employed in initial experiments. VV-MHVR encoded the complete CGM1 isoform of the receptor while VV-MHVRΔ lacked sequences encoding amino acids 10 to 122 of the mature receptor. This deleted region includes the amino-terminal immunoglobulin domain of MHVR and has been conclusively shown by Dveksler et al.[3] to be essential for receptor interaction with MHV particles; VV-MHVRΔ was therefore expected to serve as a control recombinant in studies of the virus: receptor interaction.

Immunoblot detection of the VV-derived receptors was performed using an anti-peptide antibody directed against the carboxy - terminal 16 amino acids of MHVR as a primary detection reagent. The immunoblot (Figure 1) revealed three size classes of specific product for both MHVR and MHVRΔ (arrows indicate size classes). Each size class varied in its relative sugar content. These VV-derived MHVR proteins were the only ones detected by the immunoblot procedure. All bands in lanes 1 and 2 of Figure 1 were judged non-specific as they remained regardless of the presence of anti-MHVR peptide serum. Thus while 17Cl1 cells do contain endogenous MHVR in amounts sufficient to permit infection by virus, they do not contain enough MHVR to be recognized in this assay.

Figure 1. Immunoblot detection of MHVR and MHVRΔ expressed from vaccinia recombinants. Confluent monolayers of 17Cl1 cells were left uninfected (lane 1), infected with VV-T7 (lane 2) or with VV-T7 plus the indicated VV-MHVR recombinants (lanes 3 and 4), each at multiplicities of 5 PFU per cell. At 18 hours postinfection, adherent cells were rinsed with saline and dissolved in Laemmli solubilizer [13], to 5 x 10^6 cells per ml. Aliquots representing 10^5 cells were subjected to immunoblot analysis using rabbit anti-MHVR peptide serum and alkaline phosphatase-conjugated antibody to rabbit Ig as primary and secondary antibodies, respectively. Detection of immobilized antigen was by enzymatic assay of alkaline phosphatase.

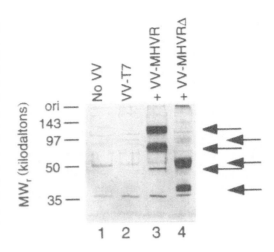

Effect of MHVR Overexpression on the Production of Infective MHV Progeny

Three hours after 17Cl1 cells were infected with VV recombinants, MHV strain A59 was inoculated on to cultures at high multiplicity. 15 hours later, supernatants and cell extracts were collected and the bulk of contaminating VV virions were removed by centrifugation. MHV A59 infectivities were then measured by plaque assay.

Infection with either VVT7 alone or VVT7 plus VV-MHVRΔ crippled the ability of the 17Cl1 cells to support MHV A59 infection by a factor of about 20. In marked contrast, expression of the complete receptor from VV-MHVR essentially eliminated all MHV A59 infectivity in both cells extracts (to < 0.005% of maximum) and in supernatants (to 0.02% of maximum).

Impairment of Spike Glycoprotein Transport in MHVR-Overexpressing Cells

To determine the fate of the MHV A59 spike glycoprotein under conditions of excessive receptor synthesis, the cell extracts were subjected to a Western immunoblotting process involving an antipeptide antibody directed against the amino-terminal 11 residues of spike cleavage product S2[14]. The results (Figure 2) demonstrated that spike glycoprotein remained uncleaved in cells expressing full - length MHVR while cleavage products S1 and S2 were detectable in cells expressing the truncated MHVRΔ. This suggested that receptor engagement with spike results in prevention of spike transport to the Golgi apparatus as this organelle contains protease(s) capable of cleaving the spike protein adjacent to the multibasic residues comprising the S1 : S2 connecting region. In addition, immunoblots showed that the total amount of accumulated spike protein was far lower in cells expressing MHVR relative to MHVRΔ. This was even more obvious when progeny virions were pelleted from supernatants, lysed and subjected to immunoblotting. The lower panel of Figure 2 shows that uncleaved S (180kd) and S2 were readily detectable in control supernates from MHVRΔ cells; however neither of these two spike proteins were present in supernates of MHVR cells.

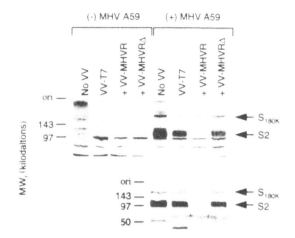

Figure 2. Inhibitory effect of MHVR on spike cleavage and secretion. Top panel: Cultures infected with the indicated vaccinia recombinants were incubated for 3 h, then mock infected (-) or infected (+) with MHV A59, at 10 PFU per cell. 15 h later, supernatants were removed and saved, then cells were lysed and aliquots were subjected to immunoblotting as described in the legend to Figure 1. Primary antibody used was directed against the amino - terminal portion of S2. Bottom Panel: Supernatant fluids from the infected cultures were clarified by centrifugation, then overlaid on 30% w/w sucrose cushions. MHV particles were pelleted by ultracentrifugation in an SW50.1 rotor spinning at 45,000 rpm for 1 h at 5°C. Virions were resuspended and their proteins were subjected to immunoblotting.

Additional experiments (data not shown) indicated that endoglycosidase H treatment specifically increased the electrophoretic mobility of the uncleaved spike present in MHVR cells, yet failed to affect spike products produced in the control cultures. Because transport of spike through the medial Golgi is required to render its covalently - linked sugars resistant to removal by endoglycosidase H[15], we conclude that spike engagement with receptor resulted in its retention and preferential degradation within a pre-medial Golgi location.

DISCUSSION

Cellular susceptibility to MHV infection after transfection of cDNA clones encoding MHVR is often measured by the cells ability to support production of infectious MHV progeny[16]. The results of this report suggest that the quantity of virus progeny produced by a given transfected cell depends on its level of MHVR expression. Expression must obviously be high enough to provide receptor surface densities recognizable by inoculum virus. However, exceedingly high MHVR synthesis rates will inhibit MHV infection through intracellular engagement of MHVR with spike protein. Thus the "optimal" receptor for MHV is one that is expressed at very low levels but accumulates and remains stable at the cell surface to provide for virus attachment.

Possible explanations for inhibition of spike transport include prevention of the further folding of either ligand by complex formation. In this regard it is worth remembering that virion - associated spike protein is able to react with denatured MHVR in blotting assays[17]; this suggests that interaction of ligands might occur before completion of folding. In addition, spike is a highly conformational molecule[18] and its folding pattern is exquisitely sensitive to the oxidizing potential of the exocytic pathway [19]; spike folding could therefore be interrupted if MHVR engagement inhibits native disulfide formation. An equally plausible explanation is that both MHVR and spike are largely folded prior to engagement, then are conformationally rearranged by the binding event This hypothesis arises from literature suggesting that viral receptors are capable of rearranging the shape of envelope glycoproteins upon binding[7][9] Extending from this model is the view that receptor - mediated induction of a fusion - active spike conformation could take place intracellularly, thereby resulting in vesiculation of endoplasmic reticulum or Golgi These issues concerning protein folding, conformational change and intracellular fusion remain to be explored

Also currently unexplored is the question of spikeless particle formation under conditions in which spike is trapped intracellularly by its receptor Early studies by Sturman et al [20] suggested that spikeless particles could form during infection in the presence of tunicamycin, and these studies have been supported and greatly extended by the recent demonstration of MHV particle formation in the absence of spike[21] Straightforward experiments intended to identify incomplete particles in MHVR - overexpressed cells are in progress

ACKNOWLEDGMENTS

Essential technical contributions to this work were made by E Sethi and B Hsiang Primary antibodies used in immunoblotting experiments were generously provided by Dr M Buchmeier This work was supported by NIH grant R29-31616 as well as by a grant from the Schweppe Foundation

REFERENCES

1 Williams, R K , Jiang, G-S , Holmes, K V (1991) Receptor for mouse hepatitis virus is a member of the carcinoembryonic antigen family of glycoproteins Proc Natl Acad Sci USA 88 5533-5536

2 Dveksler, G S , Pensiero, M N , Cardellichio, C B , Williams, R K , Jiang, G-S , Holmes, K V , Dieffenbach, C W (1991) Cloning of the mouse hepatitis virus (MHV) receptor Expression in human and hamster cell lines confers susceptibility to MHV J Virol 65 6881-6891

3 Dveksler, G S , Pensiero, M N , Dieffenbach, C W , Cardellichio, C B , Basile, A A , Elia, P E , and Holmes, K V (1993) Mouse hepatitis virus strain A59 and blocking antireceptor monoclonal antibody bind to the N-terminal domain of cellular receptor Proc Natl Acad Sci USA 90 1716-1720

4 Kubo, H , Yamada, Y K , Taguchi, F (1994) Localization of neutralizing epitopes and the receptor binding site within the amino terminal 330 amino acids of the murine coronavirus spike protein J Virol 68 5403-5410

5 Mizzen, L , Hilton, A , Cheley, S , Anderson, R (1985) Attenuation of murine coronavirus infection by ammonium chloride Virology 142 378-388

6 Gallagher, T M , Escarmis, C , Buchmeier, M J (1991) Alteration of the pH dependence of coronavirus induced cell fusion Effect of mutations in the spike glycoprotein J Virol 65 1916-1928

7 Signoret, N , Poignard, P , Blanc, D , Sattentau, Q (1993) Human and simian immunodeficiency viruses Virus - receptor interactions Trends Microbiol 1 328-333

8 Yeagle, P L The fusion of sendai virus (1993) In Viral fusion mechanisms, J Bentz, ed CRC Press

9 Meyer, W J , Johnston, R E (1993) Structural rearrangement of infecting Sindbis virions at the cell surface mapping of newly accessible epitopes J Virol 67 5114-5125

10 White, J M , Wilson, I A (1987) Anti-peptide antibodies detect steps in a protein conformational change low-pH activation of the influenza virus hemagglutinin J Cell Biol 105 2887-2896

11 Fuerst, T R , Niles, E G , Studier, F W , Moss, B (1986) Eukaryotic transient expression system based on recombinant vaccinia virus that synthesizes bacteriophage T7 RNA polymerase Proc Natl Acad Sci USA 83 8122-8126

12 Fuerst, T R , Earl, P L , Moss, B (1987) Use of a hybrid vaccinia virus T7 RNA polymerase system for expression of target genes Mol Cell Biol 7 2538-2544

13 Laemmli, U K (1970) Cleavage of structural proteins during the assembly of the head of bacteriophage T4 Nature (London) 227 680-685

14 Gallagher, T M , Parker, S E , Buchmeier, M J (1990) Neutralization-resistant variants of a neurotropic coronavirus are generated by deletions within the amino-terminal half of the spike glycoprotein J Virol 64 731-741

15 Vennema, H , Heijnen, L , Zijderveld, A , Horzinek, M C , Spaan, W J M (1990) Intracellular transport of recombinant coronavirus spike proteins implications for virus assembly J Virol 64 339-346

16 Yokomori, K , Lai, M M C (1992) The receptor for mouse hepatitis virus in the resistant mouse strain SJL is functional implications for the requirement of a second factor for viral infection J Virol 66 6931-6938

17 Boyle, J F , Weismiller, D G , Holmes, K V (1987) Genetic resistance to mouse hepatitis virus correlates with absence of virus binding activity on target tissues J Virol 61 185-189

18 Grosse, B , Siddell, S G (1994) Single amino acid changes in the S2 subunit of the MHV surface glycoprotein confer resistance to neutralization by S1 subunit-specific monoclonal antibody Virology 202 814-824

19 Opstelten, D-J E , de Groote, P , Horzinek, M C , Vennema, H , Rottier, P J M (1993) Disulfide bonds in folding and transport of mouse hepatitis coronavirus glycoproteins J Virol 67 7394-7401

20 Sturman, L S , Holmes, K V , Behnke, J (1980) Isolation of coronavirus envelope glycoproteins and interaction with the viral nucleocapsid J Virol 33 449-462

21 Vennema, H , Godeke, G-J , Horzinek, M C , Rottier, P J M (1994) Assembly of coronavirus - like particles from co - expressed structural protein genes Presented at the Sixth Intl Symposium of Corona and Related Viruses

55

MULTIPLE RECEPTOR-DEPENDENT STEPS DETERMINE THE SPECIES SPECIFICITY OF HCV-229E INFECTION

Robin Levis, Christine B. Cardellichio, Charles A. Scanga,
Susan R. Compton,[*] and Kathryn V. Holmes

Department of Pathology
Uniformed Services University of the Health Sciences
Bethesda Maryland 20814

ABSTRACT

Human coronavirus (HCV) -229E causes disease only in humans and grows in human cells and in cells of other species that express recombinant human aminopeptidase N (hAPN), the receptor for HCV-229E. We compared the species specificity of HCV-229E infection with the species specificity of virus binding using immunofluorescence, assay of virus yields, fluorescence activated cell sorting and a monoclonal antibody directed against hAPN that blocks infection. We found that HCV-229E binds to intestinal brush border membranes (BBM) and to membranes of cell lines from cats, dogs, pigs, and humans, however the virus only infects two of these species. HCV-229E will not bind to BBM or to membranes from cell lines derived from hamster or mice. Animal coronaviruses related to HCV-229E, including FIPV, CCV, and TGEV bind to cell membranes from cats, dogs, cows, pigs and humans (but not mice), while each virus infects cells from only a subset of these species. Infectious genomic HCV-229E RNA, can infect cells of all of these species. These data suggest that the species-specificity of infection for this serogroup of coronaviruses is determined at the levels of virus binding and penetration. Since binding of viral spike glycoprotein to cellular receptors is not the only limiting factor, we suggest that one or more steps associated with virus penetration may determine the species specificity of infection with the HCV-229E serogroup of coronaviruses.

---

[*] Present address: Section of Comparative Medicine, Yale University School of Medicine, New Haven, CT 06520-8019

INTRODUCTION

Coronaviruses cause disease in many different species of animals. Most coronavirus strains are highly species-specific, causing disease in only a single host[1,2]. Experimental inoculation of unnatural host species with some coronaviruses can occasionally lead to infection and antibody production. However, these infections occur primarily in neonatal animals and generally result in mild or asymptomatic disease[3].

The initial determinant for species-specific infectivity is the binding of the virus attachment protein to a host cell receptor. Three different types of receptors have been identified for coronaviruses[4-7]. A carbohydrate moiety, 9-O-acetylated neuraminic acid, serves as a receptor determinant for bovine coronavirus and hemagglutinating encephalo-myelitis virus[8]. Both the HE and S glycoproteins from BCV bind specifically to this carbohydrate moiety and require it on the cell membrane for virus infection. It is not yet clear how binding of the viruses to 9-O-acetylated neuraminic acid leads to uncoating and infection.

The murine coronaviruses, mouse hepatitis virus (MHV) use as a receptor an 110 - 120 Kda glycoprotein called MHVR[9] (or Bgp 1a) a member of the immunoglobulin superfamily. Specifically, MHVR is a biliary glycoprotein (BGP) in the carcinoembryonic antigen family[4,10]. MHV can enter cells, not only via binding to MHVR, but also via binding to several closely related BGP proteins that result from differential splicing, and BGP variants derived from several different murine genes[11].

Cells expressing recombinant human BGP, the human homolog of MHVR, are not susceptible to infection by HCV-229E or -OC43 (Williams, unpublished results). A zinc binding glycoprotein, aminopeptidase N (APN) has been identified as a receptor for HCV-229E and TGEV[6,7]. Porcine APN (pAPN) also serves as the receptor for PRCV, a porcine coronavirus related to TGEV[12]. Interestingly, even though TGEV and HCV-229E both utilize the same enzyme as a receptor and both viruses bind to purified intestinal BBM and to cell lines derived from human and porcine tissues, these viruses do not cause infection of cells from the alternate species.

HCV-229E causes upper respiratory tract infection in humans. HCV-229E is serologi-cally related to TGEV, FIPV, and CCV. Although HCV-229E is closely related to these viruses, it will only infect and cause disease in humans. HCV-229E also exhibits species-specific infectivity in cultured cell lines. To identify the steps in HCV-229E replication which determine the species-specificity of infection, we have compared the virus binding activities of intestinal BBM and cultured cells from a variety of species, and examined the suscepti-bility of the cells to infection by HCV-229E virions or infectious genomic HCV-229E viral RNA.

RESULTS

Solid phase immunoassays were done to examine the binding of HCV-229E, FIPV, CCV, TGEV, and MHV-A59 to purified BBM from their natural hosts and a variety of other species, summarized in Table 1. This analysis showed differences in the species specificity of coronavirus binding. MHV-A59 binds to BBM from MHV-susceptible adult BALB/c mice but not to BBM from MHV-resistant adult SJL/J mice, or to BBM from other species including rats, cats, dogs, pigs, or humans[13]. For the mouse coronavirus MHV, the host-range for virus binding is similar to that for MHV infection. In contrast to the narrow host range of MHV binding, coronaviruses in a different serogroup, that includes TGEV, FIPV, CCV, and HCV-229E, all bind to the BBM of their natural hosts and to the BBM from the natural

Table 1. Coronavirus binding to intestinal brush border membranes

Virus	Source of brush border membrane				
	Mouse	Human	Dog	Cat	Pig
MHV-A59	++*	–	–	–	–
HCV-229E	–	++	+	+	+
CCV	–	++	++	++	++
FIPV	–	++	++	++	++
TGEV	–	++	++	+	++

*Large print indicates binding to membranes from normal host Regular print
indicates binding to membranes from a foreign host species

hosts of the antigenically related viruses[14] However, none of these viruses bound to mouse BBM

The different binding patterns of the two serogroups of coronaviruses, the MHV- vs the HCV-229E-related viruses, suggested that the viruses recognize different types of determinants on the cell surface By the same rationale, the HCV-229E-related viruses may all recognize a similar determinant on the BBM of all of the species Even though the HCV-229E related viruses bind to the BBM of several species, they exhibit a high level of species specificity in tissue culture, with only a few instances in which a coronavirus infects cell lines derived from a different host species[3]

To characterize the species specificity of HCV-229E infectivity in more detail, we analyzed the ability of HCV-229E to bind to and productively infect cell lines derived from several different animal species, including human (WI38), feline (FCWF), canine (A72), porcine (ST fetal), and hamster (BHK) cell lines We included in these studies, BHK cells expressing recombinant human APN Except for the hamster cell lines, each of these cell lines is sensitive to infection by at least one coronavirus in the HCV-229E serogroup We analyzed the binding of HCV-229E to these cell lines using a fluorescence activated cell sorter Cells were mixed with concentrated virus, incubated at 4°C, rinsed to remove unbound virus, and fixed with paraformaldehyde Polyclonal goat antiserum directed against HCV-229E virions and R phycoerythrin-conjugated rabbit anti-goat IgG were used to detect virus bound to the cell surface

Quantitative differences in the binding of HCV-229E to the membranes of different cell lines are summarized in Table 2 The relative fluorescence intensity (RFI) showed that HCV-229E virus bound to human WI38 cells with a signal approximately ten fold above background and at a higher level, approximately fifteen-fold above background, to feline FCWF cells, and BHK cells expressing recombinant human APN (BHK-hAPN), whereas virus did not bind to BHK cells HCV-229E bound to porcine ST cells and canine A72 cells with a RFI approximately five-fold above background

While this binding assay revealed a quantitative difference in the ability of HCV-229E to bind to cell lines from different species, it is important to distinguish between the amount of receptor on each of these cell lines and the affinity of the virus for the receptor molecules present on the surface of these cell lines To correlate the levels of APN expressed on each of these cell lines with the HCV-229E binding activity of the cell lines, the levels of APN on the cell lines were measured APN activity was determined by incubating cells with substrate for varying times and measuring protease activity by colorimetric change Interestingly, the levels of APN activity, shown in Table 2, did not correlate well with the levels of HCV-229E binding to each of the cells lines The APN activity of BHK-hAPN cells

Table 2. Summary of HCV-229E binding and infectivity

| Cell lines | HCV–229E binding§ | Infection by† | | APN activity* |
		HCV–229E virions	HCV–229E RNA	
Human, WI38	++	+	+	100
Cat, FCWF	+++	+	+	64
Dog, A72	+	−	+	148
Pig, ST fetal	+	−	+	49
Hamster, BHK	−	−	+	70
Hamster + hAPN, BHK - hAPN	+++	+	+	228

§Indicates the degree of binding
to cells measured by relative
fluorescence intensity on a flow
activated cell sorter.

is two to three fold higher than the APN activity of human WI38 cells, and four fold higher than that of feline FCWF cells, yet each of these cell lines bound an equivalent amount of HCV-229E virus. Canine A72 cells which bind very low levels of virus, showed a high level of APN activity. These data suggest that the binding affinity of HCV-229E for a receptor on these cells may be an important factor in determining virus /receptor interactions. The receptor moieties to which HCV-229E binds on cells from cats, dogs, and pigs have not yet been identified, but we postulate that they may be APN or related molecules. Similarly, the receptors for FIPV and CCV in the feline or canine cells, have not been identified, although, APN or related glycoproteins are likely candidates.

In addition to HCV-229E infecting human cells and hamster cells if they express recombinant human APN[7], we find that HCV-229E readily infected the FCWF feline cell line (REF), producing a high yield of progeny virions comparable to the yield from human cells. HCV-229E may bind to feline APN or to some other unrelated receptor on feline cells. Possibly feline and human APN share determinants that bind the S glycoprotein of HCV-229E.

To determine whether the failure of HCV-229E to infect canine and porcine cells was due to a block in virus entry after binding, or to an intracellular restriction on HCV-229E translation or transcription, we transfected cells with HCV-229E genomic RNA. Twenty four hours post-transfection, viral antigens in the cells were detected by immunofluorescence. Each of the cell lines transfected with HCV-229E RNA contained cells producing viral antigens (Table 2). This indicates that the HCV-229E genomic RNA successfully established infection even across species barriers. Thus, the block in HCV-229E infectivity of canine and porcine cells probably occurs at a step after virus binding, but before viral protein synthesis.

HCV-229E RNA was also able to replicate in transfected BHK cells. This is not surprising, as HCV-229E virions can infect and replicate in BHK-hAPN cells expressing the recombinant receptor. HCV-229E virus does not bind to BHK cells in the absence of hAPN on the surface. Therefore the specificity for infection of these cells is at the level of receptor recognition.

To analyze the assembly and release of progeny HCV-229E virions from cells transfected with HCV-229E genomic RNA or with HCV-229E virions, we compared the yields of infectious virus produced in these cell lines. In general, the yield of virus, correlated well with the immunofluorescence data showing intracytoplasmic viral antigens. The amount of virus produced in HCV-229E RNA-transfected cells was low. This may be due to a low effective multiplicity of infection by HCV-229E genomic RNA. Virus released from A72, ST fetal and BHK cells was not able to reinfect these cells. For WI38 and FCWF cells, the yields of released virus from infected cells were much higher, suggesting that virus spread to other cells causing multiple cycles of infection.

Interestingly, BHK-hAPN cells transfected with HCV-229E genomic RNA or infected with HCV-229E virions expressed viral antigen as shown by immunofluorescence, but produced little or no infectious virus. Other systems show a similar pattern. Virus yields of HIV were altered in cells that overexpress CD4 receptor glycoprotein[15,16]. The low virus yields from BHK-hAPN cells, may be due to inefficient assembly of progeny virions in cells that overexpress hAPN on intracellular membranes, inefficient release of progeny virions from hAPN on the host cell plasma membrane, or inefficient penetration of virions into cells during virus infection.

DISCUSSION

All of the cell lines analyzed in this study were able to support the replication of HCV-229E genomic RNA and to produce infectious progeny virions. In contrast, the ability of HCV-229E virions to infect all of these cells is restricted, even though the virus will bind to membranes of cell lines from many species. HCV-229E virus will only infect human and feline cells, and rodent cells expressing recombinant human APN. The mechanism for virus entry into feline FCWF cells is not understood at this time. It will be interesting to analyze the receptor that HCV-229E virus is utilizing to infect feline cells and how it relates to human APN.

TGEV (and PRCV) and HCV-229E both utilize APN as a receptor for infection. Comparison of the amino acid sequences of the APN glycoproteins from these two species shows a high level of sequence similarity, however, these viruses will not infect cells from the alternate species. Delmas et al, have made recombinant chimeras between human and pig APN and have identified the region of pig APN which is essential for TGEV to infect cells[6]. This region is between amino acids 717 to 813 and is ~330 amino acids downstream from the APN active site. Several anti-hAPN monoclonal antibodies which blocked HCV-229E virus infection of human cells also inhibited enzyme activity, suggesting that the HCV-229E virus spike glycoprotein binds near the active site of hAPN[7]. Deletion of the 39 amino acids of hAPN, including the active site, also blocked infectivity. Possibly, the blocking monoclonal antibodies and the introduced mutation alter the conformation of hAPN such that HCV-229E will no longer bind even though the virus may recognize a distant region of hAPN. It will be important to further analyze human and pig APN and compare the determinants essential for infection.

The levels of APN on the surface of cell membranes from the different cell lines studied does not correlate with the level of virus binding to these cells. The initial interpretation of these data is that HCV-229E has differing affinities for the membranes of the various species tested. It will be important to determine if HCV-229E is actually binding to APN on these cells or to a different receptor molecule. APN is the receptor for the porcine coronaviruses from this serogroup and for HCV-229E, but no receptor molecule has been identified for canine and feline coronaviruses. It will be important to look at the direct binding of virus or purified S protein to cloned APNs of the other species.

Finally, the ability of the virus to bind to the cell membrane of a variety of species does not correlate with susceptibility to virus infection of these cells. Virions could only productively infect human and feline cells, but could bind to canine and porcine cells as well as to feline and human cells. Perhaps, the virus binds to a different molecule on the surface of these cell lines Alternatively, if the virus binds to APN on these cells, species specific differences in the amino acid sequence and/or conformation of the APNs may restrict receptor function Species specific accessory factors might also be required for virus uptake These factors may only interact with the viral proteins from those viruses that infect the species in nature Identification of such auxiliary factors will give important information about the restriction of host range and the evolution of these viruses in nature

In summary, cell lines from different species differ markedly in their susceptibility to infection by HCV-229E virions or by infectious HCV-229E genomic RNA We have identified three stages in the virus replicative cycle in which the species specificity of virus susceptibility may be determined First, hamster cells are resistant to infection by HCV-229E virions because the virus does not bind to any cell surface receptor, even though the cells express hamster APN Second, pig and dog cells are resistant to infection by HCV-229E virions because of restriction at an early step after virus binding and before synthesis of viral proteins, since virions bind but do not infect, while HCV-229E genomic RNA can infect these cell lines Third, hamster cells producing high levels of recombinant hAPN produce little or no virus, perhaps because of interference with the assembly or release of virions by the recombinant hAPN

ACKNOWLEDGMENTS

The authors thank Dave Wessner and Jeff Redwine for their critical review of this manuscript This work was supported by National Institutes of Health grant AI26075

The statements and assertions herein are those of the authors and do not represent the opinions of the Uniformed Services University of the Health Sciences or the Department of Defense

REFERENCES

1 Wege, H , S Siddell, and V ter Meulen The biology and pathogenesis of coronaviruses Curr Top Microbiol Immunol 1982 99 165

2 Moestl, K Coronaviridae, pathogenetic and clinical aspects An update Comp Immun Microbiol Infect Dis 1990 13 169

3 Holmes, K V and S R Compton Coronavirus receptors In, The Coronaviruses (S Siddell, Ed) Plenum Press, in press

4 Dveksler, G S , M N Pensiero, C B Cardellichio, R K Williams, G S Jiang, K V Holmes and C W Dieffenbach Cloning of the mouse hepatitis virus (MHV) receptor expression in human and hamster cell lines confers susceptibility to MHV 1991 J Virol 65 6881

5 Schultze, B and G Herrler Bovine coronavirus uses N-acetyl-9-Oacetylneuraminic acid as a receptor determinant to initiate the infection of cultured cells 1992 J Gen Virol 73 901

6 Delmas, B , J Gelfi, R L'Haridon, L K Vogel, H Sjostrom, O Noren, and H Laude Aminopeptidase N is a major receptor for the entero-pathogenic coronavirus TGEV 1992 Nature 357 417

7 Yeager, C L , R A Ashmun, R K Williams, C B Cardellichio, L H Shapiro, A T Look, and K V Holmes Human aminopeptidase N is a receptor for human coronavirus 229E 1992 Nature 357 420

8 Schultze, B ,, K Wahn, H D Klenk and G Herrler Isolated HE-protein from hemagglutinating encephalomyelitis virus and bovine coronavirus has receptor-destroying and receptor-binding acrivity 1991 Virology 180 221

9 Nedellec, P , G S Dveksler, I Daniels, c Turbide, B Chow, A A Basile, K V Holmes and N Beauchemin Bgp2, a new member of the carcinoembryonic antigen-related gene family, encodes an alternative receptor for mouse hepatitis viruses 1994 J Virol 68 4525

10 Williams, R K , G S Jiang and K V Holmes Receptor for mouse hepatitis virus is a member fo the carcinoembryonic antigen family of glycoproteins 1991 Proc Natl Acad Sci U S A 88 5533

11 Holmes, K V and G S Dveksler Specificity of coronavirus-receptor interactions In Virus Receptors (E Wimmer, Ed) Cold Spring Harbor Press In press

12 Delmas, B , J Gelfi, H Sjostrom, O Noren, and H Laude Further characterization of aminopeptidase N as a receptor for coronaviruses 1993 Adv Exp Med Biol 342 292

13 Compton, S R , C B Stephensen, S W Snyder, P G Weismiller and K V Holmes Coronavirus species specificity Murine coronavirus binds to a mouse-specific epitope on its carcinoembryonic antigen-related receptor glycoprotein 1992 J Virol 66 7420

14 Compton, S R Coronavirus attachment and replication 1988 Ph D Dissertation (unpublished) Uniformed Services University, Bethesda, MD

15 Marshall, W L , D C Diamond, M M Kowalski, and R W Finberg High level of surface CD4 prevents stable human immunodeficiency virus infection of T-cell transfectants 1992 J Virol 66 5492

16 Buonocore, L and J K Rose Blockade of human immunodeficiency virus type 1 production in CD4+ T cells by an intracellular CD4 expressed under control of the viral long terminal repeat 1993 Proc Ntl Acad Sci U S A 90 2695

56

CHARACTERIZATION OF A NEW GENE THAT ENCODES A FUNCTIONAL MHV RECEPTOR AND PROGRESS IN THE IDENTIFICATION OF THE VIRUS-BINDING SITE(S)

G. Dveksler,[1] P. Nedellec,[2] J.-H. Lu, U. Keck,[3] A. Basile,[1]
C. Cardellichio,[1] W. Zimmermann,[3] N. Beauchemin,[2] and K.V. Holmes[1]

[1] Department of Pathology
Uniformed Services University of the Health Sciences
Bethesda, Maryland 20814
[2] McGill Cancer Centre and Departments of Medicine, Biochemistry and
Oncology
McGill University
Montreal, Quebec, H3G 1Y6, Canada
[3] Institute of Immunobiology
University of Freiburg
D-79104 Freiburg, Germany

ABSTRACT

Several splice variants of the murine biliary glycoprotein 1 (Bgp 1) gene in the carcinoembryonic antigen gene superfamily serve as cellular receptors for mouse hepatitis virus. RNA PCR and immunoblot analysis of the receptor in inbred mouse strains showed that the glycoproteins expressed in SJL/J mice are encoded by an allelic variant of the Bgp 1 gene, named Bgp 1^b. We recently cloned and characterized a second gene, Bgp 2, that encodes a functional MHV receptor glycoprotein which is not recognized by anti-MHVR MAb-CC1. A third gene related to Bgp 1 was cloned and expressed and shown to encode a soluble protein called Cea-10 that differs significantly in its N-terminal domain from Bgp 1 and Bgp 2. Chimeric proteins constructed between the different murine Bgps and point mutations in the prototype MHV receptor, Bgp 1^a or MHVR, were analyzed to further characterize the MAb-CC1-binding and virus-binding domains within the N terminal domain of the receptor. Thus, the murine host for MHV expresses multiple splice variants of mRNAs encoded by several different Bgp-related genes which differ in their ability to serve as MHV receptors. The differential expression of these genes in different murine tissues may help to explain the tissue tropism of MHV strains.

Corona- and Related Viruses, Edited by P. J. Talbot and G. A. Levy
Plenum Press, New York, 1995

INTRODUCTION

Murine coronavirus MHV uses as cellular receptors several murine biliary glycoproteins in the carcinoembryonic antigen group in the immunoglobulin superfamily [1-5]. Multiple isoforms of the glycoproteins may be co-expressed in murine cells [1,6]. The MHV-resistant adult SJL/J mouse expresses the *Bgp-1[b]* gene products which include a 105 kDa glycoprotein with four external immunoglobulin (Ig)-like domains, a transmembrane and a cytoplasmic domain, and a 55 kDa glycoprotein that includes the same N-terminal and membrane-proximal Ig-like domains, transmembrane and cytoplasmic domains, but lacks the second and third Ig-like domains. The N-terminal domain common to the 2- and 4-domain Bgp1[b] glycoproteins contains 108 amino acids, of which 29 are different from the homologous 2- and 4-domain Bgp1[a] glycoproteins that are expressed in MHV-susceptible BALB/c mice. When recombinant Bgp1[a] or Bgp1[b] glycoproteins are expressed in MHV-resistant hamster cells, they serve as functional receptors for MHV-A59 and several other MHV strains [1,3]. In this paper we describe the cloning and expression of several additional genes encoding glycoproteins related to Bgp 1, and explore the elements of the N-terminal domain of Bgp1[a] that are needed for MHV receptor activity.

RESULTS

We recently cloned the cDNA of a new murine Bgp family member, named Bgp2 [7]. The Bgp2 mRNA is expressed in BALB/c, SJL/J and CD-1 mice and in the CMT 93 cell line. The two domain Bgp2-encoded protein differs markedly from the proteins encoded by Bgp1[a] and Bgp 1[b] in the sequence of the N-terminal domain (Figure 1), the 4th Ig-like domain and in the length of the cytoplasmic tail. The antireceptor antibody MAb-CC1 and two polyclonal anti-Bgp 1 antibodies failed to recognize the Bgp 2-encoded protein in immunoblots. Expression of Bgp2 in MHV-resistant hamster cells made the cells susceptible to infection with MHV-3, MHV-A59 and MHV-JHM [7]. Thus, both MHV-susceptible and MHV-resistant mouse strains express the Bgp2 glycoprotein which serves as a functional receptor for MHV when expressed in hamster cells in vitro.

To identify the components of the Bgp1[a] glycoprotein (formerly called MHVR [2,7] that are recognized by the MHV-A59 spike glycoprotein, S, a series of anchored deletion mutants of Bgp1[a] was constructed, expressed in hamster cells, and tested for the ability to bind MHV-A59 virions or blocking anti-receptor monoclonal MAb-CC1, and tested for functional receptor activity by challenging with infectious MHV-A59 and labelling the cells with antibody directed against the N-protein of MHV at 8 to 16 hours after virus challenge[8].

```
Bgp 1a(MHVR)   EVTIEAVPPQVAEDNNVLLLVHNLPLALGAFAWYKGNTTAIDKEI
Bgp 1b         EVTIEAVPPQVAEDNNVLLLVHNLPLALGAFAWYKGNPVSTNAEI
Cea 10         QVTVEAVPLQRTADNNVLLLVHNLPQTLRVFYWYKGNSGAGHNEI
Bgp 2          QVTVMAFPLHAAEGNNVILVVYNMMKGVSAFSWHKGSTTSTNAEI


ARFVPNSNMNFTGQAYSGREIIYSNGSLLFQMITMKDMGVTTLDMTDENYRRTQATVRFHV
VHFVTGTNKTTTGPAHSGRETVYSNGSLLIQRVTVKDTGVYTIEMTDENFRRTEATVQFHV
GRFVTSINRSKLGLAHSGRETIYSNGSLFFQSVTKNDEGVYTLYMLDQNFEITPISVRFHV
VRFVTGTNKTIKGPVHSGRETLYSNGSLLIQRVTMKDTGVYTIEMTDQNYRRRVLTGQFHV
```

Figure 1. Comparison of the amino acid sequence of the N-terminal domains of Bgpla- (formerly named MHVR), Bgplb- (formerly named mmCGM₂), Cea 10- and Bgp2- encoded proteins. Potential N-linked glycosylation sites are underlined.

$N_{37}-Q$ $N_{55}-Q$ $N_{70}-Q$

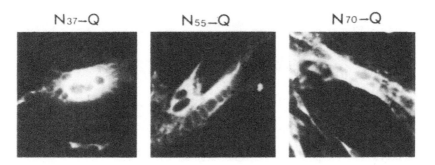

Figure 2. Effects of removal of the three protential N-linked glycosylation sites on the N-terminal domain of Bgpla (MHVR). BHK cells grown on coverslips were transfected with cDNAs encoding MHVR mutants lacking the first (N37- Q) and third (N70- Q) N-linked glycosylation sites. The cells were inoculated with MHV-A59 at 50h posttransfection, and virus antigens were detected 16h after MHV-A59 inoculation by immunofluorescence.

The N-terminal domain was essential for virus-binding, MAb-CC1-binding and virus receptor activity. However, the anchored N-terminal domain alone was not a functional receptor. Insertion of either the membrane-proximal 4th Ig domain or the second Ig domain between the N-terminal domain and the transmembrane anchor yielded functional receptors. These observations suggested that the viral S glycoprotein binds to the N-terminal domain of the receptor, but that some function of the second or fourth domain was needed for functional receptor activity. Possibly the extra domain serves as a spacer to raise the N-terminal domain above surrounding membrane glycoproteins so that virions can attach.

To further identify the elements of the N-domain of the Bgp1[a] glycoprotein that are required for receptor activity, the three potential N-glycosylation sites in the N-terminal domain were mutated by substitution of asparagine for glutamine, and the resulting recombinant proteins were expressed in hamster cells and tested for virus-binding (VOPBA), MAb-CC1 binding and virus receptor activities. Thirteen other N-glycosylation sites remained unchanged in the other Ig domains. Previous studies using tunicamycin to inhibit all N-linked glycosylation of Bgp1[a] had shown that the non-glycosylated protein had no virus-binding activity in a virus-overlay protein blot assay [9]. Studies on the glutamine substitution mutant proteins showed that removal of any one (Figure 2) or all three of the N-glycosylation sites in the N-terminal domain of Bgp1[a] did not destroy virus receptor activity, although the mutant glycoprotein lacking all three N-glycosylation sites did not bind virus in a VOPBA and bound MAb-CC1 very poorly in immunoblots. Thus, N-linked glycans in the N-terminal domain are not required for receptor function.

Recombinant chimeric proteins were constructed to aid in identification of elements of Bgp1[a] needed for receptor function (Figure 3). The recombinant Bgp1[a] protein with the N-terminal domain and the second domain anchored by the transmembrane and cytoplasmic region is a functional receptor. However, when the N-domain of Bgp1[b] was substituted for that of Bgp1[a], MAb-CC1 binding activity and MHV receptor activity were lost. A convenient BamHI site at amino acid 70 of the mature Bgp proteins was used to construct chimeras. A recombinant glycoprotein with the first 70 amino acids from Bgp1[b] and the remainder from Bgp1[a] had no receptor activity, while a recombinant glycoprotein with amino acids 71 to 108 from Bgp1[b] and the rest of the molecule from Bgp1[a] bound MAb-CC1 and served as a receptor. These observations suggested that an element of Bgp1[a] between amino acids 1 to 70 was required for good receptor activity and MAb-CC1 binding.

A CEA-related glycoprotein with an N-terminal domain markedly different from Bgp1[a], Bgp1[b] or Bgp2 was detected in mouse colon and in a C3H mouse cell line that also

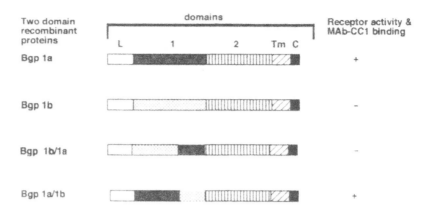

Figure 3. Schematic representation of Bgpla/Bgplb chimeras containing the leader peptide, and domains 1 and 2 followed by the transmembrane (Tm) and cytoplasmic tail (c) of Bgpla.

expressed Bgp1[a] (Figure 1; Zimmermann, personal communication; Lu, in preparation). The gene encoding this new glycoprotein, the 10th member of the murine CEA-related genes to be identified, is now called *Cea10*, according to the generally accepted nomenclature recommendations. Other genes in this family include many murine pregnancy specific glycoproteins (PSGs) [10-12] that are expressed principally in the placenta. In general, PSGs have two variable-like domains at the N-terminus, followed by a single constant Ig domain. The PSGs lack transmembrane and cytoplasmic anchor domains. Like these PSG glycoproteins, the Cea 10 glycoprotein has 2 variable-like domains and lacks an anchor sequence. The protein is secreted from murine cells in culture (Lu, in preparation). Because this CEA-related protein is secreted, it appeared possible that it might serve as a soluble receptor-homolog that might inactivate infectious virus. To test the virus-binding activity of the N-terminal domain of Cea-10, a recombinant chimeric glycoprotein was constructed with the N-terminal domain of Cea10 in place of the N-terminal domain of Bgp1[a], followed by domains 2 through 4, the transmembrane and short cytoplasmic domain of Bgp1[a]. This chimeric glycoprotein did not bind MAb-CC1 or serve as a receptor for MHV-A59. A chimeric protein with amino acids 1-70 from Cea-10 in place of the same region of Bgp1[a] in the anchored 4-domain Bgp1[a] glycoprotein also had no receptor activity. However, a construct of anchored, 4-domain Bgp1[a] with the substitution of amino acids 71-108 from Cea-10 did show virus receptor activity. Thus, these results support the data from the Bgp1[b] chimeras suggesting that a region of Bgp1[a] between amino acids 1 and 70 is required for functional receptor activity.

A series of mutant glycoproteins was constructed in which amino acids in the N-terminal domain of rat Bgp glycoproteins, which do not serve as receptors for MHV-A59, or Cea-10 were substituted for amino acids at the same location in the Bgp1[a] glycoprotein (Figure 4). Each of these mutant glycoproteins was expressed in hamster cells and tested for MHV-A59 receptor activity. None of them eliminated the receptor activity of Bgp1[a] (Dveksler, et al., in preparation). Further amino acid substitutions are in progress to identify the amino acids in the N-terminal domain of the Bgp1[a] receptor glycoprotein to which the MHV spike glycoprotein binds, leading to infection.

DISCUSSION

The CEA-related glycoproteins of mice were shown to be encoded by at least 3 genes. Several of these glycoproteins can be co-expressed in murine tissues and cell lines. Both

$$N55,F56 \longrightarrow D,P$$
$$I66,I67 \longrightarrow A,A$$
$$I66,L74 \longrightarrow T,F$$
$$V85 \longrightarrow A$$
$$E65 \longrightarrow V$$
$$T98 \longrightarrow I$$

Figure 4. Point mutations introduced in the N-domain of MHVR (Bgpla) All the mutated proteins retain virus-recptor activity

Bgp1 and Bgp2 but not Cea10 serve as receptors for MHV Thus, it will not be a simple task to examine the correlation of the expression of MHV receptor glycoproteins with the susceptibility of different tissues or cell lines to virus infection New reagents and probes to distinguish these closely related molecules will indicate what role receptors play in the pathogenesis and tissue tropism of MHV infections in vivo

Studies with recombinant mutant or chimeric receptor glycoproteins have been used to identify the domains and the particular amino acids that are important functional elements of virus receptors in the immunoglobulin superfamily including CD4, PVR, and ICAM-1 the receptors for HIV, poliovirus and the major group of rhinoviruses, respectively [13 17] The N-terminal domains of these receptors and the MHV receptor, Bgp1[a], are the sites where the virions bind At least a portion of one additional domain or of an Ig-like constant domain from a different glycoprotein is needed for function of the anchored receptor, as we have shown with the MHV receptor Bgp1[a 18] In general, because of the complex tertiary structure of the Ig domains, it appears likely that the virus-binding site of the MHV-receptor glycoproteins will include several amino acids that are not in a linear sequence, but which depend on the conformation of the receptor glycoprotein Small molecules that mimic the tertiary structure of the virus-binding site of the receptor might be effective in blocking receptor function and might thereby prevent cells or tissues from becoming infected after challenge with virulent virus In this paper we showed that one determinant for MHV binding resides in the first 70 amino acids of the N-terminal domain of Bgp1[a]

ACKNOWLEDGMENTS

This work was supported in part by NIH grants AI 25231 and AI 26075, USHS grant CO74ET, Medical Research Council of Canada grants PG-11410 and MT-12236, and a grant from the Dr Mildred Scheel Stiftung fur Krebsforschung The statements and assertions herein are those of the authors and do not represent the opinions of the Uniformed Services University of the Health Sciences or the Department of Defense

REFERENCES

1 Dveksler, G S , C W Dieffenbach, C B Cardellichio, K McCuaig, M N Pensiero, G S Jiang, N Beauchemin, and K V Holmes 1993 Several members of the mouse carcinoembryonic antigen-related glycoprotein family are functional receptors for the coronavirus mouse hepatitis virus-A59 J Virol 67 1-8

2 Dveksler, G S , M N Pensiero, C B Cardellichio, R K Williams, G S Jiang, K V Holmes, and C W Dieffenbach 1991 Cloning of the mouse hepatitis virus (MHV) receptor expression in human and hamster cell lines confers susceptibility to MHV J Virol 65 6881-6891

3 Yokomori, K and M M C Lai 1992 Mouse hepatitis virus utilizes two carcinoembryonic antigens as alternative receptors J Virol 66 6194-6199

4 Yokomorı, K and M M Laı 1992 The receptor for mouse hepatıtıs vırus ın the resıstant mouse straın SJL ıs functıonal ımplıcatıons for the requırement of a second factor for vıral ınfectıon J Vırol **66** 6931-6938

5 Wıllıams, R K , G S Jıang, and K V Holmes 1991 Receptor for mouse hepatıtıs vırus ıs a member of the carcınoembryonıc antıgen famıly of glycoproteıns Proc Natl Acad Scı U S A **88** 5533-5536

6 McCuaıg, K , M Rosenberg, P Nedellec, C Turbıde, and N Beauchemın 1993 Expressıon of the Bgp gene and characterızatıon of mouse colon bılıary glycoproteın ısoforms Gene **127** 173-183

7 Nedellec, P , G S Dveksler, E danıels, C Turbıde, B Chow, A A Basıle, K V Holmes, and N Beauchemın 1994 Bgp2, a new member of the carcınoembryonıc antıgen-related gene famıly, encodes an alternatıve receptor for mouse hepatıtıs vıruses J Vırol **68** 4525-4537

8 Dveksler, G S , M N Pensıero, C W Dıeffenbach, C B Cardellıchıo, A A Basıle, P E Elıa, and K V Holmes 1993 Mouse hepatıtıs vırus straın A59 and blockıng antıreceptor monoclonal antıbody bınd to the N-termınal domaın of cellular receptor Proc Natl Acad Scı U S A **90** 1716-1720

9 Pensıero, M N , G S Dveksler, C B Cardellıchıo, G -S Jıang, P E Elıa, C W Dıeffenbach, and K V Holmes 1992 Bındıng of mouse coronavırus MHV-A59 to ıts receptor expressed from a recombınant vaccınıa vırus depends upon post-translatıonal processıng of the receptor glycoproteın J Vırol **66** 4028-4039

10 Rudert, F , A M Saunders, S Rebstock, J A Thompson, and W Zımmermann 1992 Characterızatıon of murıne carcınoembryonıc antıgen gene famıly members Mamm Genome **3** 262-273

11 Beauchemın, N , C Turbıde, D Afar, J Bell, M Raymond, C P Stanners, and A Fuks 1989 A mouse analogue of the human carcınoembryonıc antıgen Cancer Res **49** 2017-2021

12 Nagel, G , F Grunert, T W Kuıjpers, S M Watt, J Thompson, and W Zımmermann 1993 Genomıc organızatıon, splıce varıants and expressıon of CGM1, a CD66-related member of the carcınoembryonıc antıgen gene famıly Eur J Bıochem **214** 27-35

13 Freıstadt, M S and V R Racanıello 1991 Mutatıonal analysıs of the cellular receptor for polıovırus J Vırol **65** 3873-3876

14 Wang, J H , Y W Yan, T P Garrett, J H Lıu, D W Rodgers, R L Garlıck, G E Tarr, Y Husaın, E L Reınherz, and S C Harrıson 1990 Atomıc structure of a fragment of human CD4 containing two ımmunoglobulın-lıke domaıns Nature **348** 411-418

15 Staunton, D E , M L Dustın, H P Erıckson, and T A Sprınger 1990 The arrangement of the ımmunoglobulın-lıke domaıns of ICAM-1 and the bındıng sıtes for LFA-1 and rhınovırus [publıshed erratum appears ın Cell 1990 Jun 15,61(2) 1157] Cell **61** 243-254

16 Staunton, D E , A Gaur, P Y Chan, and T A Sprınger 1992 Internalızatıon of a major group human rhınovırus does not require cytoplasmıc or transmembrane domaıns of ICAM-1 J Immunol **148(10)** 3271-3274

17 Arthos, J , K C Deen, M A Chaıkın, J A Fornwald, G Sathe, Q J Sattentau, P R Clapham, R A Weıss, J S McDougal, and C Pıetropaolo 1989 Identıfıcatıon of the resıdues ın human CD4 crıtıcal for the bındıng of HIV Cell **57** 469-481

18 Dveksler, G S , A B Basıle, C B Cardellıchıo, and K V Holmes 1995 Mouse hepatıtıs vırus receptor actıvıtıes of an MHVR/mph chımera and MHVR mutants lackıng N-lınked glycosylatıon of the N-termı-nal domaın J Vırol **69** 543-546

57

MHVR-INDEPENDENT CELL-CELL SPREAD OF MOUSE HEPATITIS VIRUS INFECTION REQUIRES NEUTRAL pH FUSION

Therese C. Nash,[1] Thomas M. Gallagher,[2] and Michael J. Buchmeier[1]

[1] Department of Neuropharmacology
The Scripps Research Institute
La Jolla, California 92037
[2] Department of Microbiology and Immunology
Loyola University Medical Center
Maywood, Illinois 60153

INTRODUCTION

Mouse hepatitis virus (MHV) attaches to susceptible cells through the interaction of MHV spike glycoprotein (S) with cellular receptors of the murine carcinoembryonic antigen (CEA) gene family[1,2]. Cells that lack murine CEA, including cells that are not of murine origin, are resistant to infection[3]; binding to MHV receptors appears to be a necessary step in the infection of cells by virions. The presence of murine CEA, however, is not required for cell-cell spread of MHV infection[4]. Cells that lack a functional MHV receptor can become productively infected through incorporation into MHV-induced syncytia. Syncytium formation is a consequence of spike-mediated cell-cell fusion. In this report, we investigated the role of spike-mediated fusion in receptor-independent, cell-associated spread of MHV infection.

RESULTS AND DISCUSSION

Previously, we demonstrated receptor-independent, cell-associated spread of MHV infection using an infectious center assay[4]. When a limited number of MHV-infected DBT cells were seeded onto a monolayer of receptor-negative BHK cells, infection spread from the overlaid cells to adjacent cells in the BHK cell monolayer, resulting in the formation of large syncytia. MHV-induced fusion is a function of the spike glycoprotein. Cell surface expression of recombinant spike glycoprotein in the absence of other MHV proteins leads to fusion of receptor-negative cells[4,5], demonstrating that receptor-binding is not required for spike-mediated cell-cells fusion. To further support the contention that cell-associated spread of MHV does not require the presence of a functional MHV receptor, and to examine

Corona- and Related Viruses, Edited by P. J. Talbot and G. A. Levy
Plenum Press, New York, 1995

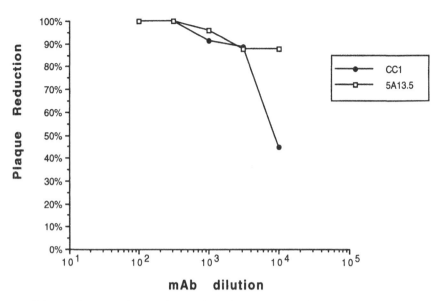

Figure 1. Inhibition of MHV-4 infection of DBT cells with anti-MHVR and anti-S monoclonal antibodies For the CC1 assay, DBT cell monolayers were incubated with the indicated dilutions of anti-MHVR monoclonal antibody CC1 for 1 hour before and 2 hours after infection with MHV-4. For the 5A13.5 assay, MHV-4 was incubated with the indicated dilution of anti-spike monoclonal antibody 5A13 15 for 1 hour at 4°C before infection of DBT cell monolayers, then 1 hour at 37°C Supernatants were then removed are replaced with an agarose overlay Plaques were enumerated after 2 days incubation

the role of the spike glycoprotein in cell-associated spread of MHV infection, infectious center assays were performed in the presence of anti-receptor and anti-spike antibodies.

Infection of susceptible cells by MHV can be prevented by either the receptor-blocking antibody CC1 or anti-spike neutralizing antibodies. The plaque reduction assay presented in Figure 1 demonstrates dose-dependent inhibition of MHV-4 infection of DBT (murine astrocytoma) cells by anti-MHVR antibody CC1[1] and anti-spike antibody 5A13.5[6]. Figure 2 shows an infectious center assay performed in the presence of antibody CC1 (1:50 dilution). In Figure 2A, isolated, MHV-4 infected cells were detected in the culture in which infected DBT cells were seeded onto coverslips alone (no monolayer). A duplicate aliquot of infected DBT cells were overlaid onto a BHK cell monolayer (Figure 2B); despite the continued presence of antibody CC1 in the culture, large syncytia developed. Therefore, anti-receptor antibody CC1 failed to prevent cell-associated spread of MHV to receptor-negative BHK cells. Antibody CC1 also did not prevent cell-associated spread of MHV to DBT cells (data not shown). Together, these results provide further evidence that cell-associated spread of MHV occurs through an MHVR-independent mechanism.

Unlike the anti-MHVR antibody, addition of anti-spike antibody 5A13.5 to the infectious center assay did prevent the cell-associated spread of MHV to the BHK cell monolayer (Figure 2C); infection remained limited to single cells. The ability of an anti-spike monoclonal antibody to prevent syncytia formation associated with receptor-independent cell-cell spread of MHV demonstrates that the spike glycoprotein mediates cell-cell spread of MHV and that this fusion activity is independent from receptor-binding activity. Several other well characterized anti-spike monoclonal antibodies were also tested for their ability to prevent cell-associated spread of MHV infection in the infectious center assay (Table 1). Only monoclonal antibodies 5A13.5 and 5B19.2 prevented receptor-independent, cell-associated spread of MHV infection. Interestingly, these two antibodies also confer protection

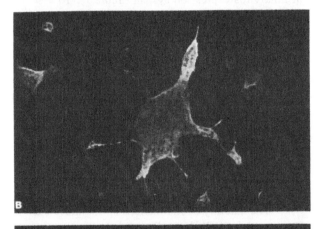

Figure 2. Infectious center assay demonstrating receptor-independent, cell-associated spread of MHV-4 infection. DBT cells were infected with MHV-4 (MOI 5.0) in suspension for 1 hr., then washed extensively to remove unbound virus. Aliquots of MHV-4 infected cells were seeded onto a blank coverslip (A), or coverslips covered with a BHK cell monolayer (B and C). All cultures had been pretreated for 1 hr. with anti-receptor antibody CC1 (1:50 dilution) and a 1:100 dilution of antibody CC1 was present throughout the course of the experiment. In (C), anti-S monoclonal antibody 5A13.5 (1:100 dilution) was also present throughout the course of the experiment. Following a 16 hour incubation at 37°C in serumfree OptiMem media (GIBCO), MHV antigen was detected by indirect immunofluorescence. Cells were fixed with 10% formalin, permeabilized with 2% NP40, blocked in 5% NGS, then incubated with a cocktail of monoclonal antibodies 4B6.2, 5A5.2, and 5B19.2, recognizing nucleocapsid, matrix, and spike proteins, respectively. Monoclonal antibodies were detected with TRITC-conjugated, sheep antimouse antibody (Accurate).

against lethal intracerebral challenge in vivo[7]. Together, these observations suggest that cell-associated spread of MHV may play an important role in dissemination of infection in vivo, and consequently, may augment viral pathogenesis.

The ability of anti-spike monoclonal antibodies to prevent cell-associated spread of infection in infectious center assays correlated with the ability to inhibit the fusion of receptor-negative BHK cells induced by recombinant spike glycoprotein in the vT7.3

Table 1. Inhibition of receptor-independent spread (RIS) and recombinant spike-mediated fusion (vv-S fusion) by anti-S and anti-MHVR monoclonal antibodies

mAb	Neutralization	*in vivo* Protection	RIS Inhibition	vv-S fusion Inhibition
5B19.2	+	+	+	+
5A13.5	+	+	+	+
4B11.6	+	–	–	–
5B93.3	–	–	–	–
5B21.5	–	–	–	–
5B207.7	–	–	nd	–
5B216.8	–	–	nd	–
CC1	+	+/–	–	–

vaccinia virus expression system (Table 1). To assess the role of fusion in receptor-independent, cell-associated spread of MHV, we examined whether the acid pH-dependent MHV variant OBLV60 was capable of receptor-independent spread. OBLV60 is a variant of MHV-4 that was isolated following a prolonged persistent infection of the murine olfactory bulb cell line, OBL21a[8]. The OBLV60 spike glycoprotein has amino acid alterations in the heptad repeat region of S2 and a concomitant inability to mediate fusion at neutral pH. Consequently, the OBLV60 spike glycoprotein is unable to mediate cell-cell membrane fusion, and infection of susceptible cells with strain OBLV60 does not lead to syncytia formation. Figure 3 shows an infectious center assay in which OBLV60-infected DBT cells were seeded with and without an underlying BHK cell monolayer. OBLV60 infection remained isolated to the single cells of the overlay and failed to spread to neighboring BHK cells. Therefore, neutral pH fusion capability of the spike glycoprotein is required for receptor-independent, cell-associated spread of MHV infection.

The inability of strain OBLV60 to induce plasma membrane fusion and the acid pH requirement for OBLV60 spike-mediated fusion suggested that OBLV60 may be restricted to an endocytic route of entry into susceptible cells. The entry of acid pH-dependent viruses into cells can be inhibited by lysosomotropic agents that prevent the acidification of endosomal vesicles. We tested the lysosomotropic base chloroquine and the carboxylic ionophore monensin for inhibitory effects on OBLV60 and JHM infection of DBT cells. Both OBLV60 and JHM were at least partly inhibited by these lysosomotropic agents; however, the reduction in the amount of infectious virus recovered from culture supernatants after overnight incubation with chloroquine or monensin was 3.2 and 4.4 $\log_{10}$ greater for OBLV60 than for JHM. In the presence of monensin, no OBLV60 was detected in the culture supernatant, which amounts to a greater than 7 $\log_{10}$ reduction compared with untreated cultures.

The effect of monensin on OBLV60 internalization was visualized by electron microscopy (Figure 4). Following a 15 minute incubation at 37°C in the presence of monensin, OBLV60 virions were found to have accumulated in large vesicles. This observation is consistent with a acid pH-dependent endosomal route for the entry of OBLV60, in

Table 2. Inhibition of MHV by lysosomotropic compounds

	Untreated	Chloroquine (50μM)	Monensin (10μM)
JHM	4.27×10^5	1.35×10^4	8.13×10^2
OBLV60	1.41×10^7	2.57×10^2	—

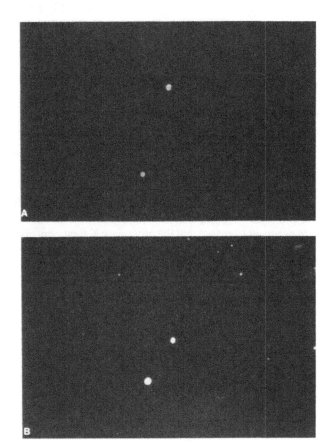

Figure 3. Infectious center assay demonstrating that OBLV60 infection does not spread through a receptor-independent, cell-associated mechanism Infectious center assay was performed as described for Figure 2, except that DBT cells were infected with OBLV60 virus OBLV60-infected DBT cells were seeded onto a blank coverslip (A) or a BHK cell monolayer (B)

which, in the presence of monensin, the pH of endosomal vesicles remains above the threshold for OBLV60 spike-mediated fusion between the viral membrane and the vesicle membrane The inhibition of OBLV60 by monensin is reversible, and the presence of intracellular infectious virus in the monensin-treated cells was demonstrated with an infectious center assay following treatment with proteinase K to remove extracellular bound virus

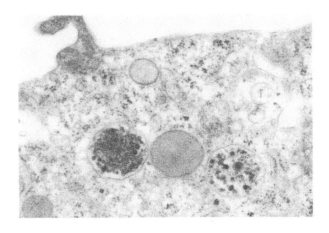

Figure 4. Electron micrograph demonstrating several OBLV60 virus particles within a large vesicle of a monensin-treated DBT cell Following absorption of OBLV60 to monensin-treated DBT cells for 1 hour at 4°C, the culture was shifted to 37°C for 15 minutes to allow internalization of bound virus, then fixed immediately with cold paraformaldehyde/glutaraldehyde and processed for electron microscopy

Table 3. Infectious center assay of proteinase K-treated DBT cells infected in the presence of lysosomotropic agents

	Untreated	Chloroquine (50μM)	Monensin (10μM)
MHV-4	1 3 x 10³	1 1 x 10³	1 9 x 10³
OBLV60	1 4 x 10⁴	7 5 x 10³	1 3 x 10⁴

(Table 3) MHV-4 or OBLV60 was absorbed to treated or untreated DBT cells at 4°C for 1 hour Cultures were incubated at 37°C for 15 minutes, then treated with 0 5mg/ml proteinase K for 45 min at 4°C in PBS to remove extracellular bound virus The proteinase K reaction was stopped by adding an equal volume of 2 mM PMSF/6% BSA in PBS The cells were then extensively washed, serially diluted and seeded onto DBT cell monolayers Cultures were overlaid with agarose and plaques were enumerated after a 2 day incubation In a control experiment to ensure that the proteinase K treatment effectively removed extracellular virus, the 15 minute incubation at 37°C was omitted from the procedure and no infectious centers were detected The data presented in Table 3 show that temporary treatment with chloroquine or monensin did not significantly affect the number of cells that were infected with MHV-4 or OBLV60

SUMMARY

Receptor-specificity is a key determinant of viral tropism In this report, however, we have demonstrated that cell-associated spread of MHV can bypass the requirement for binding to primary receptors and thereby spread to cells that are resisant to MHV infection Anti-receptor antibody CC1, which blocks infection by MHV virions, failed to prevent cell-associated spread of MHV to receptor-negative BHK cells or receptor-positive DBT cells Cell-associated MHV may be utilizing an alternative, low-affinity receptor that is inadequate for functional interaction with MHV virions Theoretically, dissemination of MHV infection through a receptor-independent, cell-associated mechanism in vivo provides the potential for broader host and tissue range, and for spread of infection despite the presence neutralizing antibodies Receptor-independent, cell-associated spread of MHV requires neutral pH fusion capability The low pH-dependent MHV variant OBLV60, which utilizes an endocytic route of entry, does not spread through a receptor-independent mechanism Additionally, antiviral antibodies that block MHV spike-mediated fusion inhibited cell-associated spread of infection

REFERENCES

1 Williams, R K , G Jiang, and K V Holmes 1991 Receptor for mouse hepatitis virus is a member of the carcinoembryonic antigen family Proc Natl Acad Sci 88 5533-5536

2 Nedellec, P , Dveksler, G S , Daniels, E , Turbide, C , Chow, B , Basile, A A , Holmes, K V , and N Beauchemin 1994 Bgp2, a new member of the carcinoembryonic antigen-related gene family, encodes an alternative receptor for mouse hepatitis virus J Virol 68 4525-4537

3 Compton, S R , C B Stephensen, S W Snyder, D G Weismiller, and K V Holmes 1992 Coronavirus species specificity murine coronavirus binds to a mouse-specific epitope on its carcinoembryonic antigen-related receptor glycoprotein J Virol 66 7420-7428

4 Gallagher, T M , M J Buchmeier, and S Perlman 1992 Cell receptor-independent infection by a neurotropic murine coronavirus Virology 191 517-522

5 Taguchi, F , Ikeda, T , H Shida 1992 Molecular cloning and expression of a spike protein of neurovirulent coronavirus JHMV variant cl-2 J Gen Virol 73 1065-1072

6 Collins, A R , R L Knobler, H Powell, and M J Buchmeier 1982 Monoclonal antibodies to murine
hepatitis virus-4 (strain JHM) define the viral glycoprotein responsible for attachment and cell-cell fusion
Virology 119 358-371

7 Buchmeier, M J , Lewicki, H A , Talbot , P J , and R L Knobler 1984 Murine hepatitis virus-4 (strains
JHM)-induced neurologic disease is modulated in vivo by monoclonal antibody Virology 132 261-270

8 Gallagher, T M , C Escarmis, and M J Buchmeier 1991 Alteration of the pH dependence of coronavirus-induced cell fusion effect of mutation in the spike glycoprotein J Virol 65 1916-1928

58

LOCALIZATION OF NEUTRALIZING EPITOPES AND RECEPTOR-BINDING SITE IN MURINE CORONAVIRUS SPIKE PROTEIN

Fumihiro Taguchi,[1] Hideyuki Kubo,[1] Hideka Suzuki,[1] and Yasuko K. Yamada[2]

[1] National Institute of Neuroscience, NCNP
4-1-1 Ogawahigashi, Kodaira
Tokyo 187, Japan
[2] National Institute of Health
4-7-1 Gakuen, Musashimurayama
Tokyo 190-12, Japan

ABSTRACT

To identify the localization of the epitopes recognized by monoclonal antibodies (MAbs) against the S1 subunit of the murine coronavirus JHMV spike protein, we have expressed the S1 proteins with different deletions from the C terminus of the S1. All of MAbs in groups A and B recognized the S1N(330) composed of 330 amino acids (aa) from the N terminus of the S1 and the larger S1 deletion mutants, but failed to react with the S1N(220) composed of 220 aa. MAbs in group C reacted only with the S1utt protein without any deletion. These results indicated that the S1N330 comprised the cluster of epitopes recognized by MAbs in groups A and B. These results together with the fact that all the MAbs in group B retained the high neutralizing activity suggested that the N terminus 330 aa are responsible for binding to the MHV-specific receptors. In pursuit of this possibility, we have expressed the receptor protein and examined the binding of each S1 deletion mutants to the receptor. It was demonstrated that the S1N(330) protein as well as other S1 deletion mutants larger than S1N(330) bound to the receptor. These results indicated that a domain composed of 330 aa at the N terminus of the S1 protein is responsible for binding to the MHV-specific receptor.

INTRODUCTION

Coronaviruses are enveloped, positive stranded RNA viruses associated with various diseases in both animals and humans[1]. The murine coronavirus (MHV) has a genome RNA of about 31 kilobases and encodes four major structural proteins: nucleocapsid protein, the integral membrane glycoprotein, the hemagglutinin-esterase glycoprotein and the spike (S) glycoprotein. Several non-structural proteins are encoded in the genome as well [1].

The spike comprises two to three molecules of the S protein of 150-200 KDa, each of which is a heterodimer consisting of two non-covalently bound S protein subunits, S1 and S2 derived from the N-terminal and C-terminal halves of the S protein[1]. The S1 subunit forms the globular head of the spike and the S2 its stalk portion[2]. The S protein of MHV is multifunctional. It attaches the virus to the cell surface by binding to MHV-specific receptors[3]. Although there are few reports concerning the analysis of the receptor-binding site on the S protein, the topologies of the S protein subunits suggest that the receptor-binding site is more likely to exist on S1 than on S2. The fusion of cultured cells infected with MHV is caused by the S protein[4,5]. The S protein is the major target of the neutralizing antibodies induced in mice after infection with MHV. It also elicits cytotoxic T cells[6]. Furthermore, the S protein is suggested to be a major determinant of viral virulence in animals.

In this study, we showed that the cluster of epitopes recognized by MAbs with neutralizing activity were localized within 330 aa from the N terminus of the S1 subunit. Furthermore, we demonstrated that this S1 domain bound to the cellular receptor protein expressed by recombinant vaccinia virus (RVV).

MATERIALS AND METHODS

Construction of the S1 Deletion Mutant Genes and Their Expression

For expression of different C terminal deletions of the S1 protein we constructed various S1 genes that lacked successively longer segments from the 3' region. The S1NM, S1N, S1N(330), and S1N(220) genes encoding 594, 453, 330 and 220 aa from the N terminus of the S1 gene were constructed by polymerase chain reaction (PCR); cl-2 S cDNA[7,8] was template, the forward primer was a positive-sense oligonucleotide corresponding to the leader sequence of JHM and the reverse primers were oligonucleotides with nucleotides TTA complementary to the stop codon at the 5' end. We inserted the PCR products into pCR™II from the TA cloning kit. Each S1 gene was then cut out from the vector and the fragment was inserted into the vaccinia virus transfer vector (VV-TV), pSF7.5EB1-B5-12 (pSF)[9]. The VV-TVs with S1utt, S1NM, S1N, S1N(330) and S1N(220) were designated pSFS1utt, pSFS1NM, pSFS1N, pSFS1N(330) and pSFS1N(220), respectively. The S1 deletion mutant proteins were expressed by the infection of wild type VV and transfection of above VV-TVs (transient expression).

Isolation of MHV-Specific Receptor Gene and Expression by RVV

The RNA extracted from BALB/c liver was reverse transcribed into cDNA with oligo(dT)12-18 as a primer[10]. And the MHV receptor gene was then amplified by a pair of primers that correspond to the published nucleotide sequences of mmCGM1 around the initiation and termination codons[11]. The amplified fragment with a length of around 900-base pairs (bp) (mL900) was incorporated either in pCR™II vector or downstream of the SRα promoter in the pcDL-SRα296 expression and cloning vector[13]. The mL900 was also inserted into the VV-TV, pSF. The RVV containing mL900 (RVV-MRe) was prepared as described previously[8].

Binding Assay of the S1 Mutant Proteins to MHV Receptor

The lysates of RK13 cells infected with RVV-MRe were electrophoresed on SDS-polyacrylamide gel and transferred onto the transfer membrane paper as described previously[13]. The membrane paper was incubated with the culture fluids from DBT cells producing S1 deletion mutants. The membrane paper was then washed and incubated with MAb No. 7. Binding of MAb No. 7 was estimated by enhanced chemiluminescence with horse radish peroxidase labeled anti-mouse IgG.

RESULTS

Expression of the S1 Mutant Proteins in DBT Cells and Their Reactivities to a Panel of MAbs

The MAbs specific for the S1 protein are classified into 3 different groups: those that react with most MHV strains (group A), those that specifically react with JHMV (group B), and those that specifically bind JHMV variants with the larger S protein (group C). All of these MAbs recognize conformational epitopes[14]. MAbs classified in group C do not react with the small S protein of sp-4 [14,15]. Cl-2 S1 protein consists of 769 aa, and that of sp-4 has a deletion of 141 aa which corresponds to aa 454 to 594 of cl-2. Based on this difference of the S protein between cl-2 and sp-4, we designated three regions of the S1 protein S1N, S1M and S1C. The S1M region is missing in the small S protein of sp-4. The S1N and S1C regions correspond to the N terminal and C terminal regions as compared with the position of S1M. We constructed deleted S1 genes that expressed the S1NM (S1N+M) and S1N proteins as well as S1N(330) and S1N(220) which respectively encoded 330 and 220 aa from the S-protein N terminus. Each S1 deletion mutant was expressed in DBT cells by the VV transient expression system. For all the constructs, indirect immunofluorescence revealed that 5 to 10% of DBT cells produced the S1 protein. Western blotting analysis using the lysates prepared from such cells showed that almost equivalent amounts of S1 deletion mutants were produced in cells transfected with various pSFs and the sizes of the expressed S1 proteins were in accord with the sizes deduced from their gene structure. DBT cells expressing these species of S1 deletion mutants were examined for their reactivities to a panel of MAbs by indirect immunofluorescence. As shown in Table 1, all of the S1-specific MAbs reacted with the S1utt which covered whole S1 protein[16] in the same way as they did with the S protein produced in DBT cells by the infection of cl-2. All MAbs classified into groups A and B reacted to S1NM, S1N and S1N(330). Group C MAbs reacted only with S1utt and not with any other S1 deletion mutants . These results indicated that the epitopes recognized by MAbs in groups A and B are clustered in the S1 domain composed of 330 aa from the N terminus. And all the proteins, with the exception of the smallest protein expressed

Table 1. Reactivities of the S1-specific MAbs to the expressed S proteins[*]

Recombinant plasmid	Reactivity													
	Group A			Group B					Group C					
	2	7	19	3	6	13	71	93	8	12	47	63	78	85
pSFS1utt	+	+	+	+	+	+	+	+	+	+	+	+	+	+
pSFS1NM	+	+	+	+	+	+	+	+	−	−	−	−	−	−
pSFS1N	+	+	+	+	+	+	+	+	−	−	−	−	−	−
pSFS1N(330)	+	+	+	+	+	+	+	+	−	−	−	−	−	−
pSFS1N(220)	−	−	−	−	−	−	−	−	−	−	−	−	−	−

*The S1 deletion mutants proteins were expressed in DBT cells infected with VV after transfection with various pSF vectors The reactivities were examined by indirect immunofluorescence

by pSFS1N(220), form the same tertiary structure as does the entire S1 protein. Epitopes recognized by the MAbs in group C were not restricted to the S1M region that is missing in the small S protein.

Binding of S1 Deletion Mutants with MHv-Specific Receptor

Most of the MAbs classified in group B showed very high neutralizing activity to cl-2[14], which may suggest that these MAbs inhibit the binding of virus to the cellular receptor. If this is the case, the epitopes recognized by such MAbs or the adjacent regions to such MAb epitopes may play an important role in the binding of the virus to the receptor. In pursuit of this possibility, we examined the binding of the S1 deletion mutants to the receptor. By reverse transcription PCR, we isolated the gene encoding MHV receptor from mouse liver. The isolated MHV receptor was identical to BgpC or MHVR(2d)[17]. The expression of this protein in BHK21 cells rendered them susceptible to MHV infection as previously described[17]. Cells infected with the RVV-MRe produced a few different species of receptor proteins; a major band was at 41 kDa and a few additional bands were around 46 kDa. We then examined the ability of these proteins to bind various S1 deletion mutants. MHV-receptor-bound membranes were incubated with the lysates prepared from cells expressing S1 with the transmembrane domain (S1tmd) or S2 with signal peptide (ssS2)[16]. The membrane paper was also incubated with culture fluids from cells producing the S1 deletion mutant proteins via a VV transient expression system; these S1 mutants lacked the transmembrane domain and therefore secreted mostly into the culture fluid. The binding of these S1 and S2 proteins were monitored with MAb No.7 and 10G, respectively. The S1 protein with a transmembrane domain bound to the 41 k and 46 k receptor protein, but binding of the S2 protein was indeterminant due to lack of reaction of MAb 10G even with the whole S protein bound to the receptor. The binding of the whole S protein could be easily detected by the MAb No.7. Fig. 1A shows that S1N(330) as well as the larger S1 proteins bound to the receptor, while S1N(220) apparently did not. S1N(220) may have bound the receptor, because MAb No. 7 used for the detection of various S1 deletion mutants failed to react with S1N(220). Also, MAb 11F recognizing the N terminal linear epitope of the S1 protein and reactive to all the S1 deletion mutant proteins failed to react with all the S1 mutants when

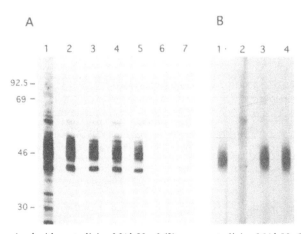

Figure 1. Receptor-binding of the S1 deletion mutants (A) and inhibition of receptor-binding of S1N(330) by MAbs (B). (A): The MHV-receptor prepared on the membrane paper by Western blotting was reacted with the culture fluids of DBT cells infected with VV and transfected with pSFS1utt (2), pSFS1NM (3), pSFS1N (4), pSFS1N(330) (5), pSFS1N(220) (6) or mock transfected cells (7). The paper was then reacted with S1-specific MAb No. 7 and subsequently with peroxidase-labeled anti-mouse IgG. Lane 1 shows the position of MHV-receptor protein detected by anti-human CEA antibody. (B): The S1N(330) was mixed with neutralizing MAb No. 3 (2), non-neutralizing MAb No.2 (3), anti-N MAb No.5 (4) or PBS (1) and incubated at RT for 60 min. These pretreated S1N(330) were monitored for their receptor-binding capacity as described in A.

these mutant S1 proteins bind to the receptor. At present, we have no way to see whether S1N(220) is capable of binding to the receptor. It was also evident from Fig. 1A that there was no striking differences in the amounts of the receptor-bound S1 mutant proteins except for S1N(220). These results clearly indicated that the region retained in the N terminal 330 aa of S1 protein is a major receptor-binding site in the S1 protein of MHV.

In order to examine the specificity of S1N(330) binding to MHV receptor, we have incubated the S1N(330) with neutralizing MAb No. 3 or non-neutralizing MAb No. 2, both of which are reactive to the S1N(330). The treated S1N(330) was then monitored for the binding to the receptor prepared on the membrane paper. As shown in Fig. 1B, the binding of the S1N(330) was perfectly inhibited by neutralizing MAb No. 3, while other non-neutralizing MAbs did not give any effect. These results clearly showed that the binding of the S1N(330) to MHV receptor was specific.

DISCUSSION

The S protein is known to be a multifunctional protein. However, very little is known about the relationship between biological functions and the structure of the S protein, except for the localization of neutralizing epitopes recognized by the MAbs[18] and fusion activity by molecularly engineered S protein[13]. The S protein has long been assumed to be involved in receptor-binding. In this paper, we showed that the clusters of antigenic epitopes exist within 330 aa from the N terminus of the S protein. Furthermore, an S1 deletion mutant protein composed of the N terminal 330 aa of S1 bound to the MHV-specific receptor. This finding agrees with the idea that S1 comprises the bulbous portion of the characteristic club-shaped spike that is located in the outer extremity of the virion[3].

Various S1 deletion mutants larger than S1N(330) bound to the receptor proteins without striking differences in intensity of binding. This excludes the possibility that some region in the S1, other than the N terminal 330 aa, is a major active site for receptor-binding. The antigenic epitopes recognized by group B MAbs are unlikely to be centers of receptor-binding sites, because these antigenic epitopes are specific for JHM strains and the receptors are known to serve as targets for JHM and A59[12,19]. The MAbs in group B probably recognize the epitopes existing in the neighboring region of the receptor-binding site on the S protein and inhibit virus binding to the receptor by steric hindrance. The receptor also served as a binding site for some other MHV strains besides JHM (unpublished data) suggesting that the receptor-binding active site is conserved within 330 aa from the S1 N terminus in most MHV strains. We are currently determining the smallest, conserved, aa sequence that functions as a receptor-binding site.

Work on different coronaviruses suggests the involvement of the S2 subunit in receptor-binding activity; MAbs specific for the S2 of MHV have neutralizing activity[4]. However, the ability of S2 antibodies to neutralize virus may be the effect of steric hindrance, since the S protein is known to have a highly conformational structure[1,3]. Moreover, viral neutralization by antibodies does not always result from the prevention of virus binding to the receptor. When the S1 subunit of avian infectious bronchitis virus (AIBV) is removed, the virion is still able to bind susceptible cells, although it is not infectious[20]. This demonstrated that the S2 subunit mediates attachment. However, this attachment did not trigger the replication of attached viruses and would not be the genuine process that occurred with infectious virions retaining the S1 subunit. In the present study, we failed to demonstrate whether or not S2 bound to the receptor, because MAb 10G and others reactive to S2 in Western blot did not work for the detection of the receptor-bound S proteins.

In picornaviruses, a small canyon formed by the outer coat protein is proposed to be the precise location of receptor-binding; this idea is the so called canyon hypothesis[21].

Members of the immunoglobulin super-family serve as the receptors of these viruses[22] The MHV receptors, mmCGM1 and mmCGM12 are also members of the immunoglobulin superfamily There is another resemblance between picornavirus receptors and MHV receptors their N terminal domains, which are similar to the immunoglobulin variable region, are critical for interaction with viral proteins [23,24] The mechanism of MHV virus-receptor binding may be similar to that of picornaviruses Analysis of the receptor-binding site of the S protein will delineate the mechanism of virus-receptor interaction, which initiates virus infection

REFERENCES

1 Spaan,W , D Cavanagh, and M C Horzinek 1988 Coronaviruses structure and genome expression J Gen Virol 69 2939-2952

2 De Groot,R J , W Luytjes, M C Horzinek, B A M Van der Zeijst, W J M Spaan, and J A Lenstra Evidence for a coiled-coil structure in the spike of coronaviruses J Mol Biol 196 963-966

3 Holmes,K V, E W Doller and J N Behnke 1981 Analysis of the function of coronavirus glycoprotein by differential inhibition of synthesis with tunicamycin Adv Exp Med Biol 142 133-142

4 Collins,A R , R L Knobler, H Powell, and M J Buchmeier 1982 Mono-clonal antibodies to murine hepatitis virus-4 (strain JHM) define the viral glycoprotein responsible for attachment and cell fusion Virology 119 358-371

5 Vennema, H , L Heijnen, A Zijderfeld, M C Horziek, and W J M Spaan 1990 Intracellular transport of recombinant coronavirus spike proteins implication for virus assembly J Virol 64 339-346

6 Kyuwa,S , and S A Stohlman 1990 Pathogenesis of a neurotropic murine coronavirus strain, JHM, in the central nervous system of mice Semin Virol 1 273-280

7 Taguchi,F , S G Siddell, H Wege, and V ter Meulen 1985 Characterization of a variant virus selected in rat brain after infection by coronavirus mouse hepatitis virus JHM J Virol 54 429-435

8 Taguchi,F , T Ikeda, and H Shida 1992 Molecular cloning and expression of a spike protein of neurovirulent murine coronavirus JHMV variant cl-2 J Gen Virol 73 1065-1072

9 Funahashi,S , S Itamura, S Iinuma, H Nerome, M Sugimoto, and H Shida 1991 Increased expression in vitro and in vivo of foreign genes directed by A-type inclusion body hybrid promoters in recombinant vaccinia viruses J Virol 65 5584-5588

10 Yamada, Y K , M Abe, A Yamada, and F Taguchi 1993 Detection of mouse hepatitis virus by the polymerase chain reaction and its application to the rapid diagnosis of infection Lab Anim Sci 43 285-290

11 Dveksler,G S , M N Pensiero, C B Cardellichio, R K Williams, G Jiang, K V Holmes, and C W Diffenbach 1991 Cloning of the mouse hepatitis virus (MHV) receptor expression in human and hamster cell lines confers susceptibility to MHV J Virol 65 6881-6891

12 Takebe Y , M Seiki, J Fujisawa, J Hoy, K Yokota, K Arai, M Yoshida, and N Arai 1988 SRa promoter an efficient and versatile mammalian cDNA expression system composed of the simian virus 40 early promoter and the R-U5 segment of human T-cell leukemia virus type 1 long terminal repeat Mol Cell Biol 8 466-472

13 Taguchi,F 1993 Fusion formation by uncleaved spike protein of murine coronavirus JHMV variant cl-2 J Virol 67 1195-1202

14 Kubo,H , S Y Takase, and F Taguchi 1993 Neutralization and fusion inhibition activities of monoclonal antibodies specific for the S1 subunit of the spike protein of neurovirulent murine coronavirus JHMV cl-2 variant J Gen Virol 74 1421-1425

15 Taguchi,F , and J O Fleming 1989 Comparison of six different murine coronavirus JHM variants by monoclonal antibodies against the E2 glycoprotein Virology 169 233-235

16 Kubo,H , and F Taguchi 1993 Expression of the S1 and S2 subunits of murine coronavirus JHMV spike protein by a vaccinia virus transient expression system J Gen Virol 74 2373-2383

17 Dveksler,G S , C W Diffenbach, C B Cardellichio, K Mccuaig, M N Pensiero, G S Jiang, N Beauchemin and K V Holmes 1993 Several members of the mouse carcinoembryonic antigen-related glycoprotein family are functional receptors for the coronavirus mouse hepatitis virus-A59 J Virol 67 1-8

18 Luytjes,W , D Geerts, W Posthumus, R Meloen, and W Spaan 1989 Amino acid sequence of a conserved neutralizing epitope of murine coronaviruses J Virol 63 1408-1412

19 Yokomori, K , and M M C Lai 1992 Mouse hepatitis virus utilizes two carcionoembryonic antigens as alternative receptors J Virol 66 6194-6199

20 Cavanagh,D , and P J Davis 1986 Coronavirus IBV removal of spike glycopeptide S1 by urea abolishes infectivity and haemagglutination but not attachment to cells J Gen Virol 67, 1442-1448

21 Rossmann,M G , E Arnold, J W Erickson, E A Frankenberger, J P Griffith, H J Hecht, J E Johnson, G Kamer, M Luo, A G Mosser, R R Rueckert, B Sherry and G Vriend 1985 Structure of a human common cold virus and functional relationship to other picornaviruses Nature 317 145-153

22 Mendelsohn,C L , E Wimmer and V R Racaniello 1989 Cellular receptor for poliovirus Molecular cloning, nucleotide sequence, and expression of a new member of the immunoglobulin superfamily Cell 56 855-865

23 Dveksler, G S , M N Pensiero, C W Dieffenbach, C B Cardellichio, A A Basile, P E Elia and K V Holmes 1993 Mouse hepatitis virus strain A59 and blocking antireceptor monoclonal antibody bind to the N-terminal domain of cellular receptor Proc Natl Acad Sci USA 90 1716-1720

24 Koike,S , I Ise, and A Nomoto 1991 Functional domains of the poliovirus receptor Proc Natl Acad Sci USA 88 4104-4108

59

ANALYSIS OF THE SIALIC ACID-BINDING ACTIVITY OF TRANSMISSIBLE GASTROENTERITIS VIRUS

B. Schultze,[1] L. Enjuanes,[2] and G. Herrler[1]

[1] Institut für Virologie
Philipps-Universität Marburg
Marburg, Germany
[2] Centro De Biologia Molecular
CSIC Universidad Autonoma
Madrid, Spain

ABSTRACT

Porcine transmissible gastroenteritis virus (TGEV) has been shown to agglutinate erythrocytes using α2,3-linked sialic acid on the cell surface as binding site. The hemagglutinating activity requires the pretreament of virus with neuraminidase. We obtained evidence that TGEV recognizes not only N-acetylneuraminic acid but also N-glycoloylneuraminic acid, a sialic acid present on many porcine cells.

INTRODUCTION

Several coronaviruses are known to be potent hemagglutinating agents: bovine coronavirus (BCV), porcine hemagglutinating encephalomyelitis virus (HEV), human coronavirus OC-43 and some murine coronaviruses[1]. Each of these viruses possess two surface glycoproteins. The S protein is a hemagglutinin and interacts with cellular receptors. It shows fusion activity and induces neutralizing antibodies. The HE-protein is an acetylesterase and functions as receptor-destroying enzyme. Like the S protein, the HE protein is a hemagglutinin and induces neutralizing antibodies. Other coronaviruses, e.g. transmissible gastroenteritis virus (TGEV) and infectious bronchitis virus (IBV) show no or only weak hemagglutinating activity[2]. These viruses possess the S-protein, but they are devoid of a receptor-destroying enzyme. It has been shown that the hemagglutinating activity of IBV and TGEV requires the pretreament of virus with neuraminidase indicating that inhibitors on the virion surface have to be inactivated in order to induce the HA-activity of these viruses[3]. The reason for the inhibitory effect of the surface-bound sialic acid on TGEV and IBV is that the virus recognizes a similar or identical sialic acid as receptor determinant.

Corona- and Related Viruses, Edited by P. J. Talbot and G. A. Levy
Plenum Press, New York, 1995

When sialic acid is removed from the surface of erythrocytes the virus can no longer agglutinate these cells. Erythrocytes which are resialylated and possess α2,3-linked N-acetylneuraminic acid on their surface are agglutinated by TGEV but to a lower extent than untreated erythrocytes. We were interested to find out whether sialic acids other than N-acetylneuraminic acid are recognized by TGEV. Therefore, we tested agglutination of erythrocytes from different species.

MATERIALS AND METHODS

Virus. The Purdue strain of TGEV was propagated in LLC-PK1 cells.

Neuraminidase treatment of virus. TGEV was harvested from the supernatant of infected LLC-PK1 cells 48 h post-infection. After clarification of the medium by low-speed centrifugation (2000xg, 10 min), virus was pelleted by ultracentrifugation for 90 min at 54,000xg. Virus was resuspended in PBS and incubated with neuraminidase from *Vibrio cholerae* (VCNA 500mU/ml) at 37°C for 30 min. Following sedimentation of the virus by ultracentrifugation for 30 min at 100,000xg, the virus was resuspended in PBS and used for hemagglutination assays.

Hemagglutination assay. Hemagglutinating activity was determined according to published procedures[4]. The hemagglutination titer (HA-titer) indicates the reciprocal value of the maximum dilution that caused complete agglutination.

Resialylation of erythrocytes. A 10% suspension of chicken erythrocytes was incubated with neuraminidase from *Vibrio cholerae* (20mU/ml) for 30 min at 37°C. Asialo cells were washed and suspended in PBS to a final concentration of 20% in a total volume of 200μl. Resialylation was accomplished by incubation with α2,3 sialyltransferase (5,2mU) and 400nmol of CMP-activated N-acetylneuraminic acid (Neu5Ac) or 200nmol of activated N-glycoloylneuraminic acid (Neu5Gc). After 2 hr at 37°C, cells were washed and used as a 1% suspension in a hemagglutination assay.

RESULTS

Agglutination of Erythrocytes from Different Species by Neuraminidase-Treated TGEV

Porcine transmissible gastroenteritis virus (TGEV) has been shown to agglutinate erythrocytes using α2,3-linked sialic acid on the cell surface as binding sites. The hemagglutinating activity requires the pretreatment of virus with neuraminidase[3]. We tested the hemagglutinating activity of TGEV with erythrocytes from several species which differ in type and amount of sialic acid on their surface. We used erythrocytes from chicken, one-day-old chicken, pig and cow in an agglutination assay with either untreated TGEV or neuraminidase-treated virus (Table 1). After pretreatment with neuraminidase TGEV was able to agglutinate erythrocytes from each of these species. In the case of bovine erythrocytes no pretreatment with neuraminidase was necessary to observe hemagglutination. Characteristic for bovine erythrocytes is that they possess a large amount (91%) of N-glycoloylneuraminic acid (Neu5Gc) on their surface. This result was a hint that Neu5Gc may be a very efficient binding component for TGEV.

Sialic acids occur mainly in the terminal position of glycoproteins and gangliosides. They represent a large family of N-and O-substituted derivatives of neuraminic acid. N-acetylneuraminic acid (Neu5Ac) and N-glycoloylneuraminic acid (Neu5Gc), respec-

Table 1. Agglutination of erythrocytes by TGEV and
neuraminidase-treated virus (TGEV-NA)

Erythrocytes	HA titer (HA units/ml)	
	TGEV	TGEV-NA
Chicken, adult	<2	1536
Chicken, 1-day-old	<2	64
Pig	<2	2048
Cow	32	768

tively, are the two most frequently occuring sialic acid species. Their structure is shown in Fig.1.

TGEV Binding to N-glycoloylneuraminic Acid on Erythrocytes

Based on the findings with bovine erythrocytes, we analyzed whether TGEV was able to recognize N-glycoloylneuraminic acid on erythrocytes more efficiently than N-acetylneuraminic acid. Asialo cells obtained by treatment with neuraminidase from *Vibrio cholerae* were resialylated with either Neu5Ac or Neu5Gc in an $\alpha2,3$-linkage. We compared the binding characteristics of neuraminidase-treated TGEV with two influenza viruses, fowl plague virus (FPV) and WSN strain for which the receptor determinant is already known, Neu5Ac, (Table 2). Erythrocytes resialylated with N-acetylneuraminic acid were agglutinated efficiently by FPV and WSN, while the hemagglutinating acitivity observed with TGEV was very low. Among the two influenza viruses only FPV was able to recognize N-glycoloylneuraminic acid. TGEV agglutinated red blood cells containing N-glycoloylneuraminic acid very efficiently. This result indicates that the preferred binding component of TGEV on erythrocytes is N-glycoloylneuraminic acid.

N-acetylneuraminic acid

N-glycoloylneuraminic acid

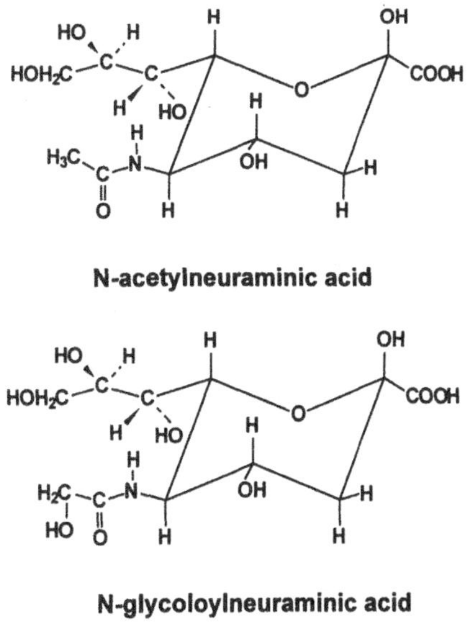

Figure 1. Structure of N-acetylneuraminic acid (Neu5Ac) and N-glycoloylneuraminic acid (Neu5Gc)

Table 2. Restoration of receptors for neuraminidase-treated TGEV
(TGEV-NA) by resialylation of asialo-erythrocytes

Erythrocytes	HA titer (HA units/ml)		
	TGEV-NA	FPV	WSN
Control	256	256	512
Asialo	<2	<2	<2
Resialylated			
Neu5Ac	4	256	256
Neu5Gc	512	128	<2

DISCUSSION

Among coronaviruses there have now been identified several potent hemagglutinating agents Members of the BCV serogroup (BCV, HCV-OC43, HEV) recognize N-acetyl-9-O-acetylneuraminic acid as receptor determinant, whereas IBV and TGEV bind to N-acetylneuraminic acid and N-glycoloylneuraminic acid, respectively They differ not only in the type of sialic acid they recognize but also in the presence or absence of a receptor-destroying enzyme on the virus particle Viruses that use sialic acid for attachment to cells have to remove this common sugar from glycoproteins and glycolipids that might act as inhibitors and prevent binding of virus to the target cell In the case of BCV the acetylesterase-activity is responsible for the lack of N-acetyl-9-O-acetylneuraminic acid from potential inhibitors IBV and TGEV are devoid of a receptor-destroying enzyme Therefore, the binding site for sialic acid is occupied by sialic acid-containing proteins Inactivation of these inhibitors by treatment with neuraminidase results in a binding activity to sialic acid The importance of the sialic acid-binding activity is not yet known

REFERENCES

1 Spaan W , Cavanagh D and Horzinek M C Coronaviruses structure and genome expression J Gen Virol 1988, 69 2939-2952

2 Schultze B , Cavanagh D and Herrler G Neuraminidase treatment of avian infectious bronchitis coronavirus reveals a hemagglutinating acitivity that is dependent on sialic acid-containing receptors on erythrocytes Virology 1992, 189 792-794

3 Schultze B , Enjuanes L , Cavanagh D and Herrler G N-acetylneuraminic acid plays a critical role for the hemagglutinating activity of avian infectious bronchitis virus and porcine transmissible gastroenteritis virus Adv Exp Med Biol 1993, 342 305-310

4 Schultze B , Gross H -J , Brossmer R , Klenk, H -D and Herrler G Hemagglutinating encephalomyelitis virus attaches to N-acetyl-9-O-acetylneuraminic acid-containing receptors on erythrocytes comparison with bovine coronavirus and influenza C virus Virus Res 1990, 16 185-194

60

ANALYSIS OF CELLULAR RECEPTORS FOR HUMAN CORONAVIRUS OC43

C. Krempl, B. Schultze, and G. Herrler

Institut für Virologie
Philipps-Universität Marburg
Marburg, Germany

ABSTRACT

Bovine coronavirus (BCV), human coronavirus OC43 (HCV-OC43) and hemagglutinating encephalomyelitis virus (HEV) are serologically related viruses that all have hemagglutinating activity. The receptor determinant for attachment to erythrocytes has been shown to be N-acetyl-9-O-acetylneuraminic acid (Neu5,9Ac$_2$). We compared the ability of the three coronaviruses to recognize 9-O-acetylated sialic acid and found that they all bind to Neu5,9Ac$_2$ attached to galactose in either A2,3 or A2,6-linkage. There are, however, some differences in the minimum amount of sialic acid that is required on the cell surface for agglutination by these viruses. Evidence is presented that HCV-OC43 uses Neu5,9Ac$_2$ as a receptor determinant not only for agglutination of erythrocytes but also for attachment to and infection of a cultured cell line, MDCK I cells.

INTRODUCTION

Bovine coronavirus (BCV) is known to use 9-O-acetylated sialic acid as a receptor determinant for attachment to cells[1,2]. In the case of erythrocytes, binding of virus results in the agglutination of cells. Studies with MDCK cells have shown that binding to 9-O-acetylated sialic acid is the initial step in the infection of cultured cells[3]. Two other coronaviruses, human coronavirus OC43 (HCV-OC43) and hemagglutinating encephalomyelitis virus (HEV) that are serologically related to BCV, are also potent hemagglutinating agents. Like BCV, HEV and HCV-OC43 require Neu5,9Ac$_2$ on the cell surface for the agglutination of erythrocytes[1,2].

We analyzed whether BCV, HEV, and HCV-OC43 differ in their ability to recognize Neu5,9Ac$_2$ present in different linkage types. Furthermore, evidence is presented that HCV-OC43 uses 9-O-acetylated sialic acid as a receptor determinant for infection of cells.

Corona- and Related Viruses, Edited by P. J. Talbot and G. A. Levy
Plenum Press, New York, 1995

MATERIALS AND METHODS

Cells. MDCK I cells, a subline of Madin-Darby canine kidney cells, were grown in minimum essential medium containing 10% fetal calf serum.

Virus. The different strains of coronaviruses were grown in MDCK I cells as described previously[2].

Resialylation of cells. Erythrocytes and MDCK I cells were treated with neuraminidase and resialylated to contain Neu5,9Ac$_2$ attached to galactose in either A2,3 or A2,6-linkage as described previously[2,3].

Hemagglutination assay. Hemagglutination titration was performed as described previously[2].

RESULTS AND DISCUSSION

Sialic acids are terminal sugars of oligosaccharides present on many glycoproteins and glycolipids. They are usually attached to galactose. Two common linkage types are SiaA2,3Gal found on O-linked oligosaccahrides and SiaA2,6Gal present on N-linked oligosaccharides.

By using CMP-activated Neu5,9Ac$_2$ and sialyltransferases specific for either linkage type, erythrocytes were modified to contain either A2,3-linked or A2,6-linked Neu5,9Ac$_2$ (Fig. 1). By varying the amount of CMP-Neu5,9Ac$_2$, batches of erythrocytes were obtained that differed in the amount of 9-O-acetylated sialic acid present on the cell surface. The erythrocytes were analyzed whether they are agglutinated or not by BCV, HEV, or HCV-OC43. As shown in Table 1, human coronavirus OC43 was most efficient in recognizing A2,6-linked Neu5,9Ac$_2$.

This virus was able to agglutinate erythrocytes that had been resialylated in the presence of as little as 0.5 nm of CMP-Neu5,9Ac$_2$, whereas HEV and BCV required a minimum amount of 4 nmol. With respect to the A2,3-linkage, the differences between

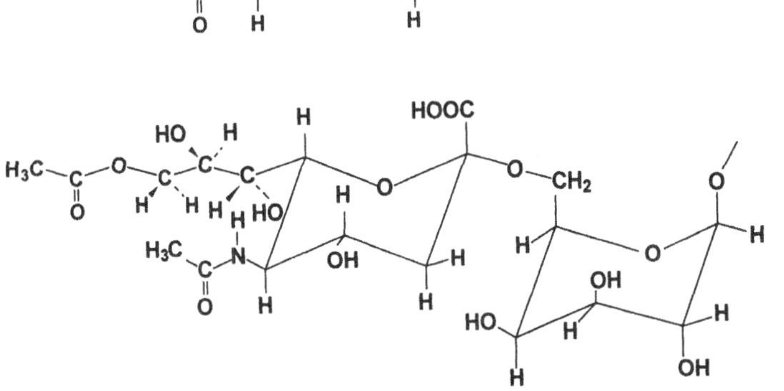

Figure 1. Structure of Neu5,9Ac$_2$ attached to galactose in A2,3-linkage (top) and A2,6-linkage (bottom).

Table 1. Comparison of the efficiency of HCV-OC43, BCV and HEV in recognizing
9-O-acetylated sialic acid as a receptor determinant on erythrocytes

	HA-activity (HA units/ml)			
Linkage Type	CMP-sialic acid	BCV	OC43	HEV
α2,3	4	256	256	128
	2	128	256	32
	1	32	< 2	< 2
	0 5	< 2	< 2	< 2
α2,6	8	64	256	256
	4	32	256	64
	2	2	< 2	64
	1	< 2	< 2	32
	0 5	< 2	< 2	5

the three strains of coronaviruses were less pronounced In this case BCV was more efficient than the two other strains agglutinating erythrocytes that had been resialylated in the presence of 1 nmol of CMP-activated sialic acid The hemagglutinating activity of HEV and HCV-OC43 was detectable only when 2 nmol or more CMP-Neu5,9Ac$_2$ was used for resialylation of the cells Thus, BCV, HEV, and HCV-OC43 are able to recognize 9-O-acetylated sialic acid both A2,3-linked and A2,6-linked to galactose, but with different efficiency

The role of Neu5,9Ac$_2$ as a receptor determinant for coronavirus in the infection of cultured cells has been demonstrated so far only with BCV[3] We analyzed whether HCV-OC43 also uses Neu5,9Ac$_2$ for binding to and infection of cultured cells For this purpose, MDCK I cells were treated with neuraminidase to remove Neu5,9Ac$_2$ from the cell surface thus inactivating endogenous receptors

As shown in Table 2, desialylated cells were resistant to infection by HCV-OC43 In contrast to untreated cells, no release of virus was detectable by hemagglutination titration in the supernatant of neuraminidase-treated cells If the asialo-cells were resialylated to contain A2,6-linked Neu5,9Ac$_2$ on the cell surface, the cells became susceptible to infection (Table 1) The virus yield obtained from the resialylated cells was lower than in the case of untreated cells indicating that the resialylated cells are infected less efficiently than the control cells One reason for this is that the supply of CMP-Neu5,9Ac$_2$ is limited and, therefore, only a limited amount could be applied in these experiments Another reason is that the resialylation time had to be kept short to avoid the regeneration of endogenous receptors The result shown in Table 2 indicates that not only BCV, but also HCV-OC43 uses 9-O-acetylated sialic acid as a receptor determinant for attachment to and infection of cultured cells

Table 2. Effect of neuraminidase treatment and
resialylation (α2,6) on the susceptibility of
MDCK I cells to infection by HCV-OC43

Treatment of cells	Virus yield (HA units/ml)
None	128
Desialylated	< 2
Resialylated, Neu5,9Ac2	16

REFERENCES

1 Vlasak R., Luytjes W., Spaan W , Palese P. Human and bovine coronavirus recognize sialic acid containing receptors similar to those of influenza C virus Proc Natl Acad Sci USA 1988; 85· 4526-4529

2 Schultze B , Gross H -J., Brossmer R., Klenk H.-D and Herrler G. Hemagglutinating encephalomyelitis virus attaches to N-acetyl-9-O-acetylneuraminic acid-containing receptors on erythrocytes: comparison with bovine coronavirus and influenza C virus Vir Res. 1990, 16: 185-194.

3 Schultze B., Herrler G. Bovine coronavirus uses N-acetyl-9-O-acetylneuraminic acid as a receptor determinant to initiate the infection of cultured cells. J Gen Virol 1992; 73: 901-906

61

POLARIZED ENTRY OF BOVINE CORONAVIRUS IN EPITHELIAL CELLS

B. Schultze and G. Herrler

Institut für Virologie
Philipps-Universität Marburg
Marburg, Germany

ABSTRACT

Epithelial cells are highly polarized cells divided into an apical and a basolateral plasma membrane. The two domains are composed of a distinct set of proteins and lipids. Concerning virus infection of epithelial cells, the polarity of host cell receptor distribution defines the domain from which infection may be mediated. We were interested to analyze the infection of polarized cells by bovine coronavirus (BCV). The entry of BCV into MDCK I cells was investigated by growing the cells on a permeable support. Cell were infected with BCV from either the apical or basolateral domain. The efficiency of infection was determined my measuring the hemaglutinating activity of the virus released into the apical compartment. Virus replication was only detectable after inoculation from the apical surface. Therefore, infection of MDCK I cells with BCV is restricted to the apical side.

INTRODUCTION

Epithelial cells line the body cavities of higher eukaryotes and, therefore, represent the primary barrier to a number of viruses infecting a host. The polarized organization of these cells involves the division of the plasma membrane into an apical and a basolateral portion that are separated by tight junctions. As the two membrane domains differ in the protein and lipid composition, they may also differ in the content of receptors for an infecting virus[1]. The presence of suitable receptors determines whether epithelial cells are infectable from the luminal (apical) or from the serosal (basolateral) surface.

We were interested to find out which way bovine coronavirus (BCV) uses to enter polarized cells. This virus may initially infect the respiratory tract, but the disease caused by BCV is due to infection of cells of the gastrointestinal tract resulting in a severe diarrhea in newborn calves. For our studies we used MDCK I cells a well characterized polarized epithelial cell line[2]. Cells grown on a filter membrane were infected with BCV from either the apical or the basolateral side. The efficiency of infection was judged by the yield of virus released into the apical medium, measured by hemagglutination titration. Infection of MDCK cells with BCV was restricted to the apical side. To rule out that the restriction was due to difficulties in penetrating through the pores of the filter membrane, MDCK cells were also infected with influenza C virus. This virus was able to infect cells from both sides indicating that the filter itself is no hindrance for virus to reach the basolateral plasma membrane.

MATERIALS AND METHODS

Cells. MDCK I cells, a subline of Madin-Darby canine kidney cells, were grown in minimum essential medium containing 10% fetal calf serum. Polycarbonate membrane filters (pore size 0.4μm, diameter 24.5mm) were purchased from Costar. Cells were grown for 3-4 days replacing the apical and basal media at daily intervals. Resistance measurements were made by using a Millicell ERS apparatus. Only cell monolayers with a resistance higher than $1000\Omega cm^2$ were used for experiments.

Virus. Strain Johannesburg/1/66 of influenza C virus was grown in 8-day-old embryonated eggs by allantoic inoculation. The allantoic fluid was harvested after incubation of the eggs for 3 days at $33°C^3$. Strain L-9 of BCV was grown in MDCK I cells as described previously[4].

Virus infection. Cells grown on a filter membrane with a resistance higher than $1000\Omega cm^2$ were infected with virus. After an adsorption time of 60 min at 37°C, cells were washed with PBS and incubated at 37°C with MEM. The efficiency of infection was judged by the yield of virus released into the medium as indicated by the hemagglutinating activity of the cell supernatant.

Hemagglutination assay. Hemagglutination titration was performed as described previously[5].

RESULTS AND DISCUSSION

The initial event in the life cycle of a virus is the attachment to specific receptors on the host cell. Epithelial cells which line the external surface of body cavities represent the primary barrier to a number of viruses during infection. Since epithelial cells are highly polarized and express distinct sets of lipids and proteins on their apical and basolateral surface, some viruses are restricted to a particular plasma membrane domain. We were interested to find out which way bovine coronavirus (BCV) uses to enter polarized epithelial cells.

For this purpose we used a filter system (Figure 1). The filter is composed of an apical chamber and a basolateral compartment separated by a microporous membrane. Cells were placed into the apical chamber and allowed to attach to the filter membrane to grow to confluency and to form a tight monolayer. Medium added to the apical and the basal compartment was replaced at daily intervals. Like in the natural environmemt, cells were fed from the basaolateral side. After three to four days, the tightness of the monolayer was determined by measuring the electrical resistance between the apical and the basolateral compartment using two electrodes and the Millicelll ERS apparatus. The extent of resistance is dependant not only on the tightness of the monolayer but also on the cell type and the number of passages. MDCK I cells usually reach an electrical resistance of more than $1000\Omega cm^2$ which is very high compared to other cell types. The tight monolayer forms a barrier between the apical and the basolateral compartment, so that substances and virus

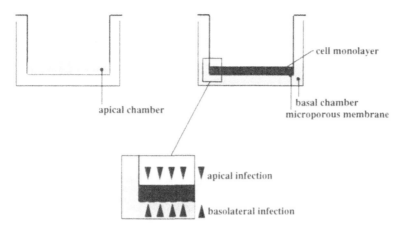

Figure 1. Model of the transwell membrane filter system.

added to one side of the cell monolayer cannot reach the other side. Therefore, cells can be infected selectively from either the apical or basolateral surface.

MDCK I cells grown on a filter membrane were infected with BCV and influenza C virus from either the apical or basolateral surface, respectively (Table 1). The efficiency of infection was judged by the yield of virus released into the apical memdium, measured by hemagglutination titration. Infection of MDCK I cells with BCV from the apical surface resulted in a high yield of virus, whereas no virus production was detectable after infection from the basolateral side. On the other hand, influenza C virus was able to infect filtergrown cells from both sides. This indicates that the filter membrane itself is no hindrance for the virus to reach the basolateral plasma membrane for infecting the cell. From these results we conclude that infection of epithelial cells with BCV occurs preferentially from the apical side.

Binding sites for viruses are a crucial component for the infectibility of a cell. The presence of suitable receptors on polarized epithelial cells may determine whether a virus can enter the cell via the apical and/or the basolateral domain of the plasma membrane. For example, SV 40 virus has been shown to infect polarized cells only from the apical side (Figure 2)[6]. This restriction correlates with the ability of the virus to bind to the apical, but not to the basolateral domain of the plasma membrane. The receptor for SV 40 has not been idientified, but it appeared to be present only on the apical surface. N-acetylneuraminic acid the receptor for influenza A and B virus is distributed on both sides of the epithelial cell; therefore, virus entry occurs from both the apical and the basolateral domain. For other viruses, e.g. vesicular stomatitis virus, canine parvovirus and vaccinia virus it has been shown that infection of epithelial cells occurs preferentially from the basolateral side[1]. As

Table 1. Infection of polarized MDCK I cells with BCV and influenza C virus from either the apical or basolateral domain. The yield of virus was determined by measuring the hemagglutinating activity of the supernatant in the apical chamber.

Virus	Infection from	Hemagglutination-titer (HA units/ml)
BCV	apical	256
	basolateral	<2
Influenza C	apical	128
	basolateral	32

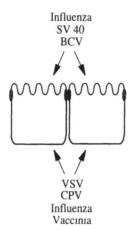

Influenza
SV 40
BCV

VSV
CPV
Influenza
Vaccinia

Figure 2. Entry of polarized epithelial cells by different viruses

far as BCV is concerned, the restriction of virus entry to the apical surface of epithelial cells is consistent with the natural infection, i e the localized infection of the intestinal epithelium

REFERENCES

1 Tucker S P and Compans R W Virus infection of polarized epithelial cells In Murphy F A , Shatkin A J , (eds) Advances in Virus Research Academic Press, San Diego, 1993, 42 187-247

2 Simmons, N L Ion transport in "tight" epithelial monolayers of MDCK cells J Membrane Biol 1981, 59 105-114

3 Herrler G and Klenk H -D The surface receptor is a major determinant of the cell tropism of influenza C virus Virology 1987 159 102-108

4 Schultze B , Gross H -J , Brossmer R , Klenk H -D and Herrler G Hemagglutinating encephalomyelitis virus attaches to N-acetyl-9-O-acetylneuraminic acid-containing receptors on erythrocytes comparison with bovine coronavirus and influenza C virus Vir Res 1990, 16 185-194

5 Herrler G , Rott R and Klenk H -D Neuraminic acid is involved in the binding of influenza C virus to erythrocytes Virology 1985, 141 144-147

6 Clayson E T and Compans R W Entry of SV 40 is restricted to apical surfaces of polarized epithelial cells Mol Cell Biol 1988, 8 3391-3396

62

OVEREXPRESSION OF TGEV CELL RECEPTOR IMPAIRS THE PRODUCTION OF VIRUS PARTICLES

B. Delmas, E. Kut, J.Gelfi, and H. Laude

Unite de Virologie et Immunologie Moléculaires
INRA
Jouy-en-Josas, France

ABSTRACT

The porcine aminopeptidase-N (pAPN) is the cellular receptor for the transmissible gastroenteritis virus (TGEV) due to the specific binding of the spike protein S to APN. In the present study, we performed both biological and biochemical experiments to analyze how the level of expression of a virus receptor can influence the viral protein biosynthesis and the virus production. We generated two swine testis cell clones overexpressing pAPN (ST-APN clones). These clones produced 10^4 less infectious virus than control ST cells. Plaque assays revealed a four-fold reduction of the diameter of the plaques in ST-APN cells compared to ST cells. Pulse-chase experiments revealed that S transport from the endoplasmic reticulum to the Golgi apparatus was not affected in ST-APN cells. Additionally, an anti-APN antibody was able to increase the virus released in the supernatant of ST-APN cells. Likewise, BHK clones expressing variable amounts of pAPN were shown to acquire TGEV susceptibility and to produce infectious particles as an inverse function of their level of pAPN expression. In contrast, MDCK clones expressing low or large amounts of pAPN failed to produce infectious particles. Taken together, these studies strongly suggest that overexpression of receptor, but also other(s) undetermined factor(s), can impair the production of viral particles.

INTRODUCTION

Aminopeptidase N (APN) has been shown to be the receptor of the porcine transmissible gastroenteritis virus on target cells by the following main criteria: i) antibodies specific of pAPN blocked efficiently the entry of TGEV in permissive cells; ii) a physical interaction between TGE virions or the spike protein S and pAPN has been evidenced, iii) pAPN expression conferred susceptibility to TGEV to non permissive cells and iv) pAPN is cointernalized with virions during the early steps of infection[1,2,3]. These observations support the view that S-pAPN complex formation at the cell surface represents the first critical step in the infection. However, little is known about the importance of the level of APN expression for the entry of virus or subsequent steps of the virus cycle. The swine testis ST cell line, which is highly susceptible to TGEV infection, provide a useful model to analyze how the level of expression of a virus receptor can influence the viral protein biosynthesis and the virus production. ST clones overexpressing pAPN were selected and were analyzed individually for S biosynthesis and capacity to produce virions. To extent the results, pAPN-expressing clones were derived from BHK and MDCK cells, two cell lines refractory to TGEV, and viral progeny was quantified as a fonction of the level of pAPN expression.

MATERIALS AND METHODS.

Virus and cell transfections. The high cell-passage Purdue-115 strain was used as a TGEV source. The cDNA encoding the pAPN was subcloned in the pTEJ4 expression vector[1]. ST, BHK-21 and MDCK cells were cotransfected with this construct and pSV2neo by lipofection or by $CaPO_4$ (three MDCK clones[1]). Cell clones resistant to the neomycin analogue G418 were selected and assayed for APN expression and TGEV susceptibility acquisition (all clones except ST clones which are naturally TGEV susceptible). APN activity was quantified as described[1].

Radiolabeling and zonal centrifugation of viruses. Confluent cell monolayers in 6 wells Costar plates were labeled 4h post-infection for 5-7hrs with 50 mCi of Tran[35]S-label (ICN) in Eagle's MEM supplemented with 5%CS. The medium was clarified and layered onto 20 to 45% (w/w) sucrose/water gradient in SW40 rotor (Beckman). After centrifugation for 4h at 25,000 r.p.m., 0.5 ml aliquots were collected and viral antigens revealed by immunoprecipitation using a mixture of anti-S, -M and -N monoclonal antibodies as described[4].

Pulse-chase labeling. ST and ST-APN cells in 12 wells Costar plates were labeled at 6.5 h p.i. for 15 min with 120 mCi Tran[35]S-label. After a period (0 to 60 min) of incubation in an excess of non-radioactive methionine and cystine, cells were solubilized by adding 0.3 ml of lysis buffer.

RESULTS

Negative Correlation Between the Level of APN Expression and the Virus Production in ST Cell Clones

ST cells were transfected with pAPN-cDNA to obtain two ST-APN clones expressing 10 to 50 fold higher amount of APN than control ST clones. Table 1 shows that ST-APN clones produced approximately 10^4 less virus than ST clones transfected only with the plasmid encoding the antibiotic resistance. To confirm this observation, plaque assays were realized in parallel with ST and ST-APN cells. We observed a five fold increase of the plaquing efficiency in APN-overexpressing cells, probably due to a higher adsorption of virions, but concomitantly a four fold reduction of the diameter of the plaques (not shown). These results suggest that overexpression of APN partially impairs the virus production. To extend this observation, we analyzed virus production by cell radioactive labeling and rate zonal centrifugation of virus particles released in the supernatant (Fig.1). Virus particles were consistently found in lower amount in ST-APN cells supernatant compared to that in ST cells. As quantified by autoradiography scanning (and in these experimental conditions), ST-APN clones produced 10 to 50 less virus than in ST cells. The relative proportion of the three main structural proteins S, M and N did not appear to be modified in virions produced in ST-APN cells.

Table 1. TGEV infectious virus production in
ST and ST-APN cell clones

Cell line	Virus production[a] (PFU/ml)
ST cells clone 2	1.3×10^8
ST cells clone 3	3.9×10^8
ST cells clone 7	4.9×10^8
ST APN cells clone 2	$<5.0 \times 10^4$
SR APN cells clone 11	$<5.0 \times 10^4$

[a]Total infectivity titer was determined 20 h post
infection by plaque assay on ST cells

S Maturation in ST-APN Cells

A possible explanation for the down-production of virus in ST-APN cells is that intracellular APN affects the biosynthesis of the S protein. The synthesis of viral antigens was analyzed by cell radioactive labeling and immunoprecipitation. Similar amounts of neosynthesized structural proteins were observed in ST and ST-APN cells (not shown). Moreover, the rate of S conversion from the 175K band (Endo H sensitive form of S) into the 220K band (Endo H resistant)[5] was not modified in the ST-APN cells (Fig. 2)

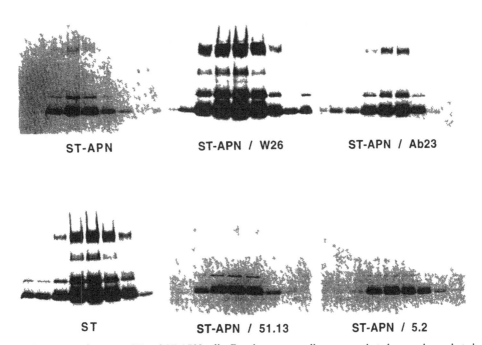

Figure 1. Virus production in ST and ST APN cells. Five hours p.i. cells were incubated or mock incubated with the indicated antibodies W26 Ab23 (anti APN) or 51 13 5 2 (anti S). Nine hours p.i. cell supernatants were fractionated by rate zonal centrifugation. Even fractions were immunoprecipitated with anti S N and M antibodies. Viral polypeptides were visualized by 8 to 15% polyacrylamide gradient SDS-PAGE and autoradiography

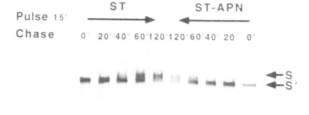

Figure 2. TGEV S protein maturation in ST and ST-APN cells. Cells were radiolabeled for 15 min, then chased 0 to 120 min. Immunoprecipitates with antibody 51.13 were resolved on an autoradiographed 8% SDS-PAGE.

An Anti-APN Antibody Restores Virus Production in ST-APN Cells

To determine if virus down-production in ST-APN cells could involve a late event in the virus maturation, we incubated infected cells with an anti-receptor antibody four hours p.i. The anti-APN antibody bound to APN is expected to be endocytosed and to reach intracellular vacuoles[6]. Interestingly, this resulted in an increased virus amount up to a level comparable to that in normal ST cells (Fig. 1). This effect was observed at concentrations of antibody as low as 1.25 µg/ml (not shown). Two anti-spike antibodies were used as controls, with one of them (51.13) having a strong neutralizing activity[4]. None of these antibodies was found to increase the amount of virus present in the cell supernatant.

Level of APN Expression and Virus Production in pAPN-BHK Cell Clones

Seven clones derived from the BHK-21 cells and expressing variable amount of recombinant pAPN were selected. All of them were susceptible to TGEV as shown by measurement of the cytopathic effect induced by the infection. Fig. 3 shows that four

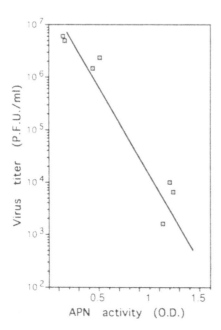

Figure 3. TGEV infectious virus production in seven BHK pAPN-expressing clones. Total infectivity titer was determined 20h p.i. by plaque assay on ST cells. APN activity was quantified by cleavage of the chromogenic substrate leucine *p*-nitroanilide.

Table 2. TGEV infectious virus production in MDCK cell clones

Cell line	APN activity (uU/ml)	Virus production[a] (PFU/ml)
MDCK-APN cl 1	158 3	5
MDCK-APN cl 2	162 5	160
MDCK-APN cl 3	128 2	35
MDCK-APN cl 12	3 5	$1\,6 \times 10^3$
MDCK-APN cl 21	3 7	$5\,9 \times 10^2$
MDCK	3 2	$4\,6 \times 10^3$
ST	2 0	5×10^7

[a]Total infectivity titer was determined 20 h post infection by plaque assay on ST cells

BHK-derived clones expressing low amount of pAPN produced infectious virions in the same range as the highly permissive swine testis cell line ST (1×10^6 to 6×10^6 p f u /ml) In contrast, the three BHK clones expressing high amount of pAPN produced few amount of particles Thus, like in ST cells, we observed an inverse correlation between the capacity of the clones to produce infectious virus and their level of pAPN expression

MDCK Cells Are Refractory to TGEV Virion Production

In a earlier study, we showed that three MDCK derived clones expressing large quantities of pAPN failed to produce virus[1] To see whether it was due to the level of APN expression, two MDCK clones transfected with pAPN-cDNA were selected on both their acquired susceptibility to TGEV by measurement of the c p e induced by the infection and on an APN activity of the same range than the parental MDCK cell line Titration of infectious viral progeny produced in the five MDCK clones shown that all of them failed to produce infectious particles compared to ST cells (Table 2) The low titer observed with the MDCK clones overexpressing pAPN was probably due to a better endocytosis and uncoating of the infectious virus, thus lowering the titer of the residual input virus This observation indicates that, in contrast to that found in ST and BHK cells, the impaired virus production is not related to an overexpression of APN

DISCUSSION

In the present study, we investigated the correlation between the level of receptor expression and the production of TGE virions in different cell systems As a main result, we observed a strong negative correlation between the level of pAPN receptor expression and the ability of the cell to produce virus particles in two different cell lines the swine testis cell line ST, naturally susceptible to TGEV and the baby hamster kidney cell line BHK, susceptible after pAPN-cDNA transfection A 10,000-fold reduction of the virus production was observed in ST clones overexpressing APN compared to that in non-transfected clones Similar observations were made with BHK clones It was also observed that the virus plaque formed in ST-APN cells were reduced in size, a finding which corroborates the decreased infectious virus production Concomitantly, the efficiency of plaquing was found to be significantly enhanced in these cells, thus indicating that the amount of receptor on the cell surface is a limiting parameter in the viral adsorption A series of experiments were carried out in an attempt to identify the step of the virus cycle potentially affected in APN-overexpressing cells. First, we tried to detect an intracellular interaction between the S spike and pAPN in pAPN-overexpressing clones without obtaining clear evidence for this (not shown). Moreover, pulse chase experiments showed that S transport (conversion from a endoH sensitive to a endoH resistant form) was identical in ST-APN and in ST cells (half time, 40mn). This suggests that, if there is a binding between APN and S inside the endoplasmic reticulum or the Golgi apparatus, it does not interfere with the spike maturation. Another possibility was that APN bound to S later in the virus cycle, during the budding or the release of the virus from the cellular membranes. The fact that the addition of an anti-pAPN antibody is able to restore a virion production nearly identical to that obtained with ST cells showed that the defectiveness of ST-APN cells to produce virion involved a late event in the virus replication cycle, such as a APN-S binding in virus secretion vesicles. To explore this hypothesis, we plan to compare the virus maturation in ST and ST-APN cells by electron microscopy.

The above observations raise the question of how TGEV can replicate and propagate efficiently in the natural target cells, the enterocytes[7], which are known to express APN at a level equivalent, if not higher, than the overexpressing cell lines here studied. An attractive explanation is that a productive infection in vivo involves a virus-driven down-regulation of the pool of receptor molecules, as recently evidenced for other viruses[8,9].

A third cell line, MDCK, was investigated in the present study. At difference with the ST and the BHK cells, MDCK cell clones expressing APN at a low level were still deficient for infectious virus production. As a confirmation, no plaque formation could be obtained in APN-MDCK cells. Nevertheless, as previously reported, infection of the cells at a high m.o.i. led to a strong cytopathic effect as well as to viral antigen synthesis[1]. Thus, the virus production is unlikely to involve an early step of the virus cycle. Such an abortive replication clearly differs of the situation reported for human APN-expressing MDCK clones infected with HCV-229E[10] or for MDCK cells transfected with feline calicivirus genomic RNA[11], where no viral synthesis could be detected. Elucidating which step of TGEV replication is severely impaired in pAPN-expressing MDCK cells might contribute to a better understanding of the coronavirus assembly or budding processes.

REFERENCES

1 Delmas, B , Gelfi, J , L'Haridon, R , Vogel, L K , Sjostrom, H , Noren, O , Laude, H Aminopeptidase N is a major receptor for the enteropathogenic coronavirus TGEV Nature 1992,357 417-420

2 Godet, M , Grosclaude, J , Delmas, B , Laude, H Major receptor-binding and neutralization determinants are located within the same domain of the coronavirus transmissible gastroenteritis virus spike protein J Virol 1994,68 (in press)

3 Hansen, G , Vogel, L K , Delmas, B , Gelfi, J , Laude, H , Noren, O , Sjostrom, H The coronavirus TGEV causes infection after endocytosis by an aminopeptidase N recruiting mechanism (manuscript in preparation)

4 Laude, H , Chapsal, J -M , Gelfi, J , Labiau, S , Grosclaude, J Antigenic structure of transmissible gastroenteritis virus I Properties of monoclonal antibodies directed against virion proteins 1986,67 119-130

5 Delmas, B , Laude, H Assembly of coronavirus spike protein into trimers and its role in epitope expression J Virol 1990,64 5367-5375

6 Louvard, D Apical membrane aminopeptidase appears at site of cell-cell contact in cultured kidney epithelial cells Proc Natl Acad Sci USA 1980,77 4132-4136

7 Pensaert, M , Callebaut, P , Cox, E Enteric coronaviruses of animals In Kapikian A Z , (ed) Viral infections of the gastrointestinal tract 2nd ed Marcel Dekker, Inc , New York 1993 pp 627-696

8 Bour, S , Geleziunas, R , Wainberg, M A The role of CD4 and its downmodulation in establishment and maintenance of HIV-1 infection Immunological Reviews 1994,140 147-171

9 Naniche, D , Wild, T F , Rabourdin-Combe, C , Gerlier, D Measles virus haemagglutinin induces down-regulation of gp57/67, a molecule involved in virus binding J Gen Virol 1993,74 1073-1079

10 Delmas, B , Gelfi, J , Kut, E , Sjostrom, H , Noren, O , Laude, H Determinants essential for the transmissible gastroententis virus-receptor interaction reside within a domain of aminopeptidase-N that is distinct from the enzymatic site J Virol 1994,68 5216-5224

11 Kreutz, L C , Seal, B S , Mengeling, W L Early interaction of feline calicivirus with cells in culture Arch Virol 1994,136 19-34

Characterization of Viral Replicase

63

IDENTIFICATION OF 120 kD AND 30 kD RECEPTORS FOR HUMAN CORONAVIRUS OC43 IN CELL MEMBRANE PREPARATIONS FROM NEWBORN MOUSE BRAIN

Arlene R. Collins

Department of Microbiology
State University of New York At Buffalo
Buffalo, New York

ABSTRACT

A biotinylated virus overlay was used to identify a 120kD virus-binding molecule in dissociated newborn mouse brain (nmb) cell suspensions after separation of the proteins by polyacrylamide gel electrophoresis, electroblotting, and blockage of non-specific binding sites. The virus-binding molecule was not detected in adult mouse brain cell suspensions. Mannose- and glucose-rich glycoproteins from nmb cell membranes were selected by ConA-Sepharose (Pharmacia) chromatography. A 30kD virus-binding molecule was eluted by 0.2 M alpha-methyl-D-mannoside. O-linked sialic acid, a receptor component, was identified in the eluate.

INTRODUCTION

OC43 is a human respiratory virus in the *Coronaviridae* family which has been adapted to grow in suckling mice and cause a lethal neurotropic infection. Infected mice display a selective tropism toward neurons[1]. Astrocytes also are susceptible to infection in vitro but do not produce infectious virus[2]. Cultured human embryo brain cells can be infected as well as the human astroglioma cell line, U87-MG[3]. A study of the molecular basis of murine neuronal cell susceptibility was initiated to find a tissue specific receptor for the virus.

MATERIALS AND METHODS

Whole brain was removed from 203 litters of euthanized, 4, 7, and 10 day old C57BL/6K mice. The tissue was dissociated by passage through a 202 µm pore size nylon mesh, washed 5 times in 0.1 M phosphate buffered saline, pH 7.2 (PBS), and lysed in lysis buffer (20 mM Tris-HCL, pH 8.0, 1 nM Na-EDTA, 150 mM NaCl, 1 mM phenylmethylsulfonylfluoride, 1% nonidet p-40, 0.02% NaAzide). Insoluble material was removed by centrifugation at 100,000 x g for 1 h. Cell lysates were stored at -70°C.

Human coronavirus OC43, obtained from G. Gerna, was propagated and assayed by plaque formation in MRC-5 cells[4]. For biotinylation, supernatant virus was harvested from RD cells, and concentrated 100-fold by ultracentrifugation and mixed with an equal volume of 0.1 mg/ml NHS-Biotin (Sigma) in PBS for 10 min at 0°C, then dialyzed overnight against 100 volumes of cold PBS, changing the buffer once. Biotinylated virus was titrated for infectivity and stored at -70°C.

Cell lysates (3-15 µm volumes) were separated on 10% SDS-Page cells and electroblotted to Immobilon (Millipore). The substrate was blocked with 5% bovine serum albumin (BSA) in 0.01 M Tris-HCL, pH 7.5, 0.15 M NaCl (TBS) for 30 min at 37°C and 0.1 M alpha-methyl-D-mannoside in PBS for 20 min at 26°C to reduce nonspecific binding of avidin to lectins. To detect virus binding proteins, the blot was probed with biotinylated OC43 virus (512 hemagglutinating units) in PBS for 1 h at 26°C and washed three times in TBS-0.1% BSA. Bound virus was visualized by incubation with avidin conjugated to horseradish peroxidase (HRP) at 1:1000 dilution in TBS-0.1% BSA and developed in TBS containing 0.5 mg/ml 4-chloro-1-naphthol and 0.03% H_2O_2. As controls for specificity of the reaction, the substrate was probed with avidin HRP alone, or unbiotinylated virus, rabbit anti-OC43 antibody (1:2000), and goat anti-rabbit IgG conjugated to HRP (1:1000), or rabbit anti-carcinoembryonic antigen (DAKO, 1:200) and the secondary antibody. Probing with antibody gave a high background.

Mouse brain cell lysate was dialyzed overnight against 200 volumes of PBS. The dialysate was mixed with an equal volume of ConA-Sepharose (25%) in binding buffer (20mM Tris-HCl, pH 7.4, p.5 M NaCl) and incubated for 1.5 h at 4°C with gentle rocking. After centrifugation at 300 x g for 1 min, the supernatant was saved and the procedure was repeated once. To elute bound glycoproteins, the ConA-Sepharose was resuspended in two volumes of PBS containing 0.2 M α-methyl-D-mannoside, and left overnight at 4°C.

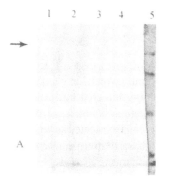 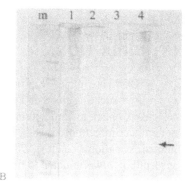

Figure 1. Identification of the OC43 receptor in mouse brain cell lysates by the solid phase virus binding assay using biotinylated virus (A) Lanes 1-4 seven-day-old mouse brain cell lysate at 15, 10, 6, and 3 ml volumes. Lane 5 molecular weight markers, from the top 97, 66, 45, and 33 kD bands respectively (B) Lane 1 molecular weight markers 116, 97, 66, 45, and 33, 21 kD Lane 2 four-day-old mouse brain lysate after dialysis against PBS, 10 ml Lane 3 and 4 First and second supernatant from ConA absorption, 10 ml Lane 5 Eluate from ConA elution with 0 2 M B a-methyl-D-mannoside, 10 ml Arrow indicates the 30kD receptor.

Table 1. Inhibition of viral esterase cleavage of
pNPA by neuronal membrane glycoproteins

Sample[a]	Rate[b] at OD405
conc OC43	0.4005
conc OC43 + eluted fraction	0.0337

[a] 20 μl of each component in 0.5 ml PBS containing 100 μg
pNPA

[b] units per min; 10 min total time O-sialic acid content was
measured by reduction in absorbance at OD405 due to
sialated glycoproteins in competition with
para-nitrophenylacetate (pNPA) for viral esterase[6]

RESULTS AND DISCUSSION

Since OC43 virus selectivity infects neurons, dissociated whole brain cell membrane lysate from 7-day-old mice was probed for biotinylated virus binding in the solid phase assay. Shown by Figure 1A, a single receptor was visible at molecular weight 120kD. When the substrate was probed with antibody to carcinoembryonic antigen, the 120kD receptor detected by unbiotinylated virus, appeared to be unrelated. This unique receptor may explain the selective tropism of the virus for neural cells. Further work must be done using embryonic or immature human neurons for comparison.

Since ConA adsorbs HIV-1 envelope glycoproteins which subsequently are able to bind antibodies to all of the conformational epitopes of gp120[5], mouse brain glycoproteins from 4-day-old mice, released after absorption to ConA-Sepharose, were probed with biotinylated virus in the solid phase assay. Figure 1B shows that the eluted fraction contained a 30kD receptor molecule. The eluate contained glycoproteins selected during adsorption in 0.5 M NaCl binding buffer. The eluted glycoproteins may be in a conformation which is more easily recognized by the virus and should be useful for molecular and immunological analysis. The dialysate prior to adsorption no longer displayed the single 120kD receptor but contained multiple bands of reactivity. This was probably due to dissociation of the receptor following removal of the detergent. Dialysis replaced Mn2+ and Ca2+ which are essential for ConA binding. No specific virus binding activity was detected in the supernatant binding buffer after ConA adsorption.

The eluate was further tested for O-sialic acid content in a substrate completion assay for the viral esterase. In the presence of eluate fraction, the rate of esterase cleavage of pNPA was reduced >90% indicating that a competing substrate was being cleaved.

The assistance of Dr. Roger K. Cunningham in collecting newborn mice and of Jeanette McGuire in preparing the manuscript is gratefully acknowledged.

REFERENCES

1. Pearson, J., Mims, C.A. Selective vulnerability of neural cells and age-related susceptibility to OC43 virus in mice. Archiv Virol 1983;77:109-118
2. Pearson, J., Mims, C A. Differential susceptibility of cultured neural cells to the human coronavirus OC43 J Virol 1985,53:1016-1019
3. Collins, A.R., Sorensen, O. Regulation of viral persistence in human glioblastoma and rhabdomyosarcoma cells infected with coronavirus OC43. Microb Path 1986;1:573-583

4 Collins, A R HLA class I antigen serves as a receptor for human coronavirus OC43 Immunol Invest 1993,22 95-103

5 Robinson, J E , Holton, S , Liu, J , McMurdo, H , Muciano, A , Gohd, R A novel enzyme-linked immunosorbent assay (ELISA) for the detection o antibodies to HIV-1 envelope glycoproteins based on immobilization of viral glycoproteins in microtiter wells coated with Con-A Immunol Meth 1990,132 63-71

6 Pfleiderer, M , Routledge, E , Siddell, S G Functional analysis of the coronavirus MHV-JHM surface glycoproteins in vaccinia virus recombinants Adv Virus Res 1990,276 21-31

64

INHIBITION OF CORONAVIRUS MHV-A59 REPLICATION BY PROTEINASE INHIBITORS

Mark R. Denison[1,2,3], James C. Kim,[1] and Theodore Ross[2]

[1] Department of Pediatrics
[2] Department of Microbiology and Immunology
[3] The Elizabeth B Lamb center for Pediatric Research
Vanderbilt University Medical School
Nashville, Tennessee 37232-2581

INTRODUCTION

Replication of mouse hepatitis virus is initiated by translation and processing of the gene 1 polyprotein. The increasing complexity of this pattern of replicase protein expression is becoming apparent. MHV-A59 encodes approximately 800 kDa of polypeptide within the two overlapping open reading frames of gene 1. In addition, at least one proteinase activity encoded by gene 1 has been identified[1,2] and two more are predicted, along with putative polymerase, helicase, NTP binding and possibly growth factor like proteins[3-5]. Only the first 380 kDa of protein products of MHV gene 1 ORF 1a have been described and characterized as to the pattern of expression and processing in virus infected cells[6]. Specifically, proteins of N-p28-65-290-C constitute the initial translation products of this region. The cleavage between p28 and p65 has been best described, both in vitro and in virus-infected cells[7]. As of yet no specific functions have been ascribed to any of these proteins. Thus a great deal remains to be learned about the processing pattern and functions of MHV gene 1 encoded proteins.

Based on sequence analysis and demonstrated activity of an ORF 1a encoded proteinase, it is assumed that most or all of the processing of the gene 1 polyprotein is effected by viral proteinases. Thus, one of the tools which has been extensively used in the investigation of gene 1 expression is the addition of proteinase inhibitors to in vitro or intracellular translation reactions, in order to determine the effect on polyprotein expression and to determine the specificity of proteolytic events involved in maturation of gene 1 products. The processing of the N-terminal protein p28 was defined based on the inhibition of appearance in the presence of leupeptin or $ZnCl_2$[8]. Subsequently, the cleavage of the p290 protein into p50 and p240 in cells was also defined in part by to its inhibition in the presence of leupeptin[7].

Leupeptin is an amino acid aldehyde (N-acetyl-leu-leu-arg-aldehyde), which inhibits most cysteine and trypsin-like serine proteinases[9]. Leupeptin acts by forming a stable, noncleavable tetrahedral intermediate in the substrate binding region of the proteinase. However leupeptin is a reversible inhibitor, and is also degradable, so we have sought more stable, irreversible inhibitors of specific serine and cysteine proteinase inhibitors to allow more careful dissection of the proteolytic pathway of the gene 1 polyprotein. One such agent which will be described in this paper is e64 (L-trans epoxysuccinyl-leucylamido(4-guanidino)butane), a specific inhibitor of cysteine (thiol) proteinases such as calpain, cathepsin B and papain[10]. E64 irreversibly binds the thiol group of cysteine catalytic residues, thereby inactivating the proteinase molecule. Thus e64 is considered an excellent active site titrant for studies of cysteine proteinases. We have utilized e64d, the uncharged diethyl ester derivative of e64, because of its excellent cell penetrability. This agent has previously been used to describe inhibition of L proteinase of the picornavirus, foot and mouth disease virus (FMDV)[11]. Leupeptin and e64d both can penetrate cells, and both are very non toxic to cell monolayers and to mice, and thus make excellent agents to study effects on virus protein processing in virus infected cells or intact animals.

During studies of polyprotein processing in MHV-A59 infected DBT cells[12], we observed that addition of either leupeptin or e64d to media at early times of infection resulted in significantly decreased viral syncytia formation. If the inhibitor is added early enough, the yield of viral nonstructural proteins at 8-10 hours is markedly decreased (Denison, unpublished data). The present studies are predicated on that observation and the hypothesis that the presence of proteinase inhibitors diminishes virus replication by acting at the level of inhibition of gene 1 polyprotein processing, an essential step in virus replication. We here show that both leupeptin and e64d inhibit MHV-A59 virus replication in DBT cells. Leupeptin and e64d both result in diminished viral RNA synthesis and infectious virus, even when added at relatively late times post infection. Thus polyprotein processing appears to be required throughout infection in order for ongoing RNA synthesis to occur.

MATERIALS AND METHODS

Virus stocks, cells, and proteinase inhibitors. Stocks of MHV-A59 were plaque purified times three and passaged at low MOI prior to use. DBT cells were used for virus growth, isolation, infected cell lysates and plaque assay[12]. Cells were maintained in DMEM 10% fetal calf serum (FCS) and infections were performed in DMEM 2% FCS. Leupeptin was obtained from Sigma, and was dissolved in water at stock concentrations of 200mM and used at final concentrations of 2mM. E64d was originally obtained from Phil Sonnett at the USDA. Subsequently e64d was obtained from Matreya Inc. E64d was dissolved in DMSO at a stock concentration of 100mg/ml and used at concentrations of 500 µg/ml in DMEM 2% FCS.

Plaque assays. Plaque assays were performed on confluent monolayers of DBT cells in six-well plates. Cells were infected for 30 minutes at 37°C with virus from media supernatants diluted in gel saline, followed by media/agar overlay for 24-30 hours and subsequent addition of agar containing 0.02% neutral red for 6-12 hours with counting of clear plaques. All plaque assays were performed in duplicate and the experiments were repeated twice.

Determination of infectious virus and cytopathic effect. Confluent monolayers of DBT cells in 60mm petri dishes were infected with MHV-A59 at an MOI of 20 PFU/cell in DMEM 2% FCS for 30 min. at 37° C. Plates were rinsed with Tris-saline pH 7.4 three times prior to addition of overlay media. For assays of infectious virus, monolayers were overlayed with 2ml of media and 100µl aliquots were removed at the time points indicated in the individual experiments. For assay of proteinase inhibitors, leupeptin (2mM) or e64d (500µg/ml) was added to the media 1 hour p.i. Monolayers were carefully observed by two independent observers who were blinded to each other, and estimates were made of the percent involvement of each monolayer by virus-induced syncytiae prior to determination of supernatant virus titers.

Determination of viral RNA synthesis. Infection was initiated exactly as described above. Actinomycin D was added to the media at a final concentration of 10μg/ml at 4 hours p.i. when determinations of RNA synthesis were to be made at 8-9 hrs p.i. [^{3}H] uridine (NEN) was added at a final concentration of 200μCi/ml at times indicated in the presence of leupeptin (2mM) or e64d (500μg/ml) added as shown in individual experiments. At the end of the labeling period, plates were washed on ice with PBS times three, lysed in hypotonic solution containing 1% SDS, and lysates were passed through a 23 gauge needle. Aliquots were taken for determination of TCA insoluble radioactivity, and were always performed in triplicate.

RESULTS AND DISCUSSION

Leupeptin Inhibition of Virus Titer and RNA Synthesis

We first investigated the effect of leupeptin on MHV-A59 virus replication in single cycle growth experiments in DBT cells by adding leupeptin at 1 hour p.i. to the overlying media, with or without supplementation every three hours throughout the infectious cycle (Figure 1).

Visually, we observed a delay in the appearance of viral syncytia in the leupeptin-treated cells. When virus titer was determined at the various time points, the eclipse phase of virus replication was markedly affected at all concentrations of leupeptin, with a delay of 4-6 hours, compared with the control infection, before increases in supernatant virus titers were observed. Virus titers in leupeptin-treated cells were diminished at all times points compared with controls, with the most marked reduction measured in those cells treated with 2mM leupeptin and supplemented every 3 hours, where 2.8 log reductions in virus titer were noted at 12 hours post infection.

We next investigated the importance of time of addition of leupeptin on replication inhibition. In the single-cycle growth experiments, leupeptin was added 1 hour after infection, and the results were therefore consistent with inhibition of an event occurring after binding and entry (Figure 2). Therefore we sought to determine whether leupeptin resulted in inhibition of viral RNA synthesis as a marker of inhibition of maturation of proteins expressed from gene 1. DBT cells were infected with MHV-A59 and 2mM leupeptin was added at various times from -1hr to 8 hrs after infection. In parallel experiments we measured either virus titer at 8 hours after infection or uptake of [^{3}H]uridine into actinomycin D resistant RNA at 8 to 9 hours after infection. When leupeptin was added at or before 2 hrs p.i., there was a greater than 90% reduction in supernatant virus titer measured at 9 hrs pi. The titer was substantially decreased when leupeptin was added as late as 7 hrs p.i., suggesting interference with an ongoing intracellular process, such as protein translation, processing or RNA synthesis.

DBT cell morphology, cell replication, and protein and RNA synthesis were unaffected by the presence of up to 10mM leupeptin, indicating that the effect on MHV-A59 replication was not due to cellular cytotoxicity(data not shown). There did not appear to be any direct antiviral effect of leupeptin on the virus, since incubation of stock virus with leupeptin did not reduce measurable virus titer. Leupeptin also did not interfere with maturation cleavage of the spike (S) surface glycoprotein. No increase in infectious virus particles was observed when MHV grown in the presence of leupeptin was incubated with trypsin at 37°C for up to two hours. These data strongly suggest that early proteolytic events,

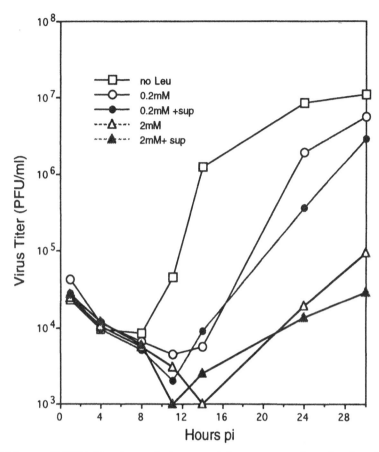

Figure 1. Inhibition of MHV-A59 replication by leupeptin. Infections were performed as described in Materials and Methods Monolayers of DBT cells were infected with MHV-A59 at an MOI of 10 PFU/ml Leupeptin was initially added at 1 hr p i to all plates except the control (no leu). For those plates which were supplemented, leupeptin was added at 3 hour intervals directly to the media. Samples were obtained at the times indicated and titer was determined by plaque assay on DBT cell monolayers

possibly those important for ORF 1a processing, may be the site of leupeptin-mediated inhibition of MHV replication.

The results of the parallel experiment, in which viral RNA was labeled with [³H] uridine in the presence of actinomycin D, are shown in Figure 2. The results are surprisingly similar to those found when measuring infectious virus titer. When leupeptin was added before 2 hrs pi, viral RNA synthesis at late times was at or below the level of mock-infected control cells in the presence of actinomycin, or essentially absent. Also similar to the virus titer experiment, addition of leupeptin up to 7 hrs p.i. resulted in a significant reduction in viral RNA synthesis at 8 to 9 hours. Together, the results of these experiments indicate that an ongoing intracelluar viral replication event is inhibited by leupeptin, and that the inhibition manifests as diminished RNA synthesis with concomitant reduction in virus titer.

E64d Inhibition of MHV-A59 RNA Synthesis and Replication

Because leupeptin is a reversible inhibitor, and because of its broad specificity for both cysteine and serine proteinases, it is not possible to state whether the need for supplementation and the decreasing inhibition when added at late times of infection is due

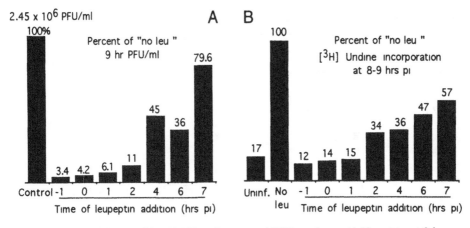

Figure 2. Leupeptin inhibition of MHV-A59 replication and RNA synthesis A) Virus titer at 9 hours p i Leupeptin was added at the times p i indicated, except for the control, to which no leupeptin was added. Supernatant virus was harvested at 9 hours post infection and titered on DBT cells B) Incorporation of [³H]uridine into viral RNA Leupeptin was added at the times p i indicated Actinomycin D was added to virus infected monolayers at 4 hrs p i and [³H]uridine was added from 8-9 hrs p i, followed by cell lysis and measurement of TCA insoluble radioactivity "Uninf" indicates uninfected cells in the presence of actinomycin D and "no leu" indicates infected cells in the presence of actinomycin D but in the absence of leupeptin

to reversibility of inhibition or synthesis of new proteinase molecules. Thus we investigated the inhibition of replication with e64d, the specific, irreversible cysteine proteinase inhibitor. This was potentially a good agent because all of the known or predicted gene 1-encoded proteinases are thought to possess cysteine catalytic residues. In addition, its irreversible inhibition is such that binding to a molecule of proteinase eliminates that molecule. Therefore any late resumption of replication can be attributed the activity of newly available proteinase. Finally, it has been shown that e64d inhibits at least one cleavage in the gene 1 polyprotein, at the carboxy terminus of p65[13]. We first repeated the single cycle growth experiments with 500μg/ml of e64d during MHV-A59 infection of DBT cells, and visually estimated percent involvement in syncytiae prior to titering supernatant virus at various time points from 0 to 24 hours (Figure 3).

E64d added at 1 hour p.i. completely blocked virus replication as measured both by titer and visible cytopathic effect when compared to control. At 10 hours post infection, there were no syncytia visible in the e64d treated plates, and virus titer was diminished greater than 6 logs compared to the control. E64d was added at only one time, 1 hr p.i.; thus it was interesting that replication as measured by both titer and syncytia estimates recommenced after a 12-16 hour lag. This indicated both that the cell monolayer was not adversely affected by the e64d and that the inhibitor was exhausted, possibly by a combination of degradation and binding to proteinase. These results were entirely consonant with those observed with leupeptin, but the inhibition was much more complete.

We next examined the effect of e64d on viral RNA synthesis, in order to assess the degree of inhibition and to determine the role of time of addition (Figure 4). Similar to leupeptin, e64d had no effect on cellular RNA synthesis in the absence of actinomycin d at 8 hours after addition, which was consistent with experience with other cell types[11].

In contrast, the addition of e64d (500μg/ml) at any time from -1 to 7 hours p.i. resulted in complete ablation of viral RNA synthesis as measured by actinomycin D resistant [³H] uridine incorporation at 8-9 hrs pi. These results show that e64d can inhibit new RNA synthesis for more than 9 hours when added at the beginning of infection, and also demonstrate that the half-life of that inhibition is short, likely less than 1 hour. This was

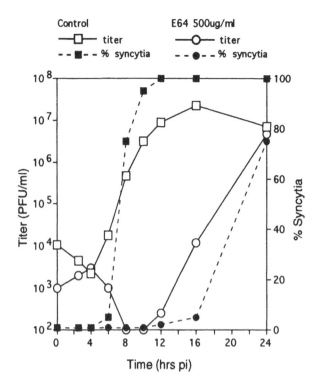

Figure 3. Inhibition of MHV-A59 virus replication by e64d Single cycle growth of MHV-A59 in DBT cells was determined The monolayer was infected at an MOI of 20 in absence proteinase inhibitor (Control), or e64d (500µg/ml in DMSO) The percent involvement of the monolayer in viral syncytia was determined as described in the Materials and Methods Aliquots of overlying media were taken at the time points indicated and titer determined. The log titer is to the left of the figure and the %syncytia to the right

consistent with the results obtained with leupeptin, but was more rapid acting and complete when added at late times of infection.

Requirement for Continuous Proteolytic Processing in MHV Replication

Both leupeptin and e64d have been used to inhibit processing of the gene 1 polyprotein. We here demonstrate that the addition of either of these agents to virus infected cells

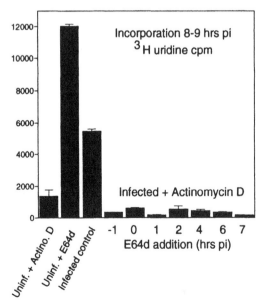

Figure 4. Time of addition of e64d and inhibition of RNA synthesis 60mm plates of confluent DBT cells were infected at an MOI of 20 Actinomycin D (10µg/ml) was added to all infected cells and one uninfected plate at 4 hrs p i E64d was added to the media of infected cells at the times indicated from -1 hr to 7 hrs p i, it was added to the uninfected cell control at 1 hr after mock infection No e64d was added to the infected control. [³H]uridine was added at 200µCi/ml at 8 hrs p.i. and cells were lysed at 9 hrs p i after a one hour labeling period TCA insoluble incorporation into RNA was determined

inhibits new viral RNA synthesis, and therefore output virus titer The effect on titer more impressive at early times of infection, whereas even at late times of addition, RNA synthesis measured shortly after addition is already diminished E64d inhibited virus replication to a greater extent than leupeptin (5 0 vs 2 8 log reduction at 8-10 hrs), and also demonstrated a longer duration of inhibition (up to 16 hrs) E64d also more completely inhibited new viral RNA synthesis (>90%) at late times of infection (7 hrs p i) than leupeptin

Together the results obtained with leupeptin and e64d suggest a model in which continued protein and expression and processing, is required for viral RNA synthesis to occur, and that interference with processing allows rapid depletion of the proteins involved in the RNA synthesis activities It has been previously demonstrated than ongoing protein synthesis is necessary for coronavirus RNA synthesis[14], and our data show that processing is required as well Early inhibition of proteolysis may result in markedly diminished RNA templates for transcription, protein synthesis, and packaging into progeny virions Our results further suggest that some of the proteins involved in RNA synthesis, whether polymerases directly or proteinases more indirectly, have a short half-life of activity during replication, rather than being continuously available throughout infection once translated and processed As the entire scheme of gene 1 polyprotein processing becomes apparent, these proteinase inhibitors will allow us to assess specifically which proteolytic events are required for ongoing RNA synthesis

ACKNOWLEDGMENTS

The help of Jim Gombold, Paul Currier and Robert Spence are greatly appreciated. This work was supported by Public Health Service grant AI-26603 from the NIAID of the NIH

REFERENCES

1 S C Baker, C -K Shieh, L H Soe, M -F Chang, D M Vannier and M M C Lai (1989) J Virol, 63 3693-3699

2 S C Baker Yokomori, K , Dong, S , Carlisle, R , Gorbalenya, A E , Koonin E V, and Lai, M M C (1993) Journal of Virology, 67 6056-6063

3 M F G Boursnell, T D K Brown, I J Foulds, P F Green, F M Tomley and M M Binns (1987) J Gen Virol , 68 57-77

4 P J Breedenbeek, C J Pachuk, A F H Noten, J Charite, W Luytjes, S R Weiss and W J M Spaan (1990) Nucleic Acids Res , 18 1825-1832

5 H-J Lee, C -K Shieh, A E Gorbalenya, E V Koonin, N LaMonica, J Tuler, A Bagdzhadhzyan and M M C Lai (1991) Virology, 180 567-582

6 M R Denison, S A Hughes and S R Weiss (1994) Virology (submitted),

7 M R Denison, P W Zoltick, S A Hughes, B Giangreco, A L Olson, S Perlman, J L Leibowitz and S R Weiss (1992) Virology, 189 274-284

8 M R Denison and S Perlman (1986) J Virol , 60 12-18

9 H Umezawa (1982) Ann Rev Microbiol , 36 75-99

10 S Mehdi (1991) TIBS, 16 150 153

11 L G Kleina and M J Grubman (1992) J Virol, 66 7168-7175

12 N Hirano, K Fujiwara and M Matumoto (1976) Japan J Micro, 20 219-225

13 J C Kim, R A Spence, P F Currier, X Lu and M R Denison (1994) (Submitted)

14 D L Sawicki and S G Sawicki (1986) J Virol , 57 328-334

65

INHIBITION OF VIRAL MULTIPLICATION IN ACUTE AND CHRONIC STAGES OF INFECTION BY RIBOZYMES TARGETED AGAINST THE POLYMERASE GENE OF MOUSE HEPATITIS VIRUS

Akihiko Maeda, Tetsuya Mizutani, Masanobu Hayashi, Kozue Ishida, Tomomasa Watanabe, and Shigeo Namioka

Department of Laboratory Animal Science
Faculty of Veterinary Medicine
Hokkaido University
Sapporo 060, Japan.

ABSTRACT

Two hammerhead ribozymes targeted against the polymerase gene of mouse hepatitis virus (MHV), which consisted of 22-nucleotide (nt) ribozyme core sequences and antisense sequences of different lengths, 243-nt (S-ribozyme) and 926-nt (L-ribozyme), were tested for their inhibitory effects on viral multiplication. Vectors that expressed the ribozymes were transfected into mouse DBT cells and several resulting cell lines constitutively expressing the ribozymes were selected and examined for intracellular MHV multiplication in acute and chronic stages of infection. The production of infectious progeny viral particles was significantly reduced in the transfected cell lines expressing either the S-ribozyme or L-ribozyme in acute infection. Although the in vitro cleavage process of the L-ribozyme was slower than that of the S-ribozyme, no difference was observed in inhibitory effects on MHV multiplication between S- and L-ribozymes in the transfected cells. In the transfected cells expressing L-ribozymes, production of viral particles was also inhibited in the chronic stage of MHV infection.

INTRODUCTION

Ribozymes, antisense agents, are catalytic RNAs that are able to cleave the phosphodiester bonds of target RNAs in a sequence-specific manner[1]. The hammerhead ribozymes contain two separable functional regions: a catalytic core region containing several conserved bases, which cleaves the target RNA, and flanking regions that direct the ribozyme core to the specific target site by nucleic acid complementarity[2,3]. By attaching the core regions to sequences complementary to those flanking the selected target site, NUX (where N is any base and X is U, C or A), ribozymes can be designed to specifically cleave any target RNA molecules. Although ribozymes are worthy of evaluation as potential antiviral agents, they have been shown to exert an effect on only a few animal viruses[4,5,6]. In the present study, we constructed two hammerhead ribozymes with antisense sequences of different lengths, which were expected to cleave the positive-stranded genomic RNA at the 5' portion and examined the effects of the ribozymes on MHVmultiplication.

MATERIALS AND METHODS

1. *Cell and virus infection:* Mouse astrocytoma DBT cells[7] were cultured in Eagle's minimum essential medium (MEM) with 5% calf serum (CS) at 37°C in an atmosphere of 5% CO_2. The JHM strain of MHV[8] was propagated in DBT cells. Cells chronically infected with MHV were established essentially according to the method of Hirano et al[9].

2. *Construction of expression vectors:* First-strand cDNA was synthesized from virion RNA using primer oligonucleotide (4) (Fig. 1a). The cDNA was used as a template for PCR amplification. To compare the cleavage efficiencies of ribozymes with different lengths, we constructed vectors which expressed ribozymes containing the 243-nt and 926-nt antisense sequences corresponding to +224 to 467-nt and +224 to +1150-nt of genomic RNA, respectively. These ribozymes are referred to as S-ribozyme and L-ribozyme, respectively. The PCR product using oligonucleotides (2) containing the catalytic domains of hammerhead ribozymes, and primer (3) was cloned into pBluescriptII KS(+). The *Pst*I fragment of this plasmid was subsequently recloned into pBluescriptII KS(+) and designated pBlue.Ribo1 (Fig. 1a). The 945-base pair (b.p.) and 267-b.p. fragments containing ribozyme sequences were cloned into the expression vector, pEF 321[10] and are referred to as pEF L-Ribo and pEF S-Ribo, respectively (Fig. 1b). A vector that expressed the 243-nt antisense RNA containing no ribozyme sequence was constructed (pEF AS). For the synthesis of S-ribozyme in vitro, a 267-b.p. *Bam*HI-fragment of pBlue.Ribo1 was recloned into pBluescriptII SK(+) and designated pBlue.Ribo2 (Fig. 1a). For construction of the vector which expressed the target RNAs, the PCR product (1308-b.p.) using oligonucleotide primers (3) and (1) was cloned into pBluescriptII SK(+) and designated pBlue.MHVt1. A 478-b.p. *Bam*HI-fragment of pBlue.MHVt1 was recloned into pBluescriptII SK(+) and designated pBlue.MHVt2.

3. *In vitro cleavage reactions:* Target and ribozyme RNAs were synthesized from linearized pBlue.MHVt2, pBlue.MHVt1, pBlue.Ribo2 and pBlue.Ribo1 DNA by T3 RNA polymerase in vitro. Target RNAs were radiolabeled with $[\alpha\text{-}^{32}P]$ UTP. Cleavage reactions were carried out under the conditions described in the legend to Fig. 2.

4. *Transfection of the cells:* DBT cells (1×10^6) were cotransfected with pEF S-Ribo, pEF L-Ribo or pEF AS, and pSV$_2$-Neo DNA by the standard calcium phosphate precipitation procedure[11]. One day after transfection, cells were selected in MEM containing 5% CS and G418 (1 mg/ml).

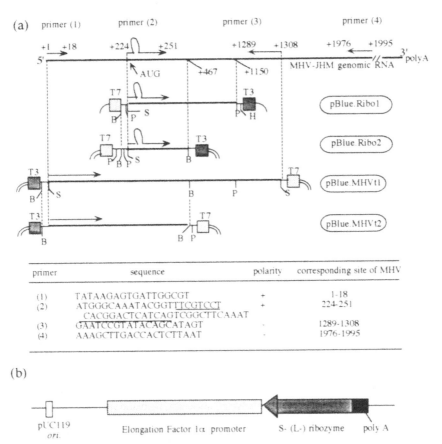

Figure 1. Construction of vectors. (a) Construction of the ribozyme against the MHV polymerase gene. Oligonucleotide (2) contained a hammerhead ribozyme core sequence (underlined). pBluescriptII SK(+) (double line) and pBluescriptII KS(+) (single line). The restriction sites, B, *Bam*HI; S, *Sma*I; P, *Pst*I; H, *Hind*III. AUG represents the putative initiation codon of the RNA polymerase gene. T3 and T7 represent the promoter regions of RNA polymerases T3 and T7, respectively. + and - polarity represent the sense and antisense directions of oligonucleotides toward the genomic RNA of MHV. (b) Construction of expression vectors of ribozymes, pEF S-Ribo and pEF L-Ribo. Poly A represents the SV40 poly A signal.

5. *Assays:* Plaque assays were performed to titrate infectious progeny viruses according to the method of Hirano et al[7]. A viral adsorption assay was carried out by the method of Asanaka and Lai[12]. The infectious center was assayed by the method of Mizzen et al[13].

RESULTS AND DISCUSSION

1. Cleavage of the Target MHV-RNAs by Ribozymes in Vitro

Hammerhead ribozymes were designed to cleave the GUC sequence at nucleotide 238 in MHV-genomic RNA encoding RNA-dependent RNA polymerase[14]. To confirm that the constructed ribozymes cleaved the target MHV-RNAs, we synthesized S- and L-ribozymes, and S-target (ST) and L-target (LT) RNA in vitro. S- and L-ribozymes contained

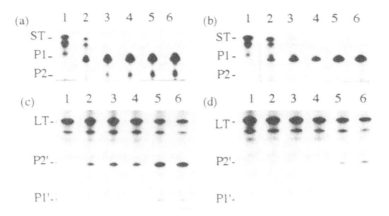

Figure 2. Time kinetics of ribozyme reactions. The reaction mixtures with equal amounts of S-ribozyme (SR) and radiolabeled S-target RNA (ST) (a), L-ribozyme (LR) and radiolabeled ST (b), SR and radiolabeled L-target (LT) (c) or LR and radiolabeled LT (d), were incubated at 37°C in the presence of 20 mM MgCl$_2$ for 0 min (Lane 1), 0.5 min (lane 2), 1 min (lane 3), 5 min (lane 4), 15 min (lane 5), or 30 min (lane 6). P1 (302-nt): 5'-cleavage product of ST, P2 (241-nt): 3'-cleavage product of ST, P1' (302-nt): 5'-cleavage product of LT, P2' (912-nt): 3'-cleavage product of LT.

243-nt and 926-nt complementary sequences against MHV genomic RNA, respectively. ST (543-nt) and LT (1214-nt) contained 467-nt (+1 to +467) and 1150-nt (+1 to +1150) MHV genomic RNA sequences, while ST and LT contained 76-nt and 64-nt plasmid sequences, respectively. The cleavage reactions of target RNAs to ribozymes in vitro were dependent on the ratio of ribozymes to target RNAs, Mg^{2+}-concentration and reaction temperature (data not shown). ST was almost completely cleaved to P1 (302-nt) and P2 (241-nt) during incubation for 1 min with S-ribozyme and for 5 min with L-ribozyme. LT was cleaved to P1' (302-nt) and P2' (912-nt) by the ribozymes, but the cleavage process of LT was significantly slower than that of ST. These results showed that the constructed ribozymes cleaved target RNAs at expected sites in vitro and that the cleavage reaction caused by L-ribozyme occurred more slowly than that caused by S-ribozyme. Therefore, the length of the ribozyme affected its cleavage efficiency in vitro. Tight hybrid formation between a ribozyme and its target RNA due to long complementary sequences may result in suppression of cyclization of the catalytic activity of ribozymes.

2. Inhibition of MHV Multiplication in the Acute Stage of Viral Infection by Ribozymes

To investigate the effects of ribozymes on MHV multiplication in the cells, transfected cell lines expressing S-ribozyme (SR-1 and SR-2) or L-ribozyme (LR-1 and LR-2) were established. These cell lines expressed ribozymes constitutively in their cytoplasm (data not shown). We also established a cell line expressing antisense RNA (AS) and a cell line which was transfected with vector pEF containing no-antisense and no-ribozyme sequences (cont.). When the cells were infected with MHV at 0.1 and 1.0 m.o.i., the yields of infectious viral particles from the cells expressing ribozymes were significantly reduced compared with those from untransfected DBT cells, AS cells and cont. cells (Table 1). In the case at 5.0 m.o.i., the production of infectious viral particles was significantly inhibited in SR-1 and LR-2. The efficiency of inhibitory effects of ribozymes on MHV multiplication may be dependent on the expression levels in the transfected cells (data not shown). When direct

Table 1. Assays of MHV-infection in acute infection

Cell	Viral adsorption assay viral titer (p f u /ml)	Plaque assay yields of viral particles (p f u /ml)		
		0 1 m o i	1 0 m o i	5 0 m o i
DBT	23 25 ± 3 63	2 74 ± 0 64 x 10^5	1 49 ± 0 03 x 10^6	1 79 ± 0 25 x 10^6
SR-1	44 00 ± 9 50	1 69 ± 0 35 x 10^4	2 15 ± 0 44 x 10^4	3 16 ± 0 22 x 10^5
SR-2	60 25 ± 19 63	1 08 ± 0 24 x 10^5	2 97 ± 0 62 x 10^5	1 61 ± 0 24 x 10^6
LR-1	36 00 ± 6 00	1 75 ± 0 25 x 10^5	4 13 ± 0 20 x 10^5	4 47 ± 0 05 x 10^6
LR-2	16 00 ± 1 00	6 22 ± 1 20 x 10^4	7 95 ± 1 28 x 10^3	2 13 ± 3 54 x 10^3
AS	26 00 ± 1 38	3 20 ± 0 61 x 10^5	1 18 ± 0 75 x 10^6	1 13 ± 0 86 x 10^6
cont	35 75 ± 1 38	2 54 ± 0 05 x 10^6	3 96 ± 0 18 x 10^6	4 08 ± 0 19 x 10^6

Each value represents the average of four separate experiments ± standard deviation (S D)

viral adsorption assays were performed, no difference was observed in the infectivity of all cell lines (Table 1) This result showed that the inhibition of MHV multiplication in the transfected cells expressing ribozymes was not due to the change of adsorbability of MHV to the cells Although the cleavage process of L-ribozyme was slower than that of S-ribozyme in the cell-free reactions, no difference between the inhibitory effects of S- and L-ribozymes on MHV multiplication in transfected cells was observed These results suggest that the length of the antisense sequence in this range may not affect the efficiency of the inhibition of viral multiplication in infected cells

Unfortunately, we could not detect the cleavage products of viral RNA in the infected cells This might have been because of the unstability of cleaved RNA products in the cytoplasm of infected cells The possibility that the antisense sequence per se inhibited the expression of the RNA polymerase gene could not be excluded However, since viral multiplication in ribozyme-expressing cells was inhibited more efficiently than in the cells expressing antisense RNA that contained no ribozyme core sequence, it is suggested that the catalytic cleavage activity in infected cells resulted in additional inhibition of MHV multiplication

3. Inhibition of MHV Multiplication in the Chronic Stage of Viral Infection by Ribozymes

Since the MHV multiplication was effectively inhibited by ribozymes in LR-2 cells (Table 1), this cell line was used for the examination of inhibitory effects of ribozymes in chronic infection by MHV As shown in Table 2, the number of infectious centers and the yield of progeny viruses in chronically infected LR-2 cells were significantly lower than in chronically infected DBT cells at 104 days postinfection (d p i) No viral particle was observed in chronically infected LR-2 cells at 200 and 250 d p i The synthesis of viral specific RNAs was significantly inhibited in chronically infected LR-2 cells compared to chronically infected DBT cells (data not shown) These results showed that MHV multiplication in the chronic stage of infection was effectively inhibited by ribozymes

In this paper, we showed that MHV multiplication was effectively inhibited by ribozymes against the RNA polymerase gene of MHV in acute and chronic stages of viral infection Thus, antisense catalytic RNA, ribozymes, might be a good antiviral therapeutic agent

Table 2. Infectious centers and yields of progeny viral particles in chronically infected DBT cells and LR-2 cells

Days postinfection (d p ı)	Cells	Infectious centers (%*)	Yields of viral particles (p f u /ml†)
104	DBT	2 97 ± 1 00	5 87 ± 0 26 x 10⁶
	LR-2	0 31 ± 0 09	0 83 ± 0 23 x 10⁶
200	DBT	0 57 ± 0 13	3 92 ± 1 49 x 10⁶
	LR-2	ND‡	ND
250	DBT	1 41 ± 0 03	1 81 ± 0 24 x 10⁶
	LR-2	ND	ND

*Percentage of plaque-forming cells
†Yields of viral particles were determined by plaque assay
‡Not detected

ACKNOWLEDGMENTS

We thank Dr Taguchı, National Institute of Neuroscience, Kodaıra, Japan, for his useful suggestions

REFERENCES

1 Cech, T R The chemistry of self-splicing RNA and RNA enzymes Science 1987, 236 1532-1539
2 Haseloff, J , Gerlach, W L Nature (Lond) Simple RNA enzymes with new and highly specific endonuclease activity 1988, 334 585-591
3 Uhlenbeck, O C A small catalytic oligoribonucleotide Nature 1987, 328 596-600
4 Sarver, N , Cantın, E M , Chang, P S , Zaıa, J A , Stephens, D A , Rossı, J J Ribozymes as potencial antı-HIV-1 therapeutic agents Science 1989, 247 1222-1225
5 Xıng, Z , Whıtton, J L An antı-lymphocytıc chorıomenıngıtıs virus ribozyme expressed in tissue culture cells diminishes viral RNA levels and leads to a reduction in infectious virus yield J Virol 1993, 67 1840-1847
6 Denman, R B , Purow, B , Rubensteın, R , Mıller, D L Hammerhead ribozyme cleavage of hamster prion pre-mRNA in complex cell-free model systems Bıochem Bıophys Res Commun 1992, 186 1171-1177
7 Hırano, N , Fujıwara, K , Hıno, S , Matumoto, M Replication and plaque formation of mouse hepatitis virus (MHV-2) in mouse cell lıne DBT culture Arch Ges Vırusforsch 1974, 44 298-302
8 Makıno, S , Stohlman, S A , Laı, M M C Leader sequences of murine coronavirus mRNAs can be freely reassorted Evidence for the role of free leader RNA in transcription Proc Natl Acad U S A 1986, 83 4204-4208
9 Hırano, N , Goto, N , Makıno, S , Fujıwara, K Persistent infection with mouse hepatitis virus JHM straın in DBT cell culture Adv Exp Med Bıol 1981, 142 301-308
10 Kım, D W , Uetsukı, T , Kajıro, Y , Yamaguchı, N , Sugano, S Use of the human elongation factor 1α promoter as a versatile and efficient expression system Gene 1990, 91 217-223
11 Graham, F L , Van der Eb, A J A new technique for the assay of infectivity of human adenovirus 5 DNA Virology 1973, 52 456-467
12 Asanaka, M , Laı, M M C Cell fusion studies identified multiple factors involved in mouse hepatitis virus entry Virology 1993, 197 732-741
13 Mızzen, L , Cheley, S , Rao, M , Wolf, R , Anderson, R Fusion resistance and decreased infectability as major host cell determinants of coronavirus persistence Virology 1983, 128 407-417
14 Brayton, P R , Laı, M M C , Patton, C D , Stohlman, S A Characterization of two RNA polymerase activities induced by mouse hepatitis virus J Virol 1982, 42 847-853

66

IDENTIFICATION OF A TRYPSIN-LIKE SERINE PROTEINASE DOMAIN ENCODED BY ORF 1a OF THE CORONAVIRUS IBV

D. X. Liu,[*] I. Brierley, and T. D. K. Brown

Division of Virology
Department of Pathology
University of Cambridge
Tennis Court Road
Cambridge CB2 1QP, United Kingdom

INTRODUCTION

Avian infectious bronchitis virus (IBV) is the prototype species of the *Coronaviridae*, a family of enveloped viruses with large positive-stranded RNA genomes. The genomic RNA of IBV is 27.6 kilobases (kb) in length and contains at least 10 distinct open reading frames (ORFs) (Boursnell et al., 1987). Available evidence suggests that five subgenomic mRNA species are produced in virus-infected cells. These mRNAs (mRNAs 2-6) together with the genome-length mRNA (mRNA1) range in length from about 2 kb to 27.6 kb, and have been shown to share a common 3'-terminus and to form a nested set structure (Stern and Kennedy, 1980ab). Three of these, mRNAs 2, 4 and 6, encode the major virion structural proteins spike (S), membrane (M), and nucleocapsid (N), respectively (Stern and Sefton, 1984). Two of the other mRNAs, mRNA 3 and mRNA 5, have recently been shown to encode three and two viral proteins respectively (Smith et al., 1990; Liu et al., 1991; Liu and Inglis, 1992).

Nucleotide sequence analysis of the genomic RNA of IBV has shown that the 5'-terminal unique region of mRNA 1 contains two large ORFs (1a and 1b), with ORF 1a having the potential to encode a polypeptide of 441K and 1b a polypeptide of 300K (Boursnell et al., 1987). The downstream ORF 1b is likely produced as a fusion protein of 741K with 1a by a ribosomal frameshift (Brierley et al., 1987, 1989). The 1a-1b fusion polyprotein is expected to be cleaved by viral or cellular proteinases to produce functional products associated with viral replication. Several putative functional domains were indeed predicted in either ORF 1a or 1b (Gorbalenya et al., 1989). For example, a picornavirus 3C-like proteinase domain was located in ORF1a between nucleotides 8937 and 9357 and an RNA-dependent-RNA polymerase domain in ORF1b between nucleotides 14100 and

---

[*] Corresponding author.

Corona- and Related Viruses, Edited by P. J. Talbot and G. A. Levy
Plenum Press, New York, 1995

14798 (Gorbalenya et al., 1989). We have recently reported the identification of a virus-specific 100 kDa polypeptide encoding by ORF 1b using a region-specific antiserum V58 (Liu et al., 1994). We show here experiments designed to identify and characterise the proteinase domain responsible for proteolytic processing of this protein from the 1a-1b polyprotein.

METHODS AND RESULTS

Expression and Processing of the 100K Polypeptide in a Transient Eukaryotic Expression System

We have recently reported the identification of a 100 kDa polypeptide in IBV-infected Vero cells using region-specific antiserum V58, which was raised in rabbits from bacterial fusion protein containing IBV sequences from nucleotides 14492 to 15520. This result suggests that the 100 kDa protein is encoded by the corresponding region of IBV ORF 1b and proteolytically cleaved from the 1a-1b polyprotein (Liu et al., 1994). To test this possibility directly, two plasmids, pIBV10 and pIBV11, which cover IBV sequences from nucleotides 8693 to 13896 and 8693 to 16980 respectively, were expressed in Vero cells using the system described by Fuerst et al (1986). Semi-confluent monolayers of Vero cells were infected with 10 pfu per cell of a recombinant vaccinia virus (vTF7-3) which expresses the T7 phage RNA polymerase, transfected with plasmid DNA from pIBV10 and pIBV11 using lipofectin (Gibco-BRL) according to the manufacturer's instructions, and then labelled with 25μCi/ml [^{35}S] methionine. The radiolabelled cells were harvested at 18 hours postinfection, and lysed with RIPA buffer (50mM Tris HCl, pH 7.5, 150mM NaCl, 1% sodium deoxycholate, 0.1% SDS). Immunoprecipitation was carried out using antiserum V58. The results of this experiment are shown in Figure 1. A polypeptide with size approximately 100 kDa was clearly detectable on immunoprecipitation of pIBV11-transfected cells with serum V58 (Figure 1), indicating that the 100 kDa polypeptide is encoded and processed by IBV sequence information within nucleotides 8693 and 16980.

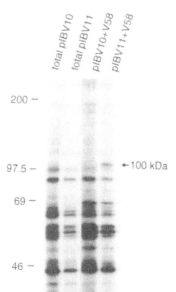

Figure 1. Analysis of transiently expressed ORFs 1a and 1b products from plasmids pIBV10 and pIBV11, using the vaccinia/T7 recombinant virus expression system. Cells were labelled with [^{35}S] methionine, lysates prepared, and polypeptides analysed directly, or immunoprecipitated with the antiserum V58. Polypeptides were separated on a 12.5% SDS-polyacrylamide gel, and detected by fluorography.

Involvement of the Putative 3C-Like Proteinase Domain in Processing of the 100 kDa Polypeptide

Since the 100 kDa polypeptide is generated when IBV sequences covering the putative 3C-like proteinase domain are present, it is tempting to speculate that this domain and its surrounding regions may be involved in processing of this polypeptide. To investigate this possibility, plasmids pIBV4 and pIBV5, which cover IBV sequences from nucleotides 10752 to 13896 and 10752 to 16980 respectively, and do not contain the 3C region, were expressed in Vero cells. As shown in Figure 2, transfection of plasmids pIBV4 and pIBV5 directed efficient synthesis of polypeptides with sizes of approximately 118 kDa (pIBV4) and 235 kDa (pIBV5). The 235 kDa polypeptide was immunoprecipitated efficiently by antiserum V58 (Figure 2). Significantly, no 100 kDa polypeptide was detected, confirming the requirement of the 3C-like domain and the surrounding regions in processing of this protein.

Determination of the C-Terminal Boundary of the 100 kDa Polypeptide

To define roughly the C-terminal boundary of the 100 kDa protein within the polyprotein encoded by IBV sequence within nucleotides 8693 and 16980, two C-terminal deletion constructs (pIBV14 and pIBV15) were made. Plasmids pIBV14 and pIBV15 were constructed by deletion of pIBV11 viral sequences from nucleotide 15537 to 16788 and 15117 to 16980 respectively. As shown in Figure 3, expression of pIBV14 in Vero cells

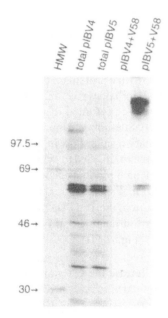

Figure 2. Analysis of transiently expressed ORFs 1a and 1b products from plasmids pIBV4 and pIBV5. Polypeptides were separated on a 12.5% SDS-polyacrylamide gel, and detected by fluorography. Lane labelled HMW represents high molecular weight markers (Amersham).

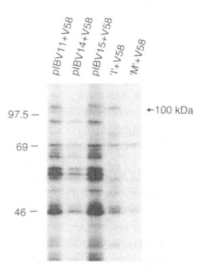

Figure 3. Analysis of transiently expressed ORFs 1a and 1b products from plasmids pIBV11, pIBV14 and pIBV15, and detection of polypeptides encoded by ORF 1b in IBV- infected and mock-infected Vero cells. Polypeptides were separated on a 12.5% SDS-polyacrylamide gel, and detected by fluorography. 'M'-mock-infected Vero cell lysate. 'I'-IBV-infected Vero cell lysate.

directed the synthesis of a polypeptide co-migrating with the 100 kDa polypeptide identified from IBV-infected and pIBV11-transfected cells. However transfection of pIBV15 yielded a polypeptide that migrated slightly more slowly than the 100 kDa polypeptide did on SDS-PAGE (Figure 3), suggesting that this deletion may block the C-terminal cleavage and therefore lead to addition of some extra amino acids to the 100 kDa polypeptide. Examination of the C-terminal sequence of pIBV15 revealed that 15 additional nucleotides derived from the vector sequence were fused with the ORF 1b frame before a stop codon was reached, resulting in synthesis of a fusion polypeptide with five extra amino acids. This suggested that the C-terminal cleavage site lies close to nucleotide position 15120.

Effect of Internal Deletion on Processing of the 100 kDa Polypeptide

To determine further the requirement of the picornavirus 3C-like domain in proteolytic processing of the 100 kDa polypeptide, sequence information covering ORF 1a from nucleotide 9911 to 12227 was deleted from pIBV14, giving plasmid pIBV14Æ1. pIBV14Æ1 was constructed by ligation of a 719 bp DNA fragment (nucleotides 12227-13046), obtained by digestion of pIBV14 with restriction enzyme SnaB1, into PvuII and SnaB1-digested pIBV14, which cut IBV sequences at 9911 and 13046 respectively . Expression of this construction in Vero cells led to synthesis of the 100 kDa polypeptide with much higher efficiency than that from pIBV14. This suggests that the deleted region is not essential for processing of this protein (Figure 4). Furthermore, this deletion reduced the size of the inserted IBV fragment from 6827 bp to 4511 bp, which would facilitate subsequent mutagenesis studies of the catalytic domain and the potential proteinase cleavage sites in the 1a-1b polyprotein.

Mutational Analyses of the Picornavirus 3C-Like Proteinase Domain and a QS Cleavage Site

Computer-aided analysis has predicted that a picornavirus 3C-like proteinase domain is located in 1a polyprotein between amino acids 2779 and 3085 (Gorbalenya et al., 1989). Three catalytically important residues, His_{2820}, Glu_{2843} and Cys_{2922}, presumed to form a catalytic triad, were identified (Figure 5A) (Gorbalenya et al., 1989). Site-directed mutagenesis was therefore carried out to test this prediction. As the preliminary results show in Figure 5B, mutation of the nucleophilic cysteine residue (Cys_{2922}) to alanine (pX17) led to synthesis of a polypeptide of approximately 180 kDa, representing the full-length product encoded by this construct. Processing of the 100 kDa protein was totally abolished. As expected, alteration of Glu_{2843} to Asp (pX16.13) did not affect the production of the 100 kDa protein (Figure 5B). Surprisingly, it was consistently observed that neither full-length nor processed products were detected after mutation of Glu_{2843} to Asn (pX16.3) (Figure 5B).

C-terminal deletion data presented above indicated that a predicted QS cleavage site encoded by nucleotides 15129 to 15135 may be responsible for release the C terminus of the 100 kDa polypeptide. To test this possibility, $Ser_{892(1b)}$ was mutated to Ala (pX20). As the results shown in Figure 5B, this mutation did not affect the processing of the 100 kDa polypeptide.

DISCUSSION

We have recently identified a 100 kDa polypeptide from IBV-infected Vero cells using a region-specific antiserum V58 (Liu et al., 1994). Our previous data suggested that

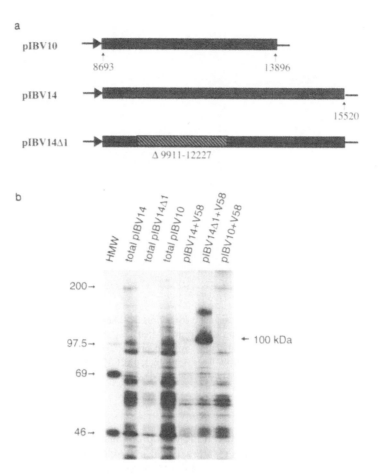

Figure 4. (a) Diagram of the structures of plasmids pIBV10, pIBV14 and pIBV14Æ1 (b) Analysis of transiently expressed ORFs 1a and 1b products from plasmids pIBV10, pIBV14 and pIBV14Æ1 Polypeptides were separated on a 10% SDS- polyacrylamide gel, and detected by fluorography

this novel protein is encoded by the 5'-portion of ORF 1b up to nucleotide 15520 and may be cleaved from the 1a-1b fusion polyprotein by the putative picornavirus 3C-like proteinase domain located in ORF1a region from nucleotide 8937 to 9357. Evidence presented here confirms that the picornavirus 3C-like proteinase domain is involved in processing of the 100 kDa polypeptide. Firstly, internal deletion of ORF 1a sequence from nucleotide 9911 to 12227 rendered no effects on the processing of the protein. Secondly, mutation of the presumed nucleophilic cysteine residue (Cys $_{2922}$) to alanine abolished the proteolytic processing of the polyprotein encoded by the mutated construct, suggesting that this residue may play an essential role in formation of the catalytic centre of the proteinase.

Cys-active-centre viral proteinases have been identified in several animal and plant viruses, such as picornaviruses, comovirus and potyviruses (Gorbaleya et al., 1986; Bazan and Fletterick, 1988). Originally, they were classified as cysteine proteinases, but they are considered now belonging to the trypsin superfamily of serine proteinase. Two recent reports on the X-ray crystal structures of the 3C proteinases from two viruses of the picornavirus family, reveal similar folding of the proteinase polypeptides, RNA-binding sites and means

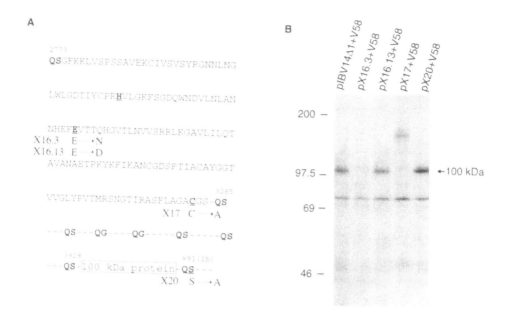

Figure 5. (A) Amino acid sequence of the presumed picornavirus 3C-like proteinase domain, including four mutants indication of the amino acids changed. The predicated QS(G) cleavage sites and the residues constituting the presumed catalytic triad are shown by bold print. (B) Immunoprecipitation analysis of transiently expressed products from plasmids pIBV14Æ1, pX16.3, pX16 13, pX17 and pX20 using antiserum V58 Polypeptides were separated on a 10% SDS-polyacrylamide gel, and detected by fluorography

for cleaving precursor polyprotein between the viral proteinases and cellular serine proteinases of trypsin or chymotrypsin class (Mattews, et al., 1994; Allaire et al., 1994).

The observation that the C-terminus of the 100K polypeptide is specified by ORF1b sequences close to nucleotide 15120 raises the possibility that this polypeptide may be generated by proteolysis at a predicted QS cleavage site encoded by nucleotides 15129 to 15135 (Gorbalenya et al., 1989). Preliminary results reported here showed that mutation of the serine residue to alanine did not alter the proteolytic processing of the 100 kDa polypeptide, suggesting that QA bond might be recognised and cleaved by the IBV serine-like proteinase. For picornavirus 3C proteinase, it was documented that most cleavages occurred between Gln-Gly peptide bond; less common cleavages were observed between Gln-Ser, Gln-Ala, Glu-Ser, or Glu-Gly pairs (Palmenberg, 1990). More dramatic changes of both Q and S residues are underway to delineate the cleavage specificities of the IBV proteinase in the 1a-1b polyprotein.

ACKNOWLEDGMENTS

This work was supported by the Agriculture and Food Research Council, U.K.

REFERENCES

Allaire, M., M. M Chernaia, B. A. Malcolm, and M. N. G. James. 1994. Picornaviral 3C cysteine proteinases have a fold similar to chymotrypsin-like serine proteinases Nature **369**:72-76

Bazan, J F , and R J Fletterick 1988 Viral cysteine proteases are homologous to the trypsin-like family of serine proteases structural and functional implications Proc Natl Acad Sci USA 85 7872-7876

Boursnell, M E G , T D K Brown, I J Foulds, P F Green, F M Tomley, and M M Binns 1987 Completion of the sequence of the genome of the coronavirus avian infectious bronchitis virus J gen Virol **68** 57-77

Brierley, I , M E G Boursnell, M M Binns, B Bilimoria, V C Blok, T D K Brown, and S C Inglis 1987 An efficient ribosomal frame-shifting signal in the polymerase-encoding region of the coronavirus IBV EMBO J **6** 3779-3785

Brierley, I , P Digard, and S C Inglis 1989 Characterisation of an efficient coronavirus ribosomal frameshifting signal requirement for an RNA pseudoknot Cell **57** 537-547

Fuerst, T R , E G Niles, F W Studier, and B Moss 1986 Eukaryotic transient-expression system based on recombinant vaccinia virus that synthesises bacteriophage T7 RNA polymerase Proc Natl Acad Sci USA **83** 8122-8127

Gorbalenya, A E , V M Blinov, and A P Donchenko 1986 Poliovirus-encoded proteinase 3C a possible evolutionary link between cellular serine and cysteine proteinase families FEBS lett **194**:253-257

Gorbalenya, A E , E V Koonin, A P Donchenko, and V M Blinov 1989 Coronavirus genome prediction of putative functional domains in the non-structural polyprotein by comparative amino acid sequence analysis Nucleic Acids Res **17** 4847-4860

Liu, D X , D Cavanagh, P Green, and S C Inglis 1991 A polycistronic mRNA specified by the coronavirus infectious bronchitis virus Virology **184**:531-544

Liu, D X , and S C Inglis 1992 Identification of two new polypeptides encoded by mRNA5 of the coronavirus infectious bronchitis virus Virology **186**:342-347

Liu, D X , Brierley, I , Tibbles, K W , and Brown, T D K (1994) A 100-kilodalton polypeptide encoded by open reading frame (ORF) 1b of the coronavirus infectious bronchitis virus is processed by ORF 1a products J Virol **68**, in press

Matthews, D A , W W Smith, R A Ferre, B Condon, G Budahazi, W Sisson, J E Villafranca, C A Janson, H E Mcelroy, C L Gribskov, and S Worland 1994 Structure of human rhinovirus 3C protease reveals a trypsin-like polypeptide fold, RNA-binding site, and means for cleaving precursor polyprotein Cell **77**:761-771

Oberst, M D , T J Collan, M Gupta, C R Peura, J D Zydlewski, P Sudarsanan, and T Glen Lawson 1993 The encephalomyocarditis virus 3C protease is rapidly degraded by an ATP-dependent proteolytic system in reticulocyte lysate Virology **193**:28-40

Palmenberg, A C 1990 Proteolytic processing of picornaviral polyprotein Annu Rev Microbiol **44**:603-623

Smith, A R , M E G Boursnell, M M Binns, T D K Brown, and S C Inglis 1990 Identification of a new membrane-associated polypeptide specified by the coronavirus infectious bronchitis virus J gen Virol **71** 3-11

Stern, D F and S I T Kennedy 1980a Coronavirus multiplication strategy I Identification and characterisation of virus specified RNA species to the genome J Virol **34**:665-674

Stern, D F and S I T Kennedy 1980b Coronavirus multiplication strategy II Mapping the avian infectious bronchitis virus intracellular RNA species to the genome J Virol **36**:440-449

Stern, D F and B M Sefton 1984 Coronavirus multiplication location of genes for virion proteins on the avian infectious bronchitis virus genome J Virol **50**:22-29

67

INVOLVEMENT OF VIRAL AND CELLULAR FACTORS IN PROCESSING OF POLYPROTEIN ENCODED BY ORF1a OF THE CORONAVIRUS IBV

D. X. Liu,[1][*] K. W. Tibbles,[1] D. Cavanagh,[2] T. D. K. Brown,[1] and I. Brierley[1]

[1] Division of Virology
Department of Pathology
University of Cambridge
Tennis Court Road, Cambridge, CB2 1QP, United Kingdom
[2] Institute for Animal Health
Compton Laboratory
Compton, Newbury, Berkshire RG16 0NN, United Kingdom

INTRODUCTION

Determination of the complete nucleotide sequence of the avian infectious bronchitis virus IBV genomic RNA (mRNA1), carried out by Boursnell *et al.* (1987), has shown that the 5′ terminal sequence of mRNA 1 contains two large ORFs, 1a and 1b, which have the potential to encode two polypeptides of molecular weights 441 kDa and 300 kDa, respectively. Several putative functional domains containing well characterised motifs and more complex homologies have been identified in the 1a or 1b regions by computer-aided techniques (Gorbalenya *et al.*, 1989; Lee *et al.*, 1991; Herold *et al.*, 1993). They include proteinase domains and viral RNA replication-related motifs commonly found in positive strand RNA virus genomes (i.e. RNA-dependent-RNA polymerase and RNA helicase motifs). For example, a papain-like and a picornavirus 3C-like proteinase domains were predicted to be located in IBV 1a (Gorbalenya *et al.*, 1989). However, in mouse hepatitis virus (MHV) and human coronavirus 229E (HCV 229E), two papain-like proteinase domains were found to be located in ORF 1a region of the genomes (Lee *et al.*, 1991; Herold *et al.*, 1993). The first of these domains in MHV, which is absent in IBV, has been identified to be responsible for proteolytic cleavage of a p28 polypeptide from the 1a polyprotein (Baker *et al.*, 1989, 1993).

---

[*] Address correspondence to Dr. Liu.

In an effort to identify viral polypeptides encoded by mRNA1, Brierley and colleagues (1990) prepared a panel of region-specific antisera, by immunising rabbits with bacterially-expressed 1a and 1b sequences (Brierley *et al.*, 1990). We describe here studies in which a number of these antisera have been used to detect viral polypetides encoded by the 5'-portion of ORF1a expressed both in virus-infected Vero cells and in an *in vitro* translation system. Two of them (V52 and V59) were able to immunoprecipitate in IBV-infected Vero cells a protein of 87 kDa. *In vitro* and *in vivo* expression and processing studies demonstrate that this 87 kDa protein is encoded by the 5'-portion of ORF 1a within the first 3000 nucleotides of the virus genome and that it appears to be cleaved from the putative polyprotein by viral and cellular proteinases.

METHODS AND RESULTS

Identification of Gene Products Encoded by the ORF 1a in IBV-Infected Cells

A set of monospecific antibodies against the predicted products encoded by the 5' portion of mRNA1 ORF1a was available in this laboratory. These antisera had been raised in rabbits using bacterially-expressed fusion proteins, containing viral sequences fused to the carboxyterminus of β-galactosidase (Brierley *et al.*, 1990). Three of these antisera, V52, V59 and V53, which was raised in rabbits with bacterial fusion protein containing IBV sequences from nucleotides 710 to 2079, 1355 to 2433 and 4398 to 4853, respectively, and have been shown to react their corresponding in vitro synthesised protein targets, were used to detect viral products in IBV-infected Vero cells. The results of this experiment are shown

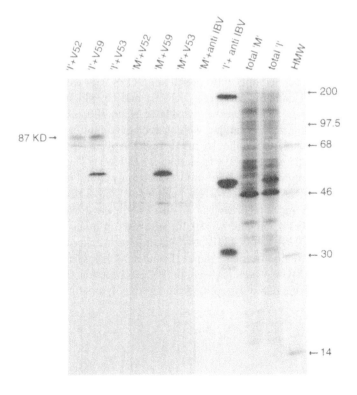

Figure 1. Detection of polypeptides encoded by ORF 1a in IBV-infected and mock-infected Vero cells by immunoprecipitation with region-specific antisera V52, V53 and V59 Confluent monolayers of Vero cells were infected with IBV, labelled with [35S]-methionine at 6 hour postinfection and after a further two hours the cells harvested Cell lysates were prepared and polypeptides analysed directly, or immunoprecipitated with the antisera indicated above each lane Polypeptides were separated on a 17 5% SDS-polyacrylamide gel, and detected by fluorography The lane labelled HMW contains high molecular weight markers (Amersham)

in Figure 1. As can be seen, antisera V52 and V59 precipitated specifically a protein with an apparent molecular weight of approximately 87 kDa from IBV-infected, but not from mock-infected Vero cell lysates; no specific protein band however was detected from the same lysates using antiserum V53.

Expression of pKT1a1 and pKT1a2 in a Transient Eukaryotic Expression System

Identification of the 87 kDa polypeptide in IBV-infected Vero cells with region-specific antisera V52 and V59 indicated that this polypeptide is encoded by the corresponding region of ORF 1a and that it is likely to be cleaved from the 1a polyprotein. To explore the possibility directly, plasmid pKT1a1, which contains IBV sequence from nucleotide 365 to nucleotide 4858 including the putative initiator AUG at position 537, was expressed in Vero cells using the system described by Fuerst et al . (1986). Vero cells were therefore infected with a recombinant vaccinia virus expressing the T7 phage RNA polymerase, and subsequently transfected with pKT1a1 DNA. The results of this experiment are shown in Figure 2A. A polypeptide migrating above the 200 kDa marker was specifically precipitated by antisera V52, V53 and V59, indicating that it may represent the full-length product encoded by pKT1a1. Once again, no processing of the 87 kDa polypeptide was observed.

Computer-assisted analysis of the predicted amino acid sequences of the IBV 1a ORF has suggested the presence of a papain-like proteinase domains located between nucleotides 4680 to 5550 (Gorbalenya et al, 1989). It is likely that this domain may be involved in cleavage of the 1a polyprotein in virus-infected cells. To investigate this possibility directly, plasmid pKT1a2, which covers IBV sequence from nucleotide 365 to 5753 and therefore contains the papain-like proteinase domain, was constructed. Expression of pKT1a2 in Vero cells using the vaccinia virus/T7 system gave rise to synthesis of a polypeptide, which

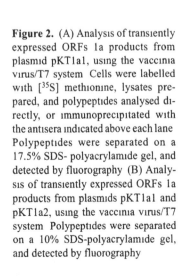
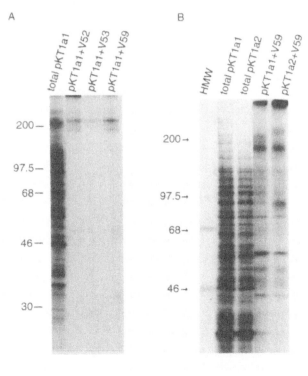

Figure 2. (A) Analysis of transiently expressed ORFs 1a products from plasmid pKT1a1, using the vaccinia virus/T7 system Cells were labelled with [^{35}S] methionine, lysates prepared, and polypeptides analysed directly, or immunoprecipitated with the antisera indicated above each lane Polypeptides were separated on a 17.5% SDS- polyacrylamide gel, and detected by fluorography (B) Analysis of transiently expressed ORFs 1a products from plasmids pKT1a1 and pKT1a2, using the vaccinia virus/T7 system Polypeptides were separated on a 10% SDS-polyacrylamide gel, and detected by fluorography

migrated more slowly on 10% SDS-PAGE than the 220 kDa polypeptide expressed from pKT1a1, and was estimated to have a molecular weight of 250 kDa (Figure 2B). This polypeptide could be specifically precipitated by antiserum V59. In addition, a polypeptide of approximately 87 kDa was also immunoprecipitated by antiserum V59, suggesting that a partial processing of the polyprotein encoded by pKT1a2 had occurred (Figure 2B).

Involvement of Cellular Factors in Processing the Polyprotein Encoded by ORF 1a

The date presented above showed that pKT1a1 expression product in Vero cells is not processed (Figure 2a) nor when it was translated in reticulocyte lysates (see Figure 3). We wished, therefore, to explore the possibility that proteinase activities could be provided *in trans* by incubating the 220 kDa *in vitro* translation product with an IBV-infected Vero cell lysate in an attempt to achieve cleavage of the *in vitro* synthesised polyprotein. For this purpose, Vero cell S10 extracts were prepared from mock-infected and IBV-infected Vero cells according to the procedures of Dorner *et al.* (1984), and incubated with the *in vitro* translation products for 1 hour at 37°C. As shown in Figure 3, incubation of the translation products from *Bam*HI-digested pKT1a1 transcripts with lysis buffer alone did not induce cleavage of either the full-length 220 kDa or the minor polypeptides arising from premature termination. However, cleavage was observed following incubation with IBV-infected Vero cell extracts and furthermore following incubation with extracts prepared from mock-infected cells. In both cases, two major products migrating at about 90 kDa and 87 kDa were observed; both products could be specifically recognised by antiserum V52 (Figure 3), indicating that they contain sequences present in the region of 1a encoding the polypeptide used to raise the antiserum. Several minor protein species migrating more slowly than the 90 kDa/87 kDa polypeptides were also observed occasionally; these might represent inter-

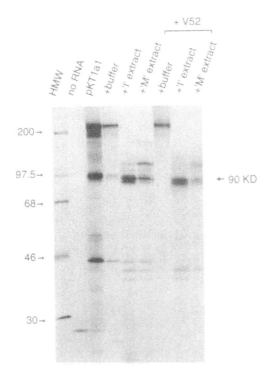

Figure 3. *In vitro* translation and processing of polypeptides encoded by pKT1a1-derived transcripts. *Bam*HI-digested pKT1a1 DNA was transcribed *in vitro* using T7 RNA polymerase, and the resulting transcript translated in the reticulocyte lysate cell-free translation system. Translation products were incubated with either buffer, IBV-infected Vero cell S10 extract or mock-infected Vero cell extract. The translated and processed materials were analysed directly, or immunoprecipitated with antiserum V52 as indicated above each lane. Polypeptides were separated on a 17.5% SDS-polyacrylamide gel, and detected by fluorography.

mediate cleavage products. In addition, a polypeptide of approximately 43 kDa, representing a premature termination product from translation of pKT1a1, was also processed to form two products with sizes differing by about 3 kDa (Figure 3).

Determination of the C-Terminal Boundary of the 87 kDa Protein

To define approximately the C-terminal boundary of the 90 kDa and 87 kDa polypeptides within the ORF1a polyprotein, pKT1a1 was linearised separately with five restriction enzymes (Figure 4A) and a set of target polypeptides with common amino-termini prepared by *in vitro* transcription and translation. These products were then tested for processing with the mock-infected Vero S10 extract. As shown in Figure 6B, translation of the RNAs derived from templates linearised by *Nci*I, *Mlu*I and *Bam*HI (which cut the IBV sequence at nucleotide positions 3002, 3997 and 4858 , respectively) resulted in synthesis of full-length products with sizes greater than 90 kDa, which were then processed down to the 90 kDa and 87 kDa forms upon incubation with the Vero cell S10 extract. No such processing was seen with the shorter translation products. This supports the idea that both the 90 kDa and the 87 kDa polypeptides are derived from within the first 3000 nucleotides of ORF1a. The C-terminal cleavage product(s) of the 220 kDa polypeptide expressed from plasmid pKT1a1 was not detected following *in vitro* processing. The reason for this is uncertain, but one possibility is that the C-terminally truncated polypeptide(s) produced by the cleavage is rapidly degraded in the presence of the S10 extract. To address this, plasmid pKT1a2 was expressed and the translation products were processed *in vitro*. As shown in Figure 4B, *in vitro* transcription and translation of pKT1a2 gave rise to a full-length translation product with an approximate size of 250 kDa (Figure 4B). This polypeptide was also processed *in vitro* to give the 90 kDa and 87 kDa polypeptides previously observed (Figure 4B). In addition to the 90 and 87 kDa polypeptides, however, a unique protein band migrating at about 60 kDa was clearly detectable. Immunoprecipitation studies indicated that it could be precipitated specifically by antiserum V53 (data not shown), suggesting that it was encoded by ORF 1a from nucleotides 3000 to 5000. The failure to detect this polypeptide in IBV-infected Vero cells (Figure 1) may be due to poor incorporation of [^{35}S]-methionine during the labelling procedure; the deduced amino acid sequence of the ORF 1a in the region between nucleotides 3300 and 4920 contains only three methionine residues.

Translation of the IBV Genomic RNA In Vitro

In order to assess the general translation pattern of mRNA1 and, more importantly, to confirm the requirement for cellular factors in the cleavage of the polyprotein encoded by the ORF1a, we translated IBV genomic RNA extracted from purified virions in the reticulocyte lysate cell-free translation system. As shown in Figure 5, translation of IBV genomic RNA gave a wide variety of polypeptide products with sizes ranging from less than 30 kDa to well over 200 kDa. Incubation of the genomic RNA translation products with Vero cell S10 extract again resulted in the appearance of cleavage products whose size was close to that of the 90 kDa/87 kDa and 60kDa polypeptides described above. The 90 kDa/87 kDa product could be immunoprecipitated specifically using antiserum V59 (Figure 5).

DISCUSSION

In this study, we have identified an 87 kDa polypeptide expressed in IBV-infected Vero cells using region-specific antisera. The evidence presented suggests that this novel

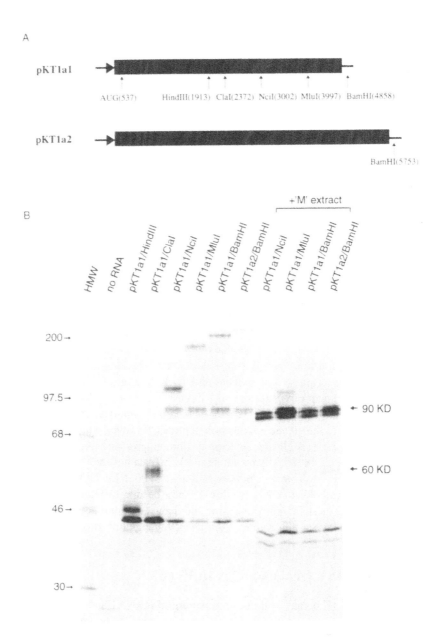

Figure 4. (A) Diagram of plasmids pKT1a1 and pKT1a2, showing the T7 promoter site and the restriction enzyme sites used to linearise the plasmids for *in vitro* transcription. Also shown are the sizes of the *in vitro* translation products expected from ORF1a RNAs transcribed from plasmids linearised at different points. (B) Analysis of cell-free translation products of mRNAs obtained by *in vitro* transcription of HindIII, ClaI, NciI, MluI and BamHI-digested pKT1a1, and from BamHI-digested pKT1a2. RNA was added to the reticulocyte lysate cell-free system, as indicated above each lane. Translation products from NciI, MluI and BamHI-digested pKT1a1, and from BamHI-digested pKT1a2 were processed by incubation with mock-infected Vero cell S10 extract. [35S] methionine-labelled translation products and the processed species were separated on a 12.5% SDS-polyacrylamide gel, and detected by fluorography.

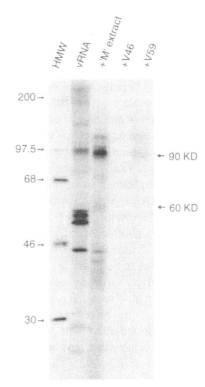

Figure 5. Analysis of the in *vitro* translation products of IBV genomic RNA. Purified virion RNA was added to the reticulocyte lysate at approximate 200µg/ml. Translation products were processed by incubation with mock-infected Vero cell S10 extracts and analysed directly, or following immunoprecipitation with the antiserum indicated above each lane. Polypeptides were separated on a 16% SDS-polyacrylamide gel, and detected by fluorography.

polypeptide is encoded by the 5'-most 3000 nucleotides of ORF 1a, and is cleaved from the 1a polyprotein by viral and cellular proteinases. Firstly, the 87 kDa polypeptide is recognised by N-terminal antisera. V52 and V59, which recognise IBV sequences from nucleotides 710 to 2079 and 1355 to 2433 respectively, and overlap by 724 nucleotides, immunoprecipitate the 87 kDa polypeptide from both virus-infected cells and from the *in vitro* translation and processing reactions. Secondly, expression of ORF 1a up to nucleotide 5763 (pKT1a2), which includes the presumed papain-like proteinase domain (Gorbalenya *et al.*, 1989), leads to synthesis of a final polypeptide product of approximately 250 kDa. This protein appears to represent the expected full-length translation product encoded by this plasmid. In addition, an inefficient processing of this polyprotein to the 87 kDa polypeptide was also observed. However, the 87 kDa polypeptide is not produced simply by translating ORF 1a *in vitro* . Translation of pKT1a1 and pKT1a2 *in vitro* in reticulocyte lysates results in synthesis of the 220 kDa and the 250 kDa polypeptides respectively, which are clearly not processed to mature products. Incubation of these *in vitro* translation products with either IBV-infected or mock-infected Vero cell lysates leads to processing of the *in vitro* synthesised polyproteins and the appearance of the 87 kDa polypeptide. This suggests strongly that cellular factors participate in processing of the 1a polyprotein.

We have recently reported that a 100 kDa polypeptide encoded by IBV ORF1b was processed by a region ORF 1a related to the picornavirus 3C-like proteinase domain (Liu *et al.*, 1994). No cleavage of the *in vitro* synthesised polyprotein was observed when IBV sequences containing the 3C-like proteinase domain were expressed *in vitro* in reticulocyte lysates. However, proteolytic processing of the 100 kDa polypeptide from the polyprotein did occur when this region was expressed in Vero cells using the vaccinia virus/T7 system (Liu *et al.*, 1994). The data presented here showed that processing of the 1a polyprotein encoded by the N-terminal 5 7 kb segment of ORF1a does not occur efficiently even when this region is expressed in intact cells using the vaccinia virus/T7 system However, processing was observed following incubation of the *in vitro* synthesised polyprotein with either IBV-infected or mock-infected Vero cell S10 extracts Currently, we are uncertain why the proteinase(s) present in the Vero cell S10 extract does not appear to function when the truncated 1a polyprotein is expressed in intact Vero cells using the vaccinia virus/T7 system One possible explanation is that infection of Vero cells with vaccinia virus may lead to inhibition of the proteinase activities required for cleavage of the polyprotein It is well documented that infection of host cells by vaccinia virus leads to shut off of host cell protein synthesis, resulting in a general reduction of host protein concentration in infected cells More specifically, three serine proteinase inhibitors (serpins) are produced during virus infection, leading to an inhibition of normal cellular proteinase activities (Smith, 1993) Alternative expression systems are currently being used to explore this possibility further

Involvement of cellular proteinases in the processing of viral polyproteins has been observed in other virus family, including *tagaviridae, flaviviridae, reoviridae, herpersviridae, poxviridae* and *retroviridae* (Dougherty and Semler, 1993) For example, cellular signal peptidases are responsible for cleavage of alpha- flavi- and pestivirus structural polypeptides from polyproteins, some cleavages of non-structural polypeptides from structural polypeptides also involve signal peptidases (Chambers *et al*, 1990) Cellular cofactors have recently been found to be required for efficient cleavage of the poliovirus 3CD polyproteins in virus-infected cells (Blair *et al*, 1993) *In vitro* translation studies also indicate that multiple proteolytic activities may be required for processing of the 1a polyprotein encoded by MHV (Denison *et al*, 1992) The cellular proteinase(s) responsible for cleavage of the IBV 1a polyprotein remains to be identified

ACKNOWLEDGMENTS

This work was supported by the Agriculture and Food Research Council, U K

REFERENCES

Baker, S C , Shieh, C-K , Soe, L H , Chang, M-F , Vannier, D M , and Lai, M M C (1989) Identification of a domain required for autoproteolytic cleavage of murine coronavirus gene A polyprotein *J Virol* **63**, 3693-3699

Baker, S C , Yolomori, K , Dong, S , Carlisle, R , Gorbalenya, A E , Koonin, E V, and Lai, M M C (1993) Identification of the catalytic sites of a papain-like cysteine proteinase of murine coronavirus *J Virol* **67**, 6056-6063

Blair, W S , Li, X , and Semler, B L (1993) A cellular cofactor facilitates efficient 3CD cleavage of the poliovirus P1 precursor *J Virol* **67**, 2339-2343

Boursnell, M E G , Brown, T D K , Foulds, I J , Green, P F , Tomley, F M , and Binns, M M (1987) Completion of the sequence of the genome of the coronavirus avian infectious bronchitis virus *J gen Virol* **68**, 57-77

Brierley, I , Boursnell, M E G , Binns, M M , Bilimoria, B , Rolley, N J , Brown, T D K , and Inglis, S C (1990) Products of the polymerase-encoding region of the coronavirus IBV *Adv exp med Biol* **276**, 275-278

Chambers, T J , Hahn, C S , Galler, R , and Rice, C M (1990) Flavivirus genome organisation, expression, and replication *Annu Rev Microbiol* **44**, 649

Denison, M R , Zoltic, P W , Hughes, S A , Giangreco, B , Olson, A L , Perlman, S , Leibowitz, J , and Weiss, S R (1992) Intracellular processing of the N-terminal ORF 1a proteins of the coronavirus MHV-A59 requires multiples proteolytic events *Virology* **189**, 274-284

Dorner, A , Semler, B L , Jackson, R J , Hanecak, R , Duprey, E , and Wimmer, E (1984) *In vitro* translation of poliovirus RNA Utilisation of internal initiation sites in reticulocyte lysate *J Virol* **50**, 507-514

Dougherty, W G , and Semler, B L (1993) Expression of virus-encoded proteinases functional and structural similarities with cellular enzymes *Microbiological Reviews* **57**, 781-822

Fuerst, T R , Niles, E G , Studier, F W , and Moss, B (1986) Eukaryotic transient-expression system based on recombinant vaccinia virus that synthesises bacteriophage T7 RNA polymerase *Proc Natl Acad Sci USA* **83**, 8122-8127

Gorbalenya, A E , Koonin, E V , Donchenko, A P , and Blinov, V M (1989) Coronavirus genome prediction of putative functional domains in the non-structural polyprotein by comparative amino acid sequence analysis *Nucleic Acids Research* **17**, 4847-4860

Herold, J , Raabe, T , Schelle-Prinz, B , and Siddell, S G (1993) Nucleotide sequence of the human coronavirus 229E RNA polymerase locus *Virology* **195**, 680-691

Lee, H-J , Shieh, C-K , Gorbalenya, A E , Koonin, E V , Monica, N L , Tuler, J , Bagdzhadzhyan, A , and Lai, M M C (1991) The complete sequence (22 kilobases) of murine coronavirus gene 1 encoding the putative proteases and RNA polymerase *Virology* **180**, 567-582

Liu, D X , Brierley, I , Tibbles, K W , and Brown, T D K (1994) A 100-kilodalton polypeptide encoded by open reading frame (ORF) 1b of the coronavirus infectious bronchitis virus is processed by ORF 1a products *J Virol* **68**, in press

Smith, G L (1993) Vaccinia virus glycoproteins and immune evasion *J gen Virol* **74**, 1725-1740

68

CHARACTERIZATION OF THE LEADER PAPAIN-LIKE PROTEASE OF MHV-A59

P. J. Bonilla, J. L. Piñón, S. Hughes, and S. R. Weiss

Department of Microbiology
University of Pennsylvania
Philadelphia, Pennsylvania 19104-6076

ABSTRACT

Sequence analysis of the mouse hepatitis virus, strain A59 (MHV-A59) genome predicts the presence of two papain-like proteases encoded within the first open reading frame of the replicase gene. The more 5′ of these domains, the leader papain-like protease, is responsible for the cleavage of the amino terminal protein, p28. We have defined the core of this protease to between amino acids 1075 and 1344 from the beginning of ORF 1a. Deletion analysis coupled with *in vitro* expression, was used to study p28 cleavage by this leader protease. Expression of a series of deletion mutants showed processing of p28, albeit at lower levels in some of them. Reduced p28 production resulting from a 0.4 kb deletion positioned between p28 and the protease domain suggests an involvement of this region in catalytic processing. Some mutants display cleavage patterns indicative of a second cleavage site. Interestingly, this newly identified cleavage site maps to a position similar to the expected cleavage site of a p65 polypeptide detected in MHV-A59 infected cells. Mutagenesis of the catalytic H1272 residue demonstrates that both cleavages observed are mediated by the leader papain- like protease encoded in ORF 1a.

INTRODUCTION

The first 21.7 kb of the MHV-A59 genome encode the putative replicase locus[1]. This gene (gene 1) encodes two overlapping open reading frames (ORF 1a and ORF 1b) predicted to possess several functional domains[1] (and references therein). Depending on the coronavirus species, one or two papain-like proteases, a picornavirus 3C-like protease and a domain of unknown function designated "X" domain are found in ORF 1a. Putative polymerase, helicase and zinc finger motifs reside in ORF 1b.

The ORFs in gene 1 potentially encode two large polypeptides (496 and 309 kDa). Presumably these are processed into mature replicase-related polypeptides by the predicted proteases encoded in ORF 1a. Some MHV-A59 ORF 1a encoded proteins have been

Corona- and Related Viruses, Edited by P. J. Talbot and G. A. Levy
Plenum Press, New York, 1995

identified including p28, p240, p290, p50 and p65 polypeptides[2-4]. The 5′ most encoded, the leader papain-like protease (PLP-1), is responsible for *in vitro* cleavage of the amino terminal protein p28[5,6].

In this report we further characterize the PLP-1 of MHV-A59. The amino and carboxy termini of the protease domain were mapped. A series of MHV-A59 ORF 1a in-frame deletion mutants was prepared. We used *in vitro* processing of p28 in a coupled transcription/translation system as a protease activity assay. A 0.4 kb deletion between the substrate and the protease down regulated processing of p28. A new cleavage site, mediated by the PLP-1 was observed during *in vitro* analysis of some of the deletion mutants. Interestingly this cleavage site corresponds in position to that which would expected to be cleaved in the processing of p65 *in vivo*.

MATERIALS AND METHODS

Plasmids. Plasmid pSPNK contains MHV-A59 sequences between nucleotides 182-4664 under the control of a T7 promoter. Neighboring *Msc* I fragments were deleted from pSPNK. The resulting plasmid ΔMsc contains a deletion of 181 amino acids (between A623-K805). Construct ΔMBst was prepared by digestion of ΔMsc with *Bst* BI followed by treatment with the Klenow fragment of DNA polymerase I, restriction with *Msc* I and self-ligation of the large fragment. ΔMBst has an in-frame deletion of 245 amino acids (between A623-K869). To prepare ΔEX the 462 bp *Eco* RI-*Xho* I fragment from pSPNK was replaced with a synthetic linker containing a 5′ end *Eco* RI overhang, the codons for amino acids C371Y372G519N520 and a 3′ end *Xho* I overhang. The 146 amino acid deletion in ΔEX is between Y372 and G519. This synthetic linker was also used to introduce the ΔEX deletion into ΔMsc, resulting in ΔEXΔMsc (total deletion of 327 amino acids). Plasmid ΔEX was cut with *Bbs* I, digested with *Bal* 31 nuclease and treated with Klenow. After the DNA was self-ligated, an in-frame clone, ΔEXΔBbs, was identified by DNA sequencing (total deletion of 278 amino acids). Next, the *Xho* I-*Kpn* I fragment from pSPNK was replaced for the corresponding fragment from ΔEXΔBbs creating ΔBbs. ΔBbs contains an in-frame deletion the 132 amino acids (between S942-A1075).

Mutagenesis. Mutagenesis of the PLP-1 catalytic H1272 was performed using recombinant PCR[7]. For the 5′ end fragment used primers FSP 3663-3683 (5′ GGCTATGAC-CAATGCTTTGTG 3′) and RMP 4035-4013 (5′ CAGCCATAGAGXGACAATCATTA 3′) where X = A, C or G. For the 3′ end fragment used FMP 4013-4035 (5′ TAATGATTGTCX-CTCTATGGCTG 3′) where X = T, C or G and RK20 (5′ GCGCTTCAACTTCCTGCAAC 3′). Both fragments were PCR-amplified with Pfu DNA polymerase. The PCR-amplified fragments were purified and used as templates for a third PCR reaction with primers FSP 3663-3683 and RK20. The 1132 bp fragment was purified, digested with *Kpn* I and *Spe* I and cloned into the corresponding sites in pSPNK. A histidine to arginine mutant, pSPNK-H1272R, was identified by DNA sequencing. The H1272R mutation was then cloned into the deletion constructs.

Antisera directed against ORF 1a polypeptides. αp28 (provided by Dr. S. Perlman) is an anti-peptide serum directed against 14 amino acids encoded between nucleotides 287-329 of the JHM strain of MHV (MHV-JHM) genome[2]. Polyclonal antiserum UP102 is directed against epitopes present in the first 593 amino acids of gene 1 of MHV-A59. The virus encoded polypeptide was expressed in *E. coli* as a viral/bacterial fusion product and used to immunize rabbits[8,9]. To generate the immunogen, an ORF 1a fragment (nucleotides 182-1989) was placed under the control of a T7 promoter in pET3B and expressed is E. coli using T7 polymerase[10]. The induced insoluble fusion protein was recovered from the pellet following bacterial cell lysis and used directly and after denaturation with 2% SDS and 5%

2-mercaptoethanol as an immunogen. Antiserum 81043 was prepared in a similar manner[2,4] from a cDNA containing gene 1 sequences between nucleotides 2819 and 4177.

In vitro transcription and translations. Expression of the plasmid DNAs was done using a coupled transcription/translation system (Promega) following manufacturer recommendations. The incorporation of [^{35}S]methionine into acid precipitable counts was used as an indicator of protein synthesis. Equal amounts of counts were used for immunoprecipitations and analyzed by SDS-PAGE as described before[11].

RESULTS

In vitro cleavage of p28. Plasmid pSPNK (Figure 1) was used to further define the properties of the coronavirus PLP-1. It contains MHV-A59 gene 1 sequences coding for p28, the PLP-1 and the "X" domain[12]. Radioabeled products resulting from *in vitro* expression of pSPNK in the presence and absence of the thiol-protease inhibitor leupeptin were immunoprecipitated with αp28 and analyzed by SDS-PAGE (Figure 2). Polypeptides of about 164 kDa and 28 kDa are immunoprecipitated with αp28. These correspond to the full-length translation product of pSPNK and the amino terminal cleavage product p28 respectively. Addition of leupeptin to the reticulocyte lysate inhibited p28 cleavage. Leupeptin-insensitive bands of intermediate size probably correspond to prematurely terminated products of the full-length protein.

Leader papain-like protease boundaries. Plasmids pSPNK and ΔBbs were used to define the amino and carboxy termini of the protease domain responsible for cleavage of p28 (Table 1). The 3' end terminus of the protease domain was defined using 3' end truncations of pSPNK. The shortest pSPNK truncation which retained p28 cleavage activity was at the *Hpa* I site (V1344). This truncation eliminates the coronavirus "X" domain (between V1342 and S1443). When *Spe* I-linearized pSPNK was used as template expression of p28 was abolished. The *Spe* I restriction site is between the catalytic residues C1121 and H1272 of the PLP-1 and removes the carboxy half of the PLP-1. ΔBbs was used to define

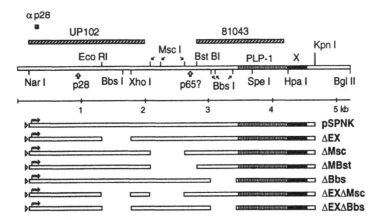

Figure 1. Map of MHV-A59 ORF 1a and diagram of plasmid constructs The PLP-1 (▥), "X" domain (▬) and significant restriction sites are shown The regions encoding the epitopes recognized by the antisera used in this study (▨) are indicated Upward pointing arrows indicate the locations of the p28 cleavage site and the putative p65 cleavage site The MHV-A59 sequences in pSPNK and its deletion mutants are shown The open triangle to the left of each construct represents the T7 promoter The zig-zag arrow on top of each construct indicates the translation initiation site

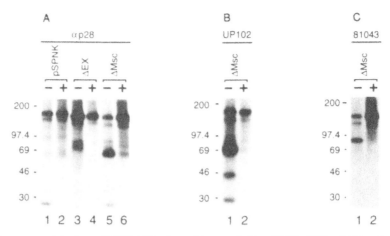

Figure 2. Immunoprecipitations of MHV-A59 polypeptides encoded in pSPNK, ΔEX and ΔMsc. Supercoiled plasmid DNA was used for *in vitro* coupled transcription/translation. Samples were analyzed by 6-18% SDS-PAGE. The (-) and (+) signs indicate the absence or presence of 2 mM leupeptin in the reaction. (A) Antiserum αp28 was used to immunoprecipitate MHV-A59 polypeptides from the [^{35}S]methionine labeled lysates. Lanes: 1-2; pSPNK, 3-4; ΔEX, 5-6; ΔMsc. (B) Immunoprecipitations with antiserum UP102. Lanes: 1-2; ΔMsc. (C) Immunoprecipitations with antiserum 81043. Lanes: 1-2; ΔMsc.

the 5' end boundary of the protease domain. *In vitro* translations reactions using full-length ΔBbs or ΔBbs truncated at the *Hpa* I site as templates, showed cleavage of p28. This defines the amino terminus of the PLP-1 to A1075, although it is possible that further deletions into the protease domain could still show p28 cleavage.

Expression of in-frame deletion mutants: detection of a second cleavage site. Processing of p28 was further studied by examining the expression of a series of in-frame deletion mutants derived from pSPNK (Figure 1). All of the deletion mutants cleaved p28 *in vitro* albeit some at reduced levels. The position of the deletion appears to determine the level of p28 processing. The ΔEX deletion, alone or in combination, dramatically decreased p28 processing. However deletions located between 2.1 kb and 3.4 kb away from the 5' end of the genome (ΔMsc, ΔMBst and ΔBbs) had little or no effect on p28 cleavage relative to pSPNK.

Some mutants displayed expression patterns indicative of a second cleavage site. The *in vitro* expression products of ΔMsc, include polypeptides of 43, 70 and 90 kDa, in addition to p28 (Figure 2). Synthesis of these polypeptides was inhibited by leupeptin suggesting that they are the products of proteolytic processing. The 43 kDa polypeptide immunoprecipitates only with UP102 antiserum indicating that this polypeptide is encoded just downstream of p28. These data, together, also suggest that the 43 kDa polypeptide results from two cleavage events. The amino terminal side of the 43 kDa polypeptide is generated by cleavage of p28 and the carboxy terminus of the 43 kDa polypeptide by a downstream cleavage. The 70 kDa polypeptide, which immunoprecipitates with antisera αp28 and UP102, results from a single cleavage at the downstream site and is a partially processed protein containing both p28 and p43 sequences. The 90 kDa polypeptide, detected only with antiserum 81043, corresponds to the cleavage product downstream of the second cleavage site and contains the protease domain. Similar, cleavage products were detected with other deletion mutants (Table 2). Interestingly expression of ΔMBst shows no evidence of a second cleavage site. We propose that ΔMBst does not carry out the second cleavage because the site at which this cleavage occurs is deleted in this construct.

Table 1. *In vitro* cleavage of p28. Either pSPNK or ΔBbs linearized at a *Pvu* I site downstream of the MHV-A59 coding region, a *Hpa* I site or the *Spe* I were used as templates for *in vitro* expression. Antiserum αp28 was used to immunoprecipitate radiolabeled polypeptides from the lysates. Samples were analyzed by SDS-PAGE. The difference in the length between plasmid templates linearized at the same restriction site arises from the 132 amino acid deletion in ΔBbs

Plasmid	Restriction site	Length (amino acids)	p28 production
pSPNK	Full-length	1485	Yes
pSPNK	*Hpa* I	1344	Yes
pSPNK	*Spe* I	1160	No
ΔBbs	Full-length	1353	Yes
ΔBbs	*Hpa* I	1212	Yes

Both cleavages are carried out by the same protease. The data described above suggest that both cleavages are carried out by the PLP-1 as it is the only protease encoded in these plasmids and both cleavages are sensitive to leupeptin, an inhibitor of this class of proteases. In order to determine the role played by this protease in the second cleavage event, mutagenesis of one of the catalytic residues of the leader protease was performed. The catalytic cysteine and histidine residues of this protease were first identified for MHV-JHM[5]. The high degree of similarity (95%) between the MHV-A59 and MHV-JHM in this region of the replicase gene[1], allowed us to identify C1121 and H1272 as the catalytic residues of the PLP-1 of MHV-A59. A H1272R mutation was introduced into pSPNK and all of the deletion mutants. Figure 3 shows the results of immunoprecipitations with UP102 after *in vitro* expression of the wild type and H1272R mutant constructs. The H1272R mutation leads to a complete inhibition of p28 processing in all of our constructs, thus confirming the importance of H1272 for the activity of the PLP-1. In addition, the presence of this mutation also inhibits the second cleavage event in ΔMsc, ΔEX, ΔExΔMsc and ΔEXΔBbs. Production of both the 43 kDa and 70 kDa polypeptides, was inhibited after introducing the arginine mutation into ΔMsc (see lanes 5 and 6). These results confirm that the PLP-1 is also responsible for the second *in vitro* cleavage event.

Table 2. Summary of *in vitro* cleavage products. Cleavage products were classified into four types. N; the p28 amino terminal cleavage product that results from processing at the p28 cleavage site and detected with both αp28 and UP102 antisera. I; internal cleavage products resulting from cleavage at both the p28 and second sites and detected only with antiserum UP102. C; PLP-1 containing carboxy terminal cleavage products obtained after cleavage at the second cleavage site and detected with antiserum 81043. N+I; fusion polypeptides containing p28 and internal products detected with antisera αp28 and UP102. No (C+I) fusion polypeptides were detected

Plasmid	N	I	C	N + I
pSPNK	28			
ΔEX	28		90	74
ΔMsc	28	43	90	70
ΔMBst	28			
ΔBbs	28			
ΔExΔMsc	28	29	90	55
ΔExΔBbs	28		60	75

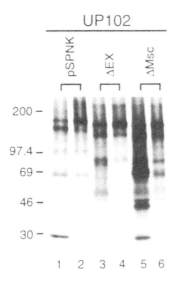

Figure 3. Effect of a mutation of the catalytic H1272 on the expression of pSPNK, ΔEX and ΔMsc. Supercoiled plasmid DNA was used for *in vitro* expression. Antiserum UP102 was used to immunoprecipitate MHV-A59 polypeptides from the [35S]methionine labeled lysates. Samples were analyzed by 6-18% gradient SDS-PAGE. Lanes: 1-2; pSPNK, 3-4; ΔEX and 5-6; ΔMsc. Lanes 1, 3 and 5 correspond to constructs containing a wild type papain-like protease (H1272). Lanes 2, 4 and 6 correspond to constructs containing a H1272R mutation in the papain-like protease.

By comparing the electrophoretic mobilities of polypeptides resulting from cleavage at the second cleavage site relative to that of molecular weight standards we have estimated the position of this second site in ORF 1a. It maps to a small area between the downstream *Msc* I and the *Bst* BI restriction sites (Figure 1). It is noteworthy that this region is deleted in ΔMBst which does not carry out the downstream cleavage.

The second cleavage site maps to a position similar to the putative cleavage site of a product observed in MHV-A59 infected cells. Our laboratory has previously examined the synthesis of ORF 1a proteins in MHV-A59 infected cells[2-4]. We have used the antisera described here to detect ORF 1a products in infected cells: p28 p65 and other high molecular weight polypeptides. The observation that p65 is detected with UP102 serum, but not with αp28 or 81043 antisera suggests it is encoded between p28 and the protease domain. However p65 is not observed during *in vitro* expression of genome RNA[13]. Assuming that the amino terminus of p65 is at the p28 cleavage site, the carboxy terminal end for a 65 kDa product would map to near K834, about 2.71 kb away from the 5′ end of the genome. Coincidentally, the second cleavage site that we have identified also maps to this same region, thus raising the possibility that these sites are identical.

DISCUSSION

Here we report a characterization of the 5′ most protease of MHV-A59. Deletion mapping was used to define the amino and carboxy termini of the PLP-1 of MHV-A59 to at least between 3.4 and 4.2 kb away from the 5′ end of the viral genome. These results map more precisely the coronavirus PLP-1 domain than previously reported[5,6]. The core of this coronavirus protease is contained within a domain of 269 amino acid residues between A1075 and V1344. Our results also show that the coronavirus "X" domain is not essential for *in vitro* cleavage of p28.

A series of deletion mutants was prepared to investigate the effect of alterations on the spacing between the substrate and the protease on p28 cleavage. All mutants showed *in vitro* processing of p28. Some however demonstrated significantly lower levels of p28 processing. The position of the deletion appears to determine the extent of p28 cleavage. Plasmids with a 0.4 kb deletion (ΔEX deletion) near the p28 cleavage site consistently produced significantly less p28. Therefore, the coding region between the *Eco* RI and *Xho* I sites appears to play a role associated with the PLP-1. This deletion may result in an altered conformation at or near the cleavage site with reduced accessibility of the scissile bond to the active site of the protease. Alternatively a protease auxiliary domain may be encoded within this region and its absence results in a basal level processing of p28. Our results resemble those observed for a flavivirus protease responsible for cleavage of nonstructural proteins. The NS3 protease of Dengue virus type 2 requires the presence of NS2B for processing of a precursor polyprotein[14]. It was previously observed that a deletion of the region between 1.1 kb and 2.0 kb of MHV-JHM abolished *in vitro* cleavage of p28 thus suggesting this region plays a role in the processing of the polyprotein. Comparison of the nucleotide sequences of the A59 and JHM strains of MHV allowed us to determine that the 0.4 kb deletion, between 1.3 kb and 1.7 kb of the polymerase gene in MHV-A59, is present within the 0.9 kb region in MHV-JHM. Thus our results refine the mapping of this putative protease auxiliary domain to between 1.3 kb and 1.7 kb of gene 1. The effects of alterations in the spacing between p28 and the protease on the cleavage of p28 was further examined in other deletion mutants. Deletions located between 2.1 kb and 3.4 kb away from the 5' end of the genome did not significantly reduce production of p28 and likely are not involved in proteolytic processing of the amino terminal portion of the genome.

Several mutants showed *in vitro* expression patterns indicating the presence of a second cleavage site downstream of the p28 cleavage site. This a novel function assigned to the coronavirus leader protease. Cleavage products derived from this downstream cleavage site were absent in reactions carried out in the presence of the protease inhibitor leupeptin. Also these products were not expressed by plasmids containing a mutation of the PLP-1 catalytic histidine. Some of the H1272R catalytic site mutants apparently produced small amounts of products derived from the second cleavage site. This could be due to a residual amount of activity from the protease itself or from another leupeptin-sensitive proteolytic activity present in the lysates. Altogether the data indicate that the PLP-1 of MHV-A59 is responsible for this second cleavage.

Cleavage products associated with the second cleavage site were absent from *in vitro* translations of pSPNK and viral genomic RNA[13]. One possibility is that the second cleavage site is unique to the deletion mutants and is not an event that occurs in MHV-A59 infected cells. However, the correlation between the predicted location of this *in vitro* cleavage site and that of the p65 product observed in MHV-infected cells may indicate otherwise. Processing of p65 in MHV-infected cells may be mediated by a cellular or viral factor, either alone or in combination with the PLP-1, that is absent from the *in vitro* translation lysates. Alternatively p65 cleavage may only occur at a specific cellular locale. An altered tertiary structure that allows access of the protease to the second cleavage site in some of our mutants may compensate for the element missing from the translation lysates.

Recently we determined the p28 cleavage site, between amino acids G247 and V248 of ORF 1a[15]. This cleavage site resembles that determined for other viral papain-like proteases[16-19]. In general cleavages by viral papain-like proteases occur between two small amino acids, the first one usually being a glycine. We examined the region to which the second cleavage site was mapped for likely cleavage sites. A G821-V822 dipeptide within this region meets the requirements of a probable candidate in agreement with our data. The predicted size for a product having V248 as the amino terminus and G821 at the carboxy end would be 63.5 kDa similar to the molecular mass of a p65 product. Experiments are underway to determine the cleavage site by direct protein sequencing.

REFERENCES

1 Bonilla, P J , Gorbalenya, A E , and Weiss, S R Mouse hepatitis virus strain A59 polymerase gene ORF 1a heterogeneity among MHV strains Virology 1994,198 736-740

2 Denison, M R , Zoltick, P W , Hughes, S A , Giangreco, B , Olson, A L , Perlman, S , Leibowitz, J L , and Weiss, S R Intracellular processing of the N-terminal ORF 1a proteins of the coronavirus MHV-A59 requires multiple proteolytic events Virology 1992,189 274-284

3 Weiss, S R , Hughes, S A , Bonilla, P J , Turner, J D , Leibowitz, J L , and Denison, M R Coronavirus polyprotein processing Arch Virol 1994,9 349-358

4 Hughes, S A , Denison, M R , Bonilla, P J , Leibowitz, J L , Baric, R S , and Weiss, S R A newly identified MHV-A59 ORF 1a polypeptide p65 is temperature sensitive in two RNA negative mutants Adv Exp Med Biol 1994,342 221-226

5 Baker, S C , Yokomori, K , Dong, S , Carlisle, R , Gorbalenya, A E , Koonin, E V , and Lai, M M C Identification of the catalytic sites of a papain-like cysteine proteinase of murine coronavirus J Virol 1993,67,6056-6063

6 Baker, S C , Shieh, C-K , Soe, L H , Chang, M-F , Vannier, D M , and Lai, M M C Identification of a domain required for autoproteolytic cleavage of murine coronavirus gene A polyprotein J Virol 1989,63, 3693-3699

7 Higuchi, R Recombinant PCR In Innis, M A , Gelfand, D H , Sninsky, J J , and White, T J (eds) PCR protocols a guide to methods and applications Academic Press, Inc , San Diego 1990 pp 177-183

8 Leibowitz, J L , Perlman, S , Weinstock, G , DeVries, J R, Budzilowicz, C , Weissemann, J M , and Weiss, S R Detection of a murine coronavirus nonstructural protein encoded in a downstream open reading frame Virology 1988,164 156-164

9 Zoltick, P W , Leibowitz, J L , DeVries, J R , Weinstock, G M , and Weiss, S R A general method for the induction and screening of antisera for cDNA-encoded polypeptides antibodies specific for a coronavirus putative polymerase-encoding gene Gene 1989,85 413-420

10 Studier, W F , Rosenberg, A H , Dunn, J J , and Dubendorf, J W Use of T7 RNA polymerase to direct the expression of cloned genes Methods in Enzymol 1990,185 60-89

11 Denison, M R , Zoltick, P W , Leibowitz, J L , Pachuk, C J , and Weiss, S R Identification of polypeptides encoded in open reading frame 1b of the putative polymerase gene of the murine coronavirus mouse hepatitis virus A59 J Virol 1991,65 3076-3082

12 Gorbalenya, A E , Koonin, E V, and Lai, M M C Putative papain-related thiol proteases of positive-strand RNA viruses FEBS Letters 1991,288 201-205

13 Denison, M R and Perlman, S Translation and processing of mouse hepatitis virus virion RNA in a cell-free system J Virol 1986,60 12-18

14 Zhang, L , Mohan, P M , and Padmanabhan, R Processing and localization of Dengue Virus Type 2 polyprotein precursor NS3-NS4A-NS4B-NS5 J Virol 1992,66 7549-7554

15 Hughes, S A , Bonilla, P J , and Weiss, S R Analysis of the MHV-A59 p28 cleavage site Elsewhere in this volume

16 Snijder, E J , Wassenaar, A L M , and Spaan, W J M The 5' end of the equine arteritis virus replicase gene encodes a papain-like cysteine protease J Virol 1992,66 7040-7048

17 Shapira, R and Nuss, D L Gene expression by a hypovirulence-associated virus of the chesnut blight fungus involves two papain-like protease activities J Biol Chem 1991,266 19419-19425

18 Carrington, J C and Herndon, K L Characterization of the potyviral HC-Pro autoproteolytic cleavage site Virology 1992,187 308-315

19 Shirako, Y and Strauss, J H Cleavage between nsP1 and nsP2 initiates the processing pathway of Sindbis virus nonstructural polyprotein P123 Virology 1990,177, 54-64

69

PROTEOLYTIC PROCESSING OF THE MHV POLYMERASE POLYPROTEIN

Identification of the P28 Cleavage Site and the Adjacent Protein, P65

Shanghong Dong, Hong-Qiang Gao, and Susan C. Baker

Department of Microbiology and Immunology
Loyola University Medical Center
2160 South First Ave., Bldg. 105
Maywood, Illinois 60153

ABSTRACT

The polymerase gene of Mouse Hepatitis Virus strain JHM (MHV-JHM) encodes a polyprotein larger than 750 kilodaltons. This polyprotein is proposed to be processed by several viral proteinases into functional subunits. The amino-terminal subunit is p28, which is cleaved by the first viral papain-like proteinase domain. In this study, we identified the cleavage site of this papain-like cysteine proteinase by amino acid sequencing of radiolabeled polypeptide adjacent to p28. Proteolysis occurs between the glycine-247 and valine-248 dipeptide bond. To determine which amino acid residues are critical for proteolysis, we preformed site-directed mutagenesis on the coding sequences surrounding the cleavage site and assayed for the efficiency of cleavage of p28 in an in vitro transcription and translation system. We report that glycine-247 and arginine-246 are the most critical residues for efficient processing of p28.

INTRODUCTION

The Mouse Hepatitis Virus (MHV) polymerase polyprotein is encoded by the 5'-most gene, gene 1. Gene 1 is 22 kilobases and has been completely cloned and sequenced[1,2]. Gene 1 encodes two overlapping open reading frames, ORF 1a and ORF 1b, which have the potential to encode a polyprotein of greater than 750 kilodaltons. ORF 1a encodes a 3C-like protease domain and two papain-like cysteine proteinases. We have previously shown that papain-like cysteine proteinase domain 1, termed PCP-1, is responsible for the autoproteolytic processing of the polymerase polyprotein to release the amino-terminal protein product, p28[3,4]. To further understand the proteolytic processing events involved in the

Corona- and Related Viruses, Edited by P. J. Talbot and G. A. Levy
Plenum Press, New York, 1995

maturation of the polymerase polyprotein, we have identified the cleavage site recognized by PCP-1 to release p28. The approach we took was to isolate the protein adjacent to the cleavage site and subject that protein to amino terminal sequencing. We then systematically mutated the residues surrounding the cleavage site to identify the residues that are essential for efficient processing of p28.

METHODS AND RESULTS

In Vitro Transcription and Translation of cDNAs Encoding p28 and PCP-1 Domains

The p28 and adjacent protein were generated by in vitro transcription and translation of a plasmid DNA which represents the 5'-region of ORF1a (Figure 1A). This construct contains the complete proteinase domain and p28 domain and has a small in-frame deletion. We have previously shown that transcription and translation of this construct results in the synthesis of a 128 kDa precursor polyprotein which is autoproteolytically processed to p28 and a 100 kDa protein[3,4]. Proteins were labeled during in vitro translation with either ^{3}H-leucine, ^{3}H-valine or ^{35}S-methionine, separated by polyacrylimide gel electrophoresis, transferred to PVDF membrane, and the isolated 100 kDa protein was subjected to amino acid cycle sequencing based on the Edman degradation reaction[5]. The results of the N-terminal sequencing of the protein labeled with ^{3}H-valine indicated that valine is in positions 1 and 7 after cleavage of p28, leucine

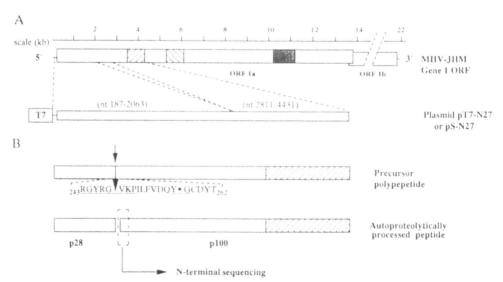

Figure 1. (A) Schematic diagram of MHV-JHM gene 1 open reading frame and the structure of the cDNA clone in plasmid pT7-S27. The two overlapping reading frames (ORFs) are shown in boxes with only ORF1a drawn to scale. Plasmid pT7-S27 contains the 5'-end of the polymerase polyprotein ORF including the authentic AUG (nucleotide 215), a small in-frame deletion (nucleotide 2063 to 2811) and the PCP-1 domain (hatched box). (B) Coupled transcription and translation of plasmid pT7-N27 DNA results in the synthesis of a polyprotein which is autoproteolytically processed to produce p28 and p100[4]. The radiolabeled C-terminal 100-kDa peptide was isolated for N-terminal microsequencing to determine the cleavage site. Partial amino acid sequence of the potential cleavage site region is shown, in which the cleavage site predicted from microsequencing analysis is indicated by a black triangle and the previously proposed cleavage site[6] is marked by an asterisk. The amino acids mutated in this study are underlined. Figure 1 reprinted with permission[7]

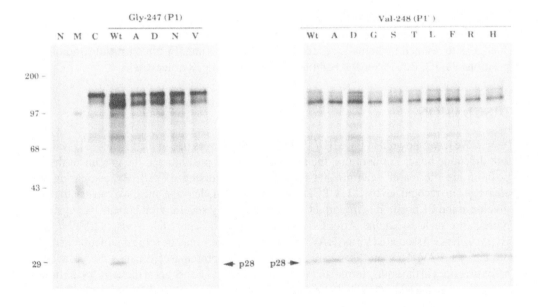

Figure 2. Schematic diagram of the p28 cleavage site. The residues from the P5 to P3' positions are indicated in boldface type. Amino acid substitutions introduced by oligonucleotide mutagenesis are indicated below each position.

was found to be in positions 5, 18 and 22 and no methionine residues were detected in the first 25 positions (data not shown). By aligning this sequence information with the deduced amino acid sequence of the region, we concluded that the cleavage site for p28 occurs between Glycine-247 and Valine-248 (Figure 1B).

To determine the amino acids which are essential for cleavage site recognition, we performed site-directed mutagenesis on the residues both upstream (P1, P2, P3, etc) and downstream (P1', P2', P3', etc) of the cleavage site (Figure 2) and determined the effect of the mutation on p28 processing in an in vitro transcription/translation system. Mutants were generated from the parent plasmid pT7-N27 using degenerate oligonucleotides[8,9]. Forty two individual mutants were then analyzed for the effect of the mutation on the processing of p28. Representative data are shown in Figure 3. Plasmid DNA encoding either the wild type

Figure 3. In vitro translation protein products of wild type and mutants at the P1 (Gly-247) and P1' (Val-248) positions. Site-specific mutations were introduced by degenerate oligonucleotide mutagenesis[8,9]. Linearized plasmids were translated in the T7 RNA polymerase coupled rabbit reticulocyte lysates system (Promega) in the presence of ^{35}S-methionine and translation products were anlyzed by 10% SDS-PAGE. Specific mutations at positions 247 and 248 are indicated at the top of the corresponding lanes, with additional lanes: M, molecular weight marker; N, no RNA; Wt, wild-type pS-N27 translation products; C, control polypeptide with inactive PCP-1 in which the catalytic residue of the proteinase, cysteine-1137, was mutated to Ser[4]. Figure 3 reprinted with permission[7].

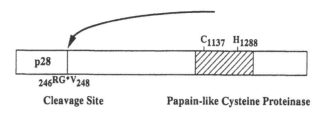

Figure 4. Schematic diagram of the proteolytic processing of the 5'-end of the MHV polymerase polyprotein The 5'-most proteinase domain, PCP-1, acts in cis to cleave the amino-terminal portion of the polyprotein, releasing p28. The catalytic residues of PCP-1, cysteine-1137 and histidine-1288, are indicated in the proteinase domain (hatched box) The p28 cleavage site, glycine-247-valine-248, is shown and the major determinants for efficient processing of p28, arginine-246 (P2) and glycine-247 (P1), are indicated

(glycine) at the P1 position or a mutant amino acid at the P1 position were transcribed and translated in rabbit reticulocyte lysates (TNT system, Promega Biotech) and the products were analyzed by polyacrylimide gel electrophoresis on a 10 % gel. As shown in Figure 3, the wild type protein is cleaved to produce p28 and the 100 kDa product. In contrast, when glycine-247 is replaced by alanine, aspartic acid, asparagine or valine, p28 processing is dramatically reduced. As a control, if the catalytic cysteine-1137 residue of the proteinase domaine is altered (Figure 3, lane C), no p28 processing occurs. Overall, these results indicate that a glycine at the P1 position is a major determinant in p28 processing.

In contrast to the P1 position, the valine at the P1' position seems to be less critical. Substitution to 9 other amino acids had little or no effect on the processing of p28 (Figure 3). Similar primary data was generated from the 42 mutants. The results of the transcription/translation experiments indicated that both the glycine at the P1 position and the arginine at the P2 position are the major determinants for the cleavage of p28 (Figure 4). Substitutions at either site resulted in dramatic reductions in p28 processing. In constrast, substitutions at positions P4, P3, P1', P2' or P3' had little or no effect on p28 processing.

DISCUSSION

We have shown that the MHV PCP-1 domain cleaves p28 at the glycine-247-valine-248 dipeptide bond. Furthermore, the arginine at the P2 position and the glycine at the P1 position are the major determinants for efficient processing of p28. By comparing the cleavage site recognized by MHV PCP-1 to other known cleavage sites, we can see that the glycine at the P1 position is important for recognition by several viral papain-like cysteine proteinases such as Equine Arteritis virus (EAV)[10], Tobacco Etch Virus (TEV)[11,12] and Hypovirulence Associated Virus (HAV)[13,14]. However, there may be some distinction in these proteinases. EAV, TEV and HAV are termed "leader" proteinases because the proteinase activity resides in the amino terminal domain and cis-cleavage occurs at the carboxy-terminal end of the proteinase[15]. MHV PCP-1 may be more similar to what Gorbalenya termed "main" proteinases[15]. The cysteine proteinase of Sindbis virus, nsp2, cleaves at multiple sites both in cis and in trans, and plays an important role in temporal regulation in Sindbis virus RNA synthesis[16]. Preliminary evidence indicates that in MHV-infected cells, the protein adjacent to p28 is a 65 kDa protein which is also proteolytically processed from the polyprotein (Gao and Baker, unpublished). However, the proteinase responsible for the cleavage of p65 is not known. Further investigation of the proteolytic processing of the MHV polymerase polyprotein is required to determine if MHV PCP-1, like Sindbis virus nsp2, cleaves at additional sites in the polyprotein.

ACKNOWLEDGMENTS

This work was supported by Public Health Service Research Grant AI32065 from the National Institutes of Health and a Junior Faculty Reasearch Award from the American Cancer Society (to S C B)

REFERENCES

1 Lee, H -J , Shieh, C -K , Gorbalenya, A E , Koonin, E V, La Monica, N , Tuler, J , Bagdzhadzhyan, A , and Lai, M M C The complete sequence (22 kilobases) of murine coronavirus gene 1 encoding the putative proteases and RNA polymerases Virology 1991, 180 567-582
2 Pachuk, C J , Bredenbeek, P J , Zoltick, P W, Spaan, W J M ,and Weiss, S R Molecular cloning of the gene encoding the putative polymerase of mouse hepatitis coronavirus, strain A59 Virology 1989, 171 141-148
3 Baker, S C , Shieh, C -K , Soe, L H , Chang, M F , Vannier, D M , and Lai, M M C Identification of a domain required for autoproteolytic cleavage of murine coronavirus gene A polyprotein J Virol 1989, 63 3693-3699
4 Baker, S C , Yokomori, K , Dong, S , Carlisle, R , Gorbalenya, A E , Koonin, E V, and Lai, M M C Identification of the catalytic sites of a papain-like cysteine proteinase of murine coronavirus J Virol 1993, 67 6056-6063
5 Matsudaira, P Sequence from picomole quantities of proteins electroblotted onto polyvinylidene difluoride membranes J Biol Chem 1987, 262 10,035-10,038
6 Soe, L H , Shieh, C -K , Baker, S C , Chang, M -F , and Lai, M M C Sequence and translation of the murine coronavirus 5'-end genomic RNA reveals the N-terminal structure of the putative RNA polymerase J Virol 1989, 61 3968-3976
7 Dong, S and Baker, S C Determinants of the p28 cleavage site recognized by the first papain-like cysteine proteinase of murine coronavirus Virology 1994, 204 541-549
8 Hutchison, C A , Phillips, S , Edgell, M H , Gilliam, S , Jahnke, P., and Smith, M Mutagenesis at a specific position in a DNA sequence J Biol Chem 1978, 253 6551-6560
9 Lewis, M K and Thompson, D V Efficient site directed in vitro mutagenesis using ampicillin selection Nucleic Acids Res 1990, 18, 3439-3443
10 Snijder, E J , Wassenaar, A L M and Spaan, W J M The 5'-end of the equine arteritis virus replicase gene encodes a papainlike cysteine protease J Virol 1992, 66 7040-7048
11 Carrington, J C , Cary, S M , Parks, T D and Dougherty, W G A second proteinase encoded by a plant potyvirus genome EMBO J 1989, 8 365-370
12 Carrington, J C and Herndon, K L Characterization of the potyviral HC-Pro autoproteolytic cleavage site Virology 1992, 187 308-315
13 Choi, G H , Shapira, R and Nuss, D L Cotranslational autoproteolysis involved in gene expression from a double-stranded RNA genetic element associated with hypovirulence of the chestnut blight fungus Proc Natl Acad Sci USA 1991a, 88 1167-1171
14 Choi, G H , Pawlyk, D M and Nuss, D L The autocatalytic proteinase p29 encoded by a hypovirulence-associated virus of the chestnut blight fungus resembles the potyvirus-encoded proteinase HC-Pro Virology 1991b, 183 747-752
15 Gorbalenya, A E , Koonin, E V, Donchenko, A P and Blinov, V M Coronavirus genome prediction of putative functional domains in the non-structural polyprotein by comparative amino acid sequence analysis Nuclic Acids Res 1989, 17 4847-4861
16 De Groot, R , Hardy, W R , Shirako, Y and Strauss, J H Cleavage-site preferences of Sindbis virus polyproteins containing the non-structural proteinase Evidence for temporal regulation of polyprotein processing in vivo EMBO J 1990, 9 2631-2638

70

ZINC-BINDING OF THE CYSTEINE-RICH DOMAIN ENCODED IN THE OPEN READING FRAME 1B OF THE RNA POLYMERASE GENE OF CORONAVIRUS

Dongwan Yoo,[1] Michael D. Parker,[2] Graham J. Cox,[1] and Lorne A. Babiuk[1]

[1] Veterinary Infectious Disease Organization
University of Saskatchewan
Saskatoon, Saskatchewan S7N 0W0 Canada
[2] Virology Annex
USAMRIID
Frederick, Maryland 21701-5000

ABSTRACT

We cloned and sequenced the second open reading frame of the RNA polymerase gene, ORF1b, of bovine coronavirus. In the region representing nucleotide positions 4919-5677 upstream from the initiation codon of the 32K non-structural protein gene, we identified two putative functional domains. One of these domains contained four leucine residues repeated exactly in every seventh position, and the other domain represented a cluster of cysteine and histidine residues. The DNA sequence representing these domains was cloned and expressed in *Escherichia coli* as fusion proteins with glutathione S-transferase from *Schistosoma japonicum*. A high level expression of the cysteine-rich domain was achieved as a fusion protein when the bacterial culture was induced with IPTG. In a solid phase zinc binding assay using the recombinant fusion protein, we found that the protein containing the cysteine-rich domain was able to bind to radioactive zinc *in vitro*, demonstrating that the polypeptide encoded by the ORF1b of coronavirus is a zinc-binding protein.

INTRODUCTION

The 5' most gene of coronavirus genome represents gene 1 encoding the putative RNA-dependent RNA polymerase. Gene 1 has been cloned and sequenced in several coronaviruses including infectious bronchitis virus[2], mouse hepatitis virus[3, 8, 10], and human coronavirus strain 229E[6]. Two large, slightly overlapping open reading frames, termed

Corona- and Related Viruses, Edited by P. J. Talbot and G. A. Levy
Plenum Press, New York, 1995

ORF1a and ORF1b, are found in coronavirus gene 1, and it has been suggested that gene 1 is able to synthesize both the ORF1a protein and the ORF1a/ORF1b fusion protein by way of ribosomal frameshifting of the two overlapping open reading frames[4]. Several putative functional domains have been suggested in the ORF1b, which include RNA polymerase, helicase-like, and cysteine-rich domains. None of these domains in coronavirus have been functionally characterized yet. We have been particularly interested in the significance of the cysteine-rich domain of the bovine coronavirus ORF1b. Modeling of the cysteine-rich domain suggests that cysteine/histidine ligands are configured as to tetrahedrally coordinate zinc ions, forming three potential finger-like structures[8]. In this study, we characterized this domain by expressing the DNA fragment as a fusion protein in *E. coli*, and demonstrated the cysteine-rich domain is able to bind to zinc.

MATERIALS AND METHODS

Protein expression in E. coli. E. coli transformants were grown in 2 ml of 2X YT (1.6% tryptone, 1% yeast extract, 0.5% NaCl) containing 2% glucose and 100 μg/ml ampicillin to an optical density of 0.5 at 600 nm. Fusion protein was expressed by adding isopropyl-ß-D-thiogalactosidase at final concentration of 2 mM, and the culture was further incubated for an additional 2 hours. Cells were pelleted in a microcentrifuge and resuspended in water. The cell suspension was mixed with an equal volume of 2X Laemmli's sample buffer, boiled for 5 min, and analysed by 12% SDS polyacrylamide gel electrophoresis. Proteins were visualized by Coommassie blue staining.

Solid phase zinc blot assay. The proteins were resolved by 12% SDS-polyacrylamide gel electrophoresis and transferred to polyvinylidone difluoride membrane (PVDF; Du Pont). The PVDF membrane was wetted in methanol for 2 min, soaked in water for 5 min, and equilibrated in transfer buffer (25 mM Tris base, 192 mM glycine, 20% methanol, pH 8.3) for 10 min before use. For the zinc binding assay[12], the membrane was briefly washed twice with renaturation buffer (100 mM Tris-HCl, pH 6.8, 50 mM NaCl, 10 mM DTT) and incubated in renaturation buffer for 1 hr with three exchanges of buffer at room temperature. The membrane was then incubated in labeling buffer (100 mM Tris-HCl, pH 6.8, 50 mM NaCl) containing 1 μCi/ml of Zn^{65} (New England Nuclear) for 1 hr with gentle shaking. At the end of incubation, the membrane was rinsed twice with washing buffer (100 mM Tris-HCl, pH 6.8, 50 mM NaCl, 1 mM DTT) and washed extensively in washing buffer with three exchanges of buffer for 1 hr at room temperature. The membrane was wrapped with Saran-Wrap and exposed to X-ray film (Reflection™, Du Pont) at -70 C.

Northwestern blot assay. The protein-RNA binding assay was performed by the procedure described previously with a minor modification[1]. Proteins were resolved by 12% SDS-PAGE and electro-transferred to PVDF membrane. The membranes were washed for 1 hr at room temperature in probe buffer containing 10 mM Tris-HCl, pH 7.5, 50 mM NaCl, 1 mM Na_2EDTA, 0.02% bovine serum albumin fraction V, 0.02% polyvinyl pyrrolidone, 0.02% Ficoll type 400 (Pharmacia), and 250 μg/ml of total *E.coli* RNA. Approximately 2 X 10^6 counts per minute of ^{32}P-labeled RNA probe was added, and the membranes were incubated for 1 hr at room temperature with continuous shaking. The membranes were washed three times for 10 min each in probe buffer and exposed to X-ray film at -70 C with a intensifying screen. The ^{32}P-labeled probe was prepared by labeling the 5' terminus of virion RNA as follows: Viral genomic RNA was dissolved in 50 mM Tris-HCl, pH 8.5, and incubated at 90 C for 30 min. The RNA was added to a 50 μl reaction containing 50 mM Tris-HCl, pH 7.6, 10 mM $MgCl_2$, 10 mM ß-mercaptoethanol, 100 μCi [gamma-^{32}P] ATP (New England Nuclear, 3,000 Ci/mmol), and 5 units of T4 polynucleotide kinase (United States Biomedical, Amersham Canada Ltd.). The reaction was incubated for 30 min at 37 C

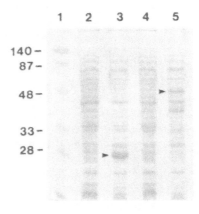

Figure 1. Expression of the GST-fusion protein in *E. coli*. The transformants were grown in 2X YT broth containing 2% glucose until OD reached 0.5 at 600 nm. IPTG was added to a final concentration of 2 mM, and the cultures were incubated for 2 additional hours. Cells were lysed in sample buffer, and the total proteins were resolved by 12% SDS-PAGE followed by Coommassie blue staining. Each lane represents approximately 150 μl of the culture. Lanes: 1, molecular weight marker; 2, uninduced GST alone; 3, IPTG-induced GST alone; 4, uninduced GST-fusion protein; 5, IPTG-induced GST-fusion protein. Arrow heads indicate the expressed proteins.

and terminated by addition of EDTA. The end-labeled RNA was separated from unincorporated radionucleotides by Sephadex G-50 column chromatography and ethanol precipitation.

RESULTS

A 759 base pair DNA fragment was PCR-cloned representing the region of nucleotide positions 4919-5677 upstream from the 32K non-structural protein gene[6] of bovine coronavirus. The cloned DNA fragment was inserted in-frame behind the glutathione S-transferase (GST) gene for expression as a fusion protein. High level expression of the GST-fusion protein was achieved under the control of a *tac* promoter using IPTG induction. Total cell lysates were prepared for SDS-PAGE, and the expressed proteins were visualized by Coommassie blue staining (Fig 1). Polypeptides with approximate molecular weights of 28 kDa and 55 kDa were identified in the IPTG-induced cultures (Fig. 1, lanes 3, 5). These proteins were not present in uninduced cultures (lanes 2, 4), and the specificity of these proteins were confirmed by immunoblotting with anti-GST antibody.

The cysteine/histidine residues present in the ORF1b region of coronavirus are similar to the configuration found in zinc-binding proteins[5]. To examine if this domain in coronavirus represents zinc-finger, a solid-phase zinc blot assay was performed. Proteins were resolved by SDS-PAGE and immobilized on the membrane under reducing conditions. The membrane was incubated with Zn[65] and exposed to X-ray film. The fusion protein containing the cysteine/histidine rich domain bound to Zn[65] (Fig. 2, lane 4), and this reaction was specific since none of the other bacterial proteins nor GST alone bound to Zn[65] (Fig. 2, lanes 1, 2, 3).

Zinc-finger motifs are often found in RNA binding proteins. To examine if the coronavirus motif was able to bind to the viral RNA, protein-RNA hybridization was carried out. This technique has been used for protein-RNA interaction studies for reovirus and coronavirus[1, 12]. Protein-bound membrane was probed with coronavirus genomic RNA labeled with [32]P. Binding of the fusion protein to coronavirus genomic RNA was not detected (Fig. 3, lane 3), whereas the BCV nucleocapsid protein efficiently bound to the viral RNA (Fig., 3, lane 1). Since the Northwestern protocol involves protein denaturation steps, we employed a filter binding assay to re-examine the RNA binding activity of the zinc protein[9]. The cell lysates were incubated with the radiolabeled genomic RNA in binding buffer, and the mixture was filtered through a nitrocellulose membrane. The membrane was washed with the binding buffer and counted for radioactivity. Approximately 80% of the input radioactivity was incorporated with the GST-fusion protein, whereas the GST alone remained only 40% (Fig. 4).

A

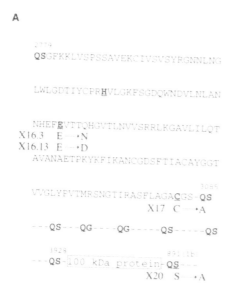

B

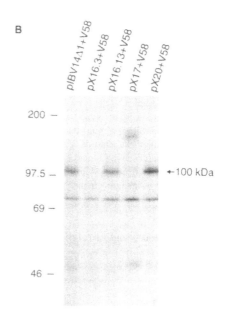

Figure 2. Zinc-binding activity of the fusion protein. Total proteins were resolved by SDS-PAGE and immobilized on PVDF membrane. The membrane was incubated in labeling buffer containing 1 μCi/ml of Zn^{65} for 1 hr. The membrane was washed twice with the labeling buffer and exposed to X-ray film. Lanes: 1, uninduced GST alone; 2, IPTG-induced GST alone; 3, uninduced GST-fusion protein; 4, IPTG-induced GST-fusion protein.

Figure 3. Northwestern blot hybridization of the fusion protein. Total cell lysates were resolved by SDS-PAGE and immobilized on PVDF membrane. Coronavirus genomic RNA was extracted from the sucrose-purified virions, and the RNA was *in vitro*-labeled with [gamma-^{32}P] using a polynucleotide kinase. The membrane was hybridized with the ^{32}P-labeled RNA probe, and exposed to X-ray film. Lanes: 1, purified coronavirus; 2, IPTG-induced GST alone; 3, IPTG-induced GST-fusion protein.

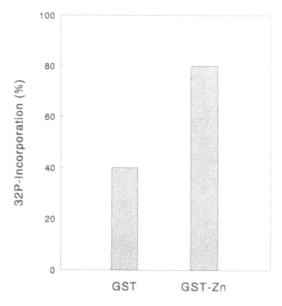

Figure 4. RNA-binding of the GST-fusion protein. The cell lysate was incubated with ^{32}P-labeled BCV RNA in 50 μl of binding buffer (17 mM HEPES-KOH, pH 7.2, 2 mM DTT, 5% glycerol, 0.5 mM magnesium acetate, 75 mM KCl), containing 100 μg/ml of bovine resum albumin. The mixture was filtered through a nitrocellulose membrane, and the filter was washed with binding buffer, air-dried, and counted in a scintillation counter. Input radioactivity was 20,000 counts per minute.

DISCUSSION

We cloned and expressed the sequence representing the cysteine-rich region of bovine coronavirus ORF1b in *E coli* This region represents nucleotide positions 4919-5677 upstream from the initiation codon of the 32K non-structural protein gene The structural arrangements in the cysteine-rich domain was highly conserved among coronaviruses, and the cysteine/histidine cluster resembles a zinc-finger motif which is commonly represented as several repeats of Cys-X-X-Cys/His, where X is any amino acid Using the GST-fusion protein expressed in *E coli*, we demonstrated that the polypeptide representing the cysteine-rich region was a zinc-binding protein The zinc-finger motif is often found in eukaryotic regulartory proteins and is able to bind to zinc[5] Cysteines and histidines form ligands co-ordinating zinc atoms, promoting sequence specific protein-DNA interactions Zinc ion is either involved in maintaining an intramolecular conformation or mediates an intermolecular linkage between two molecules resulting in the dimerization The most catalytic function of DNA-binding protein occurs in the cell nucleus However, since coronavirus replication takes place in the cytoplasm and the coronavirus zinc-finger motif is found in the putative RNA polymerase region, we speculate that the coronavirus zinc-finger may be involved in the binding of the RNA polymerase molecule to the viral RNAs, and thereby facilitates initiation of viral replication and transcription By Northwestern blot hybridization, we were not able to detect binding of the fusion protein to coronavirus RNA, while the BCV nucleocapsid protein used as control bound to the viral RNA This was probably due to the fact that the Northwestern hybridization protocol involved a protein denaturation step which altered the protein conformation neccessary for the interaction with RNA By the native filter-binding assay, we demonstrated that the fusion protein bound to the the coronavirus genomic RNA Studies to further elucidate the RNA-protein interaction and function of the zinc protein in coronavirus replication are in progress

ACKNOWLEDGMENTS

This study was supported by grants from Health Services Utilization and Research Commision of Saskatchewan, Natural Sciences and Engineering Research Council of Canada, and Medical Research Council of Canada Published as a journal series No 178 with the permission of the director of VIDO

REFERENCES

1 Boyle, L F, Holmes, K V RNA-binding proteins of bovine rotavirus 1986 J Virol 58 561-568

2 Boursnell, M E G, Brown, T D K, Foulds, I J, Green, P F, Tomley, F M, and Binns, M M Completion of the sequence of the genome of the coronavirus avian infectious bronchitis virus 1987 J Gen Virol 68 57-77

3 Bredenbeek, P J, Pachuk, C J, Noten, A F, Charite, J, Luytjes, W, Weiss, S R, Spaan, W J 1990 The primary structure and expression of the second open reading frame of the polymerase gene of the coronavirus MHV-A59 A highly conserved polymerase is expressed by an efficient frameshifting mechanism Nucl Acids Res 18 1825-1832

4 Brierley, I, Digard, P, Inglis, S C Characterization of an efficient coronavirus ribosomal frameshifting signal requirment for an RNA pseudoknot 1989 Cell 57 537-547

5 Coleman, J E Zinc proteins enzymes, storage proteins, transcription factors, and replication proteins Annu Rev Biochem 61 897-946

6 Cox, G J, Parker, M D, Babiuk, L A The sequence of cDNA of bovine coronavirus 32K nonstructural gene 1989 Nucl Acids Res 17 5847

 7 Herold, J , Raabe, T , Schelle-Prinze, B , Siddel, S G Nucleotide sequence of the human coronavirus 229E RNA polymerase locus Virology 195 680-691

 8 Lee, H J , Shieh, C K , Gorbalenya, A E , Koonin, E V , Monica, N L , Tuler, J , Bagdzhadzhyan, A , Lai, M M C The complete sequence (22 kilobases) of murine coronavirus gene 1 encoding the putative protease and RNA polymerase 1991 Virology 180 567-582

 9 Pachuk, C J , Bredenbeek, P J , Zoltik, P W , Spaan, W J M , Weiss, S Molecular cloning of the gene encoding the putative polymerase of mouse hepatitis coronavirus, strain A59 1989 Virology 171 141-148

10 Robbins, S G , Frana, M F , McGowan, J J , Boyle, J F , Holmes, K V RNA-binding proteins of MHV detection of monomeric and multimeric N protein with an RNA overlay-protein blot assay 1986 Virology 150 402-410

11 Schiff, L A , Nibert, M L , Field, B N Characterization of a zinc blotting technique evidence that a retroviral gag protein binds zinc 1988 Proc Natl Acad Sci USA 85 4195-4199

12 Methot, N , Pause, A , Hershey, J W B , Sonenberg, N The translation initiation factor eIF-4B contains an RNA-binding region that is distinct and independent from its ribonucleoprotein consensus sequence 1994 Mol Cell Biol 14 2307-2316

71

PROTEOLYTIC PROCESSING OF THE ARTERIVIRUS REPLICASE

Eric J. Snijder, Alfred L.M. Wassenaar, Johan A. Den Boon, and
Willy J. M. Spaan

Department of Virology
Institute of Medical Microbiology
Faculty of Medicine
Leiden University
PO Box 320
2300 AH Leiden, The Netherlands

INTRODUCTION

Arteriviruses are enveloped positive-stranded RNA viruses which belong to the so-called 'coronaviruslike superfamily'[1,2,3,4]. At present, the arterivirus group is comprised of equine arteritis virus (EAV, the prototype of the group), lactate dehydrogenase-elevating virus (LDV), porcine reproductive and respiratory syndrome virus (PRRSV, also known as 'Lelystad virus'), and simian haemorrhagic fever virus (SHFV). Their isometric nucleocapsid core contains a nonsegmented genome of 12.7-15.1 kb[1,5,6]. The morphological characteristics and genome size of EAV are most comparable to those of togaviruses and flaviviruses. However, the arterivirus replication strategy is similar to that of coronaviruses, which possess 25-31 kb positive-stranded RNA genomes. Among their common features are the polycistronic genome organization, the same basic gene order (5'-replicase gene-envelope protein genes-nucleocapsid protein gene-3'), and the production of a 3'-coterminal nested set of 4 to 7 subgenomic mRNAs. The 5' part of the genomes of these viruses is occupied by two large open reading frames (ORF1a and ORF1b) which encode the viral replicase[1,5-12]. Both ORF1a and ORF1b are expressed from the genomic RNA, the latter by means of ribosomal frameshifting[1,13]. The ORF1b products of various members of the coronaviruslike group contain a number of homologous domains[1,6,9] which indicate that the replicase genes of these viruses are evolutionarily related (Fig. 1).

The large coronaviruslike replicase gene product (345K-420K in arteriviruses, 740K-810K in coronaviruses) is a multidomain precursor which is posttranslationally cleaved into smaller functional units. Sequence comparison revealed the presence of multiple putative protease domains in the ORF1a polyproteins of both arteriviruses and coronaviruses[1,5,6,10,14]. One of the coronavirus proteases, responsible for the liberation of the 28K N-terminal replicase product of mouse hepatitis virus (MHV), has

Corona- and Related Viruses, Edited by P. J. Talbot and G. A. Levy
Plenum Press, New York, 1995

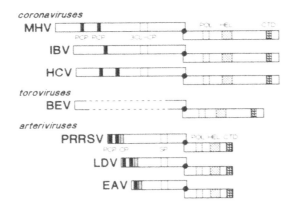

Figure 1. Location of conserved and possibly conserved domains in the replicase proteins of various members of the coronaviruslike superfamily. Abbreviations: PCP, papainlike cysteine protease; 3CL-CP, 3C-like cysteine protease; POL, polymerase motif; HEL, helicase motif; CTD, conserved C-terminal ORF1b domain; CP, cysteine protease; SP, serine protease; IBV, infectious bronchitis virus; HCV, human coronavirus 229E; BEV, Berne torovirus; other virus abbreviations: see text.

recently been characterized[15]. Furthermore, the identification of a number of (putative) coronavirus replicase cleavage products has been reported[16-19]. In the case of the arteriviruses, detailed information on the posttranslational processing of the ORF1a protein has now been obtained. Six proteolytic cleavages and four different viral proteases, all residing within the ORF1a protein, have been documented. The various arterivirus protease domains have been characterized in considerable detail. In this brief review, we will summarize our current knowledge of the complex proteolytic processing of the arterivirus replicase ORF1a protein.

PROCESSING SCHEME OF THE EAV ORF1A PROTEIN

The production of specific antisera, directed against various parts of the EAV ORF1a protein, has enabled us to study its proteolytic processing in infected cells and transient expression systems[20]. Western blotting, immunoprecipitation, and pulse-chase experiments revealed that the EAV ORF1a protein is subject to (at least) five proteolytic cleavages. These generate processing products of 29K, 61K, 22K, 31K, 41K, and 3K, which were named nonstructural protein (nsp) 1 through 6, respectively[20] (Fig. 2). Furthermore, a number of intermediate processing products was detected, of which only the smallest three have been identified so far: nsp3456, nsp34, and nsp56 (Fig. 2). After the relatively rapid cleavage of the nsp1/2, nsp2/3, and nsp4/5 junctions, the final processing steps (at the nsp3/4 and nsp5/6 sites) were found to be extremely slow[20] (Fig. 2). This suggests that, as in other viral systems, processing intermediates fulfil specific functions in the arterivirus replication cycle.

THE ARTERIVIRUS PAPAINLIKE CYSTEINE PROTEASES

Due to its activity in an in vitro translation assay, the papainlike cysteine protease in EAV nsp1, which had been predicted by comparative sequence analysis[1] (Fig. 3), could be identified and characterized before ORF1a protein-specific antisera had been obtained[21]. Residues Cys-164 and His-230 were confirmed as members of the putative catalytic dyad of the EAV nsp1 protease, which was shown to cleave the nsp1/2 site extremely rapidly and exclusively in cis[21]. A Gly-Gly dipeptide at position 260-261 was identified as the nsp1/2 cleavage site[21]. The involvement of the nsp1 protease in the nsp1/2 cleavage was recently

A)

B)

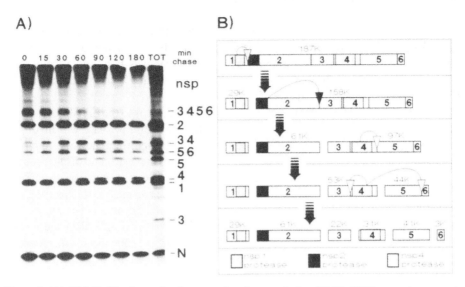

Figure 2. (A) SDS-PAGE of samples from a pulse-chase analysis of EAV ORF1a protein processing in virus-infected cells. A 15 minutes pulse-labeling was followed by chase periods of 0-180 minutes. The lane indicated with TOT shows the results of a continuous 3 hour labeling. Cells were lysed and viral nsps were immunoprecipitated using a mixture of ORF1a-protein specific antisera[20]. (B) Schematic representation of EAV ORF1a protein processing. The location of the six EAV nsps and the order of proteolytic cleavages is shown. The three EAV proteases and their (putative) relation to cleavage sites are indicated and will be discussed below.

confirmed in an in vivo expression system[20]. The nsp1 protease appears to have no other function in replicase processing than the liberation of the N-terminal ORF1a product. Therefore, it can be added to a rapidly growing family of so-called 'leader' proteases[22,23], which is comprised of papainlike autoproteases encountered in various positive-stranded RNA virus groups.

In contrast to the C-terminal half of the ORF1a protein and the ORF1b protein, the N-terminal part of the arterivirus ORF1a product is relatively variable, both in size and in sequence[6,23]. Comparative sequence analysis (Fig. 3) and the data obtained with the EAV nsp1 protease suggested that both PRRSV and LDV might contain an additional PCP domain in the N-terminus of the ORF1a protein. Recent experiments have revealed that this is indeed the case[23]. PRRSV and LDV each contain two PCP domains, provisionally named PCPα and PCPβ, each of which mediates an autoproteolytic cleavage at its own C-terminus, just downstream of the active site His residue. This means that, compared with EAV, these viruses both produce an additional ORF1a protein cleavage product. To avoid differences in nomenclature of highly conserved downstream ORF1a-encoded nsps, we have named the two N-terminal products of PRRSV and LDV nsp1α and nsp1β (Fig. 5). The sizes of these proteins differ slightly between PRRSV (20K and 27K) and LDV (22K and 26K). Both pairs of predicted active site Cys residues (Cys-76 and Cys-276 in PRRSV, Cys-76 and Cys-269 in LDV) were subjected to site-directed mutagenesis and were shown to be essential for proteolytic activity of a PCP domain[23]. Furthermore, the probable catalytic His residues of the PRRSV PCPα and PCPß proteases were identified (His-146 and His-345, respectively). Sequence alignments indicated that the PCPβ domain corresponds to the EAV nsp1 PCP domain. An equivalent of the PCPα domain may have been present in EAV[23] (Fig. 3).

THE NSP2 CYSTEINE PROTEASE

Only the EAV nsp1 protease was found to be active in an in vitro translation system[21]
We therefore continued our processing analysis using polyclonal rabbit antisera, EAV-in-
fected cells, and an in vivo expression system The latter was based on vaccinia virus
recombinant vTF7-3, which produces the T7 RNA polymerase[24], and transfection of T7
expression plasmids containing (parts of) EAV ORF1a[20]

The analysis of both deletions and site-directed mutations in the EAV ORF1a protein
indicated that, in addition to the nsp1 and nsp4 proteases, a third protease was involved in
its processing When the nsp1 and nsp4 sequences were deleted from the polyprotein[25], or
when the active site residues Cys-164 and Ser-1184 of the nsp1 and nsp4 proteases,
respectively, were mutated[25], the processing of the nsp2/3 cleavage site was not affected
Our deletion mutagenesis indicated that, if viral, the protease responsible for cleavage of the
nsp2/3 junction should be located within nsp2 or nsp3 Although we have not yet been able
to completely exclude the involvement of a host protease in the nsp2/3 cleavage, the data
available at this moment are most consistent with a proteolytic function for the conserved
N-terminus of nsp2 This 100 amino acid domain contains the only nsp2/nsp3 conserved
histidine (His-332 in EAV), a residue which is involved in catalysis in most types of
proteases A number of additional observations, including the results of site-directed mutagenesis, suggest that Cys-270 and His-332 (Fig 3) may be active site residues of a
proteolytic enzyme which represents a novel group of cysteine proteases However, our
studies of the nsp2 protease are hampered by the fact that processing of the nsp2/3 junction
appears to be very sensitive to changes in nsp2 Large or small deletions in the region which
separates the putative protease and its cleavage site all abolished cleavage of the nsp2/3 site

the arterivirus nsp1 proteases

```
                      *                    *
papain          qgscgsCWafsa 127   kvdHavaavgyn
                      *                    *
LDV    PCPα      crPgGmCWlssi  62   navHvSdesFPG
PRRSV  PCPα      ctPsGcCWlsav  61   nsmHvSdqpFPG
EAV ex-PCP?      pvPvGhkfLigw  40   stdHaSakrFPG

LDV    PCPß      dtkfskCWekif  62   yirHvsragepv
PRRSV  PCPß      dvfdgkCWlscf  60   wirHlt-ldddv
EAV    PCP       qeqdgfCWlkll  67   rawHittrsckl
                      *                    *
```

the arterivirus nsp2 proteases

```
                  *                           *
LDV    CP   gYsPPgDGaCGlhcisAmlN  42  CPsAiYkldcvnqHWtV  11  LapdClrGvC
PRRSV  CP   tYsPPtDGsCGwhvlaAimN  44  CPnAkYliklngvHWeV  10  LsreCvvGvC
EAV    CP   gYnPPgDGaCGyrcl-AfmN  39  CPnAkYamicdkqHWrV   9  LdesCfrGiC
                  *                           *
```

the arterivirus nsp4 proteases

```
                  *              *            *
trypsin     wVvsAaHcy  39  ltinnDi  86  scmGDSGgp------vvcsgklqGivswgsg
                  *              *            *
LDV    SP   vVvTAsHll  17  FKcaGDy  45  TkcGDSGSp  9  GiHTGSNkrGsgmVTTh
PRRSV  SP   tVvTAaHvl  17  FKtnGDy  46  TncGDSGSp  9  GiHTGSNklGsglVTTp
EAV    SP   vVlTAsHvv  18  FKknGDf  48  TtsGDSGSa  8  GvHTGSNtsGvayVTTp
                  *              *            *
```

Figure 3. Comparative sequence analysis of the four arterivirus protease domains Absolutely conserved
amino acid residues are shown in capitals Putative active site residues are indicated with asterisks For
comparison, the cellular proteases papain and trypsin have been included in the alignment of the nsp1 and nsp4
proteases, respectively

This was also the case when three additional conserved Cys residues in the N-terminal domain of nsp2 were substituted. On the other hand, a 10K heterologous insertion in the central part of nsp2 was tolerated[20] and a separately expressed putative protease domain was shown to be able to induce cleavage of the nsp2/3 site in trans[25]. It appears that a specific folding or posttranslational organization of the nsp2-nsp3 complex is required for proteolysis of the nsp2/3 site. Deleterious effects on the organization of the protease domain and/or the cleavage site region may explain the results obtained with many of our nsp2 mutants carrying amino acid substitutions or deletions.

Both the presence of nsp2 and the cleavage of the nsp2/3 junction appear to be extremely important for processing of the arterivirus ORF1a protein. Preliminary observations from ORF1a in vivo expression experiments indicate that the nsp4 protease is inactive when nsp2 is absent or when nsp2 contains mutations which abolish cleavage at the nsp2/3 junction. Under these conditions, only the nsp1 protease is able to function normally[26]. To increase our understanding of the role of nsp2 in ORF1a protein processing, we are currently studying two properties of the nsp2-nsp3 region which may be related to its structure. First, the conserved nsp2 N-terminus, the conserved nsp2 C-terminus, and nsp3 each contain a cluster of conserved cysteine residues[20], suggesting that disulfide bridges may be involved in the organization of this part of the ORF1a protein. Second, immunofluorescence experiments have revealed that, in infected cells, nsp2 is associated with intracellular membranes[26].

THE NSP4 SERINE PROTEASE

Comparative sequence analysis produced a convincing alignment of the arterivirus nsp4 region and the proteases that belong to the superfamily of chymotrypsinlike serine proteases and 3C-like cysteine proteases[1,6,27,28]. A putative catalytic triad of His, Asp and Ser was identified. For EAV, this prediction is now supported by the results from site-directed mutagenesis of two of these putative active site residues, His-1103 and Ser-1184. Their replacement abolishes the processing of the nsp3456 region, but does not affect the production of nsp1 and nsp2[26]. Together with data obtained from the mutagenesis of putative SP cleavage sites, these results indicate that the nsp3/4, nsp4/5, and nsp5/6 cleavages are all mediated by the nsp4 protease. Since the SP is able to cleave multiple cleavage sites, albeit with different kinetics (Fig. 2), it is the most likely candidate protease to be involved in the processing of the ORF1b product. The coronavirus 3C-like cysteine protease, which is located at a comparable relative position in the replicase, has recently been shown to be involved in processing of the coronavirus ORF1b protein[19].

CLEAVAGE SITES

Six different proteolytic cleavages have now been documented in the various arterivirus ORF1a proteins (Fig. 5). However, only one cleavage site has been determined by protein sequence analysis, the nsp1/2 site between Gly-260 and Gly-261 in EAV[21]. For the corresponding PRRSV and LDV nsp1β/2 sites, processing is predicted to occur at a Tyr-Gly dipeptide, which is present in both viruses and which can be aligned with the EAV nsp1/2 cleavage site[23] (Fig. 4). This prediction is supported by the sizes which have been observed for the PRRSV and LDV cleavage products after in vitro translation of cDNA constructs[23]. A suitable cleavage site candidate for the nsp1α/1β site in PRRSV and LDV has not been identified, although a quite accurate estimate of its location was obtained from the same in vitro translation experiments[23].

```
        nsp1  ▼  nsp2                           nsp2  ▼  nsp3
LDV     fqtrkyY GYsPPgD            LDV          glrfvsG GqIadfv
PRRSV   fgahkwY Gaagkra            PRRSV        istkttG Gasytla
EAV     lpagnyG GYnPPgD            EAV          pgfrliG Gwiygic


        nsp3  ▼  nsp4                           nsp4  ▼  nsp5                          nsp5  ▼  nsp6
LDV     nfgsvlE Gslrtrg            LDV          eslpalE Galssmq         LDV            igesdlE Aerltvd
PRRSV   avgsllE Gafrthk            PRRSV        asvpvvE Gglstvq         PRRSV          kpdnclE Aaklsle
EAV     eggmvfE Glfrspk            EAV          dglsnrE Sslsgpq         EAV            lgkgsyE Gldqdkv
```

Figure 4. Comparison of possible cleavage sites in arterivirus ORF1a proteins.

The expression of an nsp2 C-terminal deletion mutant[25] revealed that the EAV nsp2/3 site should be situated near amino acid 825 of the ORF1a protein. A Gly-Gly dipeptide (residues 831-832) appears to be the most attractive candidate (Fig. 4): it is conserved in PRRSV and LDV and it is followed directly by a very hydrophobic domain, which would in this case be the N-terminus of nsp3.

Cleavage site predictions for the arterivirus serine protease were already published before the proteolytic activity of nsp4 and the processing map of the ORF1a protein had been addressed experimentally[6]. On basis of the putative relationship of the nsp4 protease with the so-called 3C-like cysteine protease group, a set of conserved cleavage sites was proposed which is characterized by a Glu (or Asp) at the P1 position, a Gly (or Ser) at the P1' position, and a bulky hydrophobic residue (e.g. Val, Phe, or Leu) at the P3' position. For the EAV ORF1a protein, SP-directed cleavages at the C-terminal side of aa 1064 (Glu/Gly), 1268 (Glu/Ser), and 1430 (Glu/Gly) were predicted (Fig. 4). We have recently substituted the Glu residues at each of these sites with Pro. Subsequently, these mutations were tested in an ORF1a cDNA construct which was expressed in vivo. Replacement of Glu-1064 abolished cleavage of the nsp3/4 junction[26]. Since this position in the ORF1a protein sequence also corresponds well with the estimated sizes of nsp3 and nsp4, this residue may indeed be located at (or near) the nsp3/4 cleavage site. This preliminary identification of the nsp3/4 junction is supported by mutagenesis of another conserved Glu residue, located at position 1677 in EAV. When this Glu, which is followed by Gly in EAV and Ala in PRRSV/LDV, was replaced by Pro, the EAV nsp5/6 site was no longer processed[26]. The two putative candidates for the nsp4/5 cleavage site are more problematic: the estimated N-terminus of nsp5 (44K from the C-terminus of the ORF1a protein) is in the middle between the two predicted cleavage sites at residues 1268 and 1430, which are 53K and 32K from the ORF1a C-terminus, respectively. A Glu1430→Pro substitution did not have an effect on the generation of any of the currently known ORF1a protein cleavage products[26]. On the other hand, the Glu1268→Pro replacement completely abolished processing of the nsp4/5 junction[26]. If the nsp4/5 site would indeed be located at residue 1268, the sizes of both nsp4 and nsp56 would differ considerably from the size values estimated from SDS-PAGE: nsp4 would have to be 31K instead of 22K, nsp56 would not be 44K but 53K. Since the Glu1268→Pro substitution also abolished cleavage at the nsp34 junction, this mutation, which is relatively close to the SP active site, may affect the structure of the protease rather than the nsp4/5 cleavage site. Additional experiments, which will also include possible cleavage sites in the ORF1b protein, are now in progress.

ARTERIVIRUS ORF1A PROCESSING SCHEMES AND PRODUCTS

Our current knowledge of the proteolytic processing of the arterivirus ORF1a protein, and the viral proteases involved in it, is summarized in Fig. 5. The three available arterivirus

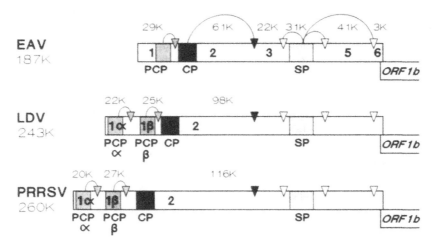

Figure 5. Comparison of the (partially putative) proteolytic processing schemes of the EAV, LDV and PRRSV replicase ORF1a proteins. The position of the four arterivirus proteases and the approximate location of cleavage sites are indicated.

replicase sequences are highly similar in the region which covers the C-terminus of nsp2, nsp3 through 6, and the ORF1b product[6,23]. We therefore expect that in this part of the replicase the proteolytic processing will be identical or largely similar for EAV, LDV, and PRRSV. The processing of one of the two relatively variable regions, the N-terminal domain of the ORF1a protein containing PCPα, has now been analyzed[23]. This leaves the variable central region of nsp2 to be studied. The high sequence similarity between the three arteriviruses abruptly ends downstream of the putative nsp2 cysteine protease domain. It is restored about 250 amino acids downstream in EAV, with a domain that contains a cluster of six invariant cysteines. However, the size (600 and 730 amino acids in LDV and PRRSV, respectively) and the sequence of the interlying central region of nsp2 are variable[6,20]. The presence of a hydrophobic domain appears to be the only common property. Obviously, we cannot exclude that the proteolytic processing of this region of the PRRSV/LDV ORF1a proteins differs from that of EAV.

The presence of hydrophobic domains at both sides of the putative main protease, either the SP domain or the 3C-like protease, is a property shared by arteriviruses and coronaviruses. Almost the entire arterivirus nsp3 sequence is hydrophobic. The only hydrophylic domain contains a cluster of conserved cysteine residues. Depending on the exact location of the nsp4/5 cleavage site (see above), a hydrophobic region downstream of the SP motif could form a C-terminal hydrophobic domain of nsp4 or an N-terminal hydrophobic region of nsp5. A more downstream hydrophobic part between amino acids 1420 and 1550 in EAV, which is again conserved in PRRSV and LDV, should definitely be within nsp5. Little can be said about the small nsp6. The extreme C terminus of the ORF1a protein is not very conserved. Although nsp6 itself may be a functional protein, the slow nsp5/6 cleavage could also serve to liberate the ORF1b-encoded proteins from the ORF1ab protein which is produced by ribosomal frameshifting[1].

CONCLUDING REMARKS

Our analysis of the processing of the arterivirus replicase has revealed the involvement of (at least) four viral proteases, an unprecedented number among positive-stranded

RNA viruses Three of these appear to be autoproteases, mediating only the liberation of nsp1α, nsp1β (or nsp1 in EAV), and nsp2 from the rest of the replicase polyprotein Although a detailed processing map of the coronavirus ORF1a protein is not yet available, a general pattern for coronaviruses and arteriviruses seems to emerge The functions encoded from the central region of ORF1a to the 3′-end of ORF1b appear to be the 'core' of the coronaviruslike replicase polyprotein the well-conserved domains (protease - polymerase - helicase - C-terminal 'unique' ORF1b domain) are within this area, and only small insertions and deletions in this part of the replicase can be detected within the coronavirus or arterivirus groups In comparable positions, the ORF1a proteins of both virus groups contain a domain belonging to the protease superfamily that comprises the chymotrypsin-like and picornavirus 3C-like proteolytic enzymes Although the predicted catalytic nucleophile of the coronavirus protease (Cys) differs from that in the arterivirus proteases (Ser), this domain may still be a remnant from a common ancestor of both virus groups The exchange of Cys for Ser at the active site of the enzyme is considered to be feasible[27 28] Furthermore, the predicted cleavage site sequences are remarkably similar for both virus groups

The N-terminal half of the ORF1a protein, on the other hand, is quite variable, suggesting that species-specific, rather than group-specific, functions have been added to the set of basic functions provided by the 'core' replicase It seems that in both corona- and arteriviruses multiple cysteine autoproteases mediate the processing of this least conserved part of the coronaviruslike replicase

ACKNOWLEDGMENTS

The experiments on the N-terminal processing of the replicase proteins of LDV and PRRSV were carried out in collaboration with Kay Faaberg and Peter Plagemann (University of Minnesota, Minneapolis, U S A) and Janneke Meulenberg (Institute for Animal Science and Health, Lelystad, the Netherlands), respectively The identification and characterization of the EAV nsp2 protease was carried out in collaboration with Alexander Gorbalenya (Institute of Poliomyelitis and Viral Encephalitides, Moscow, Russia, and Purdue University, West-Lafayette, U S A)

REFERENCES

1 Den Boon, J A , Snijder E J , Chirnside, E D , de Vries, A A F , Horzinek, M C , Spaan, W J M Equine arteritis virus is not a togavirus but belongs to the coronaviruslike superfamily J Virol 1991,65 2910-2920

2 Snijder, E J , Horzinek, M C , Spaan, W J M The coronaviruslike superfamily Adv Exp Med Biol 1993,342 235-244

3 Snijder, E J , Horzinek, M C Toroviruses replication, evolution and comparison with other members of the coronaviruslike superfamily J Gen Virol 1993,74 2305-2316

4 Plagemann, P G W , Moennig, V Lactate dehydrogenase-elevating virus, equine arteritis virus, and simian haemorrhagic fever virus a new group of positive-stranded RNA viruses Adv Virus Res 1991,91 99-192

5 Meulenberg, J J M , Hulst, M M , de Meijer, E J , Moonen, P L J M , den Besten, A , de Kluyver, E P , Wensvoort, G , Moormann, R J M Lelystad virus, the causative agent of porcine epidemic abortion and respiratory syndrome (PEARS), is related to LDV and EAV Virology 1993,192 62-72

6 Godeny, E K , Chen, L , Kumar, S N , Methven, S L , Koonin, E V, Brinton, M A Complete genomic sequence and phylogenetic analysis of the lactate dehydrogenase-elevating virus Virology 1993,194 585-596

7 Boursnell, M E G , Brown, T D K , Foulds, I J , Green, P F , Tomley, F M , Binns, M M Completion of the sequence of the genome of the coronavirus avian infectious bronchitis virus J Gen Virol 1987,68 57-77

8 Bredenbeek, P J , Pachuk, C J , Noten, J F H , Charite, J , Luytjes, W , Weiss, S R , Spaan, W J M The primary structure and expression of the second open reading frame of the polymerase gene of the coronavirus MHV-A59 Nucleic Acids Res 1990,18 1825-1832

9 Snijder, E J , den Boon, J A , Bredenbeek, P J , Horzinek, M C , Rijnbrand, R , Spaan, W J M The carboxyl-terminal part of the putative Berne virus polymerase is expressed by ribosomal frameshifting and contains sequence motifs which indicate that toro- and coronaviruses are evolutionary related Nucleic Acids Res 1990,18 4535-4542

10 Lee, H J , Shieh, C K , Gorbalenya, A E , Koonin, E V , la Monica, N , Tuler, J , Bagdzhadzhyan, A , Lai, M M C The complete sequence (22 kilobases) of murine coronavirus gene 1 encoding the putative protease and RNA polymerase Virology 1991,180 567-582

11 Herold, J , Raabe, T , Schelle-Prinz, B , Siddell, S G Nucleotide sequence of the human coronavirus 229E RNA polymerase locus Virology 1993,195 680-691

12 Bonilla, P J , Gorbalenya, A E , Weiss, S R Mouse hepatitis virus strain A59 RNA polymerase gene ORF1a heterogeneity among MHV strains Virology 1994,198 736-740

13 Brierley, I , Diggard, P , Inglis, S C Characterization of an efficient coronavirus ribosomal frameshifting signal requirement for an RNA pseudoknot Cell 1989,57 537-547

14 Gorbalenya, A E , Koonin, E V , Donchenko, A P , Blinov, V M Coronavirus genome prediction of putative functional domains in the non-structural polyprotein by comparative amino acid sequence analysis Nucleic Acids Res 1989,17 4847-4861

15 Baker, S C , Shieh, C K , Soe, L H , Chang, M F , Vannier, D M , Lai, M M C Identification of a domain required for autoproteolytic cleavage of murine coronavirus gene A polyprotein J Virol 1989,63 3693-3699

16 Brierley, I , Boursnell, M E G , Binns, M M , Bilimoria, B , Rolley, N J , Brown, T D K , Inglis, S C Products of the polymerase-encoding region of the coronavirus IBV Adv Exp Med Biol 1990,276 275-278

17 Denison, M R , Zoltick, P W , Leibowitz, J L , Pachuk, C J , Weiss, S R Identification of polypeptides encoded in open reading frame 1b of the putative polymerase gene of the murine coronavirus mouse hepatitis virus A59 J Virol 1991,65 3067-3082

18 Denison, M R , Zoltick, P W , Hughes, S A , Giangreco, B , Olson, A L , Perlman, S , Leibowitz, J L , and Weiss, S R Intracellular processing of the N-terminal ORF1a proteins of the coronavirus MHV-A59 requires multiple proteolytic events Virology 1992,189 274-284

19 Liu, D X , Brierley, I , Tibbles, K W , Brown, T D K A 100-kilodalton polypeptide encoded by open reading frame (ORF) 1b of the coronavirus infectious bronchitis virus is processed by ORF 1a products J Virol 1994,68 5772-5780

20 Snijder, E J , Wassenaar, A L M , Spaan, W J M Proteolytic processing of the replicase ORF1a protein of equine arteritis virus J Virol 1994,68 5755-5764

21 Snijder, E J , Wassenaar, A L M , Spaan, W J M The 5' end of the equine arteritis virus replicase gene encodes a papainlike cysteine protease J Virol 1992,66 7040-7048

22 Gorbalenya, A E , Koonin, E V , Lai, M M C Putative papain-related thiol proteases of positive-strand RNA viruses FEBS Lett 1991,288 201-205

23 den Boon, J A , Faaberg, K S , Meulenberg, J J M , Wassenaar, A L M , Plagemann, P G W , Gorbalenya, A E , Snijder, E J Processing of the N-terminal region of the arterivirus replicase ORF1a protein identification of a second papainlike cysteine protease Submitted

24 Fuerst, T R , Niles, E G , Studier, F W , Moss, B Eukaryotic transient-expression system based on recombinant vaccinia virus that synthesizes bacteriophage T7 RNA polymerase Proc Natl Acad Sci USA 1986,83 8122-8126

25 Snijder, E J , Wassenaar, A L M , Spaan, W J M , Gorbalenya, A E The arterivirus nsp2 protease In preparation

26 Snijder, E J , Wassenaar, A L M Unpublished observations

27 Gorbalenya, A E , Donchenko, A P , Blinov, V M , Koonin, E V Cysteine proteases of positive strand RNA viruses and chymotrypsin-like serine proteases a distinct protein superfamily with a common structural fold FEBS Lett 1989,243 103-114

28 Bazan, J F , Fletterick, R J Viral cysteine proteases are homologous to the trypsin-like family of serine proteases structural and functional implications Proc Natl Acad Sci USA 1988,85 7872-7876

72

IDENTIFICATION AND ANALYSIS OF MHV-A59 P28 CLEAVAGE SITE

Scott A Hughes, Pedro J Bonilla, and Susan R Weiss*

Department of Microbiology
University of Pennsylvania School of Medicine
Philadelphia, Pennsylvania 94104-6076

ABSTRACT

During translation of Murine hepatitis virus (MHV-A59) ORF1a, p28, the N-terminal polypeptide is cleaved from the growing polypeptide chain Amino terminal radiosequencing of the resulting downstream cleavage product demonstrated that cleavage occurs between Gly^{247} and Val^{248} Site directed mutagenesis of amino acids surrounding the p28 cleavage site revealed that substitutions of Arg^{246} (P2) and Gly^{247} (P1) nearly eliminated cleavage of p28 Single amino acid substitutions of other residues between P7 and P2' were generally permissive for cleavage although a few changes did greatly reduce proteolysis The amino acids around the p28 cleavage site represent a new sequence recognized by a virus encoded papain-like proteinase

INTRODUCTION

Murine coronavirus, mouse hepatitis virus strain A59 (MHV-A59) is an enveloped, positive stranded RNA virus Upon infection, MHV-A59 gene 1 is translated, presumably into the proteins necessary for viral RNA replication

Two overlapping open reading frames (ORF1a and ORF1b) comprise gene 1 and could potentially encode an approximately 800 kDa polyprotein[1] Sequence analysis of gene 1 predicts ORF1b to encode the RNA polymerase and several other functional domains conserved among coronaviruses, including a helicase and zinc finger like sequence[1 3] ORF1a is predicted to encode two papain-like proteinases and a poliovirus 3C-like proteinase[2 3] It is thought that the ORF1a encoded proteinases are responsible for processing both ORF1a and ORF1b encoded polypeptides into mature replicase proteins

Polypeptides encoded in ORF1a have been detected in infected cells using antibodies directed against sequences encoded in ORF1a These intracellular viral proteins include the amino terminal p28 polypeptide, p65 and a precursor polypeptide of 290 kDa which is processed into p50 and p240[4 6] During *in vitro* translation of genome RNA or of synthetic

RNAs derived from constructs containing amino terminal portions of ORF1a, only p28 and a larger downstream cleavage product are detected[7,8].

Of the predicted ORF1a encoded proteinases, only PLP-1 has been shown to have proteolytic activity. Encoded in the first 3.43-4.23 kb of ORF1a, PLP-1 is responsible for *cis* cleavage of p28 during *in vitro* translation of ORF1a[7,9]. While the catalytic cysteine and histidine residues of PLP-1 have been identified for both A59 and JHM[9,10], the p28 cleavage dipeptide is not known. Based on cleavage sites utilized by other viral papain-like proteinases, it seemed likely that cleavage would occur with a Gly or Ala in the P1 position[11-14]. In the current study, we report that the p28 cleavage site occurs between Gly^{247} (P1) and Val^{248} (P1'). Site directed mutagenesis has identified Arg^{246} (P2) and Gly247 as critical residues for efficient cleavage.

MATERIALS AND METHODS

Construction of Expression Plasmid and PCR Mutagenesis

A *Nar* I-*Hind* III fragment from cDNA ZA11 (15) was cloned into the *Eco* RV and *Hind* III sites of pSP72 (Promega) resulting in pSPZA11NH. This plasmid has MHV-A59 ORF1a sequences downstream of the bacteriophage T7 promoter. A synthetic linker containing a *Kpn* I site was inserted into the *Nde* I site of this vector, downstream of the viral sequences, to create pSPZA11NHK. pSPNK was constructed by cloning a 2.89 kb *Xho* I-*Kpn* I fragment from PCR clone X1K1[2] into the corresponding sites of pSPZA11NHK. pSPNK includes ORF1a sequences from nucleotide 182 to 4664.

Site directed mutagenesis of residues Lys^{241} to Lys^{249} in plasmid pSPNK was accomplished using the polymerase chain reaction (PCR) (16). Specific mutations were introduced into a *Bam* H1-*Eco* R1 fragment using two "inside" primers containing mismatches to the same segment of the target sequence but to the opposite strands. Recombinant PCR products were generated using inside primers in conjunction with the corresponding upstream (Fp425-444) or downstream (Rp1347-1328) outside primers. The amplified DNA was digested with *Bam* H1 and *Eco* R1 and inserted into pSPNK at the *Bam* H1 and *Eco* R1 sites (Fig 1). Arg^{246}Phe was generated by inserting a specific mutation into an *Alf* II-*Eco* R1 fragment as described above. Specific mutations were identified using the Sanger dideoxy sequencing method[17].

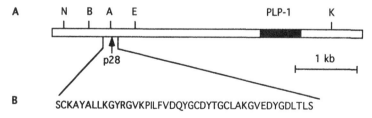

Figure 1. Schematic representation of expression construct and MHV-A59 amino acid sequence surrounding predicted p28 cleavage site. A. The 5' region of MHV-A59 gene 1 from the *Nar* I to *Kpn* I restriction sites was placed under the control of a T7 polymerase promoter The resulting plasmid encodes a 164 kDa protein that is cleaved into a 28 kDa and 136 kDa polypeptides Restiction sites *Bam* H1, *Afl* II and *Eco* R1 were utilized during PCR mutagenesis The predicted p28 cleavage region is indicated by an arrow and the PLP-1 domain is depicted as a shaded region B Amino acid sequence from Ser^{235} to Ser^{278} at and around predicted p28 cleavage site. Abbreviations. N, *Nar* I, B, *Bam* HI, A, *Afl* II, E, *Eco* RI, and K, *Kpn* I

In Vitro Transcription/Translation. in Vitro

In vitro transcriptions and translations were performed as described in Promega Technical Bulletin 126 using a coupled transcription/translation rabbit reticulocyte system. Briefly, 250ng of plasmid DNA was added per 25 μl reaction mix which included 1000 μCi/ml of [^{35}S]-methionine. Reactions were incubated for 90 minutes at 30°C. Equal cpm of samples were added to 2X Laemmli buffer, incubated for 5 minutes at 100°C and then run on a gradient polyacrylamide SDS gel. Gels were processed for fluorography in Autofluor (National Diagnostics) and exposed to X-ray film. To determine the percent cleavage, levels of polyprotein precursor and p136 cleavage product in non-flourographed gels were quantitated with a Molecular Dynamics Phosphorimager.

Radiosequencing N-Terminus of p136

For the radiosequencing of p136, *in vitro* transcription and translation of pSPNK DNA was performed as described above using either 1000 μCi/ml of [^{35}S]-cysteine (ICN) or 800 μCi [^{3}H]-leucine (ICN). Translation products were resolved by electrophoresis on a 5-12% SDS-gradient gel. Electrophoretic transfer onto a PVDF membrane (Bio-Rad) was performed at 1.0 Amp for 3 hours in 0.5X Towbin buffer[18]. Edman degradation was performed on 100 mm^2 of membrane to which p136 was immobilized. The eluate from each cycle was collected, added to scintillation cocktail (ICN) and counted in a liquid scintillation counter.

RESULTS

Identification of the p28 Cleavage Site

In order to determine the amino acid sequence surrounding the p28 cleavage site we used the polypeptides synthesized during transcription/translation of the pSPNK construct. This construct contains ORF1a sequences downstream of the T7 RNA bacteriophage polymerase promoter from nucleotides 182 through 4664 which includes p28 through PLP-1. During transcription and translation of this plasmid, a large polypeptide of 164 kDa is synthesized and quickly processed into p28 and the downstream product, p136[8,19]. Sequencing of the amino terminus of p136 would provide the sequences at the *in vitro* cleavage site of p28.

Thus, to determine the amino terminus of the downstream cleavage product, sequencing of both [^{3}H]-Leu and [^{35}S]-Cys labeled p136 was carried out. After 10 cycles of Edman degradation of [^{3}H]-Leu labeled p136, a peak was observed in cycle 5. After 20 cycles of Edman degradation of [^{35}S]-Cys labeled p136, peaks were observed in fractions 12 and 17. The profile obtained was (aa)$_4$Leu (aa)$_6$ Cys (aa)$_4$ Cys. This pattern matches with the deduced amino acid sequence of MHV-A59 ORF1a (2), in which there is Leu at residue 253 and Cys at 259 and 264. The above data are consistent only with cleavage occurring between Gly^{247} and Val^{248}. These data are not consistent with cleavage at any other sites in the region encoded by the plasmid pSPNK, that includes the first 4.6 kb of the genome.

Characterization of the p28 Cleavage Site

Identification of Gly^{247} and Val^{248} as the dipeptide cleaved by PLP-1, allows for further definition of amino acids required for efficient cleavage of p28. Eight sets of mutations from the P7 position to the P2' position were introduced into the pSPNK construct

P7	P6	P5	P4	P3	P2	P1	P1'	P2'
Leu241	**Leu242**	**Lys243**	**Gly244**	**Tyr245**	**Arg246**	**Gly247**	**Val248**	**Lys249**
Pro +	N.D.	Thr +	Val +/-	Cys +	Cys -	Val -	Asp -	Arg +
Arg +			Ala +	Ser +	Ser -	Ala -	Ala +	Met +
His +/-			Asp +	Phe ++	Gly-	Asp -	Gly ++	Thr +/-
					Leu -	Ser-	Phe+	
					His-	Cys-	Ile +/-	
					Lys -			
					Phe-			

Figure 2. Wild type amino acid sequence and substitutions around the p28 cleavage site The wild type amino acid sequence from the P7 to P2' position is indicated in boldface Corresponding substitutions are shown directly below the wild type residues The notations -, +/-, + and ++ indicate >80% inhibition, 60-80% inhibition, 40-60% inhibition and 20-40% inhibition, respectively Mutations were introduced as described in the Materials and Methods Quantitation of percentage inhibition of cleavage relative to wild type was carried out as described in the Materials and Methods Gels used to quantitate percentage inhibition are not shown The scissile bond is between the P1 and P1' residues

and the effect on cleavage was assayed using the rabbit reticulocyte coupled transcription/translation system as described above. The percentage of inhibition of cleavage was determined by calculating the ratio of uncleaved precursor p164 to p136, relative to wild type levels of p164 and p136 (see Materials and Methods). Both conservative (Ala) and non-conservative (Asp) changes of Gly[247] in the P1 position reduced cleavage by over 80% as compared to wild type levels (Figure 2). These levels of inhibition are consistent with the microsequencing analysis which identified Gly[247] as the P1 residue. Furthermore, these data show that cleavage with MHV PLP-1 is similar to cleavages carried out by other viral papain-like proteinases which require a Gly or Ala in the P1 position for efficient cleavage[3,14,20-22]. Substitution of Arg[246] in the P2 position also resulted in dramatic inhibition of p28 cleavage. Substitution of His and Lys for the positively charged Arg[246] also inhibited cleavage indicating that positive charge alone in the P2 position is not sufficient for cleavage of p28 (Figure 2) Substitution of either Gly[247] or Arg[248] resulted in levels of inhibition greater than or equal to that caused by the addition of the thiol proteinase inhibitor leupeptin which has been previously shown to be a potent inhibitor of p28 cleavage. Substitution for Val in the P1' position with either Gly or Ala did not significantly reduce cleavage although other less conservative substitutions such as Asp or Ile did reduce cleavage by over 60%. This suggests that there are some requirements for the P1' position.

Amino acid substitutions at the P7, P5, P3 and P2' positions were tolerated (Figure 2). Although cleavage was reduced somewhat in most cases, substitution of these residues did not effect cleavage to the degree seen with mutagenizing either the P2 or P1 positions. For example substitution of the aromatic Tyr[245] in the P3 position with a closely related Phe residue did not significantly effect cleavage. This is in sharp contrast to conservative substitutions which have dramatic effects in the P1 and P2 positions.

DISCUSSION

In this report we have demonstrated that p28 cleavage occurs between Gly[247] and Val[248]. This is the first identification of a cleavage site in the large polypeptide encoded in the 21 kb replicase gene (gene 1) of MHV and the first cleavage site identified for any coronavirus proteinase. It has been shown previously that this cleavage is carried out by

PLP-1 encoded in ORF1a Further characterization of the amino acids surrounding the cleavage site demonstrate that Arg[246] and Gly[247] in the P2 and P1 positions respectively are required for cleavage of p28 Val[248] in the P1' position and residues in positions P7, P6, P5, P4, P3 and P2' were not essential for cleavage, although most substitutions did result in some inhibition of cleavage

Sequence alignment analysis of MHV-A59, MHV-JHM and human coronavirus 229E reveals that the RG dipeptide is conserved in all three strains In 229E the P1' Val, which is conserved in the two MHV strains, is replaced with a Gly residue In MHV-A59 Val[248]Gly permits efficient cleavage of p28, indicating that the cleavage site has been relatively conserved among these mouse and human coronavirus strains

Based on sequence analysis, the PLP-1 of MHV-A59 appears to belong to a family of viral papain-like cysteine proteinases[23], this family contains both animal and plant viral proteinases, including the potyviral HC-proteinase, Sindbis virus nsP2, and the papain-like proteinases encoded by chestnut blight hypovirulence associated virus (HAV) and equine arteritis virus (EAV)[14 23] All of these proteinases cleave at sites which have Gly or Ala, uncharged residues with short side chains, in the P1 position It is therefore not surprising to see that PLP-1 of MHV also requires a Gly in the P1 position Our results also suggest that (Arg[246]) in the P2 position is a critical residue required for cleavage by MHV PLP-1 Interestingly the P2 position is a critical residue for cleavage by papain and some of the related cellular proteinases[24] This group of cellular enzymes usually prefers Phe in the P2 position However, these same proteinases can utilize, to different extents, substrates in which the Phe in the P2 position is replaced with Arg The structure of the PLP-1 remains to be elucidated Understanding the structure of the viral proteinases may be important in the design of antiviral strategies

REFERENCES

1 Bredenbeek, P J , Pachuk, C J , Noten, A F H , et al The primary structure and expression of the second open reading frame of the polymerase gene of the coronavirus MHV-A59, a highly conserved polymerase is expressed by an efficient ribosomal frameshifting mechanism *Nuc Acid Res* 18 1825-1832, 1990

2 Bonilla, P J , Gorbalenya, A E and Weiss, S R Mouse hepatitis virus strain A59 RNA polymerase gene ORF 1a heterogeneity among MHV strains *Virology* 198 736-740 1994

3 Lee, H J , Shieh, C K , Gorbalenya, A E , et al The complete sequence of the murine coronavirus gene 1 encoding the putative protease and RNA polymerase *Virology* 180 567-582, 1991

4 Denison, M R , Zoltick, P W , Hughes, S A , et al Intracellular processing of the N-terminal ORF 1a proteins of the coronavirus MHV-A59 requires multiple proteolytic events *Virology* 189 274-284, 1992

5 Hughes, S A , Dension, M L , Bonilla, P J , Leibowitz, J L and Weiss, S R A newly identified MHV-A59 ORF 1a polypeptide p65 is temperature sensitive in two RNA negative mutants *Advances in Experimental Medicine and Biology* 342 221-226, 1993

6 Weiss, S R , Hughes, S A , Bonilla, P J , Leibowitz, J L and Denison, M L Coronavirus polyprotein processing *Arch Virol [Suppl]* 9 349-358, 1993

7 Baker, S C , LaMonica, N , Shieh, C K and Lai, M M C Murine coronavirus gene 1 polyprotein contains an autocatlyic activity *Advances in Experimental Medicine and Biology* 276 283-290, 1990

8 RefNumSty = 8 Denison, M R and Perlman, S Translation and processing of mouse hepatitis virus virion RNA in a cell-free system *J Virol* 60 12-18, 1986

9 Bonilla, P J , Piñon, J L , Hughes, S A and Weiss, S R Characterization of the leader papain-like protease of MHV-A59 *Advances in Experimental Medicine and Biology* In press 1995

10 RefNumSty = Baker, S C , Yokomori, K , Dong, S , et al Identification of the catalytic sites of a papain-like proteinase of murine coronavirus *J Virol* 67 6056-6063, 1993

11 Carrington, J C , Cary, S M , Parks, T D and Dougherty, W G A second proteinase encoded by a plant potyviral genome *EMBO J* 8 365-370, 1989

12 Choi, G H , Shapira, R and Nuss, D L Cotranslational autoproteolysis involved in gene expression from a double stranded RNA genetic element associated with hypovirulence of the chestnut blight fungus *Proc Natl Acad Sci USA* 88 1167-1171, 1991

13 DeGroot, R , Hardy, W R , Shirako, Y and Strauss, J H Cleavage site preferences of Sindbis virus polyproteins containing the non-strucutral proteinase Evidence for temporal regulation of polyprotein processing *in vivo EMBO* 9 2631-2638, 1990

14 Snijder, E J , Wassenaar, A L M and Spaan, W J M The 5' end of the equine arteritis virus replciase gene encodes a papinlike cysteine protease *J Virol* 66 7040-7048, 1992

15 Pachuk, C J , Bredenbeek, P J , Zoltick, P W , Spaan, W J M and Weiss, S R Molecular cloning of the gene encoding the putative polymerase of mouse hepatitis virus strain A59 *Virology* 171 141-148, 1989

16 Higuchi, R , Krummel, B and Saiki, R K A general method of in vitro preparation and specific mutagenesis of DNA fragments study of protein and DNA interactions *Nucl Acids Res* 16 7351-7367, 1988

17 Sanger, F , Nicklen, S and Coulson, A P *Proc Natl Acad Sci USA* 74 5463-5467, 1977

18 Towbin, H , Staehelin, T and Gordon, J Electrophoretic transfer of proteins from polyacrylamide gels to nitrocellulose sheets procedure and some applications *Proc Natl Acad Sci USA* 76 4350-4354, 1979

19 Baker, S C , Shieh, C K , Chang, M F , Vannier, D M and Lai, M M C Identification of a domain required for autoproteolytic cleavage of murine coronavirus gene A polyprotein *J Virol* 63 3693-3699, 1989

20 Carrington, J C and Herndon, K L Characterization of the potyviral HC-Pro autoproteolytic cleavage site *Virology* 187 308-315, 1992

21 Shapira, R and Nuss, D L Gene expression by a hypovirulence-associated virus of the chestnut blight fungus involves two papain-like protease activities *J Biol Chem* 266 19419-19425, 1991

22 Shirako, Y and Strauss, J H Cleavage between nsP1 and nsP2 initiates the processing pathway of sindbis virus nonstructural polyprotein P123 *Virology* 177 54-64, 1990

23 Gorbalenya, A E , Koonin, E V and Lai, M M C Putative papain-related thiol proteases of positive strand RNA viruses *FEBS Letters* 288 201-205, 1991

24 Khouri, H E , Vernet, T , Menard, R , et al Engineering of papain selective alterations of substrate specificity by site-directed mutagenesis *Biochemistry* 30 3929-3936, 1991

73

COMPLETE GENOMIC SEQUENCE OF THE TRANSMISSIBLE GASTROENTERITIS VIRUS

J F Eleouet, D Rasschaert, P Lambert, L Levy, P Vende, and H Laude

Unite de Virologie et Immunologie Moleculaires
INRA
Jouy-En-Josas, France

We have sequenced the gene 1 of TGEV Purdue-115 strain This completes the TGEV genome sequence, which is 28579 nucleotide-long plus the poly(A) tail TGEV is the fourth coronavirus to be completely sequenced after IBV[1], MHV[2 3 4] and HCV 229E[5], and the second of a genetic subset which includes FIPV, CCV, HCV 229E and PEDV

Three cDNA libraries were constructed for sequencing of the TGEV gene 1, using TGEV-specific oligonucleotides as primers and genomic RNA purified from virions as a template A region absent of these libraries, corresponding to nucleotides 7522 - 10727, was isolated through cloning of a RT-PCR-amplified DNA fragment As for the three other coronaviruses sequenced so-far, the TGEV gene 1 contains two long ORFs, ORF1a and ORF1b Results from in vitro translation experiments indicated that translation of TGEV ORF1b also is mediated by a (-1) ribosomal frame-shifting mechanism Sequence analysis of the 5'-end of the genome revealed the presence of a three amino acid-long ORF upstream from and in frame with the initiator codon of ORF1a, instead of 8 amino acids for MHV and 11 amino acids for IBV and HCV 229E[1 2 3 5 6]

Pairwise comparison of the amino acid sequence of TGEV ORF1a with those of MHV, IBV and HCV 229E, confirmed that the predicted ORF1a polypeptide is more divergent than the predicted ORF1b product This also applies to TGEV and HCV229E, which are the two most closely related of these viruses In particular, a sequence positioned from amino acids 1435 to 1602 in HCV 229E ORF1a is not present in TGEV and largely accounts for the largest ORF1a of HCV 229E compared to TGEV Like for IBV, MHV and HCV 229E, the TGEV ORF1a predicted product contains putative functional domains including two papain-like protease domains (amino acids 1080-1273 and 1575-1770), a 3C-like protease domain (2877-3179) framed by two clearly hydrophobic regions at amino acids 2640-2980 and 3200-3450, and a growth factor/receptor-like domain (3862-4005) (Fig 1)

The TGEV ORF1b contains putative RNA polymerase (positions 530-832 from the first amino acid of ORF1b), metal ion binding domain (920-995) and helicase domain (1198-1297) (Fig 1) Multiple alignment of ORF1b showed a good conservation of the amino acid sequence all over the ORF, except in a 60-100 amino acid domain where the sequences

Corona- and Related Viruses, Edited by P J Talbot and G A Levy
Plenum Press New York, 1995

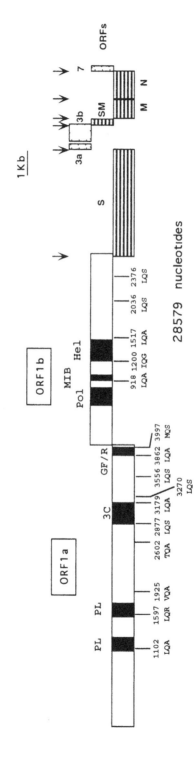

Figure 1. Organization of the TGEV coding sequences. The major ORFs are represented as boxes in the 0, -1 and -2 reading frames. Black boxes indicate putative functional domains. Streaked boxes corresond to structural genes. Vertical arrows indicate functional intergenic consensus sequences. Amino acid positions are given from the beginning of the ORF. PL, papain-like protease, 3C, 3C-like protease, GF/R, growth factor/receptor-like domain, Pol, polymerase, MIB, metal ion binding site, Hel, helicase. Putative 3C-like cleavage sites are also indicated on figure 1.

are much more variable (nucleotides 18715-19007) This region contains a MHV packaging signal[7]

REFERENCES

1 Boursnell M E G , Brown, T D K , Foulds, I J , Green, P F , Tomley, F M , and Binns, M M Completion of the sequence of the genome of the coronavirus avian infectious bronchitis virus J Gen Virol 1987, 68 57-77

2 Soe L H , Shieh, C -K , Baker, S C , Chang, M -F , and Lai, M M C Sequence and translation of the murine coronavirus 5'-end genomic RNA reveals the N-terminal structure of the putative RNA polymerase J Virol 1987, 61 3968-3976

3 Pachuk C J , Bredenbeek, P J , Zoltick, P W , Spaan, W J M , and Weiss, S Molecular cloning of the gene encoding the putative polymerase of mouse hepatitis coronavirus, strain A59 Virology 1989, 171 141-148

4 Lee H -J , Shieh, C -K , Gorbalenya, A E , Koonin, E V , La Monica, N , Tuler, J , Bagdzhadzhyan, A , and Lai, M M C The complete sequence (22 kilobases) of murine coronavirus gene 1 encoding the putative proteasses and RNA polymerase Virology 1991, 180 567-582

5 Herold J , Raabe,T , Schelle-Prinz, B , and Siddell, S G Nucleotide sequence of the human coronavirus 229E RNA polymerase locus Virology 1993, 195 680-691

6 Brown, J D K , Boursnell, M E G , Binns, M M , and Tomley, F M Cloning and sequencing of the 5' terminal sequences from avian infectious bronchitis virus genomic RNA J Gen Virol 1986, 67 221-228

7 Makino, S , Yokomori, K , and Lai, M M C Analysis of efficiently packaged defective interfering RNAs of murine coronavirus localization of a possible RNA-packaging signal J Virol 1990, 64 6045-6053

8 Gorbalenya A E , Koonin, E V , Donchenko, A P , and Blinov, V M Coronavirus genome prediction of putative functional domains in the non-structural polyprotein by comparative amino acid sequence analysis Nucleic Acid Res 1989, 17 4846-4861

Transcription and Replication of Viral RNA

74

TRANSCRIPTION, REPLICATION, RECOMBINATION, AND ENGINEERING OF CORONAVIRUS GENES

Michael M. C. Lai

Howard Hughes Medical Institute and
Department of Molecular Microbiology
University of Southern California School of Medicine
Los Angeles, California 90033

INTRODUCTION

When one discusses coronavirus RNA synthesis, several unique features of coronavirus RNAs immediately come to mind[1]: the extraordinary length (27-31 kb) of the RNA genome, the subgenomic mRNAs with a leader RNA derived from the 5'-end of the genome, and the presence of common sequences between the 3'-end of the leader RNA and the sequence (intergenic sequence, IG) preceding each gene (Figure 1). These characteristics indicate that coronavirus RNA synthesis has to be discontinuous, so that the leader RNA can be joined with mRNAs. This feature, coupled with several unique phenomena associated with coronavirus replication, e.g., high frequency of RNA recombination[2] and rapid generation and evolution of defective interfering (DI) RNA[3], suggests that coronavirus RNA polymerase is probably nonprocessive, being able to jump between different RNA regions and RNA molecules during synthesis. This discontinuous nature of coronavirus RNA synthesis appears to contrast with another requisite function of RNA polymerase, i.e., faithful replication of the long RNA genome. These seemingly conflicting demands on coronavirus RNA polymerase suggest that the coronavirus polymerase is very versatile and the mechanism of coronavirus RNA synthesis is unique.

The first macromolecular event after virus entry is the synthesis of a viral RNA polymerase from the incoming viral genome, which is then copied into a (-)-strand RNA. Subsequently, the (-) strand RNA serves as a template for the synthesis of mRNAs and genomic RNA. Thus, the discussion of coronavirus RNA synthesis should cover four separate but intertwined issues:

1. Synthesis, processing and function of RNA polymerase
2. Synthesis of (-)-strand RNA
3. Synthesis of subgenomic mRNAs (transcription)
4. Synthesis of genomic RNA (replication).

Corona- and Related Viruses, Edited by P J Talbot and G A Levy
Plenum Press, New York, 1995

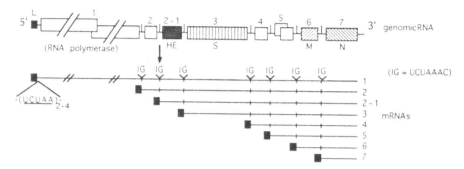

Figure 1. The structure of MHV genomic RNA and subgenomic mRNAs.

In this review, I will discuss only the last three issues. The studies of the synthesis and processing of polymerases have just begun to come of age, and will be extensively reviewed elsewhere in the book.

MODELS OF CORONAVIRUS SUBGENOMIC mRNA SYNTHESIS

Several potential models for explaining coronavirus subgenomic mRNA synthesis have been proposed, which can be classified into continuous and discontinuous mechanisms (Figure 2). One continuous mechanism proposes that the joining of mRNAs with the leader sequence occurs by a continuous transcription using a looped-out template. This may occur during either (+)- or (-)-strand RNA synthesis. Alternatively, the mRNAs are proposed to be amplified from the virion-packaged subgenomic mRNAs (termed "subgenomic replicons")[4] by a faithful continuous transcription. The discontinuous transcription mechanisms include three models: The first is leader-primed transcription, in which the leader RNA is synthesized from the 3'-end of the full-length, (-)-strand RNA template, dissociates from the template, and then rejoins the template at the IG sites to serve as a primer for mRNA transcription[5]. In this model, IG sequences serve as promoter and transcription initiation sites. The second model proposes that discontinuous transcription occurs during (-)-strand RNA synthesis[6]; thus, IG sites serve as transcription terminators. By an unknown mechanism, the (-)-strand, subgenomic transcript then joins the leader RNA and continues to transcribe an antigenomic complement of the leader sequence. The antigenomic, subgenomic RNAs serve as the templates for faithful transcription of the subgenomic mRNAs. The third model is the classical *trans*-splicing mechanism[7], in which subgenomic mRNAs and leader RNAs are transcribed separately and then joined together by a post-transcriptional *trans*-splicing mechanism. This may occur during either (+)- or (-)-strand RNA synthesis. Although the continuous transcription models cannot be rigorously ruled out, the existing evidence favors the discontinuous models because of recent evidence showing that the subgenomic mRNAs and leader RNAs are derived from two separate RNA molecules[8,9]. The different discontinuous transcription models, however, have not been clearly distinguished. The two favored models, the leader-primed transcription and discontinuous (-)-strand RNA synthesis models, are both consistent with some of the experimental data. It has thus been proposed that both models might operate during two different stages of transcription, so-called primary transcription and secondary transcription[10]. Although these two models differ in the stage of transcription at which discontinuous synthesis occurs, i.e., either during (+)- or (-)-strand RNA synthesis, both models have a common feature, i.e., both require that the IG sequence

CONTINUOUS

1) Continuous RNA synthesis on the "looped-out" template

2) Amplification of virion-packaged subgenomic mRNAs (subgenomic "replicon")

DISCONTINUOUS

1) Leader-primed transcription during positive-strand synthesis

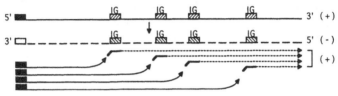

2) Discontinuous transcription during negative-strand synthesis

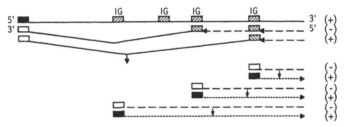

3) *Trans*-splicing

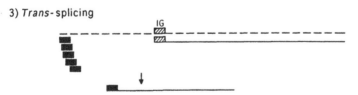

Figure 2. Models of coronavirus subgenomic mRNA transcription.

interacts with the 3′-end of the leader sequence, so that the leader RNA can join the mRNAs at precise sites. These two models cannot yet be unequivocally distinguished.

The following discussion will focus on the regulation of RNA synthesis, including the *trans*- and *cis*-acting regulatory sequences for the synthesis of various RNA species, and the interaction between the leader and IG sequences. These features are critical to coronavirus RNA synthesis regardless of the actual mechanism of RNA transcription.

THE REGULATORY RNA SEQUENCES FOR CORONAVIRUS mRNA TRANSCRIPTION AND GENOMIC RNA REPLICATION

The discontinuous transcription models predict that mRNA transcription is regulated by the leader and IG sequences. The importance of these regulatory elements has been examined extensively, using a DI RNA as an experimental model, in the past several years[11,12]. The DI RNA can be considered a mini-genome, because it contains both the 5'- and 3'-ends of the genomic RNA and can replicate autonomously in the presence of a helper virus, which provides RNA polymerases. The replication signals for coronavirus RNA have been determined using these DI RNAs, and localized to approximately 400 nucleotides at both ends[13,14]. For some DI RNA species, a short stretch of internal sequences is also needed for RNA replication, probably to maintain proper RNA conformation[13,14]. By inserting an intergenic sequence into these DI RNA vectors, it has been unequivocally established that IG sequences are the necessary regulatory elements for initiating RNA synthesis[11,12]. As few as seven nucleotides (UCUAAAC) are sufficient[15]. More recently published reports further established that the upstream sequences at the 5'-end of the genome are also required for mRNA transcription initiating from the IG sequence[16]. These upstream sequences include the leader sequence and some ill-defined sequences located at the 5'-end of the genomic RNAs. The most surprising finding is that the leader sequence is not sufficient. If the 5'-end sequences immediately downstream of the leader are replaced with exogenous sequences or even the sequences of subgenomic mRNAs, no mRNA transcription occurs despite the presence of both the leader and IG sequences[16]. These results indicate that certain sequences from the genomic RNAs are required for mRNA transcription, and that only the genomic RNA can be used for mRNA synthesis. These findings thus suggest that mRNAs are not likely the results of amplification of the virion-packaged subgenomic mRNAs[4]. The leader RNA also provides an enhancer-like function for mRNA transcription[16]. Thus, the leader RNA has dual functions: providing the leader RNA for mRNAs and enhancing transcription. Most interestingly, recent data indicate unequivocally that the leader RNA of mRNAs is derived from the helper virus RNA, thus representing a *bona fide trans*-acting leader RNA[8,9,16]. Furthermore, the efficiency of transcription is affected by the nature of the leader sequence not only of the DI RNA, but also of the helper viral RNA[9]. Therefore, the leader RNA can regulate transcription, not only *in cis* but also *in trans*, and the extent of the influence of the leader RNA on transcription varies with the nature of the IG sequence[9,16]. These studies thus establish that mRNA transcription is regulated by three components: IG, the *cis*-acting leader and a *trans*-acting leader.

THE POSSIBLE MECHANISM OF THE LEADER-IG INTERACTIONS

Because of the presence of homologous sequences between the IG sites and 3'-end of the leader RNA, it was previously proposed that the leader RNA can bind to the intergenic sequence via this stretch of sequences, which are complementary between the (+)-strand leader and the (-)-strand IG[5]. Site-specific mutagenesis of IG sequences indeed caused alteration of the transcription efficiencies from the IG site[12,17]. However, there is no reasonable correlation between the transcription efficiency and the extent of sequence homology between the IG site and the leader[17]. In fact, in most coronaviruses (one exception being MHV), the amounts of subgenomic mRNAs do not correlate with the extents of homology. Therefore, direct RNA-RNA interaction cannot explain satisfactorily the transcriptional regulation of mRNAs. The possibility that the coronaviral transcriptional regulation is mediated by factors other than direct RNA-RNA interaction recently has been suggested by a couple of studies, which demonstrated that the leader-mRNA fusion occurs outside of these complementary sequences[17,18]. The most dramatic example is the transcription by a murine coronavirus, MHV-2C. This virus transcribes mRNAs which contain unusually aberrant leader-fusion sites; there is no sequence homology between the leader and these sites[18]. These data strongly suggest that direct RNA-RNA interaction may not be the driving force behind coronavirus mRNA transcription.

An alternative possibility is that the transcriptional regulation is mediated by RNA-protein and protein-protein interactions. It is likely that certain cellular or viral proteins bind to the RNA regulatory regions; these proteins then interact with each other to bring the essential RNA components together to form a transcription or replication complex[18]. Thus, the selection of the transcriptional initiation sites may be mediated by the specificity of these RNA-binding proteins, which can be considered transcription factors. This model is similar to DNA-dependent RNA transcription, in which transcription is regulated mainly by the interactions of transcription factors. Some of the cellular proteins interacting with the leader RNA and intergenic sequences of both strands have been detected[19]. These cellular proteins conceivably are the transcription factors which regulate coronavirus mRNA transcription.

THE REGULATION OF (-)-STRAND RNA SYNTHESIS

The (-)-strand RNA is the template used for the synthesis of the subgenomic mRNAs and genomic RNA. Thus, it controls the synthesis of all the major viral RNA species. Several features of coronavirus (-)-strand RNAs are already known: (1) Both subgenomic- and genomic-sized (-)-strand RNAs are present in the infected cells[4,6]. (2) (-)-strand RNAs contain both an antigenomic-sense leader RNA at the 3'-end and poly(U) sequences at the 5'-end, thus representing complementary copies of the (+)-strand RNAs[20,21]. (3) (-)-strand RNA synthesis peaks slightly ahead of (+)-strand RNA synthesis and gradually declines during the viral replication cycle[22]; however, the ability to synthesize (-)-strand RNA persists throughout the infection[10]. (4) (-)-strand RNAs in the cells are present in the double-strand (ds) RNA forms[22,23]; thus, the ds RNA may be the actual template for RNA transcription and replication.

This possibility has not been rigorously tested. The regulation of (-)-strand RNA synthesis has not been as well studied as (+)-strand RNA because the amount of (-)-strand RNA is very small in the infected cells. Recently, our laboratory has optimized a ribonuclease protection method and utilized a transfected DI RNA to study the mechanism of the regulation of (-)-strand RNA synthesis[24]. The results demonstrated that the *cis*-acting signal for (-)-strand RNA synthesis resides entirely within the 55 nucleotides at the 3'-end, plus poly(A) sequences. No specific 5' upstream sequence is required[24]. This sequence requirement is considerably less stringent than those required for RNA transcription and replication. Most notably, coronavirus RNA replication requires more than 400 nucleotides of virus-specific sequences at the 3'-end[13,14]. Why RNA replication requires a longer stretch of specific sequences at the 3'-end of the genomic RNA than that required for (-)-strand RNA synthesis is of considerable interest. This finding, therefore, suggests that the 3'-end sequences are required not only for (-)-strand RNA synthesis but also for (+)-strand RNA synthesis. (+)-strand RNA synthesis is initiated from the 3'-end of the (-)-strand template (corresponding to the 5'-end of the genomic RNA), and the signal for RNA replication would be expected to be at the 5'-end of genomic RNA. Thus, the presence of the replication signal at the 3'-end of the genome raises an interesting possibility that (+)-strand RNA synthesis requires the interaction between the 5'- and 3'-end sequences of the template RNA. Furthermore, we have also shown that transcription from an internal IG site of the DI RNA inhibits (-)-strand RNA synthesis of the very RNA, again suggesting that transcription requires certain RNA interactions, which interfere with (-)-strand RNA synthesis[24]. The interaction between the 5'- and 3'-end sequences has previously been shown to be required for RNA synthesis of some other RNA viruses[25]. Thus, this interaction may be a common feature of RNA-dependent RNA synthesis.

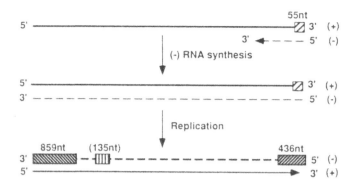

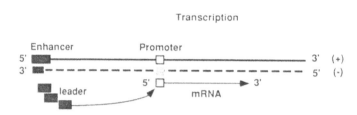

Figure 3. *Cis*-acting signals for (-)-strand RNA synthesis, RNA replication and subgenomic mRNA transcription. The boxed regions are the required sequences for the indicated steps of RNA synthesis. The figure is reproduced from Reference 29, with permission.

Another surprising feature of (-)-strand RNA synthesis is that RNAs which transcribe a subgenomic mRNA make smaller amounts of (-)-strand RNA than nontranscribing RNAs[24]. Therefore, the amounts of mRNA transcription do not parallel the quantities of (-)-strand RNA templates available, and that the quantitative regulation of RNA synthesis likely occurs mainly at the step of (+)-strand RNA synthesis rather than (-)-strand RNA synthesis. When the *cis*-acting signals for (-)-strand RNA, mRNA and genomic RNA synthesis are compared (Figure 3), it appears that the signal for (-)-strand RNA synthesis is strikingly less complex than that for the other two RNA species. This is conceptually logical, since (-)-strand RNA has to be synthesized directly from the incoming genomic RNA at the time when very few viral proteins have been synthesized. Since the 5'-end sequences of the genomic RNA are not required for (-)-strand RNA synthesis, any RNAs which contain the 55 nucleotides from the 3'-end of the genome and poly(A) will be able to initiate (-)-strand RNA synthesis. This is consistent with the detection of the subgenomic (-)-strand RNA species which correspond to all of the subgenomic mRNAs in the virus-infected cells[4,6]. In contrast, both RNA replication and transcription require longer stretches of sequences at both the 5'- and 3'-ends (Lin and Lai, unpublished observation), suggesting a complex interaction between various RNA regions.

SUMMARY OF THE MECHANISM OF CORONAVIRUS mRNA SYNTHESIS

Although the precise models of coronavirus mRNA transcription cannot yet be unequivocally deciphered, several features of mRNA synthesis have already been established: (1) mRNA transcription is discontinuous, since the leader sequences on most mRNAs are derived *in trans* from a different RNA molecule[8,9,16]. (2) mRNA synthesis can be regulated *in trans* by RNA elements located on different RNA molecules[9]. (3) leader-mRNA fusion may not be precise, suggesting that complementary RNA sequences between the leader and IG sequences are not necessarily directly involved in leader fusion[17,18]. (4) the RNA regulatory regions are complexed with cellular and, probably, viral proteins[19]. These proteins might mediate RNA- and protein-protein interactions to form transcription or replication complexes. (5) The amounts of (-)-strand RNA do not correspond to the amounts of mRNAs or genomic RNAs synthesized, suggesting that the quantitative regulation of RNAs is not at (-)-strand RNA synthesis, but rather at (+)-strand RNA synthesis[24]. (6) Subgenomic (-)-strand RNAs have been shown to be derived from subgenomic replicative form RNA, indicating that subgenomic (-)-strand RNA can function as a template[6]. However, it is not clear what the predominant mechanism for mRNA transcription is. The most crucial question is at what step, either (+)-strand or (-)-strand synthesis, discontinuous transcription occurs. This issue will require future studies. It is possible that these different mechanisms proposed are not mutually exclusive and may operate simultaneously or at different stages of the viral replication cycle.

RNA RECOMBINATION AND GENETIC ENGINEERING

The studies of RNA viruses in general have been aided tremendously by the availability of infectious viral RNA or cDNA. This approach, unfortunately, has not been available to coronaviruses. The task of achieving such a technical feat is also formidable, because of the extremely large size of the viral RNA. However, progress has been made in the genetic engineering of coronavirus RNA by taking advantage of the fact that coronavirus RNA undergoes a very high frequency of recombination. RNA recombination now has been shown to occur not only in MHV, but also in several other coronaviruses, although at a much lower frequency (see this volume). It can occur between two viral RNAs, and also between viral RNA and a transfected RNA fragment[26,27] or defective-interfering RNA[28]. This capacity to undergo RNA recombination has been harnessed to replace the 5'- and 3'-ends of viral genomic RNA with desired sequence[27,28]. However, the success of this approach so far has been limited to both ends of the genomic RNA; it has not yet been possible to replace the internal sequences or insert foreign genes, probably because multiple cross-overs are required. We have recently developed a DI RNA vector for expressing foreign genes. This DI vector is made to contain either an internal ribosomal entry site (IRES)[13] or IG sequence[16]. When a viral gene is inserted behind either IG or IRES sequences, it can be expressed, and the protein product can be incorporated into the virions, thus generating a pseudotyped virus (Liao and Lai, unpublished). Furthermore, DI RNA can be incorporated into virions, thus generating a pseudo-recombinant virus, which contains both the wild-type viral genome and a defective-interfering RNA containing a foreign gene. This represents a first step toward the genetic engineering of coronavirus. This system can potentially be used to express a foreign gene of interest; for instance, it can be used to express cytokines or cytokine antagonists to study the effects of these molecules on viral infections. This is particularly powerful, since these molecules will be expressed only in virus-infected cells. It can be used to express spike proteins of different viruses and, thereby, alter the host range of the virus. Finally, it can be used to express a desired protein to induce antibodies, thus serving as a virus vaccine. Therefore, even though an infectious full-length cDNA clone is not yet available, the era of genetic engineering of coronaviruses may be at hand.

CONCLUSIONS

The large size of coronavirus RNA genome and the discontinuous nature of its mRNA synthesis are the unique features of coronavirus. Although the detailed mechanism of RNA synthesis has not yet been unequivocally established, the presently available evidence suggests that it has many interesting features, enabling coronavirus RNA to serve as a paradigm for the studies of other viral RNAs Progress in the studies of RNA synthesis also has made possible the genetic engineering of coronavirus. However, many unanswered questions remain, e.g., what is the nature of coronavirus RNA polymerase? What are the transcription factors? How are the various steps of RNA synthesis regulated? What is the nature of the templates for RNA synthesis? These and other questions will demand further studies in the future.

REFERENCES

1 Lai M M C Coronavirus Organization, Replication and Expression of Genome Ann Rev Microb 1990,44 303-333

2 Makino S , Keck J G , Stohlman S A , Lai M M C High-frequency RNA recombination of murine coronaviruses J Virol 1986,57 729-737

3 Makino S , Fujioka N , Fujiwara K Structure of the intracellular defective viral RNAs of defective interfering particles of mouse hepatitis virus J Virol 1985,54 329-36

4 Sethna P B , Hung S L , Brian D A Coronavirus subgenomic minus-strand RNAs and the potential for mRNA replicons Proc Natl Acad Sci USA 1989,86 5626-5630

5 Lai M M C Coronavirus leader RNA-primed transcription An alternative mechanism to RNA splicing BioEssays 1989,5 257-260

6 Sawicki S G , Sawicki D L Coronavirus transcription Subgenomic mouse hepatitis virus replicative intermediates function in RNA synthesis Proc Natl Acad Sci USA 1990,64 1050-1056

7 Sutton R E , Boothroyd J C Evidence for trans splicing in trypanosome Cell 1986,47 527-535

8 Jeong Y S , Makino S Evidence for coronavirus discontinuous transcription J Virol 1994,68 2615-2623

9 Zhang X , Lai M M C Coronavirus leader RNA regulates and initiates subgenomic mRNA transcription, both in trans and in cis J Virol 1994,68 4738-4746

10 Jeong Y S , Makino S Mechanism of coronavirus transcription Duration of primary transcription initiation activity and effects of subgenomic RNA transcription on RNA replication J Virol 1992,66 3339-3346

11 Makino S , Joo M , Makino J K A system for study of coronavirus mRNA synthesis A regulated, expressed subgenomic defective-interfering RNA results from intergenic site insertion J Virol 1991,65 6031-6041

12 Joo M , Makino S Mutagenic analysis of the coronavirus intergenic consensus sequence J Virol 1992,66 6330-6337

13 Lin Y-J , Lai M M C Deletion mapping of a mouse hepatitis virus defective-interfering RNA reveals the requirement of an internal and discontiguous sequence for replication J Virol 1993,67 6110-6118

14 Kim Y-N , Jeong Ys , Makino S Analysis of cis-acting sequences essential for coronavirus defective interfering RNA replication Virology 1993,197 53-63

15 Makino S , Joo M Effect of intergenic consensus sequence flanking sequences on coronavirus transcription J Virol 1993,67 3304-3311

16 Liao C -L , Lai M M C Requirement of the 5'-end genomic sequence as an upstream cis-acting element for coronavirus subgenomic mRNA transcription J Virol 1994,68 4727-4737

17 Van der Most R G , de Groot R J , Spaan W J M Subgenomic RNA synthesis directed by a synthetic defective interfering RNA of mouse hepatitis virus A study of coronavirus transcription initiation J Virol 1994,68 3656-3666

18 Zhang X , Lai M M C Unusual heterogeneity of leader-mRNA fusion in a murine coronavirus Implications for the mechanism of RNA transcription and recombination J Virol 1994,68 6626-6633

19 Furuya T , Lai M M C Three different cellular proteins bind to the complementary sites on the 5'-end positive- and 3'-end negative-strands of mouse hepatitis virus RNA J Virol 1993,67 7215-7222

20 Hofmann M A , Brian D A The 5'-end of coronavirus minus-strand RNAs contains a short poly(U) tract J Virol 1991,65 6331-6333

21 Sethna P B , Hofmann M A , Brian D A Minus-strand copies of replicating coronavirus mRNAs contain antileaders J Virol 1991,65 320-325

22 Sawicki S G , Sawicki D L Coronavirus minus-strand RNA synthesis and effect of cycloheximide on coronavirus RNA synthesis J Virol 1986,57 328-334

23 Perlman S , Ries D , Bolger E , Chang L J , Stoltzfus C M MHV nucleocapsid synthesis in the presence of cyclohexamide and accumulation of negative-strand MHV RNA Virus Res 1986,6 261-272

24 Lin Y -J , Liao C L , Lai M M C Identification of the cis-acting signal for minus-strand RNA synthesis of a murine coronavirus Implications for the role of minus-strand RNA in RNA replication and transcription J Virol 1994,(In press)

25 Luo G , Luytjes W , Enami M , Palese P The polyadenylation signal of influenza virus RNA involves a stretch of uridines followed by the RNA duplex of the panhandle structure J Virol 1991,65 2861-2867

26 Liao C -L , Lai M M C RNA recombination in a coronavirus Recombination between viral genomic RNA and transfected RNA fragments J Virol 1992,66 6117-6124

27 Koetzner C A , Parker M M , Ricard C S , Sturman L S , Masters P S Repair and mutagenesis of the genome of a deletion mutant of the coronavirus mouse hepatitis virus by targeted RNA recombination J Virol 1992,66 1841-1848

28 Van der Most R G , Heijnen L , Spaan W J M , de Groot R J Homologous RNA recombination allows efficient introduction of site-specific mutations into the genome of coronavirus MHV-A59 via synthetic co-replicating RNA Nucl Acids Res 1992,20 3375-3381

29 Lai M M C , Liao C -L , Lin Y -J , Zhang X Coronavirus How a large RNA viral genome is replicated and transcribed Infect Agents and Dis 1994,3 98-105

75

ANALYSIS OF CORONAVIRUS TRANSCRIPTION REGULATION

Myungsoo Joo and Shinji Makino

Department of Microbiology
The University of Texas
Austin, Texas

ABSTRACT

Insertion of an intergenic region from murine coronavirus mouse hepatitis virus (MHV) into an MHV defective interfering (DI) RNA led to transcription of subgenomic DI RNA in helper virus-infected cells Using this system we studied how two intergenic regions positioned in close proximity affected subgenomic RNA synthesis When two intergenic regions were separated by more than 100 nt, slightly less of the larger subgenomic DI RNA (synthesized from the upstream intergenic region) was made, this difference was significant when the intergenic region separation was less than about 35 nucleotides Deletion of sequences flanking the two intergenic regions inserted in close proximity did not affect transcription No significant change in the ratio of the two subgenomic DI RNAs was observed when the sequence between the two intergenic regions was altered Removal of the downstream intergenic region restored transcription of the larger subgenomic DI RNA These results demonstrated the downstream intergenic sequence was suppressing subgenomic DI RNA synthesis from the upstream intergenic region

INTRODUCTION

Mouse hepatitis virus (MHV) is the coronavirus prototype MHV contains a single-stranded, positive sense RNA of approximately 31 kb[1 3] In MHV-infected cells, seven to eight species of virus-specific mRNAs are synthesized, they are named mRNA 1 to 7 according to decreasing order of size[4 5] The 5' end of the MHV genomic RNA and subgenomic mRNAs starts with a 72-77 nucleotide-long leader sequence[6 7] The mRNA body sequences begin from a consensus sequence (UCUAAAC, or a very similar sequence) in the intergenic region, which is located upstream of each MHV gene[6 7] The intergenic region preceding gene 7 (gene 7 encodes the nucleocapsid protein) carries the same 18 nucleotide-long sequence (AAUCUAAUCUAAACUUUA) found at the 3′ region of the genomic leader sequence[8].

A system that exploits defective interfering (DI) RNAs of MHV for the study of coronavirus transcription is established[9]. One study using this system demonstrated that the sequences flanking the intergenic region preceding gene 7 do not play a role in subgenomic DI RNA transcription[10]. However, studies of an MHV mutant virus[11] and bovine coronavirus (BCV) subgenomic mRNAs[12] raise the possibility that sequence(s) outside of the intergenic consensus sequence may affect coronavirus transcription. From these studies we hypothesized that two coronavirus intergenic consensus sequences that are located in close proximity may interact in such a way that the presence of a downstream consensus sequence may inhibit transcription of subgenomic mRNA from an upstream consensus sequence. We examined this possibility, and present new aspects of coronavirus transcription regulation.

MATERIALS AND METHODS

Viruses and Cells

The plaque-cloned A59 strain of MHV (MHV-A59) was used as a helper virus. Mouse DBT cells were used for growth of viruses.

DNA Construction

A procedure based on recombinant polymerase chain reaction (PCR) was employed for site-directed mutagenesis[13]. Construction of various plasmids will be described elsewhere.

RNA Transcription and Transfection

Plasmid DNAs were linearized by Xba I digestion and transcribed with T7 RNA polymerase as previously described[14]. The lipofection procedure was used for RNA transfection as previously described[9].

Primer Extension

The oligonucleotides were 5′-end labeled with [γ-^{32}P] ATP with polynucleotide kinase[15]. Poly (A) containing RNAs were used for primer extension analysis as described previously[16]. Reaction products were analyzed on 6% polyacrylamide gels containing 7M urea.

PCR and Direct Sequencing of the PCR Products

Primer extension products were purified from the gel and amplified by PCR under the same condition described above. The gel-purified RT-PCR products were separated by agarose gel electrophoresis. Direct PCR sequencing was performed according to the procedure established by Winship[17].

RESULTS

The Effect of Two Proximally Inserted Intergenic Regions on MHV Subgenomic DI RNA Transcription

In an attempt to gauge the effect of a downstream intergenic region on its upstream neighbor, we measured the synthesis of two subgenomic DI RNAs that were transcribed from two proximally inserted intergenic consensus sequences. We constructed a series of MHV DI cDNAs, each of which contained two intergenic regions in the same parental clone, MHV DI RNA-derived cDNA clone MT1/174[10]. All clones contained an insertion upstream of the 18 nucleotide-long intergenic region of MT1/174. The inserted sequence consisted of the 18 nt-long intergenic region preceding gene 7 attached to varying lengths of downstream sequence and a few nucleotides from the 3'-end, which were generated by construction procedures. Therefore, these newly constructed DI cDNA clones bore two 18 nt-long intergenic regions; most of the sequences between the two intergenic regions derived from downstream of the intergenic region between genes 6 and 7. We named these DI cDNAs according to the distance between the first nucleotide of the upstream intergenic region and the first nucleotide of the downstream intergenic region. In MS124, the first nucleotide of the downstream intergenic region is located 124 nt from the first nucleotide of the upstream intergenic region.

We used primer extension analysis to examine the quantities of the two subgenomic DI RNAs in the mutants with similar size subgenomic DI RNAs. The 5'-end labeled oligonucleotide, which specifically binds with genomic and subgenomic DI RNAs, but not to helper virus mRNAs, hybridized with intracellular RNA species and the hybridized primer was extended by reverse transcriptase. Primer extension products were then analyzed by electrophoresis on sequencing gels (Fig. 1). Large amounts of a primer extension product corresponding to a smaller subgenomic DI RNA, synthesized from the downstream intergenic region, were made in MS23-, MS29- and MS36-replicating cells (Fig. 1A). We frequently observed a minor primer extension product, which migrated as if it were 5 nucleotides longer than the major primer extension product (Fig. 1, shown by the open triangle). The amount of a primer extension product corresponding to a larger subgenomic

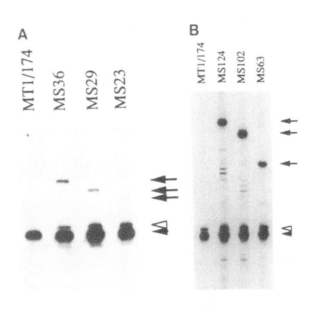

Figure 1. Primer extension analysis of subgenomic DI RNAs. The 5'-end labeled oligonucleotide was hybridized with intracellular RNAs and extended with reverse transcriptase. The products were electrophoresed on sequencing gels. Panels A and B represent two independently electrophoresed gels The arrows and arrowheads indicate the larger subgenomic DI RNA primer extension products and the smaller subgenomic DI RNA primer extension products, respectively. The minor bands indicated by the open triangles are the smaller subgenomic DI RNAs which contained three UCUAA repeats at the leader-body junction.

DI RNA, produced from the upstream intergenic region, was significantly reduced (Fig. 1A). Direct sequencing of PCR products demonstrated that the larger and the smaller primer extension products contained the expected structure of the larger and the smaller subgenomic DI RNA, respectively. All the subgenomic DI RNAs contained two repeats of UCUAA at the leader-body junction site. The minor primer extension products, which migrated slightly slower than the major primer extension product of the smaller subgenomic DI RNAs, exhibited structures that were similar to the smaller subgenomic DI RNAs, except that they contained three UCUAA repeats. These analyses demonstrated that transcription of the larger subgenomic DI RNA was inhibited when two intergenic regions were inserted within 23 to 36 nucleotides of each other.

We examined the effect of the length of the "gap" between the two intergenic sites on inhibition of the larger subgenomic DI RNA transcription. The amounts of the larger and the smaller subgenomic DI RNAs of MS124, MS102 and MS63 were compared (Fig. 1B). Primer extension analysis and direct sequencing of PCR products of the primer extension products demonstrated that more of the larger subgenomic DI RNAs (shown by the arrows in Fig. 1B) were synthesized in MS124-, MS102 and MS63-replicating cells than had been made in MS36-, MS29- and MS23- replicating cells (Fig. 1A). When two intergenic regions were separated by about 60 nucleotides, a distinct inhibition of transcription of the larger subgenomic DI RNA occurred. Of the DI RNAs that we analyzed, the degree of inhibition increased as the distance between the two intergenic regions decreased; MS23, with the shortest sequence between the two intergenic sites, demonstrated the greatest inhibition of the larger subgenomic DI RNA transcription.

The Effect of Sequences Flanking the Two Inserted Intergenic Regions on Subgenomic DI RNA Transcription

We looked for an effect of the sequences flanking the two inserted intergenic regions on subgenomic DI RNA transcription by analyzing two MS23-derived DI cDNAs, MS23ΔDF and MS23ΔUF. Downstream of the MS23 intergenic region, we deleted 156 nt to create MS23ΔDF. Upstream of the MS23 intergenic region we removed 0.8 kb to make MS23ΔUF. Primer extension analysis and direct sequencing of the PCR products of the primer extension products demonstrated that the transcription of the large subgenomic DI RNA was inhibited in both DI RNAs (Fig. 2A). These studies demonstrated that the flanking sequences of the two inserted intergenic regions did not affect the inhibition of the larger subgenomic DI RNA's transcription.

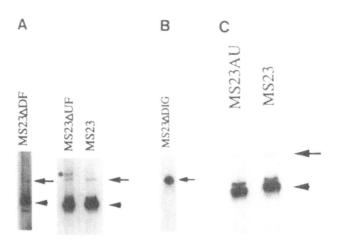

Figure 2. Primer extension analysis of subgenomic DI RNAs. The arrows and arrowheads indicate the the larger subgenomic DI RNA primer extension products and the smaller subgenomic DI RNA primer extension products, respectively. An asterisk most likely indicates a premature termination product of the genomic DI RNA primer extension product.

Investigation of a Possible Effect of the Downstream Intergenic Region on Inhibition of Transcription from the Upstream Intergenic Consensus Sequence

To directly test the effect of the downstream intergenic region on inhibition of transcription from the upstream intergenic region we used a new construct, MS23ΔDIG. This MS23-derived DI cDNA lacked the entire 18 nucleotide-long downstream intergenic region; in its place, MS23ΔDIG carried an 8 nucleotide insertion from a non-MHV sequence, which was generated through the DNA construction procedure. Northern blot analysis of MS23ΔDIG intracellular RNA species demonstrated that the amount of MS23ΔDIG subgenomic DI RNA was comparable to that of MT1/174 (data not shown), which synthesizes a high level of subgenomic DI RNA. Primer extension analysis of MS23ΔDIG showed an abundance of product corresponding to the larger subgenomic DI RNA, whereas no primer extension product made from the smaller subgenomic DI RNA was seen (Fig. 2B). This data clearly demonstrated that the presence of a downstream intergenic region inhibited transcription from the upstream intergenic region.

The actual sequence of the nucleotides between the two intergenic regions might affect transcriptional efficiency of the larger subgenomic DI RNA; we tested this by changing the nucleotides between the two repeated UCUAA sequences of the intergenic regions in clone MS23, making clone MS23AU. Primer extension analysis of MS23AU revealed that transcription of the larger subgenomic DI RNA was inhibited (Fig. 2C). This experiment indicated that inhibition of transcription from the upstream intergenic region did not depend on the nucleotide sequence between the two intergenic regions, because there was significant inhibition after mutation of the "gap" between the intergenic sites. That mutation did have an effect, however, which we saw as a small increase in the inhibition of transcription from the upstream intergenic site.

DISCUSSION

In the present study, transcription of the larger of two subgenomic DI RNAs, which was synthesized from the upstream intergenic region of two closely inserted intergenic regions, was inhibited; its inhibition was caused by the presence of the downstream intergenic sequence. In the case of the gene 6-7 intergenic sequence, transcription efficiency is not affected by sequences flanking the intergenic sequence[10]. A contrasting exception to that, presented here, was when the flanking sequence was itself an intergenic region, and affected transcription by inhibiting the intergenic region that it flanked. Recently, we placed a transcription consensus sequence in the middle of a 0.4 kb fragment, which was located at a fixed position but derived from various regions of MHV, and found that transcription of subgenomic DI RNAs varied among the DI RNA constructs (unpublished data). This data indicated that flanking sequences of the inserted intergenic region affected subgenomic DI RNA transcription efficiency. It is possible that each of the naturally occurring MHV intergenic sites is regulated in an analogous but slightly different way that depends upon flanking sequences.

Our finding explains well why a BCV subgenomic mRNA is not synthesized from the predicted intergenic consensus sequence, but is synthesized from another sequence, located 15 nucleotides downstream of the predicted intergenic consensus sequence[12]. The conclusions from our study also provide a reason for why it is that in MHV-S No. 8-infected cells the amount of the larger mRNA 7, which is synthesized from the upstream consensus sequence, is significantly lower than of that of the smaller mRNA 7, which is synthesized from the downstream consensus sequence[11] The transcriptional regulation of MHV DI RNAs is most likely governed by the same mechanism as coronavirus transcription, and the data shown in the present study, therefore, sheds light directly on the understanding of the actual coronavirus transcription mechanism

ACKNOWLEDGMENTS

We thank Young-Nam Kim for performing some of the preliminary work This work was supported by Public Health Service grants AI29984 and AI32591 from the National Institutes of Health, U S A

REFERENCES

1 Lai, M M C, and Stohlman, S A RNA of mouse hepatitis virus J Virol 1978,26 236-242

2 Lee, H -J, Shieh, C -K, Gorbalenya, A E, Eugene, E V, La Monica, N, Tuler, J, Bagdzhadzhyan, A, and Lai, M M C The complete sequence (22 kilobases) of murine coronavirus gene 1 encoding the putative proteases and RNA polymerase Virology 1991,180 567-582

3 Pachuk, C J, Bredenbeek, P J Zoltick, P W, Spaan, W J M, and Weiss, S R Molecular cloning of the gene encoding the putative polymerase of mouse hepatitis virus, strain A59 Virology 1989,171 141-148

4 Lai, M M C, P R Brayton, P R, Armen, R C, Patton, C D, Pugh, C, and Stohlman, S A Mouse hepatitis virus A59 mRNA structure and genetic localization of the sequence divergence from hepatotropic strain MHV-3 J Virol 1981,39 823-834

5 Leibowitz, J L, Wilhelmsen, K C, and Bond, C W The virus-specific intracellular RNA species of two murine coronaviruses MHV-A59 and MHV-JHM Virology 1981,114 39-51

6 Lai, M M C, Baric, R S, Brayton, P R, and Stohlman, S A Characterization of leader RNA sequences on the virion and mRNAs of mouse hepatitis virus, a cytoplasmic RNA virus Proc Natl Acad Sci USA 1984,81 3626-3630

7 Spaan, W, Delius H Skinner, M, Armstrong, J, Rottier, P, Smeekens, S, van der Zeijst, B A M, and Siddell,S G Coronavirus mRNA synthesis involves fusion of non-contiguous sequences EMBO J 1983,2 1939-1944

8 Shieh, C -K, Soe, L H, Makino, S, Chang, M -F, Stohlman, S A, and Lai, M M C The 5'-end sequence of the murine coronavirus genome implications for multiple fusion sites in leader-primed transcription Virology 1987,156 321-330

9 Makino, S, Joo, M, and Makino, J K A system for study of coronavirus mRNA synthesis a regulated, expressed subgenomic defective interfering RNA results from intergenic site insertion J Virol 1991,65 6031-6041

10 Makino, S, and Joo, M Effect of intergenic consensus sequence flanking sequences on coronavirus transcription J Virol 1993,67 3304-3311

11 Taguchi, F, Ikeda, T, Makino, S, and Yoshikura, H A murine coronavirus MHV-S isolate from persistently infected cells has a leader and two consensus sequences between the M and N genes Virology 1994,198 355-359

12 Hofmann, M A, Chang, R -Y, Ku, S, and Brian, D A Leader-mRNA junction sequences are unique for each subgenomic mRNA species in the bovine coronavirus and remain so throughout persistent infection Virology 1993,196 163-171

13 Higuchi, R In Innis, M A, Gelfand, D H, Sninsky, J J, White, T J (eds) PCR protocols Academic Press, San Diego 1990 pp177-183

14 Makino, S, and Lai, M M C High-frequency leader sequence switching during coronavirus defective interfering RNA replication J Virol 1989,63 5285-5292

15 Sambrook, J, Fritsch, E F and Maniatis, T Molecular cloning Cold Spring Harbor Laboratory, Cold Spring Harbor, N Y, 1989

16 Makino, S, Shieh, C -K, Soe, L H, Baker, S C, and Lai, M M C Primary structure and translation of a defective interfering RNA of murine coronavirus Virology 1988,166 550-560

17 Winship, P R An improved method for directly sequencing PCR material using dimethyl sulfoxide Nucleic Acids Res 1989,17 1266

76

EXPRESSION OF MURINE CORONAVIRUS GENES 1 AND 7 IS SUFFICIENT FOR VIRAL RNA SYNTHESIS

Kyongmin Hwang Kim and Shinji Makino

Department of Microbiology
The University of Texas at Austin
Austin, Texas

INTRODUCTION

The genome of mouse hepatitis virus (MHV) is known to contain eight or nine genes that are encoded by mRNAs 1, 2, 2-1, 3, 4, 5 (for both 5a and 5b), 6 and 7. The gene 2, 2-1, 4 and 5a products are not essential for MHV replication at least in tissue culture[1-3]. The 5'-most MHV gene, gene 1 most probably encodes virus RNA polymerase and proteases, the activities that are necessary for MHV RNA synthesis[4]. Gene 3 encodes S protein which is responsible for binding to cellular receptor[5] and for induction of cell fusion[6]. MHV 5b protein is present as a virus structural protein[7]. Gene 6 encodes M protein which is believed to be essential for virus assembly. The most 3' region of the MHV genome, gene 7, encodes the N protein. N protein binds to MHV genomic RNA forming a helical nucleocapsid. Anti-N antibody inhibits MHV RNA replication in vitro[8], indicating that N protein is necessary for MHV RNA replication. Whether the remaining proteins, S , ns5b and M are necessary for MHV RNA is not known.

An MHV-JHM defective-interfering (DI) RNA, DIssA, which is nearly genomic in size[9], replicates by itself in the absence of helper virus infection[10] and is efficiently packaged into MHV particles[9, 11]. Almost all MHV mRNA synthesis is strongly inhibited in DIssA-replicating cells, whereas synthesis of mRNA 7 and its product N protein is not inhibited[9]. Oligonucleotide T1 fingerprinting analysis of DIssA suggested that gene 1 and gene 7 of DIssA are essentially intact, whereas multiple deletions are present from genes 2 to 6[9].

In the present study, we examined possibilities that mRNA 7 is synthesized form DIssA template RNA, but not from helper virus template RNA, and that the gene 1 products and N protein are sufficient for the MHV RNA synthesis. Our study demonstrated that these possibilities are, in fact, the case.

MATERIALS AND METHODS

Viruses and Cells

The MHV-A59 temperature-sensitive mutant, LA 16[12], the plaque-cloned MHV-JHM, and virus sample obtained after 19 undiluted passage of original plaque-cloned MHV-JHM (JHM19th)[9] were used. Mouse DBT cells were used for RNA transfection and propagation of viruses.

Radiolabeling of Viral RNAs and Agarose Gel Electrophoresis

Virus-specific RNAs in virus-infected cells were labeled with ^{32}Pi as previously described[13] and separated by electrophoresis on 1% urea-agarose gels as described previously[9].

Preparation of Virus-Specific Intracellular RNA and Northern Blotting

Virus-specific RNAs were extracted from virus-infected cells[13]. For each sample, 1.5 μg of intracellular RNA was denatured and electrophoresed through a 1% agarose gel containing formaldehyde, and the separated RNA was blotted onto nylon filters [14]. The RNA on the filters was hybridized with ^{32}P-labeled probes specific for the various region of MHV RNA[14].

Isolation of Clones Containing DIssA-Specific Sequence

For the amplification of a DIssA-related subgenomic RNA, cDNA was first synthesized from intracellular RNA as previously described[15], using as a primer oligonucleotide 1116 (5'-CTGAAACTCTTTTCCCT-3'), which binds to positive-strand MHV mRNA 7 at nucleotides 250 to 267 from the 5'-end of mRNA 7. MHV-specific cDNA was then incubated with oligonucleotide 78[15], which binds to antileader sequence of MHV RNA, in PCR buffer as described previously[15]. DIssA subgenomic RNA-specific reverse transcriptase (RT)-PCR products were examined by Southern blot analysis in which RT-PCR products were separated by agarose gel electrophoresis and hybridized with a probe which corresponds to 1.5 to 1.7 kb from the 3'-end of MHV genomic RNA. The identified 1.2 kb-long DIssA subgenomic RNA-specific RT-PCR product was eluted from the preparative gel and cloned into TA cloning vector (Invitrogen). Clones containing DIssA-specific sequence were isolated by colony hybridization using the same probe which was used for Southern blot analysis.

RESULTS

Strategy for Analysis of DIssA-Related RNAs

We wanted to determine whether mRNA 7 detected in DIssA RNA-replicating cells was derived from the DIssA template or from helper virus genomic template. We established experimental conditions so that DIssA-derived RNA, but not helper virus-derived RNA, was efficiently synthesized. Two parental viruses were used: JHM19th and LA16, which is an MHV-A59 temperature sensitive mutant with an RNA$^-$ phenotype. In JHM19th-infected cells DIssA, mRNA 7 and the 2.2 kb-long DIssE[9] are synthesized, whereas synthesis of the helper virus-derived mRNAs is strongly inhibited at nonpermissive temperature[9]. DBT cells were infected with either JHM19th or LA16 alone, or were coinfected with both virus samples and cultured at 32.5°C, which is the permissive temperature for LA16. After overnight

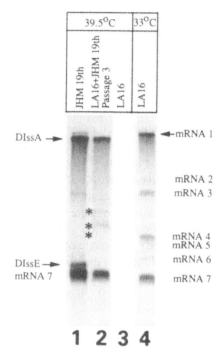

Figure 1. Urea-agarose gel electrophoresis of MHV-specific intracellular RNA species. DBT cells were infected with JHM19th (lane 1), P3 virus sample from JHM19th and LA 16-coinfected cells (lane 2), and LA 16 (lanes 3 and 4). Viruses were grown at 39.5°C (lanes 1-3) or 32.5°C (lane 4) and virus-specific RNA was labeled with ^{32}Pi in the presence of actinomycin D. Extracted RNA was electrophoresed on a 1% urea-agarose gel. The three minor bands marked with asterisks (lane 2) were newly synthesized DI RNA species generated during virus passage.

incubation, culture fluid was collected and virus samples were named passage 0 (P0) samples and these virus samples were further passaged on DBT cells several times at the permissive temperature. Because MHV-A59 usually grows about 10 times better than MHV-JHM, we expected that LA16 would grow better than MHV-JHM in coinfected cells and that during passage, MHV-JHM would be eliminated from the virus samples.

We examined MHV-specific intracellular RNA species of these passaged virus sample at the nonpermissive temperature. Infected cells were cultured at 39.5°C or 32.5°C, and virus-specific RNAs were labeled with ^{32}Pi in the presence of actinomycin D. Extracted intracellular virus-specific RNA was electrophoresed on a 1% urea-agarose gel (Fig. 1). Synthesis of DIssA and mRNA 7 was clearly apparent in the cells infected with the P3 sample from coinfected cells, whereas synthesis of helper virus mRNAs and DIssE RNA was not detected (Fig. 1, lane 2). In addition to DIssA and mRNA 7, other minor MHV-specific RNA species were also detected (Fig. 1, lane 2 asterisks); these bands were most likely newly generated DI RNAs. Helper virus-derived mRNAs, 1 through 6, were not evident at the nonpermissive temperature in the cells infected with the P3 sample from JHM19th and LA 16-coinfected cells, therefore, the mRNA 7 shown in Fig. 1 lane 2 was most likely derived from the DIssA template RNA.

We studied the structure of the mRNA 7 by RNase T$_1$ 1-D oligonucleotide fingerprinting. The mRNA 7 fingerprint patterns from P1 and P3 samples of co-infected cells were very similar to that of JHM, but not to LA16 (data not shown). This data clearly demonstrated that mRNA 7 was indeed synthesized from the MHV-JHM-derived DIssA template RNA.

Primary Structure of DIssA

In addition to mRNA 7 subgenomic RNA, DIssA may synthesize small amounts of other DIssA-specific subgenomic RNAs, and these putative DIssA-specific RNAs might be amplified by RT-PCR. Sequence analysis of cloned RT-PCR products of putative DIssA-specific subgenomic RNAs should reveal the primary structure of part of DIssA. MHV-specific cDNA was synthesized using oligonucleotide 1116. This oligonucleotide should hybridize with all MHV mRNAs and DIssA. The RT-PCR products were synthesized by incubating the cDNA with oligonucleotide 1116 and oligonucleotide 78. RT-PCR products were then examined by Southern blot analysis with a probe that specifically hybridizes to the 5'-region of gene 7. We detected a 1.2 kb-long RT-PCR product in JHM19th-infected cells, whereas we did not find this PCR product in all other samples, indicating that this RT-PCR product was synthesized from a DIssA-specific subgenomic RNA. We isolated this PR-PCR product and cloned it into a plasmid vector. We identified several clones by colony hybridization, and completely sequenced one 1.2 kb-long clone, TA23 clone.

Sequence analysis revealed that structure of TA23 was similar to MHV-JHM mRNA 2-1, but that it contained a large internal deletion spanning from nucleotide 876 in the transcription initiation site of gene 2-1 (or HE gene) to the 3'-region of gene 6. The leader-body fusion site of TA23 was the same as MHV-JHM mRNA 2-1[2]. This sequence analysis suggested that TA23 represented a cloned cDNA of a DIssA-specific subgenomic RNA. Because coronavirus mRNA forms a 3'-coterminal nested structure, the structure of TA23 indicated that DIssA lacks all of genes 3, 4, 5, most of gene 6 and the 3'-one third of the HE gene.

We further confirmed the structure of DIssA by Northern blot analysis by using a probe corresponding to 362 - 670 nucleotides from the 5'-transcription initiation site of the HE gene. With this probe, DIssA and three minor other RNAs, approximately 5.4 kb, 3.4 kb and 2.6 kb in length, were detected in the DIssA-replicating cells (data not shown). According to the size of RNAs, these three additional RNAs most probably represented the DIssA-specific subgenomic RNAs. The size of TA23 indicated that it most likely derived from the smallest 2.6 kb RNA. Based on the sizes of the other two RNAs, the 3.4 kb RNA probably had the same structure as MHV-JHM mRNA 2, which includes a large internal deletion like that of TA23; probably the largest 5.4 kb RNA, with the same deletion, started about 2 kb from the 3'-region of DIssA gene 1. This Northern blot analysis demonstrated that DIssA synthesized three minor RNA species, in addition to mRNA 7.

Gene Products from DIssA-related RNAs Supported Replication and Transcription of Another DI RNA

Our data indicated that the gene 1 products, ns2 protein, part of the HE protein, and the N protein were expressed in the DIssA-replicating cells and that their expression was sufficient for replication and transcription of these DIssA-related RNAs. Would these gene products support replication and transcription of another DI RNA? We next examined whether these gene products supported RNA synthesis of other MHV RNA molecules in trans. MHV DIssF-derived MHV DI cDNA clone, MT 1/24 was used to examine this possibility. MT 1/24 DI RNA contains an inserted intergenic region preceding gene 7 and a subgenomic DI RNA is synthesized in MT 1/24-replicating, MHV-infected cells[16]. MT1/24 DI RNA was transfected into monolayers of DBT cells infected with LA16 1 hr prior to transfection[16]. After incubation of virus-infected cells at 32.5°C for 16 h, P0 virus sample was obtained and further passaged to generate P1 virus sample. DBT cells were coinfected with this P1 virus sample and DIssA containing LA16 virus sample and cultured at nonpermissive temperature. Intracellular RNA was extracted from virus-infected cells and examined by Northern blot analysis (Fig. 2). Replication of MT 1/24 genomic DI RNA and transcription of MT 1/24 subgenomic DI RNA were observed in coinfected cells, demon-

Figure 2. Northern blot analysis of intracellular RNA species. DBT cells were coinfected with two virus samples, P2 virus sample from LA 16 and JHM19th coinfected cells and P1 virus sample from MT 1/24-transfected, LA 16-infected cells, and cultured at 32.5°C (lanes 1 and 5) and 39.5°C (lanes 2-4 and 6). Intracellular RNA was extracted, and 1.5 μg of cytoplasmic RNA was separated on 1% formaldehyde-agarose gels and transferred to nylon filters. MHV RNA species were detected using a probe corresponding to the 3'-end of genomic RNA. An arrowhead and an arrow represent MT 1/24 genomic DI RNA and subgenomic DI RNA, respectively. Lanes 1-4 were exposed for the same length of time. Lanes 5 and 6 are shorter exposure of lanes 1 and 2.

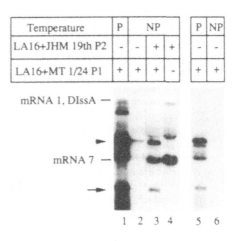

Temperature	P	NP			P	NP
LA16+JHM 19th P2	-	-	+	+	-	-
LA16+MT 1/24 P1	+	+	+	-	+	+

mRNA 1, DIssA —

mRNA 7 —

1   2   3   4   5   6

strating that gene 1 products and N protein supported RNA synthesis of other MHV RNA molecules in trans.

The ns2 protein and the HE protein are not necessary for MHV replication[1, 2]. Although DIssA-specific subgenomic RNA appeared to encode the ns2 protein and part of the HE protein, the 3.4 kb and 2.6 kb RNAs are most probably not necessary for synthesis of DIssA RNA and its related RNAs or for MT 1/24 DI RNA. We concluded that expression of gene 1 products and N protein was sufficient for MHV RNA replication and transcription.

DISCUSSION

The present study and an in vitro MHV RNA replication study[8] suggested that the N protein is necessary for MHV RNA synthesis. Coronavirus nucleocapsid is not required for initiation of coronavirus RNA synthesis, because coronavirus genomic RNA extracted from purified virions is infectious[17, 18]. Possibly a subsequent step in coronavirus RNA synthesis, e.g., genomic RNA replication, may require a nucleocapsid template. How coronavirus N protein functions in coronavirus RNA synthesis is fundamental for the understanding of coronavirus RNA synthesis.

We found that the molar ratios of mRNA 1 to mRNA 7 in the helper virus-infected cells, and of DIssA and DIssA-derived mRNA 7 in the DIssA-replicating cells were both basically the same (Fig. 1). Also no significant difference was found between the molar ratios of MT 1/24 genomic RNA to subgenomic RNA in LA 16-infected cells and in DIssA-replicating cells (Fig. 2). These data indicated that MHV transcription efficiency was not altered by the expression of the MHV S protein, ns4 protein, ns5a protein, ns5b protein, or M protein. This situation is different from influenza virus and vesicular stomatitis virus. In these viruses M protein inhibits viral RNA transcription in vitro[19, 20].

ACKNOWLEDGMENTS

We thank Michael Lai and Stephen Stohlman for LA 16. We thank John Repass and Gwen Giles for excellent technical help. This work was supported by Public Health Service grants AI29984 and AI32591 from the National Institutes of Health.

REFERENCES

1 Schwartz, B , Routledge, E , and Siddell, S G Murine coronavirus nonstructural protein ns2 is not essential for viral replication in transformed cells J Virol 1990,64 4784-4791

2 Shieh, C -K , Lee, H J , Yokomori, K , La Monica, N , Makino, S , and Lai, M M C Identification of a new transcriptional initiation site and the corresponding functional gene 2b in the murine coronavirus RNA genome J Virol 1989,63 3729-3736

3 Yokomori, K , and Lai, M M C Mouse hepatitis virus S sequence reveals that nonstructural proteins ns4 and ns5a are not essential for murine coronavirus replication J Virol 1991,65 5605-5608

4 Lee, H -J , Shieh, C -K , Gorbalenya,A E , Eugene, E V , La Monica, N , Tuler, J , Bagdzhadzhyan, A , and Lai, M M C The complete sequence (22 kilobases) of murine coronavirus gene 1 encoding the putative proteases and RNA polymerase Virology 1991,180 567-582

5 Dveksler, G S , Pensiero, M N , Cardellicjio, C B , Williams, R K , Jiang, G -S Holmes, K V, and Dieffenbach, C W Cloning of the mouse hepatitis virus (MHV) receptor expression in human and hamster cell lines confers susceptibility to MHV J Virol 1991,65 6881-6891

6 Collins, A R , Knobler, R L , Powell, H , and Buchmeier, M J Monoclonal antibodies to murine hepatitis virus-4 (strain JHM) define the viral glycoprotein responsible for attachment and cell fusion Virology 1982,119 358- 371

7 Yu, X , Bi, W , Weiss, S R , and Leibowitz, J L Mouse hepatitis virus gene 5b protein is a new virion envelope protein Virology 1994,202 1018-1023

8 Compton, S R , Rogers, D B , Holmes, K V , Fertsch, D , Remenick, J , and McGowan, J J In vitro replication of mouse hepatitis virus strain A59 J Virol 1987,61 1814-1820

9 Makino, S , Fujioka, N , and Fujiwara, K Structure of the intracellular defective viral RNAs of defective interfering particles of mouse hepatitis virus J Virol 1985,54 329-336

10 Makino, S , Shieh, C -K , Keck, J G , and Lai, M M C Defective-interfering particles of murine coronavirus mechanism of synthesis of defective viral RNAs Virology 1988,163 104-111

11 Makino, S , Taguchi, F , and Fujiwara, K Defective interfering particles of mouse hepatitis virus Virology 1984,133 9-17

12 Baric, R S , Fu, K , Schaad, M C , and Stohlman, S A Establishing a genetic recombination map for murine coronavirus strain A59 complementation groups Virology 1990,177 646-656

13 Makino, S , Taguchi, F , Hirano, N , and Fujiwara, K Analysis of genomic and intracellular viral RNAs of small plaque mutants of mouse hepatitis virus, JHM strain Virology 1984,139 138-151

14 Jeong Y S , and Makino, S Mechanism of coronavirus transcription duration of primary transcription initiation activity and effect of subgenomic RNA transcription on RNA replication J Virol 1992, 66 3339-3346

15 Makino, S , Joo,M , and Makino, J K A system for study of coronavirus mRNA synthesis a regulated, expressed subgenomic defective interfering RNA results from intergenic site insertion J Virol 1991,65 6031-6041

16 Makino, S , and Joo, M Effect of intergenic consensus sequence flanking sequences on coronavirus transcription J Virol 1993,67 3304-3311

17 Lamniczi, B Biological properties of avian coronavirus RNA J Gen Virol 1977,36 531-533

18 Schochetman, G , Stevens, R H , and Simpson, R W Presence of infectious polyadenylated RNA in the coronavirus avian infectious bronchitis virus Virology 1977,77 772-782

19 Caroll, A R , and Wagner, R R Role of the membrane (M) protein in endogenous inhibition of in vitro transcription by vesicular stomatitis virus J Virol 1979,29 134-142

20 Zvonarjev, A Y , and Ghendon, Y Z Influence of membrane (M) protein on influenza A virus transcriptase activity in vitro and its susceptibility to rimantadine J Virol 1980,33 583-586

77

DEPHOSPHORYLATION OF THE NUCLEOCAPSID PROTEIN OF INOCULUM JHMV MAY BE ESSENTIAL FOR INITIATING REPLICATION[*]

Kishna Kalicharran[†] and Samuel Dales

Cytobiology Group
Department of Microbiology and Immunology
Health Sciences Centre
University of Western Ontario
London, Ontario
Canada, N6A 5C1

INTRODUCTION

The type of neurological disease induced by JHMV in the central nervous system of rats is related to the post partum age at challenge[1]. Resistance to the demyelinating form of disease in the white matter sets in at the time myelination is completed, when the progenitor number of oligodendrocyte lineage cells becomes minimal[1,2]. In primary rat glial cultures, mature oligodendrocytes including dbcAMP-treated progenitors are non-permissive for JHMV[3]. Permissiveness was restricted to a discrete stage in their differentiation pathway[4]. Viral restriction was shown to occur at an early stage after penetration but before viral gene expression, presumably mediated by a cellular component(s)[3]. Interestingly, an endosomal-associated phosphoprotein phosphatase (PPPase) activity isolated from L cells, a glioma cell line and primary cerebral culture has specificity for dephosphorylating JHMV nucleocapsid protein (N)[5]. Moreover, the type 1 regulatory subunit (R1) of protein kinase A, which is specifically upregulated when primary progenitor oligodendrocytes are treated with the cAMP analog dbcAMP, could inhibit the endosomal PPPase activity in a cell-free system[3,6]. This study addresses the role of dephosphorylation and modulation of N during the early stages of viral replication.

[*] Supported by the Medical Research Council of Canada.

[†] Recipient of a studentship from the Multiple Sclerosis Society of Canada.

RESULTS AND DISCUSSION

JHMV maintains tropism for explanted rat oligodendrocytes although infectability is confined to a discrete stage in the differentiation process, between the mitotically active progenitors (O-2A cells) and the terminally differentiated oligodendrocytes[4]. Permissiveness seems to be determined by an intracellular factor(s) beyond viral absorption which is not affected in the non-permissive cells[3]. To elaborate on our previous results demonstrating that the restriction occurs at an early step prior to onset of transcription, isolated JHMV genomic RNA was transfected into primary telencephalic cultures established from neonatal Wistar Furth pups as previously described[4]. After 8-10 days in culture, the more loosely adherent O-2A lineage cells were released by sharply tapping the culture flask, plated into fresh growth medium, then treated with 1 mM dbcAMP 24 hours before inoculation with JHMV or 2 days postinoculation. As shown in Table 1, pretreatment of cultures to induce maturation completely blocks replication, whereas post-treatment does not when compared with untreated cells. To circumvent events in penetration and uncoating which virions must undergo, cells were transfected with RNA. Genomic RNA was isolated, combined with Lipofectin (Gibco) and added to the cultures. The normally non-permissive oligodendrocytes pretreated with dbcAMP supported JHM replication when compared to untreated controls (Table 1), demonstrating that the restriction does indeed involve an early event as previously presumed[3]. The efficiency of initiating infection by RNA transfection is, however, much lower, as evident with both L cells and the primary oligodendrocytes (Table 1).

The N protein of JHMV is phosphorylated on serine residues[7, 8]. Work in our laboratory has described an endosomal-associated PPPase that specifically dephosphorylates N when assayed in vitro[5]. To determine whether dephosphorylation of N is essential during an early stage of replication, the reversible inhibitor of types 1 and 2A PPPase, calyculin A,[9] was applied. L cells were treated for 30 minutes either before inoculation or 2 hours postinfection and viral titers determined 10 hours postinfection. As shown in Figure 1, there was a greater than 2 $\log_{10}$ reduction in PFU/mL produced when 100 nM calyculin was added prior to inoculation as compared to a much lesser reduction when calyculin was added 2 hours postinfection. These data are in line with the concept that a PPPase activity is a crucial requirement early in viral replication. Since N is the only phosphorylated JHM virion protein, the effect of calyculin A on the state of N phosphorylation inside host cells was tested. L cells were treated with inhibitor as described, then 2 hours postinfection, samples were taken and prepared for an immunoblot probing with anti-N antibodies. As shown in Figure 2, increasing concentrations of calyculin A inhibited conversion from the 56kDa to 50kDa form of N, consistent with a restriction in the dephosphorylation of N. These results indicate that

Table 1. Transfection of non-permissive oligodendrocytes with genomic RNA

Treatment of cells	Infection (PFU/mL)		Transfection (PFU/mL)	
	48 hr postinfection	72 hr postinfection	48 hr postinfection	72 hr postinfection
Untreated	1,300	4,680	30	50
1 mM dbcAMP postinfection	3,340	10,000	ND	ND
1 mM dbcAMP before infection	0	0	10	30
L cells*	ND	ND	38,000	ND

*Titre from L cells infection >10[7] PFU/mL
Virus production on primary rat oligodendrocytes after JHMV infection or RNA transfection Enriched oligodendrocyte cultures were treated with dbcAMP either 24 hr before inoculation and addition of RNA or 48 after inoculation Supernatant virus was titrated by plaque assay on L cells

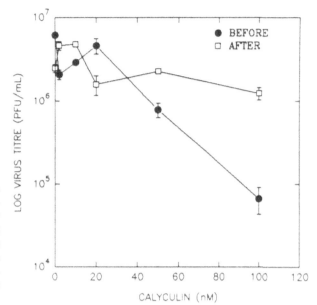

Figure 1. Effects of calyculin A on virus replication Various concentrations of calyculin A were added either 30 minutes before inoculation or 2 hours post inoculation Supernatants were sampled at 10 hours for plaque assay on L cells Pretreated solid circles Treated after inoculation open squares

dephosphorylation of N takes place early during virus-cell interaction and is crucial for replication to commence.

In brain tissue the cAMP-dependent protein kinase II is present but the type I enzyme is missing. After treatment of the oligodendrocytic lineage cells with dbcAMP the normal activation of the cAMP-dependent protein kinase was uncoupled, resulting in a selective upregulation of the type 1 regulatory subunit (R1)[3, 10]. Several studies demonstrated the capacity of the R subunit, isolated from various sources, to suppress by itself PPPase activity in a time-related and dose-dependent manner[11, 12, 13]. Neither the catalytic subunit nor the holoenzyme could exert such an effect[11, 12]. In the context of this study, previous work from our lab showed that isolated R1 can suppress the endosomal PPPase activity[6]. Therefore, to determine whether R1 overexpression in host cells modulates JHMV replication, L cells were transfected with plasmids encoding the rat R1 gene under the control of the metallothionein promoter[14]. Stable clones were established and selected in the presence of G418, and the transfected clones and normal L cells used as controls were treated with 8 µM cadmium chloride to induce the R1 gene and levels of expression were determined. Levels of R1 induced in two different producing clones were compared with untransformed cells by immunoblotting employing anti-R1 antibodies. The results are illustrated in Figure 3C. The influence of R1 concentration on viral replication was determined by titration of JHMV

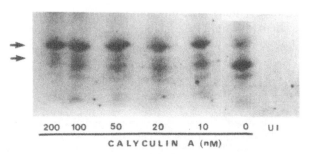

Figure 2. Immunoblots demonstrating inhibition of processing of N from 56 kDa to 50 kDa (ie dephosphorylation) in the presence of calyculin A. L cells were treated for with various concentrations of calyculin A (0, 10, 20, 50, 100, 200 nM) 30 minutes prior to inoculation Two hours after absorption, cell samples were processed for immunoblotting with anti-N antibodies

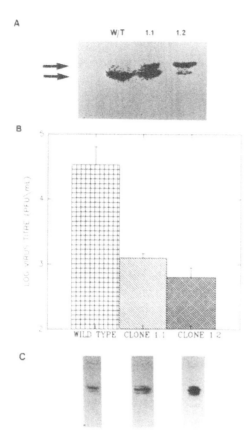

Figure 3. Influence of R1 on JHMV replication when overexpressed in transfected clones. Stable transfected clones of L cells overexpressing the R1 subunit were identified. (A) Clones 1.1, 1.2, and normal L cells were infected with JHMV. Two hours post infection, samples were processed for immunoblotting with anti-N antibodies: lane 1, normal L cells; lane 2, clone 1.1; lane 3, clone 1.2. (B) Supernatants collected 8 hours post infection and the cells were assayed for virus produced. Lanes as in A. (C) Equivalent amount of cells were processed after overnight induction of the metallothionein promoter with 8 μM cadmium chloride. Samples were prepared for immunoblotting with anti-R1 subunit antibodies as above. Lanes as in A.

present in supernatants at 8 hours postinfection. As shown in Figure 3B, the titers produced in cells overexpressing R1 were reduced by over a $\log_{10}$ compared to those generated in untransfected L cells. Furthermore, the inhibition was more profound in the higher expressing clone 1.2. To examine the in-vivo effect of R1 on dephosphorylation of N in inoculum virions, infected cells were taken 2 hours postinoculation for immunoblotting with anti-N antibodies. It is evident in Figure 3A that processing from 56 kDa to 50 kDa was inhibited in the transfected cells. Furthermore, a greater restriction was evident in clone 1.2 which express R1 more abundantly (Figure 3C). Our results are consistent with previous data from cell-free assays in demonstrating that R1 inhibits dephosphorylation of N and indicates that the inhibition of processing from 56 kDa to 50 kDa occurs at an early stage of replication .

In summary, the non-permissiveness of mature oligodendrocytes to JHMV infection can be circumvented by means of transfection using genomic RNA. Dephosphorylation of N at early times after inoculation may be essential for initiating the replication cycle, as evident with the PPPase inhibitor calyculin A and the influence of R1 of the cAMP-dependent protein kinase when it is overexpressed in host L cells.

REFERENCES

1. Beushausen, S., and S. Dales. In vivo and in vitro models of demyelinating disease XI. tropism and differentiation regulate the infectious process of coronaviruses in primary explants of the rat CNS. Virology. 1985; 141:89-101.

2 Wilson, G A R , S Beushausen and S Dales In vivo and in vitro models of demyelinating disease XV differentiation influences the regulation of coronavirus infection in primary explants of mouse CNS Virology 1986, 151 253-264

3 Beushausen, S , S Narindrasorasak, B D Sanwal and S Dales In vivo and in vitro models of demyelinating disease activation of the adenylate cyclase system influences JHM virus expression in explanted rat oligodendrocytes J Virol 1987, 61 3795-3803

4 Pasick, J M M , and S Dales Infection by coronavirus JHM of rat neurons and ligodendrocyte type-2 astrocyte lineage cells during distinct developmental stages J Virol 1991, 65 5013-5028

5 Mohandas, D V , and S Dales Endosomal association of a protein phosphatase with high dephosphorylating activity against a coronavirus nucleocapsid protein FEBS Lett 1991,282 419-424

6 6 Wilson, G A R , D V Mohandas, and S Dales In vivo and in vitro models of demyelinating disease Possible relationship between induction of regulatory subunit from cAMP dependent protein kinase and inhibition of JHMV replication in cultured oligodendrocytes In D Cavanagh and T K Brown (eds) Coronaviruses and their disease Plenum Press New York 1990 261-266

7 Stohlman, S A , J O Fleming, C D Patton, amd M M C Lai Synthesis and subcellular localization of the murine coronavirus nucleocapsid protein Virology 1983, 130 527-532

8 Wilbur, S M , G W Nelson, M M C Lai, M McMillan, and S A Stohlman 1986 Phosphorylation of the mouse hepatitis virus nucleocapsid protein Biochem Biophys Res Commun 1986, 141 7-12

9 Ishihara, H , B L Martin, D L Brautigan, H Karaki, H Ozaki, Y Kato, N Fusetani, S Watabe, K Hashimoto, D Uemura, D J Hartshorne Calyculin A and okadaic acid inhibitors of protein phosphatase activity Biochem Biophys Res Commun 1989, 159 871-877

10 Loffler, F , S Lohman, B Walckhof, U Walter, and B Hamprecht Selective increase of R1 subunit of cyclic AMP-dependent protein kinase in glia-rich primary cultures upon treatment with dibutyryl cyclic AMP Brain Res 1985, 344 322-328

11 Khatra, B S , R Printz, C E Cobb, and J D Corbin Regulatory subunit of cAMP-dependent protein kinase inhibits phosphoprotein phosphatase Biochem Biophys Commun 1985, 130 567-573

12 Jurgensen, S R , P B Chock, S Taylor, J R Vandenheede Inhibition of the Mg(II) ATP-dependent phosphoprotein phosphatase by the regulatory subunit of cAMP-dependent protein kinase Pro Natl Acad Sci US 1985, 82 7565-7569

13 Srivasta, A K , R L Khandewal, J L Chiasson, and A Haman Inhibitory effect of the regulatory subunit of type 1 cAMP-dependent protein kinase on phosphoprotein phosphatase Biochem Int 1988, 16 303-310

14 Correll, L A , T A Woodford, J D Corbin, P L Mellon, and G S McKnight Functional characterization of a cAMP-binding mutations in type 1 protein kinase J Biol Chem 1989, 264 16672-16678

78

EVIDENCE THAT MHV SUBGENOMIC NEGATIVE STRANDS ARE FUNCTIONAL TEMPLATES

Ralph S. Baric and Mary C. Schaad

Department of Epidemiology
University of North Carolina
Chapel Hill, North Carolina 27599-7400

INTRODUCTION

The genome of mouse hepatitis virus (MHV) is transcribed into a full length (~32 kb) and 6-8 subgenomic length mRNAs upon entry into susceptible cells[1]. Several discontinuous transcription models have been proposed to explain the presence of leader RNAs on positive-stranded mRNAs and antileader RNAs on full-length and subgenomic-length negative-stranded RNAs. One model proposes that subgenomic mRNAs are initially synthesized from a full-length negative-stranded RNA template by a leader-priming mechanism followed by mRNA amplification through subgenomic negative-stranded RNA intermediates[2,3]. Alternative models propose that subgenomic-length negative strands are synthesized directly from the genomic template by either a looping-out or transcription attenuation, or are functionally unimportant dead-end transcriptional products[5]. In this study, we demonstrate that the MHV group C1 mutants regulate negative strand synthesis and that the subgenomic length negative-stranded RNAs are the predominant template for mRNA synthesis late in infection.

MATERIALS AND METHODS

Virus, Cell Lines, and Preparation of RNA

MHV A59 group C *ts* mutants were propagated in 17CL1 or DBT cells as previously described[6]. Viral mRNA and RF RNA were radiolabeled, isolated and analyzed according to published protocols[4], and quantified by the AMBIS Radioanalytic Imaging System.

Corona- and Related Viruses, Edited by P. J. Talbot and G. A. Levy
Plenum Press, New York, 1995

RESULTS

Phenotypic Characterization of the Group C Mutants

We have previously shown that at least one group C mutant (LA9) probably encodes an early function regulating negative strand synthesis[6]. To determine if other group C mutants have similar phenotypes, cultures of cells were infected with different mutants, shifted to the restrictive temperature at 5 h postinfection, and radiolabeled with 200 μCi/ml ^{3}H-Uridine for 3 h (Figure 1). The subgroup C1 mutants (NC3 and LA9), but not the group C2 mutants (NC1, NC10) continue to synthesize significant quantities of mRNA and infectious virus after temperature shift. Using strand specific RNA probes that specifically bind to negative-stranded RNA[6], overall negative strand synthesis was reduced following temperature shift with the group C1 mutants (Figure 2).

Effect of Early Temperature Shift on NC3 mRNA and RF RNA Synthesis

If the group C1 mutants were defective in negative-stranded RNA synthesis after shift, a corresponding reduction in RF RNA should be seen in cultures shifted early in infection. Moreover if subgenomic negative-stranded RNAs are dead end transcription products, a decrease in subgenomic-length, but not full length RF RNA, should occur over time as pre-existing subgenomic templates transcribe a single positive strand RNA and terminate[5]. Following temperature shift at 5.5 h, NC3 full-length as well as subgenomic-length mRNA and RF RNA were prevented from increasing to maximal levels as detected at the permissive temperature (Figure 3 a,b). Relative to controls, AMBIS scans indicated that NC3 total RF RNA and mRNA levels were reduced by about 83% and 49%, respectively,

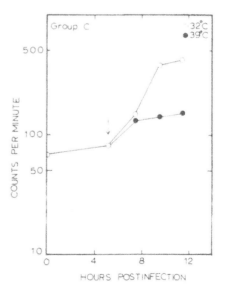

Figure 1. RNA synthesis in the group C mutants. Panel A: 32°C, Panel B: 39.5°C Lanes 1,2,3, LA9, NC10, NC3

Figure 2. Negative strand synthesis following temperature shift LA9 intracellular RNA was bound to nitrocellulose filters and probed with negative strand specific RNA probes and counted[6] 0-32°C, 0-39 5° C.

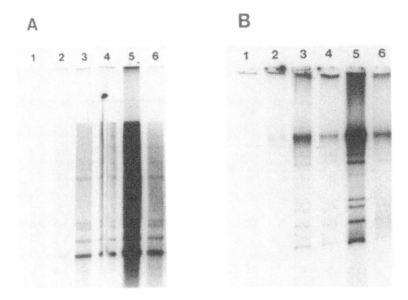

Figure 3. Viral mRNA and RF RNA synthesis after temperature shift NC3-infected 17Cl1 cells were shifted to the restrictive temperature at 5 5 h and radiolabeled with ^{32}Pi for 1 hr Panel A NC3 mRNA, Panel B NC3 RF RNA Lane 1 4 hpi, Lane 2 5 hpi, Lane 3 7 hpi, 32°C, Lane 4 7 hpi, 39 5°C, Lane 5 8 hpi, 32°C, Lane 6 8 hpi, 39 5°C

but that both full length and subgenomic length RF RNAs were transcriptionally active after shift Relative percent molar ratios of mRNA and RF RNA closely paralleled each other at permissive and restrictive temperature, also showing that the subgenomic negative strand templates were not depleted (Table 1)

Time Required to Transcribe NC3 mRNA after Temperature Shift

The defect in NC3 transcription resulting in a reduction in the amount of radiolabeled RF RNA after shift could result from a decrease in negative and/or positive strand RNA synthesis To distinguish between these possibilities, we calculated polymerase rates from a "same sized" subgenomic-length RF template using the formula used for Sindbis virus (Figure 4a)[7] Alternatively, the "leader-primed" transcription model dictates that all viral mRNAs will

Table 1. Relative percent molar ratio of mRNA and RF RNA after temperature shift

| RNA species | Percent molar ratio | | | | Mol. wts. ($\times 10^{-6}$) |
| | mRNA | | RF RNA | | |
	32°C	39.5°C	32°C	39.5°C	
1	0.2	0.9	8.1	7.8	9.79
2	3.3	4.4	9.4	4.4	3.04
3	6.3	5.5	9.2	2.2	2.40
4	12.1	9.8	6.5	3.0	1.15
5	15.5	9.6	9.4	7.8	1.00
6	19.0	14.3	17.4	33.6	0.81

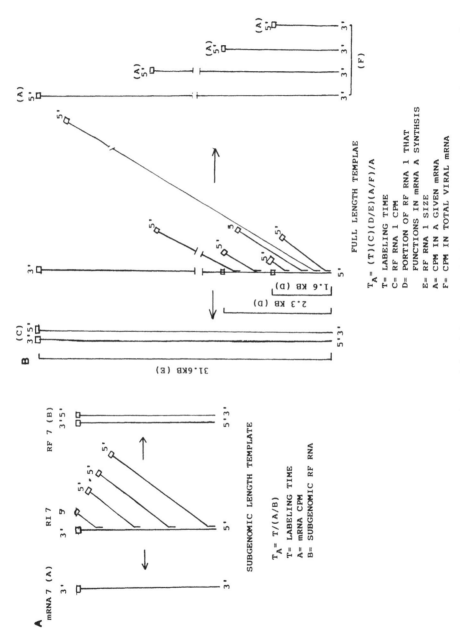

Figure 4. Formula for rate calculations from subgenomic length (Panel A) or full length (Panel B) negative-stranded RNA templates.

Table 2. Putative times required to transcribe mRNA from a same-sized subgenomic length replicative intermediate

NC3 mRNA species	Permissive temperature		Restrictive temperature	
	6-7 hr	7-8 hr	6-7 hr	7-8 hr
mRNA7	1.45 min	1.92 min	1.09 min	1.01 min
mRNA6	2.86 min	2.75 min	1.93 min	2.75 min
mRNA3	3.32 min	2.80 min	1.61 min	1.53 min

be transcribed from a full length RI; predicting that only that portion of the full length RI that is equivalent to the size of a given mRNA functions in its synthesis, and that each mRNAs nascent positive strands are present on the full length RI (Figure 4b)[2]. Thus, we modified the B in the original formula to correct for: (1) the portion of the full-length RF that could be contributing to the synthesis of a particular subgenomic mRNA, (C)(D/E) and; (2) the ratio of nascent positive strands synthesized from this portion of the RF (A/F)(Figure 4b).

Since our previous results indicated that subgenomic negative strands were functional templates throughout infection[3,4], we calculated the rates of mRNAs 3, 6 and 7 synthesis from their equivalent RF RNA template (Table 2). Appropriately, the times required to transcribe mRNA from a corresponding subgenomic-length template increased with increasing size of the mRNA. Less time was required to synthesize equivalently-sized mRNA at 39.5°C as compared to 32°C. We also calculated rates of mRNA synthesis assuming that only the full-length template was functional. Rates of mRNA synthesis were also increased at 39.5°C as compared to 32°C. However, the time required to synthesize any mRNA from the full-length template was drastically reduced as compared to that from subgenomic-length templates. RNA polymerase rates were calculated at $1.34 \pm 0.7 \times 10^3$ nucleotides/min (npm) and $1.85 \pm 1.30 \times 10^3$ npm (p<.05) from "same-sized" RF RNA templates at 32°C and 39.5°C, respectively, and $14.68 \pm 1.90 \times 10^3$ npm and $26.38 \pm 5.8 \times 10^3$ npm from a full length RF RNA template, respectively, (p<.001) demonstrating that the group C1 defect did not decrease the rate of mRNA synthesis.

DISCUSSION

The group C mutants are tightly linked by complementation and standard genetic recombination mapping techniques (data not shown), yet the group C1 mutants synthesize mRNA and produce infectious virus after temperature shift whereas the group C2 mutants can not, suggesting the presence of two noncomplementable functions. A variety of data support our hypothesis that the group C1 mutation regulates full-length and subgenomic-length negative strand RNA synthesis[6]. Compared to control cultures, group C1 negative

Table 3. Putative times required to transcribe mRNAs from a full-length negative strand RNA template by leader-primed transcription

NC3 mRNA species	Permissive temperature		Restrictive temperature	
	6-7 hr	7-8 hr	6-7 hr	7-8 hr
mRNA 7	0.13 min	0.17 min	0.07 min	0.05 min
mRNA 6	0.17 min	0.16 min	0.09 min	0.08 min
mRNA 3	0.56 min	0.52 min	0.29 min	0.26 min

strand RNA synthesis was reduced by about 80%. Total full-length and subgenomic-length RF RNA synthesis was also reduced by about 83-92% after shift to 39.5°C. Since the rates of positive strand synthesis at 39.5°C were at least equal to, or increased, as compared to the rates at 32°C, the most logical conclusion was that the MHV group C1 defect regulated negative-stranded RNA synthesis.

Support for the idea that subgenomic negative strands were functional templates for positive strand synthesis was also demonstrated by temperature shift experiments early in infection. In the absence of significant amounts of new negative strand RNA, full-length and subgenomic-length RF RNA were labeled at a constant rate after shift indicating that the previously transcribed full-length and subgenomic-length RIs were stable and transcriptionally active throughout infection. If subgenomic negative strands are dead-end transcriptional intermediates, a reduction in the amount of radiolabel incorporated at 39.5°C into the subgenomic RFs, but not full-length RFs, should occur over time in cultures infected with *ts* mutants transcribing little new negative-stranded RNA. Since this is clearly not the case, we conclude that subgenomic-length negative strands are the predominant templates for the synthesis of positive-stranded RNA throughout infection as proposed previously[3,4]. Not surprisingly, the relative percent molar ratio of mRNA and RF RNA remain equivalent at permissive and restrictive temperatures readily explaining the concentrations of each viral mRNA transcribed during infection.

Consonant with these findings, the times required for the transcription of mRNA from a subgenomic length template are more in line with the times required to synthesize SB 26S mRNA as well as other previously published eukaryotic RNA polymerase rates of approximately 1,800 nt/min[7,8]. In contrast, if leader-primed transcription from full-length negative-stranded RNA is the only mechanism by which subgenomic mRNAs are synthesized, it would have to occur at a remarkable rate and efficiency.

Careful examination of the literature demonstrates a wealth of data supporting an early discontinuous transcription mechanism by either the leader-primed, looping out or transcription attenuation models[1,2,5] as well as the synthesis of mRNA from subgenomic negative strands containing antileaders[3,4]. Since these mechanisms are not mutually exclusive and provide coronaviruses with considerable regulatory control over gene expression, the most likely interpretation is that both function during MHV infection. If mRNA is initially transcribed from a full length negative-stranded RNA by leader-primed transcription, then these mRNA must be subsequently transcribed into subgenomic negative strands; the predominant template for mRNA synthesis. If discontinuous transcription occurs by looping out or transcription attenuation early in infection to directly synthesize subgenomic negative strands from a genome length RNA, these subgenomic negative strands could immediately function in the synthesis of subgenomic mRNAs, in the absence of a leader-primed transcription model.

ACKNOWLEDGMENTS

This research was supported by the American Heart Association (AHA 90-1112), NIH (AI-23946) and was performed during the tenure of an Established Investigator from the AHA (AHA 89-0193) (RSB).

REFERENCES

1. Lai, M.M.C. 1990. Coronavirus: organization, replication and expression of genome. Annu. Rev. Microbiol. 44, 303-333.

2 Baric, R S , Stohlman, S A and Lai, M M C 1983 Characterization of replicative intermediate RNA of mouse hepatitis virus presence of leader RNA sequences on nascent chains J Virol **48:**633-640

3 Sethna, P B , Hofmann, M A and Brian, D A 1991 Minus-strand copies of replicating coronavirus mRNAs contain antileaders J Virol **65:**320-325

4 Sawicki, S G and Sawicki, D L 1990 Coronavirus transcription subgenomic mouse hepatitis virus replicative intermediates function in RNA synthesis J Virol **64:**1050-1056

5 Jeong, Y S and Makino, S 1992 Mechanism of coronavirus transcription duration of primary transcription initiation activity and effects of subgenomic RNA transcription on RNA replication J Virol **66:**3339-3346

6 Schaad, M C , Stohlman, S A , Egbert, J , Lum, K , Fu, K , Wei, Jr , T and Baric, R S 1990 Genetics of mouse hepatitis virus transcription identification of cistrons which may function in positive and negative strand RNA synthesis Virology **177:**634-645

7 Simmons, D T and Strauss, J H 1972 Replication of sindbis virus II Multiple forms of double-stranded RNA isolated from infected cells J Mol Biol **71:**615-631

8 Chambon, P 1974 Eucaryotic RNA polymerases, p 261-331 *In* Boyer, P D (ed) The Enzymes, vol 10 Academic Press, Inc New York

79

CORONAVIRUSES USE DISCONTINUOUS EXTENSION FOR SYNTHESIS OF SUBGENOME-LENGTH NEGATIVE STRANDS

Stanley G. Sawicki and Dorothea L. Sawicki

Department of Microbiology
Medical College of Ohio
P.O. Box 10008
Toledo, Ohio 43699

ABSTRACT

We have developed a new model for coronavirus transcription, which we call discontinuous extension, to explain how subgenome-length negatives stands are derived directly from the genome. The current model called leader-primed transcription, which states that subgenomic mRNA is transcribed directly from genome-length negative-strands, cannot explain many of the recent experimental findings. For instance, subgenomic mRNAs are transcribed directly via transcription intermediates that contain subgenome-length negative-strand templates; however subgenomic mRNA does not appear to be copied directly into negatives strands. In our model the subgenome-length negative strands would be derived using the genome as a template. After the polymerase had copied the 3'-end of the genome, it would detach at any one of the several intergenic sequences and reattach to the sequence immediately downstream of the leader sequence at the 5'-end of genome RNA. Base pairing between the 3'-end of the nascent subgenome-length negative strands, which would be complementary to the intergenic sequence at the end of the leader sequence at the 5'-end of genome, would serve to align the nascent negative strand to the genome and permit the completion of synthesis, i.e., discontinuous extension of the 3'-end of the negative strand. Thus, subgenome-length negative-strands would arise by discontinuous synthesis, but of negative strands, not of positive strands as proposed originally by the leader-primed transcription model.

Coronaviruses contain an unusually long (27-32,000 ribonucleotides) positive-sense RNA genome (1, 2) that is polyadenylated at the 3' end and capped at the 5' end. During the coronavirus replication cycle, genome and subgenomic mRNAs are produced that together comprise a 3' co-terminal nested set. At their 5' ends they also all possess an identical sequence, called the leader RNA, of 60-80 nucleotides. The original proposal (3) that transcription of subgenomic mRNA was discontinuous was based on the published report (4) that only genome-length negative-strand templates existed in infected cells. Splicing of the genome had been ruled out because the rate of inactivation by ultraviolet light (UV) of the synthesis of genomic and subgenomic mRNA was proportional to the length of the RNA (5, 6); therefore, it appeared that the leader RNA was joined to the body of subgenomic mRNA during positive strand synthesis. The leader-primed transcription model was proposed (7) because only genome-length negative strands were reported to be found in replication intermediates containing nascent subgenomic mRNA that already had acquired their leader RNA. Taken together the experimental evidence suggested RNA synthesis initiated at the 3' end of the negative strand copy of the genome and then proceeded either continuously to produce genomic RNA or discontinuously to produce subgenomic mRNA, in which case the leader RNA together with the viral transcriptase detached or jumped after copying the leader RNA and started again further downstream at specific intergenic sequence (IS) elements. The evidence for leader-primed transcription is indirect (reviewed in 1, 2) and based mostly on the fact that the IS elements possess a consensus sequence, which is found also at the end of the leader RNA, and that the IS elements determine subgenomic mRNA synthesis (8, 9, 10, 11).

Several years after the leader-primed transcription model had been proposed and generally accepted as correct, David Brian's laboratory (12) demonstrated that transmissible gastroenteritis virus (TGEV) infected cells contained subgenome-length negative strands. Shortly afterward we demonstrated (13) that replication intermediates (RIs) containing subgenome-length negative strands were transcriptionally active in mouse hepatitis virus (MHV) infected cells. Finally, Brian's laboratory demonstrated (14) that the subgenome-length negative strands contained at their 3'-end a sequence that was complementary to the leader RNA. Taken together, these observations showed unequivocally that subgenomic-length templates were utilized for the production of subgenomic mRNA. Figure 1 illustrates one of the approaches that we are using currently to study the synthesis of genome-length and subgenome-length negative strands in MHV infected cells.

Oligodeoxyribonucleotides were used as primers to first copy with reverse transcriptase a defined region of either positive or negative strand MHV RNA into cDNA. The cDNA was then amplified with PCR. To detect negative strands, we used during reverse transcription a primer that was identical to a sequence very near the 5' end of the leader RNA and that would anneal to genome-length negative strands as well as subgenome-length negative strands if they both possessed at their 3' ends a complementary copy of the leader RNA. To detect positive strands, we used two primers for reverse transcription: one that was complementary to a unique sequence 550 nucleotides downstream of the 5' end of the genome RNA and a second primer that was complementary to a sequence 370 nucleotides downstream of the 5' end of mRNA-7. Because the positive strands form a 3' nested set, the second primer would recognize also sequences that are present not only in RNA-7 but also in the other subgenomic mRNAs as well as in the genomic RNA. However, because we used only 5 min of incubation for the reverse transcriptase reaction, mostly RNA-7 was copied into cDNA. During the PCR amplification all three primers were present. Figure 1 shows the results when we used RNA isolated from cells seven hours after infection with MHV. The primer pairs for the genome and the negative strand copy of the genome gave a PCR product of the expected length of 550 nucleotides; and the primer pairs for RNA-7 and the negative-strand copy of RNA-7 gave a PCR product of the expected length of 370 nucleotides. Therefore, MHV infected cells possess both subgenome-length and genome-length negative strands that have a complementary copy of the leader RNA at their 3' ends. If we mix the primer pairs appropriately, we can detect in infected cells the genome and subgenomic mRNAs and their negative strand copies. Figure 2 shows that uninfected cells do not produce a PCR product using these primers.

Cells infected with MHV in the cold (4°C) and harvested after removing unadsorbed virus contained only genomic RNA and not the negative strand copy of the genome.

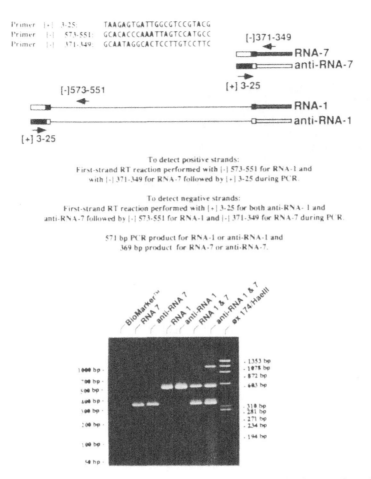

Figure 1. Detection by PCR of genomes and subgenomic mRNA and their negative strand templates. Super-script-II reverse transcriptase (Life Technologies, Bethesda, MD) for first strand synthesis and Taq DNA polymerase (Boehringer-Mannheim, Indianapolis, IN) for PCR were used according to the manufacturer's directions. After denaturation at 95°C for 5 min and annealing with the first strand primer, reverse transcription reactions were carried out for 5 min at 42°C. After heating to 99°C for 2 min and cooling to 4°C, the second strand primer(s) were added with the Taq polymerase and DNA amplification was performed for 35 cycles: 1 min at 65°C for annealing, 5 min at 72°C for extension and 1 min at 95°C for denaturation.

Interestingly, they did not have either RNA-7 or the negative strand copy of RNA-7, although RNA-7, but not its negative strand copy, was present in preparations of purified virions. The virion RNA was obtained from virions that had been purified twice by isopycnic tartrate-gradient centrifugation. Although subgenomic mRNAs were present in particles that have the same density as virions, they were not adsorbed and, therefore, were not in the same particles as the genome. By seven hours post-infection, the infected cell possessed all of the aforementioned RNA molecules. These experiments confirm those from Brian's laboratory (14) and extend their findings to MHV: MHV infected cells contain negative strand copies of subgenomic mRNA including a sequence at their 3' end that is complementary to the leader RNA. Furthermore, these experiments demonstrate directly that subgenomic mRNA and subgenome-length negative strands arise from genomic RNA because that was the only RNA that was in the virions that adsorbed to the cells.

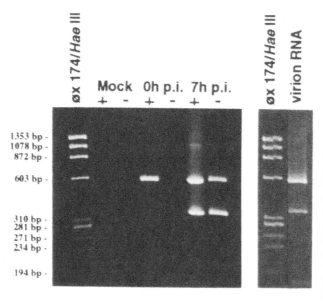

Figure 2. Only virions containing genome RNA adsorb to 17Cl-1 cells and cause the synthesis of genomes and subgenomic mRNA and their negative strand templates. The conditions for reverse transcription and DNA amplification by PCR were as described in Figure 1. Virion RNA was obtained from virus that had been purified twice by isopycnic tartrate-gradient centrifugation (15).

Previously, we published (13) experiments that demonstrated that ^{3}H-uridine accumulated rapidly and at the same rate in the smallest RI, which would produce mRNA-7, as in the largest RI, which would produce genomes. Figure 3 shows the results of this type of experiment.

These experiments demonstrated conclusively that RIs with subgenome-length templates were transcriptionally active in mRNA synthesis. These RIs were synthesizing positive strand RNAs because negative strand synthesis had virtually ceased at the time of labeling

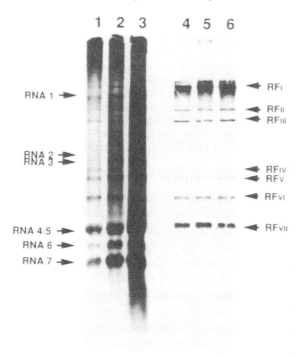

Figure 3. Transcriptionally active MHV subgenome-length RIs. 17Cl-1 cells (3 x 10⁶) in 60 mm petri dishes were infected with MHV at 20 pfu/cell and labeled with 1 mCi of ^{3}H-uridine/ml for 2 min [lanes 1 and 4], with 0.25 mCi/ml for 5 min [lanes 2 and 5], or with 0.1 mCi/ml for 20 min [lanes 3 and 6]. The cells were solubilized, extracted with phenol and chloroform, and ethanol precipitated. Extracts from 5 x 10⁵ cells were treated with DNase and either were not treated [lanes 1-3] or treated [lanes 4-6] with 1 µg of RNase A/ml before electrophoresis on an 0.8% agarose gel in TBE buffer. [Reproduced from Sawicki and Sawicki (13) by courtesy of ASM.]

(13, 15). We also noted that the relative amount of ^{3}H-uridine incorporated into each RI did not vary with the length of the labeling period and that the relative amount of label in each species of RI RNA was proportional to the synthesis of genome and subgenomic mRNAs. Therefore, it appeared from the data that all the subgenomic mRNAs were produced from RIs containing subgenome-length templates; and the RIs containing genome-length templates were producing exclusively genomes. This conclusion was consistent with the UV inactivation studies (5, 6) and the observations from Brian's laboratory (12, 14, 16) and our own laboratory (15) that negative strands accumulated during the early phase of the replication cycle when viral RNA synthesis was increasing at an exponential rate and stopped accumulating when the rate of viral RNA synthesis became constant. Thus, contrary to the recent suggestion (17) that subgenome-length negative strands were dead-end products, the experimental data (13) clearly demonstrated they were active as templates for subgenomic mRNA synthesis.

How do subgenome-length templates arise? Sethna, Hofmann and Brian (14) proposed that subgenomic mRNAs would be replicated either after they were produced by leader-primed transcription or directly from subgenomic mRNAs that entered the cells because they were packaged within the virions. They reasoned the coronavirus replicase would be able to recognize the subgenomic mRNAs because they possessed the same 3' and 5' ends as genomic RNA. However, subsequently it was shown that infected cells did not replicate transfected subgenomic mRNAs or subgenomic DI-RNAs (8, 9, 10). Subgenomic mRNAs themselves, therefore, appear unable to participate directly in replication. Based on this observation, it appears that sequences downstream of the leader RNA, which are present in genomes and DI RNAs, are required to form replication complexes. If subgenomic mRNAs are incapable of acting as templates to initiate the formation of replication complexes, it would appear that discontinuous transcription of subgenomic mRNA or leader-primed transcription would not have a role to play in coronavirus transcription as originally proposed.

If subgenome-length negative-strand templates are not created by copying subgenomic mRNAs, then they must arise directly from the genome as we had proposed (13). Based on the fact that subgenome-genome length negative-strand templates are transcriptionally active, we propose a model, called 3'-discontinuous extension of negative-strand templates, to explain the discontinuous synthesis of subgenome-length negative stands. This model is depicted in Figure 4.

The major points of the model are that the viral replicase begins synthesis at the 3' end of the genome and pauses at the IS elements; it then either elongates through the IS elements or discontinues synthesis and switches to the 5' end of the genome and completes synthesis by copying the leader RNA. The result of 3' discontinuous extension during negative strand synthesis is that the subgenome-length negative strands acquire a complementary copy at their 3' end of the leader RNA. The replication complex that synthesized a subgenome-length negative strand would retain it as a template and go on to transcribe continuously, not discontinuously, subgenomic mRNA. The replication complexes of coronaviruses, like alphaviruses, would retain the negative strand template they created by copying the genome and use it as the preferred template (18). If the viral negative strand replicase elongated continuously to the 5' end of the genome, the replication complex would become loaded with a genome-length negative strand template and synthesize genomic RNA. Leader switching (19) would occur during negative strand synthesis if the incomplete or nascent subgenome negative strand switched templates and copied the 5' end of a different positive strand RNA. The extent of leader switching would be governed by the efficiency at which a particular genome was replicated, by the nature and location of the IS elements and by the sequence at the end of the leader RNA, i.e. whether it is a 2-repeat or 3- repeat element or has insertions or deletions. The coronavirus negative strand replicase might copy the

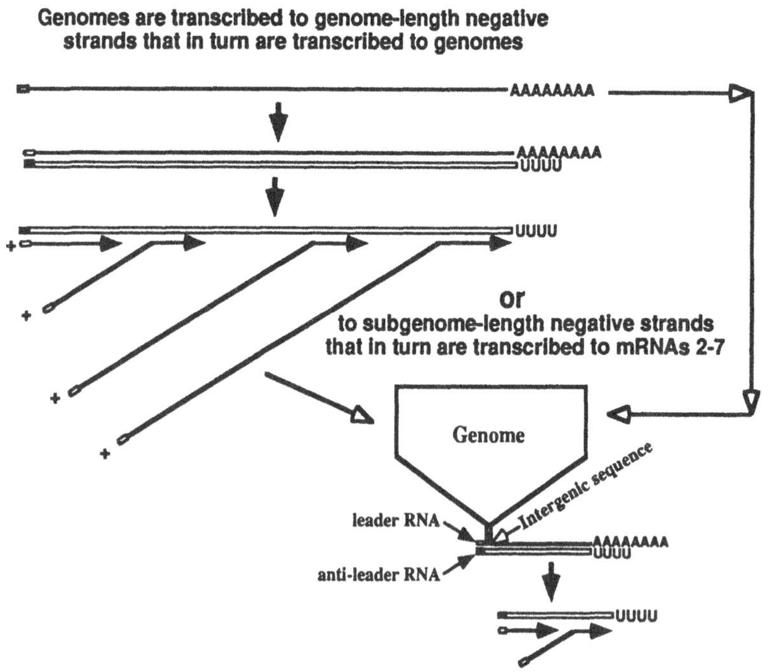

**Genomes are transcribed to genome-length negative
strands that in turn are transcribed to genomes**

Figure 4. A model for coronavirus negative strand synthesis: 3'-discontinuous extension of negative-strand RNA synthesis.

template RNA in a fashion analogous to DNA-dependent RNA polymerases that are capable of retraction (20) and remain associated with the growing nascent strand rather then with the template (21): If retraction occurred after the viral replicase had copied the IS element, the exposed 3' end of the nascent negative strand would relocate and align precisely to complementary sequences at the 3' side of the leader RNA at the 5' end of the genome and complete synthesis of the negative strand. Therefore, the alignment for joining the body of the mRNA with the leader RNA would occur during the formation of the negative-strand template for the subgenomic mRNAs and not during mRNA synthesis. The interaction of the nascent negative strand with the 5' end of the genome might be mediated by protein:protein interaction between the replicase attached to the growing negative strand and a protein associated with the 5' end of the genome. Only positive strand templates that possess the promoter for negative strand synthesis at their 3' end and the appropriate sequences at the 5' end would be capable of forming active replication complexes. Therefore, positive strands might get copied into negative strands that would be incapable of producing positive strands because they would lack the promoter element at their 3' end.

Discontinuous extension of the 3' end of the negative strands has many attractive features. It would explain for instance why shorter mRNAs are overproduced relative to the longer mRNAs. van der Most et al (11) have shown that if the IS element for mRNA-3 was moved close to the 3' end of the genome, it became as active as the IS element for mRNA-7. The negative-strand template for the shortest mRNA would have the greatest probability of being produced since its IS element is promoter proximal. The sequence of the IS element also influenced the abundance of subgenomic mRNA (11). Therefore, position effect would not be the only factor governing the relative rate of synthesis of a given species of subgenomic mRNA. The open reading frame encoding the viral replicase might be translated simultaneously on the genome serving as a template for subgenome-length negative-strand synthesis. If a viral protein were to bind to the 5' end of the genome and determine the

3'-discontinuous extension of subgenome-length negative strands, its absence during the initial round of replication would guarantee the genome was copied first continuously into a genome-length negative strand template. Because coronavirus replication and transcription requires continuous translation (15), presumably of ORF1, subgenomic mRNA synthesis would be dependent on genome RNA synthesis because a constant supply of viral polymerase proteins would be needed to fuel transcription. Thus, the danger of the subgenome-length templates acting as DI RNA would be mitigated by several factors: Firstly, subgenome-length negative strand templates would arise from the genome, not from subgenomic mRNA; and secondly, the production of subgenomic mRNA would be dependent on the presence of enough genomic mRNA to supply the viral nonstructural proteins that would be required for transcription. To our knowledge, there are no experimental data that disprove such a model. Furthermore, it has many features that make it an attractive and informative model for the elucidation of the strategy that coronaviruses use to express their genetic information and replicate.

ACKNOWLEDGMENTS

We wish to acknowledge Ryan Lee who performed the experiments shown in Figures 1 and 2. Support for these studies was derived from Public Health Service grant AI-28506 from the National Institutes of Health.

REFERENCES

1 Spaan, W, D Cavanagh, and M C Horzinek 1988 Coronaviruses structure and genome expression J Gen Virol 69 2939-2952

2 Lai, M M C 1990 Coronavirus-organization, replication and expression of genome Annu Rev Microbiol 44 303-333

3 Spaan, W, H Delius, M Skinner, J Armstrong, P Rottier, S Smeekens, B A M van der Zeijst, and S G Siddell 1983 Coronavirus mRNA synthesis involves fusion of noncontiguous sequences EMBO J 2 1939

4 Lai, M M C, C D Patton, and S A Stohlman 1982 Replication of mouse hepatitis virus negative-stranded RNA and replicative form RNA are of genome length J Virol 44 487

5 Jacobs, L, W J M Spaan, M C Horzinek, and B A M van der Zeijst 1981 Synthesis of subgenomic mRNAs of mouse hepatitis virus is initiated independently evidence from UV transcription mapping J Virol 39 401-406

6 Stern, D F, and B M Sefton 1982 Synthesis of coronavirus mRNAs kinetics of inactivation of IBV RNA synthesis by UV light J Virol 42 755-759

7 Baric, R S, S A Stohlman, and M M C Lai 1983 Characterization of replicative intermediate RNA of mouse hepatitis virus presence of leader RNA sequences on nascent chains J Virol 48 633

8 Brian, D A, R -Y Chang, M A Hofmann, and P B Sethna 1994 Role of subgenomic minus-strand RNA in coronavirus replication Arch Virol (Suppl) 9 173-180

9 Makino, S, M Joo, and J K Makino 1991 A system for study of coronavirus mRNA synthesis a regulated, expressed subgenomic defective interfering RNA results from intergenic site insertion J Virol 65 6031-6041

10 Masters, P S, C A Koetzner, C A Kerr, and Y Heo 1994 Optimization of targeted RNA recombination and mapping of a novel nucleocapsid gene mutation in the coronavirus mouse hepatitis J Virol 68 328-337

11 van der Most, R G, R J de Groot, and W J W Spaan 1994 Subgenomic RNA synthesis directed by a synthetic defective interfering RNA of mouse hepatitis virus a study of coronavirus transcription initiation J Virol 68 3656-3666

12 Sethna, P B S-L Hung, and D A Brian 1989 Coronavirus subgenomic minus-strand RNAs and the potential for mRNA replicons Proc Natl Acad Sci USA 86 5626-5630

13 Sawicki, S G, and D L Sawicki 1990 Coronavirus transcription subgenomic mouse hepatitis virus replicative intermediates function in RNA synthesis J Virol 64 1050-1056

14 Sethna, P B M A Hofmann, and D A Brian 1991 Minus-strand copies of replication coronavirus mRNAs contain antileaders J Virol 65 320-325

15 Sawicki, S G , and D L Sawicki 1986 Coronavirus minus-strand RNA synthesis and the effect of cycloheximide on coronavirus RNA synthesis J Virol 25 19-27

16 Hofmann, M A , P B Sethna, and D A Brian 1990 Bovine coronavirus mRNA replication continues throughout persistent infection in cell culture J Virol 64 4108-4114

17 Jeong, Y S , and S Makino 1992 Mechanism of coronavirus transcription duration of primary transcription initiation activity and effects on subgenomic RNA transcription on RNA replication J Virol 66 3339-3346

18 Sawicki, S G, and D L Sawicki 1994 Alphavirus positive and negative strand RNA synthesis and the role of polyproteins in formation of viral replication complexes Arch Virol (Suppl) 9 393-405

19 Jeong, Y S , and S Makino 1994 Evidence for coronavirus discontinuous transcription J Virol 68 2615-2623

20 Kassavetis, G A , and E P Geiduschek 1993 RNA polymerase marching backward Science 259 944-945

21 Johnson, T L , and M J Chamberlin 1994 Complexes of yeast RNA polymerase II and RNA are substrates for TFIIS-induced RNA cleavage Cell 77 217-224

80

REGULATION OF TRANSCRIPTION OF CORONAVIRUSES

Guido van Marle, Robbert G. van der Most, Tahar van der Straaten,
Willem Luytjes, and Willy J.M. Spaan

Department of Virology
Faculty of Medicine
Leiden University
PO Box 320
2300 AH Leiden, The Netherlands

ABSTRACT

To study factors involved in regulation of transcription of coronaviruses, we constructed defective interfering (DI) RNAs containing sg RNA promoters at multiple positions. Analysis of the amounts of sg DI RNA produced by these DIs resulted in the following observations: (i) a downstream promoter downregulates an upstream promoter; (ii) an upstream promoter has little or no effect on the activity of a downstream promoter. Our data suggest that attenuation of upstream promoter activities by downstream promoter sequences plays an important role in regulating the amounts of sg RNAs produced by coronaviruses. Our observations are in accordance with the models proposed by Konings et al. (8) and Sawicki and Sawicki (16).

Coronaviruses produce a 3'-coterminal nested set of subgenomic (sg) mRNAs. All sg mRNAs contain a common leader sequence derived from the 5' end of the genome. For mouse hepatitis virus MHV this leader sequence is 72 nucleotides (nt) in length (20). The joining of the 5' leader RNA to the mRNA is believed to be a discontinuous transcription process (10,19), since the results of UV transcription mapping argue against RNA splicing (5,23). On the genome the transcription units for the mRNAs are preceded by the intergenic sequence (IS) (10,19). For MHV every IS contains a sequence element related to the consensus 5' AAUCUAAAC 3' (2,8,18). These IS elements function, on the negative stranded RNA template, as promoters for sg mRNA synthesis (14,22). On the negative strand the IS promoter elements are called intergenic promoter sequence (IPS).

The mechanism of coronavirus sg mRNA synthesis is a subject of considerable debate. In earlier experiments only genome length negative strands were found (11) and it was believed that genomic negative strands were the exclusive templates for the synthesis of sg mRNAs. To explain the synthesis of leader containing sg mRNAs, it has been proposed that short leader RNA species act as primers (9,20). According to this leader-primed

Corona- and Related Viruses, Edited by P. J. Talbot and G. A. Levy
Plenum Press, New York, 1995

transcription model, the leader RNAs are transcribed from the 3' end of the genome, translocated to the several IPSs on the negative stranded template and then extended to form leader containing sg RNAs. The key observation that supports the priming of the leader during transcription initiation is the fact that the leader RNA includes an IS that allows base pairing between the 3' end of the leader and the IPS (2).

The discovery in recent years of sg negative strands (4,6,16,17) has had consequences for the leader primed transcription model. The sg negative strands seem to be involved actively in the synthesis of sg mRNAs (3,16), although it has been argued that they are merely dead-end products synthesized from the sg mRNAs (6). Additional models have now been proposed for coronavirus transcription, in which the sg mRNAs are transcribed from sg negative stranded templates.

Sethna et al. (17) speculated that sg mRNAs produced in the classic leader-primed fashion, are amplified from negative stranded counterparts as replicons. However, to date all attempts to obtain direct evidence for mRNA replication have failed. Transfecting synthetic mRNAs into coronavirus infected cells did not result in replication of the sg RNA (1,12,14). However, it could well be that transfected sg RNAs are not suitable templates for replication.

Sawicki and Sawicki (16) proposed an alternative model. They suggested that sg negative strands are synthesized first to serve as templates for the synthesis of the corresponding mRNA and not vice versa. In this model transcription should be regulated on the level of negative strand synthesis.

Many of the basic features of coronavirus transcription are unclear because of the lack of an appropriate experimental system. Recently, it has been shown that full length cDNA clones of defective interfering (DI) RNAs can be used to study MHV mRNA transcription (7,13-16,22). Inserting an IS into the genome of a synthetic MHV DI-RNA and transfecting this DI-RNA into MHV infected cells gives rise to a DI derived sg RNA. To study transcription, we use a DI RNA vector based on a full length cDNA clone of a natural occuring 5.5 kb DI RNA of MHV-A59, pMIDIC (21,22).

Coronavirus mRNAs are, in general, synthesized in amounts that are inversely related to their size. Previously, we have proposed that the generation of this gradient of sg mRNA arizes because larger RNA molecules are more prone to premature transcription termination and therefore produced less abundantly than smaller RNAs (8). There are two stages in which transcription termination can occur. In one case transcription initiation events on downstream promoters on the negative strands are attenuating factors for positive strand synthesis (8). Alternatively, premature termination could occur during negative strand synthesis (16). This is based on the model in which a nested set of sg negative strands is synthesized first. In this case larger negative strands are produced in lower quantities because they encounter more transcription attenuating antipromoters on the positive strand during their synthesis then smaller ones.

To test the hypothesis of attenuation we inserted wildtype (wt) as well as mutant sg RNA3 promoters at different positions of our DI-RNA vector (Fig. 1). The mutant RNA3 promoter is inactive due to a single point mutation. DI-RNAs containing the wt and the mutant promoters replicated efficiently and produced sg DI-RNAs of the expected length. The DI-RNA constructs containing a wt RNA3 promoter at position A or C (Fig. 1), produced equal amounts of sg DI-RNA. However, analysis of the sg RNA synthesis of the DI-RNA containing wt promoters at positions A and C showed a difference in promoter activities. The activity of the promoter at position A was reduced by the presence of the wt promoter at position C, while the activity of the promoter at position C remained the same. The presence of an additional wt promoter at position B, downstream of A (Fig. 1), reduced the activity of the wt promoter at position A to an almost undetectable level. The activity of the promoter at position C was not affected by the presence of an additional upstream promoter.

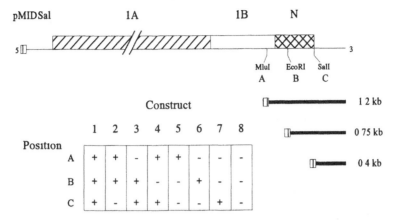

Figure 1. Schematic representation of the constructs containing RNA3 promoters the *Mlu*I, *Eco*RI and *Sal*I site of pMIDSal (positions A, B and C on the DI RNA) The black bars represent the subgenomic DI RNAs produced by the DIs The table shows which constructs contain which combinations of wildtype and mutant RNA3 promoters (+ = wildtype (UAAUCUAAAC), - = mutant (UAAUGUAAAC))

From these data we concluded that a downstream promoter attenuates the amounts of sg RNA generated by upstream promoter and not vice versa Our observations are in agreement with the models of Konings et al (8) and Sawicki and Sawicki (16) However, our data can not discriminate between attenuation during positive or negative strand synthesis We also observed that a wt RNA3 promoter at position B gave rise to more sg DI-RNA then at positions A and C Even in the presence of wt promoters at positions A and C, the sg DI-RNA produced by the promoter at position B is the most abundant This suggests that there are additional factors that regulate the accumulation of sg RNAs It could be possible that the polymerase initiates or terminates (depending on the model one prefers) transcription more efficiently at the promoter at position B Protein binding domains or the RNA secondary structure could play a role in this preference for the promoter at position B This could explain why for the coronaviruses TGEV and FIPV the smallest sg mRNA is not the most abundant as the gradient would predict Nevertheless, our data suggest that attenuation of promoter activities by promoter sequences is important in regulating the amounts of sg mRNAs of coronaviruses

REFERENCES

1 Brian, D A , R -Y Chang, M A Hofmann, and P B Sethna 1994 Role of subgenomic minus-strand RNA in coronavirus replication Arch Virol 9 173-180

2 Budzilowicz, C J , S P Wilczynski, and S R Weiss 1985 Three intergenic regions of coronavirus mouse hepatitis virus strain A59 genome RNA contain a common nucleotide sequence that is homologous to the 3' end of the viral mRNA leader sequence J Virol 53 834-840

3 Hofmann, M A , R -Y Chang, S Ku, and D A Brian 1993 Leader-mRNA junction sequences are unique for each subgenomic mRNA species in the bovine coronavirus and remain so throughout persistent infection Virology 196 163171

4 Hofmann, M A , P B Sethna, and D A Brian 1990 Bovine coronavirus mRNA replication continues throughout persistent infection in cell culture J Virol 64 4108-4114

5 Jacobs, L , W J M Spaan, M C Horzinek, and B A M van der Zeijst 1981 Synthesis of subgenomic mRNAs of mouse hepatitis virus is initiated independently Evidence from UV transcription mapping J Virol 39 401-406

6   Jeong, Y S and S Makino 1992 Mechanism of coronavirus transcription duration of primary transcription initiation activity and effects of subgenomic RNA transcription on RNA replication J Virol 66 3339-3346

7   Joo, M and S Makino 1992 Mutagenic analysis of the coronavirus intergenic consensus sequence J Virol 66 6330-6337

8   Konings, D A M , P J Bredenbeek, J F H Noten, P Hogeweg, and W J M Spaan 1988 Differential premature termination of transcription as a proposed mechanism for the regulation of coronavirus gene expression Nucl Acids Res 16 10849-10860

9   Lai, M M , R S Baric, P R Brayton, and S A Stohlman 1984 Characterization of leader RNA sequences on the virion and mRNAs of mouse hepatitis virus, a cytoplasmic RNA virus Proc Natl Acad Sci USA 81 3626-3630

10 Lai, M M C 1990 Coronavirus - organization, replication and expression of genome Ann Rev Microbiol 44 303-333

11 Lai, M M C , C D Patton, and S A Stohlman 1982 Replication of mouse hepatitis virus negative-stranded RNA and replicative form RNA are of genome length J Virol 44 487-492

12 Luytjes, W , H Gerritsma, and W J M Spaan 1994 Unpublished results

13 Makino, S and M Joo 1993 Effect of intergenic consensus sequence flanking sequences on coronavirus transcription J Virol 67 3304-3311

14 Makino, S , M Joo, and J K Makino 1991 A system for study of coronavirus mRNA synthesis a regulated, expressed subgenomic defective interfering RNA results from intergenic site insertion J Virol 65 6031-6041

15 Makino, S , K Yokomori, and M M Lai 1990 Analysis of efficiently packaged defective interfering RNAs of murine coronavirus localization of a possible RNA-packaging signal J Virol 64 6045-6053

16 Sawicki, S G and D L Sawicki 1990 Coronavirus transcription subgenomic mouse hepatitis virus replicative intermediates function in RNA synthesis J Virol 64 1050-1056

17 Sethna, P B , S L Hung, and D A Brian 1989 Coronavirus subgenomic minus-strand RNAs and the potential for mRNA replicons Proc Natl Acad Sci USA 86 5626-5630

18 Shieh, C-K , H J Lee, K Yokomori, N La Monica, S Makino, and M M C Lai 1989 Identification of a new transcriptional initiation site and the corresponding functional gene 2b in the murine coronavirus RNA genome J Virol 63 3729-3736 @RefNum = 19
    Spaan, W , D Cavanagh, and M C Horzinek 1988 Coronaviruses Structure and genome expression J Gen Virol 69 2939-2952

20 Spaan, W , H Delius, M Skinner, J Armstrong, P Rottier, S Smeekens, B A M van der Zeijst, and S G Siddel 1983 Coronavirus mRNA synthesis involves fusion of non-contiguous sequences EMBO J 2 1839-1844

21 Van der Most, R G , P J Bredenbeek, and W J M Spaan 1991 A domain at the 3' end of the polymerase gene is essential for the encapsidation of coronavirus Defective Interfering RNAs J Virol 65 3219-3226

22 Van der Most, R G , R J de Groot, and W J M Spaan 1994 Subgenomic RNA synthesis directed by a synthetic defective interfering RNA of mouse hepatitis virus a study of coronavirus transcription initiation J Virol 68 3656-3666

23 Yokomori, K , L R Banner, and M M Lai 1992 Coronavirus mRNA transcription UV light transcriptional mapping studies suggest an early requirement for a genomic length template J Virol 66 4671-4678

81

EVIDENCE FOR A PSEUDOKNOT IN THE 3' UNTRANSLATED REGION OF THE BOVINE CORONAVIRUS GENOME

G. D. Williams,[1] R. -Y. Chang,[2] and D. A. Brian[2]

[1] Program in Biotechnology
[2] Department of Microbiology
University of Tennessee
Knoxville, Tennessee 37996-0845

ABSTRACT

A potential pseudoknot was found in the 3' untranslated region of the bovine coronavirus genome beginning 63 nt downstream from the stop codon of the N gene. Mutation analysis of the pseudoknot in a cloned defective interfering RNA indicated that this structural element is necessary for defective interfering RNA replication.

INTRODUCTION

The 291 nt 3' untranslated region (UTR) of the bovine coronavirus (BCV) likely contains sequences that serve as a promoter for minus-strand synthesis during RNA replication. To explore the structural features that might be important for this function, a computer program was used to predict thermodynamically stable secondary structural elements. This analysis identified a potential pseudoknot beginning 63 nt downstream from the termination codon of the N gene having two stems with a high negative free energy. A phylogenetic comparison with other coronaviruses indicated conservation at the secondary and tertiary levels.

To investigate the functional significance of the proposed pseudoknot in viral RNA replication, site-directed mutagenesis of a cloned 2.2 kb reporter containing defective interfering RNA (pDrep1) was undertaken to disrupt and restore base pairing in the first stem of the pseudoknot. Mutants were then tested for replication. These experiments strongly suggest a role for the pseudoknotted structure in viral RNA synthesis.

MATERIALS AND METHODS

Cloning and construction of the bovine coronavirus defective-interfering RNA which carries a reporter sequence of 30 nt (pDrep1) have been described[1]. Synthesis of pDrep1

RNA transcripts and assay for pDrep1 replication using Northern analysis were done as previously described[1]. Site-directed mutations of pDrep1 3' UTR were done with a previously published PCR-based mutagenesis procedure[2]. For these, oligodeoxynucleotides UCUAAAC and AGAUUUG were used. For computer analysis, the Microgenie program (Beckman Instruments) which employs the Tinoco algorithm for analysis of RNA secondary structure[3] was used.

RESULTS

A computer-assisted analysis of RNA secondary structures in the BCV 3' UTR showed the existence of two stem regions between nucleotides -173 and -226 from the base of the poly(A) tail. The downstream stem, stem 2, has a predicted stability of -10.2 kcal/mole, and the upstream stem, stem 1, which involves base-pairing between upstream sequences and the loop of stem 2, has a predicted stability of -8.6 kcal/mole (Fig. 1). The potential structure formed would be considered an H(hairpin)-type pseudoknot in which stems 1 and 2 of the folded structure would become adjacent to each other to cause coaxial stacking of the two stems and formation of a quasi-continuous double helix[4].

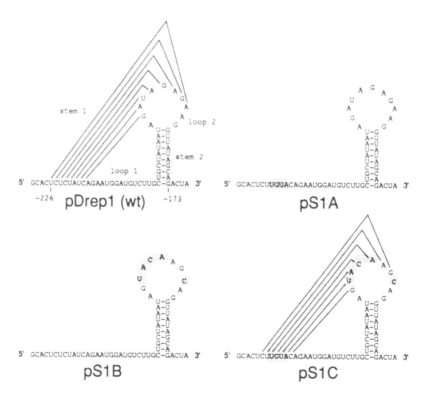

Figure 1. Proposed pseudoknot in the bovine coronavirus 3' UTR and the mutations used to test its role in RNA replication A Wild-type pseudoknot in the genome 3' UTR It is shown here as part of the 3' UTR in the 2 2 kb reporter-containing defective interfering RNA of BCV (pDrep1) B Mutant pS1A in which nucleotides CUAU in stem 1 were changed to UGUA C Mutant pS1B in which AUAGnnA in stem 1 were changed to UACAnnC D Mutant pS1C, the double mutant, in which the changes in pS1A and pS1B were combined to reform stem 1 which now has a stability of -5 6 kcal/mole

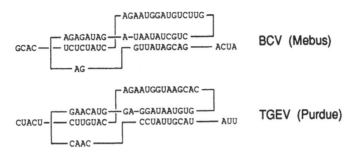

Figure 2. Comparison of the phylogenetically conserved pseudoknots between the bovine coronavirus and the porcine transmissible gastroenteritis coronavirus. Stem 1 is shown on the left, and loop 1 is shown on the right. Note the sequence divergence in the two stem regions.

To determine whether the pseudoknotted structure is phylogenetically conserved among coronaviruses, the 3′ UTR of the porcine transmissible gastroenteritis virus (TGEV) and ten other mammalian coronaviruses was likewise examined. Despite differences in nucleotide sequences, a pseudoknoted structure was found in the same relative position in TGEV (Fig 2) and the other coronaviruses examined (data not shown). Between BCV and TGEV, nucleotide similarities were 44% for stem 1 and 55% for stem 2. Yet in each case compensatory changes were found in TGEV that maintained stable stem structures. These were predicted to be -5 8 and -10 kcal/mole for stems 1 and 2, respectively. A similar picture was found for the other coronaviruses.

These observations led to the hypothesis that the pseudoknot must have evolved to serve a biological function. To test the idea that it plays a role in RNA replication, mutations were made to disrupt the pseudoknot in pDrep1, and the effect of the mutations on replication of the DI RNA were tested. pS1A and pS1B are constructs that have mutations on one side of stem 1. Point mutations were chosen to create mismatches predicted to destroy the thermodynamic stability of the stem. The double mutant pS1C incorporates both sets of mutations and mimics the analogous stem in TGEV. The three mutants were tested in a replication assay (Fig 3).

A Northern analysis showed that pDrep1 with the wild-type pseudoknot replicated as evidenced by an increase in abundance and by having been passaged in virions (Fig 3, lanes 5, 6 and 7). Disruption of the stem in the single mutants pS1A and pS1B abolished replication when assessed by the same analysis (Fig 3, lanes 10, 11 and 12, and 15, 16 and 17). On the other hand, the double mutant pS1C with the compensated stem replicated at or near wild-type levels and became packaged into virions (Fig 3, lanes 20, 21 and 22). These experiments demonstrate the functional importance of this stem and the pseudoknot in RNA replication.

DISCUSSION

To date, few phylogenetically conserved structural elements have been described in the 3′ untranslated region of coronaviruses that would suggest functional elements involved in RNA replication. In addition to the pseudoknot reported here, an octameric consensus sequence GGAAGAGC is found approximately 70 nt upstream from the poly (A) tail. This conserved element, however, has not yet been shown to play a role in RNA synthesis.

How the disrupted pseudoknot is preventing RNA replication in our studies is not known. Since the pseudoknot and the octameric consensus sequence are in the 3′ UTR of genomic RNA, they might function as polymerase recognition or promoter sites for minus-strand RNA synthesis. This, however, remains to be shown. Both primary and secondary

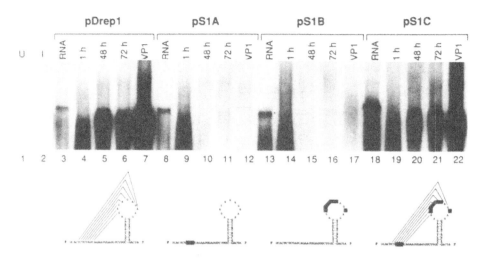

Figure 3. Importance of stem 1 in DI RNA replication. Synthetic transcripts of pDrep1 and of the three mutants pS1A, pS1B, and pS1C were transfected into BCV helper virus-infected cells, and cytoplasmic RNA was extracted at various times posttransfection, or at 48h after the first virus passage (VP1). Extracted RNA was separated by electrophoresis on a formaldehyde-agarose gel and probed in a Northern blot with radiolabeled probe for detection of the plus strand of the reporter sequence. Note the presence of the DI RNA in lanes identified pDrep1 and pSIC, but not in lanes identified as pS1A and pS1B.

structures have been shown to be important for RNA-dependent RNA polymerase promoter activity[5]. Another mechanistic possibility is that the disrupted pseudoknot may be preventing translation of the pDrep1 open reading frame and secondarily preventing replication. A dependency on the open reading frame for pDrep1 replication has been shown[1], and a strong preference for maintenance of the open reading frame in the replication of MHV DI RNAs has been demonstrated[6,7]. Precedent for a pseudoknot in the 3′ UTR of a plant virus being important for translation of the positive stranded genome has recently been reported[8].

Further studies are needed to determine the mechanistic role of the pseudoknot in coronavirus RNA replication.

REFERENCES

1. Chang R.-Y., Hofmann M.A., Sethna P.B., Brian D.A. A *cis*-acting function for the coronavirus leader in defective interfering RNA replication. J Virol 1994;68:8223-8231.
2. Horton R.M., Cai Z., Ho S.N., Pease L.R. Gene splicing by overlap extension: tailor made genes using the polymerase chain reaction. BioTechniques 1990;8:525-535.
3. Tinoco I., Borer P.N., Dengler B., Levine M.D., Uhlenbeck O.C., Crothers D.M., Gralla J. Improved estimation of secondary structure in ribonucleic acids. Nature (London) New Biol 1973;246:40-41.
4. ten Dam E., Pleij K., Draper D. Structural and functional aspects of RNA pseudoknots. Biochemistry 1992;31:1665-1676.
5. Dreher T.W., Hall T.C. Mutational analysis of the sequence and structural requirements in Brome Mosaic Virus RNA fro minus strand promoter activity. J Mol Biol 1988;201:31-40.
6. De Groot R.J., Van der Most R.G., Spaan W.J.M. The fitness of defective interfering murine coronavirus DI-a and its derivatives is described by nonsense and frameshift mutations. J Virol 1992;66:5898-5905.
7. Kim Y.-N., Lai M.M.C., Makino S. Generation and selection of coronavirus defective interfering RNA with large open reading frame by RNA recombination and possible editing. Virology 1993;194:244-253.
8. Leathers V., Tanguay R., Kobayashi M., Gallie D.R. A phylogenetically conserved sequence within viral 3′ untranslated RNA pseudoknots regulates translation. Mol Cell Biol 1993;13:5331-5347.

82

REGULATION OF CORONAVIRUS RNA TRANSCRIPTION IS LIKELY MEDIATED BY PROTEIN-RNA INTERACTIONS

X M Zhang[1] and M M C Lai[1][2][3]

[1] Department of Neurology
[2] Howard Hughes Medical Institute
[3] Department of Microbiology
University of Southern California School of Medicine
Los Angeles, California

ABSTRACT

Coronavirus mRNA transcription was thought to be regulated by the interaction between the leader RNA and the intergenic (IG) sequence, probably involving direct RNA-RNA interactions between complementary sequences In this study, we found that a 9-nucleotide sequence immediately downstream of the leader RNA up-regulated mRNA transcription and that a particular strain of mouse hepatitis virus (MHV) lacking this 9-nucleotide transcribed subgenomic mRNA species containing unusually heterogeneous leader-fusion sites These results suggest that the sequence complementarity between the leader and IG is not necessarily required for mRNA transcription UV cross-linking experiments using cytoplasmic extracts of uninfected cells and the IG sequence showed that three different cellular proteins bound to IG of the template RNA Deletion analyses and site-directed mutagenesis of IG further demonstrated a correlation between protein-binding and transcription efficiency, suggesting that these RNA-binding proteins are involved in the regulation of coronavirus mRNA transcription We propose that coronavirus transcription is regulated by RNA-protein and protein-protein interactions

INTRODUCTION

Coronavirus mRNA synthesis involves a unique mechanism of discontinuous transcription, generating subgenomic mRNAs which contain a leader RNA fused to a distant RNA sequence[4] The precise mechanism of this RNA synthesis is not yet fully

Corona and Related Viruses Edited by P J Talbot and G A Levy
Plenum Press New York 1995

understood. Regardless of the mechanism of mRNA transcription, the regulation of coronavirus mRNA synthesis has to involve the interaction between the leader RNA, which is present at the 5'-end of the genomic RNA and is incorporated into subgenomic mRNAs, and the intergenic (IG) sequence, which precedes every gene[5]. In this study, we use mouse hepatitis virus (MHV) as a model system to study the mechanism of the transcriptional regulation. The 3'-end of the leader on MHV genomic RNA generally contains 2 to 4 copies of a pentanucleotide sequence (UCUAA)[8]. The IG sequence also contains a consensus UCUAAAC or similar sequence[10]. The UCUAAAC sequence is necessary and sufficient for the initiation of mRNA synthesis[3]. Fusion of the leader sequence with the mRNA body sequence usually occurs between the UCUAA repeats of the leader and the consensus IG sequence[9]. It has thus been hypothesized that mRNA initiation is mediated by direct RNA-RNA interaction between the leader RNA and consensus IG[4]. The copy number of UCUAA at the 3' end of the subgenomic mRNA leader may differ from that of the genomic RNA leader, presumably because of imprecise alignment between the leader and the intergenic regions[4,9]. The 5'-end of genomic RNA, including the leader, also acts as an enhancer-like element for mRNA synthesis[6]. Furthermore, the sequence of IG significantly affects the pattern of transcriptional regulation. For instance, the efficiency of transcription initiation from the IG of mRNA2-1, varies with the leader·RNA containing either 2 or 3 UCUAA copies[10], while the IG of mRNA7 is constitutively transcribed. In addition, a 9-nucleotide (nt) sequence (UUUAUAAAC) immediately downstream of the leader influenced the frequency of incorporation of the leader RNA into mRNAs[13]. Deletion of this sequence also prevented leader-switching during RNA replication[8]. Thus, coronavirus mRNA transcription is regulated by direct or indirect interactions of multiple *cis*- and *trans*-acting RNA sequences.

In this study, we found that a particular strain of MHV, JHM2c, which has a deletion of the 9-nt sequence immediately downstream of the leader, generated subgenomic mRNAs with very heterogeneous leader-fusion, which did not directly involve complementary RNA sequences. These results strongly suggest that leader-mRNA fusion in coronavirus transcription does not require direct RNA-RNA interaction between complementary sequences. Ultraviolet (UV) light cross-linking studies with cytoplasmic extracts further identified several cellular proteins binding to the IG sequence and/or the leader RNA, both being the RNA regulatory elements. Furthermore, site-directed mutagenesis of IG7 demonstrated a correlation between the protein-binding and transcription efficiencies. This study thus provides the first evidence that cellular RNA-binding proteins are involved in coronavirus RNA synthesis, suggesting that these RNA-binding proteins may be the transcription factors.

MATERIALS AND METHODS

Viruses and cells: MHV strains JHM(3) and JHM2c(3) contain three UCUAA repeats, and JHM2c(4) contains four UCUAA repeats at the 3'-end of the leader RNA. Both JHM2c(4) and JHM2c(3) have a deletion of a 9-nt sequence immediately downstream of the leader. The properties of these viruses have been described previously[10]. DBT cells[2] were used for infections and transfections.

UV cross-linking of RNA-protein complex: Cytoplasmic extracts were prepared from uninfected DBT cells and UV cross-linking experiments were performed as described previously[1].

RESULTS AND DISCUSSION

Effects of the Nine-Nucleotide Sequence Immediately Downstream of the Leader RNA on mRNA Transcription

By using an MHV DI RNA containing a CAT gene behind an IG, we have previously demonstrated that deletion of the 9-nt sequence (UUUAUAAAC) in DI RNA reduced the efficiency of transcription from the IG, suggesting the involvement of this sequence in the regulation of subgenomic mRNA transcription[13]. Furthermore, the deletion of the 9-nt sequence inhibited the ability of the leader RNA of the DI RNA to be incorporated into subgenomic mRNAs, suggesting that this sequence might be required as a transcriptional terminator to synthesize a free leader RNA[13]. Thus, an intriguing question arises: if an MHV genome (e.g. JHM2c) lacks this 9-nt sequence, how are viral subgenomic mRNAs transcribed? To answer this question, we constructed several DI RNA-reporter plasmids which contain a CAT gene behind an IG of mRNA7 or mRNA2-1[6,13], in the presence or absence of the 9-nt sequence immediately downstream of the leader (Fig.1). These RNAs were transfected into cells infected with JHM2c, which lacks the 9-nt sequence in the genomic RNA. The transcriptional efficiencies from each IG site were determined by CAT assays, which accurately reflect transcriptional efficiencies[6]. As shown in Table 1, in the presence of JHM2c(3) and JHM2c(4), both of which lack the 9-nt sequence, CAT activities were approximately 3-fold higher from DIs with the 9-nt sequence than those from DIs without the 9-nt sequence. For example, in JHM2c(3)-infected cells, CAT activity was 3-fold higher in 25CAT+9nt than 25CAT, and 7-fold higher in DECAT2-1+9nt than DECAT2-1. Similarly, in JHM2c(4)-infected cells, CAT activity of DECAT-2-1+9nt was approximately 3.5-fold higher than in DECAT2-1. DECAT2-1(2)m RNA, which is identical to DECAT2-1+9nt except that it contains only 2 UCUAA repeats (Fig.1), has a transcriptional efficiency similar to that of DECAT2-1+9nt in both infected cells, suggesting that the differences in transcriptional efficiencies among different DI constructs were mainly due to the presence or absence of the 9-nt sequence, and not the copy number of UCUAA or other differences within the leader sequence. However, in JHM2c(4)-infected cells, 25CAT and 25CAT+9nt exhibited similar transcriptional efficiencies. These results indicate that the 9-nt sequence in *cis* (i.e. in DI RNA) could affect CAT expression, depending on the context of the IG and helper virus. This result is consistent with the previous finding that MHV transcription is regulated by both *trans*- and *cis*-acting leaders and the IG sequence[13]. Nevertheless, subgenomic mRNAs were transcribed from almost all of the DI RNA constructs, even though both the helper virus and DI RNA lack the 9-nt sequence. Therefore, we conclude that this 9-nt sequence is not absolutely necessary for transcription. When JHM(3), which contains the 9-nt sequence, was used as a helper virus, CAT activities in all of the RNA constructs were generally higher than in the corresponding RNAs in JHM2c(3)-infected cells (Table 1). These results together suggest that the 9-nt sequence in either helper viral RNAs (in *trans*) or DI RNA (in *cis*) could enhance transcription of subgenomic mRNAs. However, the magnitude of these effects varied with respect to the context of other

Figure 1. Structures of the cDNA clones of DI RNAs. Only the DI-derived regions, which are placed behind a T7 RNA polymerase promoter, are shown. *Xba*I was used for digestion of constructions for in vitro run-off transcription. Each unlabeled small open rectangle in the leader and IG represents a copy of UCUAA. The V-shaped breaks in the lines represent a 9-nucleotide deletion.

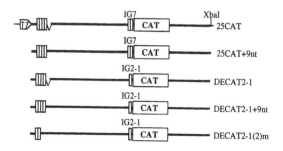

Table 1. Effect of the nine-nucleotide sequence on CAT expression

Transfected RNA	CAT activity (fold increase)* with		
	JHM2c(3)	JHM2c(4)	JHM(3)
25CAT	67×	252 ×	291 ×
25CAT+9nt	200 ×	225 ×	277 ×
DECAT2-1	1 ×	7 ×	11 ×
DECAT2-1+9nt	7 ×	24 ×	36 ×
DECAT2-1(2)m	6 ×	19 ×	70 ×

*The data represent the averages of five separate infection and transfection experiments in DBT cells. CAT activity is indicated as fold increase against the background values for a culture in which either virus for infection or RNA for transfection alone was used as a negative control, which was set at 1×.

sequences, such as the number of UCUAA repeats in the leader and IG (either IG7 or IG2-1) in both the DI RNA and helper virus. Since the 9-nt sequence is not part of the leader RNA and is not incorporated into mRNAs, this result further indicates that these are *cis*-acting upstream regulatory sequences for mRNA transcription.

Heterogeneity of Leader-mRNA Fusion Sites

We further determined the structure of subgenomic mRNAs from cells infected with JHM2c(3), and transfected with 25CAT, 25CAT+9nt or DECAT2-1(2)m RNA by sequencing. We found that, when JHM2c(3), which lacks the 9-nt sequence, was used as a helper virus, the leader-mRNA fusion sites in subgenomic mRNAs were extremely heterogeneous. Similar heterogeneity was also detected in mRNA2-1 of JHM2c(3)-infected cells[12]. Most notably, in 10 of 53 clones sequenced, the leader is fused with mRNAs at a site either upstream or downstream of the consensus IG. These sites bear little or no sequence homology with the leader. As a result, mRNAs contain sequences upstream, or have a sequence deletion downstream, of the consensus UCUAAAC. These heterogeneous leader-mRNA fusions suggest that coronavirus mRNA transcription does not require a complementary sequence between the leader and IG, implying that the RNA regulatory elements must be brought together by other factors such as cellular proteins which bind to these RNAs.

Interactions between the Cytoplasmic Proteins and the Intergenic Sequence of MHV RNA

We have previously shown that some cellular proteins, most notably p35/38, bind to the template strand of the leader RNA. Another protein p55 binds to the leader RNA per se[1]. To determine whether any cellular proteins interact with the third component of the RNA regulatory sequences, i.e. IG sequence, of MHV RNA, we performed UV cross-linking experiments using cytoplasmic extracts of uninfected DBT cells and the template strand of IG7. We found that four major proteins p110, p70, p48, and p35/38 bound to IG7-SF135(-) RNA. To determine the binding specificity of these proteins, we used an unlabeled homologous RNA and a nonspecific vector RNA to perform competition assays. When increasing amounts of the unlabeled homologous RNA (2- to 50-fold molar excess) were preincubated with a fixed amount of cytoplasmic extracts (20 μg of cellular proteins), most of the protein bands were competed away, except that p110 was not efficiently competed. In contrast, the nonspecific vector sequence (25- to 50-fold molar excess) did not affect the amount of protein-binding(data not shown) We conclude that at least three cytoplasmic proteins (p70, p48, and p35/38) specifically bound to IG7-SF135(-) RNA, while p110 binding was probably not specific Using a series of deletion constructs, we found that p35/38 bound only to the fragments which contain the consensus sequence of IG7, but not RNAs without the consensus sequence We thus conclude that the consensus sequence of IG7 is essential for p35/38 binding In contrast, the binding of p70 and p48 was only slightly decreased in some of the deletion clones, suggesting that the sequence requirement for p70 and p48 was not very specific or that these two proteins bound to a longer stretch of RNA sequence

Site-Specific Mutagenesis of IG7 Reveals That RNA-Protein Interaction Correlated with the Transcription Efficiency of Subgenomic mRNAs

To test the hypothesis that cellular proteins may be involved in the regulation of mRNA transcription, we constructed five different IG mutants (Fig 2A), which have previously been shown to result in altered transcriptional efficiencies[3 7], for UV cross-linking studies with cytoplasmic extracts The results showed that the binding efficiencies of these IG mutants with p35/38 correlated approximately with their transcriptional efficiencies,

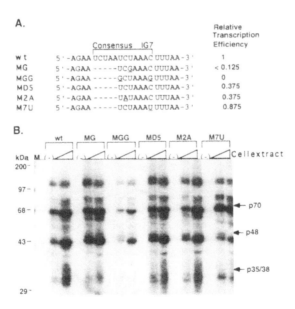

Figure 2. Site-specific mutagenesis of the consensus sequence of IG7 with regards to their protein binding properties (A) Sequences (in positive sense) of the wild-type (wt) IG7 and the mutants used The consensus sequence of the wt IG7 is indicated at the top of the sequence Nucleotide mutations are underlined, and deletions are indicated by dashes The transcription efficiencies of the mutant IGs indicated are relative to that of the wt IG, which is set as 100% These data are recalculated from the published reports[3 7] (B) UV cross-linking of cellular proteins with negative strand RNAs of the wt and mutant IG7 Cytoplasmic extracts (2 and 20 µg proteins, respectively) were mixed with a fixed amount of ^{32}P-labeled RNAs and UV-irradiated In (-) lanes, no cell extract was added M molecular mass standard in kDa Protein bands p70 p48 and p35/38 are indicated with arrows on the right side

except for mutant M7U (Fig. 2A and B). Specifically, mutant MGG, which has a deletion of a UCUAA sequence and two additional single-nucleotide mutations, has been shown to lack the transcriptional activity. Correspondingly, MGG(-) RNA did not bind p35/38, and bound slightly less p48 and p70. Mutant MG, which has a deletion of a UCUAA sequence and a C-_G mutation, had retained less than 12.5% of transcription efficiency of the wild-type IG7. Correspondingly, this IG mutant bound significantly less p35/38 than the wild-type RNA. Mutants M2A and M7U, which had similar transcription efficiencies, also had similar binding activities with p35/38. These results demonstrated that the binding efficiencies of p35/38 to IG7 correlated roughly with the transcription efficiencies of these mutants, suggesting that their binding to IG7 is required for subgenomic mRNA transcription, Interestingly, a C->U mutation in mutant M7U, which had retained most of the transcription activity, bound a relatively small amount of p35/38, but it bound an increased amount of p70 (Fig. 2). Thus, p70 binding may substitute for p35/38 in restoring the transcription activity. These results suggest that these cellular RNA-binding proteins may be transcription-associated factors for coronavirus RNA synthesis.

Recently, we have suggested that protein-RNA and protein-protein interactions rather than direct RNA-RNA interactions are the driving force for the initiation of coronavirus RNA transcription[13]. This model proposes that viral and cellular proteins first bind to the RNA regulatory elements through protein-RNA interactions; these components are then brought together with viral RNA polymerase and other viral and cellular factors to form a transcription initiation complex through protein-protein interactions (Fig.3). In this model, the sequence complementarity between leader and IG is not absolutely required; they may only

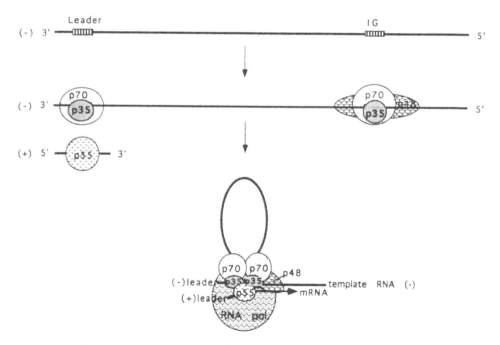

Figure 3. Model of coronavirus RNA transcription through protein-RNA and protein-protein interactions The striped boxes indicate the UCUAA repeats at the 3'-end of the leader and the consensus IG The proteins bound to the different regions were identified in reference[1] and this study The presence of the putative viral RNA polymerase in the complex is hypothetical In this putative complex, the negative-strand, genome-length RNA serves as a template, and the leader RNA (positive sense) *in trans* binds to the negative-strand IG RNA and serves as a primer The subgenomic mRNA transcription starts at the IG region inside the complex

provide the recognition signals for viral and cellular proteins Any alterations of the RNA elements and protein components within this complex could possibly affect the initiation and efficiency of subgenomic mRNA transcription The identification of the unusual heterogeneity in leader-mRNA fusion and cellular RNA-binding proteins involving in transcriptional regulation in this study strongly supports this notion

ACKNOWLEDGMENTS

This work was supported by a Public Health Service Research Grant AI16144 from the National Institutes of Health M M C L is an Investigator of the Howard Hughes Medical Institute

REFERENCES

1 Furuya T , Lai M M C Three different cellular proteins bind to complementary sites on the 5'-end-positive and 3'-end-negative strands of mouse hepatitis virus RNA J Virol 1993,67 7215-7222

2 Hirano N , Fujiwara K , Hino S, Matsumoto M Replication and plaque formation of mouse hepatitis virus (MHV-2) in mouse cell line DBT culture Arch Gesamte Virusforsch 1974,44 298-302

3 Joo M , Makino S Mutagenic analysis of the coronavirus intergenic consensus sequence J Virol 1992,66 6330-6337

4 Lai M M C Coronavirus organization, replication and expression of genome Annu Rev Microbiol 1990,44 303-333

5 Lai M M C , Baric R S , Brayton P R , Stohlman S A Characterization of leader RNA sequences on the virion and mRNAs of mouse hepatitis virus, a cytoplasmic RNA virus Proc Natl Acad Sci USA 1984,81 3626-3630

6 Liao C L , Lai M M C Requirement of the 5'-end genomic sequence as an upstream cis-acting element for coronavirus subgenomic mRNA transcription J Virol 1994,68 4727-4737

7 Makino S , Joo M Effect of intergenic consensus sequence flanking sequences on coronavirus transcription J Virol 1993,67 3304-3311

8 Makino, S , Lai M M C High-frequency leader sequence switching during coronavirus defective interfering RNA replication J Virol 1989,63 5285-5292

9 Makino S , Soe L H , Shieh C K , Lai M M C Discontinuous transcription generates heterogeneity at the leader fusion sites of coronavirus mRNAs J Virol 1988,62 3870-3873

10 Shieh C K , Lee H J , Yokomori K , La Monica N , Makino S , Lai M M C Identification of a new transcriptional initiation site and the corresponding functional gene 2b in the murine coronavirus RNA genome J Virol 1989,63 3729-3736

11 Shieh C K , Soe L H , Makino S , Chang M F , Stohlman S A , Lai M M C The 5'-end sequence of the murine coronavirus genome implications for multiple fusion sites in leader-primed transcription Virology 1987,156 321-330

12 Zhang X M , Lai M M C Unusual heterogeneity of leader-mRNA fusion in a murine coronavirus Implications for the mechanism of RNA transcription and recombination J Virol (in press)

13 Zhang X M , Liao C L , Lai M M C Coronavirus leader RNA regulates and initiates subgenomic mRNA transcription both in trans and in cis J Virol 1994,68 4738-4746

83

INTERACTIONS BETWEEN THE IBV NUCLEOCAPSID PROTEIN AND RNA SEQUENCES SPECIFIC FOR THE 3' END OF THE GENOME

Ellen W. Collisson[1], Anna K. Williams[2], Shan-Ing Chung,[3] and Minglong Zhou[1]

[1] Department of Veterinary Pathobiology, Texas A&M University
College Station, Texas 77843-4467
[2] The Scripps Research Institute
La Jolla, California 92037
[3] Molecular Biology Division, Development Center for Biotechnology
Taipei, Taiwan

ABSTRACT

The infectious bronchitis virus (IBV) nucleocapsid protein was expressed as a fusion protein in bacteria. The coding sequence differed from the native protein only in the addition of six histidine residues at the amino terminus which were used for enrichment with a nickel affinity column. In gel shift assays, the mobility of labelled G RNA was decreased with increasing concentrations of the fusion protein. Competitive gel shift assays with labelled G RNA indicated that the protein interacted with relatively high avidities to several unlabelled RNAs representing sequences at the 3' noncoding end of the IBV genome. Cache Valley virus (a bunyavirus) mRNA transcribed from the smaller segment cDNA also inhibited the interaction with IBV G RNA to the same extent as homologous unlabelled G RNA. In contrast, interactions of the fusion protein with a region from 99 to 249 bases from the 3' terminus of the IBV genome and bovine liver RNA were relatively weak. The binding of IBV nucleocapsid protein with RNA probably requires specific sequences and/or structures that are present at a number of sites on the genome, and may represent a common mechanism used by similar viral proteins whose functions depend on binding to RNA.

INTRODUCTION

The nucleocapsid protein of the avian coronavirus, infectious bronchitis virus (IBV), is a highly basic, phosphorylated structural protein of 409 residues or about 50kD[1,2]. A highly basic region in the IBV protein between residues 238 and 293 is identical among the strains that have been examined[2]. The corresponding region of mouse hepatitis virus (MHV) has been shown to bind to RNA[3,4]. The MHV nucleocapsid protein specifically binds to small leader-containing RNAs, and within the cytosol of MHV infected cells, it can interact with membrane-bound small leader RNAs in transcription complexes[5]. The amount of nucleo-capsid protein from MHV found associated with the genome and the putative functions of this protein suggest that it could readily associate with additional regions of coronavirus RNA. Because transcription of the negative template must initiate at the 3' end of genomic RNA and that nucleocapsid protein may be involved, it is likely that the 3' noncoding region will also readily associate with the nucleocapsid protein. This study examined the association of a recombinant IBV nucleocapsid protein with the 3' end of the IBV genome. Quantities of enriched nucleocapsid recombinant protein were obtained with an *Escherichia coli* expression vector. Competitive binding gel shift assays were used to determine the relative avidities of interactions of this recombinant protein with RNA fragments corresponding to the 3' terminus of the IBV genome.

MATERIALS AND METHODS

Preparation of the Recombinant Nucleocapsid Protein

The nucleocapsid gene of the IBV Gray strain was excised from a pCR1000 recombinant plasmid[2], transfected into M15/pRep4 cells (6; BTX, San Diego, CA; Qiagen, Chatsworth, CA). The fusion protein was induced as described in the Qiagen manual (Qiagen, Chatsworth, CA). Western blot analysis was used to compare the fusion protein with native nucleocapsid protein prepared from purified Gray strain IBV grown in em-bryonating chicken eggs[7,8].

Preparation of RNA Transcripts

The *in vitro* transcripts were produced from the pCR1000 and the pGem3Z vectors with the T7 promoter (Promega, Madison, WI). Inserts specific for selected fragments of the 3' end of the Gray genome were obtained by cDNA cloning of PCR products that corre-sponded to the sequences of interest. Cache Valley virus (CVV) RNA used as competitor was produced by transcribing *Bam*H1 digested S segment cDNA that had been cloned into pGEM3Z[9]. After linearizing with restriction enzymes, transcription was carried out as described in the Promega Protocol and Applications Guide (Promega, Madison, WI) with T7 RNA polymerase and G was labelled with ^{32}P-CTP. The RNA products were quantitated by comparing the ethidium bromide stained samples with known standards of yeast tRNA (Promega, Madison, WI).

RNA Protein Binding Assays

Protein-RNA interactions were analyzed by a modified gel-shift assay[10]. RNA and nucleocapsid protein were co-incubated for 20min at room temperature in 10μl of gel-shift buffer. Following the addition of sample buffer, the reaction mixtures were loaded onto a 0.5% agarose gel and electrophoresed at 60V in 1XTris-borate-EDTA[6]. Gels were then dried and autoradiographed. The 25% inhibition (I_{25}) was defined as the concentration needed to inhibit 25% of the mobility resulting from the interaction of labelled G RNA with nucleo-capsid protein.

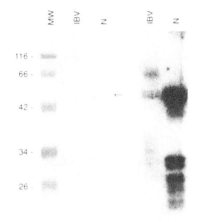

Figure 1. Recombinant protein eluted from a Ni^{2+}-column. The left panel is the Coomassie blue stained gel and the right panel shows the corresponding western blot. IBV represents protein from virus; N recovered recombinant protein and M, the molecular weight marker proteins. The arrow indicates the nucleocapsid protein band.

RESULTS

Recombinant Nucleocapsid Protein and RNA Transcripts

Nucleocapsid protein from the Gray strain of IBV was expressed in a pQE8 bacterial expression vector in order to obtain quantities of enriched protein used for studying interactions with RNA. The fusion protein with tandem histidine residues at the amino terminus was enriched from the soluble cytoplasmic fraction by Ni^{2+}-NTA affinity chromatography. A Coomassie blue stain of the eluted protein on an SDS-polyacylamide gel shows that the protein migrates similar to the native nucleocapsid protein and the western blot shows that the recombinant protein also reacted with anti-IBV serum in a Western blot assay (Fig. 1). Anti-IBV antibody did not react with bacterial proteins.

Positive sense RNAs specific for the 3′ noncoding end of the genome were synthesized as targets for the synthetic nucleocapsid protein binding studies (Fig. 2). Clone G (1832nt) was used for expression of the whole nucleocapsid gene and 3′ noncoding region, and I for transcription of the 3′ HVR (266nt). The RNA fragments AB$^+$ (181nt), CD$^+$ (155nt) and EF$^+$ (151nt), which partially overlapped with AB$^+$ and CD$^+$, were used as templates for the synthesis of RNAs representing conserved regions extending from the HVR to the 3′ terminus of the genome. The following gel shift studies were done using 0.15M sodium chloride.

Gel Shift Analyses

Gel shift assays were used to further examine protein-RNA interactions to determine relative avidities of the recombinant nucleocapsid protein for various species of RNA, including transcripts corresponding to the 3′ terminus of the IBV genome. Interactions

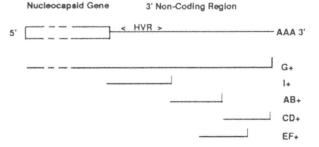

Figure 2. Diagrammatic representation of DNA used for *in vitro* transcription of the generated IBV-specific RNAs. | represents the restriction enzyme recognition sites (*Hind*III for G, I, AB, & EF; *Bam*H1 for CD) used for linearization of plasmid DNA Western blot.

20
10
5
2.5
1.25
0.62
0.31
0.15
0.075
0

Figure 3. Gel-shift assay demonstrating the binding of ^{32}P-labeled G RNA to the synthetic nucleocapsid protein. The amount of protein in μg is indicated above each lane.

between G and the fusion protein could be identified by the decrease in mobility of the labelled RNA (Fig. 3).

No shift in mobility could be observed when labelled RNA was mixed with bovine serum albumin. The effects of protein concentration on formation of G RNA-protein complexes on gel shift assays were first determined in the absence of competing RNA. The mobility decreased, that is, the size of the complexes increased, with increasing concentrations of nucleocapsid protein. The greatest concentrations of protein resulted in formation of additional large, megacomplexes that did not migrate into the gel. No shift in mobility of the labelled RNA was observed when reacted with an equivalent concentration of either egg allantoic proteins or bovine serum albumin.

Competitive Gel Shift Analyses

The interactions of this recombinant protein with each RNA fragment was examined with a gel shift assay and the relative avidities of these interactions were calculated. Radiolabelled G was mixed with each competing RNA species before reacting with the nucleocapsid protein. Differences could be seen in the efficiency of inhibition of labelled G RNA interaction with the nucleocapsid protein by each RNA species. Fig. 4 shows the inhibition of G binding to the protein in the presence of unlabelled G, bovine liver RNA and CVV RNA. Unlabelled positive sense G, AB, CD, I and CVV RNA bound relatively efficiently with the nucleocapsid protein, whereas the protein interactions with EF^+ and bovine liver RNA were relatively inefficient (Table 1). The calculated I_{25}'s determined from 5 concentrations of inhibiting RNA indicated that the IBV fragments from the 3' end reacted with the recombinant nucleocapsid protein with relatively high avidity ($I_{25} = 10$ to 26) except the EF fragment which corresponded to nt 99 to 249 from the 3' end of the genome. The

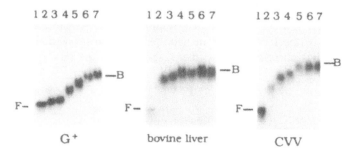

Figure 4. Competition gel-shift assay in which the interaction of the recombinant nucleocapsid protein with ^{32}P-labelled G RNA is inhibited with varying concentrations of the unlabeled RNA. The corresponding competitor RNA used is indicated below each autoradiograph. Lanes 1 represent free labeled G RNA in the absence of protein and competitor, lanes 2 to 6, labelled G with protein and two fold dilutions of competitor beginning with 100ng in 2 and lane 7 represents labeled G in the presence of protein and no competitor.

Table 1.

RNA	Nucleotide length	I_{25}
G^+	1800	10
CD^+	155	11
CVV	800	11
AB^+	181	26
I^+	266	24
EF^+	151	100
BL^*	1900	88
	4700	

*Bovine liver.

binding efficiency as indicated by the 100 I_{25} for EF was at least as poor as that of bovine liver RNA with an I_{25} of 88. These inhibition studies suggested that CD bound with greater avidity to the fusion protein than AB or I. The I_{25} of 11 resulting from competition with CVV RNA was similar to the of I_{25} of 10 observed with the homologous G RNA. No correlation could be found between the relative avidities and the size of the interacting RNA fragment.

DISCUSSION

In this paper, we have shown that the nucleocapsid protein of IBV is able to bind with relatively high affinity to RNA representing specific regions at the 3' end of the IBV genome. Competitive binding analyses indicated that the nucleocapsid protein bound with similar specificity to three positive sense transcripts which spanned the length of the entire 3' noncoding region but the recombinant fusion protein did not bind to an overlapping fragment that was 249 to 99 nts from the 3' terminus. Sequences that promoted minimum nucleocapsid protein binding to IBV RNA under conditions used in these studies were not present or were sufficiently altered in EF^+ and bovine liver RNA.

The bacterial expressed fusion protein would not be expected to be identical to the viral protein synthesized in the eukaryotic host. Differences probably exist in the phosphorylation status of the native and synthetic proteins. However, it was obvious that this recombinant protein was at least functional with respect to its basic interaction with RNA and should provide a valuable source for further characterization of the functions of the nucleocapsid in IBV replication.

The nature of the binding between nucleocapsid protein and RNA should involve cooperative binding among multivalent RNAs and protein molecules. In addition to the presumed binding of viral protein at a packaging signal sequence, packaging of genomic RNA may dictate the necessity of nucleocapsid protein binding to repeated sites along the molecule. This interaction appears to require sequences or structures that are not ubiquitous even on the IBV genome.

ACKNOWLEDGEMENTS

This research was supported by Southeastern Poultry & Egg Association, #29 and #260, USDA National Competitive Research Initiative Grants Program #RF-93-762, USDA Formula Animal Health (section 1433) #TEXO-6824, and the Texas Advance Technology Program #99902-063.

REFERENCES

1 Boursnell, M E G , Binns, M M , Foulds, I J , and Brown, T D K (1985) Sequences of the nucleocapsid genes from two strains of avian infectious bronchitis virus *J Gen Virol* **66**, 573-580

2 Williams, A K , Wang, L, Sneed, L W, and Collisson, E W (1992) Comparative analyses of the nucleocapsid genes of several strains of infectious bronchitis virus and other coronaviruses *Virus Res* **25**, 213-222

3 Masters, P S (1992) Localization of an RNA-binding domain in the nucleocapsid protein of the coronavirus mouse hepatitis virus *Archives of Virol,* **125**, 141-160

4 Nelson, G W, and Stohlman, S A (1993) Localization of the RNA-binding domain of mouse hepatitis virus nucleocapsid protein *J Gen Virol* **74**, 1975-1979

5 Stohlman, S A , Baric, R S , Nelson, G N , Soe, L H , Welter, L M , and Deans, R J (1988) Specific interaction between coronavirus leader RNA and nucleocapsid protein *J Virol* **62**(11), 4288-4295

6 Sambrook, S , Fritsch, E F, and Maniatis, T (1989) "Molecular cloning- a laboratory manual " 2nd edition Cold Spring Harbor Laboratory Press

7 Sneed, L W , Butcher, G D , Parr, R , Wang, L , and Collisson, E W (1989) Comparisons of the structural proteins of avian infectious bronchitis virus as determined by western blot analysis *Viral Immunol* **2**(3), 221-227

8 Parr, R L , and Collisson, E W 1993 Epitopes on the spike protein of a nephropathogenic strain of infectious bronchitis virus Arch Virol **133**, 369-383

9 Chung, S -I (1992) Molecular and Pathogenesis Studies of Cache Valley Virus A Common Teratogenic Agent in Sheep *Dissertation Thesis* Department of Veterinary Microbiology, Texas A&M University, College Station, TX

10 Chodosh, L A (1992) Chapter 12 2 DNA-Protein interactions Mobility shift DNA-binding assay using gel electrophoresis *Current Protocols, Supplement 3* Editors F M Ausubel, R Brent, R E Kingston, D D Moore, J G Seidman, S A Smith & K Struhl Wiley & Sons, New York

84

CHARACTERIZATION OF THE TRANSMISSIBLE GASTROENTERITIS VIRUS (TGEV) TRANSCRIPTION INITIATION SEQUENCE

Characterization of TGEV TIS

Julian A. Hiscox, Karen L. Mawditt, David Cavanagh, and Paul Britton[*]

Division of Molecular Biology
Institute for Animal Health
Compton, Newbury
Berkshire, RG16 0NN, United Kingdom

ABSTRACT

The ability of the TGEV transcription initiation sequence (TIS) to produce subgenomic RNAs was investigated by placing a reporter gene, chloramphenicol acetyltransferase (CAT) under the control of either the mRNA 6 or the mRNA 7 TISs. Both constructs only produced CAT in TGEV infected cells and the amount of CAT produced from the mRNA 7 TIS was less than from the mRNA 6 TIS. Mutations were made within and around the TISs and the effect on CAT production assayed. The results showed that the TGEV TIS acted as an initiator of transcription for CAT, though the degree of base pairing between the TIS and leader RNA was not the only factor implicated in the control of transcription.

Introduction

A number of mechanisms have been proposed for the transcription of coronavirus subgenomic mRNAs. These include, leader primed transcription (1), differential premature termination of transcription (2) and the synthesis of subgenomic negative strand RNAs (3). The mechanism of coronavirus transcription has not been elucidated, although to date the leader priming hypothesis has been predominantly used to explain the origin of subgenomic mRNAs. Leader primed transcription postulates that a *trans* acting leader RNA, transcribed

---

[*] To whom all correspondence should be sent.

from the 3' end of a negative copy of the genomic RNA, binds to conserved regions on the negative strand template(s) and is extended to produce the subgenomic mRNAs. These conserved sequences act as TISs and are found 5' to coronavirus genes, though their sequence varies depending on the coronavirus. The mouse hepatitis virus (MHV) subgenomic mRNAs follow a gradational relationship in which shorter RNAs are produced in larger amounts than longer species. However, in the TGEV group of coronaviruses this pattern is not observed, the smallest mRNAs of TGEV (4, 5), porcine respiratory coronavirus (6), feline infectious peritonitis virus (7) and canine coronavirus (8) being produced in lower amounts than the next larger mRNA. Sequence analysis of TGEV indicated that the minimum TIS is CUAAAC (9, 10), but the degree of base pairing between the TIS and the leader RNA in the leader polymerase complex (LPC) varies and this may be responsible for the different amounts of each of the mRNAs. To investigate interactions between the TIS and leader RNA we produced a series of constructs to analyse the role of nucleotides (nts) flanking the CUAAAC sequence involved in the synthesis of TGEV mRNAs 6 and 7.

MATERIALS AND METHODS

Virus and Cells

TGEV, FS772/70, was grown in LLC-PK$_1$ cells and the recombinant vaccinia virus, vTF7-3, in HTK$^-$ cells. Both viruses were titrated by plaque assay.

Construction of reporter genes

Various modified TISs (Table 1) were fused to the CAT gene, by PCR using the pCAT Basic Vector (Promega) as template, to produce a series of gene cassettes (Fig 1). A final elongation step in the PCR was excluded to reduce the addition of an extra 3' adenine residue to the PCR products, resulting in 3' overhangs, which have been reported to produce extraneous T7 RNA polymerase-generated transcripts (11).

Table 1. 5'-end sequences of reporter gene cassettes

Reporter Gene Cassette	TGEV Sequence of TIS Region
E6CAT	ACATATGGTATAACTAAACTTCTAAATG
6JAH4	ACATATGGTATAACTAAAC*gag*ATG
6JAH5	ACATATGGTAT*cg*AACTAAACTTCTAAATG
6JAH6	ACATATGGTAT*tt*CTAAACTTCTAAATG
6CAT	GGTATAACTAAACTTCTAAATG
6SCAT	ACATATGGTAT*ttagctga*TTCTAAATG
7CAT	TAACGAACTAAACGAGATG
7JAH1	TAA*tc*AACTAAACGAGATG
7JAH3	TAACGAACTAAAC*ttctaa*ATG
7SCAT	TAA*tcgtaccgtt*GAGATG

Note: Potential TISs are double underlined; nt changes in lowercase bold italics; CAT initiation codon is underlined; prefix 6 indicates mRNA 6- and prefix 7 indicates mRNA 7-based constructs.

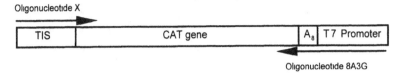

Figure 1. Schematic representation of the CAT reporter gene under the control of TGEV TISs Constructs were produced by PCR using oligonucleotide X which contained the sequences shown in Table 1 and the first 20 nts of the CAT gene Oligonucleotide 8A3G contained the T7 RNA polymerase promoter, a tract of eight T residues and 20 nts complementary to the 3'-end of the CAT gene

Virus Infection and Transfection protocol

LLC-PK$_1$ cells were grown to 95% confluence and infected with TGEV at an MOI of 10, for 1 h at 37°C, the cells were washed, incubated for 1 h in fresh medium, infected with vTF7-3 at an MOI of 10 and incubated for 2 h. The cells were washed and incubated for 2 h in fresh medium, transfected for 5 h with the PCR products, prepared by adding 10 µg of the appropriate PCR product with 45 µl of LIPOFECTAMINE™ (Gibco/BRL) in 3 ml of Opti-MEM (Gibco/BRL) at room temperature for 30 min, and incubated in fresh medium for 19 h. To account for potential variations in CAT synthesis, duplicate cell sheets were used in two experiments and triplicate cell sheets in one experiment.

Detection and Quantification of CAT Protein

Cells were lysed by freeze/thawing and CAT detected by CAT ELISA (Boehringer Mannheim), with an initial incubation time of 2 h, and analysed in a Titertek Multiscan®MCC/340 MK11 spectrophotometer for quantitation.

RESULTS

Expression of CAT from TGEV mRNA 6 and 7 TISs

A reporter gene system was used to investigate the mechanism of coronavirus subgenomic mRNA synthesis. A series of gene cassettes, consisting of the reporter gene, CAT, under the control of potential TGEV TISs followed by a short poly(A) tail and a copy of the T7 promoter were produced (Fig. 1). The cassettes were designed so that negative RNA copies of the CAT gene, with a negative copy of a TGEV TIS or a modified TIS at the 3'-end, could be synthesised in cells by T7 RNA polymerase. Transfection of the cassettes into cells infected with TGEV and vTF7-3 would allow the negative RNAs, produced by the T7 RNA polymerase from vTF7-3, to be tested as potential templates for the TGEV LPC. Recognition of the RNAs by the TGEV LPC would result in the transcription of RNAs, with the TGEV leader attached at the 5'-end, capable of acting as mRNAs for the translation of CAT. The RNAs produced by the T7 RNA polymerase would contain a negative copy of the CAT gene and therefore could not be translated into CAT. The only way a positive sense RNA could be generated would be from the recognition of the RNA, via the TIS, by the TGEV LPC in an analogous manner to the transcription of TGEV subgenomic mRNAs.

Constructs, 6CAT and 7CAT (Table 1), were produced such that the initiation codon of the CAT gene was preceded by the same nts as are present upstream of the initiation codons of the nucleoprotein (N; gene 6) and ORF-7 (gene 7) genes (12). TGEV mRNAs 6 and 7 TISs were chosen because previous studies had indicated that mRNA 6 was more abundant than mRNA 7 in TGEV infected cells. The amount of CAT synthesised by each construct was analysed and the mean values and standard deviations for the amounts of CAT synthesised were 39 ± 5.1 and 24 ± 3.7 pg/cell extract for 6CAT and 7CAT, indicating that 6CAT produced significantly more CAT than 7CAT with a confidence interval of 95%. Constructs, 6SCAT and 7SCAT, similar to 6CAT and 7CAT except that the TISs were scrambled (Table 1) did not generate any detectable CAT, indicating that the TISs in 6CAT and 7CAT had been utilised for the transcription of CAT mRNA.

Differential Affects of Nucleotide Changes on the Expression of CAT

Results on the expression of CAT using mRNA 6 and 7 TISs indicated that the system could be used for analysing the role of various nts in the synthesis of subgenomic mRNAs. Several constructs were generated in which modified versions of the TGEV mRNA 6 or mRNA 7 TISs were placed 5' to the CAT gene (Table 1). Essentially, various nts within or flanking the mRNA 6 and 7 TISs were interchanged, e.g. the three nts between the mRNA 7 TIS and ORF-7 AUG were exchanged with the six nts of mRNA 6 (Table 1). Data was analysed by the statistical analysis package, GLIM (13), to determine whether there was any significant variation in the amount of CAT produced from the modified mRNA 6 constructs when compared to E6CAT and the modified mRNA 7 constructs when compared to 7CAT. GLIM analysis indicated that there was no significant variation between the three experiments at the 5% level therefore the data was pooled for a more comprehensive analysis (Fig 2). Results showed that changing the nts between the mRNA 7 TIS and the CAT AUG codon from GAG (as in mRNA 7), construct 7CAT, to TTCTAA (as in mRNA 6), construct 7JAH3, significantly increased the amount of CAT produced (Fig 2B). However, decreasing the number of nts involved in base pairing between the TIS and the LPC from ten, CGAAC-TAAAC (as in mRNA 7), construct 7CAT, to eight, AACTAAAC (as in mRNA 6), construct 7JAH1, did not significantly affect the amount of CAT produced (Fig 2B).

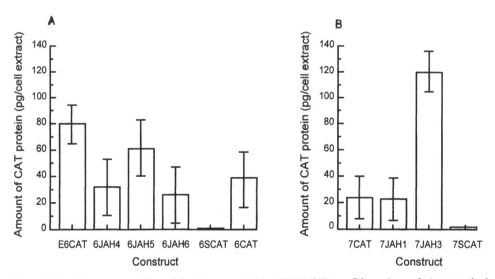

Figure 2. Bar chart representation of the data generated by GLIM, 95% confidence intervals (not standard error) are indicated by the vertical lines. Mean values are significantly different when the confidence interval lines do not overlap. (A) Modified mRNA 6 constructs were compared with E6CAT and (B) modified mRNA 7 constructs were compared with 7CAT. Comparisons can only be made within the gene 6 TIS group or within the gene 7 TIS group and not between the two groups.

Increasing the number of nts, found upstream of the mRNA 6 TIS on the genomic sequence, from five to eleven nts, construct E6CAT, significantly increased the amount of CAT produced when compared to 6CAT (Fig 2B). Reducing the number of nts, between the TIS and the AUG of the CAT gene, from six to three as found for mRNA 7, construct 6JAH4, resulted in a significant decrease in the amount of CAT produced when compared to E6CAT (Fig 2A). Increasing the base pairing between the TIS and the LPC from eight to ten nts as found in the mRNA 7 TIS, construct 6JAH5, did not significantly affect the amount of CAT produced when compared to E6CAT (Fig 2A). Reducing the mRNA 6 TIS from eight (AACTAAAC) to six (CTAAAC) nts, the minimum potential TIS, construct 6JAH6, significantly decreased the amount of CAT produced when compared to E6CAT (Fig 2A).

DISCUSSION

This study has demonstrated the expression of a reporter gene under the control of the TGEV mRNA 6 and 7 TISs. The constructs were produced so that the AUG of the CAT gene was in the same position with respect to the number and type of nts from the TISs as for the TGEV N and ORF-7 genes in mRNAs 6 and 7. Similar constructs in which the TISs were scrambled did not produce any detectable CAT indicating that nts within the TISs are essential for the transcription of subgenomic RNAs. The reporter gene constructs only contained coronavirus derived sequences at their 5' ends indicating that the TISs acted as sites of transcription for the synthesis of the subgenomic mRNAs rather than as attenuators as proposed in one model (2). The observation that the TGEV TISs could produce mRNAs capable of translation allowed us to use the system for studying the role of various nts, within and about the TISs, in the transcription of TGEV mRNAs. The use of TISs used for the synthesis of TGEV mRNAs 6 and 7 indicated that the amount of CAT produced from the mRNA 7-derived construct, 7CAT, was significantly less than for the mRNA 6 derived construct 6CAT. This confirmed previous observations in which Northern blot analysis of TGEV infected cells indicated that mRNA 7 was produced in much lower amounts than mRNA 6 (4, 5). The amount of CAT detected from 7CAT when compared to 6CAT was higher than expected from the comparison of the amounts of mRNAs 6 and 7 in TGEV infected cell indicating that other sequences or nts may be involved in the regulation of mRNA 6 or 7 synthesis. Increasing the number of nts upstream of the TIS sequence in 6CAT from five to eleven nts, using nts derived from the genomic sequence, construct E6CAT, significantly increased the production of CAT. This indicated that sequences upstream of the TIS, and not expected to interact with the leader RNA, may influence the control of transcription. This observation might result from potential differences in the secondary structure of the RNAs in which the TIS of E6CAT is more available for interaction with the LPC than the TIS of 6CAT, causing an increased level of CAT mRNA.

Changing the nts between the TIS and the AUG of the CAT gene in E6CAT from TTCTAA to GAG, thereby converting the sequences downstream of the gene 6 TIS for those of gene 7, significantly reduced the amount of CAT produced. In contrast, introducing the TTCTAA sequence in place of GAG in 7CAT significantly increased the amount of CAT produced. This observation indicated that the number of nts between the TIS and the AUG of the gene might influence transcription, whether this is by altering the potential secondary structure of the RNA or the alteration of some, as yet unknown, control element is not known. An alternative explanation may arise from the fact that translation of the CAT mRNA was used as an indicator of transcription levels and that alterations to the RNA sequence may affect translation. Thus the changes in the amount of CAT detected may not be due to changes in the levels of transcription. Potential alterations to the Kozak sequence about the CAT AUG may influence translation of CAT. According to Kozak (14, 15) the optimal sequence for initiation by eukaryotic ribosomes is ACC*ATG*G in which a purine at position -3 has a dominant effect and if replaced by a pyrimidine becomes less favourable for initiation of translation The Kozak sequence was changed from TAA*ATG*G (E6CAT) to GAG*ATG*G in 6JAH4 and from GAG*ATG*G (7CAT) to TAA*ATG*G in 7JAH3 Thus E6CAT to 6JAH4 represented a T → G change at the -3 position, indicating a more favourable Kozak sequence, however, the amount of CAT detected from 6JAH4 was less than from E6CAT 7CAT to 7JAH3 represented a G → T change at the -3 position, indicating a less favourable Kozak sequence, however, the amount of CAT detected from 7JAH3 was more than from 7CAT These results were the opposite to those that would have been expected if the levels of translation observed had depended on the Kozak sequences, favouring a change in the level of transcription as an explanation

Increasing the base pairing between the TIS and 3'-end of the leader RNA from eight (AACTAAAC) to ten (CGAACTAAAC) but keeping the nts 3' of the TIS as for mRNA 6 or decreasing the base pairing of the TIS for mRNA 7 to eight nts with the 3' nts the same as for mRNA 7, did not significantly affect the amount of CAT detected These observations indicated that increasing the potential base pairing between the TIS and LPC beyond that achieved with the sequence AACUAAAC did not enhance transcription However, reducing the base pairing from eight to six nucleotides (CTAAAC), construct 6JAH6, so that the TIS site was similar in length to the TGEV mRNA 4 TIS, a low abundance mRNA species, significantly decreased the production of CAT, indicating some base pairing is essential

Similar findings have recently been described in which an MHV defective interfering (DI) RNA was used to study transcription (16) The authors inserted sequences corresponding to the TISs of MHV mRNAs 3 and 7 into a synthetic DI Although the MHV mRNA 3 and 7 derived TISs have base pairings of 10 and 17 nts with the MHV leader RNA, the amounts of a subgenomic DI RNA generated were similar The authors concluded that base pairing, between the TISs and the leader RNA, does not control mRNA abundance, though some residual base pairing was essential and that transcription initiation could require recognition of the TIS by the transcriptase

ACKNOWLEDGMENTS

J A H was the recipient of a BBSRC studentship We would like to thank Rob van Es for help with the statistical analysis and Sarah Duggan for preparation of the cells used in this study This work was supported by the Ministry of Agriculture, Fisheries and Food, UK

REFERENCES

1 Lai, M M C (1986) Coronavirus leader-RNA-primed transcription An alternative mechanism to RNA splicing BioEssays **5**, 257-260

2 Konings, D A M , Bredenbeek, P J , Noten, J F H , Hogeweg, P and Spaan, W J M (1988) Differential premature termination of transcription as a proposed mechanism for the regulation of coronavirus gene expression Nucl Acids Res **16**, 10849-10860

3 Sawicki, S G and Sawicki, D L (1990) Coronavirus transcription subgenomic mouse hepatitis virus replicative intermediates function in RNA synthesis J Virol **64**, 1050-1056

4 Jacobs, L , van der Zeijst, B A M and Horzinek, M C (1986) Characterization and translation of transmissible gastroenteritis virus mRNAs J Virol **57**, 1010-1015

5 Page, K W , Britton, P and Boursnell, M E G (1990) Sequence analysis of the leader RNA of two porcine coronaviruses Transmissible gastroenteritis virus and porcine respiratory coronavirus Virus Genes **4**, 289-301

6 Page, K W, Mawditt, K L and Britton, P (1991) Sequence comparison of the 5' end of mRNA 3 from transmissible gastroenteritis virus and porcine respiratory coronavirus J Gen Virol **72**, 579-587

7 de Groot, R J , ter Haar, R J , Horzinek, M C and van der Zeijst, B A M (1987) Intracellular RNAs of the feline infectious peritonitis coronavirus strain 79-1146 J Gen Virol **68**, 995-1002

8 Horsburgh, B C , Brierley, I and Brown, T D K (1992) Analysis of a 9 6 kb sequence from the 3' end of canine coronavirus genomic RNA J Gen Virol **73**, 2849-2862

9 Rasschaert, D , Gelfi, J and Laude, H (1987) Enteric coronavirus TGEV - partial sequence of the genomic RNA, its organization and expression Biochimie **69**, 591-600

10 Britton, P , Lopez Otin, C , Martin Alonso, J M and Parra, F (1989) Sequence of the coding regions from the 3 0 kb and 3 9 kb mRNA subgenomic species from a virulent isolate of transmissible gastroenteritis virus Arch Virol **105**, 165-178

11 Schenborn, E T and Mierendorf Jr, R C (1985) A novel transcription property of SP6 and T7 RNA polymerases dependence on template structure Nucl Acids Res **13**, 6223-6236

12 Britton, P , Carmenes, R S , Page, K W , Garwes, D J and Parra, F (1988) Sequence of the nucleoprotein from a virulent British field isolate of transmissible gastroenteritis virus and its expression in *Saccharomyces cerevisiae* Molec Microbiol **2**, 89-99

13 GLIM, version 3 77, update 0 The Royal Statistical Society, London

14 Kozak, M (1983) Comparison of initiation of protein synthesis in prokaryotes, eukaryotes and organelles Microbiol Rev **47**, 1-45

15 Kozak, M (1986) Point mutations define a sequence flanking the AUG initiator codon that modulates translation by eukaryotic ribosomes Cell **44**, 283-292

16 van der Most, R G , de Groot, R J and Spaan, W J M (1994) Subgenomic RNA synthesis directed by a synthetic defective interfering RNA of mouse hepatitis virus a study of coronavirus transcription initiation J Virol **68**, 3656-3666

85

IN VIVO AND IN VITRO TRANSCRIPTION OF SMALL mRNAS CONTAINING A LEADER SEQUENCE FROM MOUSE HEPATITIS VIRUS STRAIN JHM

M. Hayashi, T. Mizutani, K. Ishida, A. Maeda, T. Watanabe, and
S. Namioka

Department of Laboratory Animal Science
Faculty of Veterinary Medicine
Hokkaido University
Sapporo 060, Japan

Mouse hepatitis virus (MHV) genomes are divided into at least seven coding regions. Recently, it has been shown that one or two additional small mRNAs (mRNA8 and 9) are synthesized in DBT cells infected with MHV strains A59, -1 and -S[1]. It is suggested that the transcription may occur via a leader-priming mechanism whereby a trans-acting leader RNA binds to highly conserved intergenic sequences, UC(U/C)AAAC, on the full-length negative-stranded template to prime transcription of subgenomic mRNAs[2,3]. mRNA8 is initiated from a perfectly conserved intergenic sequence, UCCAAAC, at 828 nt of the nucleocapsid (N) protein gene of MHV-A59. mRNA9 is initiated from a nearly perfect sequence, UCUAAAU, at 982 nt. However, whether mRNA8 and 9 are synthesized in MHV-infected mice and whether the products from mRNA8 and 9 play a biological role in infected cells remain unknown. In this experiment we studied the course of synthesis of these two small mRNAs in DBT cells and in mouse organs.

Total RNA was extracted from DBT cells infected with MHV-JHM at a multiplication of infection (m. o. i.) of 2.0 as described previously[4] and cDNA was synthesized using reverse transcriptase (RT) and primer-R (5' TGCCGACATAGGATTCATTCTCT 3') corresponding to 1368-1390 nt of the N protein gene of JHM. The synthesized cDNA was amplified by polymerase chain reaction (PCR) using primer-F1 (5' TATAAGAGTGATTGGCGTCCG 3') and -F2 (5'CTCTAAAACTCTTGTAGTTT 3') corresponding to 1-21 nt and 37-56 nt in the leader RNA, respectively, as a forward pimer. Primer-R was used as a reverse primer. The amplified products were analyzed by Southern blot hybridization using cDNA of mRNA7[5], which contains sequences of mRNA 8 and 9, as a probe as described previously[4].

Cellular RNA was extracted from DBT cells infected with JHM and A59 strains of MHV at 37°C at 6 h p.i. The amplified products derived from mRNA7 and 2 small mRNAs were observed in both strains using primer-F1 and -R (Fig. 1a). Sequence data of the PCR

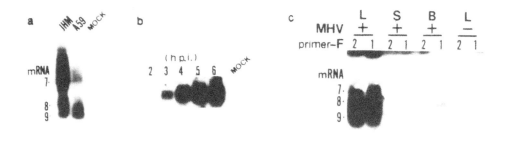

Figure 1. Small mRNAs in DBT cells and the mouse tissues infected with JHM-MHV. The cellular RNA was extracted from DBT cells infected with MHV-JHM and -A59 at 6 h p. i. (a), and from cells infected with MHV-JHM at 2, 3, 4, 5 and 6 h p. i. (b). RNA was extracted from the liver (L), spleen (S) and brain (B) of the infected mouse at 5 days p. i. and amplified by RT-PCR using primer-F1 or -F2 and -R (c). + and - represent the infected and uninfected mouse, respectively. The RT-PCR products were analyzed by Southern blot hybridization using cDNA of mRNA7 as a probe.

products showed that these small mRNAs corresponded to the mRNA8 and 9 reported by Schaad & Baric[1] (data not shown). These small mRNAs from JHM-MHV were detected at 3 h p. i. by analysis using RT-PCR (Fig. 1b). Since ORF1a gene products of MHV that contain RNA-dependent RNA polymerases are translated from genomic RNA at 1.5-3 h p. i.[6], mRNA8 and 9 were synthesized in the DBT cells infected with JHM-MHV at the early stage of the infection. This result suggests that the products from mRNA8 and 9 of MHV may play a role in the early stage of the viral replication cycle.

When a C57BL/6 mouse was inoculated i.p. with 3×10^5 PFU MHV-JHM. The mRNA8 and 9 were observed in the liver and brain of the infected mouse at 5 days p.i. by RT-PCR amplifications of the RNAs (Fig. 1c). This result showed that mRNA8 and 9 were synthesized in the liver and brain, which are target tissues of MHV-JHM. Therefore, the products of mRNA8 and 9 may play a role in MHV infection in vivo. The mRNA8 and 9 synthesized in vivo contained the leader sequence (data not shown). Although mRNA8 is synthesized from a perfectly conserved sequence, mRNA9 is from an imperfect sequence[1]. In the infected cells, it is suggested that mRNAs of MHV are synthesized by an imprecise trans-acting leader-primed transcription mechnism[2,7,8]. This study suggested that the imperfect intergenic sequence also can serve as an initiation site for the leader-primed mRNA synthesis of MHV in vivo.

REFERENCES

1. Schaad, M. C., Baric, R. S. Evidence for new transcriptional units encoded at the 3'-end of the mouse hepatitis virus genome. Virol. 1993; 196: 190-198.
2. Baric, R. S., Stohlman, S. A., Lai, M. M. C. Characterization of replicative intermediate RNA of mouse hepatitis virus: Presence of leader RNA sequences on nascent chains. J. Virol. 1983; 48: 633-604.
3. Makino, S., Joo, M., Makino, J. K. A system for study of coronavirus mRNA synthesis: a regulated, expressed subgenomic defective interfering RNA results from intergenic insertions. J. Virol. 1991; 65: 6031-6041.
4. Mizutani, T., Hayashi, M., Maeda, A., Yamashita, T., Isogai, H., Namioka S. Inhibition of mouse hepatitis virus multiplication by antisense oligonucleotide complementary to the leader RNA. J. Vet. Med. Sci. 1992; 54: 465-472.
5. Skinner, M.A., Siddell, S.G. Coronavirus JHM; nucleotide sequence of the mRNA that encodes nucleocapsid protein. Nucleic Acids Res. 1983; 11: 5045-5054.

6 Dennison, M R , Zoltick, P W , Hughes, S A , Giangreco, B , Olson, A L , Perlman, S , Leibowitz, J L , Weiss, S R Intracellular processing of the N-terminal ORF1a proteins of the coronavirus MHV-A59 requires multiple proteolytic events Virol 1992, 189 274-284

7 Baric, R S , Stohlman, S A Razavi, M K , Lai, M M C Characterization of leader-related small RNAs in coronavirus infected cells Further evidence for leader-primed mechanism of transcription Virus Res 1985, 3 19-33

8 Makino, S , Lai, M M C Evolution of the 5' end of genomic RNA of murine coronaviruses during passages in vitro Virol 1989, 169 227-232

86

PEDV LEADER SEQUENCE AND JUNCTION SITES

Kurt Tobler and Mathias Ackermann

Institute of Virology
Vet. -med. Faculty
University of Zurich
Zurich, Switzerland

The leader sequence of coronaviruses plays an important role in the replication of the coronaviruses. One model for the mRNA synthesis describes the leader as the primer which binds to the intergenic sequences (IS) on the negative copy of the genome.[1] Recently it was reported that the leader sequence regulates the expression of the viral proteins.[2]

The leader sequence of porcine epidemic diarrhoea virus (PEDV) CV777 was determined on the mRNA encoding the N protein by reverse transcription from within its open reading frame, followed by head-to-tail ligation of the cDNA, PCR amplification over the ligation sites of these concatemers and finally cloning and sequencing.[3] Fig. 1 shows 71 nt of the 5′ end including the IS of the N mRNA aligned to the leader sequence of HCV 229E.[4]

The leader sequence was used to design a leader specific primer (oligo 117). This oligonucleotide and primers within the known open reading frames[5,6] were used to amplify and subsequently to determine the sequence of the junction sites of the subgenomic mRNA species. The junction sites of the mRNAs encoding the structural proteins are depicted in Fig. 2. The leader junction sites of the ORF3 mRNA from the cell culture adapted (ca) PEDV and its parent wild type (wt) PEDV are shown in Fig. 3.

Previous sequencing results of ca PEDV revealed a polymorphic open reading frame for the putative ORF3 protein.[6] On cDNA clones derived from RNA isolated from infected pigs (wt PEDV CV777) an open reading frame without any deletions was found.[7]

The results established thus far question the existence of a functional mRNA for the expression of the ORF3 protein in the ca PEDV. In contrast, the absence of the deletions and the presence of an mRNA species which has the capacity to encode for the 224 amino acids protein in the wt PEDV support the existence of the product in wt PEDV. The virus appears to have lost the ability to synthesize the ORF3 product in Vero cells. The postulated loss of

```
ACTTAAAAAGATTTTCTATCTACGGATAGTTAGCTCTTTTTCTAGACCTTGTCTACTCAATTCAACTAAAC
 |||||  |   ||  ||||||||||  ||||  |  |   ||| ||||| ||||     | |||   |||||
-CTTAAGTACCTTATCTATCTACAAATAGAAAAGTTGCTTTTTAGACTTTGT--GTCTACTTC---TAAAC
```

Figure 1. Alignment of the leader sequence from PEDV CV777 (upper line) and HCV 229E (lower line). The sequence of the leader specific primer (oligo 117) is underlined.

Corona- and Related Viruses, Edited by P. J. Talbot and G. A. Levy
Plenum Press, New York, 1995

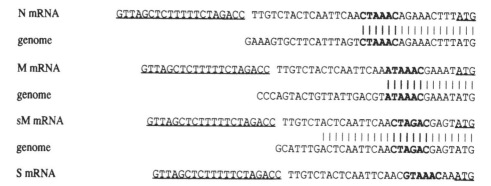

Figure 2. Leader-mRNA junction for four PEDV subgenomic mRNA The sequences are shown in alignment with the genomic RNA (except for the S mRNA, because this part of the genome is not sequenced yet) The intergenic sequences are marked as bold letters The leader specific primer 117 and the initiation codons for the corresponding genes are underlined

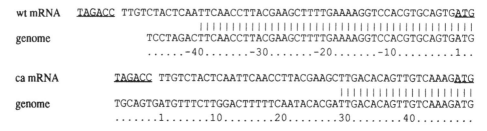

Figure 3. Differences found on the 5' end of the ORF3 mRNA between the ca PEDV and the wt PEDV The 3' end of the leader sequence specific primer and the initiation codons are underlined The numbering of the nucleotides are in both parts of the figure identical, the A of the initiation codon from the 224 amino acids open reading frame is set as 1 on the genomic RNA The ca PEDV mRNA showed a deletion covering the initiation codon for the putative 224 amino acid ORF3 protein The wt PEDV mRNA isolated from infected pigs possessed the expected initiation codon for the expression of the ORF3 product

the ORF3 product demonstrates an unexpected feature caused by adaption to replication in cell culture We hypothesize that the ORF3 product is incompatible with replication of PEDV in cell culture. In contrast, the observation that a correct ORF3 mRNA is made in wt virus indicates that ORF3 may be of importance *in vivo*

ACKNOWLEDGMENTS

These studies were supported by the Swiss National Science Foundation, grant #31-37418 93

REFERENCES

1 M M C Lai, Annu Rev Microbiol 1990,44 303-333
2 S M Tahara, T A Dietlin, C C Bergmann, et al , Virology 1994,202 621-630
3 M A Hofmann and D A Brian, PCR Methods and Appplications 1991,1 43-45
4 S S Schreiber, T Kamahora and M M C Lai, Virology, 1989,169 142-151
5 A Bridgen, M Duarte, K Tobler, H Laude and M Ackermann, J General Virol 1993,74 1795-1804
6 M Duarte, K Tobler, A Bridgen, D Rasschert, M Ackermann and H Laude, Virology 1994,198 466-476
7 K Tobler and M Ackermann, (unpublished observation)

Recombination and Mutation of Viral RNA

87

MUTAGENESIS OF THE GENOME OF MOUSE HEPATITIS VIRUS BY TARGETED RNA RECOMBINATION

Paul S. Masters, Ding Peng, and Françoise Fischer

David Axelrod Institute
Wadsworth Center for Laboratories and Research
New York State Department of Health
Albany, New York 12201-2002

ABSTRACT

Our laboratory has described a method for introducing site-specific mutations into the genome of the coronavirus mouse hepatitis virus (MHV) by RNA recombination between cotransfected genomic RNA and a synthetic subgenomic mRNA. By using a thermolabile N protein mutant of MHV as the recipient virus and synthetic RNA7 (the mRNA for the nucleocapsid protein N) as the donor, engineered recombinant viruses were selected as heat-stable progeny resulting from cotransfection. We have recently reported an optimization of the efficiency of targeted recombination in this process by using a synthetic defective interfering (DI) RNA in place of RNA7. The frequency of recombination is sufficiently high that recombinants can often be directly identified without employing a thermal selection. We present here a progress report on our use of this system to map MHV mutants and to construct N gene mutants which include (1) a mutant in which the internal open reading frame within the N gene (the I gene) has been disrupted, and (2) a series of recombinants in which portions of the MHV N gene have been replaced by the homologous regions from the N gene of bovine coronavirus. We also report on some mutants we have not been able to construct.

INTRODUCTION

Although the tools of molecular biology have been widely applied to the study of separate components of RNA viruses, the direct engineering of these viruses has lagged behind that of other biological entities because of the unique composition of their genetic material. The genomes of coronaviruses, like those of all other positive-sense RNA viruses, are, in isolation, fully infectious when transfected into appropriate host cells. This property ought to allow them to be manipulated by the same methodology that has been used to carry

Corona- and Related Viruses, Edited by P. J. Talbot and G. A. Levy
Plenum Press, New York, 1995

out site-directed mutagenesis of other positive-sense RNA viruses, that is, by the transcription of infectious RNA from a cDNA clone of the viral genome. This technique was pioneered with poliovirus[1] and has been applied successfully to a number of viruses, the largest thus far being Sindbis virus.[2] Coronaviruses, however, have the largest known genomes of all the RNA viruses (26 to 31 kb), approaching four times the size of the typical picornavirus genome or three times the size of the Sindbis virus genome. Thus, the construction of a full-length, error-free coronavirus cDNA would appear to be a major technical feat and may yield a product that would be too cumbersome for practical manipulation.

An alternative approach has been to exploit the high frequency of RNA recombination[3] that occurs in coronaviruses, in a manner analogous to that used for large DNA viruses. Our laboratory[4] and the Spaan laboratory[5] concurrently reported a method to target site-specific mutations to the nucleocapsid (N) gene and the 3' untranslated region (UTR) of the genome of mouse hepatitis virus (MHV). We describe here our recent improvements on this system and its application to particular research questions.

METHODS

The methods used in these studies have been described in detail elsewhere.[4,6] In brief, transcription vectors encoding MHV subgenomic RNA7 or a synthetic DI RNA were constructed by standard techniques, and mutations or substitutions were made in these by a variety of current PCR-based methods. Recombination experiments were performed by either cotransfection of genomic and synthetic RNA or infection followed by transfection. Transfections were carried out by electroporation of mouse L2 cells, which were then plated onto monolayers of mouse 17 clone 1 cells. Progeny virus, harvested at 24-30 h post-infection, were titered on mouse L2 cells at 39°C, and candidate recombinants able to form large plaques were purified and analyzed further. Initial characterization of recombinants typically involved RT-PCR analysis or sequencing of RNA from infected cells; final verification of recombinants was accomplished through sequencing of genomic RNA from purified virions.

RESULTS AND DISCUSSION

We originally reported a system for the mutagenesis of MHV via targeted recombination between cotransfected genomic RNA and a synthetic copy of RNA7, the smallest of the subgenomic transcripts of MHV.[4] The recipient virus used in this work was a temperature-sensitive and thermolabile N gene deletion mutant, Alb4. Alb4 produces tiny plaques at the nonpermissive temperature (39°C), and its virions are orders of magnitude more susceptible to heat inactivation than wild-type virions. Recombinants were selected as thermally stable progeny of the cotransfection. The efficacy of the technique was demonstrated by the incorporation of a 5 nt insertion into the 3' UTR of the virus, a short distance beyond the stop codon of the N gene, at the point where a BstEII site occurs in the cDNA copy of the 3' UTR. Thus, repair of the Alb4 deletion, presumably by a template switching event involving the cotransfected RNA7, could be used to cotransduce a linked mutation into the MHV genome. The efficiency of recovery of recombinants, however, was very low, roughly on the order of 10^{-5}, and was dependent on the powerful counterselection afforded by the thermolability of Alb4.

Van der Most et al.[5] also generated site-specific mutations in MHV, but at a much higher frequency, using Alb4 as the recipient virus and a defective interfering (DI) RNA as the donor species. On the basis of this observation, we sought to increase the efficiency of our recombination system by constructing a vector that would act as template for a synthetic

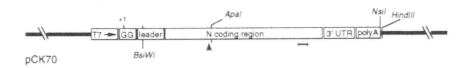

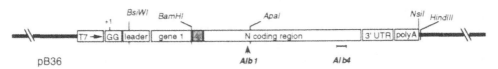

Figure 1. Transcription vectors used for synthesis of donor RNAs for targeted recombination Plasmid pCK70 serves as the template for MHV subgenomic RNA7 Plasmid pB36 serves as the template for a synthetic DI RNA The start of transcription is denoted by +1 The loci of the point mutation in Alb1 and the deletion in Alb4 are indicated

DI RNA. The vector we made, pB36, was patterned after a naturally occurring DI RNA of bovine coronavirus (BCV), and contained the first 467 nt of the MHV genome, linked, via a spacer of 48 heterologous nt, to the entire MHV N gene and 3' UTR (Fig. 1). We demonstrated by metabolic labeling that RNA produced by runoff transcription of pB36 was robustly replicated in MHV-infected cells.[6] This is noteworthy, since this synthetic DI RNA comprises only MHV sequences from the 5' and 3' termini of the genome, and it does not contain any internal discontiguous sequences reported to be necessary for MHV DI RNA replication.[7,8]

In confirmation of the results of van der Most et al.,[5] we observed that use of the DI RNA as the donor species in recombination experiments with Alb4 led to an increase in the frequency of recombination of at least 100-fold in comparison to that obtained with RNA7 as the donor species. A practical consequence of this was that it became possible to identify recombinants without the need for prior thermal selection. We have subsequently used this system for three general types of applications: (1) mapping of mutants; (2) site-directed mutagenesis to produce a defined product; and (3) domain exchange to produce multiple possible products that remain undefined until analyzed.

In the first type of experiment, pB36-derived RNA was used to determine if any of various temperature-sensitive mutants mapped to the N gene. The best candidates to screen in this fashion were mutants that were also thermolabile, since thermolability is likely to indicate a lesion in a viral structural gene. One such mutant, Alb1, yielded recombinants with transfected DI RNA against an undetectably low frequency of reversion, clearly establishing it as an N gene mutant.[6] An additional mutant, Alb11, did not give rise to recombinants under these conditions, and thus did not map in N. A third mutant, Alb25, did not exhibit a frequency of recombination significantly higher than its reversion frequency. Thus, its status is presently undetermined. Alb1 was found to have two closely liked point mutations in the first third of the N protein, in a region that is highly conserved among all coronavirus N proteins. One of these two mutations was shown by reversion analysis to be the relevant lesion, while the other was found to be phenotypically silent. The sites of the Alb1 mutation and the previously mapped Alb4 deletion are shown in Fig. 1. These results demonstrated that targeted RNA recombination can be used to provide definitive genetic proof for the locus of a given mutation. In principle, it should be possible to map any mutation (with a sufficiently low reversion rate) using a set of DI RNAs constructed to span the MHV genome.

The second application of targeted RNA recombination, site-directed mutagenesis, presents a powerful tool for the examination of questions of structure and function of coronavirus proteins. There are obviously huge advantages to examining the effects of mutational alterations *in situ*, rather than indirectly by means of an expression system. Equally obvious, however, is the major drawback of this technique: the mutations one designs cannot be lethal. One specific mutant that we have made is a virus in which the internal open reading frame (I ORF or I gene) contained within the N gene has been disrupted. Many coronavirus genomes contain a second ORF, in the +1 reading frame, embedded entirely within the first half of the N gene. The I ORF encodes a largely hydrophobic protein that could potentially be expressed via a leaky scanning translational strategy. Indeed, for BCV it has been shown that the I protein is expressed in infected cells, and furthermore, that hyperimmune serum from BCV-infected calves recognizes I protein.[9] The function of the I protein, if any, is unknown. To determine whether the MHV counterpart of this protein is an essential gene product, we engineered a mutant in which the start codon of the I gene was replaced by a threonine codon, and the fourth codon of the ORF was changed from a serine codon to a stop codon (Fig. 2). Both of these mutations are silent in the N reading frame. The resulting double mutant was found to be viable at all temperatures tested and formed plaques only slightly smaller than those of its wild-type counterpart. Moreover, preliminary experiments have revealed no difference between the I gene mutant and wild-type MHV in their ability to infect mice. Therefore, we conclude that the I protein is not essential to MHV infection, either in tissue culture or in the natural host. One reason for our interest in this question was that it was essential to our analysis of any mutants we might construct in the first half of the N gene. If the I gene had been essential to MHV, then genetic analysis of the region in which N and I overlap would have been greatly complicated.

The third application of targeted RNA recombination, domain exchange, is an approach we have used to determine the extent to which portions of a heterologous N protein (that of BCV) could be substituted for the corresponding portions of the MHV N protein. Initially, a small defined substitution was made in the highly variable spacer B region[10] near the carboxy terminus. This putative spacer between functional domains shows considerable sequence heterogeneity among different strains of MHV, and so it was not unexpected that part of it could be replaced by a more than 50% divergent sequence from BCV without producing any phenotypic alterations in the resulting recombinant virus. This first chimeric BCV/MHV N protein is shown schematically at the top of Fig. 3.

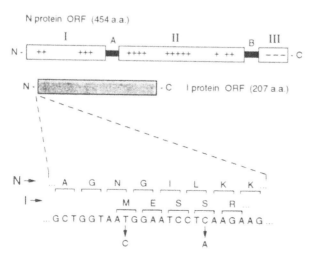

Figure 2. Mutagenesis of the MHV I gene. The relative positions of the ORFs for the N and I proteins are indicated above. Shown below are the two nucleotide changes introduced into a pB36-derived plasmid in order to disrupt the I ORF without changing the N ORF.

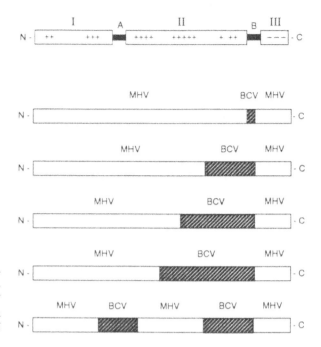

Figure 3. Independently isolated mutants containing chimeric N genes composed of portions of the BCV and MHV N genes are shown in the context of a three domain model of the MHV N protein,[10] indicated above.

Taking this as a starting point to proceed in the amino-terminal direction, we then sought to obtain recombinants containing larger extents of the BCV N gene in place of that of MHV. This was accomplished by constructing a derivative of pB36 in which the entirety of domains I and II, as well as spacer A and the above described portion of spacer B were substituted by their BCV counterparts. Overall, the nucleotide and amino acid sequence divergence between BCV and MHV in these regions is about 30%.[11] This vector was used to obtain a number of independent recombinants. We assumed that if these arose via a single crossover event, then for each resulting chimeric virus, the right junction of the BCV insert would be that determined by the junction constructed in the spacer B region of the vector. The left junction, however, could vary from recombinant to recombinant, and would be determined by the locus of the crossover event that gave rise to the recombinant. This turned out to be the case, and we found that resulting recombinants had left junctions that fell in three regions, represented by the second, third and fourth chimeric N proteins shown in Fig. 3. We were unable to obtain recombinants whose left junction entered or went beyond the serine- and arginine-rich cluster that is found near the amino end of domain II. This region is one of the most variable between BCV and MHV N, and apparently is not functionally interchangeable between the two even though most of the rest of domain II can be interchanged. A second pB36-derived vector was prepared which replaced this region in the earlier chimeric vector with the corresponding sequence from MHV. This allowed the construction of new BCV/MHV chimeric viruses, containing still greater amino-terminal extents of BCV sequence, although a boundary was reached in domain I beyond which recombination did not occur, probably because of the extent of sequence divergence between the two N proteins beyond this point (see the final chimeric N protein in Fig. 3). These studies have shown what portions of the N protein have been functionally conserved across different species (most of domain II, the RNA-binding portion of N[12]) as well as what portions have diverged too far to be functionally equivalent (the amino terminus and the SR-rich region). We have thus identified regions to focus upon in future mutagenesis experiments. In addition, these studies clearly demonstrate that targeted recombination can be used to make extensive

A. Cysteine replacement mutants

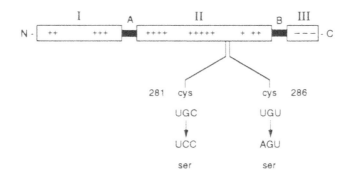

B. Insertion of an extra gene into the 3' UTR

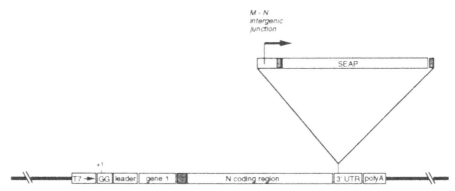

Figure 4. Mutants we have not been able to construct. (A) N protein mutants in which one or both of the two cysteine residues of N have been replaced by serine. (B) A mutant containing a heterologous gene (secreted alkaline phosphatase, SEAP) inserted into the upstream portion of the 3' UTR.

substitutions in the coronavirus genome, generating recombinants that could not be made otherwise between two viruses separated by a species barrier.

Finally, some negative results deserve mention. First, we have attempted to replace one or both of the two cysteine residues of the MHV N protein by serine (Fig. 4). This was done to test the biological significance of the disulfide-linked multimers of N, thought to be trimers, that have been observed in vitro.[13,14] Our inability to construct cysteine-deficient mutants of N implies, but by no means proves, that such mutations are lethal and that these intermolecular bonds may be important in nucleocapsid structure. However, further evidence would have to be obtained to corroborate this argument.

A second mutant we have been unable to obtain, despite a number of attempts, is one in which an extra gene has been inserted into the upstream portion of the 3' UTR of MHV (Fig. 4). We have speculated that this region of the genome ought to tolerate such an insertion because (a) a number of other coronaviruses contain one or more extra transcription units downstream of the N gene; (b) this region of the 3' UTR is the least conserved among coronavirus 3' UTRs; and (c) the original 5 nt insertion into the 3' UTR was made at this locus. Our failure to obtain this mutant may imply that such an insertion is lethal at this point in the genome, for as yet unknown reasons, or it may mean that there are limitations on the size of nonhomologous material that can participate in the recombination process Much further work remains to sort out these and other possibilities

ACKNOWLEDGMENTS

We are grateful to Dr Lawrence Sturman for providing mutants, advice, and encouragement This work was supported in part by Public Health Service grant AI31622 from the National Institutes of Health

REFERENCES

1 V R Racaniello, and D Baltimore Cloned poliovirus cDNA is infectious in mammalian cells *Science* 214 916 (1981)

2 C M Rice, R Levis, J H Strauss, and H V Huang Production of infectious RNA transcripts from Sindbis virus cDNA clones mapping of lethal mutations, rescue of a temperature-sensitive marker, and in vitro mutagenesis to generate defined mutations *J Virol* 61 3809 (1987)

3 M M C Lai RNA recombination in animal and plant viruses *Microbiol Reviews* 56 61 (1992)

4 C A Koetzner, M M Parker, C S Ricard, L S Sturman, and P S Masters Repair and mutagenesis of the genome of a deletion mutant of the coronavirus mouse hepatitis virus by targeted RNA recombination *J Virol* 66 1841 (1992)

5 R G van der Most, L Heijnen, W J M Spaan, and R J de Groot Homologous RNA recombination allows efficient introduction of site-specific mutations into the genome of coronavirus MHV-A59 via synthetic co-replicating RNAs *Nucl Acids Res* 20 3375 (1992)

6 P S Masters, C A Koetzner, C A Kerr, and Y Heo Optimization of targeted RNA recombination and mapping of a novel nucleocapsid gene mutation in the coronavirus mouse hepatitis virus *J Virol* 68 328 (1994)

7 Y-N Kim, Y S Jeong, and S Makino Analysis of cis-acting sequences essential for coronavirus defective interfering RNA replication *Virology* 197 53 (1993)

8 Y-J Lin, and M M C Lai Deletion mapping of a mouse hepatitis virus defective interfering RNA reveals the requirement of an internal and discontiguous sequence for replication *J Virol* 67 6110 (1993)

9 S D Senanayake, M A Hofmann, J L Maki, and D A Brian The nucleocapsid protein gene of bovine coronavirus is bicistronic *J Virol* 66 5277 (1992)

10 M M Parker, and P S Masters Sequence comparison of the N genes of five strains of the coronavirus mouse hepatitis virus suggests a three domain structure for the nucleocapsid protein *Virology* 179 463 (1990)

11 W Lapps, B G Hogue, and D A Brian Sequence analysis of the bovine coronavirus nucleocapsid and matrix protein genes *Virology* 157 47 (1987)

12 P S Masters Localization of an RNA-binding domain in the nucleocapsid protein of the coronavirus mouse hepatitis virus *Arch Virol* 125 141 (1992)

13 S G Robbins, M F Frana, J J McGowan, J F Boyle, and K V Holmes RNA-binding proteins of coronavirus MHV detection of monomeric and multimeric N proteins with an RNA overlay-protein blot assay *Virology* 150 402 (1986)

14 B G Hogue, B King, and D A Brian Antigenic relationships among proteins of bovine coronavirus, human respiratory coronavirus OC43, and mouse hepatitis coronavirus A59 *J Virol* 51 384 (1984)

88

FIRST EXPERIMENTAL EVIDENCE OF RECOMBINATION IN INFECTIOUS BRONCHITIS VIRUS

Recombination in IBV

Sanneke A. Kottier, David Cavanagh, and Paul Britton[*]

Division of Molecular Biology
Institute for Animal Health
Compton, Newbury
Berkshire, RG16 ONN, United Kingdom

ABSTRACT

A high frequency of recombination has been shown to occur during replication of the coronavirus mouse hepatitis virus (MHV) *in vitro* as well as *in vivo*. Although sequencing of field strains of coronavirus infectious bronchitis virus (IBV) has indicated that IBV strains also undergo recombination, there has been no experimental evidence to support this deduction. To investigate whether recombination occurs in IBV, embryonated eggs were coinfected with IBV-Beaudette and IBV-M41. Potential recombinants were detected by strain-specific polymerase chain reaction (PCR) amplifications, using oligonucleotides corresponding to regions in the 3' end of the genome. Sequencing of the PCR products confirmed that a number of recombinations had occurred between the two strains.

INTRODUCTION

RNA-RNA recombination involves the exchange of genetic information between non-segmented viruses and has been detected at a high frequency in picornaviruses and in coronavirus MHV. In MHV, recombination sites were observed over almost the entire genome and several recombinants resulted from multiple cross-over events (1). A recombination map for MHV (2) suggested the frequency of recombination to be roughly 1% per 1300 nucleotides, which represents 25% for the entire genome, assuming that recombination occurs randomly. Recombination has also been shown to occur between coronavirus

---

[*] To whom all correspondence should be sent.

Corona- and Related Viruses, Edited by P. J. Talbot and G. A. Levy
Plenum Press, New York, 1995

genomic RNA and transfected RNA, enabling site-specific mutagenesis (3,4,5). However, such recombinants have only been sought and detected in the 5′ and the 3′ ends of the genome.

Studies on the mechanisms of recombination in picornaviruses showed that recombination was more likely to occur by template-switching than by breaking and rejoining of RNA and recombination frequencies were found to be higher between more closely related strains (6). The mechanism of recombination in coronaviruses has been suggested to occur by a mechanism similar to that described for poliovirus (7). Although these findings have contributed greatly to the understanding of RNA recombination in viruses, MHV remains the only coronavirus in which recombination has been experimentally demonstrated. Sequencing studies on isolates of IBV have suggested that recombination does occur in the field (8,9,10) but there is no direct evidence of recombination in coronaviruses other than MHV.

In this paper recombination was tested in the coronavirus IBV using embryonated eggs, coinfected with IBV-M41 and IBV-Beaudette. IBV-M41 and IBV-Beaudette are both of the Massachusetts serotype. The entire genome of Beaudette has been sequenced (11) and several regions of M41 have been sequenced (as reviewed in 12). Beaudette has 184 nucleotides inserted in the noncoding region at the 3′ end of the genome when compared with M41. Otherwise, the two strains have a high degree of sequence identity, increasing the chance of recombination between the two strains. Strain-specific oligonucleotides were designed, from sequences within the 3′ ends of both IBV strains, for detection of potential recombinants. Potential recombinants were detected by RT-PCR using a negative sense Beaudette-specific primer and a positive sense M41-specific primer or vice versa. PCR products from potential recombinants were sequenced and analyzed.

METHODS

Coinfection of Embryos with IBV M41 and Beaudette

Eleven-day-old specified pathogen free embryonated Rhode Island Red (RIR) eggs were inoculated, in quintuplicate, in the allantoic fluid, with 7×10^6 egg infectious dose $_{50}$ (EID_{50}) IBV-Beaudette, or with 3×10^7 EID_{50} IBV-M41, or coinfected with 7×10^6 EID_{50} IBV-Beaudette and 3×10^7 EID_{50} IBV-M41 (inoculum X1), or with 7×10^6 EID_{50} IBV-Beaudette and 3×10^6 EID_{50} IBV-M41 (inoculum X2), or with 7×10^6 EID_{50} IBV-Beaudette and 3×10^5 EID_{50} M41 (inoculum X3) or inoculated with PBSa. Eggs were sealed with collodium and incubated at 37°C for 24 h. After overnight incubation at 4°C the allantoic fluid was harvested.

Reverse Transcription (RT)

Virions were pelleted from the allantoic fluid and RNA was extracted from virions, using the guanidinium isothiocynate method (13). The final pellets were dissolved in 50 µl of nuclease-free water (Sigma). For first strand cDNA synthesis 3 µl of the extracted RNA was denatured for 10 min at 60°C in the presence of 0.21 nmol of oligonucleotide 100 (CAGGATATCGCTCTAACTCTATACTAGCCT), complementary to position 27587-27607 at the 3′ end of the IBV-genome. The mixture was immediately cooled on ice and incubated for 2 h at 42°C in a 35 µl reaction mixture containing; 250 U of M-MLV reverse transcriptase (Promega), 50 mM Tris-HCl (pH 8.3), 75 mM KCl, 3 mM $MgCl_2$, 7 mM DTT and 1.5 mM of dNTPs.

Polymerase Chain Reaction (PCR)

From the first strand cDNA reaction mixture, 5 μl was taken and mixed in a total volume of 50 μl containing, 1 5 U Taq DNA polymerase, 50 mM KCl, 10 mM Tris-HCl (pH 9 0), 0 1% Triton-X100, 1 5 mM MgCl$_2$, 0 2 mM dNTPs and 0 3 nmol of positive and negative sense primers (see results) The mixture was overlayered with 50 μl of mineral oil and subjected to 25 cycles on a Hybaid Omnigene heating block at 94°C for 1 min, 42°C for 2 min and 72°C for 3 min, followed by a 9 min elongation step at 72°C

Cloning and sEquencing

PCR products, derived from potential recombinants, were cloned into the pCR™ vector (Invitrogen) according to the manufacturer's protocol and sequenced on a 373A DNA Sequencer (Applied Biosystems Inc) using the PRISM™ Ready Reaction DyeDeoxy Terminator cycle sequencing kit according to the manufacturer's instructions (Applied Biosystems Inc)

RESULTS

Strain-specific PCR has been demonstrated to be a powerful technique for studying RNA recombination in the absence of selection pressure (14, 15) Strain-specific primers were designed from the 3' end of the genomes of IBV-Beaudette and M41 with as many nucleotide differences as possible, at the 3' end of the oligonucleotides (Fig 1) Recombination was detected in MHV within a region of 330 nucleotides (16),therefore IBV strain-specific primers over 300 nucleotides apart from each other were chosen The IBV-Beaudette specific primers were of positive sense, primer B5 (CCTGATAACGAAAACATT), corresponding to position 26041-26058 and of negative sense, primer B7 (GCACAGCAA-CAATACAAT), complementary to position 27233-27250 of the IBV-Beaudette genome The IBV-M41 specific primers were of positive sense, primer M6 (CCTGATAAT-GAAAATCTA), corresponding to position 26041-26058 and of negative sense, primer M8 (AGGTCAATGCTTTATCCA), complementary to position 26990-2707 The nucleotides positions in M41 referred to equivalent regions on the IBV-Beaudette genome

RT-PCR was carried out on virion RNA obtained from the infected eggs as described in the methods As a control to show that no PCR artifacts were produced under the chosen conditions, equal amounts of RNA from eggs infected only with Beaudette and only with M41 were mixed for one of the RT-PCR amplifications PCR amplifications were carried out on all cDNA samples with the Beaudette-specific primers, B5 and B7 (Fig 2A), the M41-specific primers, M6 and M8 (Fig 2B) and to detect potential recombinants, with the primer combinations, B5 and M8 (Fig 2C) and M6 and B7 (Fig 2D) No PCR products were

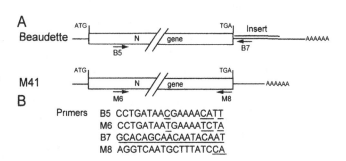

Figure 1. Strategy for detection of IBV-Beaudette-M41 recombinants (A) shows the position of the strain-specific primers in the two genomes (B) shows the oligonucleotide sequences with the mismatched nucleotides underlined

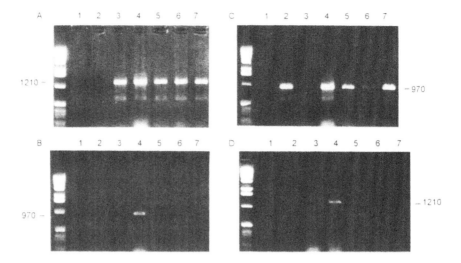

Figure 2. Detection of recombinants with strain-specific PCR, using different oligonucleotide primer combinations; (A) Beaudette specific B5+B7, (B) M41 specific M6+M8, (C) recombinant specific B5+M8, (D) recombinant specific M6+B7. For all primer combinations; lane 1, mock infected eggs; lane 2, M41-infected eggs; lane 3, Beaudette-infected eggs; lane 4-6 (respectively inoculations X1, X2, X3), M41- and Beaudette-coinfected eggs; lane 7, mixed M41 and Beaudette RNA. The marker used was the 1kb ladder from Boehringer.

detected with any primer combinations from the cDNA derived from eggs inoculated with PBSa, confirming that the PCR products were not derived from egg mRNAs. Furthermore, no PCR products were detected from cDNA derived from eggs infected with only IBV-M41 when the IBV-Beaudette specific primer combination was used (Fig. 2A, lane 2), whereas cDNA derived from eggs infected with Beaudette led to the expected PCR product of 1210 bp (Fig. 2A, lanes 3-7). In addition, no PCR products were detected from cDNA derived from eggs infected with only IBV-Beaudette when the IBV-M41 specific primer combination was used (Fig. 2B, lane 3), whereas cDNA derived from eggs infected with IBV-M41 led to the expected PCR product of 970 bp (Fig. 2B, lanes 2, 4-7), confirming that the primers were strain-specific. The primer combinations used to detect potential recombinants did not result in PCR products for cDNA derived from eggs infected with only IBV-M41 (Fig. 2C and 2D, lane 2) or infected only with IBV-Beaudette (Fig. 2C and 2D, lane 3) and did not give rise to any detectable products when cDNA derived from a mixture of Beaudette and M41 RNA was used (Fig. 2C and 2D, lane 7), indicating that no mispriming and no PCR artifacts had arisen under the chosen conditions. However, both recombinant primer combinations showed a product of the expected sizes (970 bp for primer combination B5+M8 and 1210 bp for primer combination M6+B7) when cDNA derived from coinfection X1 was used (Fig. 2C and 2D, lane 4), indicating that recombination had occurred between the two strains. These results were confirmed by three more independent experiments.

Although recombination was detected in one of the coinfections, recombination could not be detected in two other coinfections. The difference between these coinfections was the ratio of the amount of infectious Beaudette and M41 used to infect the eggs. The mixed infection in which recombination was detected had a ratio of Beaudette and M41 of approximally 1:5. However, if more Beaudette than M41 was used, recombination was not detected. This might be explained by the fact that IBV Beaudette has been adapted to grow in embryonated eggs and appears to grow much quicker in eggs than M41. These results showed that it is important to use an appropriate ratio of viruses in a coinfection for detection of recombinants.

In order to confirm that the PCR products, derived from potential recombinants, consisted of both M41 and Beaudette sequences, they were cloned and eight clones were sequenced. Each recombinant contained sequences characteristic of both Beaudette and M41, showing that the DNA obtained had been derived from recombinants. Moreover, none of the recombinant DNAs had the same cross-over site, indicating that the PCR products were derived from several recombination events.

DISCUSSION

We have demonstrated experimentally that IBV strains undergo recombination during mixed infection. Although previous sequencing studies have provided circumstantial evidence that recombination occurs in IBV in the field (8,9,10), there was no direct experimental evidence that recombination did occur in IBV. This paper, however, describes the first experimental evidence of RNA-RNA recombination in IBV.

Strain-specific polymerase chain reactions were used for the detection of IBV recombinants because the technique is very sensitive and is extremely powerful for distinguishing between closely related strains. Further, it opens the possibility to study recombination in the absence of selection. All primers designed for the detection of IBV recombinants were shown to be able to differentiate between Beaudette and M41 and could therefore be used for detection of recombinants. Although oligonucleotide M8 had only two nucleotide mismatches in relation to Beaudette RNA, the primer was specific for M41. Previously, PCR had been shown to be strain-specific using an oligonucleotide with only 3 mismatches at the 3' end (16).

To investigate the possibility of template-switching by the Taq-polymerase, RT-PCRs were conducted on a mixture of M41 and Beaudette RNA. PCR amplifications with primer combinations of one M41-specific primer and one Beaudette-specific primer did not result in a PCR product, indicating that template-switching of the Taq-polymerase was not detectable. This demonstrates that the observed PCR products from coinfected extracts did indeed arise from a recombination event by the coronavirus polymerase.

Sequencing of cloned PCR products confirmed that they were derived from IBV recombinants. All sequenced clones showed different recombination sites, suggesting that a number of recombination events had occurred. These recombination events seemed to occur randomly between the selected primers. Although initially it had been suggested that there might be favoured recombination sites in the coronavirus genome (7), it is now believed that the apparent clustering of recombination sites resulted from the use of selection pressure in the experiments (14,16). Banner et al.(14) demonstrated that recombination sites were randomly distributed after a mixed infection of two strains of MHV. However, when this mixed population was further passaged, recombination sites seemed to be clustered. To detect recombination in IBV, we used RT-PCR from pelleted virions from the mixed infections, without passaging to increases the chances of detecting recombination events, with as little selection pressure as possible, including recombinants with selective disadvantages.

Although PCR is a very powerful technique, it does have limitations. For example, the strain-specific PCR described in this paper can only detect recombinants that have an odd number of recombination events between the two selected primers. Further, recombination was detected in an area of less than 1/20th of the genome. Nevertheless, it is most likely that recombination also occurred in other parts of the genome, considering that recombination in MHV was detected in most areas of the genome.

The experimental evidence of recombination in IBV opens new possibilities for IBV research As in MHV, recombination between IBV genomic- and synthetic RNA could be examined, allowing the production of site-specific mutants For detection of this kind of recombination using synthetic RNA of the 3' end of the genome, the same strain-specific primers can be used The possibility of creating various site-specific IBV recombinants by making use of the natural ability of IBV to recombine is an exciting prospect and might be useful for examining differences in pathogenicity among IBV strains

ACKNOWLEDGMENTS

We would like to thank Miss Karen Mawditt for synthesising the oligonucleotides This work was supported by the Ministry of Agriculture, Fisheries and Food, UK

REFERENCES

1 Lai, M M C (1992) RNA recombination in Animal and Plant Viruses Microbiology Reviews 56, 61-79

2 Baric, R S, Fu, K, Schaad, M C & Stohlman, S A (1990) Establishing a genetic-recombination map for murine coronavirus strain A59 complementation groups Virology 177, 646-656

3 Koetzner, C A, Parker, M M, Ricard, C S, Sturman, L S & Masters, P S (1992) Repair and mutagenesis of the genome of a deletion mutant of the coronavirus mouse hepatitis virus by targeted RNA recombination Journal of Virology 66, 1841-1848

4 van der Most, R G, Heijnen, L, Spaan, W J M & de Groot, R J (1992) Homologous RNA recombination allows efficient introduction of site-specific mutations into the genome of coronavirus MHV-A59 via synthetic co-replicating RNAs Nucleic Acids Research 20, 3375-3381

5 Masters, P S, Koetzner, C A, Kerr, C A & Heo, Y (1994) Optimization of targeted RNA recombination and mapping of a novel nucleocapsid gene mutation in the coronavirus mouse hepatitis virus Journal of Virology 68, 328-337

6 Kirkegaard, K & Baltimore, D (1986) The mechanism of RNA recombination in poliovirus Cell 47, 433-443

7 Makino, S, Keck, J G, Stohlman, S A & Lai, M M C (1986) High frequency RNA recombination of murine coronaviruses Journal of Virology 57, 729-737

8 Kusters, J G, Jager, E J, Niesters, H G M & van der Zeijst, B A M (1990) Sequence evidence for RNA recombination in field isolates of avian coronavirus infectious bronchitis virus Vaccine 8, 605-608

9 Cavanagh, D, Davis, P J & Cook, J K A (1992) Infectious bronchitis virus evidence for recombination within the Massachusetts serotype Avian Pathology 21, 401-408

10 Wang, L, Junker, D, Collisson, E W (1993) Evidence of natural recombination within the S1 gene of infectious bronchitis virus Virology 192, 710 716

11 Boursnell, M E G, Brown, T D K, Foulds, I J, Green, P F, Tomley, F M & Binns, M M (1987) Completion of the sequence of the genome of the coronavirus avian infectious bronchitis virus Journal of General Virology 68, 57-77

12 Spaan, W, Cavanagh, D & Horzinek, M C (1988) Coronaviruses structure and genome expression Journal of General Virology 69, 2939-2952

13 Chomczynski, P & Sacchi, N (1987) Single step method of RNA isolation by acid guanidinium thiocyanate phenol chloroform extraction Analytical Biochemistry 162, 156-159

14 Banner, L R & Lai, M M C (1991) Random nature of coronavirus RNA recombination in the absence of selection pressure Virology 185, 441-445

15 Jarvis, T C & Kirkegaard, K (1992) Poliovirus RNA recombination mechanistic studies in the absence of selection EMBO Journal 11, 3135-3145

16 Liao, C -L & Lai, M M C (1992) RNA recombination in a coronavirus recombination between viral genomic RNA and transfected RNA fragments Journal of Virology 66, 6117-6124

89

MOLECULAR BASES OF TROPISM IN THE PUR46 CLUSTER OF TRANSMISSIBLE GASTROENTERITIS CORONAVIRUSES

M. L. Ballesteros, C. M. Sánchez, J. Martín-Caballero, and L. Enjuanes

Department of Molecular and Cellular Biology
Centro Nacional de Biotecnología, CSIC
Campus Univesidad Autónoma
Cantoblanco, 28049 Madrid, Spain

ABSTRACT

Transmissible gastroenteritis coronavirus (TGEV) infects both, the enteric and the respiratory tract of swine. S protein, that is recognized by the cellular receptor, has been proposed that plays an essential role in controlling the dominant tropism. The genetic relationship of S gene from different enteric strains and non-enteropathogenic porcine respiratory coronaviruses (PRCVs) was determined. A correlation between tropism and the genetic structure of the S gene was established. PRCVs, derived from enteric isolates have a large deletion at the N-terminus of the S protein. Interestingly, two respiratory isolates, attenuated Purdue type virus (PTV-ATT) and Toyama (TOY56) have a full-length S gene. PTV-ATT has two specific amino acid differences with the S protein of the enteric viruses. One is located around position 219, within the deleted area, suggesting that alterations around this amino acid may result in the loss of enteric tropism.

To study the role of different genes in tropism, a cluster of viruses closely related to PUR46 strain was analyzed. All of them have been originated by accumulating point mutations from a common, virulent isolate which infected the enteric tract. During their evolution these viruses have lost, virulence first, and then, enteric tropism. Sequencing analysis proved that enteric tropism could be lost without changes in ORFs 3a, 3b, 4, 6, and 7, and in 3'-end untranslated regions (3'-UTR). To study the role of the S protein in tropism recombinants were obtained between an enteric and a respiratory virus of this cluster. Analysis of the recombinants supported the hypothesis on the role in tropism of S protein domain around position 219.

INTRODUCTION

TGEV, infects both, the enteric and the respiratory tract of swine. TGEV must attach to host cells through the S glycoprotein, since monoclonal antibodies (MAbs) specific for the S glycoprotein, but not MAbs specific for the N or M proteins inhibit the binding of the virus to ST cells[16]. Then it should be expected that S protein plays an important role in the control of dominant tropism. In fact, there are data from different laboratories establishing a correlation between the S protein gene structure and tropism. PRCVs have been originated, independently, in Europe[12, 3] and in North America[18], from enteric isolates[14]. PRCVs show a large deletion at the N-terminus of the S protein, including the antigenic sites C and B[14]. Interestingly, there are two respiratory isolates, PTV-ATT and TOY56, with a full-length S gene, without deletion. The S genes of these viruses have been sequenced and compared to S genes from the enteric isolates. In PTV-ATT, only two nucleotide differences, leading to two specific amino acid changes were found. One was located around amino acid 219, within the deleted area, while the other was outside[14]. Our laboratory has proposed that alterations around position 219 may be responsible for the loss of enteric tropism in these viruses.

However, it cannot be excluded that other genes could be involved in tropism. ORF3 has accumulated a large number of nucleotide changes among enteric isolates and PRCVs, including both, insertions and deletions[13, 2, 17, 9]. Most important, ORF3a is expressed in the enteric isolates but not in PRCVs, due to a nucleotide change in the ORF3a consensus region. In contrast, in the nearest genes 3b and 4, there are no specific mutations between enteric viruses and PRCVs. To analyze the role of different genes in tropism a cluster of viruses closely related to the PUR46 strain was studied.

MATERIALS AND METHODS

Cells and Viruses

All viruses were grown on swine testis (ST) cells[11]. The characteristics of TGEV strains: PUR46-SW11[6], provided to us by M. Pensaert and PUR46 [1, 15, 14], have been described. PTV-ATT, a Purdue type virus, was previously named NEB72. Due to the relationship between NEB72 and other TGEVs in the epidemiological tree developed[14] and to its sequence homology with the PUR46 isolate we have renamed this virus strain. Both, PTV-ATT and a temperature sensitive mutant, PTV-ATTts, derived from it, were kindly provided by M. Welter and L. Welter. PTVtsdmar was derived from PTV-ATTts. This virus is temperature sensitive and has modified the antigenic subsites Aa and Ab of the S protein.

Recombinants were obtained by coinfection with two parental strains: respiratory PTVtsdmar and enteric PUR46. The progeny was selected using MAbs and restrictive temperature.

RNA Analysis

RNA was sequenced by different procedures: directly from viral RNA[5] or deriving cDNA fragments by RT-PCR; cDNA fragments were sequenced using *fmol* system (Promega) or cloned into pBluescript to be sequenced using the *Sequenase* kit (USB).

To analyze a point mutation in nucleotide 655 of S gene, cDNA fragments containing the mutated position were RT-PCR derived and their susceptibility to *Bsm*AI restriction endonuclease enzyme studied.

Virus Tropism

Tropism was studied in conventional, non-colostrum-deprived, newborn mini-swine. Piglets were orally inoculated by stomach tube and slaughtered at 24, 48, and 72 hours post-infection. Animal room was kept at 22°C. Lungs and small intestine were collected and homogenized in PBS. A sample was separated and virus titer determined by plaque-assay[7].

RESULTS

To study the role of different genes in tropism a cluster of closely related viruses was studied (Fig. 1). All of them have been originated, by accumulating point mutations, from a common and virulent, ancestor PUR46-SW11 that infects the enteric tract. All these isolates have, as a trade mark, a small deletion of six nucleotides in the S gene. These viruses have evolved, losing their virulence (PUR46) and their enteric tropism (PTV-ATT and PTVtsdmar).

ORFs 2, 3a, 3b, 4, 6, and 7 and the 3'-UTR were sequenced in PUR46, PTV-ATT and PTVtsdmar. Sequence comparison of the genomes of the enteric isolate PUR46 and its respiratory derivative PTV-ATT showed that no nucleotide was changed in 3'-UTR and in ORFs 7, 6, and 4. In ORFs 3a and 3b only two nucleotide differences were found. One was located in the non-coding region between the two genes, and the other in ORF3b. Viruses of this cluster do not express ORF3b because they have a point mutation in the consensus region. These data show that, in contrast to what has been previously suggested[17], enteric tropism may be lost without changes in ORF3a.

Between the S genes of PUR46 and PTV-ATT, only two nucleotides have mutated leading to two specific amino acid changes. One was nucleotide 655, located within the deleted area of the PRCVs. The other was nucleotide 2098. To determine if these positions were involved in tropism, ST cells were coinfected with the enteric PUR46 and the respiratory PTVtsdmar. Recombinants were obtained having either, both or only one nucleotide changed. The recombinant isolation frequency was estimated to be very low

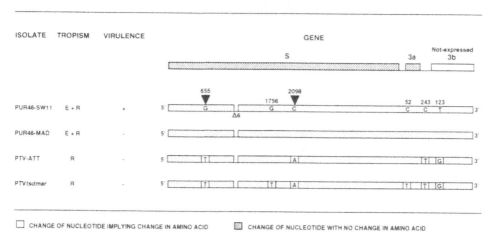

Figure 1. Summary of nucleotide changes in genes S, 3a, and 3b of PUR46-related viruses. Genes 2, 3a, and 3b were sequenced and compared in the cluster of PUR46-related viruses, including PUR46-SW11, PUR46-MAD, PTV-ATT, and PTVtsdmar isolates. Bars represent the studied genes. White and dotted squares indicate the position of nucleotide causing a change or no modification, respectively, in the amino acid sequence. Letters represent nucleotides. Numbers above these letters indicate the location of nucleotides referring to the ATG of each gene. Δ6 indicates a deletion of 6 nucleotides, in relation to the sequence of the MIL65 strain. Virus tropism is indicated with E+R when virus infects both, the enteric and the respiratory tract, and with R when virus only infects the respiratory tract. Virulence is designed by + if virus kills piglets or by - when not, in the assay conditions described in Materials and Methods.

Table 1. Phenotypic characterization of potential recombinants between enteric
PUR46 and respiratory PTV*tsdmar*

Recombinant group	Neutralization index (MAbs)	Inactivation index (temperature)	Number of clones	%
PUR46*wt*	3.7±0.3	0.5±0.3	0	—
PTV*tsdmar*	0.5±0.3	3.0±0.2	3	9.7
Group 1	3.7±0.3	3.0±0.2	0	—
Group 2	0.5±0.3	0.5±0.3	15	48.4
Group 3	1.2>NI>0.5	2.4>II>0.5	13	41.9

($<1.0 \times 10^{-6}$) in this cluster of viruses. The progeny phenotype was determined (Table 1). Recombinants were classified in three groups according to their resistance to MAbs and restrictive temperature. Selection was performed against group 1 (sensitive to both, neutralization by MAbs and inactivation at restrictive temperature) and, in fact, no virus with this phenotype was isolated. Group 3 contained viruses having an intermediate phenotype between the two parental strains. Almost 50% of the progeny showed the expected recombinant phenotype (group 2), resistant to both, neutralization by MAbs and inactivation at restrictive temperature.

To genotypically characterize group 2 recombinants two nucleotide differences between the S genes of the two parental viruses were used. According to the position of the crossing-over, recombinants were split up into two groups (Fig. 2). Group 2A had recombined in the S gene, between the two nucleotide differences, taking position 655 from the enteric parental (PUR46), while nucleotide 2098 from the respiratory one (PTV*tsdmar*). The mutation responsible for the lack of antigenic subsites Aa and Ab had been previously mapped at nucleotide 1756 of ORF2. The crossing-over of this group of recombinants was located between nucleotides 655 and 1756. A high number of the analyzed recombinants, more than 40%, showed this structure. Group 2B had recombine 5' upstream of the S gene, although the exact position of the crossing-over has not yet been mapped. Group 2B

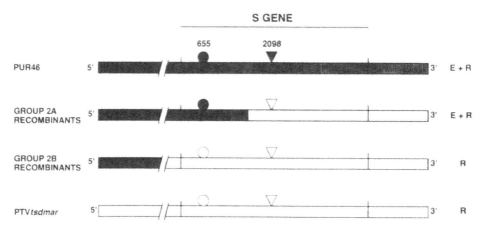

Figure 2. Genetic structure of the recombinants obtained by crossing the enteric PUR46 and the respiratory PTV*tsdmar*. Full and empty bars indicate sequences coming from PUR46 or PTV*tsdmar*, respectively. The nucleotide difference of position 655 is marked by a circle and the difference at position 2098 by a triangle. Two Kb of ORF1b 3'-end were sequenced.

recombinants has taken the entire protein from the respiratory parental virus. If the predicted role of the S protein domain around amino acid 219 in tropism is correct, group 2A recombinants should infect the enteric tract, while those of group 2B should be respiratory. The tropism of PUR46-cluster of viruses, as well as group 2A of recombinants was tested. All these viruses could efficiently replicate in lungs. PUR46-SW11 and PUR46 could be isolated in the enteric tract, while PTV-ATT and PTV*tsdmar* were not, as expected. Recombinants of group 2A were able to infect the small intestine. These results support the hypothesis that the area around position 219 of the S protein, taking in this group from the enteric parental strain, is involved in tropism.

DISCUSSION

PUR46-cluster of viruses does recombine. The estimated frequency of recombinants isolation (<1.0 x 10^{-6} per 10^3 nucleotides) is lower than that described for murine coronaviruses[10]. Due to this low frequency, selective pressure has to be used to eliminate parental viruses in order to isolate recombinants. The use of selective markers mapping in distal parts of the genome increases the distance to recombine. The markers that were used to select the recombinants (*dmar* and *ts* mutations) were located, at least, at 4 kb from each other. Nevertheless, 40% of the isolates that were analyzed have recombined within nucleotides 655 and 1756 of the S protein. This concentration of recombinants having the crossing-over in this region could be due either, to a high frequency of recombination in the area, or to a selective advantage of viruses having this genome.

The respiratory PTV-ATT was originated from the enteric PUR46. Sequencing analysis has shown that enteric tropism has been lost in PTV-ATT without changes neither in 3'-UTR nor in the proteins encoded by ORFs 3a, 3b, 4, 6, and 7, when compared to PUR46 strain. Although in this cluster of viruses enteric tropism could be lost without changes in these genes, it does not imply that mutations affecting ORFs 7, 6, 4, 3b, and 3a could not lead to a change of tropism. In fact, several laboratories have reported data suggesting a possible involvement of ORF3a in TGEV enteropathogenicity[9] and attenuation.

The tropism of group 2A recombinants strongly suggests that the area around nucleotide 655 of the S gene is involved in the enteric tropism of this cluster of viruses. To make a definitive statement group 2B recombinants has to be further characterized. S protein plays an essential role in the attachment of virions[16, 14] and in the fusion of viral and cellular membranes[9]. Aminopeptidase N (APN) is known to act as a major receptor for TGEV in cell culture[4]. MAbs specific for the antigenic sites A and D are the best inhibitors of virus binding to ST cells, suggesting that the domain recognized by the cellular receptor on ST cells must be located spatially close to these sites[16]. PRCVs, that have a large deletion at the 5'-end of the S gene including the area around nucleotide 655, may also use APN to entry into cells[9]. In fact, binding of APN to both TGEV and PRCVs is mediated by residues located between amino acids 500-800 of S protein[8]. All these data indicate that the viruses of this cluster having nucleotide 655 mutated which do not infect the enteric tract still recognize APN. Different mechanisms could be proposed to explain how the area around position 219 of the S protein is involved in tropism. One possibility is the presence of a second cellular receptor needed for *in vivo* virus entry into cells of the enteric tract. The receptor binding-site would map around position 219 in the S protein. Another possibility is, as proposed for MHV[19], the need of a spike protein-dependent cellular factor, other than a virus receptor, required for a productive virus cycle in the enteric tract. The cellular factor would interact with the spike protein in the area around amino acid 219. The analysis of group 2 recombinants will be very helpful to explain the role of S protein in tropism.

ACKNOWLEDGMENTS

We are grateful to M Welter and L Welter (Ambico, Dallas Center, IA) for providing to us the PTV-ATT and the *ts* mutant derived from it This work has been supported by grants from Consejo Superior de Investigaciones Científicas, the Comision Interministerial de Ciencia y Tecnologia, Instituto Nacional de Investigaciones Agrarias, La Consejeria de Educacion y Cultura de la Comunidad de Madrid, from Spain, and the European Communities (Projects Science and Biotech) M L Ballesteros received a fellowship from Consejo Superior de Investigaciones Científicas

REFERENCES

1 Bohl, E H Antibody in serum, colostrum, and milk of swine after infection or vaccination with transmissible gastroenteritis virus Infect Immun 1972, 6 289-301

2 Britton, P, Page, K W, Mawditt, K, Pocock, D H Sequence comparison of porcine transmissible gastroenteritis virus (TGEV) with porcine respiratory coronavirus VIIIth International Congress of Virology 1990 IUMS, Berlin pp P6-018

3 Callebaut, P, Correa, I, Pensaert, M, Jimenez, G, Enjuanes, L Antigenic differentiation between transmissible gastroenteritis virus of swine and a related porcine respiratory coronavirus J Gen Virol 1988, 69 1725-1730

4 Delmas, B, Gelfi, J, L'Haridon, R, Vogel, L K, Noren, O, Laude, H Aminopeptidase N is a major receptor for the enteropathogenic coronavirus TGEV Nature 1992, 357 417-420

5 Fichot, O, Girard, M An improved method for sequencing of RNA templates Nucleic Acids Res 1990, 18 6162

6 Haelterman, E O, Pensaert, M B Pathogenesis of transmissible gastroenteritis of swine Proc 18th World Vet Congress 1967 Paris Vol 2, pp 569-572

7 Jimenez, G, Correa, I, Melgosa, M P, Bullido, M J, Enjuanes, L Critical epitopes in transmissible gastroenteritis virus neutralization J Virol 1986, 60 131-139

8 Laude, H, Godet, M, Bernard, S, Gelfi, J, Duarte, M, Delmas, B Functional domains in the spike protein of transmissible gastroenteritis virus VI International symposium on corona- and related viruses 1994 Quebec, Canada

9 Laude, H, Vanreeth, K, Pensaert, M Porcine respiratory coronavirus - molecular features and virus host interactions Vet Res 1993, 24 125-150

10 Makino, S, Keck, J G, Stohlman, S A, Lai, M M C High-frequency RNA recombination of murine coronaviruses J Virol 1986, 57 729-737

11 McClurkin, A W, Norman, J O Studies on transmissible gastroenteritis of swine II Selected characteristics of a cytopathogenic virus common to five isolates from transmissible gastroenteritis Can J Comp Vet Sci 1966, 30 190-198

12 Pensaert, M, Callebaut, P, Vergote, J Isolation of a porcine respiratory, non-enteric coronavirus related to transmissible gastroenteritis Vet Q 1986, 8 257-260

13 Rasschaert, D, Duarte, M, Laude, H Porcine respiratory coronavirus differs from transmissible gastroenteritis virus by a few genomic deletions J Gen Virol 1990, 71 2599-2607

14 Sanchez, C M, Gebauer, F, Suñe, C, Mendez, A, Dopazo, J, Enjuanes, L Genetic evolution and tropism of transmissible gastroenteritis coronaviruses Virology 1992, 190 92-105

15 Sanchez, C M, Jimenez, G, Laviada, M D, Correa, I, Suñe, C, Bullido, M J, Gebauer, F, Smerdou, C, Callebaut, P, Escribano, J M, Enjuanes, L Antigenic homology among coronaviruses related to transmissible gastroenteritis virus Virology 1990, 174 410-417

16 Suñe, C, Jimenez, G, Correa, I, Bullido, M J, Gebauer, F, Smerdou, C, Enjuanes, L Mechanisms of transmissible gastroenteritis coronavirus neutralization Virology 1990, 177 559-569

17 Wesley, R D, Woods, R D, Cheung, A K Genetic analysis of porcine respiratory coronavirus, an attenuated variant of transmissible gastroenteritis virus J Virol 1991, 65 3369-3373

18 Wesley, R D, Woods, R D, Hill, H T, Biwer, J D Evidence for a porcine respiratory coronavirus, antigenically similar to transmissible gastroenteritis virus, in the United States J Vet Diagn Invest 1990, 2 312-317

19 Yokomori, K, Asanaka, M, Stohlman, S A, Lai, M M C A spike protein-dependent cellular factor other than the viral receptor is required for mouse hepatitis virus entry Virology 1993, 196 45-56

90

GENERATION OF A DEFECTIVE RNA OF AVIAN CORONAVIRUS INFECTIOUS BRONCHITIS VIRUS (IBV)[*]

Defective RNA of Coronavirus IBV

Zoltan Penzes,[1] Kefford W. Tibbles,[2] Kathy Shaw,[1] Paul Britton,[1] T. David K. Brown,[2] and David Cavanagh[1†]

[1] Division of Molecular Biology
Institute for Animal Health
Compton, Newbury, Berkshire RG16 ONN, United Kingdom
[2] Division of Virology, Department of Pathology
University of Cambridge
Cambridge, Cambridgeshire, CB2 1QP, United Kingdom

ABSTRACT

The Beaudette strain of IBV was passaged 16 times in chick kidney (CK) cells. Total cellular RNA was analyzed by Northern hybridization and was probed with ^{32}P-labeled cDNA probes corresponding to the first 2 kb of the 5′ end of the genome, but excluding the leader, and to the last 1 8 kb of the 3′ end of the genome. A new, defective IBV RNA species (CD-91) was detected at passage six. The defective RNA, present in total cell extract RNA and in oligo-$(dT)_{30}$-selected RNA from passage 15, was amplified by the reverse transcription-polymerase chain reaction (RT-PCR) to give four fragments. The oligonucleotides used were selected such that CD-91 RNA, but not the genomic RNA, would be amplified. Cloning and sequencing of the PCR products showed that CD-91 comprises 9 1 kb and has three regions of the genome. It contains 1133 nucleotides from the 5′ end of the genome, 6322 from gene 1b corresponding to position 12423 to 18744 in the IBV genome and 1626 from the 3′ end of the genome. At position 749 one nucleotide, an adenine residue, was absent from CD-91 RNA. By Northern hybridization CD-91 RNA was detected in virions in higher amounts than the subgenomic mRNAs.

---

[*] Sequence data from this article have been deposited with the EMBL, GenBank and DDJB Nucleotide Sequence Databases under Accession No Z30541

[†] To whom all correspondence should be addressed

INTRODUCTION

In infected cells the IBV-encoded RNA-dependent RNA polymerase replicates the genome into a minus-sense RNA, which then serves as template for synthesis of both the genome RNA (gRNA) and transcription of the five subgenomic mRNAs. During transcription and replication the RNA polymerase may pause, fall off and then rejoin the original or another RNA template. The discontinuous and non-processive nature of transcription may give rise to incomplete RNA intermediates (1, 2) defective RNAs (D-RNAs) (7, 12, 17) and recombinants (4, 10). To date, coronavirus D-RNAs have been reported only for MHV (12, 17). In this article we report the cloning, sequencing and characterization of a naturally occurring replicating and packaged D-RNA, CD-91, of coronavirus IBV.

MATERIALS AND METHODS

Virus and Cells, Undiluted Passage of IBV-Beaudette

Beaudette-US, an egg-adapted strain of IBV, chick kidney (CK) cells prepared from one-week-old Rhode Island Red chicks and Vero cells were used throughout the experiments. IBV-Beaudette was passaged undiluted in confluent CK and Vero cells at 37°C for 24 h.

Preparation of Purified Virions, Cell Extract and Viral RNAs

Cell-associated RNA (CK-cells) and RNA in pelleted virions was extracted using the guanidinium isothiocyanate method. Some of the total cellular RNA was selected with paramagnetic oligo-(dT)$_{30}$ particles according to the manufacturer's instructions (Scigen Ltd., Sittingbourne, UK).

Northern Blot Analysis

Viral and cell-extract RNA was analyzed and radiolabeled probes ([α-^{32}P]dCTP) were prepared as described previously (18). The filters were probed with various radiolabeled cDNA probes covering different parts of the IBV-Beaudette genome.

RT-PCR Amplification, Cloning and Sequencing of CD-91 RNA

CD-91 RNA present in total cellular RNA and in oligo-(dT)$_{30}$-selected RNA from passage 15 was amplified by RT-PCR in four separate fragments. The cDNA synthesis (Superscript RT, BRL) of the four fragments was primed with oligonucleotides 21 (position in the IBV-Beaudette genome: 12733-12714), 93/118 (15650-15631), 93/104 (26092-26074) and oligo-(dT)$_{18}$-*Not*I. Four µl of the resulting cDNA was then amplified by the PCR (*Taq* polymerase, Promega) with oligonucleotides 43 (1-22) and 21; 93/116 (1111-1131) and 93/118; 93/117 (15578-15597) and 93/104 and with 35 (16785-16803) and oligo-(dT)$_{18}$-*Not*I to obtain the four fragments of CD-91 RNA.

The amplified CD-91 fragments were cloned into the *Sma*I site of pBluescript II SK(+) (Stratagene) using standard cloning procedures or cloned into pCR™ vector (Invitrogen) according to the manufacturer's instructions. The clones were sequenced on a 373A DNA Sequencer (Applied Biosystems Inc.) using the PRISM™ Ready Reaction DyeDeoxy Terminator cycle sequencing kit (Applied Biosystems Inc.).

RESULTS

Detection of a Defective IBV RNA in CK Cells

IBV Beaudette-US was passaged undiluted in CK cells 16 times. At 24 h post infection (p.i.) total cellular RNA was extracted and analyzed by Northern hybridization using ^{32}P-labeled cDNA probes corresponding to the 5' (excluding the leader) and the 3' end of the genome. A new RNA species, CD-91, appeared at passage 6 which was larger than mRNA 2 (Fig. 1). TCID$_{50}$ analysis showed no fluctuation of infectious virus titer, all passages having a TCID$_{50}$ between $10^{7.5}$ - $10^{8.0}$.

Mapping and Sequencing of Defective RNA CD-91

In order to determine which genomic sequences were present in the CD-91 RNA, total RNA from IBV infected cells was probed with various ^{32}P-labeled cDNA probes which covered most of the Beaudette genome. The results indicated that in addition to the genomic 5' and 3' sequences, CD-91 RNA contained an internal region, corresponding to approximately the first 6 kb of gene 1b. Gene 1a was found to be almost absent and none of the structural genes, except some of N gene were present (Fig. 2A).

CD-91 RNA was amplified by RT-PCR such that the 5' and 3' ends of CD-91 RNA and the middle region, corresponding to gene 1b, were amplified separately with four different oligonucleotide pairs, as described in Materials and Methods. Each of the four overlapping PCR fragments (5' to 3': 1.5 kb, 3.2 kb, 3.2 kb and 3.6 kb) contained one of the two putative rearrangement sites of CD-91 RNA. Cloning and sequencing of the PCR products confirmed that CD-91 RNA was composed of three regions from the IBV genome, as shown in Fig. 2A. With one exception, the sequence of CD-91 RNA corresponded to the equivalent region of the published IBV-Beaudette sequence (3): at position 749, near the 5' end, an adenine residue was absent from CD-91 RNA. The adenine residue deletion in CD-91 RNA was confirmed on RNA from passages 1 (see below), 2, 7 and 15. Using total RNA from infected cells (passage 7) the corresponding region of the gRNA was also sequenced and the results confirmed the presence of the adenine residue in the gRNA.

CD-91 RNA has one long open reading frame (ORF) nucleotides 996 to 7463, corresponding to nucleotides 997-1133, 12423-18744 and 25983-25990 in the IBV genome. Due to the adenine residue deletion, this frame is 467 nucleotides shorter at its 5' end when compared to IBV gene 1a. This CD-91 RNA ORF stops after the 3' rearrangement site at nucleotide 7463 on CD-91 RNA. A second ORF corresponding to the 3' half of the N gene

Figure 1. Northern blot of IBV Beaudette-US RNAs from CK cells. Beaudette-US was passaged undiluted in CK cells (passage 1 to 15) and the total cell extract RNA was separated in denaturing agarose gels and probed with a ^{32}P-labeled genomic 3' end 1.8 kb probe. Lane M is IBV-Beaudette RNA extracted from purified virions which also contain subgenomic mRNAs, used as a marker. Lanes 1-15 show the undiluted passages of Beaudette-US, where a new RNA species, CD-91, appeared at passage six and persisted in high amounts during subsequent passages.

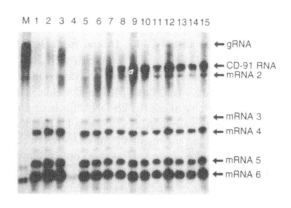

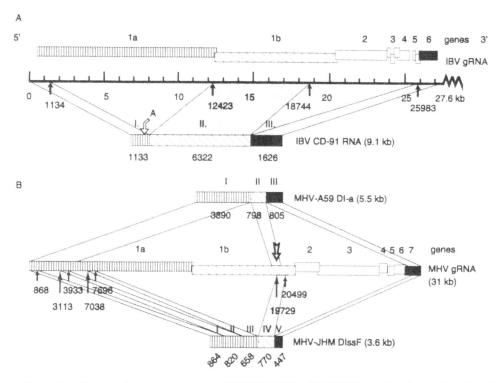

Figure 2. A. Diagram showing the structure of IBV CD-91 RNA. CD-91 RNA contains three regions from the IBV genome: 1133 nucleotides from the 5' end of the IBV-Beaudette genome (region I), 6322 nucleotides from gene 1b (region II) and 1626 nucleotides from the 3' end of the genome (region III). At position 749 an adenine residue deletion was found in CD-91 RNA (marked with unfilled arrow). **B.** Diagram showing the regions of MHV-A59 DI-a (17) and MHV-JHM DIssF (15). The MHV packaging signal is marked with an unfilled arrow (6). The solid arrows indicate the positions of the rearrangement sites of the different D-RNAs on the IBV and MHV genomes.

is also present in CD-91 RNA. It comprises nucleotides 8009 to 8575, corresponding to nucleotides 26536-27102 in the genome.

Detection of CD-91 RNA in Early Passages in CK Cells

To investigate whether CD-91 RNA was present earlier than passage 6, in which it was first detected by Northern hybridization, the 5' 1a-1b rearrangement site of CD-91 was selectively amplified by RT-PCR from total RNA from infected CK cells from passage 0 (starting inoculum), 1-7, 15, 16 and from Beaudette-US Vero passage 2. Use of oligonucleotides 93/102 and 21 yielded the expected 1.3 kb product for CD-91 RNA starting from passage 4. Then, each PCR product was re-amplified with nested oligonucleotides 93/106 (674-692) and ST4 (12490-12471) and yielded a 0.5 kb DNA fragment starting from passage 1. The specificity of the 0.5 kb DNA has been confirmed by cloning and sequencing of the PCR products from passages 1, 2, 7 and 15. All the clones from passages 2, 7 and 15 contained the 5' 1a-1b rearrangement site of CD-91 RNA and lacked the adenine residue at position 749 (CD-91-like clones). However, out of ten clones sequenced from passage 1, only a minority of them was CD-91-like. Four different types of clones were identified with a 1a-1b rearrangement site which was shifted about 100 to 400 nucleotides towards the 5' end of the genome, when compared to the CD-91 RNA 5′ rearrangement site. None of these different types of clones lacked the adenine residue at position 749.

Packaging of CD-91 RNA

To study whether CD-91 RNA was packaged into virions, RNAs extracted from purified virions at passage 0 and 15 were analyzed by Northern hybridization. The results showed that only virus preparations from passage 15 contained CD-91 RNA.

DISCUSSION

This study reports the detection, cloning, sequencing and characterization of CD-91 RNA, a 9.1 kb defective RNA of coronavirus IBV. CD-91 RNA of IBV Beaudette-US was detected after undiluted passage in CK but not Vero cells. The defective RNA was not detected in CK cell passage 0 or in Vero cell passage 2, even by nested-set PCR, whereas CD-91 was detected by this method starting at CK cell passage 1. The finding that CD-91 RNA was not detected in the Vero cell passages and that heterogeneous defective RNAs were detected in CK cell passage 1 suggests that CD-91 RNA may have been generated in CK passage 1 and subsequently became the predominant D-RNA species. However, the possibility that a pool of minute amounts of D-RNAs were already present in passage 0 cannot be excluded.

Cloning and sequencing of CD-91 RNA revealed that it comprised three discontinuous regions of the IBV genome (Fig. 2A); it is therefore generally similar to MHV defective interfering RNAs (DIs) (13, 14, 15, 17). However, CD-91 RNA is considerably larger (9.1 kb) than the MHV DIssF (3.6 kb) or DI-a or DI-b (5.5 kb and 6.5 kb (Fig. 2B). CD-91 lacks the 3′ end of gene 1b that is present in the MHV DIs and where a 61 nucleotide MHV packaging signal has been identified (6). If CD-91 RNA does contain a packaging signal then it has a different location than in MHV (Fig. 2B). The 5′ end gene 1a region of CD-91 RNA is smaller (1.1 kb) than the MHV gene 1a region present in the MHV DIs (1.6 to 3.9 kb) and lacks the 0.2 kb region of the MHV gene 1a that is located about 3.2 kb from the 5′ end of the MHV genome and which has been found to be necessary for MHV DI replication (9, 11). Since CD-91 RNA is replicated efficiently in CK cells, we conclude that this region may not be required for the replication of CD-91 RNA, or that a homologous region may be located elsewhere.

CD-91 RNA does not have one long ORF that spans the whole sequence. This is in contrast to DIssE and DI-a in which the regions of the genome comprising the DIs are joined in one frame. The ORF of DI-a has been shown to be essential for replication (5). Due to the 'A' deletion at 749, confirmed in passages 1, 2, 7 and 15, the IBV gene 1a ORF is truncated, spanning only nucleotides 529 to 765. However, due to the in frame 1a-1b junction and the long (6.3 kb) gene 1b region (Fig. 2A), CD-91 RNA does have a long ORF, potentially encoding a protein of deduced M_r 244,000 that would include most of gene 1b. Since the CD-91 RNA 3′ 1b-N junction is out of frame, only the carboxy-terminal half of the N protein would possibly be translated from a separate ORF. At present it is not known which, if any, of these ORFs is essential for the replication of CD-91 RNA.

We did not observe a decrease in virus titer or in the amounts of subgenomic RNAs during undiluted virus passage. This was probably because we were unable to achieve high multiplicity of infection. Zhao et al. (18) have calculated that approximately 1 in 24 virus particles of IBV-Beaudette contained mRNA 6. We have estimated that virus preparations contained approximately 2.3-fold more CD-91 than mRNA 6, i.e. about 1 in 10 virus particles would contain CD-91 RNA. Even if all the CD-91 molecules were present in virus particles that also contained gRNA, most cells would have been infected with virions containing gRNA but not CD-91. Hence the interfering property of CD-91 RNA would be masked by normal replication of gRNA in the majority of cells.

The amount of CD-91 in CK cell extracts (passage 9 onwards) was somewhat greater than that of mRNA 4 and slightly less than mRNA 6 (Fig. 1). Zhao *et al* (18) calculated that the molar ratios of mRNA 4/gRNA and mRNA 6/gRNA in cell extracts is approximately 2.2 and 7.7, respectively, indicating that in the cell extracts as a whole there was about 5-fold more CD-91 RNA than gRNA. This observation indicates that CD-91 was replicated more efficiently than gRNA, presumably because it is only one-third of the size of gRNA.

There is a great deal of evidence to support the view that discontinuous, leader-primed transcription is the primary and major mechanism whereby coronavirus subgenomic mRNAs are generated (1, 2, 8). However, there may also be a secondary mechanism for generating subgenomic mRNAs. Negative-sense RNAs containing anti-leader sequence have been demonstrated in cells infected with coronavirus transmissible gastroenteritis virus (16). This has led to the view that coronavirus subgenomic mRNAs can be replicated. In our experiments cell extracts contained almost as much CD-91 RNA as mRNA 6, the most abundant mRNA, even though only a minority of infected cells would have contained CD-91 whereas the mRNAs would have been present in all the infected cells. If only a minority of the subgenomic mRNAs are generated by replication, our results suggest that CD-91 RNA replicates much more efficiently than the subgenomic mRNAs. This suggests that parts of the gene 1a and/or 1b sequences present in CD-91 RNA are required for efficient IBV RNA replication.

ACKNOWLEDGMENTS

This work was supported by the Ministry of Agriculture, Fisheries and Food, UK and an Agricultural and Food Research Council Linked Research Group Award.

REFERENCES

1 Baric, R S , Stohlman, S A , Razavi, M K , and Lai, M M C (1985) Characterisation of leader related small RNAs in coronavirus infected cells Further evidence for leader primed mechanism transcription *Virus Res* **3**, 19-33

2 Baric, R S , Shieh, C , Stohlman, S A , and Lai, M M C (1987) Analysis of intracellular small RNAs of mouse hepatitis virus evidence for discontinuous transcription *Virology* **156**, 342-354

3 Boursnell, M E G , Brown, T D K , Foulds, I J , Green, P F , Tomley, F M , and Binns, M M (1987) Completion of the sequence of the genome of the coronavirus avian infectious bronchitis virus *J Gen Virol* **68**, 57-77

4 Cavanagh, D , Davis, P J , and Cook, J K A (1992) Infectious bronchitis virus evidence for recombination within the Massachusetts serotype *Avian Pathol* **21**, 401-408

5 de Groot, R J , van der Most, R G , and Spaan, W J M (1992) The fitness of defective interfering murine coronavirus DI-a and its derivatives is decreased by nonsense and frameshift mutations *J Virol* **66**, 5898-5905

6 Fosmire, J A , Hwang, K , and Makino, S (1992) Identification and characterization of a coronavirus packaging signal *J Virol* **66**, 3522-3530

7 Furuya, T , Macnaughton, T B , La Monica, N , and Lai, M M C (1993) Natural evolution of coronavirus defective-interfering RNA involves RNA recombination *Virology* **194**, 408-413

8 Jeong, Y S , and Makino, S (1994) Evidence for coronavirus discontinuous transcription *J Virol* **68**, 2615-2623

9 Kim, Y , Jeong, Y , and Makino, S (1993) Analysis of cis-acting sequences essential for coronavirus defective interfering RNA replication *Virology* **197**, 53-63

10 Lai, M M C , Baric, R S , Makino, S , Keck, J G , Egbert, J , Leibowitz, J L , and Stohlman, S A
    (1985) Recombination between nonsegmented RNA genomes of murine coronaviruses *J Virol* **56,**
    449-456

11 Lin, Y , and Lai, M M C (1993) Deletion mapping of a mouse hepatitis virus defective interfering RNA
    reveals the requirement of an internal and discontiguous sequence for replication *J Virol* **67,** 6110-6118

12 Makino, S , Taguchi, F , and Fujiwara, K (1984) Defective interfering particles of mouse hepatitis virus
    Virology **133,** 9-17

13 Makino, S , Fujioka, N , and Fujiwara, K (1985) Structure of the intracellular viral RNAs of defective
    interfering particles of mouse hepatitis virus *J Virol* **54,** 329-336

14 Makino, S , Shieh, C , Soe, L , Baker, S C , and Lai, M M C (1988) Primary structure and translation
    of a defective interfering RNA of murine coronavirus *Virology* **166,** 550-560

15 Makino, S , Yokomori, K , and Lai, M M C (1990) Analysis of efficiently packaged defective interfering
    RNAs of murine coronavirus localization of a possible RNA-packaging signal *J Virol* **64,** 6045-6053

16 Sethna, P B , Hofmann, M A , and Brian, D A (1991) Minus-strand copies of replicating coronavirus
    mRNAs contain antileaders *J Virol* **65,** 320-325

17 van der Most, R G , Bredenbeek, P J , and Spaan, W J M (1991) A domain at the 3' end of the
    polymerase gene is essential for encapsidation of coronavirus defective interfering RNAs *J Virol* **65,**
    3219-3226

18 Zhao, X , Shaw, K , and Cavanagh, D (1993) Presence of subgenomic mRNAs in virions of coronavirus
    IBV *Virology* **196, 172-178.**

91

HIGH RECOMBINATION AND MUTATION RATES IN MOUSE HEPATITIS VIRUS SUGGEST THAT CORONAVIRUSES MAY BE POTENTIALLY IMPORTANT EMERGING VIRUSES

Ralph S. Baric, Kaisong Fu, Wan Chen, and Boyd Yount

Department of Epidemiology
University of North Carolina
Chapel Hill, North Carolina 27599-7400

INTRODUCTION

Coronaviruses are common respiratory and gastrointestinal pathogens of mammals and birds. Not only do they cause about 15-20% of the common colds in humans, they are also occasionally associated with infections of the lower respiratory tract and central nervous system[1]. The prototype, mouse hepatitis virus (MHV), contains a 32 kb genomic RNA which encodes two large orfs at the 5' end, designated orf 1a and orf 1b. Orf 1b contains highly conserved polymerase, helicase and metal binding motifs typical of viral RNA polymerases while orf 1a contains membrane and cysteine rich domains, and serine-and poliovirus 3c-like protease motifs[1]. The large size of the genome coupled with it's unique replication strategy and high recombination frequencies during mixed infection predict a considerable capacity to evolve[1,2,3].

The majority of emerging RNA viruses are probably zoonotic pathogens that bridge the species barrier and spread into the human population. The probable emergence of HIV, Hantaan, and influenza viruses from zoonotic hosts suggests that this is a natural, almost predictable phenomenon, yet we know little about the molecular mechanisms mediating virus spread between species[3]. To address this question in coronaviruses, we calculated MHV polymerase error rates and RNA recombination frequencies. Further MHV's capacity to bridge the species barrier was confirmed by isolating variants which grow efficiently on nonpermissive baby hamster kidney cells in vitro.

Corona- and Related Viruses, Edited by P. J. Talbot and G. A. Levy
Plenum Press, New York, 1995

MATERIAL AND METHODS

Temperature sensitive (*ts*) mutants from the group F RNA$^+$ mutants (NC6, NC16), group E (LA18) and group C (NC3) RNA$^-$ mutants were used in this study[2,3]. MHV-A59/JHM was adapted to baby hamster kidney cells (BHK) by serial passage in progressively increasing ratio's of BHK/DBT mixed cultures.

RESULTS

Sequence Analysis of TS and Revertent Viruses

Identification of the mutant allele in *ts* virus in MHV is complicated by the large size of the viral genome including the ~22 kb polymerase region encoding at least 5 or more genetic functions[2]. To simplify this problem, we focused on the RNA$^+$ mutants which mapped in the S gene of MHV (NC6, NC16) and the RNA$^-$ mutants which mapped in the n-(NC3) or c-termini (LA18) of orf 1b[2,3]. Since infectious vectors are not available, revertent viruses were isolated to assist in the identification of the mutant allele. Revertent viruses had similar titers and were of the RNA$^+$ phenotype when assayed at both permissive and restrictive temperatures (Table 1).

Overlapping primer pairs were developed to clone 1.0-1.5 kb stretches of viral RNA in the MHV orf 1b and S genes using PCR[5]. To obviate PCR-induced mutations and circumvent high RNA polymerase error rates[4], the PCR product was reamplified by asymmetric PCR, and sequenced directly. For the RNA$^+$ mutants (NC6, NC16), the entire S glycoprotein gene was sequenced in *ts* and revertent viruses. For the RNA$^-$ mutants (LA18, NC3), the entire orf 1b region and the c-terminal 1-2 kb of orf 1a was sequenced.

NC6 contained a single nucleotide substitution at the 5' end of the S glycoprotein gene which resulted in an A to G transition at position 630 (Thr to Ala; amino acid 207). NC16 contained a single nucleotide change involving a C to G transversion at nt 2511 (Asp to Glu) in the S2 domain of the S glycoprotein gene. In revertents of NC16, the mutation had reverted back to wildtype sequence (Table 2).

Among the group E RNA$^-$ mutants, LA18 contained a single nucleotide substitution at nt 7100 in orf 1b. The G to A alteration led to a Arg to Lys change at amino acid 2286. In ten LA18 revertants, the mutation had reverted to the wildtype sequence. The group C mutant NC3 contained two alterations at nt 171 and 172; located upstream from the ribosomal

Table 1. *Ts* and revertant virus titers

Ts mutant	Titer at Temperature (°C)		
	32	39.5	39.5/32
LA18	6.1×10^7	2.4×10^3	4.0×10^{-5}
LA18R1	2.3×10^7	9.5×10^7	4.0×10^{-1}
NC3	2.4×10^8	5.5×10^4	2.3×10^{-4}
NC3R1	2.0×10^8	2.0×10^8	1.0×10^0
NC6	2.8×10^7	3.5×10^3	1.2×10^{-4}
NC6R1	2.3×10^7	2.4×10^7	1.0×10^0
NC16	8.5×10^7	2.3×10^4	2.7×10^{-4}

Table 2. Nucleotide locations of the MHV-A59 *ts* alleles

TS mutant	Location in genome	nt site	nt change wt[a]	nt change ts[b]/Rev[c]	AA change wt	AA change ts(Rev)
Group F						
NC6	S gene	620	A	G	Tyr	Cys
NC6R1	—	—	—	A	—	Tyr
NC16		2502	C	G	Asp	Glu
NC16R1	—	—	—	C	—	Asp
Group C						
NC3	ORF1a	171,172	GC	CG	Gly	Ala
NC3R1	—	—	—	GC	—	Gly
Group E						
LA18	ORF1b	7100	G	A	Arg	Lys
LA18R1	—	—	—	G	—	Arg

[a]wildtype.
[b]temperature sensitive mutant.
[c]revertant.

frameshifting site in orf 1a. The alteration (GC to CG) resulted in a Gly to Ala at amino acid 4438. In revertant virus, reversion to wildtype sequence had occurred.

Additional nucleotide changes were also found in *ts* and revertant viruses which were not tightly linked to the *ts* phenotype (data not shown).

Establishing Precise Recombination Rates in the MHV A59 Genome

All mutants were crossed 3-5 X to the other mutants used in the study and standardized to the standard cross (Table 3)[2,3]. To establish recombination frequencies between highly defined alleles, the nucleotide distances between individual mutants was divided by the recombination frequencies between each particular cross. Within the S gene (NC6 x NC16), the average recombination frequency was 1% RF/629±331 bp as compared to 1%/8979±1191 bp in the polymerase gene (LA18 x NC3). Between the 5' end of the S glycoprotein gene (NC6) and the 3' end of the polymerase gene (LA18), the average recombination frequency was about 1%/1054±139 bp, indicating that progressively increasing recombination frequencies were evident from the 5' to 3' end of the genome[3].

RNA Polymerase Error Rates during MHV Infection

MHV RNA polymerase error rates were calculated from the average reversion frequencies of several *ts* mutants and approached $1.32\pm0.89 \times 10^{-4}$ or about 2.4 mutations (range 1.5-7.2) per genome round of replication.

Table 3. Recombination frequencies between *ts* mutants[a]

Mutant	NC3	LA18	NC6	NC16
NC3	—	0.84 ± 0.21	5.50 ± 2.10	ND
LA18		—	0.90 ± 0.20	4.6 ± 0.60
NC6			—	2.8 ± 0.90
NC16				—

[a]mean ± standard deviation. ND, not done.

Table 4. Characterization of MHV BHK-adapted virus isolates

	Virus titers[a]	
	DBT	BHK
Virus stocks		
MHV-A59	7.3×10^8	0
MHV-H1	3.5×10^7	3.0×10^7
MHV-H2	1.0×10^8	7.5×10^7
Virus Growth[b]		
MHV-A59 (0h)	2.6×10^4	4.8×10^4
MHV-A59 (24h)	7.0×10^7	1.7×10^3
MHV-H2 (0h)	1.0×10^4	2.5×10^4
MHV-H2 (24h)	1.0×10^6	1.1×10^8

[a]virus stocks titered on DBT or BHK cells.
[b]cultures of DBT or BHK cells were infected with virus and titer on
DBT cells.

MHV Evolution in Vitro

The high rate of recombination coupled with a high RNA polymerase error rate
suggested that MHV has considerable potential to evolve rapidly in vitro. Since most
emerging viruses are thought to be zoonotic pathogens that bridge the species barrier[4], we

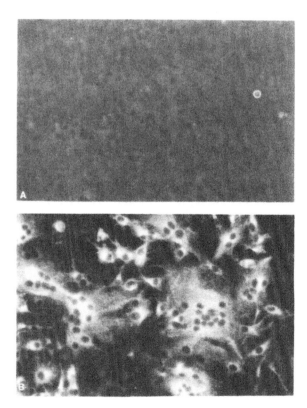

Figure 1. Expression of MHV-A59 (Panel A) and MHV-A59H2 (Panel B) viral antigen in BHK cells.

examined the capacity of MHV to bridge the species barrier and grow in nonpermissive BHK cells. Cultures of mixed BHK/DBT cells were infected with MHV and the progeny serially passaged in progressively increasing ratios of BHK to DBT cells (1 to 10, to 5 to 1). Following additional selection, two MHV variants (MHV-H1, MHV-H2) were isolated which replicated, expressed viral antigen, fused BHK cells, and plaque in BHK and DBT cells efficiently (Figure 1) (Table 4). Under identical conditions, MHV-A59 wild type virus did not replicate or express significant quantities of viral protein in BHK cells.

DISCUSSION

Ts and revertant viruses were sequenced to identify specific mutations which correlated with the *ts* phenotype, map the location of the MHV complementation groups, determine RNA polymerase error rates and precise RNA recombination frequencies throughout the genome. While this approach suffers from the practical limits of sequencing the entire 32 kb genome, each mutation was uniquely present in *ts* but not in wildtype or revertent viruses, and mapped to a domain as predicted by standard genetic recombination mapping techniques[2]. Although more *ts* mutants must be sequenced, these data suggest that the group F RNA+ mutants map within the S glycoprotein gene of MHV while the group E RNA- mutants, which are defective in positive strand synthesis after temperature shift, reside near the c-terminus of orf 1b. The group C mutants, which are defective in negative but not positive strand RNA synthesis, probably map at the orf 1a/orf 1b junction.

It is well documented that viral RNA-dependent RNA polymerases do not contain proofreading activities and have high mutation rates approaching 10^{-3} to 10^{-5} per round of replication[4]. The resulting mixed virus populations, quasi-species, is probably an important mechanism of RNA virus evolution and spread between species. From the reversion frequencies of highly plaque purified *ts* mutants, we determined the average MHV RNA polymerase error rate to approach $1.32 \pm 0.89 \times 10^{-4}$, or about 2.4 mutations per genome round of replication (range, 1.5-7.4). High mutation rates in a genome that is 3-4 X larger than most other positive-stranded RNA viruses suggests that coronaviruses probably exist in large quasispecies populations, providing vast reservoirs of natural virus variants.

Previous studies in our laboratory and others have suggested that RNA recombination occurs at high frequency and varies in different portions of the MHV genome[1,2,3]. The physical map locations of several MHV *ts* mutants provides accurate estimates of the homologous RNA recombination frequencies in the MHV genome which approach 1%/8979±1191 bp or about 4.9% throughout the 22 kb polymerase region (assuming reciprocal crosses). Between the polymerase and S glycoprotein genes, the recombination frequency is significantly increased, and approaches 1%/1054±139 bp (~3.8% in the 2.0 kb p30/HE genes). In the 4.0 kb S glycoprotein gene, RNA recombination rates approach 1%/629±331 bp (~12.7%). If recombination frequencies at the 3' end of the genome approaches or exceeds rates measured in the S gene, the overall recombination rate in MHV probably exceeds 25%. Currently, the most likely model to explain the high, progressively 5' to 3' increased recombination rate in MHV is from: (1) the large size of the genome, (2) discontinuous transcription, and (3) the presence of transcriptionally active full and subgenomic length positive- and negative-stranded RNAs which increase the amount of template for strand switching[2,3].

The high rate of coronavirus RNA recombination coupled with high RNA polymerase error rates provides coronaviruses with a natural mechanism for rapid antigenic variation and evolution, especially within the highly immunogenic structural genes. It is clear that MHV can rapidly alter its species specificity and infect rats and primates; the resulting virus variants are associated with demyelinating diseases in these alternative species[5]. Newly recognized animal coronaviruses include porcine epidemic diarrhea coronaviruses which probably evolved by mutation and recombination from the human coronavirus 229E[6]. We have also isolated MHV variants which efficiently replicate not only in BHK cells, but to a lesser extent, have retained their natural ability to replicate in mouse cells in vitro. These data further demonstrate the capacity of coronaviruses to evolve rapidly and bridge the species barrier in vitro. Using these variants, we can map determinants which mediate species specificity, map MHV virulence markers in the hamster and/or mouse, determine if the BHK CE homologue or other cellular proteins acts as a receptor in alternative hosts, and determine evolution rates in structural and nonstrucural genes. The data presented in this paper demonstrates that coronaviruses can evolve rapidly and transverse the species barrier in vitro.

ACKNOWLEDGMENTS

This research was supported by grants from the American Heart Association (AHA 90-1112), the NIH (AI-23946) and was performed during the tenure of an Established Investigator Award from the American Heart Association (RSB) (AHA 89-0193).

REFERENCES

1. Lai, M.M.C. 1990. Coronavirus: organization, replication and expression of genome. Annu. Rev. Microbiol. 44, 303-333.
2. Baric, R.S. Fu, K.S., Schaad, M.C. and Stohlman, S.A. 1990. Establishing a genetic recombination map for MHV-A59 complementation groups. Virology 177: 646-656.
3. Fu, K. and Baric, R.S. 1992. Evidence for variable rates of recombination in the MHV genome. Virology 189:88-102.
4. Morse, S.S. 1994. The viruses of the future? Emerging viruses and evolution. In: Morse, S.S. (eds), The Evolutionary Biology of Viruses, Raven Press, Ltd, New York.
5. Murray, R.S., et al., 1992. Coronavirus infects and causes demyelination in the primate central nervous system. Virology 188: 274-284.
6. Brigden, A., M. Duarte, K. Tobber, H. Laude, and M. Ackermann. 1993. The nucleocapsid protein gene of the porcine epidemic diarrhoea virus confirms that this virus is a coronavirus related to HCV 229E and TGEV. J. Gen. Virol. 74:1795-1804.

92

HEPATITIS MUTANTS OF MOUSE HEPATITIS VIRUS STRAIN A59

S. T. Hingley[1], J. L. Gombold[2], E. Lavi,[3] and S. R. Weiss[4]

[1] Department of Microbiology and Immunology
Philadelphia College of Osteopathic Medicine
Philadelphia, Pennsylvania
[2] Department of Microbiology and Immunology
Louisiana State University Medical Center
Shreveport, Louisiana
[3] Department of Pathology
[4] Department of Microbiology
University of Pennsylvania
Philadelphia, Pennsylvania

ABSTRACT

MHV-A59 causes acute meningoencephalitis and hepatitis in susceptible mice, and a persistent productive, but nonlytic, infection of cultured glial cells. We have shown previously that viruses isolated from persistently infected glial cell cultures have a fusion-defective phenotype and were impaired in their abilities to cause hepatitis compared to wild-type MHV-A59. Two mutants chosen for detailed study, B11 and C12, display two distinct hepatitis phenotypes. The ability of B11 to replicate in the liver was dependent on infectious dose and route of inoculation, while C12 consistently displayed decreased liver titers regardless of dose and route of inoculation. Sequence analysis of wild-type, mutant and revertant S proteins indicates that 1) a mutation in the N terminal subunit of S, resulting in a glutamine to leucine amino acid substitution (Q159L), may affect ability to cause hepatitis and 2) a cleavage site mutation (H716D) which determines fusogenicity is not responsible for the altered hepatitis phenotype. Sequence analysis indicated that hepatitis-producing revertants did not revert at mutation Q159L, although it is possible that a mutation in the heptad repeat domain of S2 may compensate for the mutation in S1. Since B11, C12 and a nonattenuated fusion mutant (B12) have identical S protein sequences, there must be additional mutations outside of S which influence both virulence and ability to replicate in the liver.

Corona- and Related Viruses, Edited by P. J. Talbot and G. A. Levy
Plenum Press, New York, 1995

INTRODUCTION

The murine coronavirus, mouse hepatitis virus strain A59 (MHV-A59), produces both hepatitis and neurological disease in susceptible mice. Neurological disease includes both acute meningoencephalitis and chronic demyelinating disease. *In vitro*, MHV-A59 causes a lytic infection in mouse fibroblast cells, but a persistent productive, nonlytic, infection of cultured glial cels.

We have previously characterized fusion-defective mutants of MHV-A59 which were derived from persistently infected glial cell cultures[1]. The fusion defect was shown to result from the substitution of an aspartic acid residue for a histidine at the cleavage site of the MHV fusion glycoprotein, or spike (S) protein. This point mutation prevented cleavage of the S protein during the maturation of virions, and correlated with lack of wild-type syncytia formation during infection of mouse fibroblast cells[1]. The fusion mutants were shown to be attenuated *in vivo* and have diminished capacity to cause hepatitis[2]. Interestingly, two distinct hepatitis phenotypes were observed. The mutant B11 appeared to be blocked in its spread from the central nervous system (CNS) to the liver, while the mutant C12 was inhibited in its ability to replicate in hepatocytes[2].

The S protein is responsible for viral attachment, and mediates binding to members of the carcinoembryonic antigen gene family[3,4]. Since the S protein would presumably be involved in determination of viral tropism, the sequence of wild-type (WT) and mutant S protein was compared to determine whether mutations in this glycoprotein might account for inhibition of viral replication in the liver. Aside from the cleavage site mutation (H716D), which did not affect hepatotropism, the mutants contained one additional point mutation in S1 at amino acid 159, which produced a glutamine to leucine amino acid substitution[1] (Q159L). An association between the Q159L mutation and impaired growth in the liver supports the hypothesis that the mutation in the S1 subunit may play a role in determining ability to produce hepatitis.

METHODS

S Gene Sequences

Sequencing was performed using the fmol sequencing system (Promega) to sequence overlapping DNA fragments of the S gene obtained by reverse transcriptase-PCR[1].

Viral Replication in Infected Tissue

C57BL/6 mice were infected either intracranially (ic) or intrahepatically (ih) with 5,000 plaque forming units (PFU) of virus as described previously[2]. Virus in homogenized brain or liver tissue was titered, in duplicate, by plaque assay on L2 cell monolayers in 6-well microtiter plates. Results were expressed as the log of the PFU/g of tissue. The limit of detection in this assay was 2.7 log PFU/g.

Isolation of Hepatitis-Producing Revertants

Revertants were plaque purified three times from a liver homogenate of a mouse infected with virus obtained after eight *in vivo* passages of C12 through the livers of C57Bl/6 mice. A revertant hepatitis phenotype was confirmed by measuring replication in the liver following both ic and ih inoculation.

RESULTS AND DISCUSSION

The Q159L mutation first appeared in virus plaque purified from two independent persistently infected glial cell cultures at 6 weeks post infection (pi), while the H716D mutation did not appear until 12 weeks pi (Table 1). The appearance of this point mutation coincided with the inability to replicate in the liver[2].

The hepatitis phenotype, S gene sequence, and fusogenicity of different MHV-A59 mutants are listed in Table 2. With the exception of B11, which replicates in the liver following intrahepatic (ih) inoculation but not intracranial (ic) inoculation, a loss of hepatotropism is associated with a leucine residue at amino acid position 159. As demonstrated previously, the cleavage site mutation at amino acid position 716 correlates with fusogenicity but not inhibition of growth in the liver[1,2]. It is not known what site(s) account for the two distinct hepatitis phenotypes exhibited by B11 and C12. However, the difference between these two mutants does not map to the S gene, since B11 and C12 have identical S gene sequences. Additional experiments with early B mutants, such as B7, will determine whether their hepatitis phenotypes resemble B11, or B10 and B12.

To try to substantiate the hypothesis that the mutation in S1 helps to determine tissue specificity for MHV, we derived hepatitis-producing revertants of the mutant C12. Three such revertants (LR1, LR2, and LR3) demonstrated varying degrees of growth in the liver in mice inoculated both ic and ih. As shown in Table 3, LR2 and LR3 titers from ic- and

Table 1. Amino acid sequence at positions 159 and 716
of B and C culture mutants

Virus	Weeks pi	Codon 159	Codon 716
B1	4	Q	H
B2		Q	H
B3		Q	H
B4	8	L	H
B6		L	H
B7	12	L	H
B8		L	D
B9		L	D
B10	16	L	D
B11		L	D
B12		L	D
C1	4	Q	H
C2		Q	H
C3		Q	H
C4	8	L	H
C5		L	H
C6		L	H
C7	12	L	D
C8		L	D
C9		L	D

Table 2. S sequence, fusogenicity and growth in the liver of variants
of MHV-A59

Virus	aa 159	aa 716	Cell Fusion[a]	Growth in Liver[b]	
				ic	ih
MHV-A59	Q	H	+	+	+
B1	Q	H	+	+	ND
B7	L	H	+	−	ND
B10	L	D	−	−	−
B11	L	D	−	−	+
B12	L	D	−	−	−
B11R5	L	Y	+	−	ND
C3	Q	H	+	ND	+
C5	L	H	+	ND	−
C8	L	D	−	ND	−
C12	L	D	−	−	−
C12R1	L	H	+	−	ND
C12R3	L	G	+	−	ND

[a]Cell fusion was scored "+" if the virus produced syncytia formation with
wild-type kinetics, and "−" if the virus displayed delayed fusion kinetics
[b]The ability to replicate in the liver was assessed after ic and/or ih
inoculation If viral titers were within 1 log of WT levels, the virus was
scored "+" for growth in liver, if titers resembled C12 levels (2 logs less
than WT), the virus was scored "−" for growth in liver.
ND, not determined

ih-inoculated mice, taken at 4 days post inoculation (dpi), were comparable to WT levels;
viral titers from LR1-infected animals were intermediate between C12 and WT levels.

The sequence of the S proteins of LR1, LR2, and LR3 was compared to that of C12
and MHV-A59 (Table 4). All three LR viruses retained the mutations present in the C12 S
gene at amino acid residues 159 and 716. LR1 differed from C12 at positions 839 (a proline
to leucine substitution) and 1035 (a glutamic acid to asparagine substitution). LR2 and LR3
both contained a conservative arginine to histidine substitution at position 654 and, like LR1,
the glutamic acid to asparagine substitution as position 1035. In addition, LR3 substituted a
serine for a phenylalanine at amino acid residue 371. The positions of these point mutations

Table 3. *In vivo* replication of hepatitis-producing
revertants of C12

Virus	ic inoculation[a]		ih inoculation[a]
	Brain	Liver	Liver
WT	7 0	5 0	6 2
LR1	6 7	3 6	5 5
LR2	6 8	5 2	6.9
LR3	7 4	6.5	7 4
C12	7 8	<2.7	4.1

[a]Viral titers from ic- or ih-infected mice taken at 4 dpi Each
titer is the mean from duplicate mice and expressed as the
$\log_{10}$pfu/g

Table 4. S gene sequence of hepatitis-producing revertants of C12

Virus	Amino acid codon					
	159	371	654	716	839	1035
A59	GLN	PHE	ARG	HIS	PRO	GLU
C12	LEU			ASP		
LR1	LEU			ASP	LEU	ASN
LR2	LEU		HIS	ASP		ASN
LR3	LEU	SER	HIS	ASP		ASN

in S relative to the defined hypervariable domain, cleavage site, and heptad repeat domains, is shown schematically in Figure 1

Although none of the three LR viruses contained a mutation at or near amino acid residue 159, the site proposed to be involved in determining growth in the liver, they all contained the same mutation at position 1035, in heptad repeat domain 1 It is conceivable that a mutation in the heptad repeat region could compensate for a mutation in S1 by affecting the conformation of S

The premise that there is an interaction between regions in S1 and S2 has been proposed by Daniel et al [5], whose data suggest that an immunodominant functional domain in the S2 subunit is part of a complex three-dimensional structure, possibly involving sites in S1 Furthermore, it has been noted that monoclonal antibodies (MAbs) which can compete with each other have been mapped to different portions of the S protein[6], and that mutations in antibody escape mutants have been mapped to sites outside of the binding site of the MAb used to generate these mutants[7 8] These observations support the suggestion that epitopes in S1 and S2 subunits are adjacent to each other and interact with each other in the native protein

Therefore, the possibility exists that sites in both S1 and the heptad repeat domain of S2 could determine MHV tissue tropism Pertinent to this hypothesis is the observation by Grosse and Siddell[7] that monoclonal antibody-resistant mutants of a MAb with specificity for an epitope in S1 had point mutations which mapped adjacent to the second heptad repeat domain These investigators suggest that the proper spatial arrangement of the S1 and S2 subunits is crucial for the biologic functions of the S protein To directly assess the potential role of these two sites in producing hepatitis requires a system for generating recombinant viruses with defined mutations in the S genes Such a system for engineering recombinant viruses using defective interfering particles as vectors for recombination is being developed in several laboratories[9 10] If successful, it would be interesting to construct MHV-A59 and C12 variants with mutations at amino acid positions 159 and 1035, and look for an affect on viral replication in the liver

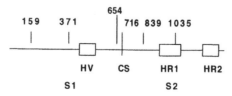

Figure 1. Schematic of S showing the approximate position of the point mutations of the LR viruses relative to the hypervariable region (HV) cleavage site (CS) and heptad repeat domains (HR1 and HR2) of S

REFERENCES

1 Gombold, J L , Hingley, S T , Weiss, S R Fusion-defective mutants of mouse hepatits virus A59 contain a mutation in the spike protein cleavage signal J Virol 1993,67 4504-4512

2 Hingley, S T , Gombold, J L , Lavi, E , Weiss, S R MHV-A59 fusion mutants are attenuated and display altered hepatotropism Virology 1994,200 1-10

3 Dveksler,G S , Dieffenbach, C W , Cardellichio, C B , McCuaig,K , Pensiero, M N , Jiang, G -S , Beauchemin, N , Holmes, K V Several members of the mouse carcinoembryonic antigen-related glyco-protein family are functional receptors for the coronavirus mouse hepatitis virus-A59 J Virol 1993,67 1-8

4 Williams, R K , Jiang, G -S , Holmes, K V Receptor for mouse hepatitis virus is a member of the carcinoembryonic antigen family of glycoproteins Proc Natl Acad Sci USA 1991,88 5533-5536

5 Daniel, C , Anderson, R , Buchmeier, M J , Fleming, J O , Spaan, W J M , Wege, H , Talbot, P J Identifi-cation of an immunodominant linear neutralization domain on the S2 portion of the murine coronavirus spike glycoprotein and evidence that it forms part of a complex tridimensional structure J Virol 1993,67 1185-1194

6 Wege, H , Dorries,R Wege, H Hybridoma antibodies to the murine coronavirus JHM characterization of epitopes on the peplomer protein (E2) J Gen Virol 1984,65 1931-1942

7 Grosse, B , Siddell, S G Single amino acid changes in the S2 subunit of the MHV surface glycoprotein confer resistence to neutralization by S1 subunit-specific monoclonal antibody Virology 1994,202 814-824

8 Wang, F I , Fleming, J O , Lai, M M C Sequence analysis of the spike protein gene of murine coronavirus variants study of genetic sites affecting neuropathogenicity Virology 1992,186 742-749

9 Koetzner, C A , Parker M M , Ricard, C S , Sturman, L S , Masters, P S Repair and mutagenesis of the genome of a deletion mutant of the coronavirus mouse hepatitis virus by targented RNA recombination J Virol 1992,66 1841-1848

10 van der Most, R G , Heijnen, L , Spaan, W J M , de Groot R J Homologous RNA recombination allows efficient introduction of site -specific mutation into the genome of coronavirus MHV-A59 via synthetic co-replicating RNAs Nucleic Acids Res 1992,20 3375-3381

93

STRUCTURE AND ENCAPSIDATION OF TRANSMISSIBLE GASTROENTERITIS CORONAVIRUS (TGEV) DEFECTIVE INTERFERING GENOMES

Ana Méndez, Cristian Smerdou, Fátima Gebauer, Ander Izeta, and Luis Enjuanes

Department of Molecular and Cellular Biology
Centro Nacional de Biotecnología
Campus Universidad Autónoma
Canto Blanco
Madrid 28049, Spain

ABSTRACT

Serial undiluted passages were performed with the PUR46 strain of TGEV in swine testis (ST) cells. Total cellular RNA was analyzed at different passages after orthophosphate metabolic labeling. Three new defective RNA species of 24, 10.5, and 9.5 kb (DI-A, DI-B, and DI-C respectively) were detected at passage 30, which were highly stable and significantly interfered with helper mRNA synthesis in subsequent passages. By Northern hybridization DIs A, B, and C were detected in purified virions at amounts similar to those of helper RNA. Standard and defective TGEV virions could be sorted in sucrose gradients, indicating that defective and full-length genomes are independently packaged. cDNA synthesis of DI-B and DI-C RNAs was performed by the reverse transcription-polymerase chain reaction (RT-PCR) to give four fragments in each case. Cloning and sequencing of the DI-C PCR products showed that the smallest DI particle comprises 9.5 kb and has 4 discontinuous regions of the genome. It contains 2.1 kb from the 5'-end of the genome, about 7 kb from gene 1b, the first 24 nucleotides of the S gene, 12 nucleotides of ORF 7, and the 0.4 kb of the UTR at the 3'-end.

INTRODUCTION

In recent years, the assembly of full-length, infectious cDNA clones of a number of positive-strand RNA viruses has greatly facilitated molecular genetic analyses of viral proteins and *cis* regulatory elements[9]. The large size of coronavirus genomes (26 to 31 kb),

Corona- and Related Viruses, Edited by P. J. Talbot and G. A. Levy
Plenum Press, New York, 1995

however, has been a constraint on this approach. The generation of coronavirus defective RNAs, containing the *cis* signals required for replication but dependent on viral replicase functions to be supplied in *trans*, was exploited to circumvent this limitation. cDNA constructs of small Mouse Hepatitis Virus (MHV) defective RNAs have been successfully employed to define *cis*-acting signals required for replication[2, 4], transcription[2, 4, 11, 13] and encapsidation of the virus[6]. Their manipulation has also permitted the introduction of site-specific mutations into the 3'-end of the genome by homologous RNA recombination[6, 12, 7], and the engineering of chimeric viruses expressing heterologous genes[4, 3]. Defective interfering genomes have, in addition, a growing potential in protection against viral disease.

To date, coronavirus defective RNAs have been reported only for MHV[5] and Infectious Bronchitis Virus (IBV)[8]. In this chapter we report the characterization, cloning and sequencing of naturally occurring defective RNAs of TGEV which are replicated and packaged into virus particles.

MATERIALS AND METHODS

Virus and Cells

The plaque-cloned PUR46 strain of TGEV was used to perform the serial undiluted passages in ST cells.

Radiolabeling of Intracellular Viral RNA and Agarose Gel Electrophoresis

Briefly, ST cell monolayers were infected with viruses from different passages at a multiplicity of infection of 10 and labeled with ^{32}Pi from 6 to 9 hours post infection in the presence of 2.5 µg/ml of actinomycin D. Cytoplasmic extracts were prepared by lysing the cells in TSM buffer (0.15M NaCl, 0.01M Tris-hydrochloride [pH 7.6], 5mM MgCl$_2$) containing 0.2% Nonidet P-40, and pelleting nuclei by centrifugation at 13000 g, 30 sec. RNA was isolated by the addition of urea-SDS lysis buffer (1.5% SDS, 15mM EDTA, 0.24M NaCl, 0.04M Tris-hydrochloride [pH 7.6], 8M urea) and phenol-chloroform extraction. Gel electrophoresis was conducted after denaturation of RNA with formaldehyde.

Purification of Virions through Sucrose Layers

Virus was harvested at 12 hpi from 12 roller bottles which had been infected with PUR46p41 at m.o.i. 10. Virus was clarified, and divided into two portions. One of them was underlaid with 15% sucrose and the other with 31% sucrose. Samples were centrifuged at 84 000 g for 2 hours. RNA was extracted from pelleted virions in each case.

Separation of Virions in a Linear Sucrose Gradient for RNA Extraction

Virus was harvested at 12 hpi from 12 roller bottles which had been infected with PUR46p41 at m.o.i. 10. Virus harvests were clarified and concentrated by pelleting the virus through 15% sucrose. Pelleted virus was resuspended in TEN-Tween 20 (0.01 M Tris-HCl pH=7.4, 1 mM EDTA, 1M NaCl, 0.2% v:v Tween 20) and placed on top of a 15-42% linear sucrose gradient. Sample was centrifuged at 84000 g for two hours. After fractionation of the gradient, each fraction was diluted and virions were pelleted by centrifugation in the above conditions. RNA from virions in each fraction was extracted and analyzed by Northern blot.

RT-PCR Amplification of DI-B and DI-C RNAs

DI-B and DI-C RNAs from purified virions from passage 41 were amplified by RT-PCR in four separate fragments. As the complete sequence of TGEV genome was not known, primers were derived from high homology sequence regions among HCV 229E and FIPV. Virus sequences were kindly provided by S.Siddel and R. de Groot, respectively.

RESULTS

Detection of Defective RNAs in ST Cells

PUR46 strain of TGEV was passaged undiluted 45 times in ST cells. RNAs induced by TGEV infection at several passage levels were analyzed in a denaturing gel after [^{32}P] orthophosphate metabolic labeling. At passage 30 three new subgenomic RNAs of 24, 10.5, and 9.5 kb (DI-A, DI-B and DI-C) were detected (Fig. 1). These defective RNAs persisted in increasing amounts during 15 subsequent passages in ST cells, significantly interfering with helper genomic and subgenomic RNA synthesis. When ten more undiluted passages

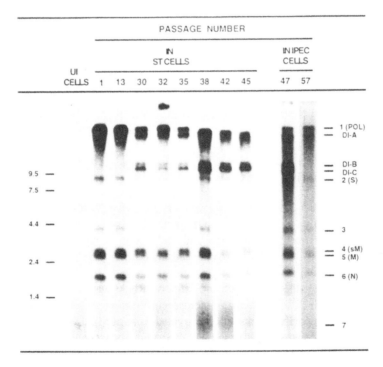

Figure 1. Electrophoretic analysis of metabolically labeled TGEV RNAs Cytoplasmic RNA at different passage levels was extracted and separated in denaturing agarose gels after [^{32}P] orthophosphate metabolic labeling Passage numbers are indicated at the top of each lane Bars on the right show TGEV mRNAs and DIs, and bars on the left the position of RNA molecular weight markers (GibcoBRL) Three new RNA species of 24, 10 5 and 9.5 kb (DI-A, B, and C) were detected at passage 30 and persisted at least after 15 additional passages in ST cells DI-B and DI-C were lost after 10 undiluted passages in a different cell line (IPEC)

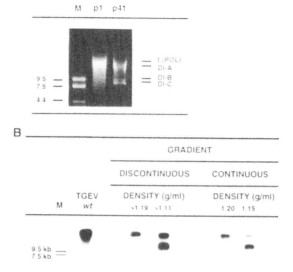

Figure 2. Analysis of RNA from passage 41 purified virions. **A.** Virions from passages 1 and 41 were purified by centrifugation through 15%w/v sucrose. RNA was extracted and analyzed in an ethidium bromide stained agarose gel. At passage 41 defective RNAs A, B, and C were detected in addition to *wt* RNA. **B.** Virions were purified by centrifugation through 31% w/v or 15% w/v sucrose cushions, or through a continuous sucrose gradient (15%-42%). RNA from pelleted virions was extracted, separated in denaturing agarose gels, and hybridized to a leader specific probe. Only *wt* genomes were detected in virions with densities >1.19 g/ml, but defective RNAs were also present in virions with lower densities (>1.11 g/ml). Both types of virions, standard and defective, could be sorted in a continuous gradient. RNA from virions at passage 1 was used as a marker.

were performed on intestinal epithelial cells (IPEC), only DI-A persisted, whereas DIs B and C were lost. This suggests that host cell factors may affect the stability of defective particles.

To determine whether defective RNAs had the 5' and 3' end of the genome Northern hybridization was performed with [^{32}P]-labeled oligonucleotides complementary to the leader and the 3'-UTR sequences. Both oligonucleotides hybridized to all *wt* mRNAs and also detected DIs A, B, and C. As a first approach to study which genomic sequences were present in the DIs, total RNA from ST-infected cells was also probed with oligonucleotides mapping in S, M, and N structural protein genes. None of them detected the DIs. These results indicate that DIs A, B, and C contain genomic 5' and 3' sequences, but lack structural genes (data not shown).

Packaging of Defective RNAs

To study whether defective RNAs were encapsidated, virions from passages 1 and 41 were partially purified by centrifugation through a 15% w/v sucrose cushion. RNA from pelleted virions was extracted in each case and analyzed in an ethidium bromide gel. (Fig. 2A). Defective RNAs A, B, and C were detected in addition to full-length RNA in virions from passage 41.

To determine whether full-length and defective RNAs were independently packaged, centrifugation through different density sucrose cushions as well as continuous sucrose gradients were used to purify virions. RNA extracted from purified virions was analyzed by Northern blot using a leader specific probe (Fig. 2B). Only the *wt* genome was detected when 31% w/v sucrose (1.19 g/ml) was used. However when the sucrose density was decreased to 15% w/v (1.11 g/ml) both the full-length and the defective RNAs were present. These data indicate that DI RNAs are packaged in defective virions, which differ in density from the standard ones. An almost complete separation of *wt* and defective virions was achieved in continuous sucrose gradients (15%-42%). Lower fractions (density 1.20 g/ml) were enriched in standard virions, while defective virions were concentrated in upper fractions (density 1.15 g/ml).

Cloning and Sequencing of DIs B and C

DIs B and C RNA from passage 41 purified virions were used together as templates for cDNA synthesis by RT-PCR. Due to their large size DIs could not be amplified using primers specific for the ends of the genome. Four different sets of primers were designed to amplify DI-B and C RNAs in four overlapping fragments (I, II, III, and IV from the 5' to the 3' end). PCR products were cloned. Fragments I and II were common to DIs B and C, but clones of two different sizes were found for fragments III and IV which account for the difference in length of DIs B and C (data not shown). The DI-C four fragments were sequenced from at least two independent RT-PCR reaction clones.

The genetic structure of DI-C was determined and is summarized in Fig. 3. It has about one third of the genome length and includes: i) 2.1 kb from the 5'-end; ii) ORF1b almost complete but with a small internal deletion; iii) the first 24 nucleotides of S gene; iv) the last 12 nucleotides of the ORF 7; and, v) the 3'-end UTRs. Two main deletions remove most of ORF 1a and the structural and 3' non structural protein genes. Preliminary data indicate that DI-B has a similar genetic structure than DI-C, but differs in the size and the precise location of the main 3'-end deletion and the internal deletion in ORF 1b (data not shown).

DISCUSSION

The frequency of DI particle generation in TGEV is lower than in MHV. While in MHV defective RNAs are easily detected after a few passages, in TGEV novel RNA species were only identified at late passages in one out of four isolates assayed. TGEV strains PUR46-MAD, HOL87, and PUR46*mar*1CC12 were passaged 35 times in ST cells at a multiplicity of infection close to 100 PFU/cell. An uncloned isolate, PUR46SW37CC4, was serially passaged in IPEC cells [10, 1]. Defective RNAs were only detected in PUR46-MAD

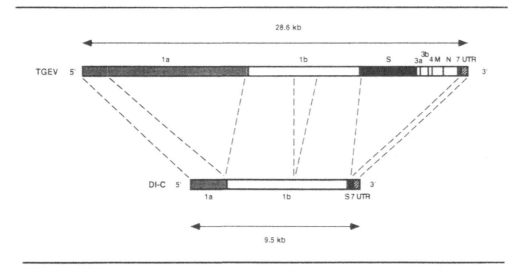

Figure 3. Diagram summarizing the structure of TGEV DI-C. DI-C, the smallest TGEV defective RNA isolated, contains 4 discontinuous regions of the TGEV genome. It comprises 2.1 kb from the 5' end, almost complete ORF 1b including the 1a-1b overlapping region, the beginning of S gene, incomplete ORF 7, and the UTRs.

strain after 30 passages. This is in contrast to what has been described in MHV, where seven JHM defective RNAs of different size were characterized from passage 3 through 22 in DBT cells.

In addition, MHV defective RNAs naturally evolve early after generation, whereas TGEV defective RNAs, once generated, were highly stable. DIs A, B, and C persisted along 15 passages in ST cells, and no changes in size or the generation of novel RNA species were observed. This difference in the rate of generation and evolution of naturally occurring DIs in both viruses may be due to a more accurate replication of TGEV RNA.

Interference of standard TGEV-PUR46 replication by DI particles was observed after passage 30. The levels of genomic and subgenomic RNAs were reduced from passage 32 to 45, as the defective RNAs were enriched. DI-C (9.5 Kb) is the most abundant RNA at late passages. One possible mechanism of interference is through competition between standard virus RNA and DI RNA for the viral polymerase. There may be other reasons for the selective advantage of DIs, such as secondary or tertiary RNA structure, or encapsidation efficiency.

The present study clearly indicates that DI and standard RNAs are independently encapsidated. It can not be excluded that more than one DI genome is encapsidated per virion, although the clear density difference between virions with full length or defective genomes suggests that only one or two RNA molecules are present in defective virions.

Cloning and sequencing of DIs B and C revealed that they comprise 4 discontinuous regions of the TGEV genome. They are similar in primary structure to both MHV and IBV defective interfering RNAs, but their length and structure make them more comparable to CD-91, a defective genome of 9.1 kb isolated from IBV. However, DIs B and C include most of gene 1b, including the 3' end of gene 1b that DI CD-91 from IBV lacks. In this region of MHV a 61-nt packaging signal has been identified. The precise location of TGEV RNA encapsidation signal is not known.

The 5' end of gene 1a region of DIs B and C (2.1 kb) is larger than the IBV gene 1a region present in DI CD-91 (1.1 kb), but also lacks the 0.2-kb region of the MHV gene 1a that is located about 3.2 kb from the 5' end of the MHV genome and which has been found essential for MHV DI replication in some strains. This region may not be required for TGEV replication, or a homologous region may be located elsewhere. The ORF 1a-1b overlapping region was maintained in both DI B and C, including the pseudoknot motif necessary for ribosomal frameshift. At present, it is not known if this motif may contribute to the fitness of these long DIs.

ACKNOWLEDGMENTS

This work has been supported by grants from the Consejo Superior de Investigaciones Científicas, the Comisión Interministerial de Ciencia y Tecnología, Instituto Nacional de Investigaciones Agrarias, La Consejería de Educación y Cultura de la Comunidad de Madrid, and Laboratorios Sobrino (Cyanamid) from Spain, and the European Communities (Projects Science and Biotech). Ana Méndez and Ander Izeta received fellowships from the Comunidad de Madrid and Gobierno Vasco.

REFERENCES

1 Gebauer, F., Posthumus, W A P , Correa, I , Suñe, C., Sanchez, C M., Smerdou, C , Lenstra, J. A., Meloen, R., Enjuanes, L Residues involved in the formation of the antigenic sites of the S protein of transmissible gastroenteritis coronavirus Virology 1991, 183: 225-238

2 Kim, Y N , Lai, M M C , Makino, S Generation and selection of coronavirus defective interfering RNA with large open reading frame by RNA recombination and possible editing Virology 1993, 194 244-253

3 Liao, C -L , Lai, M M C The requirement of 5′-end genomic sequence as an upstream cis-acting element for coronavirus subgenomic mRNA transcription J Virol 1994, 68 4727-4737

4 Lin, Y J , Lai, M M C Deletion mapping of a mouse hepatitis virus defective interfering RNA reveals the requirement of an internal and discontiguous sequence for replication J Virol 1993, 67 6110-6118

5 Makino, S , Fujioka, N , Fujiwara, K Structure of the intracellular viral RNAs of defective interfering particles of mouse hepatitis virus J Virol 1985, 54 329-336

6 Makino, S , Yokomori, K , Lai, M M C Analysis of efficiently packaged defective interfering RNAs of murine coronavirus - localization of a possible RNA-packaging signal J Virol 1990, 64 6045-6053

7 Masters, P S , Koetzner, C A , Kerr, C A , Heo, Y Optimization of targeted RNA recombination and mapping of a novel nucleocapsid gene mutation in the coronavirus mouse hepatitis virus J Virol 1994, 68 328-337

8 Penzes, Z , Tibbles, K , Shaw, K , Britton, P , Brown, T D K , Cavanagh, D Characterization of a replicating and packaged defective RNA of avian coronavirus infectious bronchitis virus Virology 1994, 203 286-293

9 Racaniello, V R , Baltimore, D Cloned poliovirus cDNA is infectious in mammalian cells Science 1981, 214 916-919

10 Sanchez, C M , Jimenez, G , Laviada, M D , Correa, I , Suñe, C , Bullido, M J , Gebauer, F , Smerdou, C , Callebaut, P , Escribano, J M , Enjuanes, L Antigenic homology among coronaviruses related to transmissible gastroenteritis virus Virology 1990, 174 410-417

11 Van der Most, R G , De Groot, R J , Spaan, W J M Subgenomic RNA synthesis directed by a synthetic defective interfering RNA of mouse Hepatitis virus a study of Coronavirus transcription initiation J Virol 1994, 68 3656-3666

12 Van der Most, R G , Heijnen, L , Spaan, W J M , Degroot, R J Homologous RNA recombination allows efficient introduction of site-specific mutations into the genome of coronavirus MHV-A59 via synthetic coreplicating RNAs Nuc Ac Res 1992, 20 3375-3381

13 Zhang, X , Liao, C L , Lai, M M C Coronavirus leader RNA regulates and initiates subgenomic mRNA transcription both in *trans* and in *cis* J Virol 1994, 68 4738-4746

94

MUTATIONS ASSOCIATED WITH VIRAL SEQUENCES ISOLATED FROM MICE PERSISTENTLY INFECTED WITH MHV-JHM

J. O. Fleming,[1,2] C. Adami,[3] J. Pooley,[1] J. Glomb,[3] E. Stecker,[1] F. Fazal,[3] and S. C. Baker[3]

[1] The Departments of Neurology and Medical Microbiology and
    Immunology
University of Wisconsin School of Medicine
Madison, Wisconsin
[2] William S. Middleton Memorial Veterans Hospital
Madison, Wisconsin
[3] Department of Microbiology and Immunology
Loyola University of Chicago Stritch School of Medicine
Maywood, Illinois

ABSTRACT

Mouse hepatitis virus JHM (JHMV or MHV-4) induces subacute and chronic demyelination in rodents and has been studied as a model of human demyelinating diseases, such as multiple sclerosis. However, despite intensive investigation, the state of JHMV during chronic disease is poorly understood. Using reverse transcription-polymerase chain reaction amplification (RT-PCR) to "rescue" viral RNA, we have found that JHMV-specific sequences persist for at least 787 days after intracerebral inoculation of experimental mice. Analysis of persisting viral RNA reveals that it is extensively mutated, and we hypothesize that the mutations observed reflect adaptation of the viral quasispecies to low-level intracellular replication during chronic disease.

INTRODUCTION

Central nervous system (CNS) disease induced by JHMV in rodents typically is biphasic (Figure 1). During the acute phase of infection, abundant infectious virus is present in brain and spinal cord, and there may be clinical signs of encephalitis. After the acute phase, infectious virus declines to low or non-detectable levels; however, during this phase, inflammatory infiltrates and multifocal plaques of primary demyelination resembling the

Corona- and Related Viruses, Edited by P. J. Talbot and G. A. Levy
Plenum Press, New York, 1995

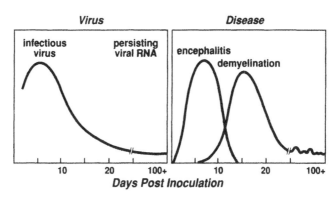

Figure 1. Pathogenesis of coronavirus JHMV in the murine central nervous system.

lesions of multiple sclerosis are found[1]. At very late times, there may be fluctuations or relapses of disease, as represented by the wavy line in Figure 1.

Recent evidence suggests that this pattern - of rapid viral clearance, followed by prolonged persistence of low-levels of viral nucleic acid - may characterize many or most viral infections. In this regard, semiquantitative studies of JHMV persistence by ourselves or other viruses such as Sindbis by Diane Griffin's laboratory[2] have indicated that the level of persisting viral nucleic acid is approximately 10,000-100,000 less than the level during acute viral replication. Thus it is likely that the biology of the virus during persistence will be different in kind from that during acute, productive infection.

As indicated, the pathology of the JHMV infection in the CNS mimics that of multiple sclerosis. Additionally, in a more general context, JHMV infection of the CNS may be an instance in which a virus serves as the initiator or trigger for chronic degenerative or autoimmune disease, as been hypothesized for many human diseases. Nonetheless, the relationship of persistence of viral nucleic acid to chronic disease is poorly understood. JHMV infections of rodents may be one system in which this important phenomenon can be studied experimentally. In a first attempt to understand JHMV persistence in the murine CNS, we have used RT-PCR to probe for mutations in RNA found in the brains of experimental mice.

MATERIALS AND METHODS

Male C57BL/6J mice, 6 weeks of age, were obtained from the Jackson Laboratory (Bar Harbor, ME); selected mice were screened by enzyme-linked immunosorbent assay and found to be seronegative for murine coronaviruses. Mice were kept in microisolator cages and were inoculated with 10^3 plaque forming units of JHMV, variant 2.2-V-1[3] intracerebrally 72 hours after arrival. At the indicated times, mice were sacrificed, and total brain RNA was extracted and subjected to RT-PCR as previously described[4]. Random hexamers were used to prime the RT step, and oligonucleotide primers recognizing either the S1 region of the spike gene (nucleotides 1084-1916) or the 3' half of the nucleocapsid gene (nucleotides 672-1237) were used in the PCR reaction, which consisted of 35 cycles of amplification, with each cycle consisting of 94°C for 1 min, 60°C for 1 min, and 72°C for 1 min, with an additional 72°C extension step at the end of the procedure. PCR was performed with *Taq* polymerase (Promega, Madison, WI), utilizing conditions to minimize nucleotide incorporation errors by this enzyme.[5] Amplified PCR products were ligated into pGEM-T vectors (Promega, Madison, WI), and the ligation products were used to transform competent *E.*

coli. Bacterial cultures were grown from individual colonies, and the plasmid DNA was purified from these cultures and sequenced by the dideoxy chain termination method (Sequenase 2.0, U.S. Biochemical, Cleveland, Ohio). In a pilot study, stock virus was added to uninfected mouse brain, and RNA was extracted and subjected to RT-PCR and sequencing as indicated above. Analysis of 8 clones corresponding to the S1 region showed the consensus or "wild type" sequence in 7 clones and one point mutation in the eighth clone, indicating that the error rate of the procedure and/or the mutation rate in individual viral genomes is in the order of 1/6,000 nucleotides analyzed.

RESULTS

The frequencies with which mutations were found in S1 and N regions are shown in Figure 2. (Please note that the oligonucleotide primers used in RT-PCR recognize these sequences in either JHMV genome or mRNA.) The numbers of clones analyzed at the S1 region were as follows: 8 at day 0 PI (input virus, pilot experiment above), 20 at day 4 PI, 4 at day 8 PI, 4 at day 13 PI, 4 at day 20 PI, and 20 at day 42 PI; for N region analysis, 20 clones were analyzed at day 4 PI, 4 at day 20 PI, and 20 at day 42 PI. Since the majority of our studies focussed on day 4 PI and 42 PI, the frequency of mutations is known with most precision at these timepoints.

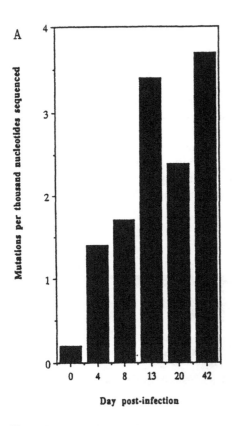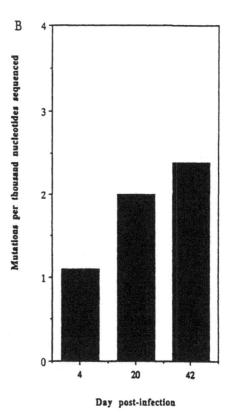

Figure 2. Mutations detected in (A) the S-1 region of the spike protein gene or (B) nucleocapsid gene during persistent infection of murine brain by coronavirus JHMV. The frequency of mutations observed are expressed as mutations per thousand nucleotides sequenced.

Analysis of individual mutations showed that they were approximately evenly distributed over the regions studied; that is, there were no apparent "hot spots" of clustered mutations in either the S1 or N regions analyzed. Of 144 mutations observed, 137 were point mutations, and the majority (74%) of these mutations were non-synonymous or amino acid-changing. Five mutations resulted in termination codons, and 2 mutations were deletions. Surprisingly, 63% of the mutations were U to C or A to G transitions, reminiscent of the biased hypermutational patterns observed during persistent CNS infections by measles virus. In parallel experiments, infectious virus could not be isolated from the brains of mice beyond day 13 PI, although RNA corresponding to JHMV could be detected at day 787 PI, the latest timepoint studied (data not shown).

DISCUSSION

Our present conception of JHMV persistence in the CNS is shown in Figure 3. In this hypothesis, persistence of viral RNA depends on both host and virus. We think it unlikely that persistence depends on a unique viral mutation, that is, an altered viral genome which would cause persistence in a naive animal; rather, the viral RNA we observed is the product of a process in which microbe and host mutually adapt to each other after the clearance of acute, productive infection.

Furthermore, we think that our findings may best be understood in the context of concepts such as those introduced by Eigen[6], Domingo[7], Holland[8], and others to characterize the evolution of viral populations. Thus, RNA viral populations have been shown to consist of quasispecies, or a distribution of mutants centered around a hypothetical consensus or "master sequence" not actually present in any existing virion. Under natural conditions, these quasispecies frequently undergo drastic constrictions or bottlenecks, for example during transmission or immune pressure.

In these circumstances, founder effects or the properties of individual mutant genomes may become important or even dominate the viral population.

The application of evolutionary concepts to JHMV persistence in the CNS is shown in Figure 3. Initially the virus population expands geometrically; soon, however, an active immune response to JHMV is generated and clears most virus. Nevertheless, a small fraction of the JHMV population escapes immune clearance, possibly because of altered antigenic determinants or, more likely, restricted and altered replication. Possibly, selection favors

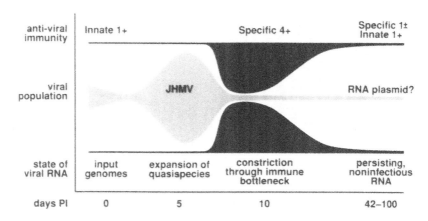

Figure 3. Evolution of coronavirus JHMV RNA during persistent infection of murine brain.

JHMV genomes which are adapted to intracellular replication rather than to the production of infectious, extracellular virions . Such RNA genomes are well-known in plant virology and have been described under the provocative term "RNA plasmids"[9,10].

In pilot experiments, we have performed strand-specific RT-PCR assays for negative sense JHMV RNA and have detected these species as late as 90 days PI (the latest timepoint assayed), providing evidence for the continued replication of JHMV RNA during persistence (data not shown). In addition, we have evidence that JHMV RNA which persists in the CNS of mice predisposes such mice to recurrent paralysis, mimicking one of the most prominent clinical features of multiple sclerosis, the tendency to relapses and remissions[11]. Thus, persisting JHMV RNA is not an inconsequential remnant of acute infection but has important biological consequences for the host. We anticipate that further study of this experimental model will yield insights into the mechanisms by which RNA viruses persist, and these insights hopefully will be relevant to the study of viruses as possible etiologic agents of chronic diseases in humans.

REFERENCES

1 Kyuwa S , Stohlman S A Pathogenesis of a neurotropic murine coronavirus, strain JHM in the central nervous system of mice Semin Virol 1990,1 273-280

2 Levine B , Griffin D E Persistence of viral RNA in mouse brains after recovery from acute alphavirus encephalitis J Virol 1993,66 6429-6436

3 Fleming J O , Trousdale M D , El-Zaatari F A K , Stohlman S A , Weiner L P Pathogenicity of antigenic variants of murine coronavirus JHM selected with monoclonal antibodies J Virol 1986,58.869-875

4 Fleming J O , Houtman J J , Alaca H , Hinze H C , McKenzie D , Aiken J , Bleasdale T , Baker S Persistence of viral RNA in the central nervous system of mice inoculated with MHV-4 Adv Exp Med Biol 1993,327-332

5 Chang S Y, Shih A , Kwok S Detection of variability in natural populations of viruses by polymerase chain reaction Methods Enzymol 1993,224 428-432

6 Eigen M Steps Towards Life A Perspective on Evolution Oxford University Press, Oxford, 1992

7 Domingo E , Escarmis C , Martinez M A , Martinez-Salas E , Mateu M G Foot-and-mouth disease virus populations are quasispecies Curr Top Microbiol Immunol 1992,176 33-47

8 Holland J J Replication error, quasispecies populations, and extreme evolution rates of RNA viruses In Morse S S (ed) Emerging Viruses Oxford University Press Oxford 1993 pp 203-218

9 Brown G G , Finnegan P M RNA plasmids Int Rev Cytol 1989,117 1-55

10 Atkins J F Contemporary RNA genomes In Gesteland R F , Atkins J F (eds) The RNA World Cold Spring Harbor Laboratory Press, Plainview 1993 pp 535-556

11 Fleming J O , Houtman J J , Hinze H C Experimentally produced relapses of demyelination in murine coronavirus-infected mice Neurology 1994,44(Suppl 2) A332

Participants in the Sixth International Symposium on Corona and Related Viruses, held August 27–September 1, 1994, in Laval, Quebec, Canada

PARTICIPANTS

Dr. Mathias Ackermann
Institut fuer Virologie
  Veterinärmedizinische Fakultät
Universität Zürich
Winterhurerstrasse 266a
CH-8057 Zürich
Switzerland
Ph.: 01 365 15 34
Fax: 01 363 01 40
ma@vetvir.unizh.ch

Ms. Nathalie Arbour
Centre de recherche en virologie
Institut Armand-Frappier, Université du
  Québec
531, boulevard des Prairies
Laval (Québec), Canada
H7N 4Z3
Ph.: 514-687-5010 x4282 or 4407
Fax: 514-686-5626

Dr. Jean-Christophe Audonnet
Rhône Mérieux
254, rue Marcel Mérieux
69342 Lyon Cedex 07
France
Ph.: 33 72 72 33 79
Fax: 33 72 72 34 97

Dr. Susan C. Baker
Department of Microbiology and
  Immunology
Loyola University School of Medicine
2160 South First Avenue, Bldg 105
Maywood, Illinois 60153
Ph.: 708-216-6910
Fax: 708-216-9574

Mr. Michael Baker
Department of Microbiology and
  Immunology
Loyola University School of Medicine
2160 South First Avenue, Bldg 105
Maywood, Illinois 60153
Ph.: 708-216-6910
Fax: 708-216-9574

Ms. Maria L. Ballesteros
Centro Nacional de Biotecnologia (CNB),
  CSIC
Campus Universidad Autonoma,
  Cantoblanco
28049 Madrid
Spain
Ph.: 34-1-5854526
Fax: 34-1-5854506
mballesteros@samba.cnb.uam.es

Dr. Ralph S. Baric
Program in Infectious Diseases
Department of Epidemiology
University of North Carolina
Chapel Hill, North Carolina 27599-7400
Ph.: 919-966-3895
Fax: 919-966-2089

Mr. Pierre Baudoux
Institut national de recherche agronomique
  (I.N.R.A.)
Laboratoires de virologie et immunologie
  moléculaires
Domaine de Vilvert
78352 Jouy-en-Josas
France
Ph.: 33 1 34 65 26 06
Fax: 33 1 34 65 26 21

Dr. David A. Benfield
Department of Veterinary Science
P.O. Box 2175, North Campus Drive
South Dakota State University
Brookings, South Dakota 57007-1396
Ph.: 605-688-4317
Fax: 605-688-6003

Dr. Weizhen Bi
The University of Texas Health Science
  Center at Houston
Medical School MSB 2.136
6431 Fannin
Houston, Texas 77030
Ph.: 713-792-5200 x3100
Fax: 713-794-4149

Mr. Aurelio Bonavia
Centre de recherche en virologie
Institut Armand-Frappier, Université du
  Québec
531, boulevard des Prairies
Laval (Québec), Canada
H7N 4Z3
Ph.: 514-687-5010 x4282 or 4407
Fax: 514-686-5626

Dr. Pedro J. Bonilla
209 Johnson Pavilion, Department of
  Microbiology
36th & Hamilton Walk
University of Pennsylvania
Philadelphia, Pennsylvania 19104-6076
Ph.: 215-898-4672
Fax: 215-898-9557
bonilla@pobox.upenn.edu

Ms. Evelyne Bos
Department of Virology, AZL, Building 1,
  L4-V
Institute of Medical Microbiology, Fac. of
  Medicine
Leiden University, P.O. Box 320
2300 AH, Leiden
The Netherlands
Ph.: 71-261654
Fax: 71-263645
azruviro@rulcri.leidenuniv.nl

Dr. David A. Brian
Department of Microbiology
University of Tennessee
Knoxville, Tennessee 37996-0845
Ph.: 615-974-4030
Fax: 615-974-4007

Dr. Paul Britton
Institute for Animal Health
Compton Laboratory
Compton, Newbury, Berkshire
RG16 0NN
United Kingdom
Ph.: 44 635 577268
Fax: 44 635 577263
britton@afrc.ac.uk

Dr. Michael J. Buchmeier
Department of Neuropharmacology
  (CVN-8)
The Scripps Research Institute
10666 North Torrey Pines Road
La Jolla, California 92037
Ph.: 619-554-7056
Fax: 619-554-6470

Dr. Gary F. Cabirac
Rocky Mountain MS Center
501 East Hampden Avenue
Englewood, Colorado 80110
Ph.: 303-788-4051
Fax: 303-788-5418
cabirac@csn.org

Dr. Paul E. Callebaut
Laboratory of Virology
Faculty of Veterinary Medicine, University
 of Gent
Casinoplein 24
B-9000 Gent
Belgium
Ph.: 9 223 37 65
Fax: 9 233 22 34

Mr. Raymond F. Castro
University of Iowa
246 Med Labs
Iowa City, Iowa 52242
Ph.: 319-335-7576
Fax: 319-356-4855

Dr. Dave Cavanagh
Institute for Animal Health
Compton Laboratory
Compton, Newbury, Berkshire
RG16 0NN
United Kingdom
Ph.: 0635 578411
Fax: 0635 577263
cavanagh@afrc.ac.uk

Ms. Fanny Chagnon
Centre de recherche en virologie
Institut Armand-Frappier, Université du
 Québec
531, boulevard des Prairies
Laval (Québec), Canada
H7N 4Z3
Ph.: 514-687-5010 x4282 or 4407
Fax: 514-686-5626

Dr. Wan Chen
Department of Epidemiology
School of Public Health
University of North Carolina at Chapel Hill
Chapel Hill, North Carolina 27599-7400
Ph.: 919-966-3881
Fax: 919-966-2089

Dr. Jane Christopher-Hennings
Department of Veterinary Science
P.O. Box 2175, North Campus Drive
South Dakota State University
Brookings, South Dakota 57007-1396
Ph.: 605-688-5171
Fax: 605-688-6003

Dr. Arlene R. J. Collins
Department of Microbiology
138 Farber Hall
State University of New York at Buffalo
Buffalo, New York 14214
Ph.: 716-829-2161
Fax: 716-829-2158
acollins@ubvms.edu

Dr. Ellen W. Collisson
Department of Veterinary Pathobiology
College of Veterinary Medicine
Texas A&M University
College Station, Texas 77843-4467
U.S.A
Ph.: 409-845-4122
Fax: 409-845-9972

Dr. Susan R. Compton
Section of Comparative Medicine
Yale University School of Medicine
P.O. Box 208016
New Haven, Connecticut 06520-8016
Ph.: 203-785-6733
Fax: 203-785-7499
compton@biomed.med.yale.edu

Ms. Geneviève Côté
Centre de recherche en virologie
Institut Armand-Frappier, Université du
 Québec
531, boulevard des Prairies
Laval (Québec), Canada
H7N 4Z3
Ph.: 514-687-5010 x4282 or 4407
Fax: 514-686-5626

Dr. Jean-Paul J. A. Coutelier
Unit of Experimental Medicine
ICP, UCL
Avenue Hippocrate, 74
1200 Brussels
Belgium
Ph.: 32 2 764 7437
Fax: 32 2 764 7430
coutelier@mexp.ucl.ac.be

Dr. Samuel Dales
Department of Microbiology and
    Immunology
Health Science Centre, Univ. of Western
    Ontario
London (Ontario)
Canada
N6A 5C1
Ph.: 519-661-3448
Fax: 519-661-3499

Dr. Raoul J. de Groot
Institute of Virology
Faculty of Veterinary Medicine
Utrecht University, Yalelaan 1
3584 CL Utrecht
The NetherlandS
Ph.: 31-30-532460
Fax: 31-30-536723

Dr. Serge Dea
Centre de recherche en virologie
Institut Armand-Frappier, Université du
    Québec
531, boulevard des Prairies
Laval (Québec), Canada
H7N 4Z3
Ph.: 514-686-5303

Dr. Bernard Delmas
Institut national de recherche agronomique
    (I.N.R.A.)
Laboratoires de virologie et immunologie
    moléculaires
Domaine de Vilvert
78352 Jouy-en-Josas
France
Ph.: 33 1 34 65 26 14
Fax: 33 1 34 65 26 21

Dr. Mark R. Denison
Vanderbilt University Medical Center
D-7235 Medical Center North
1611 - 21st Avenue South
Nashville, Tennessee 37232-2581
Ph.: 615-322-2250
Fax: 615-343-9723
denison@ctrvax.vanderbilt.edu

Dr. Brian Derbyshire
Department of Vet. Microbiology and
    Immunology
University of Guelph
Guelph (Ontario)
Canada
N1G 2W1
Ph.: 519-823-8800
Fax: 519-767-0809

Dr. Ram B. C. Dessau
Department of Clinical Microbiology
Herlev Hospital
75, Herlev Ringvej
DK-2730
Denmark
Ph.: 45 44 53 53 00 ext. 3799 or 45 31 65
    55 21
Fax: 45 44 53 53 32
ram@biobase.aau.dk

Dr. Jean-François Eleouet
Institut national de recherche agronomique
    (I.N.R.A.)
Laboratoires de virologie et immunologie
    moléculaires
Domaine de Vilvert
78352 Jouy-en-Josas
France
Ph.: 33 1 34 65 26 06
Fax: 33 1 34 65 26 21

Dr. Luis Enjuanes
Centro Nacional de Biotecnologia (CNB),
    CSIC
Campus Universidad Autonoma,
    Cantoblanco
28049 Madrid
Spain
Ph.: 341-585-4555
Fax: 341-585-4506
lenjuanes@samba.cnb.uam.es

Dr. Robert Fingerote
c/o Dr. Gary Levy, The Toronto Hospital
621 University Avenue, NU-10-151
Toronto (Ontario)
Canada
M5G 2C4
Ph.: 416-340-5166
Fax: 416-340-3492

Dr. John O. Fleming
Department of Neurology, University of
  Wisconsin
H6/550 CSC Neurology
600 Highland Avenue
Madison, Wisconsin 53792
Ph.: 608-263-5420
Fax: 608-263-0412
fleming@neurology.wisc.edu

Dr. Kosaku Fujiwara
D-806,
1241-4 Okamoto
Kamakura 247
Japan
Ph.: 81-467-43-1818
Fax: 81-466-82-1310

Ms. Atsuko Fujiwara
D-806,
1241-4 Okamoto
Kamakura 247
Japan
Ph.: 81-467-43-1818
Fax: 81-466-82-1310

Dr. Laisum Fung
c/o Dr. Gary Levy, The Toronto Hospital
621 University Avenue, NU-10-151
Toronto (Ontario)
Canada
M5G 2C4
Ph.: 416-340-5166
Fax: 416-340-3492

Dr. Thomas M. Gallagher
Loyola University Medical Center
Department of Microbiology and
  Immunology
2160 South First Avenue
Maywood, Illinois 60153
Ph.: 708-216-4850
Fax: 708-216-9574
tgallag@lucpug.it.lyc.edu

Dr. Russell R. Gettig
Virogenetics Corporation
465 Jordan Road
Troy, New York 12180
U.S.A
Ph.: 518-283-8389
Fax: 518-283-0936

Dr. Jim Gombold
Department of Microbiology and
  Immunology
Louisiana State University Medical Center
1501 Kings Highway
Shreveport, Louisiana 71130
Ph.: 318-675-6684
Fax: 318-675-5764

Dr. Masanobu Hayashi
Department of Laboratory Animal Science
Faculty of Veterinary Medicine
Hokkaido University, Kita 18-Joe, Nishi
  9-Chome
Kita-Ku, Sapporo 060
Japan
Ph.: 011-716-2111 x5107
Fax: 011-717-7569

Mr. Jens Herold
Institute für Virologie
Universität Würzburg
Versbacher Str. 7
97078 Würzburg
Germany
Ph.: 931 201 3966
Fax: 931 201 3934

Mr. Arnold A. P.M. Herrewegh
Institute of Virology
Faculty of Veterinary Medicine
Utrecht University, Yalelaan 1
3584 CL Utrecht
The Netherlands
Ph.: 31-30-534195
Fax: 31-30-536723

Dr. Georg Herrler
Institut für Virologie
Philipps-Universität Marburg
Robert-Koch-Str. 17
35037 Marburg
Germany
Ph.: 06421-28-5360
Fax: 06421-28-5482
herrler@convex.hrz.uni-marburg.de

Dr. Susan T. Hingley
7 Rose Lane
West Chester, Pennsylvania 19380
Ph.: 215-898-3551
Fax: 215-898-9557

Dr. Norio Hirano
Department of Veterinary Microbiology
Iwate University
Morioka 020
Japan
Ph.: 0196-23-5171 x2572
Fax: 0196-52-6945

Mr. Julian A. Hiscox
Institute for Animal Health
Department of Molecular Biology
Newbury, Berkshire
RG16 0NN
United Kingdom
Ph.: 44 635 578411 x2657
Fax: 44 635 577263
hiscox@afrc.ac.uk

Dr. Brenda G. Hogue
Baylor College of Medicine
One Baylor Plaza
Department of Microbiology and
    Immunology
Houston, Texas 77030
Ph.: 713-798-6412
Fax: 713-798-7375
bhogue@bcm.tmc.edu

Dr. Kathryn V. Holmes
Department of Pathology
Uniformed Services Univ. of the Health
    Sciences
(USUHS), 4301 Jones Bridge Road
Bethesda, Maryland 20814-4799
Ph.: 301-295-3456
Fax: 301-295-1640

Ms. Jacqueline J. Houtman
Department of Medical Microbiology and
    Immunology
University of Wisconsin-Madison
1300 University Avenue
Madison, Wisconsin 53706
Ph.: 608-262-6930
Fax: 608-262-8418
jhoutman@macc.wisc.edu

Mr. Scott A. Hughes
209 Johnson Pavilion, Department of
    Microbiology
36th & Hamilton Walk
University of Pennsylvania
Philadelphia, Pennsylvania 19104-6076
Ph.: 215-898-4672
Fax: 215-898-9557

Dr. Jagoda Ignjatovic
CSIRO Division of Animal Health
Private Bag No 1
Parkville 3052
Australia
Ph.: 61-3-342 9759
Fax: 61-3-347 4042
jagoda@mel.dah.csiro.au (AARNET)

Dr. Mark W. Jackwood
Department of Avian Medicine
College of Veterinary Medicine, Univ. of
  Georgia
953 College Station Road
Athens, Georgia 30602-4875
Ph.: 706-542-5475
Fax: 706-542-5630
mjackwood@uga.cc.uga.edu

Dr. Daral J. Jackwood
The Ohio State University
Ohio Agricultural Research and
  Development Center
1680 Madison Avenue
Wooster, Ohio 44691
Ph.: 216-263-3964
Fax: 216-263-3677

Dr. Gundula Jäger
Max von Pettenkoferinstitut
Pettenkoferstr. 9a
80336 München
Germany
Ph.: 49-89-5160-5233
Fax: 49-89-538-0584

Ms. Dara Jamieson
Division of Neuropathology
University of Pennsylvania School of
  Medicine
449 Johnson Pavilion
Philadelphia, Pennsylvania 19104-6079
Ph.: 215-662-6653
ehud_lavi@path1a.med.upenn.edu

Dr. Jeymohan Joseph
Jefferson Medical College of Jefferson
  University
1025 Walnut Street, Suite 511
Philadelphia, Pennsylvania 19107
Ph.: 215-955-5208
Fax: 215-923-6792

Mr. Kishna Kalicharran
Department of Microbiology and
  Immunology
Health Science Centre, Univ. of Western
  Ontario
London (Ontario)
Canada
N6A 5C1
Ph.: 519-661-6628
Fax: 519-661-3797

Ms. Gertrud Kern-Siddell
Institute für Virologie
Universität Würzburg
Versbacher Str. 7
97078 Würzburg
Germany
Ph.: 931 201 3896
Fax: 931 201 3934
siddell@vax.r3.uni-wuerzburg.d400.de

Dr. Robert L. Knobler
Jefferson Medical College of Jefferson
  University
1025 Walnut Street, Suite 511
Philadelphia, Pennsylvania 19107
Ph.: 215-955-7365
Fax: 215-923-6792

Ms. Sanneke A. Kottier
Institute for Animal Health
Compton Laboratory
Compton, Newbury, Berkshire
RG16 0NN
United Kingdom
Ph.: 0635 578411
Fax: 0635 577263

Dr. Shigeru Kyuwa
Department of Animal Pathology
Institute of Medical Science
University of Tokyo
4-6-1 Shirokanedai, Minato-ku, Tokyo 108
Japan
Ph.: 81-3-3443-8111 x332
Fax: 81-3-3443-6819

Dr. Michael M. C. Lai
Department of Microbiology
Univ. of Southern California, School of
  Medicine
2011 Zonal Avenue
Los Angeles, California 90033
Ph.: 213-342-1748
Fax: 213-342-9555

Dr. Nadira Lakdawalla
405-1100 Sheppard Avenue, East
Willowdale (Ontario)
Canada
M2P 1R4

Mr. Alain Lamarre
Centre de recherche en virologie
Institut Armand-Frappier, Université du
  Québec
531, boulevard des Prairies
Laval (Québec), Canada
H7N 4Z3
Ph.: 514-687-5010 x4282 or 4407
Fax: 514-686-5626

Dr. Lucie Lamontagne
Département des sciences biologiques
Université du Québec à Montréal,
C.P. 8888, Succursale A
Montréal (Québec), Canada
H3C 3P8
Ph.: 514-987-3184
Fax: 514-987-4647

Dr. Hubert Laude
Institut national de recherche agronomique
  (I.N.R.A.)
Laboratoires de virologie et immunologie
  moléculaires
Domaine de Vilvert
78352 Jouy-en-Josas
France
Ph.: 33 1 34 65 26 13
Fax: 33 1 34 65 26 21
laude@biotec.jouy.inra.fr

Dr. Ehud Lavi
Division of Neuropathology
University of Pennsylvania School of
  Medicine
449 Johnson Pavilion
Philadelphia, Pennsylvania 19104-6079
Ph.: 215-662-6653
ehud_lavi@path1a.med.upenn.edu

Dr. Sang-Koo Lee
305-380
Lucky Research Park, Biotechnology
Science Town
Dae Jeon
Korea
Ph.: 03-3812-2111 x5401

Dr. Julian Leibowitz
The University of Texas Health Science
  Center
at Houston, Medical School MSB 2.136
6431 Fannin
Houston, Texas 77030
Ph.: 713-792-8360
Fax: 713-794-4149
leib@casper.med.uth.tmc.edu

Dr. Robin Levis
Department of Pathology
Uniformed Services Univ. of the Health
  Sciences
(USUHS), 4301 Jones Bridge Road
Bethesda, Maryland 20814
Ph.: 301-295-3499
Fax: 301-295-1640

Dr. Gary A. Levy
The Toronto Hospital
621 University Avenue, NU-10-151
Toronto (Ontario)
Canada
M5G 2C4
Ph.: 416-340-5166
Fax: 416-340-3492

Dr. Dingxiang Liu
Division of Virology, Department of
  Pathology
University of Cambridge
Tennis Court Road
Cambridge CB2 1QP
United Kingdom
Ph.: 0223 336918
Fax: 0223 336926
dxl10@uk.ac.cambridge.phoenix

Dr. Willem Luytjes
Department of Virology, AZL, Building 1,
  L4-V
Institute of Medical Microbiology, Fac. of
  Medicine
Leiden University, P.O. Box 320
2300 AH, Leiden
The Netherlands
Ph.: 71-261654
Fax: 71-263645
azruviro@rulcri.leidenuniv.nl

Dr. Knud G. Madsen
State Veterinary Institute for Virus Research
Lindholm
DK-4771 Kalvehave
Denmark
Ph.: 45 55 81 45 23
Fax: 45 55 81 17 66
kgmgen@biobase.aau.dk

Mr. Akihiko Maeda
Department of Laboratory Animal Science
Faculty of Veterinary Medicine
Hokkaido University, Kita 18-Joe, Nishi
  9-Chome
Kita-Ku, Sapporo 060
Japan
Ph.: 011-716-2111 x5107
Fax: 011-717-7569

Dr. Ronald Magar
Laboratoire d'Hygiène Vétérinaire et
  Alimentaire
Agriculture Canada, 3400 Casavant Ouest
St-Hyacinthe (Québec)
Canada
J2S 8E3
Ph.: 514-773-7730
Fax: 514-773-8152

Dr. Shinji Makino
Department of Microbiology
The University of Texas at Austin
24th at Speedway, ESB 304
Austin, Texas 78712
Ph.: 512-471-6876
Fax: 512-471-7088
makino@emx.cc.utexas.edu

Dr. Paul S. Masters
Wadsworth Center for Laboratories and
  Research
New York State Department of Health
New Scotland Avenue, P.O. Box 22002
Albany, New York 12201-2002
Ph.: 518-474-1283
Fax: 518-473-1326

Ms. Ana Mendez
Centro Nacional de Biotecnologia (CNB),
  CSIC
Campus Universidad Autonoma,
  Cantoblanco
28049 Madrid
Spain
Ph.: 34-1-5854526
Fax: 34-1-5854506
amendez@samba.cnb.uam.es

Dr. Janneke J. M. Meulenberg
Central Veterinary Institute
Department of Virology
Moutribweg 39
8221 RA Lelystad
The Netherlands
Ph.: 31-320076805
Fax: 31-320042804

Dr. Ghania Milane
Centre de recherche en virologie
Institut Armand-Frappier, Université du
  Québec
531, boulevard des Prairies
Laval (Québec), Canada
H7N 4Z3
Ph.: 514-687-5010 x4380 or 4389

Dr. Timothy J. Miller
Smithkline Beecham Animal Health
601 West Cornhusker Highway
Lincoln, Nebraska 68521
Ph.: 402-441-2254 or 441-2732
Fax: 402-441-2530

Ms. Charmaine D. Mohamed
c/o Dr. Gary Levy, The Toronto Hospital
621 University Avenue, NU-10-151
Toronto (Ontario)
Canada
M5G 2C4
Ph.: 416-340-5166
Fax: 416-340-3492

Dr. Samir Mounir
Centre de recherche en virologie
Institut Armand-Frappier, Université du
  Québec
531, boulevard des Prairies
Laval (Québec), Canada
H7N 4Z3
Ph.: 514-682-0486

Dr. Vibeke Moving
Statens Veterinärmedicinska (SVA)
Dag Hammarskjöldsväg 32 B
75183 Uppsala
Sweden
Ph.: 18-506353
Fax: 18-554868

Dr. Ronald S. Murray
Rocky Mountain MS Center
501 East Hampden Avenue
Englewood, Colorado 80110
Ph.: 303-788-4210
Fax: 303-788-5488

Dr. Therese C. Nash
Department of Neuropharmacology
  (CVN-8)
The Scripps Research Institute
10666 North Torrey Pines Road
La Jolla, California 92037
Ph.: 619-554-7161
Fax: 619-554-6470
tnash@pluto.scripps.edu

Dr. Eric A. Nelson
Department of Veterinary Science
P.O. Box 2175, North Campus Drive
South Dakota State University
Brookings, South Dakota 57007-1396
Ph.: 605-688-5171
Fax: 605-688-6003

Mr. Ryuji Nomura
Department of Veterinary Microbiology
Iwate University
Morioka 020
Japan
Ph.: 0196-23-5171 x2572
Fax: 0196-52-6945

Mr. Dirk-Jan E. Opstelten
Institute of Virology
Faculty of Veterinary Medicine
Utrecht University, Yalelaan 1
3584 CL Utrecht
The Netherlands
Ph.: 31-30-533337
Fax: 31-30-536723

Mr. Jean-Sébastien Paquette
Centre de recherche en virologie
Institut Armand-Frappier, Université du
  Québec
531, boulevard des Prairies
Laval (Québec), Canada
H7N 4Z3
Ph.: 514-687-5010 x4282 or 4407
Fax: 514-686-5626

Dr. Becky Parr
The University of Texas Health Science
  Center at Houston
Medical School MSB 2.136
6431 Fannin
Houston, Texas 77030
Ph.: 713-792-5200 x3061
Fax: 713-794-4149
parr@casper.med.uth.tmc.edu

Mr. Zoltan Penzes
Institute for Animal Health
Compton Laboratory
Compton, Newbury, Berkshire
RG16 0NN
United Kingdom
Ph.: 44 635 578411 x2657
Fax: 44 635 577263
penzes@afrc.ac.uk

Dr. Stanley Perlman
Department of Pediatrics
University of Iowa
Med Labs 207
Iowa City, Iowa 52242
U.S.A
Ph.: 319-335-8549
Fax: 319-356-4855
stanley_perlman@umaxc.uiowa.edu

Dr. Marc Pope
c/o Dr. Gary Levy, The Toronto Hospital
621 University Avenue, NU-10-151
Toronto (Ontario)
Canada
M5G 2C4
Ph.: 416-340-5166
Fax: 416-340-3492

Dr. Dragan R. Rogan
Vetrepharm Research
4077 Breen Road, RR #1, P.O. Box 19
Putnam (Ontario)
Canada
N0L 2B0
Ph.: 519-485-4731
Fax: 519-485-4749

Mr. John W. A. Rossen
Institute of Virology
Faculty of Veterinary Medicine
Utrecht University, Yalelaan 1
3584 CL Utrecht
The Netherlands
Ph.: 31-30-533337
Fax: 31-30-536723

Dr. Peter J. M. Rottier
Institute of Virology
Faculty of Veterinary Medicine
Utrecht University, Yalelaan 1
3584 CL Utrecht
The Netherlands
Ph.: 31-30-532462
Fax: 31-30-536723

Ms. Francine Ruskin
Department of Microbiology and
  Immunology
Health Science Centre, Univ. of Western
  Ontario
London (Ontario)
Canada
N6A 5C1
Ph.: 519-661-3448
Fax: 519-661-3499

Dr. Linda J. Saif
Food Animal Health Research Program
OARDC, 1680 Madison Avenue
The Ohio State University
Wooster, Ohio 44691
Ph.: 216-263-3744
Fax: 216-263-3677

Mr. Yannic Salvas
Centre de recherche en virologie
Institut Armand-Frappier, Université du
  Québec
531, boulevard des Prairies
Laval (Québec), Canada
H7N 4Z3
Ph.: 514-687-5010 x4282 or 4407
Fax: 514-686-5626

Dr. Stanley G. Sawicki
Department of Microbiology
Medical College of Ohio
3000 Arlington Avenue
Toledo, Ohio 43614
Ph.: 419-381-3921
Fax: 419-381-3002
sawicks%opus@cutter.mco.edu

Dr. Dorothea L. Sawicki
Department of Microbiology
Medical College of Ohio
3000 Arlington Avenue
Toledo, Ohio 43614
Ph.: 419-381-4337
Fax: 419-381-3002
sawickd%opus@cutter.mco.edu

Dr. Beate Schultze
Institut für Virologie
Philipps-Universität Marburg
Robert-Koch-Str. 17
35037 Marburg
Germany
Ph.: 06421-285146
Fax: 06421-285482

Mr. Shinwa Shibata
Department of Veterinary Pathology
Faculty of Agriculture
The University of Tokyo
1-1-1 Yayoi, Bunkyo-ku, Tokyo 113
Japan
Ph.: 03-3812-2111 x5401
Fax: 03-5800-6919

Dr. Stuart G. Siddell
Institute für Virologie
Universität Würzburg
Versbacher Str. 7
97078 Würzburg
Germany
Ph.: 931 201 3896
Fax: 931 201 3934
siddell@vax.r3.uni-wuerzburg.d400.de

Dr. Emil Skamene
Montreal General Hospital
1650 Cedar Avenue, Room B7.118
Montréal (Québec)
Canada
H3G 1A4
Ph.: 514-934-8038
Fax: 514-933-7146

Dr. Helen E. Smith
Langford Cyanamid
131 Malcolm Road
Guelph (Ontario)
Canada
N1K 1A8
Ph.: 519-823-5490 x439
Fax: 519-823-8303

Dr. Willy J. M. Spaan
Department of Virology, AZL, Building 1,
  L4-V
Institute of Medical Microbiology, Fac. of
  Medicine
Leiden University, P.O. Box 320
2300 AH, Leiden
The Netherlands
Ph.: 71-261652
Fax: 71-263645
azruviro@rulcri.leidenuniv.nl

Dr. John S. Spencer
Rocky Mountain MS Center
501 East Hampden Avenue
Englewood, Colorado 80110
Ph.: 303-788-4016
Fax: 303-788-5418
jspencer@csn.org

Dr. Steve A. Stohlman
Department of Neurology
University of Southern California
2025 Zonal Avenue, MCH 142
Los Angeles, California 90033
Ph.: 213-342-1063
Fax: 213-225-2369

Dr. Bertel S. Strandbygaard
State Veterinary Institute for Virus Research
Lindholm
DK-4771 Kalvehave
Denmark
Ph.: 45 55 81 45 23
Fax: 45 55 81 17 66

Dr. Fumihiro Taguchi
National Institute of Neuroscience, NCNP
4-1-1 Ogawahigashi
Kodaira
Tokyo 187
Japan
Ph.: 0423-41-2711 x5272
Fax: 0423-46-1754
taguchi@ncnpja.ncnp.go.jp

Dr. Pierre J. Talbot
Centre de recherche en virologie
Institut Armand-Frappier, Université du
  Québec
531, boulevard des Prairies
Laval (Québec), Canada
H7N 4Z3
Ph.: 514-687-5010 x4406 or 4369
Fax: 514-686-5626
pierre_talbot@iaf.uquebec.ca

Mr. Takafumi Tawara
Department of Veterinary Microbiology
Iwate University
Morioka 020
Japan
Ph.: 0196-23-5171 x2572
Fax: 0196-52-6945

Mr. Kurt Tobler
Institut fuer Virologie
  Veterinärmedizinische Fakultät
Universität Zürich
Winterhurerstrasse 266a
CH-8057 Zürich
Switzerland
Ph.: 01 365 15 20
Fax: 01 363 01 40

Dr. Koji Uetsuka
Department of Veterinary Pathology
Faculty of Agriculture
The University of Tokyo
1-1-1 Yayoi, Bunkyo-ku, Tokyo 113
Japan
Ph.: 03-3812-2111 x5401
Fax: 03-5800-6919

Dr. Anna Utiger
Institut fuer Virologie
  Veterinärmedizinische Fakultät
Universität Zürich
Winterhurerstrasse 266a
CH-8057 Zürich
Switzerland
Ph.: 01 365 15 38
Fax: 01 363 01 40

Mr. Guido van Marle
Department of Virology, AZL, Building 1,
  L4-V
Institute of Medical Microbiology, Fac. of
  Medicine
Leiden University, P.O. Box 320
2300 AH, Leiden
The Netherlands
Ph.: 71-261654
Fax: 71-263645
azruviro@rulcri.leidenuniv.nl

Dr. Harry Vennema
Department of Surgical and Radiological
  Sciences
School of Veterinary Medicine
University of California at Davis
Davis, California 95616
Ph.: 916-752-1064
Fax: 916-752-6042
hxvennema@ucdavis.edu

Dr. Elke Vieler
Institut für Hygiene und
  Infektionskrankheiten der Tiere der
  Justus-Liebig-Universität Giessen
Frankfurter Str. 89-91
D-35392 Giessen
Germany
Ph.: 49-641-702 4886
Fax: 49-641-702 4876

Dr. Terri L. Wasmoen
Fort Dodge Laboratories
800 Fifth Street N.W.
Fort Dodge, Iowa 50501
Ph.: 515-955-4600 x2388
Fax: 515-955-9189

Mr. Fred Wassenaar
Department of Virology, AZL, Building 1,
  L4-V
Institute of Medical Microbiology, Fac. of
  Medicine
Leiden University, P.O. Box 320
2300 AH, Leiden
The Netherlands
Ph.: 71-261652
Fax: 71-263645
azruviro@rulcri.leidenuniv.nl

Dr. Hana M. Weingartl
Institut national de recherche agronomique
  (I.N.R.A.)
Laboratoires de virologie et immunologie
  moléculaires
Domaine de Vilvert
78352 Jouy-en-Josas
France
Ph.: 33 1 34652121
Fax: 33 1 34652273 or 33 1 34652621

Dr. Lisa M. Welter
Ambico West Inc.
2011 Zonal Avenue, HMR 603
Los Angeles, California 90033
Ph.: 213-342-3364 or 800-544-1565
Fax: 213-342-3364

Dr. Dongwan Yoo
Veterinary Infectious Disease Organization
  (VIDO)
124 Veterinary Road
Saskatoon (Saskatchewan)
Canada
S7N 0W0
Ph.: 306-966-7485
Fax: 306-966-7478
yoo@sask.usask.ca

Mr. Wei Yu
The University of Texas Health Science
  Center at Houston
Medical School MSB 2.136
6431 Fannin
Houston, Texas 77030
Ph.: 713-792-5200 x3100
Fax: 713-794-4149
weiyu@casper.med.uth.tmc.edu

Ms. Mathilde Yu
Centre de recherche en virologie
Institut Armand-Frappier, Université du
  Québec
531, boulevard des Prairies
Laval (Québec), Canada
H7N 4Z3
Ph.: 514-687-5010 x4282 or 4407
Fax: 514-686-5626

Dr. Xuming Zhang
Department of Microbiology
University of Southern California, School
  of Medicine
2011 Zonal Avenue, HMR-504
Los Angeles, California 90033
Ph.: 213-342-3505
Fax: 213-342-9555

INDEX

Made in the USA
Las Vegas, NV
26 October 2024

10376724R10359